MASSEY FERG

T0256165

SHOP MANUAL MF-201

Model ■ MF65

Models ■ MF85 ■ MF88 ■ MF Super 90 ■ MF Super 90 WR

Models ■ MF1080 ■ MF1085

Models ■ MF1100 ■ MF1130

Model ■ MF1150

Models ■ MF1105 ■ MF1135 ■ MF1155

I&T
&
SHOP MANUALS

Information and Instructions

This shop manual contains several sections each covering a specific group of wheel type tractors. The Tab Index on the preceding page can be used to locate the section pertaining to each group of tractors. Each section contains the necessary specifications and the brief but terse procedural data needed by a mechanic when repairing a tractor on which he has had no previous actual experience.

Within each section, the material is arranged in a systematic order beginning with an index which is followed immediately by a Table of Condensed Service Specifications. These specifications include dimensions, fits, clearances and timing instructions. Next in order of arrangement is the procedures paragraphs.

In the procedures paragraphs, the order of presentation starts with the front axle system and steering and proceeding toward the rear axle. The last paragraphs are devoted to the power take-off and power lift sys-

tems. Interspersed where needed are additional tabular specifications pertaining to wear limits, torquing, etc.

HOW TO USE THE INDEX

Suppose you want to know the procedure for R&R (remove and reinstall) of the engine camshaft. Your first step is to look in the index under the main heading of ENGINE until you find the entry "Camshaft." Now read to the right where under the column covering the tractor you are repairing, you will find a number which indicates the beginning paragraph pertaining to the camshaft. To locate this wanted paragraph in the manual, turn the pages until the running index appearing on the top outside corner of each page contains the number you are seeking. In this paragraph you will find the information concerning the removal of the camshaft.

More information available at Clymer.com
Phone: 805-498-6703

Haynes Publishing Group
Sparkford Nr Yeovil
Somerset BA22 7JJ England

Haynes North America, Inc
859 Lawrence Drive
Newbury Park
California 91320 USA

ISBN-10: 0-87288-376-0
ISBN-13: 978-0-87288-376-5

MASSEY FERGUSON

Model ■ MF65

Previously contained in I&T Shop Manual No. MF-19

SHOP MANUAL
MASSEY-FERGUSON

MODEL MF 65

Tractor serial number stamped on instrument panel name plate.
Engine serial number stamped on side of engine.

INDEX (By Starting Paragraph)

CONDENSED SERVICE DATA

GENERAL	NON-DIESEL		DIESEL
Torque recommendations	See End of Shop Manual		
Engine Make	Cont'l	Perkins	Perkins
Engine Model	G176	4A-203	AD4-203
No. Cylinders	4	4	4
Bore—Inches	3.58	3.6	3.6
Stroke—Inches	4.38	5.0	5.0
Displacement—Cubic inches	176	203.5	203.5
Compression Ratio, Gasoline	7.1:1	......	......
Compression Ratio, LP-Gas	8.1:1	......	......
Compression Ratio, Diesel	......	17.4:1	18.5:1
Pistons Removed From?	Above	Above	Above
Main Bearings, Number of?	3	5	5
Cylinder Sleeves	Wet	Dry	Dry
Forward Speeds, No. of?	6*	6	12
Reverse Speeds, No. of?	2*	2	4
Generator & Starter Make	D-R	D-R	D-R

*12 forward, 4 reverse; if equipped with "Multi-Power" transmission.

TUNE-UP			
Firing order	1-3-4-2	1-3-4-2	1-3-4-2
Intake Valve Tappet Gap (Hot)	0.016	0.010	0.010
Exhaust Valve Tappet Gap (Hot)	0.018	0.010	0.010
Inlet Valve Face Angle	30°	44°	44°
Inlet Valve Seat Angle	30°	45°	45°
Exhaust Valve Face Angle	44°	44°	44°
Exhaust Valve Seat Angle	45°	45°	45°
Ignition Distributor Model	1112583	......	......
Breaker Contact Gap	0.022	......	......
Plug Electrode Gap	0.025-0.027	......	......
Carburetor, Gasoline	M-S, TSX695	......	......
Carburetor, LP-Gas	Zenith	......	......
Float Setting, M-S	1/4 inch	......	......
Injection Pump Make	......	C.A.V.	C.A.V.
Injection Pump Model	......	DPA	DPA

TUNE-UP (Cont.)	NON-DIESEL		DIESEL
Injection Pump Timing	......	18°BTC	24°BTC
Injector Nozzle Make	......	C.A.V.	C.A.V.
Engine Low Idle Rpm	450	500	600
Engine Load Rpm	2000	2000	2000
Engine High Idle Rpm	2150	2200	2175
Pto Load Rpm	720*	720	636
Pto No Load Rpm	774*	792	691

*Loaded rpm 636; no-load rpm 684, if equipped with "Multi-Power" transmission.

SIZES—CAPACITIES—CLEARANCES

(Clearances in thousandths)

Crankshaft Journal Diameter	2.375	2.749	2.749
Crankpin Diameter	2.062	2.249	2.249
Camshaft Journal Diameter:			
Front	1.937	1.870	1.870
Center	1.750	1.860	1.860
Rear	1.687	1.840	1.840
Piston Pin Diameter	1.125	1.250	1.250
Piston Ring Width:			
Compression	0.125	0.093	0.093
Oil	0.188	0.250	0.250
Main Bearing, Diameter Clearance	0.5-2.7	2.5-4.5	3.0-5.0
Rod Bearing, Diameter Clearance	0.7-2.7	2-3.5	2.5-4.0
Crankshaft End Play	4-8	2-10	2-14
Camshaft Bearing, Diameter Clearance	3-4.5	4-8	4-8
Camshaft End Play	3-7	3-6	3-6
Cooling System—Quarts	10.5	10	10
Crankcase Oil—Quarts	5	6	7
Transmission, Differential and Hydraulic Lift—Gallons	8	8	8
Power Steering Reservoir—Quarts	2/3	1-2/3	1-2/3
Steering Gear Housing—Pints	2	2	2

FRONT SYSTEM

ting hole, drill a small depression in top wall of bushing bore. Install the new bushing with grease hole in bushing and support aligned, then stake bushing in place by inserting a slim punch through grease fitting hole and indenting bushing into previously drilled depression.

When installing the axle center member, reverse the removal procedure and vary the number of shims (15) to obtain a center member end play of 0.002-0.008 when checked between rear face of pivot bracket (13) and front face of center member (11). Shims are available in thicknesses of 0.002, 0.005 and 0.010.

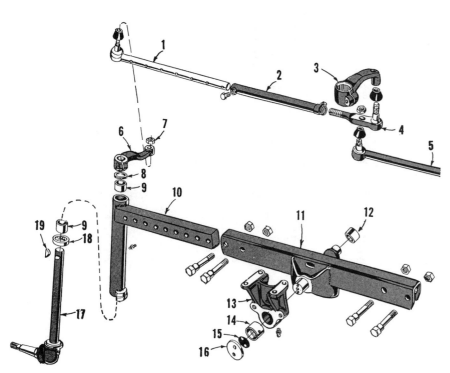

Fig. MF1—Standard and Hi-arch front axle used on model MF65. Recommended axle end play of 0.002-0.008 is adjusted with shims (15). Toe-in adjustment is accomplished by turning tie rod tube (2).

1. Right tie rod	8. Dust seal	14. Front pivot bushing
2. Tie rod tube	9. Spindle bushings	15. Shims
3. Steering crank arm	10. Axle extension	16. Thrust plate
4. Tie rod end	11. Axle center member	17. Spindle and knuckle
5. Left tie rod tube	12. Rear pivot bushing	18. Thrust bearing
6. Spindle steering arm	13. Pivot bracket	19. Woodruff key

SPINDLE BUSHINGS

2. Each axle extension contains two renewable bushings (9) which require final sizing after installation. Recommended clearance between spindles and bushings is 0.0035-0.005. Nominal spindle diameter is 1¼-inches for adjustable axle models; and 1½-inches for Utility models.

TOE-IN, TIE-RODS AND/OR DRAG LINKS

3. Tie-rod ends are of the automotive type and are not adjustable to compensate for wear.

Recommended toe-in of 0-¼ inch is adjusted by varying the length of the right hand tie-rod. This is accomplished by loosening the tube set screw and clamp bolt and turning the tube (2—Fig. MF1) either way as required. Be sure the tube set screw is in the forward position before tightening.

AXLE ASSEMBLY

1. To remove the axle main (center) member (11—Fig. MF1), support tractor under engine, remove the grille lower panel, disconnect tie rods from the spindle steering arms and remove both axle extensions and wheel assemblies from the center member. Unbolt and remove the axle pivot thrust plate (16) and save shims (15) for reinstallation. Remove the cap screws retaining pivot bracket (13) to the front support casting and remove the pivot bracket. Pull center member (11) forward and out of rear pivot bushing (12) which is located in the front support casting.

The non-adjustable axle assembly used on Utility models is renewed in the same manner except spindles must be withdrawn from the one-piece axle member.

Inside diameter of new pivot bushings is 1.877-1.879 for rear bushing (12) and 2.002-2.004 for front bushing (14). Pivot pin diameter for a new center member is 1.874-1.876 for the rear pin and 1.999-2.001 for the front pin. Renew center member and/or bushings if running clearance is excessive. Replacement bushings are pre-sized and will not require reaming if carefully installed with a suitable arbor. Make certain, however, that lubrication holes in bushings are in register with similar holes in pivot bracket and front support.

NOTE: On some early models, the rear pivot bushing is not staked in place in front support. When renewing the rear bushing, refer to Fig. MF2 and proceed as follows: Remove the old bushing and grease fitting. Use a long 7/32-inch drill bit and, working up through the grease fit-

Fig. MF2 — Axle pivot pin rear bushing installation.

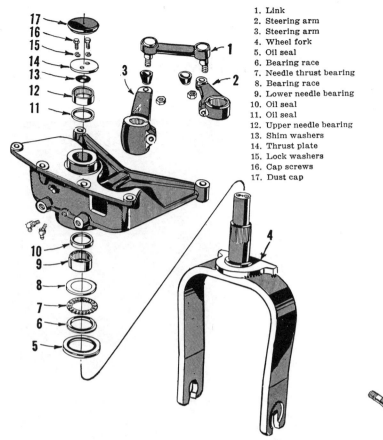

1. Link
2. Steering arm
3. Steering arm
4. Wheel fork
5. Oil seal
6. Bearing race
7. Needle thrust bearing
8. Bearing race
9. Lower needle bearing
10. Oil seal
11. Oil seal
12. Upper needle bearing
13. Shim washers
14. Thrust plate
15. Lock washers
16. Cap screws
17. Dust cap

Fig. MF3 — Exploded view of the single wheel fork, support casting and associated parts used on the model MF65. End play of wheel fork is controlled by shim washers (13).

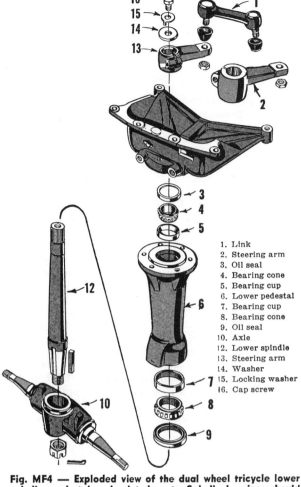

1. Link
2. Steering arm
3. Oil seal
4. Bearing cone
5. Bearing cup
6. Lower pedestal
7. Bearing cup
8. Bearing cone
9. Oil seal
10. Axle
12. Lower spindle
13. Steering arm
14. Washer
15. Locking washer
16. Cap screw

Fig. MF4 — Exploded view of the dual wheel tricycle lower spindle, pedestal and related parts. Spindle bearings should be adjusted to remove all end play without binding.

SINGLE WHEEL & FORK

4. The fork mounted single front wheel is carried in taper roller bearings which should be adjusted to provide a very slight rotational drag.

To remove the wheel fork, support tractor under engine and remove the grille lower panel. Open the grille door and remove the sheet metal dust cap (17—Fig. MF3). Remove the cap screws (16) retaining the thrust plate (14) to upper end of wheel fork and withdraw thrust plate (14) and shim washers (13). Working through opening in the support casting, loosen the clamp bolt retaining steering arm (3) to wheel fork. Raise front of tractor and at the same time, withdraw the wheel fork from below. CAUTION: Wheel fork must not be cocked during removal or seals and needle bearings may be damaged.

Examine needle bearings (9 and 12), seals (5, 10 and 11) and needle thrust bearing (6, 7 and 8) and renew any questionable parts. When install-

ing the caged needle bearings (9 and 12), be sure to align the oil hole in the bearings with the oil feed holes in the support casting.

Install the wheel fork by reversing the removal procedure and vary the number of shim washers (13) to provide the wheel fork with an up and down end play of 0.002-0.008.

NOTE: The wheel fork and steering arm (3) have a blind spline to facilitate correct installation.

DUAL WHEELS, LOWER SPINDLE & LOWER PEDESTAL

5. Each of the dual wheels is mounted on taper roller bearings which should be adjusted to provide a very slight rotational drag.

To remove the lower spindle and wheels assembly, support tractor under engine and remove the grille lower panel. Open the grille door, un-

lock and remove cap screw (16—Fig. MF4), lock washer (15) and flat washer (14). Working through opening in the support casting, loosen the clamp bolt retaining steering arm (13) to the spindle. Raise front of tractor and at the same time, withdraw the spindle, axle and wheels assembly from below.

Lower pedestal (6) can be removed from the support casting at this time.

Examine the bearing cones (4 and 8), cups (5 and 7) and seals (3 and 9) and renew any questionable parts.

Install the spindle and lower pedestal by reversing the removal procedure and tighten cap screw (16) to remove all spindle end play without causing any bearing drag.

NOTE: The lower spindle (12) and steering arm (13) have a blind spline to facilitate correct installation.

MANUAL STEERING SYSTEM

For the purposes of this section, the model MF65 manual steering system will include the cam and lever type gear unit, the housing of which is integral with the transmission top cover; and the upper pedestal assembly which is mounted on the tractor front support. Refer to Fig. MF5.

6. ADJUST GEAR UNIT. Before attempting to adjust the steering gear unit, first make certain that the gear housing is properly filled with lubricant, then disconnect the drag link from the pitman arm to remove load from the gear unit.

7. CAMSHAFT BEARINGS. To check and/or adjust the camshaft bearings, first loosen the adjusting screw locknut (17—Fig. MF6) and back-off the lever shaft adjusting screw (16) at least two full turns. Pull up and push down on the steering wheel to check for camshaft end play. If end play exists, unbolt the steering column from gear housing

and raise the column; then, split and remove sufficient quantity of shims (4—Fig. MF5) to allow camshaft to turn with a barely perceptible drag. Shims are available in thicknesses of 0.002, 0.003 and 0.010.

8. LEVER BACKLASH. With the camshaft bearings adjusted as outlined in the preceding paragraph, turn the steering gear to the mid or straight ahead position (pitman arm in vertical position) and turn adjusting screw (16—Fig. MF6) **in** to obtain a very slight drag when steering gear is rotated through this mid-position. The gear unit should turn freely in all other positions. Tighten lock nut (17) when adjustment is complete.

9. REMOVE AND REINSTALL GEAR UNIT. To remove the steering gear housing assembly, drain cooling system, remove the hood, side panels and grille assembly and disconnect the heat indicator sending unit from

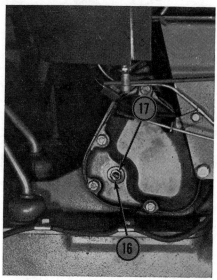

Fig. MF6 — The steering gear lever shaft should be adjusted with screw (16) to provide a slight drag when gear unit is rotated through the mid-position.

16. Adjusting screw
17. Lock nut

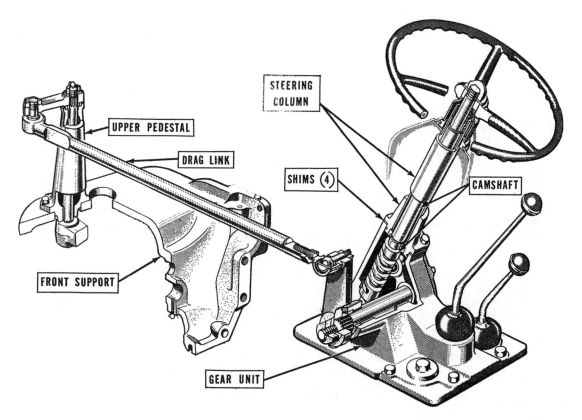

Fig. MF5—Cut-away view of a typical cam and lever type steering gear unit. The pedestal unit shown in this illustration is used on models with manual steering only. The gear unit, however, is the same on models with or without power steering.

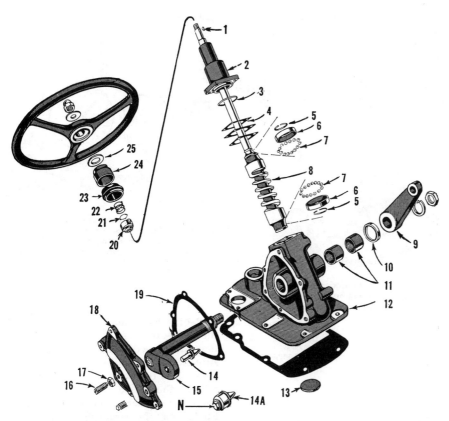

Fig. MF7—Exploded view of the cam and lever type steering gear unit. Camshaft bearings and lever backlash are adjustable.

N. Adjusting Nut	9. Pitman arm	17. Lock nut
1. Woodrufff key	10. Oil seal	18. Side cover
2. Steering column	11. Bushings	19. Gasket
and cover	12. Housing	20. Bearing assembly
3. "O" ring	13. Housing plug	21. Bearing spring
4. Shims	14. Lever stud	seat
5. Retainer ring	(tricycle models)	22. Bearing spring
6. Ball cup	14A. Lever stud	23. Seal
7. Bearing balls	(axle models)	24. Steering column
8. Camshaft	15. Lever shaft	cap
assembly	16. Adjusting screw	25. Felt seal

to the steering gear housing and lift the instrument panel and battery platform assembly from tractor.

10. Disconnect drag link from pitman arm, remove the cap screws retaining the transmission cover to transmission housing and lift the steering gear and transmission cover unit from tractor.

11. **OVERHAUL GEAR UNIT.** The steering gear unit can be overhauled without removing the gear housing assembly from tractor by removing the instrument panel and battery platform assembly as outlined in paragraph 9.

Remove pitman arm from lever shaft and remove the gear housing side cover (18—Fig. MF7). Withdraw the lever shaft and stud assembly. Examine the stud and renew same if damaged or excessively worn.

Note: On axle type tractors, the stud (14A) is mounted in roller bearings, but component parts of the stud assembly are not sold separately. When installing a new lever stud (14A), tighten nut (N) until a rolling torque of 6-8 Inch-Lbs. is required to turn the stud in its bearings.

Remove the steering column and gear housing cover (2) and save shims (4) for reinstallation. Withdraw the camshaft, remove snap rings (5), cups (6) and bearing balls (7). Examine all parts and renew any which are questionable.

New lever shaft bushings (11) should be reamed after installation, if necessary, to provide an inside diameter of 1.6235-1.625. The lever shaft (15) should have a clearance of 0.0005-0.003 in the bushings.

the water outlet elbow. Disconnect cable from starting motor and wires from coil, headlights and generator. Disconnect tractormeter cable from generator, choke rod from carburetor and oil gage line from right side of cylinder block. Shut off the fuel and remove the fuel line. Unbolt the fuel tank from its rear support, loosen the fuel tank front support bolts and block-up between fuel tank and rocker cover.

Note: Some mechanics prefer to completely remove the fuel tank. Disconnect throttle rod and rear ball joint. Disconnect battery cables and remove battery. Disconnect tail light wires and wires from starter safety switch. Unbolt battery platform from engine and remove steering wheel, steering wheel Woodruff key, felt washer (25—Fig. MF7), spring (22), chrome cap (24) and rubber seal (23).

Remove the cap screws retaining the instrument panel and battery platform

Fig. MF8—Typical upper pedestal and drag link installation on all models equipped with manual steering and adjustable type front axle. The installation on tricycle models is similar except for details of the lower steering arm.

When reassembling the unit, adjust the camshaft bearings as in paragraph 7 and the lever backlash as in paragraph 8.

12. R&R AND OVERHAUL UPPER PEDESTAL. To remove the upper pedestal, first remove the grille screens and the left side panel as shown in Fig. MF8. Loosen clamp bolt (B) retaining steering arm to lower end of pedestal shaft and disconnect the drag link at forward end. Remove the cap screws retaining upper pedestal to front support casting and remove pedestal assembly from tractor.

Remove the nut (26—Fig. MF9) retaining steering arm (29) to pedestal shaft and remove the steering arm. Remove pedestal shaft (30), examine all parts and renew any which are damaged or show wear. Ream new bushings (32), after installation, to an inside diameter of 1.5005-1.5015. Pedestal shaft (30) should have a clearance of 0.0005-0.002 in the bushings.

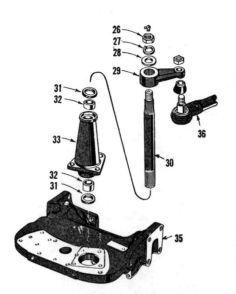

Fig. MF9 — Exploded view of the upper pedestal and components used on model MF65 with manual steering.

26. Nut
27. Lock washer
28. Flat washer
29. Upper steering arm
30. Pedestal shaft
31. Dust seals
32. Bushings
33. Pedestal
35. Front support

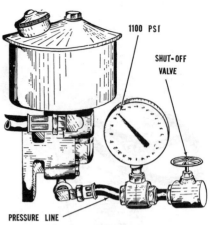

Fig. MF10 — Pressure gage and shut-off valve installation for checking the power steering system operating pressure. At 2000 engine rpm, the pump output pressure should be 1100 psi.

POWER STEERING SYSTEM

NOTE: The maintenance of absolute cleanliness of all parts is of utmost importance in the operation and servicing of the hydraulic power steering system. Of equal importance is the avoidance of nicks or burrs on any of the working parts.

FILLING AND BLEEDING

13. Fluid capacity for the complete power steering system is ⅔ qt. for non-diesel models and 1⅔ qt. for diesel models.

Only automatic transmission fluid, Type A is recommended for use in the power steering system. Reservoir fluid level and the paper filter element in reservoir should be checked every 750 hours of operation or more often in severe dust conditions. Reservoir fluid level should be maintained ¼-½- inch above the filter element.

To bleed the system, fill reservoir, start engine and turn the steering wheel full right and full left several times to bleed air from the system;

then, refill reservoir to the proper level.

TROUBLE SHOOTING

14. The accompanying table lists troubles which may be encountered in the operation of the power steering system. The procedure for correcting most of the troubles is evident; for those not readily remedied, refer to the appropriate subsequent paragraphs.

SYSTEM OPERATING PRESSURE AND RELIEF VALVE

15. A pressure test of the hydraulic circuit will disclose whether the pump, relief valve or some other unit in the system is malfunctioning. To make such a test, proceed as follows:

Connect a pressure test gage and shut-off valve in series with the pump pressure line as shown in Fig. MF10. Note that the pressure gage is connected in the circuit between the

shut-off valve and the pump. Open the shut-off valve and run engine at low idle speed until oil is warmed. Advance the engine speed to 2,000 rpm, close the shut-off valve and retain in the closed position only long enough to observe the gage reading. Pump may be seriously damaged if valve is left in the closed position for more than 10 seconds. If gage reading is 1,100 psi with the shut-off valve closed, the pump and relief valve are O. K. and any trouble is located in the control valve, power cylinder and/ or connections.

If the pump output pressure is more than 1,100 psi, the relief valve is either improperly adjusted or stuck in the closed position. If the output pressure is less than 1,100 psi, either the relief valve is improperly adjusted or the pump requires overhauling. In any event, the first step in eliminating trouble is to adjust the relief valve. This may be accomplished by removing the reservoir cover and filter element on non-diesel models; or cap plug on top of pump (diesels), and turning the relief valve adjusting screw (20—Fig. MF11) either way as required. One turn of the adjusting screw will increase the output pressure approximately 300 psi. Be sure to stake the screw after adjustment is complete. If relief valve adjustment will not restore the pressure with the shut-off valve closed, renew the pump.

PUMP

16. Component parts are not available for the hydraulic pump. If trouble develops therein, renew the pump as an assembly.

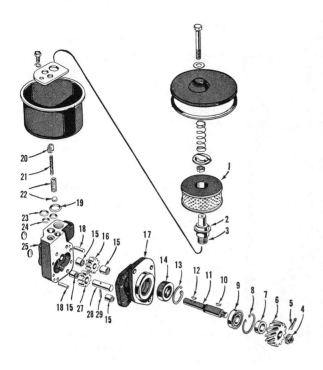

1. Filter element
2. Element seat cup
3. Stud
4. Nut
5. Cotter pin
6. Drive gear
7. Bearing spacer
8. Snap ring
9. Ball bearing
10. Woodruff key
11. Drive shaft
12. Woodruff key
13. Snap ring
14. Oil seal
15. Needle bearings
16. Driven gear
17. Pump body
18. Dowel pin
19. "O" ring
20. Adjusting screw
21. Spring
22. Relief valve spring and ball
23. "O" ring
24. "O" ring
25. Gear housing
27. Follower gear
28. Follower gear shaft
29. Shear pin

the valve components from the body. Thoroughly clean all parts and renew any which are questionable. When reassembling, be sure to renew all "O" rings and seals and install a new seal retainer (6).

18. After the unit is installed on tractor, fill and bleed the system as in paragraph 13 and adjust the linkage as follows: With the engine running at 1500 rpm and pin (P—Fig. MF12) removed, loosen jam nut (N) and tighten the adjusting pin (S) until it bottoms. Loosen the jam nuts on the adjusting link and turn sleeve (B) either way as required until pin (P) can be freely installed. Install cotter pins in pin (P) and tighten the linkage jam nuts. Back-off the adjusting pin (S) seven full turns and tighten jam nut (N).

Now, turn the steering wheel to both extreme positions and make certain the system pressure relief valve opens. If valve will not open when wheel is turned to the right, lengthen the linkage by turning sleeve (B). If valve will not open when wheel is turned to the left, shorten the linkage.

STEERING VALVES

All Models Prior 650 369

17. **R&R AND OVERHAUL.** To remove the power steering valves, open the grille door and disconnect the hose lines from the valve unit. Remove pin (P—Fig. MF12), unbolt and remove the valve unit from cylinder.

To disassemble the unit, disconnect the adjusting linkage from link (10—Fig. MF13) and pry out the dust seal retainer (6). Extract the dust seal (7), remove snap ring (8) and withdraw

All Models After 650 368

18A. As shown in Fig. MF13A, the valve housing is integral with the cylinder end cap (2). To overhaul the unit, remove the grille screens, remove pin (P) and disconnect the hy-

POWER STEERING SYSTEM TROUBLE-SHOOTING CHART

	Loss of Power Assistance	Power Assistance in One Direction Only	Unequal Turning Radius	Erratic Steering Control	Fluid Foaming Out of Reservoir	Unequal Turning Effort
Binding, worn or bent mechanical linkage	★		★	★		
Insufficient fluid in reservoir	★					
Low pump pressure	★					
Sticking or binding valve spool	★	★		★		★
Damaged or restricted hose or tubing	★	★			★	
Wrong fluid in system	★			★	★	
Improperly adjusted tie rods or drag links			★	★		
Steering arms not positioned properly			★			
Air in system				★	★	
Plugged filter element					★	
Internal leak in valve			★		★	★
Faulty cylinder	★			★		
Faulty linkage adjustment			★	★		★
Improper adjusting pin clearance	★			★		

Fig. MF12 — Typical power steering valves, cylinder and pedestal installation on model MF65 prior to Ser. No. 650369. The power steering system incorporates the same cam and lever steering gear as models with manual steering. Refer to text.

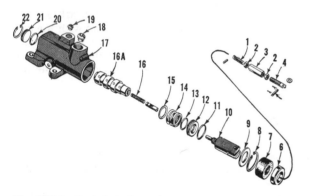

1. Valve link end
2. Jam nut
3. Adjusting sleeve
4. Link end
6. Dust seal retainer
7. Dust seal
8. Snap ring
9. Washer
10. Valve link
11. "O" ring
12. Valve guide
13. "O" ring
14. Outer spring
15. Washer
16. Inner spring
16A. Valve
17. Valve body
18. Connector
19. Connector
20. "O" ring
21. End plug
22. Snap ring

Fig. MF13—Exploded view of the power steering valve unit used on models prior to Ser. No. 650369.

draulic lines. Bump the pin (39) down to release the cylinder rod end and remove the cap screws clamping pin (49) in the cylinder tube end cap (6). Swing the cylinder and valve unit outward, remove pin (49) and withdraw the cylinder and valve unit from tractor.

Remove Allen screws (SS) then insert a small punch into one of the screw holes and into a hole in spool (15). Then, while preventing the spool from turning with the punch, unscrew and remove valve end (18). Remove cap (24) and push the spool (15) out of valve body. Extract seal (22), snap ring (21) and items (19 and 20).

Thoroughly clean all parts and renew any which are questionable. When reassembling, be sure to renew all "O" rings and seals.

18B. After the unit is installed on tractor, fill and bleed the system as in paragraph 13 and adjust the linkage as follows: With engine running at 1500 rpm, and pin (P) removed, loosen jam nut (N) and tighten the adjusting pin (S) until it bottoms. Loosen the jam nuts on the adjusting link and turn sleeve (B) either way as required until pin (P) can be freely installed. Install cotter pins in pin (P) and tighten the linkage jam nuts.

Back-off the adjusting pin (S) seven full turns and tighten jam nut (N).

Now, turn the steering wheel to both extreme positions and make certain the system pressure relief valve opens. If valve will not open when wheel is turned to the right, lengthen the linkage by turning sleeve (B). If valve will not open where wheel is turned to the left, shorten the linkage.

STEERING CYLINDER

All Models Prior 650369

19. R&R AND OVERHAUL. To remove the power steering cylinder, remove the grille screens, remove the cap screws retaining the valve assembly (Fig. MF12) to top of cylinder and move the valve unit out of way. Bump the pin (X) down to release the cylinder rod end and remove the cap screws (CS). Swing the cylinder outward and as far toward the left as possible. Remove pin (Y) and withdraw the cylinder.

Remove ring (2—Fig. MF14) and spacer (3). Push the tube head (8) inward and remove retainer ring (4). Withdraw the rod, piston and head assembly. Remove tube head from rod. When reassembling, renew all "O" rings and seals and be sure retainer rings are properly seated in their grooves. Note: Removal of piston (12) from rod (1) is not recommended.

Before installing the cylinder, remove the rod end bracket (Fig. MF12), extract pin (X), then install the bracket and cylinder. Bleed the power steering system as in paragraph 13.

All Models After 650 368

19A. R&R AND OVERHAUL. To remove the power steering cylinder, remove the grille screens, remove pin (P—Fig. MF13A) and disconnect the hydraulic lines. Bump pin (39) down to release the cylinder rod end and remove the cap screws clamping pin (49) in the cylinder tube end cap (6). Swing cylinder and valve unit outward, remove pin (49) and withdraw the cylinder and valve unit from tractor. Remove tie bolts (33) and disassemble the remaining parts. Thoroughly clean all parts and renew any which are damaged or worn. Be sure to renew all "O" rings and seals.

Before installing the cylinder, remove the rod end bracket (38), extract pin (39), then install the bracket and cylinder. Bleed the power steering system as in paragraph 13.

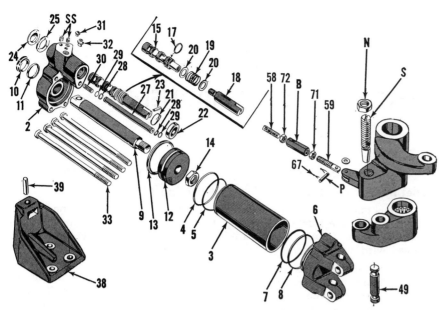

Fig. MF13A — Power steering control valve and cylinder used on models after Serial No. 650368.

2. End cap	14. Nut	28. "O" ring
3. Cylinder tube	15. Spool	29. Back-up washer
4. "O" ring	17. "O" ring	30. Spring
5. Back-up washer	18. Valve end rod	31. Seat
6. Tube end cap	19. Valve compression	32. Seat
7. "O" ring	spring	33. Tie bolts
8. Back-up washer	20. Washer	38. Support
9. Piston rod	21. Snap ring	39. Pin
10. Dust seal	22. Seal	49. Pin
11. "T" ring	23. Pin	58. Sleeve end
12. Piston	24. Cap	59. Sleeve end
13. "T" ring	25. Dust seal	67. Cotter pins
	27. Oil tube	

Fig. MF15 — Power steering pedestal and components.

15. Snap ring	23. Pin
16. Pedestal shaft	24. Bushing
17. Bushing	25. Bushing
18. Pedestal	26. Adjusting pin
19. Dust seal	27. Nut
20. Steering arm	28. Bushing
21. Bushing	29. Valve connecting
22. Cylinder	arm
connecting arm	

STEERING PEDESTAL

20. R&R AND OVERHAUL. The power steering pedestal is shown installed in Fig. MF12. To remove the pedestal, first remove the grille screens and the left side panel. Disconnect drag link from pedestal arm and remove pin (P).

Remove the cap screws (CS) and bump pin (X) downward to release the cylinder rod end. Swing the cylinder and valve assembly outward and

as far toward left as possible. Remove pin (Y) and lay cylinder and valve assembly out of way. Loosen the cap screw retaining the steering arm to lower end of pedestal shaft, unbolt pedestal from the front support casting and lift the pedestal assembly from tractor.

NOTE: There are no blind splines to facilitate correct assembly of the cylinder connecting arm (22—Fig. MF15) with respect to pedestal shaft (16). It

is important, therefore, that the relative position of the two parts be marked or otherwise identified before disassembly. Remove snap ring (15) and press the pedestal shaft (16) down and out of pedestal. Be sure to renew any parts which show an excessive amount of wear. All bushings are pre-sized and available for service installation.

After pedestal is installed on tractor, adjust the linkage as outlined in paragraph 18 or 18B.

GEAR UNIT

21. The gear unit used on models with power steering is the same as the unit used on models without power steering. The procedure for adjusting the unit is outlined in paragraphs 6, 7 and 8. Removal, reinstallation and overhaul of the unit is outlined in paragraphs 9, 10 and 11.

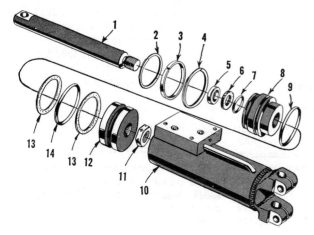

Fig. MF14 — Exploded view of model MF65 power steering cylinder used on models prior to Serial No. 650369.

1. Piston rod
2. Retainer ring
3. Spacer
4. Retainer ring
5. Wiper seal
6. Leather washer
7. "O" ring
8. Tube head
9. "O" ring
10. Cylinder
11. Nut
12. Piston
13. Back-up washer
14. "O" ring

NON-DIESEL ENGINE AND COMPONENTS

R&R ENGINE WITH CLUTCH

Model MF-65 Non-Diesel

22. To remove the engine and clutch, first drain cooling system and if engine is to be disassembled, drain oil pan. Disconnect head light wires, remove hood and on models so equipped, disconnect radius rod and/or drag link and power steering oil lines. Disconnect radiator hoses. Support tractor under transmission housing and unbolt front support casting from engine. Roll front axle, support and radiator as an assembly away from tractor. Shut-off fuel and remove fuel tank. Disconnect the heat indicator sending unit from water outlet elbow and cable from starting motor. Disconnect wires from coil and generator, tractormeter cable from generator, choke rod from carburetor and oil gage line from right side of cylinder block. Remove the air cleaner pipe, loosen the front end of throttle rod and disconnect exhaust pipe from manifold. Disconnect battery cables and remove battery. Support engine in a hoist and unbolt engine from transmission case.

CYLINDER HEAD

Model MF-65 Non-Diesel

Gasoline and LP-Gas cylinder heads are not interchangeable.

23. **REMOVE AND REINSTALL.** To remove the cylinder head, drain cooling system, shut off fuel and remove fuel tank. Disconnect governor-to-carburetor rod and loosen front end of throttle rod. Disconnect choke rod from carburetor, loosen carburetor to air cleaner pipe hose and unbolt exhaust pipe from manifold. Unbolt manifold from cylinder head and remove manifold and carburetor assembly. Remove upper radiator hose, rocker arm cover and rocker arms assembly. Unbolt and remove cylinder head from tractor.

When reinstalling the cylinder head, tighten the stud nuts from the center cutward and to a torque value of 70-75 Ft.-Lbs. Valve tappet gap is given in paragraph 24.

VALVES AND SEATS

Model MF-65 Non-Diesel

24. Valve tappet gap should be set hot to 0.016 for the intake and 0.018 for the exhaust. All valves are equipped with stem caps.

Intake valves seat directly in cylinder head and the valve stems are equipped with neoprene oil guards. Exhaust valves are equipped with renewable type seat inserts and on gasoline models, the exhaust valves are fitted with positive type rotators. Replacement exhaust valve seat inserts are 0.010 oversize. Desired interference fit of new inserts is 0.003.

Intake valves have a face and seat angle of 30 degrees with a desired seat width of $\frac{3}{32}$ inch. Exhaust valves have a face angle of 44 degrees and a seat angle of 45 degrees. Desired exhaust valve seat width is $\frac{3}{32}$ inch. Seats can be narrowed, using 15 and 75 degree stones.

Valve stem diameter is 0.3141-0.3149 for the intake, 0.3124-0.3132 for the exhaust.

VALVE GUIDES

Model MF-65 Non-Diesel

25. The pre-sized intake and exhaust valve guides are interchangeable and can be driven from cylinder head if renewal is required. Guides should be pressed into the cylinder head, using a piloted drift 0.002 smaller than bore of guide, until port end of guide is $\frac{11}{16}$ inches from spring seat in cylinder head.

Desired valve stem clearance in guides is 0.0006-0.0029 for the intake, 0.0030-0.0053 for the exhaust.

VALVE SPRINGS

Model MF-65 Non-Diesel

26. Intake and exhaust valve springs are interchangeable. Renew any spring which is rusted, discolored or does not meet the test specifications which follow:

Free length .... 2 5/64"
Lbs. test @ $1\frac{21}{32}$".......... 41- 47
Lbs. test @ $1\frac{7}{32}$".......... 103-110

VALVE ROTATORS

Gasoline Models

27. Normal servicing of the positive type exhaust valve rotators ("Rotocaps") consists of renewing the units. It is important, however, to observe the valve action after engine is started. The valve rotator action can be considered satisfactory if the valve rotates a slight amount each time the valve opens.

VALVE TAPPETS

Model MF-65 Non-Diesel

28. Mushroom type tappets (cam followers) operate directly in machined bores of the cylinder block. Refer to Fig. MF18. The 0.5615-0.5620 diameter tappets are furnished in standard size only and should have a clearance of 0.0005-0.002 in the block bores. Tappets can be removed after removing camshaft as outlined in paragraph 33.

Refer to paragraph 24 for valve tappet gap.

ROCKER ARMS

Model MF-65 Non-Diesel

29. Rocker arms and shaft assembly can be removed after removing fuel tank and rocker arm cover. The rocker arms, being right and left hand assemblies, are not interchangeable. Refer to Fig. MF18. Desired clearance between new rocker arms and new rocker arm shaft is 0.0002-0.0014. Renew shaft and/or rocker arms if clearance exceeds 0.003. Diameter of new rocker arm shaft is 0.9678-0.9685. Oil holes in rocker arm shaft face toward valve spring.

Refer to paragraph 24 for valve tappet gap.

VALVE TIMING

Model MF-65 Non-Diesel

30. Valves are properly timed when single punch marked tooth on crankshaft gear is meshed with double punch marked tooth space on camshaft gear. Refer to Fig. MF19.

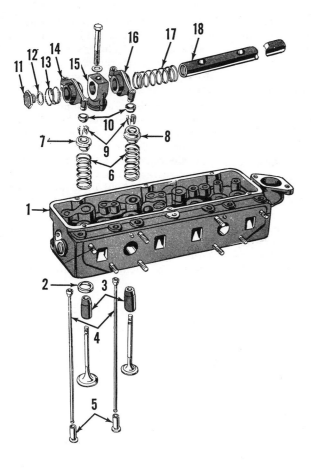

Fig. MF18 — Exploded view of MF65 non-diesel cylinder head, valves and rocker arms assembly.

1. Cylinder head
2. Exhaust valve seat insert
3. Valve guide
4. Push rod
5. Valve tappet
6. Valve spring
7. Roto-Cap assembly
8. Intake valve spring seat
9. Valve stem lock
10. Ball socket
11. Shaft plug
12. Gasket
13. Spring
14. L. H. rocker arm
15. Rocker arm support
16. R. H. rocker arm
17. Rocker arm spring
18. Rocker arm shaft

dersize to facilitate obtaining the desired backlash. Gear markings are as follows: Standard, "S"; Undersize, "U"; Oversize, "O".

During installation, mesh the single punch marked tooth on crankshaft gear with the double punch marked tooth space on camshaft gear and when camshaft gear is being drifted on, it is advisable to remove oil pan and buck up the shaft with a heavy bar.

Tighten the camshaft gear retaining nut to a torque of 65-70 Ft.-Lbs.

CAMSHAFT

Model MF-65 Non-Diesel

33. To remove the camshaft, first remove the camshaft timing gear as outlined in paragraph 32. Remove fuel tank, rocker arm cover, rocker arms and shaft assembly and push rods. Remove the ignition distributor and oil pan and block-up or support tappets (cam followers). Remove the camshaft thrust plate (12—Fig. MF20) and withdraw camshaft from front of engine.

The three main camshaft journals ride directly in machined bores in cylinder block. Outboard bushing (7) is renewable. Camshaft bearing journals have a normal running clearance of 0.003-0.0045. If running clearance exceeds 0.0055, renew camshaft and/or cylinder block or bushing.

Camshaft journal diameters are as follows:

Front	1.9370-1.9375
Center	1.7495-1.7500
Rear	1.6870-1.6875

To check valve timing when engine is assembled, first rotate crankshaft until both valves of No. 1 cylinder are closed (top of compression stroke). Set No. 1 exhaust valve tappet gap to 0.025 and No. 1 intake valve tappet gap to 0.017. Continue rotating crankshaft until No. 1 exhaust opens fully and then starts to close. Further movement of crankshaft will close exhaust valve completely; backing up will close intake valve. At this point ONLY, when both valves are open (further ascertained by tightening of push rods), the "TDC" mark on flywheel should be in register (within five degrees) with groove in timing hole on left side of engine.

Twelve degrees early or late indicates the timing is off one gear tooth.

Reset valves to the proper tappet gap given in paragraph 24.

TIMING GEAR COVER

Model MF-65 Non-Diesel

31. To remove the timing gear cover, drain cooling system, disconnect head light wires and remove hood assembly. Disconnect radiator hoses and on models so equipped, disconnect the radius rods and/or drag link and

power steering oil lines. Support tractor under transmission housing, unbolt the front axle support from engine and roll the front axle, support and radiator as an assembly away from tractor. Remove fan blades and disconnect spring and rod from governor lever. Remove starting jaw and crankshaft pulley. Remove the cap screws retaining oil pan to timing gear cover and loosen the remaining oil pan cap screws. Unbolt and remove timing gear cover from engine.

The crankshaft front oil seal (26—Fig. MF20) can be renewed at this time and should be installed with lip facing the engine.

TIMING GEARS

Model MF-65 Non-Diesel

32. Timing gears can be renewed after removing timing gear cover as outlined in paragraph 31. Remove the governor assembly (38—Fig. MF20) and using a suitable puller, remove the crankshaft timing gear (27). The procedure for removing the camshaft gear is evident.

Recommended timing gear backlash is 0.001-0.003. Gears are available in sizes from 0.006 oversize to 0.006 un-

Fig. MF19—MF65 non-diesel timing gear train. Mesh single punched marked tooth on crankshaft gear with double punch marked tooth space on camshaft gear.

To install camshaft, reverse the removal procedure.

Desired camshaft end play of 0.003-0.007 is controlled by thrust plate (12). Renew the plate if end play exceeds 0.010. The extreme front end of camshaft rides in bushing (7) which is located in support (6). Renew the bushing if it shows any wear.

ROD AND PISTON UNITS

Model MF-65 Non-Diesel

34. Connecting rod and piston units are removed from above after removing cylinder head and oil pan. Pistons and rods are installed with the rod correlation marks facing the camshaft.

Replacement rods are not marked and should be installed with the oil spray hole (lower end of rod) facing away from camshaft side of engine. Tops of pistons are installed with arrow marked front facing toward front of engine.

Connecting rod bolt torque is 35-40 Ft.-Lbs.

PISTONS, SLEEVES AND RINGS

Model MF-65 Non-Diesel

35. Aluminum alloy, cam ground pistons are supplied only in the standard size and are available only in a sleeve and piston kit consisting of pistons, pins, rings and sleeves. Recommended piston skirt clearance is 0.002.

Piston clearance is checked with a spring scale pull of 5-10 lbs., using a 0.002 x ½ inch feeler gage. Cylinder sleeves must be renewed when the difference between the worn and unworn portions exceeds 0.008.

With the piston and connecting rod assembly removed from the cylinder block, use a suitable puller to remove the sleeve. Before installing the wet type sleeve, clean all cylinder block sealing surfaces. The top of the sleeve should extend 0.001-0.003 above the top surface of the cylinder block. If this standout is in excess of 0.003, check for foreign material under sleeve flange. Excessive standout will cause water leakage at cylinder head gasket. To facilitate installation of

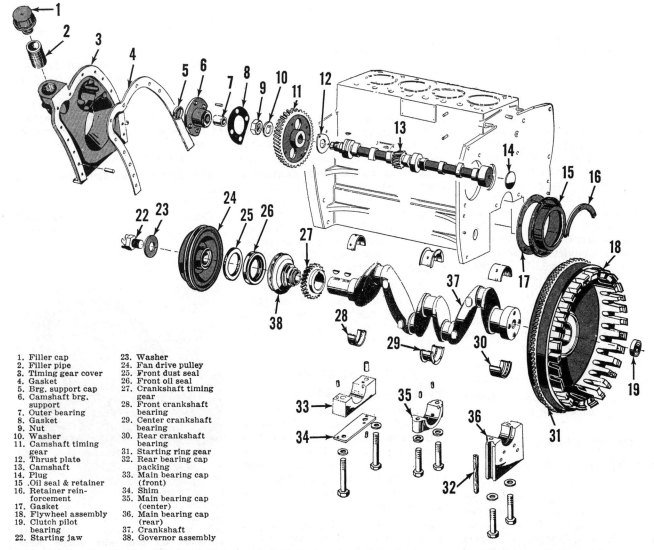

1. Filler cap
2. Filler pipe
3. Timing gear cover
4. Gasket
5. Brg. support cap
6. Camshaft brg. support
7. Outer bearing
8. Gasket
9. Nut
10. Washer
11. Camshaft timing gear
12. Thrust plate
13. Camshaft
14. Plug
15. Oil seal & retainer
16. Retainer reinforcement
17. Gasket
18. Flywheel assembly
19. Clutch pilot bearing
22. Starting jaw
23. Washer
24. Fan drive pulley
25. Front dust seal
26. Front oil seal
27. Crankshaft timing gear
28. Front crankshaft bearing
29. Center crankshaft bearing
30. Rear crankshaft bearing
31. Starting ring gear
32. Rear bearing cap packing
33. Main bearing cap (front)
34. Shim
35. Main bearing cap (center)
36. Main bearing cap (rear)
37. Crankshaft
38. Governor assembly

Fig. MF20—Non-diesel camshaft, crankshaft and timing gears. Crankshaft end play is controlled by the flanged center bearing.

sleeves, use a lubricant (palm oil or vaseline) on the two neoprene sealing rings.

There are three ⅛ inch wide compression rings and one $\frac{3}{16}$ inch wide oil control ring per piston. Recommended end gap is 0.010-0.020 for the oil ring and compression rings. Recommended side clearance for the top compression ring is 0.003-0.0045; for 2nd and 3rd compression rings, 0.002-0.0035; for oil control ring, 0.001-0.0025.

Standard size cylinder bore is 3.578.

PISTON PINS
Model MF-65 Non-Diesel

36. The 1.125-1.1252 diameter floating type piston pins are retained in piston bosses by snap rings and are available in standard size only. The bushing in the upper end of connecting rod has a very thin wall and if sized by reaming, do so only with a fluted type, taking very light cuts. Be sure oil hole in bushing registers with oil hole in top end of the connecting rod and clean same thoroughly after sizing the bushing. Pin should be fitted to a 0.0001-0.0005 clearance in the rod and should be a thumb press fit in the piston when same is heated to 160 degrees.

CONNECTING RODS AND BEARINGS
Model MF-65 Non-Diesel

37. Connecting rod bearings are of the shimless, non-adjustable, slip-in precision type renewable from below after removing oil pan. When installing new bearing shells, be sure that

the projection engages milled slot in rod and cap and the rod and rod cap correlation marks are in register. Replacement rods are not marked and should be installed with the oil spray hole facing away from camshaft side of engine. Bearings are available in 0.002, 0.010 and 0.020 undersize, as well as standard.

Crankpin diameter	2.0615-2.0625
Running clearance	0.0007-0.0027
Side clearance	0.006 -0.010
Rod length C to C	6.998-7.002
Cap screw torque	35-40 Ft.-Lbs.

CRANKSHAFT AND BEARINGS
Model MF-65 Non-Diesel

38. Crankshaft is supported on 3 shimless, non-adjustable, slip-in, precision type main bearings, renewable from below without removing the crankshaft. The rear main bearing cap contains packing (32—Fig. MF20) on each side of the filler block to prevent oil leakage into the clutch housing. This packing is in addition to the separate crankshaft rear seal (15). To remove the rear main bearing cap and filler block, first remove the two cap screws which retain the crankshaft rear oil seal retainer to the bearing cap and then remove two bearing cap retaining screws.

Bearing inserts are available in standard, 0.002, 0.010 and 0.020 undersizes.

Normal crankshaft end play of 0.004-0.008 is controlled by the flanged center main bearing.

To remove crankshaft it is necessary to remove engine, clutch, flywheel, rear oil seal, timing gear cover, oil pan, and main bearing caps.

Check the crankshaft journals for wear, scoring and out-of-round condition against the values listed below:

Journal diameter	2.374-2.375
Running clearance	0.0005-0.0027
Cap screw torque	85-95 Ft.-Lbs.

CRANKSHAFT REAR OIL SEAL
Model MF-65 Non-Diesel

39. Crankshaft rear oil seal (15—Fig. MF20) is contained in a one piece retainer and serviced only as an assembly. To renew the seal, first separate the engine from the transmission

case as outlined in paragraph 122, and remove the flywheel. Remove the three seal retainer to crankcase cap screws and the two seal retainer to rear main bearing retaining cap screws.

FLYWHEEL
Model MF-65 Non-Diesel

40. To remove flywheel, separate the engine from transmission case as outlined in paragraph 122, then remove the clutch unit from flywheel. The starter ring gear can be renewed after removing the flywheel. To install a new ring gear, heat same to 500 deg. F. and install on flywheel with beveled end of teeth facing timing gear end of engine. One flywheel mounting bolt hole is off-center.

OIL PUMP
Model MF-65 Non-Diesel

41. Gear type pump, shown in Fig. MF21, is bolted to bottom of number one main bearing cap and is gear driven from crankshaft pinion. Pump is accessible after oil pan is removed. Shims interposed between pump body and main bearing cap are varied to obtain the desired 0.010-0.012 backlash between the crankshaft pinion and the oil pump driven gear.

Check the pump internal gears for wear or scoring and renew any damaged parts. Pump shaft bushings are not available for service. If shaft running clearance exceeds 0.005, renew entire pump.

RELIEF VALVE
Model MF-65 Non-Diesel

42. Plunger type relief valve (10—Fig. MF21) is located in oil pump body and can be adjusted with spacers inserted under the spring so as to maintain a pressure of 20-30 psi at 2200 engine rpm. At idle speed, pressure should be not lower than 15 psi. The piston of the relief valve should have a slide fit in its bore. The relief valve spring should have a free length of 2 inches and should check 7¾-8¼ pounds at a working length of 1⅜ inches.

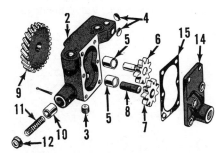

Fig. MF21—Exploded view of non-diesel oil pump.

2. Pump body	10. Pressure relief
3. Plug	valve
4. Expansion plug	11. Pressure relief
5. Bushing	valve spring
6. Drive gear	12. Pressure spring
7. Follow gear	retainer
8. Follow gear shaft	14. Pump cover
9. Pump drive gear	15. Gasket

DIESEL ENGINE AND COMPONENTS

Perkins diesel engines are used in all diesel models. All engines have thin-walled cylinder sleeves and a piston displacement of 203.5 cubic inches. During the 1961 production, the engine design was changed and, although the general appearance is similar, many of the parts are not interchangeable.

The early engine (Model 4A-203) utilized an indirect injection system, with the injected fuel going half into a spherical combustion chamber located in the cylinder head, and the other half into the cylinder area above the piston. A two orifice injection nozzle was used.

The late engine (Model AD4-203) features a direct injection system in which the entire fuel charge is directed into the cylinder area above the piston, the combustion chamber being formed by an identation in top of piston.

For the purposes of this manual, the two engines will be identified in the manner used by Massey-Ferguson, Inc., in their parts books and literature; the early engine being designated as "Regular Diesel"; and the late engine as "Direct Injection Diesel."

R&R ENGINE WITH CLUTCH

All Diesel Models

43. To remove the engine with clutch, first drain the cooling system and if engine is to be disassembled, drain oil pan. Disconnect the headlight wires and remove the hood and side panels. Disconnect drag link and on tractors so equipped, unbolt power steering reservoir bracket from cylinder head and fan shroud, and power

steering pump from timing gear case. Lay pump and reservoir in grille enclosure and disconnect air cleaner and radiator hoses. Place a support under transmission case and support front end assembly from top of steering pedestal. Remove bolts securing front support to engine and roll front end assembly away from engine as a unit. Note: Generator will need to be removed for access to right upper support bolt.

Disconnect and remove batteries and battery clamps. Shut off fuel then unbolt and remove fuel tank. Unbolt and remove starting motor and disconnect wiring harness from cold starting element. Disconnect heat indicator sending unit from outlet elbow and tractormeter cable from engine block. Unbolt and remove exhaust pipe. Disconnect fuel pressure and return lines from secondary filter and injector pump linkage from pump. Support engine in a hoist and unbolt engine from transmission case.

CYLINDER HEAD

All Diesel Models

44. **REMOVE AND REINSTALL.** To remove the cylinder head, first remove hood and side panels. Remove the batteries and battery clamps, shut off fuel, then unbolt and remove fuel tank. Drain the cooling system, disconnect upper radiator hose and remove the heat indicator sending unit from the water outlet elbow. Disconnect injector lines from injectors and

pump adapters and remove injectors and lines. Disconnect the exhaust pipe from the manifold. Disconnect the air inlet hose from intake manifold, wire from cold starting unit, and external oil feed line from head and block. Remove the rocker arm cover and rocker arms assembly, then unbolt and remove the cylinder head.

Mushroom type cam followers operate directly in machined bores in the cylinder head and are prevented from falling out as the cylinder head is raised by the tappet adjusting screw lock nuts.

The center head stud on regular diesel engine is machined to a larger diameter and must always be installed in the center position. The purpose of this stud is to prevent installation of the Perkins P-4 head gasket which looks similar.

When installing cylinder head, coat both sides of gasket with a suitable gasket compound. Gasket is marked "Top Front" for proper installation. Tighten cylinder head nuts to a torque of 55-60 Ft.-Lbs., using the sequence shown in Fig. MF23 for regular diesel engines; or Fig. MF24 for direct injection diesel engines. Head should be re-torqued and valves adjusted after engine is warm.

VALVES AND SEATS

All Diesel Models

45. Intake and exhaust valves are not interchangeable and seat directly in the cylinder head. Valve heads and

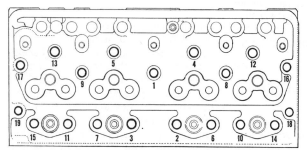

Fig. MF23 — On regular diesel models tighten the cylinder head nuts to a torque of 55-60 Ft.-Lbs. in the sequence shown. Head studs along left side also serve as injector mounting studs, so a deep socket is required.

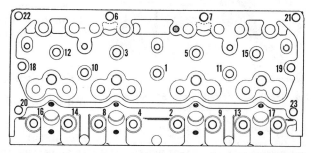

Fig. MF24—On direct injection diesels, tighten the cylinder head stud nuts in the sequence shown. Recommended tightening torque is 55-60 Ft.-Lbs.

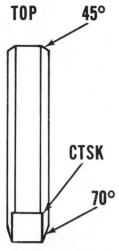

TOP　　45°

CTSK

70°

Fig. MF25 — Cross sectional view of the shoulderless valve guide used in late models and available as a service replacement for early engines. Install guide with countersunk end to gasket surface of head as shown.

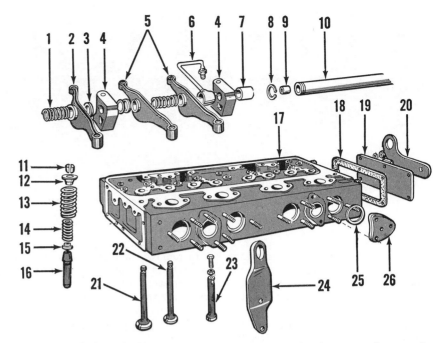

Fig. MF26A — Cylinder head, rocker arms, valves and associated parts used on regular diesel models.

1. Spring	10. Rocker shaft	15. Lower spring seat
2. RH rocker	11. Split cone keepers	16. Valve guide
3. Spacer	12. Spring retainer	17. Cylinder head
4. Bracket	13. Outer valve spring	18. Gasket
5. LH rocker	14. Inner valve spring	19. End plate
6. Oil feed pipe		20. Engine lifting plate
7. Spacer		21. Exhaust valve
8. Snap ring		
9. Plug		

22. Intake valve
23. Tappet
24. Engine lifting plate
25. Combustion chamber gasket
26. Combustion chamber cap

seat locations are numbered consecutively from the front of the engine. Any replacement valves should be so marked prior to installation. Intake and exhaust valves have a face angle of 44 degrees, a seat angle of 45 degrees and a desired seat width of $\frac{1}{16}$-inch. Seats can be narrowed, using 20 and 70 degree stones.

Valve heads should be recessed a specified amount into the cylinder head. Clearance can be measured using a straight edge and feeler gage, by placing the straight edge across the clean gasket surface of head, then measuring to top of valve head. Production clearances are held within the limits of 0.061-0.074. A maximum clearance of 0.140 is permissible before renewing valve or seat providing other conditions of satisfactory valve service are met. Inlet and ex-

haust valve tappet gap should be set to 0.010, hot.

Valve stem diameter is 0.311-0.312 for both the intake and exhaust.

VALVE GUIDES

All Diesel Models

46. The pre-sized intake and exhaust valve guides are interchangeable and can be pressed or driven from the cylinder head if renewal is required.

Engines before engine serial num-

ber 2814187 were equipped with a shouldered type valve guide. Install by pressing into cylinder head until machined shoulder contacts the head.

Regular diesel engines after serial number 2814186, and all direct injection diesels use a shoulderless guide which can also be used as a service replacement for the shouldered guides provided the correct spring seat (12 —Fig. MF26) is used. A cross section of the shoulderless guide is shown in Fig. MF25. Guides should be installed with counterbored end toward head of valve, until top of guide extends 0.584-0.594 above surface of head.

Desired valve stem diametral clearance in the guide is 0.002-0.0045 for both the intake and exhaust.

VALVE SPRINGS

All Diesel Models

47. Springs, retainers and locks are interchangeable for the intake and exhaust valves. The valves are fitted with an inner and outer spring as shown in Fig. MF26. Springs may be installed on the valve with either end up. The inner spring has a shorter assembled length than the outer due to the seating washer (15) and the ma-

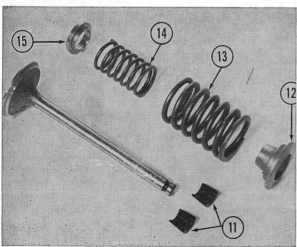

Fig. MF26—Valve, spring and retainer arrangement used on diesel engines. Spring seat (15) and retainer (12) have milled steps to apply proper tension to shorter inner spring.

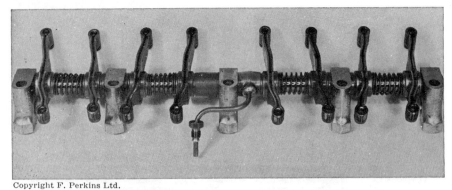

Copyright F. Perkins Ltd.

Fig. MF27——Diesel rocker shaft showing correct assembly of components.

chined step on the spring retainer (12). Due to the difference in construction of the shouldered and shoulderless guides, the correct seating washer (15) must be used.

Renew the springs if they are distorted, discolored, or fail to meet the test specifications which follow:

INNER SPRING:
Approx. free length......1⅜ inches
Lbs. test @ 1³⁄₁₆ inches.......8 min.
Lbs. test @ ²⁷⁄₃₂ inches.........21-25

OUTER SPRING:
Approx. free length.....1²⁵⁄₃₂ inches
Lbs. test @ 1½ inches.......22-23
Lbs. test @ 1⁵⁄₃₂ inches.......48-52

VALVE TAPPETS

All Diesel Models

48. The mushroom type tappets (cam followers) operate directly in machined bores in the cylinder head. The 0.6223-0.6238 diameter tappets are furnished in standard size only and should have a diametral clearance of 0.0008-0.0035 in the cylinder head bores.

Fig. MF28—Rear view of rocker shaft used on diesel engines. During original assembly, punch mark (P) is placed on rear support bracket to indicate position of slot (S) in shaft for proper oil supply. Maximum lubrication is provided when slot is horizontal.

To remove the tappets after the cylinder head is removed, first remove the adjusting screw and lock nut, then withdraw the tappet from its bore.

ROCKER ARMS

All Diesel Models

49. Rocker arms and shaft assembly can be removed after removing the fuel tank and rocker arm cover. The rocker arms are right and left hand assemblies and should be installed on the shaft as shown in Fig. MF27. Desired diametral clearance between new rocker arms and new rocker arm shaft is 0.0008-0.0035. Renew shaft and/or rocker arm if clearance is excessive.

The amount of oil circulating to the rocker arms is regulated by the rotational position of the rocker shaft in the support brackets. This position is indicated by a slot in the rear end of the rocker shaft as shown in Fig. MF28. When the slot is positioned horizontally the maximum oil circulation is obtained. In production, this slot is positioned 30 degrees from the vertical, and the position indicated by a punch mark on the rear support bracket. When reassembling, position the rocker shaft slot as indicated by the punch mark and check the rocker arms assembly for ample but not excessive lubrication.

VALVE TIMING

All Diesel Models

50. To check the valve timing when engine is assembled, set the intake valve clearance on No. 4 cylinder to 0.012, insert a 0.002 feeler gage in No. 4 intake valve tappet gap and turn the engine in the normal direction of rotation until a distinct drag is felt on the feeler gage. At this time the crankshaft should be at 13 degrees before top dead center; or 1⅝ inches before the TDC mark on the flywheel.

Fig. MF29 — Front view of direct injection engine showing assembled timing gear cover. Lower portion (4) is removed with oil pan.

1. Upper cover
2. Inspection cover
3. Front seal retainer
4. Lower cover

Fig. MF30—Lower cover, including the two clamp screws (C), should be loosened when removing oil seal retainer (3).

TIMING GEAR COVER

Regular Diesel

51. To remove the timing gear cover, drain cooling system, disconnect headlight wires and remove the hood and side panels. Disconnect drag link, and on tractors so equipped unbolt power steering reservoir bracket from cylinder head and fan shroud and

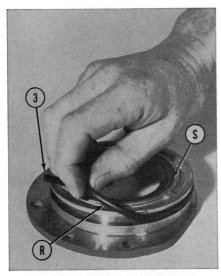

Fig. MF31—Oil leakage around outside of retainer (3) is prevented by "O" ring seal (R). Lip seal (S) should be installed from front, with sealing lip toward rear.

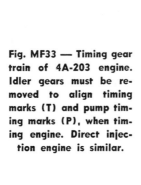

Fig. MF33 — Timing gear train of 4A-203 engine. Idler gears must be removed to align timing marks (T) and pump timing marks (P), when timing engine. Direct injection engine is similar.

Copyright F. Perkins Ltd.

Fig. MF32—Use the crankshaft pulley (P) as a pilot when installing front seal retainer (3). Lower timing gear cover should be loosened to prevent damage to "O" ring (R).

power steering pump from timing gear case. Lay pump and reservoir in grille enclosure and disconnect air cleaner and radiator hoses. Unbolt and remove generator. Support tractor under transmission housing and front end assembly from top of steering pedestal. Unbolt and remove front support and front end assembly as a unit.

Remove water pump and fan assembly and unstake and remove starting jaw and crankshaft pulley. Remove the 22 cap screws retaining the front cover to the timing gear housing and remove the cover. Crankshaft front oil seal may be renewed at this time.

Direct Injection Diesel

52. The two-piece timing gear cover also contains a separate injection pump gear cover and front oil seal retainer as shown in Fig. MF29. To remove the timing gear cover or front oil seal, it is first necessary to remove the entire front end as a unit as outlined in paragraph 51 for earlier model tractors.

If oil seal only is to be renewed, it will not be necessary to remove the main timing gear cover. To renew the oil seal, remove the cap screw from front end of crankshaft and withdraw the crankshaft pulley. Loosen the two cap screws (C—Fig. MF30), and remove the remainder of the cap screws attaching lower cover (4—Fig. MF29) to oil pan and the screws attaching oil seal retainer (3) to upper cover. The front oil seal retainer can now be removed as shown in Fig. MF30.

Remove and discard the "O" ring (R—Fig. MF31). Press the seal (S) into retainer with seal lip toward rear, until front of seal is $\frac{3}{32}$-inch from front face of bore.

When reinstalling the retainer, use the crankshaft pulley as a pilot as shown in Fig. MF32. Carefully work

the "O" ring into bore of upper cover. Loosely install the cap screws retaining the lower cover and the retainer, then tighten the two clamping cap screws (C—Fig. MF30). The remainder of the assembly may now be completed.

The lower cover (4—Fig. MF29) may be removed separately to examine the oil pump drive gears; or with the oil pan as outlined in paragraph 71, without disturbing the remainder of the covers.

The main timing gear cover (1) may be removed after removing the seal retainer (3) as previously outlined, then removing the water pump. Lower cover (4) must be removed for main cover installation to prevent damage to the seal retainer "O" ring (R—Fig. MF31).

TIMING GEARS

All Diesel Models

53. Timing gears can be renewed after removing timing gear cover as outlined in paragraph 51 or 52. Also, remove the rocker arms assembly before removing any of the timing gears, to prevent valve and piston interference.

19

Copyright F. Perkins Ltd.

Fig. MF34 — One cap-screw hole in camshaft flange is marked with a stamped letter "D" as shown. One attaching hole in gear is similarly marked. Align the two marked holes when installing the camshaft gear.

Copyright F. Perkins Ltd.

Fig. MF37—Injection pump drive used on 4A 203 engine. Align timing marks (P) when installing drive gear (D). Master spline (S) on pump drive shaft enters a mating spline in adapter (E). Snap ring (C) retains the assembly on mounting boss.

Note: Due to the odd number of teeth in the two idler gears, the complete set of timing gear marks will align only once in more than 2000 engine revolutions. Therefore, if any of the gears in the timing gear train are to be removed, or if retiming is indicated, both idler gears must be removed.

Recommended timing gear backlash is 0.003-0.006 between any two gears in the timing gear train. The complete timing gear train with timing marks aligned is shown in Fig. MF33.

54. CAMSHAFT GEAR. The cap screw holes in the camshaft and camshaft timing gear are evenly spaced and the timing gear can be installed in three positions; only one of which is correct. A letter "D" is stamped on the camshaft hub flange and gear

web near one of the cap screw holes as shown in Fig. MF34. The camshaft gear is correctly installed when the two "D" marks are aligned.

55. INJECTION PUMP DRIVE. On regular diesel models, the injection pump driving mechanism is a three-piece unit as shown in Figs. MF35, 36 and 37. Boss (A—Fig. MF35) is an integral part of the injection pump carrier plate and can be renewed by removing the injection pump and installing a new carrier. Gear hub (B) is retained to the boss by snap ring (C—Fig. MF36). Install drive gear (D—Fig. MF37) and adapter (E) by aligning the master spline (S) in the adapter with the pump drive shaft; then, turn adapter so that pump

timing marks (P) align. Install and tighten retaining cap screws. Note: Timing marks (P) on adapter and gear face are not scribed on parts stock items; therefore, when renewing these parts, align and time the injection pump as outlined in paragraph 94, then scribe the timing marks to simplify future service.

56. On direct injection diesel models, the injection pump drive gear (G—Fig. MF38) is attached directly to the injection pump drive shaft (P) with cap screws. The locating dowel pin (D) aligns with machined slot (S) to facilitate timing. Timing marks of ALL the gears in timing gear train must be aligned as shown in Fig. MF33, when installing the injection pump drive gear.

57. CRANKSHAFT GEAR. The crankshaft gear does not normally have a tight fit on the shaft and can usually be removed by using a small pry bar behind the gear. If gear cannot be removed by using pry bars, it will be necessary to use a suitable puller.

58. IDLER GEAR AND HUBS. Idler gears should have 0.001-0.0035 diametral clearance on the idler hubs. Idler hub retainer plates and the flanged bushings in the lower idler are renewable.

Idler hubs are a light press fit in the engine block and may be pried out if renewal is indicated. When renewing the hubs, make sure the oil feed hole

Copyright F. Perkins Ltd.

Fig. MF35—On early models with 4A 203 engine, the injection pump timing gear boss (A) is an extension of pump carrier plate. Hub (B) has three unevenly spaced, threaded holes for attaching drive gear.

Copyright F. Perkins Ltd.

Fig. MF36 — Attach hub to carrier plate boss with snap ring (C).

Fig. MF38—Injection pump drive gear arrangement used on direct injection engines. Dowel (D) in pump drive gear (G) fits in milled slot (S) in injection pump shaft (P) for correct pump timing. Timing marks (T) in the entire timing gear train must be aligned as shown in Fig. MF33, when installing gear.

in same aligns with the oil passage in the block as shown in Fig. MF39. Also, make sure that oil holes (A and B—Fig. MF40) are open and clean.

59. **TIMING.** With the camshaft gear, crankshaft gear and injection pump drive gear installed, install the idlers and time the engine as follows: Turn the crankshaft until the gear timing mark aligns with the center line of the lower idler hub and install the lower idler so that one timing mark aligns with the mark on the crankshaft gear and the second mark aligns with the center line of the upper idler hub. Turn the camshaft gear and pump drive gear so that the timing marks align with the upper idler gear hub and install upper idler so that the three timing marks align with the marks on the camshaft gear, pump drive gear and lower idler. See Fig. MF33 for a view of the timing gear train with the timing marks in proper alignment.

In addition to the oil holes shown in Fig. MF40, timing gear teeth on models before engine serial number 2812139 are spray lubricated by the oil feed line leading from the lower idler hub as shown in Fig. MF41.

Make certain oil passages are open when timing gears are serviced. Install rocker arms assembly and adjust tappet gap.

TIMING GEAR HOUSING

All Diesel Models

60. After removal of the engine timing gears, the timing gear housing may be removed from the engine block. On the block front face, cup type expansion plug (A—Fig. MF42) closes the water jacket and a small cup type

Copyright F. Perkins Ltd.

Fig. MF40 — Oil holes (A and B) through gear align during rotation, with hole in idler hub stud to feed oil to gear teeth.

plug (B) blocks the front opening of the engine oil gallery. A one-piece gasket seals the timing gear housing to the engine block and oil pan.

CAMSHAFT

All Diesel Models

61. To remove the camshaft, first remove the timing gear cover as outlined in paragraph 51 or 52, remove fuel tank, rocker arm cover and rocker arms assembly. Secure the valve tappets in their upper-most position, re-

Copyright F. Perkins Ltd.

Fig. MF39 — Upper idler hub removed showing oil passage (A, B and C) which must be aligned in installation. Hub extension (D) is a light press fit in block recess. (E). Lower idler hub is identical.

Fig. MF41 — Timing gears on some models are spray lubricated through oil hole in tube from lower idler stud. Outer end of tube is sealed by rubber plug in timing gear housing.

Fig. MF42 — Front face of diesel engine block with timing gear housing removed, showing (A) water jacket expansion plug and (B) oil gallery plug.

Copyright F. Perkins Ltd.

Fig. MF43 — Camshaft thrust washer (A) is positioned in the engine block by locating pin (B).

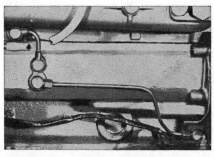

Fig. MF44 — Long external oil line leads from main oil gallery to center camshaft bearing. Short external oil line feeds metered oil to cylinder head for valve rocker lubrication.

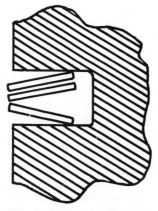

Fig. MF45 — Cross sectional view of the segmented compression ring showing correct method of installation. Refer to text for details.

move the fuel lift pump, and withdraw the camshaft and gear as a unit.

Camshaft end thrust of 0.003-0.006 is controlled by a machined boss on the timing gear cover and a renewable thrust washer (A—Fig. MF43) located in the front of the engine block and positioned with a locating pin (B).

The camshaft runs in three journal bores machined directly in the engine block. The front and rear camshaft journals are gravity lubricated by the return oil from the rocker arms assembly. The center camshaft journal is pressure lubricated by the external oil feed line shown in Fig. MF44. The center journal, in turn, meters oil which is fed to the rocker shaft through the second short oil feed line.

Camshaft bearing journals have a normal diametral clearance of 0.004-0.008 in all three bearing bores. Journal diameters are as follows:

Front 1.869-1.870
Center 1.859-1.860
Rear 1.839-1.840

ROD AND PISTON UNITS

All Diesel Models

62. Connecting rod and piston units are removed from above after removing cylinder head, oil pan and rod bearing caps. Cylinder numbers are stamped on the connecting rod and cap. When reinstalling rod and piston units, make certain the correlation numbers are in register and face away from the camshaft side of the engine.

Connecting rod nut torque is 70-80 Ft.-Lbs.

PISTONS, SLEEVES AND RINGS

All Diesel Models

63. Aluminum alloy pistons are supplied in standard size only, and are available in a kit, consisting of piston, pin, rings and sleeve. Note: Piston skirt clearance specifications are not supplied by the manufacturer. As a general rule, ring groove wear will determine the necessity for piston renewal; however, piston renewal due to skirt wear or collapse must be left to the discretion of the servicing mechanic.

All regular diesels and some direct injection diesels were originally equipped with a chrome plated, thin steel cylinder sleeve. Some late models use a spun cast iron cylinder sleeve which is available for service on all models. Sleeve dimensions are identical and overhaul procedures the same, for both the chrome and cast iron sleeves. The cylinder sleeves should be renewed when visible wear is present at the top of the ring travel, or if the sleeve is damaged in any other way.

The 0.0425 thick cylinder sleeves fit 0.001 loose to 0.001 tight in the 3.6875-3.6885 block bores. When installing sleeves, make certain that the block bore including the counterbore for the sleeve flange, and the cylinder sleeve is absolutely clean and free of any particles of carbon or dirt which might cause distortion of the thin walled sleeves. Also, examine the sleeves for nicks, burrs or damage. Apply a light coating of engine oil to O.D. of sleeve and press same into cylinder bore until the flange is fully seated in the block counterbore. After allowing a few minutes for sleeve to conform to the shape of the cylinder wall, measure for concentricity of sleeve in cylinder bore. Sleeve should measure 3.6005-3.6030 when measured in two directions at top, center and bottom of sleeve. Top flange should be 0.001-0.009 below top surface of the block.

On regular diesel engines, the pistons are not marked and may be installed either way on the connecting rods. On direct injection diesel engines, the combustion chamber cavity in top of piston crown is slightly offset, and piston is marked "FRONT" for proper installation.

All piston rings should have an end gap of 00.009-0.013. All one-piece rings have a recommended side clearance of 0.002-0.004 in piston grooves.

On all engines, the top ring is parallel faced, and may be installed either side up. The second compression ring on all engines; and second and third rings on early engines; are taper faced and marked "Top" for correct installation. Late engines with chrome sleeves use a three-piece, segmented compression ring in the third groove. The segmented ring may be used as a service replacement for the solid ring on early models.

The segmented, third compression ring, correctly installed, is shown in Fig. MF45. End gaps of upper and lower segment should be aligned with one end of piston pin, with the end gap of center segment aligned with opposite end of pin. NOTE: Ring ends of lower segment will point downward, and ends of two upper segments point upward, when segment is compressed before installation.

The two oil control rings are interchangeable and may be installed either side up.

A re-ring set is available for regular diesel engines which consists of three identical compression rings which contain a cutout groove on inside, top surface; and two identical oil control rings with beveled lands, which must be installed with bevels up.

Fig. FM47 — Rear main bearing showing lower thrust washer (A) with locating tab (B) which is installed in cap recess (C) to prevent rotation. Upper thrust washer (D) is held in place by lower washer half.

Copyright F. Perkins Ltd.

PISTON PINS

All Diesel Models

64. The 1.24975-1.2500 diameter floating type piston pins are retained in piston bosses by snap rings and are available in standard size only. The renewable connecting rod bushing must be final sized after installation to provide a diametral clearance of 0.0005-0.001 for the pin. Be sure the pre-drilled oil hole in the bushing is properly aligned with the hole in the top of the connecting rod when installing new bushings. The piston pin should have a thumb press fit in the piston after same is heated to 160 degrees F.

CONNECTING RODS AND BEARINGS

All Diesel Models

65. Connecting rod bearings are precision type, renewable from below after removing oil pan and bearing caps. When renewing bearing shells, be sure that the projection engages milled slot in rod and cap and that the correlation marks are in register and face away from the camshaft side of the engine. Replacement rods should be marked with the cylinder number in which they are installed. Bearings are available in standard, as well as undersizes of 0.010, 0.020 and 0.030.

On regular diesel engines, tighten connecting rod nuts to a torque of 70-80 Ft.-Lbs. On direct injection diesel engines, tighten unplated connecting rod nuts to a torque of 65-70 Ft.-Lbs.; or cadmium plated nuts to a torque of 45-50 Ft.-Lbs.

Crankpin diameter 2.2485-2.2490
Diametral clearance
   Regular diesel ....0.002 -0.0035
   Direct injection 0.0025-0.004
Side clearance 0.0095-0.0133

Copyright F. Perkins Ltd.

Fig. MF46 — Block lower surface with crankshaft in place. Matched block and bearing caps bear serial number (A) and bearing caps are numbered front to rear (B) as shown.

CRANKSHAFT AND BEARINGS

All Diesel Models

66. The crankshaft is supported in precision type main bearings. The front bearing cap, containing the oil pump, and the three intermediate bearing caps may be removed after removing the oil pan as outlined in paragraph 70 or 71. To remove the rear main bearing cap, it is first necessary to remove the engine (paragraph 43), flywheel (paragraph 69), adapter plate (paragraph 67) and the crankshaft rear oil seal retainer plate. Upper and lower bearing liners are interchangeable on all bearings except the front main bearing. When renewing the front main bearing, make sure the correct bearing half is installed in the engine block bore. Bearing inserts are available in standard, and undersizes of 0.010, 0.020 and 0.030.

Main bearing caps are numbered from front to rear as shown at (B— Fig. MF46). Block and cap bores are machined with caps in place and caps cannot be interchanged. Block and caps are stamped with a serial number (A) when the machining operation is performed. When installing caps, cap serial number must be on same side of block as the block serial numbers. Bearing caps are aligned to block by ring dowels as shown at (A—Fig. MF-48).

On regular diesel engines, tighten main bearing cap screws to a torque of 110-120 Ft.-Lbs. Use the same torque for direct injection diesels un-less shim washers are installed between cap screw heads and the locking tab washer. If shims are used, tighten main bearing cap screws to a torque of 90-95 Ft.-Lbs.

Crankshaft end play is controlled by renewable thrust washers (A and D—Fig. MF47) at the front and rear of the rear main bearing. The lower thrust washer half is prevented from rotation by the locking tab (B) which fits in recess (C) in the bearing cap. Upper half (D) is held in place by lower half (A). After bearing cap is removed, upper half of thrust washer can be "rolled" out without R&R of crankshaft. Crankshaft end play should be 0.002-0.010 on regular diesel engines; or 0.002-0.014 on direct injection diesels.

When renewing rear main bearing, refer to paragraph 68 for rear oil seal installation.

Check the crankshaft journals against the values listed below:
Journal diameter 2.7485-2.7490
Diametral clearance
  Regular diesel 0.0025-0.0045
  Direct injection . . . . . .0.003 -0.005

ENGINE ADAPTER PLATE

All Diesel Models

67. The engine flywheel is housed in a cast iron adapter plate which is located to the engine block by two dowels and secured to the block rear face by six cap screws.

To obtain access to the rear main bearing, crankshaft, or rear oil seal, it is first necessary to remove the adapter plate as follows:

Split tractor between engine and transmission and remove clutch and flywheel. After the flywheel is off, the plate may be removed by removing the six cap screws and tapping the plate free of the locating dowels.

CRANKSHAFT REAR OIL SEAL

All Diesel Models

68. The asbestos rope type rear oil seal (C—Fig. MF48) is contained in a two piece seal retainer (B), which

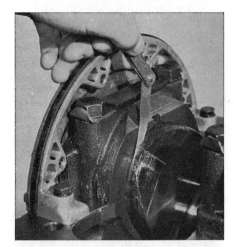

Fig. MF49—Crankshaft end play should be 0.002-0.010 on regular diesel; or 0.002-0.014 on direct injection diesel. Shown is the AD4-203 direct injection engine.

Copyright F. Perkins Ltd.

Fig. MF48 — Rear main bearing cap removed, showing ring dowels (A) which locate bearing caps. Seal retainer (B) is in two pieces, containing rope-type asbestos seal (C).

Fig. MF50—To renew the crankshaft rear oil seal (S), the oil pan, flywheel and engine adapter plate must be removed. Shown is the direct injection engine.

Fig. MF51 — When installing rear oil seal housing on direct injection diesels, make sure rear face is flush with rear of block as shown.

can be removed after removing the engine adapter plate as outlined in paragraph 67.

On regular diesel engines, lower half of oil seal retainer attaches to rear main bearing cap as shown in Fig. MF48. On direct injection diesel engines, lower half attaches to a separate seal housing which contains the cork oil pan gasket as shown in Figs. MF49 and MF50. On these models, the seal housing must be removed for removal of rear main bearing cap, and must be reinstalled with rear face of housing and block aligned as shown in Fig. MF51.

The rope type crankshaft seal is precision cut as to length, and must be installed in retainer halves so that

0.010-0.020 of the seal projects from each end of the retainer. To install the seal, clamp each half of retainer in a vise as shown in Fig. MF52. Make sure the seal groove is clean, and start each end of seal in its groove with the fingers, the specified amount of free end extending. Allow the seal rope to buckle up in the center until about an inch of each end is bedded in seal groove, work center of seal into position, then roll with a round bar as shown. Repeat the process on the other half of seal.

Position the retainer halves around the crankshaft and install the retaining cap screws loosely, then install and tighten the long clamping bolts or cap screws. Complete the assembly by reversing the disassembly procedure.

FLYWHEEL

All Diesel Models

CAUTION: Flywheel is only slightly piloted to crankshaft. Use caution when unbolting flywheel to prevent same from falling and causing possible personal injury.

69. To remove flywheel, first separate the engine from transmission case then remove the clutch. The starter ring gear can be renewed after removing the flywheel. To install a new ring gear heat same to 475-500 deg. F. and install on flywheel with beveled end of teeth facing timing gear end of engine.

Flywheel is secured to crankshaft flange by six evenly spaced cap screws. To properly time flywheel to engine during installation, be sure that unused hole in flywheel aligns with untapped hole in crankshaft flange. Tighten retaining cap screws to a torque of 75 Ft.-Lbs. Maximum allowable flywheel runout is 0.008.

OIL PAN

Regular Diesel

70. The cast-iron oil pan is fitted with a wedge-shaped adapter, located between the pan and the flywheel housing as shown in Fig. MF53. Oil pan and adapter are matched, and should be renewed as a unit. To remove the oil pan, first unbolt and remove the wedge-shaped adapter. Two oil pan retaining cap screws, located at the rear corners underneath the adapter, are fitted with copper sealing washers. After removing the adapter, place a jack under oil pan and remove the cap screws retaining the pan to the engine block and timing gear housing. Lower the oil pan slightly and carefully separate front edge of oil pan from timing gear housing gasket without tearing gasket. A one piece gasket seals timing gear housing at engine block and oil pan joint and renewing gasket involves considerable extra work.

When reinstalling oil pan, use cup grease, rather than gasket sealer, on gasket between oil pan and timing gear housing. Install oil pan to block cap screws finger tight; then install and securely tighten oil pan to timing gear housing cap screws. Now, tighten cap screws retaining oil pan to block and reinstall wedge-shaped adapter.

Direct Injection Diesel

71. To remove the oil pan, first drain the cooling system and oil pan, disconnect headlight wires and remove the hood and side panels. Disconnect drag link, and on tractors so equipped, unbolt power steering reservoir bracket from cylinder head and fan shroud and power steering pump from timing gear case. Lay pump and reservoir in grille enclosure and disconnect air cleaner and

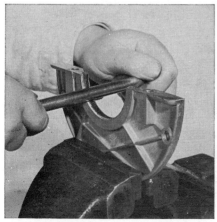

Fig. MF52—Using a round bar to bed the asbestos rope seal in retainer half. Refer to text for details.

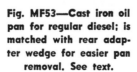

Fig. MF53—Cast iron oil pan for regular diesel; is matched with rear adapter wedge for easier pan removal. See text.

Fig. MF54 — Before removing oil pan on direct injection engine, it is necessary to remove front end unit and the stud nut (N) from each side of timing gear housing.

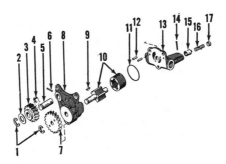

Fig. MF56 — Exploded view of diesel engine oil pump showing component parts.

1. Snap ring	11. "O" ring
2. Washer	12. Dowel pins
3. Idler gear	13. End plate
4. Idler bushing	14. Cotter pin
5. Idler shaft	15. Relief valve
6. Retaining pin	plunger
7. Pump drive gear	16. Relief valve
8. Pump body	spring
9. Woodruff key	17. Spring seat
10. Rotor assembly	

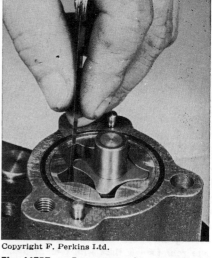

Fig. MF57 — Rotor gear clearance should not exceed 0.006 when measured with a feeler gage as shown.

radiator hoses. Support tractor under transmission housing and front end assembly from top of steering pedestal. Unbolt and remove front support, radiator and front end assembly as a unit.

Remove the stud nut (N—Fig. MF-54) at left front corner of oil pan, and similar nut on right side which is located behind the front support spacer (S—Fig. MF55). Remove the two cap screws (C) from timing gear cover, remove remainder of oil pan retaining cap screws; support pan on a rolling floor jack, and remove the oil pan with lower timing gear cover (L) attached.

Before reinstalling the oil pan, remove lower timing gear cover (L), then reinstall finger tight using a new gasket. Examine the "O" ring on front oil seal retainer and, if in good condition reinstall and tighten pan retaining cap screws; install and tighten the two cap screws (C), then tighten the screws retaining lower timing gear housing (L) to the oil pan.

OIL PUMP

All Diesel Models

72. A rotor type oil pump (Fig. MF56) is mounted on the front main bearing cap and driven from the crankshaft timing gear through idler gear (3). Pump may be removed, together with the front bearing cap, after the oil pan is removed as outlined in paragraph 70 or 71.

Pump rotors (10) are available as a matched pair only. All other pump parts are available individually. Check the rotor clearance (Fig. MF57) with a feeler gage. Radial clearance should not exceed 0.006. Check rotor to body clearance (Fig. MF58). Radial clearance should not exceed 0.010. Check rotor end clearance with straight edge and feeler gage (Fig. MF59); clearance should not exceed 0.003. Renew rotor assembly and/or body if clearances are excessive.

When reinstalling pump, tighten main bearing cap screws to a torque of 110-120 Ft.-Lbs. Note: When renewing timing gears, pump drive gear and idler may be removed from the front without disturbing pump, if renewal is required.

Fig. MF58 — Rotor to body clearance should not exceed 0.010 when measured with feeler gage as shown.

Fig. MF59 — Rotor end clearance should not exceed 0.003 when measured with straight edge and feeler gage as shown.

Fig. MF55—Right front view of direct injection engine assembly. Before removing oil pan, the front support spacer (S) must be removed for access to hidden stud nut. Remove cap screw (C) from each side of timing gear cover and allow lower cover (L) to remain on oil pan.

RELIEF VALVE

All Diesel Models

73. Plunger type relief valve (15 —Fig. MF56) is located in oil pump body and can be adjusted by renewing spring (16) to maintain a relief pressure of 50-65 psi at operating speed.

To adjust the pressure on regular diesel, drain engine oil and remove pressed steel strainer housing from bottom of oil pan. Working through access hole in bottom of pan, remove cotter pin (14) and extract spring seat (17), spring (16) and plunger (15).

On direct injection diesel, the oil pan will need to be removed as outlined in paragraph 71, before relief valve can be disassembled.

CARBURETOR

(Except LP-Gas)

74. Marvel-Schebler TSX-695 carburetor is used on gasoline models of the MF65 tractor. Specifications are as follows:

Float setting .... ¼-inch
Repair kit 286-1235
Gasket set 16-654
Inlet needle and seat........ 233-543

LP-GAS SYSTEM

Two types of factory installed LP-Gas systems have been used, both utilizing the liquid withdrawal principle. On tractors prior to serial number 673300, a secondary regulator is mounted on the vaporizer assembly and fuel enters the carburetor at approximately atmospheric pressure. On tractors after serial number 673299 a pressure regulating carburetor is used and fuel enters the carburetor at the 10 psi vaporizer outlet pressure.

CAUTION: LP-Gas expands readily with any decided increase in temperature. If tractor must be taken into a warm shop to be worked on during extremely cold weather, make certain that fuel tank is not overly full of fuel. Unburned LP-Gas fuel should never be released inside a building or in an unventilated place.

CARBURETOR

Early Type

LP-Gas tractors before serial number 673300 are equipped with a Zenith, Model GO-12256 carburetor.

75. **ADJUSTMENT.** Refer to Fig. MF61. Clockwise rotation of the idle mixture adjustment screw (1) or the main fuel adjustment (2), leans the mixture. Initial idle needle setting is one turn open from the closed position. Initial setting for the main adjustment is 1¾ turns open. To adjust

the carburetor, start engine and bring to operating temperature then turn the throttle stop screw (3) either way to provide an engine slow idle speed of 450-475 rpm. Turn the idle adjusting needle (1) clockwise until engine begins to "stall," then turn counter-clockwise until engine runs smoothly. To adjust the main fuel adjustment (2), bring engine up to operating temperature, remove any three spark plug wires, open the throttle to full throttle setting and adjust the main needle until highest engine operating speed is obtained. Turn main needle clockwise until engine speed just begins to drop, then turn needle counter-clockwise just enough to bring the engine speed back to top rpm.

76. **OVERHAUL.** Except for the removal of air-borne dirt and resealing of throttle and choke shafts, overhaul is not generally required. Removal of venturi is rarely necessary. If venturi is to be removed, first remove throttle valve and shaft and invert carburetor body; then remove the venturi locking screw located above the choke valve and allow venturi to drop out top of carburetor. Install venturi so that step in counterbore faces upward. Remove burrs in choke and throttle shaft bores caused by previous staking; install new seals and retainers, and stake new seal retainers ⅛-inch below outside of body bores.

Fig. MF60 — Side view of LP-Gas engine showing location of system components.

Fig. MF61 — Installation view of LP-Gas carburetor. Refer to text for adjustment procedure.

1. Idling mixture adjustment
2. Main fuel adjustment
3. Idling speed adjustment

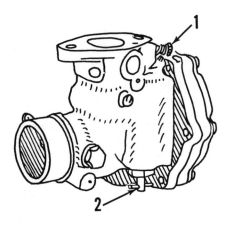

Fig. MF62—A Zenith PC, pressure regulating carburetor is used on late LP-Gas models.

1. Idle mixture needle
2. Main fuel adjustment

Late Type

LP-Gas tractors after serial number 673299 use a Zenith PC Type, pressure regulating carburetor.

77. ADJUSTMENT. Clockwise rotation of the idle mixture adjustment screw (1—Fig. MF62) or the main fuel adjustment mixture screw (2), leans the mixture. Normal initial setting for the idle mixture needle is 1½-2 turns open from the closed position. Adjust the main fuel mixture screw 2-3 turns open. To make the operational adjustment, proceed as follows: Start engine and bring to operating temperature. Adjust to a slow idle speed of 450-475 rpm and turn the idle mixture needle (1) clockwise until engine begins to stall; then counter-clockwise until engine

runs smoothly. Readjust idle speed if necessary. Engine must be under load for proper adjustment of main adjustment needle (2). Load can be applied in the field; or in shop by use of a dynamometer. In the absence of a dynamometer, a close approximation can be obtained by removing and grounding any three spark plug wires; then adjusting the main needle (2) to the leanest setting at which maximum engine speed can be obtained.

78. OVERHAUL. Mixing chamber components are similar to those used on early models, and overhaul procedures are similar. The carburetor incorporates a dual diaphragm pressure regulator which reduces the fuel inlet pressure to slightly less than atmospheric pressure for efficient and economical operation. Carburetor is completely sealed.

Inlet valve seat (10—Fig. MF63) is locked in position by nylon plug (15) and set screw (16). The position of the seat controls the opening position of diaphragm lever (17). When assembling or adjusting the carburetor, loosen the locking plug (16) and thread the seat (10) in or out until the lower end of valve lever (17) is flush with gasket surface of carburetor body (4).

Inspect the diaphragms (19 & 21) for pin holes or cracks, and install so that cupped spacer on inner diaphragm (19) faces toward the steel center plate of outer diaphragm (21). Make sure the passage holes in diaphragms, housings and gasket are aligned when the carburetor is assembled.

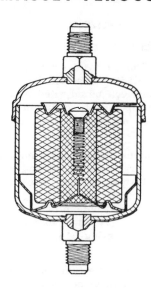

Fig. MF64 — Cross-sectional view of A-P Controls Model 408 fuel filter used on the MF65 LP-Gas engine.

RELIEF VALVE

All LP-Gas Models

79. The safety relief valve, located at the top forward end of tank, is pre-set to operate at 312 psi pressure. If tank internal pressure rises above 312 psi, the relief valve will open and allow fuel to escape until the pressure is reduced below 312 psi.

FILTER

All LP-Gas Models

80. The A-P Controls Division Model 408 filter used in this system is installed in the fuel line between the fuel tank and vaporizer and positioned between the fuel withdrawal valves. It is of the "throw-away" type, containing a fibrous filter element as shown in Fig. MF64.

To renew the filter, close both fuel withdrawal valves and allow engine to operate until all fuel in lines, vaporizer, regulator and carburetor is burned out and engine stops. Turn off ignition switch.

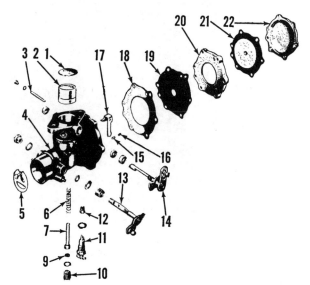

Fig. MF63 — Exploded view of Zenith pressure regulating carburetor.

1. Throttle valve
2. Venturi
3. Lever shaft
4. Body
5. Choke valve
6. Valve spring
7. Valve plunger
9. Valve
10. Valve seat
11. Main adjustment needle
12. Main jet
13. Choke shaft
14. Throttle shaft
15. Nylon plug
16. Lock screw
17. Diaphragm lever
18. Gasket
19. Inner diaphragm
20. Housing
21. Outer diaphragm
22. Diaphragm cover

Fig. MF65 — Fuel filter installation on the MF65 LP-Gas tractor.

Remove three sheet metal screws retaining left side panel to instrument panel and swing out slightly to provide working clearance. Remove two flare fittings at the "tee" in the filter top and a single fitting at the bottom. Refer to Fig. MF65. Remove the two

"tee" fittings from the filter and install them on a new filter.

Installation is the reverse of removal procedure.

VAPORIZER

All LP-Gas Models

81. **R&R AND OVERHAUL.** Before disconnecting any lines, be sure that by closing the tank withdrawal valves all fuel is burned out of the lines, vaporizer, regulator and carburetor and allowing engine to run until it stops. Turn off the ignition switch. Raise hood, drain cooling system completely and remove left side panel. Disconnect the vaporizer to regulator connection and the tank to vaporizer hose. Remove the vaporizer retaining capscrew at right side of tractor, which holds the vaporizer in the water outlet housing and withdraw the vaporizer assembly.

Refer to Fig. MF66 and further disassemble as follows: Remove heat exchanger (25) from vaporizer body by removing retainer screw (27). Fibre washer (26) and "O" rings (23 and 24) can be renewed at this time.

Remove inlet orifice retainer (37) and washer (20). Remove the inlet orifice (38) by loosening the lock nut (2) and turning pressure adjustment (1) down as far as it will go by hand; then, threading a ¼ x 20 inch standard screw into the inlet orifice, grasp the screw body with pliers and tap the inlet orifice out of vaporizer body as shown in Fig. MF67. Note: Do not attempt to turn while withdrawing, since this part is located with a dowel pin, which may shear. Inspect the fuel inlet orifice tip (38—Fig. MF66) for nicks or scratches. The tip must be smooth and the seat contact surface parallel with the orifice body. Remove fibre washer (22) with a scribe or wire hook.

Back-out pressure adjustment screw (1) to release spring tension, then remove from diaphragm cover (5). Remove regulator spring (3). Remove the diaphragm cover screws and the diaphragm cover (5).

Separate the edge of the diaphragm (9) from the vaporizer body (16). Grasping the diaphragm and diaphragm plate (8), twist back and forth gently until piston (10), "O" ring (11), diaphragm and diaphragm plate can be removed from vaporizer body as an assembly. Refer to Fig. MF68.

Note: Do not loosen diaphragm assem-

bly screw (6—Fig. MF66) unless diaphragm is defective and needs renewing.

If diaphragm (9) does show signs of deterioration or cracking, renew as stall new diaphragm, but leave assembly screw loose enough for diaphragm to turn on piston. Push piston (10) in follows: Loosen and remove diaphragm assembly screw (6), then remove diaphragm from piston (10). Into position in the vaporizer body with the opening in the piston skirt parallel with the inlet orifice channel. With a small straight edge inserted in the channel, hold the piston in this position. Rotate the diaphragm (9) until the holes align with those in the vaporizer body (16); then tighten the assembly screw (6).

With the piston (10) removed from the body, valve seat (13), retainer (12) and valve spring (15) may be removed. If retainer (12) sticks in body, use a small wire hook to pull it out. Note: Do not lose spring button (14).

When completely disassembled, clean metal parts thoroughly using an approved solvent. Dry with an air hose, making sure all drilled passages and vents (31 and 39) are open and clean. Use new "O" rings, seals and fibre washers.

82. To reassemble the vaporizer, proceed as follows: Assemble the seat retainer (12), spring (15) and spring button (14) at the retainer end. Refer to Fig. MF69. Slide the parts into the vaporizer body until spring (15— MF66) rests in the spring base spool which is not a removable part.

Align the seat retainer (12) within the vaporizer body so that one flat surface faces the inlet orifice channel. This position allows spring pressure to be distributed equally to all four corners of the seat retainer by the mating surfaces of the piston (10).

Install the assembled piston and diaphragm in the correct relation as noted in paragraph 81.

Place the diaphragm cover (5) over the diaphragm (9) and install the six cover screws. Note: Ascertain that the vent hole (39) in the diaphragm cover is toward bottom of tractor.

Install and center the regulator spring (3) on the diaphragm assembly screw (6). Install the pressure adjustment (1) and lock nut (2) as an assembly, by compressing the spring (3) slightly and threading the assembly into the cover (5). **Screw the**

Fig. MF66—Exploded view of the vaporizer assembly.

1. Screw	16. Body
2. Lock nut	20. Washer
3. Regulator spring	22. Fibre washer
5. Diaphragm cover	23. "O" ring
6. Assembly screw	24. "O" ring
8. Diaphragm plate	25. Heat exchanger
9. Diaphragm	shell
10. Piston	26. Fibre washer
11. "O" ring	27. Retainer screw
12. Seat retainer	31. Vent
13. Inlet valve seat	37. Fuel inlet
14. Spring button	38. Inlet orifice
15. Retainer spring	39. Vent

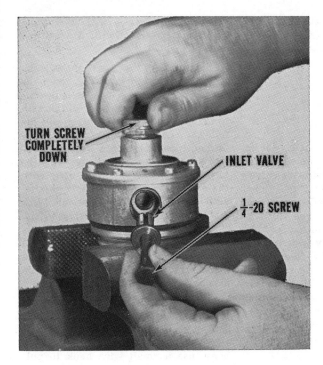

Fig. MF67 — Using a ¼-20 screw to remove the vaporizer inlet orifice. Pressure adjusting screw must be turned completely down during this operation.

adjustment down 6 to 8 turns until the inlet valve seat (13) has moved below the level of the inlet orifice channel.

Install the inlet orifice (38) and a new washer (22), using the ¼-20 inch threaded screw used in disassembly as a handle. The dowel pin in the body and the slot in the inlet orifice must be in alignment.

Install washer (20) and tighten the orifice retainer (37). Install new "O" ring (24) on heat exchanger (25) and place in position. Install retainer screw (27) and fibre washer (26) and tighten securely.

83. VAPORIZER TEST AND ADJUSTMENT. Before installing vaporizer into the water jacket, proceed as follows: Plug fuel outlet with a suitable pipe fitting to which a shut-off valve is attached. Connect a 30-lb. pressure gage into the test connection port in the vaporizer body. Refer to Fig. MF70. Connect the fuel inlet to a compressed air line having approximately 75 lbs. pressure. Back regulator adjustment out as shown. Turn on air pressure and screw in pressure adjustment gradually. The vaporizer should hold each increase without rising or dropping. At several points in the check, release air by opening the shut-valve in the fuel outlet fitting. The vaporizer should recover the pressure to the original setting as observed on the test gage.

CAUTION: Make certain test air is dry. Do not permit water or vapor to enter vaporizer assembly.

Final adjustment is made by screwing adjustment clockwise until 10 psi is observed on test gage. Tighten lock nut (2—Fig. MF66) and remove test apparatus. Install vaporizer into water jacket housing, then install and tighten the retaining capscrew securely. Make certain "O" ring (23) is in good condition to prevent water leakage.

Connect the vaporizer to regulator connection and the tank to vaporizer hose. The remainder of assembly is evident.

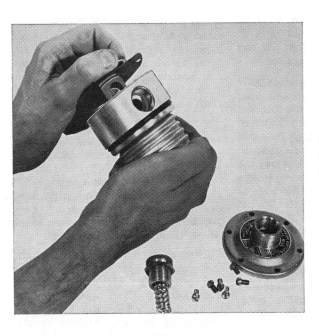

Fig. MF68 — Removing the diaphragm and piston assembly from the vaporizer body. Refer to text.

REGULATOR

Early Type

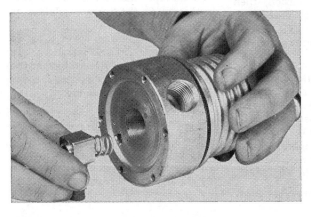

Fig. MF69 — Installing the valve seat retainer, valve seat and spring into the LP-Gas vaporizer.

84. R&R AND OVERHAUL. Before disconnecting any lines, be sure all fuel is burned out of the lines, vaporizer, regulator and carburetor by closing the tank withdrawal valves and allowing engine to run until it stops. Turn off ignition switch. Raise hood, drain cooling system completely and remove left side panel. Disconnect

Fig. MF70 — Using an air hose and pressure gage to test and adjust the LP-Gas vaporizer. Working pressure is set to 10 psi.

Fig. MF73 — Installing the regulator pressure adjusting screw. Stem of valve must enter the hole in the screw.

Fig. MF71 — Exploded view of the LP-Gas regulator.

1. Diaphragm vent screen
2. Diaphragm and cover
3. Gasket
4. Regulator body
5. Control valve block and spring
6. Regulator valve seat
7. Fuel inlet orifice
8. Leaf spring lock screw
9. "O" ring
10. Pressure adjusting screw

(6) is smooth and even. Check vent screens (1) for cleanliness; remove any dirt or obstructions.

85. To reassemble, insert valve block and leaf spring assembly (5) into regulator body (4). Refer to Fig. MF72. Make sure that ends of leaf spring enter their respective slots in the regulator body and are visible through holes for leaf spring lock screws (8). Install and tighten the inlet orifice (7).

vaporizer to regulator connection and regulator to carburetor hose. Remove the regulator assembly.

Referring to Fig. MF71, remove the diaphragm and cover assemblies (2), using Zenith Service Tool C161-190

to facilitate removal. Remove fuel inlet orifice (7) and pressure adjusting screw (10). Remove both leaf spring lock screws (8). Remove control valve block and leaf spring assembly (5).

Clean and inspect all parts and renew any which are questionable. "O" ring (9) and gasket (3) can be renewed at this time. Ascertain that contact surface of regulator valve seat

With "O" ring (9) slightly lubricated, insert pressure adjusting screw (10) into regulator body. Refer to Fig. MF73. Note: If resistance is felt, round valve stem may not be centered in the adjusting screw. Remove and repeat until adjusting screw can be turned by hand and slotted head is almost flush with body. Then lightly

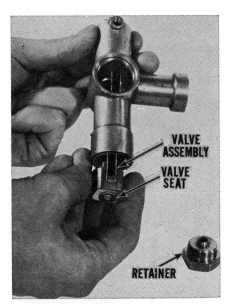

Fig. MF72 — Installing valve assembly in the regulator housing. Leaf springs must enter slots at opposite end of bore.

Fig. MF74—Correctly installing the leaf spring assembly using Zenith Gage C161-189. Refer to text.

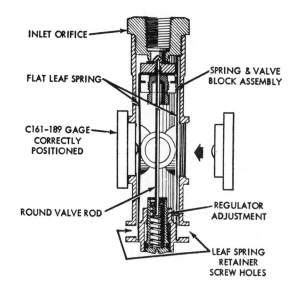

seat the pressure adjusting screw, turning clockwise. Back out approximately 3 turns.

Insert Zenith Part C161-189 leaf spring gauge set into diaphragm cover bores as shown in Fig. MF74. This will position the leaf springs correctly. Then install and tighten the leaf spring lock screws (8—Fig. MF71).

Install and tighten thoroughly by hand both diaphragm and cover assemblies (2). Install the regulator on engine and connect the regulator to carburetor hose and vaporizer to regulator connection. Remainder of assembly is obvious after examination.

86. **ADJUSTMENT.** The regulator must be adjusted for slightly below atmospheric pressure regulation as follows: Connect an air hose to the inlet side of the vaporizer unit which has been previously adjusted to 10 psi as outlined in paragraph 83. Apply air pressure to the unit, cover the regulator outlet with soapy film solution and slowly turn the pressure adjusting screw (10—Fig. MF71) counter-clockwise until a soap bubble begins to form; then turn the adjusting screw clockwise until the bubble can be maintained without breaking or enlarging. The soap bubble can then be removed and another applied which should lie flat. If bubble lies flat, then turn the adjusting screw ½-turn more clockwise. Add more solution. If the bubble enlarges rather than lying flat, recheck the regulator for improper parts or incorrect assembly; then repeat the procedure.

Fig. MF75—Diaphragm type primary fuel pump (B) is mounted on right side of engine block and driven by engine camshaft. Shut-off valve (A) is renewable as a unit. Sediment bowl (C) serves as a water trap and should be cleaned periodically.

DIESEL SYSTEM

The Massey-Ferguson MF-65 diesel fuel system consists of the fuel filters, injection pump and injection nozzles. When servicing any unit associated with the fuel system, the maintenance of absolute cleanliness is of utmost importance. Of equal importance is the avoidance of nicks or burrs on any of the working parts.

Probably the most important precaution that service personnel can impart to owners of diesel powered tractors, is to urge them to use an approved fuel that is absolutely clean and free from foreign material. Extra precaution should be taken to make certain that no water enters the fuel storage tanks. This last precaution is based on the fact that all diesel fuels contain some sulphur. When water is mixed with sulphur, sulphuric acid is formed and the acid will quickly erode the closely fitting parts of the injection pump and nozzles.

87. **QUICK CHECKS - UNITS ON TRACTOR.** If the diesel engine does not run properly, and the diesel fuel system is suspected as the source of trouble, refer to the Diesel System Trouble Shooting Chart and locate points which require further checking. Many of the chart items are self-explanatory; however, if the difficulty points to the fuel filters, injection nozzles and/or injection pump, refer to the appropriate paragraphs which follow.

DIESEL SYSTEM TROUBLE-SHOOTING CHART

	Sudden Stopping of Engine	Lack of Power	Engine Hard to Start	Irregular Engine Operation	Engine Knocks	Excessive Fuel Consumption
Lack of fuel	★	★	★	★		
Water or dirt in fuel	★	★	★	★		
Clogged fuel lines	★	★	★	★		
Inferior fuel	★	★	★	★		
Faulty primary pump	★	★	★	★		
Faulty injection pump timing		★	★	★	★	★
Air traps in system	★	★	★	★		
Clogged fuel filters		★	★	★		
Deteriorated fuel lines	★					★
Faulty nozzle				★	★	★
Faulty injection pump		★	★	★	★	★

FUEL FILTERS

All Diesel Models

88. **CIRCUIT DESCRIPTION AND MAINTENANCE.** Fuel is drawn from the fuel tank and through the shutoff valve (A—Fig. MF75) and the sediment bowl (C) by the primary pump (B), which is driven by a lobe on the engine camshaft. Fuel then passes through the first and second stage filters and on to the injection pump. The primary pump incorporates a hand priming lever to be used in bleeding the system. The sediment bowl should be inspected frequently and if water and/or particles of foreign matter are observed, the bowl should be removed and cleaned. The primary pump can be disassembled for renewal of gaskets, diaphragm assembly or valve assemblies.

The first stage filter (A—Fig. MF 76) contains a replaceable cartridge type element. The filter should be drained at not more than 100 hour intervals and cartridge renewed at not more than 500 hour intervals. When renewing the filter element, thoroughly clean filter interior before installing new element. A bleed screw is located at the top of the filter head for air bleeding the first stage filter.

The second stage filter (B) further filters the fuel and also contains a renewable cartridge type element. Filter should be drained at 100 hour intervals and element renewed with every other renewal of first stage element. A bleed screw is located in the top of the filter body for air bleeding the filter.

The fuel now enters the injection pump where proper metering and timing sequence occurs.

BLEEDING

All Diesel Models

89. To bleed the system, proceed as follows: Open shut-off valve and loosen the air vent on the first stage filter. Operate the manual lever on primary fuel pump up and down fully until air-free fuel comes out filter vent; then, tighten the vent plug securely. Loosen vent on the second stage filter and continue pumping until air-free fuel comes out the vent; then tighten plug.

Loosen the top vent plug on the injection pump and repeat pumping procedure until air-free fuel flows out the port. Tighten the vent plug and repeat the procedure with the lower vent plug opened.

Operate the primary pump lever ten extra strokes to provide fuel in bleed back lines to the second stage filter.

Loosen the injector pressure lines at the injectors, make sure the fuel shut-off rod is in the operating position and turn the engine over with the starting motor until fuel escapes from the line ends.

Tighten the pressure line connections and start the engine. If the engine fails to start, it may be necessary to air bleed the system again.

INJECTOR NOZZLES

WARNING: Fuel leaves the injector nozzle with sufficient pressure to penetrate the skin. When testing, keep unprotected parts of your body clear of the nozzle spray.

All MF-65 diesel engines are equipped with C. A. V. multiple hole nozzle assemblies. In early regular diesel models, fuel leaves the nozzle in two fine sprays, one of which is directed into the spherical combustion chamber, and the other into the cylinder area above the piston.

Late models with direct injection system are equipped with a four-orifice nozzle tip which extends through the cylinder head to inject the fuel charge into the combustion chamber located in top of piston.

90. **TESTING AND LOCATING A FAULTY NOZZLE.** If the engine does not run properly, and the quick checks outlined in paragraph 87 point to a faulty injector, locate the faulty unit as follows:

If one engine cylinder is misfiring, it is reasonable to suspect a faulty injector. Generally, a faulty injector can be located by loosening the high pressure line fitting to each injector nozzle in turn, thereby allowing fuel to escape at the union rather than enter the cylinder. As in checking spark plugs in a spark ignition engine, the faulty unit is the one which, when its line is loosened, least affects the running of the engine.

Remove the suspected injector unit from the engine as outlined in paragraph 92. If a suitable nozzle tester is available, check the unit as in paragraph 91. If a tester is not available, reconnect the fuel line to the injector assembly and with the nozzle tip directed where it will do no harm, crank the engine with the starting motor and observe the nozzle spray pattern.

If the spray patterns are ragged, unduly wet, streaky and/or not symmetrical, or if nozzle dribbles, the nozzle valve is not seating properly. Send the complete nozzle and holder assembly to an authorized diesel service station for overhaul.

91. **NOZZLE TESTER.** A complete job of testing and adjusting the injector requires the use of a special test rig. The nozzle should be tested for opening pressure, seat leakage and spray pattern as follows:

Note: Only clean, approved testing oil should be used in the tester tank.

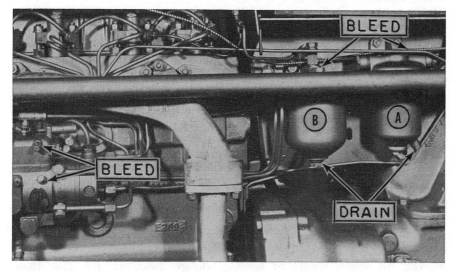

Fig. MF76 — First stage filter (A) and second stage filter (B) contain renewable cartridges. Filter bodies are equipped with drain screws, and pump and filters are equipped with bleed screws. See text for procedures.

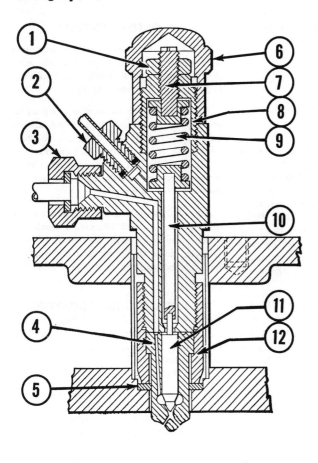

Fig. MF77 — Sectional view of typical C. A. V. injector showing component parts. Injector used in MF-65 Diesel differs in some respects.

1. Lock nut
2. Leak-off connection
3. Inlet connection
4. Nozzle body
5. Sealing washer
6. Cap nut
7. Adjusting screw
8. Spring cap nut
9. Spring
10. Valve spindle
11. Nozzle valve
12. Nozzle retaining nut

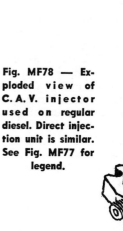

Fig. MF78 — Exploded view of C. A. V. injector used on regular diesel. Direct injection unit is similar. See Fig. MF77 for legend.

Operate the tester lever until oil flows and attach the injector assembly.

Close the tester valve and apply a few quick strokes to the lever. If undue pressure is required to operate the lever, the nozzle valve is plugged and should be serviced as outlined in paragraph 93.

While operating the tester handle, observe the gage pressure at which the spray occurs. On the two-hole nozzle, the gage pressure should be 1760 psi. The four-hole nozzle used in direct injection engines should open at 2500 psi. If the pressure is not as specified, remove the injector cap (6—Fig. MF78), back off lock nut (1) and turn adjusting screw (7) either way as required to correct the opening pressure.

The nozzle valve should not leak at a pressure 150 psi below nozzle opening pressure. To check for leakage, actuate the tester handle slowly and as the gage needle approaches 1600 psi (early type); or 2350 psi (late type), observe the nozzle tip for drops of fuel. If drops of fuel collect at pressures less than the specified pressure, the nozzle valve is not seating properly and should be serviced as outlined in paragraph 93.

If leakage test as outlined above was satisfactory, check nozzle valve and body for excessive wear or scoring by actuating the tester handle as above until the gage needle reaches 1600 psi and observe the length of time gage needle requires to drop to 500 psi. If time is less than six seconds, and no tip leakage occurs, the nozzle assembly should be renewed. NOTE: Leakage of the tester check valve or lines can cause an excessively fast pressure drop. This fact should be taken into consideration, and if all injectors tested show a fast rate of pressure drop, the tester, rather than the injectors should be suspected.

If leakage and pressure tests are satisfactory, operate the tester handle at approximately 100 strokes per minute and observe the nozzle spray pattern. On early type, two symmetrical sprays should emerge from the nozzle tip. Each should be a misty spray spreading to about three inches in diameter at about one foot away from the nozzle tip. On late type, four finely atomized conical sprays should emerge. All sprays should be symmetrical and equal in appearance. If the spray patterns are unduly wet, streaky or ragged, the nozzle tip

should be serviced as outlined in paragraph 93.

92. **REMOVE AND REINSTALL.** The fuel tank bottom on regular diesel models is provided with an indented channel over the injector assemblies enabling them to be withdrawn without removing the fuel tank. Extreme care should be taken, however, to avoid particles of dirt or carbon lodging on the block seating surface when the injectors are installed. A better practice for satisfactory injector service is to remove the hood, side panels and fuel tank so the seating surface can be examined and cleaned before installing the injector assemblies.

Before loosening any fuel lines, thoroughly clean the lines, connections, injectors and block, with air pressure and a solvent spray. Discon-

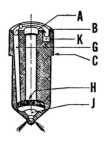

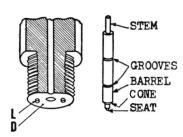

nect the high pressure and leak off lines and immediately cap the open ends to avoid dirt entry. Remove the injector retaining nuts and withdraw the injector from the cylinder head, being careful not to strike the tip end against any hard surface.

Thoroughly clean the nozzle recess in the cylinder head before reinstalling the injector assembly. It is important that the seating surface be free of even the smallest particle of carbon or dirt which could cause the unit to be cocked and result in blow-by. No hard or sharp tools should be used in cleaning. A piece of wood dowel or brass stock properly shaped is very effective. Do not reuse the copper sealing washer (5—Fig. MF 77) located between the injector assembly and head, always install a new one. Tighten the retaining nuts evenly to a torque of 18-20 Ft.-Lbs. After engine is started examine injectors for blow-by, making the necessary corrections before releasing tractor for service.

93. OVERHAUL. Hard or sharp tools, emery cloth, crocus cloth, grinding compounds or abrasives of any kind should NEVER be used in the cleaning of nozzles.

Wipe all dirt and loose carbon from the injector assembly with a clean, lint free cloth. Carefully clamp injector assembly in a soft jawed vise or injector fixture and remove the protecting cap (6—Fig. MF77). Loosen the jam nut (1) and back off the adjusting screw (7) enough to relieve the load from spring (9). Remove the nozzle cap nut (12) and nozzle body (4). Normally, the nozzle valve (11) can easily be withdrawn from the nozzle body. If it cannot, soak the assembly in fuel oil, acetone, carbon tetrachloride or similar carbon solvent to facilitate removal. Be careful not to permit the valve or body to come in contact with any hard surface.

If more than one injector is being serviced, keep the component parts of each injector separate from the others by placing them in a clean compartmented pan covered with fuel oil or solvent. Examine the nozzle body and remove any carbon deposits from exterior surfaces using a brass wire brush. The nozzle body must be in good condition and not blued due to overheating.

All polished surfaces should be relatively bright without scratches or dull patches. Pressure surfaces (A, B and D—Fig. MF79) must be absolutely clean and free from nicks, scratches or foreign material, as these surfaces must register together to form a high pressure joint.

Clean out the small fuel feed channels (G), using a small diameter wire as shown in Fig. MF80. Insert the special groove scraper into nozzle body until nose of scraper locates in the fuel gallery. Press nose of scraper hard against side of cavity and rotate scraper to clean all carbon deposits from the gallery as shown in Fig. MF81. Using seat scraper, clean all carbon from valve seat (J—Fig. MF 79) by rotating and pressing on the scraper as shown in Fig. MF82.

Using the pin vise with a suitable size probe, thoroughly clean the spray holes in the nozzle body end. A 0.013 (0.35 mm) cleaning needle should be used for early, two-hole nozzles; and a 0.009 (0.24 mm) cleaning needle for the four-hole nozzle used in direct injection engines.

Examine the stem and seat end of the nozzle valve and remove any carbon deposit using a clean, lint free cloth. Use extreme care, however, as any burr or small scratch may cause valve leakage or spray pattern distortion. If valve seat has a dull circumferential ring indicating wear or pitting, or if valve is blued, the valve and body should be turned over to an authorized diesel service station for possible overhaul.

Before reassembling, thoroughly rinse all parts in clean diesel fuel and make certain that all carbon is re-

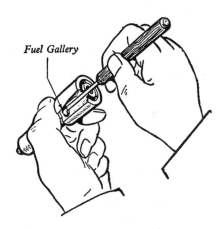

Fig. FM80 — Cleaning fuel feed channels in nozzle body.

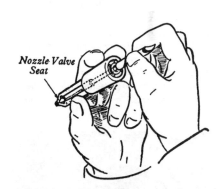

Fuel Gallery

Fig. MF81 — Cleaning fuel gallery in nozzle body.

Nozzle Valve Seat

Fig. MF82 — Cleaning valve seat in nozzle body.

moved from the nozzle holder nut. Examine body (4—Fig. MF77) and holder. On some nozzles, the body is located by a small dowel pin. On others, there are scribed lines on side of body and holder which must register when reassembling. Install nozzle body and holder nut, making certain that the valve stem is located in the hole of the holder body. Tighten the holder nut.

NOTE: Over-tightening may cause distortion and subsequent seizure of the nozzle valve.

Test the injector as outlined in paragraph 91. If the nozzle will not pass the tests, renew the nozzle valve and seat, which are available only as a matched set, or send the injector to an authorized C. A. V. diesel service station for a complete overhaul which includes reseating the nozzle valve cone and seat.

Overhaul injector body as follows: Remove injector jam nut (1—Fig. MF77) and adjusting screw (7) and withdraw spring (9) and spindle (10). Thoroughly wash all parts in clean diesel fuel and examine the end of the spindle which contacts the nozzle stem for any irregularities. If the contact surface is pitted or rough, renew the spindle. Renew any other parts which are questionable.

Reassemble the nozzle holder and leave the adjusting screw locknut loose until after the nozzle pressure has been adjusted as outlined in paragraph 91.

INJECTION PUMP

All Diesel Models

The injection pump is a completely sealed unit. No service work of any kind can be accomplished on the pump or governor without the use of special, costly pump testing equipment. The only adjustment which can be made without the use of a special test stand is the adjustment of the engine idle speed. If additional service work is required, the pump should be turned over to an authorized C.A.V. service station for overhaul. Inexperienced service personnel should never attempt to overhaul a diesel injection pump.

94. PUMP TIMING TO ENGINE (REGULAR DIESEL). The injection pump and drive gear adapter are equipped with a master spline. As long as the pump drive gear is in proper relation to the engine timing gear train as outlined in paragraph 53, the pump may be installed at any time without regard to crankshaft or timing mark location. When a new injection pump, pump drive gear or drive gear adapter is installed, or when incorrect pump timing is suspected, it will be necessary to check the pump timing as follows: Shut off fuel and remove the pump drive gear inspection cover (Fig. MF83), flywheel timing cover (Fig. MF84), and injection pump timing cover (Fig. MF85). Slowly crank engine until number one piston is coming up on the compression stroke and the flywheel "SPILL 18" timing mark is centered in the flywheel timing cover hole as shown in Fig. MF84. Check the timing marks on the timing gear housing, pump carrier plate and pump body flange as shown in Fig. MF86. Mounting holes in pump carrier plate and pump body flange are slotted to provide slight adjustment. If alignment is required, loosen mounting bolts, align the units and retighten bolts. With the flywheel spill mark centered in timing hole as shown in Fig. MF84, the "C" scribed

line on injection pump rotor should align with the scribed line at the lower hole of the pump snap ring as shown in Fig. MF85. If the pump timing marks are not in line, loosen the three cap screws (A—Fig. MF83) and rotate the drive hub and adapter on the drive gear until the pump timing marks are properly aligned. After timing pump, scribe new timing marks (TM) on the gear and adapter flange as shown in Fig. MF83 to assist in future service operations.

95. PUMP TIMING TO ENGINE (DIRECT INJECTION DIESEL). The injection pump drive shaft is fitted with a milled slot (S—Fig. MF87) in forward end which engages dowel pin (D) in pump drive gear. Thus injection pump can be removed and reinstalled without regard to timing position. NOTE: Injection pump drive

Copyright F. Perkins Ltd.

Fig. MF85 — Injection pump with timing cover removed showing timing marks. The "C" timing mark is aligned with scribe mark on snap ring as shown, when timing regular diesel. On direct injection diesel, the "B" timing mark is used.

Copyright F. Perkins Ltd.

Fig. MF86 — Left side view of engine showing timing marks on pump housing, carrier plate and timing gear housing. Master spline on pump drive shaft and adapter enables pump to be removed and installed without checking crankshaft location.

Fig. MF83 — Regular diesel pump drive gear with cover removed showing timing marks (TM). Adapter and hub can be moved in slotted holes in gear after loosening cap screws (A).

Copyright F. Perkins Ltd.

Fig. MF84—Flywheel timing cover removed showing injection timing mark used for regular diesel engines. Injection timing for direct injection diesel is 24°BTDC.

Fig. MF87—Disassembled view of injection pump drive used on direct injection diesels.

D. Locating dowel
G. Drive gear
P. Pump shaft
S. Locating slot
T. Timing marks

Fig. MF88—To remove the injection pump on late models, it is first necessary to remove the inspection cover and three gear retaining cap screws. Dowel (Arrow), times the drive gear to pump shaft.

Fig. MF89 — Installing the injection pump on a late model tractor.

C. Timing cover
S. Locating slot

gear cannot become unmeshed from idler gear when timing cover is installed; therefore, gear cover can be removed as shown in Fig. MF88, and pump removed and installed without altering pump timing.

To check the injection pump timing, shut off the fuel, remove the pump timing window (C—Fig. MF89) and the flywheel timing window on left front side of engine adapter plate. Injection timing is correct when 24° BTDC or "SPILL" timing mark on flywheel is aligned with scribe mark on timing window; the No. 1 piston is coming up on compression stroke; and the "B" timing mark cn pump rotor is aligned with scribed line on lower end of snap ring. The mounting holes in pump flange are elongated to allow for minor timing variations. If timing marks cannot be properly aligned by shifting pump on mounting studs, the timing gear cover must be removed as outlined in paragraph 52, and the gears retimed.

96. REMOVE AND REINSTALL INJECTION PUMP. Before attempting to remove the injection pump, thoroughly wash the pump and connections with clean diesel fuel. Disconnect injection lines from pump adapters and inlet line, outlet line and controls from pump.

On regular diesel models, remove the three cap screws securing pump to mounting plate and withdraw the pump.

On direct injection diesel models, remove the pump gear cover from front of timing gear cover, remove the three cap screws securing drive gear to pump shaft as shown in Fig. MF88; then, unbolt and remove pump.

To install the pump on regular diesel models, align the master spline on pump drive shaft and adapter hub, insert pump, install mounting bolts and align timing marks as shown in Fig. MF86. Tighten the mounting bolts securely and bleed the system as outlined in paragraph 89.

To install the pump on direct injection diesels, align the milled slot (S—Fig. MF89) in pump shaft with the locating dowel (Arrow—Fig. MF88) and insert the pump. Align timing marks as shown in Fig. MF90 and install and tighten the retaining stud nuts. Install the three cap screws securing drive gear to pump shaft and reinstall drive gear cover. Connect fuel lines and controls and bleed the system as outlined in paragraph 89.

On all models, if injection pump or drive components are renewed, timing should be checked as outlined in paragraph 94 or 95.

97. GOVERNOR. The mechanical governor is an integral part of the injection pump and no adjustments are recommended except adjustment of engine low idle speed and linkage adjustment as follows:

Engine low idle speed (see CONDENSED SERVICE DATA table) is controlled by stop screw (2—Fig. MF91). The recommended engine high idle speed is controlled by stop screw (4). This screw is factory adjusted and sealed and should not need adjustment in the field.

To adjust throttle linkage on regular diesel models, refer to Fig. MF92, and loosen nut on control lever clamp bolt (B). Pull hand throttle lever (1) down until cam (C) strikes stop pin

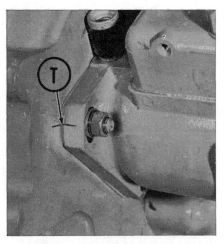

Fig. MF90 — Direct injection diesel tractor showing injection pump timing marks aligned. Refer to text.

Fig. MF91 — Injector pump, showing linkage and speed adjusting screws.

1. Shut-off control rod
2. Idle speed stop screw
3. Throttle control rod
4. Maximum speed stop screw

37

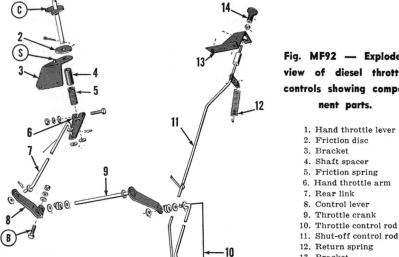

Fig. MF92 — Exploded view of diesel throttle controls showing component parts.

1. Hand throttle lever
2. Friction disc
3. Bracket
4. Shaft spacer
5. Friction spring
6. Hand throttle arm
7. Rear link
8. Control lever
9. Throttle crank
10. Throttle control rod
11. Shut-off control rod
12. Return spring
13. Bracket
14. Stop button

is conventional. Exhaust manifold must be removed to service the two center chambers. Always renew copper sealing washers when pre-combustion chambers are removed.

COLD WEATHER STARTING AID

All Diesel Models

99. The cold weather starting aid consists of a valve, solenoid and heating element located in the forward portion of the intake manifold on the right side of the tractor. Turning the starter switch to the "Heat" position activates the heating element and the fuel valve solenoid. When the fuel valve opens, fuel from the auxiliary fuel tank flows into coils and becomes vaporized. Further turning of the starter switch to "Heat-Start" position activates the starter and draws outside air into the manifold. This air mixes with the previously vaporized fuel, becomes ignited and pre-heats the incoming air for easier starting.

The starting aid is a self-contained unit. If starting aid does not function correctly, disconnect fuel line and electrical lead wire, remove the unit and install a new one.

Note: The starting aid must be installed with the stamped arrow in the direction of manifold air flow (toward rear of tractor) to insure correct operation.

(S) in support bracket; then, reach around to left side of steering housing and push throttle crank (9) forward until governor throttle arm on injection pump contacts high speed stop screw. While holding crank in this position, tighten nut on clamping bolt (B) securely. On direct injection diesels, arm (8) is pinned to crank (9) and rear link (7) is adjustable.

PRE-COMBUSTION CHAMBERS
Regular Diesel

98. Service or cleaning of the pre-combustion chambers is rarely necessary. Inspection and cleaning is advised when excessive smoking exists, fuel economy or power output drops, or when badly carbon fouled injectors are encountered. Removal and service

NON-DIESEL GOVERNOR

Except for slight structural differences between gasoline and LP-Gas models of the MF65 tractor, care and maintenance of the governor are the same. Refer to Fig. MF93.

100. **MINOR ADJUSTMENT.** Warm up the engine and adjust carburetor mixture. Disconnect governor to carburetor rod (28—Fig. MF93) at carburetor and adjust engine idle speed stop screw to obtain 400-450 rpm. Reconnect governor to carburetor rod. Set desired engine speed of 2100-2200 rpm as in the following paragraph.

Stop the engine and open hand throttle to wide open position. Loosen and relocate the control rod ball joint (16) on the governor control rod (15). Moving the ball joint forward (shortening) on the control rod increases engine rpm; backward (lengthening) decreases engine rpm. Retighten ball joint lock nut. Start engine and check for desired 2100-2200 rpm. If still not within limits, stop engine and repeat until correct engine rpm is obtained.

With engine operating at 1000 rpm, check for surging or unsteady running.

If surging exists, check for and remove any binding in the operating linkage.

Note: Engine speeds can be observed on tractormeter.

101. **MAJOR ADJUSTMENT.** This adjustment will cover the control linkage if same has been disassembled, or renewed.

With the engine stopped and the carburetor rod (28) disconnected from the governor arm (36), loosen the governor control rod ball joint (16).

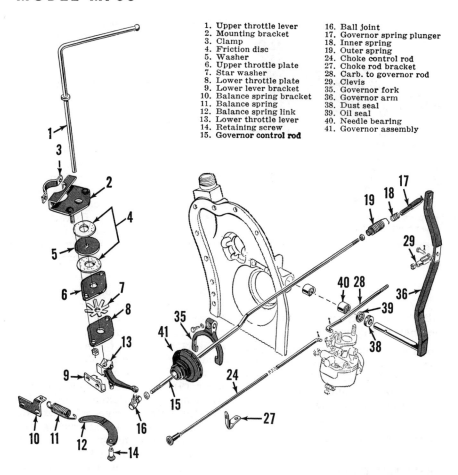

1.	Upper throttle lever	16.	Ball joint
2.	Mounting bracket	17.	Governor spring plunger
3.	Clamp	18.	Inner spring
4.	Friction disc	19.	Outer spring
5.	Washer	24.	Choke control rod
6.	Upper throttle plate	27.	Choke rod bracket
7.	Star washer	28.	Carb. to governor rod
8.	Lower throttle plate	29.	Clevis
9.	Lower lever bracket	35.	Governor fork
10.	Balance spring bracket	36.	Governor arm
11.	Balance spring	38.	Dust seal
12.	Balance spring link	39.	Oil seal
13.	Lower throttle lever	40.	Needle bearing
14.	Retaining screw	41.	Governor assembly
15.	**Governor control rod**		

Fig. MF93—Exploded view of non-diesel throttle, choke and governor controls and linkage.

Pull the carburetor control rod (28) rearward against the carburetor stop; pull the governor arm (36) to its rearmost position. Adjust the carburetor control rod clevis (29) until it is necessary to move the rod forward approximately $\frac{1}{32}$ inch to insert the clevis pin through the control rod and governor arm. Lock the pin and tighten the clevis lock nut.

With the governor arm in the forward (idle) position and the throttle hand lever in the upper (idle) position, tighten the governor control rod ball joint lock nut.

Make minor adjustment for engine speed as listed in paragraph 100. If the hand throttle tends to creep, prevent same by tightening the two friction plate lock nuts on the lower throttle friction plate (8), located under the instrument panel. If this does not remove the throttle creep, renew the cork washer.

NOTE: Correct throttle lever movement should not exceed 50° travel from lower to upper stop for the MF65 gasoline and LP-Gas tractor.

102. **R&R AND OVERHAUL.** First step in the removal of the governor is to remove the timing gear cover as outlined in paragraph 31. The governor fork and lever assembly, bearings and/or oil seal, Fig. MF93, located in the timing gear cover can be renewed at this time.

Remove the Woodruff key and withdraw the governor weight unit from the crankshaft.

To disassemble the governor weight unit remove the snap rings (49—Fig.

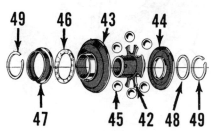

Fig. MF94 — Exploded view of flyball type governor assembly.

42.	Governor driver	46.	Thrust bearing
43.	Inner race	47.	Fork base
44.	Outer race	48.	Thrust washer
45.	Ball	49.	Snap ring

MF94). Renew the governor balls (45), the inner race (43) or outer race (44) if they are worn, scored or discolored.

Reassembly is the reverse of the disassembly procedure. When installing the timing gear cover, make certain the governor fork falls behind the governor weight unit.

COOLING SYSTEM

RADIATOR

Model MF-65 Non-Diesel

103. Radiator filler neck is fitted with a 9-11 pound pressure cap which raises the coolant boiling point 27-33 degrees.

To remove the radiator, first drain cooling system. Remove hood and grille assembly and disconnect upper and lower radiator hose connections. Remove radiator support rod and loosen the two cap screws retaining radiator to front axle support. Remove radiator by withdrawing same forward.

Model MF-65 Diesel

104. The radiator is equipped with a 10 psi pressure cap which raises the coolant boiling point approximately 28 degrees F.

To remove the radiator, first drain cooling system, remove hood and side panels, air cleaner and grille. On power steering equipped tractors, disconnect power steering reservoir bracket from fan shroud. Disconnect radiator hose connections, and support rod, remove two cap screws retaining radiator to front support and lift radiator and fan shroud forward off tractor.

THERMOSTAT

Model MF-65 Non-Diesel

105. Thermostat is located inside the engine water outlet casting and the renewal procedure is evident.

hose and pump outlet and bypass hoses. Unbolt and remove pump by withdrawing same from right side of tractor.

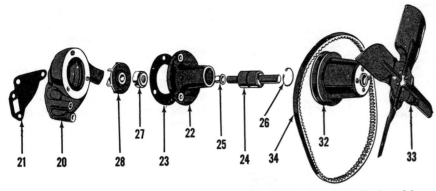

Fig. MF95—Exploded view of non-diesel water pump and fan assembly. Shaft and bearing (24) are available as an assembled unit only.

20. Pump body	23. Gasket	26. Snap ring	32. Drive pulley
21. Gasket	24. Pump shaft	27. Seal assembly	33. Fan
22. Pump support	25. Water slinger	28. Impeller	34. Fan belt

The lower right hand mounting bolt must remain in the housing until the pulley is removed and must be inserted before pulley is installed. To disassemble pump, first remove self-locking nut (1—Fig. MF96) and washer; then, remove pulley, using a suitable puller. Remove snap ring (3) from front of housing and press shaft (4) and bearings (5) forward out of housing and impeller. Extract felt bearing seal (8), retainers (7 & 9) and pump seal (10) from pump housing. Press bearings from shaft if renewal is indicated.

When assembling pump, press bearings (5) and spacer (6) on shaft so that open ends of the two bearings are together. Fill the bearings and the area between, half full of a high melting point grease before installing shaft in housing. Install pump seal (10), retainer (9), felt seal (8) and retainer (7) in housing. Retainer (7) must be installed with concave side next to bearing. Press shaft and bearing assembly into housing and install snap ring (3). Insert bolt (B) in housing and press pulley on shaft until nut (1) can be installed. Tighten nut to a torque of 55-60 Ft.-Lbs. Press impeller on shaft so that 0.015-0.025 clearance is maintained between inner edge of impeller blades and pump body. Impeller clearance is correct when back face of impeller is even with pump mounting flange rear surface. The two shorter cap screws (top right and lower left) are sealed with copper washers when reinstalling pump.

Model MF-65 Diesel

106. Thermostat is located inside the cylinder head outlet housing (15—Fig. MF96) and retained by the water outlet connection (17) which also serves as a mounting base for the fuel tank. To remove the thermostat, remove the hood and side panels, loosen fuel tank mountings and block up tank sufficiently to permit removal of outlet connection (17). Thermostat begins to open at 157-162 degrees F. and is fully open at 182 degrees F.

WATER PUMP AND FAN

Model MF-65 Non-Diesel

107. To remove the water pump, first drain cooling system and remove upper and lower radiator hose.

Remove the radiator. Remove fan blades and fan belt. Remove the three nuts retaining drive shaft support (22—Fig. MF95) to pump body and remove assembly (seal, impeller and

shaft). Pump body can be unbolted from engine if desired.

To disassemble pump, remove impeller and pump seal assembly. Remove snap ring (26) and press shaft and bearing assembly (24) toward front and out of shaft support.

Shaft and bearings are available as an assembled unit only.

Model MF-65 Diesel

108. To remove the water pump, first drain cooling system, loosen fan belt and unbolt fan blades from pump pulley. Remove hood and right side panel, then disconnect lower radiator

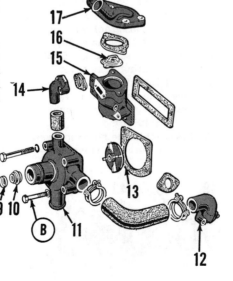

Fig. MF96 — Exploded view of diesel water pump, thermostat housing and connections.

1. Nut	5. Pump bearing	10. Pump seal	15. Thermostat housing
2. Pulley	6. Spacer	11. Pump body	16. Thermostat
3. Snap ring	7. Retainer	12. Block inlet	17. Outlet housing
4. Pump shaft	8. Felt seal	13. Impeller	
	9. Retainer	14. Bypass elbow	

Fig. MF97 — Timing mark on the MF65 gasoline tractor is 6° BTC as shown. On LP-Gas tractors, use the 4° BTC mark.

IGNITION AND ELECTRICAL SYSTEM

DISTRIBUTOR

Model MF65 Gasoline

109. A Delco-Remy distributor 1112583 is used. Specification data follows:

Breaker contact gap 0.022
Breaker arm spring tension . . 17-21 oz.
Cam angle . . . . 25-34°

Advance data is in distributor degrees and distributor rpm.

Start advance . . . . 0-2 @ 325
Intermediate advance . . . 5-7 @ 700
Maximum advance . . . . . 9-11 @ 1000

Model MF65 LP-Gas

110. Delco-Remy distributor model number 1112589 is used. Specification data follows:

Breaker contact gap. 0.022
Breaker arm spring tension . . 17-21 oz.
Cam angle 25-34

Advance data is in distributor degrees and distributor rpm.

Start advance . . . . 0-2 @ 300
Intermediate advance . . . . 6-8 @ 650
Maximum advance . . . . . . 9-11 @ 825

All Models

111. **TIMING.** To time the distributor, first set the ignition breaker contact gap to 0.022. Crank engine until the number one piston is coming up on compression stroke and continue cranking until the mark indicating 6 degrees B.T.D.C. for the gasoline model or 4 degrees B.T.D.C. for the LP-Gas model is in register with the groove in the timing hole as shown in Fig. MF97. Loosen the distributor clamp screw and turn distributor until breaker contacts just start to open and tighten the clamp screw.

Running spark timing can be checked with a neon timing light. With engine running at 2000 rpm, maximum advance is 26-30 degrees B.T.C. for the gasoline model and 24-28 degrees B.T.C. for the LP-Gas model.

GENERATOR, REGULATOR AND STARTING MOTOR

All Models

112. Delco-Remy electrical units are used. Test specifications are as follows:

Generator—D-R 1100362
Brush spring tension (oz.) 28
Field draw
  Volts 12.0
  Amperes 1.58-1.67
Output (cold)
  Maximum amperes 20
  Volts 14.0
  RPM 2300

Generator—D-R 1100998
Brush spring tension (oz.) 16
Field draw
  Volts 12.0
  Amperes 2.0-2.14
Output (hot)
  Maximum amperes 10-12
  Volts 14.0
  RPM 2800

Generator—D-R 1101012
Brush spring tension (oz.) 24
Field draw
  Volts 12.0
  Amperes 2.0-2.14
Output (hot)
  Maximum amperes 10-12
  Volts 14.0
  RPM 2800

Regulator—D-R 1118979
Cutout relay
  Air gap 0.020
  Point gap 0.020
  Closing voltage (range) 11.8-14.0
  Adjust to 12.8
Voltage regulator
  Air gap 0.075
  Voltage range 13.6-14.5
  Adjust to 14.0

Regulator—D-R 1118981
Cutout relay
  Air gap 0.020
  Point gap 0.020
  Closing voltage (range) 11.8-14.0
  Adjust to 12.8
Voltage regulator
  Air gap 0.075
  Voltage range 13.6-14.5
  Adjust to 14.0

Starting Motor—D-R 1107226
Brush spring tension (oz. min.) . . . 35
No-load test
  Volts 10.3
  Amperes 75
  RPM 6900
Lock test
  Volts 5.8
  Amperes 435
  Torque (ft.-lbs.) 10.5

Starting Motor—D-R 1107503
Brush spring tension (oz. min.) 24
No-load test (Including solenoid)
  Volts 11.8
  Amperes 55-85
  RPM (Min.) 6300
  RPM (Max.) 8700
Resistance test (Including solenoid)
  Volts 5.0
  Amperes (Min.) 700
  Amperes (Max.) 825

Starting Motor—D-R 1107508
Brush spring tension (oz. min.) . . . 35
No-load test
  Volts 10.6
  Amperes 75-95
  RPM (Min.) 6400
  RPM (Max.) 9500
Resistance test
  Volts 3.5
  Amperes (Min.) 545
  Amperes (Max.) 600

Starting Motor—D-R 1107512
Brush spring tension (min. oz.) 35
No-load test (Including solenoid)
  Volts 10.6
  Amperes 75-95
  RPM (Min.) 6400
  RPM (Max.) 9500
Resistance test (Including solenoid)
  Volts 3.5
  Amperes (Min.) 520
  Amperes (Max.) 590

Starting Motor—D-R 1107654
Brush spring tension (oz., min.) 35
No-load test
  Volts 10.3
  Amperes 75
  RPM 6900
Lock test
  Volts 5.8
  Amperes 435
  Torque (ft.-lbs.) 10.5

Starting Motor—D-R 1108662
Brush spring tension (oz., min.) 24
No-load test
  Volts 11.8
  Amperes 70
  RPM 6800
Lock test
  Volts 5.8
  Amperes 615
  Torque (ft.-lbs.) 29

CLUTCH

The MF65 tractor is equipped with a combination coil and Belleville spring dual clutch assembly, Auburn model 100183-2. Depressing the clutch pedal through its first stage retracts the primary pressure plate from its disc and interrupts the power flow to the transmission. Further depressing the clutch pedal through its second stage interrupts the power flow to the hydraulic pump and power take-off

113. PEDAL FREE PLAY ADJUST-MENT. To insure full clutch engagement and protect the clutch release bearing, it is important that adequate clearance exists between the clutch release fingers (Inset—Fig. MF99) and the clutch release bearing (25). This clearance is evident externally as free play of the clutch pedal and is checked by depressing the clutch pedal with the hand until the engagement of the release bearing with the release levers is felt. The forward edge of the clutch release shaft arm should be ⅛ inch from the transmission case as shown in Fig. MF100.

If the clutch pedal free play is not correct, adjust as follows: Loosen the clamp bolt (2) in the release shaft arm (1) and place a rod through the hole provided in the extended end of the clutch release shaft (3). Rotate the shaft clockwise until engagement with the clutch fingers is felt. Hold the release shaft in this position; then move the release shaft arm in relation to the shaft until the arm is ⅛ inch from the transmission case.

Tighten the clamp bolt (2) securely and recheck the clearance.

1. Transmission case
2. Clutch shaft bushing
3. Brake shaft bushing
4. Release shaft arm
5. Bottom cover
6. Gasket
7. Drain plug
21. Release shaft
22. Release pivot shaft
23. Release bearing fork
24. Release bearing carrier
25. Release bearing
26. Carrier spring
27. Fork screw
33. Clutch cover
34. Pressure plate (9")
35. Pressure plate (11")
36. Bellville spring
37. Driven disc assy.
39. Release lever
41. Upper lever pin
42. Dowel pin
46. Adjusting bolt
47. Lock nut
48. Lever link
49. Link pin
50. Torsion spring
51. Bracket bolt
52. Lock nut
54. Lower link pin
55. Pressure spring
56. Washer
57. Drive disc assy.
60. Air ring

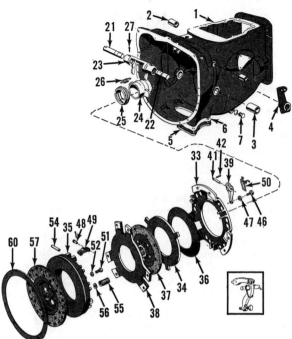

Fig. MF99 — Exploded view of the transmission case and clutch assembly used on the MF65 tractor.

114. CLUTCH LINKAGE ADJUST-MENT. Adjustment to compensate for lining wear is accomplished by adjusting the clutch pedal linkage, NOT by adjusting the position of the release levers on the clutch cover assembly.

To make the adjustment, remove the tractor pto cap and shift the pto lever to "Engine pto." Place a 1-inch block under the clutch pedal as shown in Fig. MF98. Start the tractor engine and allow it to run at idle speed; then slowly depress the clutch pedal and note the action of the pto shaft. If the pto shaft continues to rotate after the pedal reaches the block, remove the clutch pull rod clevis pin (1) and shorten the rod by rotating the clevis and recheck. If the pto shaft stops before the clutch pedal reaches the block, lengthen the rod slightly by turning the clevis and recheck.

When the correct adjustment is obtained, insert and secure the clevis pin and tighten the clevis lock nut.

115. REMOVE AND REINSTALL. To remove the clutch, first split tractor as outlined in paragraph 122 and proceed as follows: Make-up three special "T" bolts by welding a cross bar to ¼x6-inch screws, then add lock nuts. Install the special "T" bolts through the three holes in the clutch cover (33—Fig. MF99) to hold the

assembly together while removing it from the flywheel and tighten the "T" bolt lock nuts to compress the pressure springs. Note: In order to maintain the clutch balance when reassembling, punch assembly marks on the clutch cover (33), the 9-inch pressure plate (34), the flywheel plate (38) and the 11-inch pressure plate (35). Remove the six cap screws securing the cover assembly to the flywheel and remove the clutch as shown in Fig. MF102, using care not to damage or distort the air ring between the clutch assembly and the ventilated flywheel.

116. To install the dual clutch assembly, install the air ring on the flywheel using two $\frac{5}{16}$ x 3-inch guide

Fig. MF98—Adjusting the clutch pedal linkage. Refer to paragraph 114.

Fig. MF100 — Adjusting the clutch pedal free play. Refer to paragraph 113.

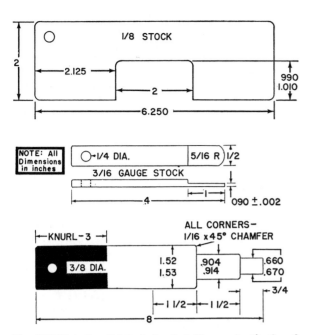

Fig. MF101 — Special tools for installing and adjusting the dual clutch used on the MF 65 tractor. Tools can be made, using the dimensions shown.

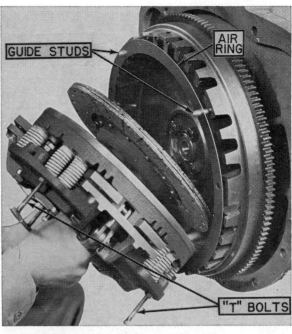

Fig. MF102—Install guide studs as shown.

studs as shown in Fig. MF102. Then proceed as follows:

Insert the pilot tool (Fig. MF101) through the driven discs and install the clutch cover into the flywheel.

Note: Be sure that the pressure plate Bellville spring (36—Fig. MF99) is centered in the cover assembly (33) when loosening the special "T" bolt lock nuts.

Remove the pilot tool and the three special "T" bolts. If a new clutch cover assembly is to be installed, it is also necessary to remove three cap screws which are used to secure the assembly together when received from stock.

117. **OVERHAUL.** NOTE: Dimensions of special tools necessary for servicing the clutch are shown in Fig. MF101. In addition, three special "T" bolts can be fabricated by welding a crossbar to a ¼ x 6-inch bolt, then adding a lock nut. To disassemble a removed clutch, proceed as follows: Unhook the three torsion springs (50 —Fig. MF99) from the clutch release levers (39). Loosen the adjustment cap screw lock nuts (52) and tighten the cap screws (51) until they bottom against the 11-inch pressure plate (35). Back off the special "T" bolts until it is possible to drive the roll pin (42) down sufficiently to remove the upper release lever pivot pin (41).

Fig. MF103 — Removing the upper release lever pivot pin.

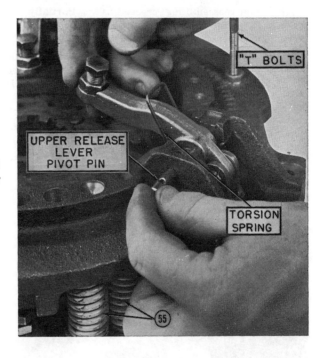

Note: Do not attempt to drive roll pins (42) completely out of the cover assembly. To do so will damage the cover assembly or Bellville spring. The pins can be removed very easily once the cover has been removed.

Remove the three upper release lever pivot pins and torsion springs (Fig. MF103); loosen the lock nuts on the special "T" bolts evenly until the coil springs (55—Fig. MF99) are fully extended and the clutch cover is free. Then remove the special "T" bolts.

The remainder of the disassembly is evident. The roll pins (42—Fig. MF99) can be removed at this time.

Coil spring specifications are as follows:

Color ....Lavender
Free length ....2 21/64 inches
Length @
   test load .....1½ in. @ 111-113 lbs.

Thoroughly clean and examine all other parts and renew any which are damaged or worn. Replacement linings are available for both driven plates.

Fig. MF104 — Adjusting the clutch release lever height. Refer to paragraph 120.

118. When reassembling, proceed as follows: Place the clutch cover upside down on a bench; then center the Bellville spring (36) in the cover groove with the convex side up. Place the pressure plate (34) on the spring, aligning the previously affixed punch marks. Install the driven disc (37) with the hub side down; place the smooth side of the flywheel plate (38) down on the disc and align the punch marks. Bolt the assembly together temporarily with three $\frac{5}{16}$ x 1½-inch bolts through the cover to flywheel plate mounting holes.

Partially insert the lower release lever pins (54) and coat them with "Lubriplate."

Place the eleven inch pressure plate (35) with friction face down on a bench and install the insulating washers (56) and pressure springs (55). Install the previously assembled clutch cover over the springs, aligning the punch marks. Install the three special "T" bolts until they bottom in the pressure plate (35). Tighten the "T" bolt lock nuts evenly until lower release lever links (49) can clear the

flywheel plate sufficiently to permit inserting of the lower release lever pins (54). Then tighten the "T" bolt lock nuts further to allow insertion of the upper release lever pins (41) through the clutch release levers (39) and torsion springs (50). Secure the upper release lever pivot pins with the roll pins (42) and hook the torsion spring ends. Finally, remove the three temporary 5/16 x 1½ inch bolts fastening the clutch cover and flywheel plate together. Adjust the unit as follows:

119. INITIAL ADJUSTMENTS. After the clutch assembly has been installed on the engine flywheel, two adjustments are necessary for proper clutch operation:

120. Clutch Release Lever Height. This adjustment is necessary to insure that the clutch pedal travel will fully disengage the primary and secondary clutch discs. Proceed as follows: When the clutch assembly has been installed, with a NEW primary disc, the release lever height is checked with the special gage shown in Fig. MF101. The release lever height is measured from the machined surface of the clutch cover to the bearing contacting surface on the adjusting screws as shown in Fig. MF104. Loosen the lock nuts and turn the adjusting screws in or out to obtain the proper height. Tighten the lock nut and recheck the adjustment.

Note: Clutch release lever height must be adjusted with a NEW primary disc installed. If it is desired to use a partially worn disc, the adjustment should first be made with a NEW disc, then the partially worn disc can be subsequently installed without changing the adjustment.

121. Primary Pressure Plate Free-Play. This adjustment determines the point in the clutch pedal travel where the secondary pressure plate begins to release. Proceed as follows: With the release levers properly adjusted as previously outlined, loosen the lock nuts (52—Fig. MF99) and turn each of the adjusting screws (51) in or out until the 0.090 end feeler gage (Fig. MF101) can just be inserted between the cap screw head and the secondary pressure plate as shown in Fig. MF-105. Tighten the lock nuts when adjustment is complete. Note: This adjustment can be made with either a new or used, but still serviceable, secondary lined disc in position.

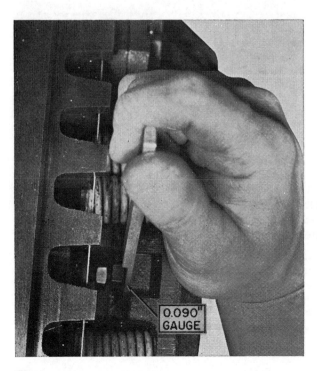

Fig. MF105 — Using the 0.090-inch gage to set the primary pressure plate free-play.

SPLITTING TRACTOR

All Models

A general procedure for splitting the tractor is given in the following paragraph. LP-Gas models will differ from gasoline models in obvious details due to differences in tank and lines.

122. To detach engine from transmission, proceed as follows: Drain cooling system and remove hood and side panels. Disconnect drag link from pitman arm and exhaust pipe from manifold. Disconnect heat indicator sending unit from engine and tractor-meter cable at front end. Disconnect oil gage line from cylinder block.

On gasoline models, shut off fuel and remove fuel line. Unbolt fuel tank from its rear support, loosen fuel tank front support bolts, and block up between fuel tank and rocker arm cover. Disconnect wires from coil and generator, choke rod from carburetor and governor spring plunger assembly from front end of throttle rod. Disconnect battery cables and remove battery.

On diesel models, shut off fuel and disconnect pressure line from primary fuel pump, fuel feed and return lines from secondary filter, disconnect fuel tank rear support and block up between fuel tank and rocker arm cover. Disconnect wiring harness from generator and cold starting unit. Disconnect battery cables and remove batteries, unbolt and remove starter and disconnect injector pump throttle and stop controls.

On all models unbolt battery platform from engine, support engine in a hoist and place a rolling floor jack under transmission case. Unbolt engine from transmission and separate the tractor halves.

DUAL RANGE TRANSMISSION

123. **REMOVE AND REINSTALL.** To remove the complete transmission housing assembly, first detach (split) engine from transmission as outlined in paragraph 122 and proceed as follows: Disconnect tail light wires and wires from starter safety switch. Unbolt and remove steering gear housing and transmission top cover assembly. Disconnect brake rods. Support transmission housing and rear axle center housing separately, then unbolt and separate the units.

124. **R&R TOP COVER.** The transmission top cover is integral with the steering gear housing. To remove the unit, proceed as outlined in paragraph 9.

125. **OVERHAUL.** Data on overhauling the various transmission components are as follows:

126. MAIN DRIVE SHAFT AND GEAR (CLUTCH SHAFT). To remove the main drive shaft and gear (29—Fig. MF107) first detach (split) engine from transmission housing as outlined in paragraph 122. Remove set screws retaining the clutch release fork to the shafts, pull the release fork shafts outward and remove the release fork. Remove the front pto housing bearing cap (41—Fig. MF106) and extract snap ring (28) from forward end of the live power take-off shaft. Use two puller cap screws in the tapped holes provided in the pto shaft front bearing housing (44) and remove the housing and bearing assembly. Remove thrust washer (46). Remove the internal snap ring (47) from the pto main drive gear and withdraw the front pto main drive shaft from transmission case. This allows the drive gear to drop down and clear the transmission main drive shaft retainer.

Unbolt the main drive shaft retainer from transmission case and withdraw the drive shaft and retainer as an assembly.

To disassemble the unit, remove snap ring (30—Fig. MF106) and withdraw the main drive gear and shaft. Remove snap ring (33) and bump the pto main drive shaft from the housing. Oil seals (37) in housing and (32) in the pto main drive shaft can be renewed at this time. The need and procedure for further disassembly is evident.

When reassembling, reverse the disassembly procedure and use a protector sleeve or shim stock to avoid damaging the seals when shafts are installed. Shoulder of thrust washer (46) goes toward front.

127. SHIFTER RAILS AND FORKS. To remove the transmission shifter rails and forks, first remove the transmission top cover and steering gear housing assembly and proceed as follows:

Disconnect the brake rods, step plates and exhaust pipe. Support both halves of tractor, unbolt transmission from rear axle center housing and separate the units.

Unwire and remove the set screws retaining the selector and shifter forks to the rails. Remove the detent springs and plungers. Unbolt and remove stop plate (16—Fig. MF108) and withdraw the shifter rails and forks from transmission case.

When reassembling, reverse the disassembly procedure. Forks (21 & 21A) are interchangeable, but rails for same are not. Rails should be installed with the milled flat toward rear and on top and with the selector lock grooves to the center.

128. MAIN (SLIDING GEAR) SHAFT. To remove the transmission mainshaft (24—Fig. MF107), first remove the transmission assembly (paragraph 123), main drive shaft and gear (paragraph 126) and the shifter rails and forks (paragraph 127).

Remove the four cap screws retaining the planetary unit to the transmission case and withdraw the planetary rear cover plate (60—Fig. MF106), thrust washer (61) and planet carrier (62). Using two screw drivers, work the planetary ring gear (69) and dowels from locating holes in case and remove the planetary front cover (70) and shim (71).

Remove snap ring (28) from front of mainshaft, block-up between the mainshaft sliding gears and carefully bump mainshaft rearward and out of transmission case. Rear bearing can be removed from shaft and front bearing can be removed from case at this time.

When reinstalling the shaft, proceed as follows: Place the mainshaft sliding gears in position; larger gear goes to the front with the fork groove toward rear and larger gear of cluster gear goes toward rear. Install the mainshaft and rear bearing assembly. Install the bronze thrust washer (61) in recess of planetary ring gear. Install cover (70) on ring gear dowels with oil grooves of cover toward the bronze thrust washer. Install shim (71) on dowels and use a lead hammer to bump dowels into transmission case, making certain that thrust washer is in ring gear recess and is free to rotate. Install planet carrier, other thrust washer (61) and rear cover (60) with oil grooves of cover toward the bronze thrust washer. Install and tighten cap screws retaining planetary unit to transmission case.

Remove snap ring (49) and pull the countershaft drive gear (50) from countershaft, thereby providing sufficient room to install the mainshaft front bearing. Use a suitable piece of pipe and drive the mainshaft front bearing into position and install the snap ring. Reinstall the countershaft drive gear (50) and its retaining snap ring (49).

129. COUNTERSHAFT. To remove the countershaft (53—Fig. MF107), first remove the mainshaft as outlined in paragraph 128 and proceed as follows: Remove snap ring (22—Fig. MF106) from rear end of countershaft and snap ring (49) and gear (50) from front end of shaft. Carefully bump the countershaft forward and out of case. Front bearing (51) can be removed from shaft and rear bearing (23) can be removed from case at this time. Renew the needle bearing (54) in rear end of shaft if bearing is damaged.

When reassembling, install the countershaft gears with the small gear to rear and with long hubs of gears together. Push the countershaft in as far as it will go and wedge a wooden block in the front compartment of the transmission housing to prevent the countershaft from moving forward while rear bearing is being drifted on. Using a suitable piece of pipe, drive the countershaft rear bearing into position, install the snap ring and remove the wooden wedge.

130. REVERSE IDLER. The reverse idler gear and shaft can be removed after removing the mainshaft as outlined in paragraph 128 and either before or after removing the countershaft. Remove cap screw and stop (77—Fig. MF106) and pull the idler shaft rearward.

The reverse idler gear contains needle bearings which can be renewed at this time. Install the reverse idler gear cluster with larger gear to rear and with a bronze thrust washer (76) on each side of gear.

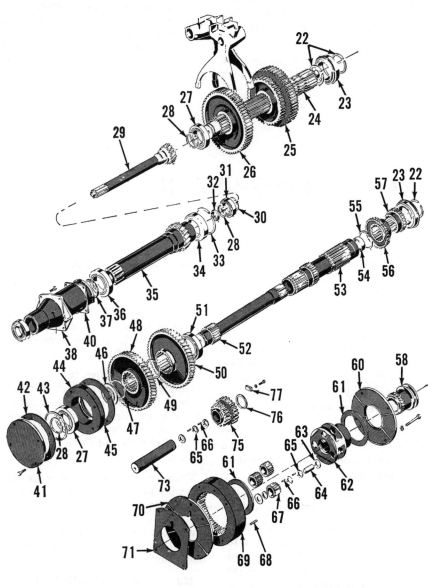

Fig. MF106 — Exploded view of transmission shafts and gears used on models without multipower. Many parts are interchangeable with multipower units.

22. Snap ring	40. Gasket	58. Planetary shift coupler
23. Ball bearing	41. Cap	60. Planetary rear cover plate
24. Mainshaft	42. Gasket	
25. Intermediate and high sliding gear	43. Snap ring	61. Thrust washer
	44. Pto shaft front bearing housing	62. Planet carrier
26. Low speed sliding gear	45. Gasket	63. Planetary pinion thrust washer
27. Ball bearing	46. Thrust washer	64. Shaft
28. Snap ring	47. Snap ring	65. Washer
29. Main drive shaft pinion	48. Pto main drive gear	66. Needle rollers
	49. Snap ring	67. Planetary pinion
30. Snap ring	50. Countershaft drive gear	68. Dowel pin
31. Ball bearing	51. Ball bearing	69. Planetary ring gear
32. Oil seal	52. Pto front main drive shaft	70. Planetary front cover
34. Ball bearing	53. Countershaft	71. Shim
35. Pto main drive shaft pinion	54. Needle bearing	73. Reverse shaft
36. Ball bearing	56. High speed pinion	75. Reverse cluster gear
37. Oil seal	57. Intermediate speed pinion	76. Washer
38. Main drive shaft retainer		77. Stop

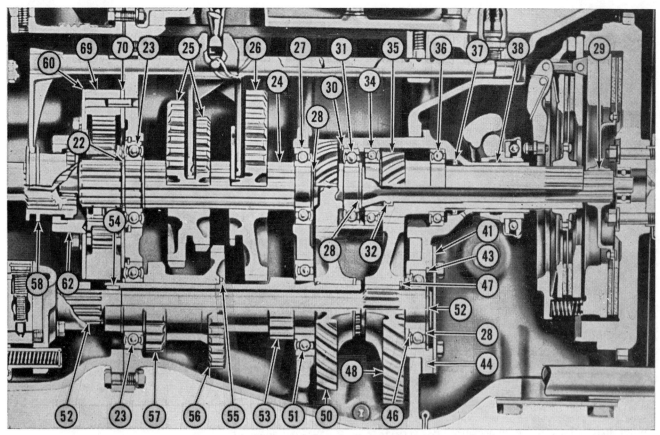

Fig. MF107—Cross sectional view of transmission shafts and gears used on tractors without multipower. Tractors with multipower are similar except for dual drive. Refer to Fig. MF106 for legend.

131. PLANETARY UNIT. The planetary unit can be removed after detaching the transmission from the rear axle center housing; or, if the planetary unit is the only unit to be serviced, it can be removed through the top of the rear axle center housing as follows:

Remove the hydraulic lift cover as outlined in paragraph 168 and proceed as follows:

Working through the top opening in the rear axle center housing, remove the large cotter pin, collapse and remove the rear drive shaft assembly. Remove set screw (5—Fig. MF108) and withdraw the planetary shifter fork and coupler (58—Fig. MF106). Remove the four cap screws retaining the planetary unit to the transmission case and withdraw the planetary rear cover plate (60), thrust washer (61) and planet carrier (62).

Using two screw drivers or equivalent, work the planetary ring gear (69) and dowels from locating holes in case and remove the front cover (70) and shim (71).

Using a small punch and hammer, drive the planetary pinion shafts (64) out of planet carrier and remove the planetary pinions and washers.

Thoroughly clean all parts and examine them for excessive wear. Renew any questionable parts. Heavy grease will facilitate installation of the planetary pinions and rollers.

To install the planetary unit, proceed as follows: Install one of the bronze thrust washers (61) in recess of planetary ring gear. Install cover (70) on ring gear dowels with oil grooves of cover toward the bronze thrust washer. Install shim (71) on dowels and use a lead hammer to bump dowels into transmission case, making certain that thrust washer is in ring gear recess and is free to rotate. Install planet carrier, other thrust washer (61) and rear cover (60) with oil grooves of cover toward the bronze thrust washer. Install and tighten the cap screws. Install the planetary shifter fork and coupler (58). When installing the hydraulic lift cover, refer to paragraph 168.

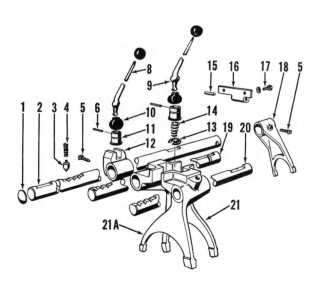

Fig. MF108 — Exploded view of shifter rails and forks. Forks (21 and 21A) are interchangeable, but rails (19 and 20) are not.

1. Expansion plug
2. Planetary rail
3. Detent plunger
4. Spring
5. Fork set screw
6. Pin
8. Planetary shift lever
9. Shift lever
10. Shift lever cover
11. Shift lever cup
12. **Planetary shift rail selector**
13. Spring seat
14. **Spring**
15. Selector lock pin
16. Lever stop plate
17. Cap screw
18. Planetary shifter fork
19. Low and reverse rail
20. Intermediate and high rail
21 & 22A. Shifter forks

MULTIPOWER TRANSMISSION

The Massey-Ferguson "Multipower" transmission is a modification of the standard dual-range, three speed transmission, providing an adidtional hydraulically operated high-low range, thus making available a total of twelve forward and four reverse gear speeds. The "Multipower" unit may be shifted while the tractor is moving, without disengaging the transmission clutch. Power for the "Multipower" clutch is supplied by a separate hydraulic pump which is mounted in the rear axle center housing on top of the regular tractor hydraulic pump.

SPLITTING TRACTOR

132. To detach (split) engine from transmission, follow the general procedure outlined in paragraph 122. NOTE: Do not attempt to split tractor by leaving steering gear attached to front unit, on tractors equipped with "Multipower" transmission.

Service on the "Multipower" control valve or relief valve; or renewal of input shaft oil seals, can be accomplished after the split. Any other service requires removal of the transmission as outlined in paragraph 135.

133. **CONTROL VALVE.** To remove the control valve, first detach (split) engine from transmission housing as outlined in paragraph 122. Remove clutch release bearing, release fork and shafts, and the brake cross shaft. Refer to Fig. MF110. Disconnect shift linkage (1) and remove shift bracket (3) and pressure tube. Remove the

cap screws securing the pto input shaft and retainer to front wall of transmission housing, and withdraw the pto input shaft, retainer and control valve as a unit from transmission housing.

When detaching control valve (5) from input shaft retainer, note that front Allen head cap screw is sealed with a copper washer. This cap screw retains the control valve spool. Use care not to drop or damage spool when this screw is removed.

The regulating valve (21 — Fig. MF115) is set to provide a pressure of 150 psi to the "Multipower" clutch and is not adjustable. Examine valve housing, control valve and regulating valve for scoring or excessive wear, and renew as indicated. Always renew the quad-ring seal on control valve spool when unit is disassembled; and make sure that copper sealing washer is installed on the screw which retains valve spool. Refer to Fig. MF115 for an exploded view of pump and control valve.

134. **PTO DRIVE SHAFT AND INPUT SHAFT SEALS.** To renew the input shaft oil seals, first remove the pto input shaft, retainer and "Multipower" control valve unit as outlined in paragraph 133; then remove the retaining snap ring (44—Fig. MF113). Withdraw the pto input shaft (46) and bearing (47) as a unit from the retainer.

Fig. MF110 — Front view of transmission housing showing control valve and associated parts.

1. Control rod	4. Clutch fork
2. Actuating lever	5. Control valve
3. Bracket	6. Input shaft

The outer seal (51) is provided with an external retainer which must be staked in place on front end of retainer. Inner seal (45) seats against a shoulder on inside of pto input shaft, and can be removed using the special Nuday puller (MFN 850). Install seal until it bottoms, with lip toward rear (gear) end of shaft.

Fig. MF109—On "Multipower" transmission, a range unit section is added ahead of main transmission unit. Power is transmitted through the two overdrive gears (O) when the hydraulically activated multiple disc clutch is engaged; or through the direct drive gears (D) and jaw-type overrunning clutch (J) when clutch is disengaged. Items (33 & 43) are snap rings which must be released during countershaft removal.

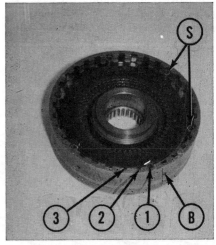

Fig. MF111 — Assembled view of "Multipower" clutch unit.

1. Assembly position, 1st clutch plate	3. Assembly position, 3rd clutch plate
2. Asembly position, 2nd clutch plate	B. Bleed holes
	S. Clutch return springs

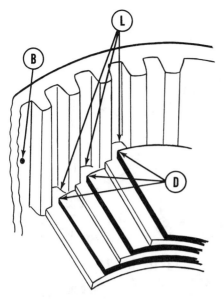

Fig. MF112 — Cross sectional view of "Multipower" clutch showing recommended method of assembly.

B. Bleed hole
D. Clutch discs
L. Drive lugs on clutch plates

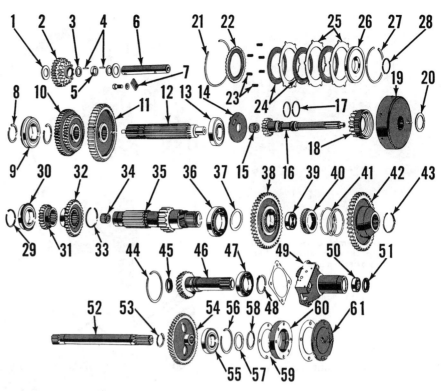

Fig. MF113—Exploded view of "Multipower" transmission shafts and gears. The range type planetary unit which attaches to rear of transmission (not shown) is identical to that used on other models.

1. Thrust washer	22. Retaining plate	42. Overdrive gear
2. Reverse cluster gear	23. Piston return	43. Snap ring
3. Spacer	springs	44. Snap ring
4. Needle roller	24. Clutch discs	45. Oil seal
5. Spacer	25. Clutch plates	46. PTO input shaft
6. Reverse shaft	26. Clutch piston	47. Ball bearing
7. Stop	27. Piston ring	48. Snap ring
8. Snap ring	28. Piston ring	49. Input shaft
9. Ball bearing	29. Snap ring	retainer
10. Sliding gear	30. Ball bearing	50. Oil seal
11. Sliding gear	31. Intermediate speed	51. Retainer
12. Rear mainshaft	pinion	52. PTO countershaft
13. Ball bearing	32. High speed pinion	53. Snap ring
14. Spacer	33. Snap ring	54. PTO countershaft
15. Pilot bearing	34. Pilot bearing	gear
16. Transmission input	35. Countershaft	55. Ball bearing
shaft	36. Ball bearing	56. Snap ring
17. Sealing washers	37. Thrust washer	57. Spacer
18. Overdrive pinion	38. Direct drive gear	58. Snap ring
19. Clutch drum	39. Bushing, drive gear	59. Gasket
20. Thrust washer	40. Drive clutch	60. Bearing housing
21. Snap ring	41. Clutch spring	61. Cap

Operating fluid for the "Multipower" clutch is directed through drilled passages in pto input shaft (46) and retainer (49), to the drilled transmission input shaft (16). This pressure passage is sealed by cast iron sealing rings (17) on transmission and pto input shafts. Examine sealing rings for breakage or wear; and seating surfaces on inside of pto input shaft and retainer for wear or scoring. Use care when assembling, to prevent damage to sealing rings and seating surfaces. Make sure that thrust washer (20) is installed brass side forward.

TRANSMISSION R&R

135. To remove the "Multipower" transmission, first drain transmission and rear axle center housing; then split transmission from engine as outlined in paragraph 122. Disconnect rear light wires, wires from starter safety switch, and the "Multipower" shift linkage; then unbolt and remove steering gear housing and transmission top cover assembly. Disconnect brake rods and the "Multipower" pressure hose at front end. Support transmission housing and rear axle center housing separately and unbolt and move transmission housing straight forward away from rear unit.

Reinstall by reversing the removal procedure. After transmission top cover assembly is installed, move the "Multipower" shift lever fully up-

ward to "High" position and, with clamping set screw loose, push control rod on top of transmission case down as far as possible. Tighten the clamping screw to secure the adjustment.

OVERHAUL

136. **SHIFTER RAILS AND FORKS.** Shifter mechanism is identical to that used on tractors without "Multipower" transmission, and can be serviced as outlined in paragraph 127.

137. **PLANETARY UNIT.** Service on the planetary unit is identical to that outlined in paragraph 131, except that detaching transmission from rear axle center housing is the recommended removal procedure. NOTE:

Before making the rear split, it will be necessary to remove the left inspection cover and disconnect the "Multipower" pressure line at the rear connection.

138. **MAINSHAFT.** To remove the transmission mainshaft (output shaft), first remove the transmission assembly. Remove rear planetary unit and shifter rails and forks. Remove snap ring (8—Fig. MF113) from shaft at rear of rear bearing (9), and drive the bearing rearward out of housing, using a punch. Jar the front bearing (13) from the shaft by sliding shaft rearward several times through the gears. Withdraw main shaft (12) from

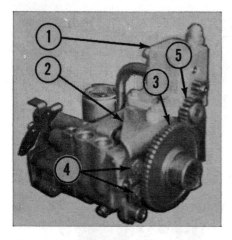

Fig. MF114 — Assembled view of tractor hydraulic pump showing attached "Multipower" pump.

1. "Multipower" pump
2. Mounting bracket
3. Drive gear
4. Through bolts
5. Idler gear

rear of housing while lifting gears (10 and 11) and front bearing (13) out the top.

When reinstalling, place the cluster gear (10) on shaft first with larger gear to the rear. Single gear (11) is installed forward of cluster gear with fork groove to the rear. Use a thin spacer plate as a support to bump front bearing (13) into position on shaft.

139. TRANSMISSION INPUT SHAFT AND "MULTIPOWER" CLUTCH. To remove the transmission input shaft (16—Fig. MF113), first remove the mainshaft as outlined in paragraph 138, and the pto input shaft as outlined in paragraph 133. Lift out the spacer (14) at rear of input shaft, then carefully withdraw input shaft rearward out of "Multipower" clutch (19) and overdrive pinion (18). Lift out the clutch unit through top of housing.

Input shaft contains a needle roller bearing (15), which pilots on front end of mainshaft (12), and cast iron sealing rings (17) for oil passages. Examine sealing rings and needle bearing, as well as inspecting the polished bearing and sealing surfaces of shaft for wear or other damage. Renew parts as required, and reassemble by reversing the disassembly procedure.

140. "MULTIPOWER" CLUTCH. To disassemble the removed "Multipower" clutch assembly (19), place unit on a clean bench with overdrive

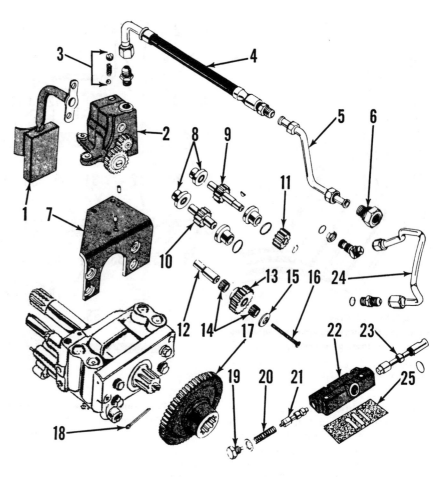

Fig. MF115—Exploded view of "Multipower" pump, control valve and associated parts.

1. Inlet strainer	9. Pump gear	18. Cotter pin
2. Pump assembly	10. Pump gear	19. Body plug
3. Relief valve	11. Driven gear	20. Regulator spring
4. Pressure hose	12. Idler shaft	21. Regulator valve
5. Pressure tube	13. Idler gear	22. Valve body
6. Connector	14. Idler bearings	23. Control valve
7. Mounting bracket	15. Washer	24. Pressure tube
8. Pump bearing	16. Retaining bolt	25. Gasket
	17. Drive gear	

pinion (18) up. Apply slight pressure to clutch retainer plate (22) and remove snap ring (21) with a narrow blade screwdriver. Completely disassemble clutch and examine the component parts for wear or scoring.

When reassembling, refer to Figs. MF111 and MF112. Note that the three clutch plates each contain six external driving lugs which are unevenly spaced around outside of plate. Lug spacing corresponds with the location of bleed holes (B—Fig. MF111) in clutch drum. With clutch piston installed, place the first clutch plate in drum with the six driving lugs one spline clockwise from bleed holes (B) as indicated by (1). Install an internally splined clutch disc; then the second plate with each driving lug one spline clockwise from lugs of first plate as shown at (2). Install a second clutch disc; then the third plate with driving lugs one spline

clockwise (3) from those of center plate. Install the last clutch disc, then place the six piston return springs (S) on the driving lugs of the lower plate as shown. When assembly is complete, the external driving lugs should appear staggered as shown in Fig. MF112, and the bleed holes (B) will be open and unobstructed. Install the overdrive pinion in clutch hub and secure with retainer plate and snap ring.

141. COUNTERSHAFT ASSEMBLY. To remove the countershaft (35 —Fig. MF113), first remove the transmission input shaft as outlined in paragraph 139. Remove the cap screws retaining the pto countershaft front bearing cover plate (61); remove snap ring (58) and spacer washer (57) from front of pto countershaft (52), then slide shaft out rear of transmission. Use two ⅜-inch NC cap screws as forcing screws and remove the bearing housing (60). The pto countershaft gear (54) can now be lifted out through top opening of housing.

Use the special Nuday clamping tool (MFN 830) or a small "C" clamp to secure the countershaft drive gears (38 and 42) in position, then remove snap rings (29 and 43) from front and rear of countershaft. Insert a step plate of proper size in rear end of countershaft; then bump shaft forward slightly, unseat snap ring (33) in front of high speed pinion and move snap ring forward on shaft.

Insert the step plate in forward end of countershaft, and drift shaft rearward until the countershaft driving gears (38 & 42) and overrunning clutch assembly can be slipped off forward end of shaft and removed. Drive countershaft forward until free of rear bearing (30); then remove shaft by working it forward while lifting the contained gears out top of housing.

Assemble by reversing the disassembly procedure. The two pinions (31 & 32) must be installed with hubs together, and with the larger (high speed) pinion forward. The countershaft driving gears (38 & 42) and overrunning clutch assembly may be installed as a unit (by using Nuday Tool MFN 830); or individually. If Nuday tool is used, leave clamping screws slightly loose until splines are engaged, then tighten clamp and leave in place until all snap rings are seated to minimize bouncing of overdrive gear against snap ring.

142. REVERSE IDLER. The reverse idler gear (2—Fig. MF113) and shaft (6) can be removed after removing the transmission main shaft, and either before or after removing countershaft. Remove the retaining cap screw and stop (7) at rear of reverse idler gear shaft, and slide shaft rearward out of gear assembly. The gear (2) contains 56 loose needle bearing rollers (4), arranged in a double row separated by a spacer washer (5).

Install reverse idler with larger gear to rear with a bronze thrust washer (1) on each side of gear.

"MULTIPOWER" PUMP

Power for the "Multipower" disc clutch is supplied by a separate gear-type pump which mounts on top of the regular hydraulic pump in the rear axle center housing. The "Multipower" pump is driven by a gear (3—Fig. MF114) which is attached to the front of the main hydraulic pump camshaft by a cotter pin, and which serves as the pto countershaft coupling.

143. REMOVE & REINSTALL. To remove the "Multipower" pump, first remove the hydralic lift cover as outlined in paragraph 168. Disconnect the "Multipower" pressure line from top of pump; remove the three cap screws securing pump (1—Fig. MF114) to mounting bracket (2) and lift the pump from tractor. The regular tractor hydralic pump can now be removed for service as outlined in paragraph 166.

The "Multipower" pump mounting bracket (2) is attached to front of regular tractor hydraulic pump by the four through-bolts (4) which secure the pump assembly. Cap screw holes in bracket are slightly oversize to permit adjustment of backlash between pump drive gear (3) and idler gear (5). The recommended backlash is 0.002-0.005. If any of the parts are renewed, or if tractor hydraulic pump is disassembled, the backlash must be checked and readjusted as follows:

Assemble the tractor hydraulic pump, leaving the four through-bolts (4) loose. Temporarily install drive gear (3) and "Multipower" pump (1). Shift the "Multipower" pump and bracket assembly to remove all gear backlash without binding; then tighten the two accessible through-bolts to prevent shifting of bracket. Remove drive gear (3) and securely tighten all through-bolts.

Remove the "Multipower" pump from mounting bracket, reinstall drive gear (3) and secure with cotter pin; then install tractor hydraulic pump assembly in tractor as outlined in paragraph 166. After hydraulic pump has been aligned and secured, reinstall the "Multipower" pump, using shims, if necessary, between pump and mounting bracket, to obtain the necessary 0.002-0.005 backlash.

NOTE: Backlash cannot be correctly established until hydraulic pump unit is positioned in tractor, because of loose fit of drive gear on pump shaft splines.

144. OVERHAUL. Refer to Fig. MF115. End bearings (8) are pressure loaded and balanced to automatically control gear end clearance. Note location of bushings (8) and gears (9 and 10) upon disassembly, and return them to same relative positions when reassembling. All component parts are available individually.

DIFFERENTIAL, BEVEL GEARS & FINAL DRIVE

DIFFERENTIAL

145. REMOVE & REINSTALL. The ring gear and differential unit can be removed after removing the complete left final drive unit as outlined in paragraph 152.

Prior to 1962, no **provision was** made for adjustment of differential carrier bearing pre-load or main drive bevel gear backlash other than the renewal of parts. At about the beginning of 1962, a selective thickness shield (22—Fig. MF117) was made available to control differential bearing pre-load on models equipped with differential lock. A 0.005 thickness shim (Part No. 892 180 M1) was made available to be inserted between bearing cone (8—Fig. MF116) and Carrier plate (6—Fig. MF119) to be used as required to control bearing pre-load on models without differential lock. At the same time, a shim was provided for installation between main drive bevel gear (11—Fig. MF116) and the flange of differential case (2) where required for backlash adjustment. When renewing any of the parts carefully measure

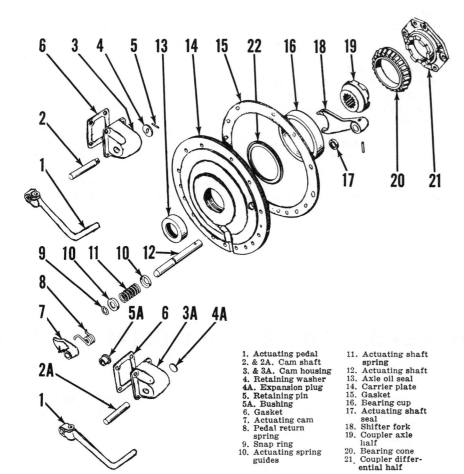

2. Differential case assembly
3. Bearing cup
4. Bearing cone
5. Outer differential case
6. Case adapter assy.
6A. Coupling half (differential lock)
7. Bearing cup
8. Bearing cone
8A. Cone (differential lock)

11. Ring gear
12. Bevel pinion
16. Differential side gear
17. Thrust washer
18. Differential spider
19. Differential pinion
20. Thrust washer
25. Pilot bearing
26. Snap ring
27. Sleeve assembly
28. Dowel pin
29. Bearing cup
30. Bearing cone
31. Drive gear hub
32. Lock washer
33. Adjustment nut
34. Drive gear (pto)
35. Snap ring

Fig. MF116—Exploded view of main drive bevel gears, differential and associated parts.

1. Actuating pedal
2. & 2A. Cam shaft
3. & 3A. Cam housing
4. Retaining washer
4A. Expansion plug
5. Retaining pin
5A. Bushing
6. Gasket
7. Actuating cam
8. Pedal return spring
9. Snap ring
10. Actuating spring guides
11. Actuating shaft spring
12. Actuating shaft
13. Axle oil seal
14. Carrier plate
15. Gasket
16. Bearing cup
17. Actuating shaft seal
18. Shifter fork
19. Coupler axle half
20. Bearing cone
21. Coupler differential half

Fig. MF117—Exploded view of right differential carrier plate and differential lock components. Items (1 through 5) in upper left hand corner were superceded in later production by alternate type shown.

the installed parts and renew with parts of the same thickness.

Factory recommendations call for a pre-load of 0.006-0.018 for the carrier bearings of tractors equipped with differential lock; or 0.006-0.016 for tractors without differential lock. Bevel gear backlash should be 0.008-0.012 on all models.

146. OVERHAUL. To disassemble the removed differential unit, first place correlation marks on both halves of the differential case to insure correct reassembly. Remove the eight retaining bolts and separate the case halves. Differential pinions (19—Fig. MF116), spider (18) and side gears (16) can now be removed. Recommended backlash of 0.003-0.008 between the differential pinions and side gears is controlled by the side gear thrust washers (17) and/or pinion thrust washers (20).

The main drive bevel ring gear is retained to the differential case by bolts or rivets. Recommended procedure for removing the rivets is by drilling.

Where ring gear is bolted to differential case, nuts are secured by using LOCTITE, grade AV. When removing the bolted on ring gear, heat the nuts to approximately 480° F, using an open flame. When reinstalling, old sealant should be thoroughly removed from bolts or preferably, new attaching parts used. Nuts and bolts must be thoroughly cleaned and degreased using trichlorethylene or similar solvent, then two drops of grade AV, LOCTITE applied to end threads of each bolt. Tighten the nuts to a torque of 110-120 Ft.-Lbs. New bolts and nuts are included with new gear assembly, but not with differential case.

Ring gear runout should not exceed 0.002.

Tooth contact (mesh pattern) of the main drive bevel gears is fixed and non-adjustable.

DIFFERENTIAL LOCK

147. OPERATION. The mechanically actuated differential lock assembly is furnished as standard equipment on the MF65 Diesel tractor and as a factory installed option on all other MF65 models. The assembly consists of a foot pedal, actuating mechanism and differential lock coupler.

When the pedal is depressed, the axle half of the coupler assembly is moved inward to contact the differential case half of the coupler. If slippage is occurring at one wheel, depressing the pedal will cause the coupler dogs to contact each other and lock the differential case to the right axle. The differential case and both drive wheels will then rotate together as a unit. As soon as contact is made by the coupler dogs, coupler contact pressure will keep the differential lock engaged and the foot pedal may be released. When ground traction on both drive wheels again becomes equal, coupler contact pressure will be relieved and the lock shaft spring will disengage the coupler.

See Fig. MF117 for a view of the component parts of the differential lock assembly. Rear axle, final drive housing and differential assembly are common to all tractor models, whether or not differential lock is installed. Carrier plate seal (13) is identical to seal used in differential carrier plate on models without differential lock.

148. ADJUSTMENT. The differential lock should be fully engaged when the pedal contacts the tractor step plate. To adjust, loosen the pedal clamp bolt slightly so that pedal can be moved on pedal shaft. Bolt must be tight enough that downward movement will engage coupler. Shift pedal on shaft by pulling same upward beyond its normal position; then, engage coupler by pressing pedal down until fully engaged. Continue pushing pedal down until it contacts step plate and tighten clamp bolt.

Note: If axle and differential halves of coupler are so turned that ends of coupler dogs contact, coupler assembly cannot be engaged. To check, depress pedal before starting to make the adjustment. If pedal contacts step plate, coupler is properly aligned. If pedal is obstructed from movement some distance from step plate, start tractor and pull forward or backward in a circle while depressing differential lock pedal. Stop tractor at the instant further pedal movement indicates that coupler halves are properly aligned, and adjust as indicated above.

149. REMOVE AND REINSTALL. To remove differential lock coupler halves, first drain the transmission and hydraulic reservoir, block up under rear axle center housing and remove right fender and rear tire and wheel assembly. Remove right lower link and disconnect right brake linkage. Support right final drive assembly in

a hoist and remove retaining nuts. Slide right final drive as a unit away from the rear axle center housing.

Remove the bearing cone and roller assembly (20—Fig. MF117) from the differential coupling half, remove safety wire and unbolt and remove the coupler half (21). To remove the axle half of the coupler, first remove the two Allen head screws from the carrier plate (14) and the roll pin from the shifter fork (18); then, withdraw the fork, coupler half and carrier plate from the final drive assembly as a unit. Be careful while removing plate not to damage axle seal (13) or shaft seal (17).

To reassemble, reverse the disassembly procedure using care not to damage the seals.

To remove the differential lock actuating shaft (12) and spring (11), first remove the right brake lever support from the final drive housing, unbolt and remove the cam housing (3 or 3A) with its attached parts and drive the roll pin from the shifter fork (18). Working through the hole in the final drive housing, unseat the snap ring (9) and withdraw the shaft from the differential end of the final drive housing. Spring (11) and spring guides (10) may be renewed at this time. Procedure and need for further disassembly of the cam housing assembly is evident. Refer to Fig. MF117.

MAIN DRIVE BEVEL GEARS

150. BEVEL PINION. The main drive bevel pinion is available only in a matched kit which also includes bevel ring gear and attaching bolts and nuts.

To remove the bevel pinion, first remove the hydraulic lift cover as outlined in paragraph 168 and proceed as follows:

Working through the top opening in the rear axle center housing, remove the large cotter pin, collapse and remove the rear drive shaft assembly. Unbolt pinion bearing sleeve (27—Fig. MF116) from case wall and using two jack screws in the tapped holes provided in the sleeve flange, pull the pinion and sleeve assembly forward and out of case bore.

To disassemble the unit, remove snap ring (35) and gear (34). Unlock and remove nut (33) and bump pinion out of sleeve (27). If bearing cups in sleeve are damaged, renew same.

When reassembling, tighten nut (33) to provide a perceptible bearing preload, or when 6-8 inch pounds of torque are required to rotate the pinion shaft in its bearings.

Bearing cups (29) are an integral part of sleeve (27), and are not available separately. Two different sleeve assemblies have been used. When renewal is required, make sure the correct parts are installed. NOTE: Each transmission assembly bears a serial number which is located next to safety switch opening on top of housing. Bearing change was made at Transmission Serial Number CL2686-IL2.

Install the pinion assembly by reversing the removal procedure and when installing the hydraulic lift cover, refer to paragraph 168.

151. BEVEL RING GEAR. The main drive bevel ring gear is only available in a matched kit which includes ring gear, pinion and attaching bolts and nuts.

To remove the main drive bevel ring gear, first remove differential assembly as outlined in paragraph 145. On early models where ring gear is riveted to differential case, rivets should be drilled for removal. When bolts and nuts are used for attaching ring gear, nuts are secured by using LOCTITE, grade AV. Before removing, heat nuts to approximately 480° F., using an open flame.

Make sure mating faces of ring gear and differential case are thoroughly clean and free from nicks or burrs. Degrease the new attaching bolts and nuts by using trichlorethylene or similar solvent; and apply two drops of Grade AV, LOCTITE to outer threads of bolts before installing nuts. Tighten the nuts to a torque of 110-120 Ft.-Lbs. and check the runout of ring gear, which should not exceed 0.002.

Mesh position of main drive bevel gears is fixed and non-adjustable. Differential carrier bearing pre-load is not adjustable except that a selective thickness bearing shield is used between bearing cup and carrier plate on right side of models equipped with differential lock. Shield is available in five thicknesses varying from 0.034 to 0.060. If any of the parts are renewed, it is essential that correct thickness shield be installed as outlined in paragraph 145.

FINAL DRIVE

The final drive unit used on all models consists of a long axle shaft which is splined into the differential side gears, plus a planetary gear reduction unit mounted on outer end of axle housing as shown in Fig. MF118. Effective with Lo-Clearance tractor serial number 662236; or Hi-Clearance tractor serial number 660007, the wheel axle inner bearing was changed from a taper roller bearing (16 and 17—Fig. MF119) to axial and thrust needle bearings (1 through 6—Fig. MF120). Wheel axle shaft and long axle shaft were changed at the same time; and drive cover (22—Fig. MF119) was modified to include a drain plug.

152. REMOVE AND REINSTALL. To remove either one of the axle shaft housings and components as an assembly, jack up under differential housing and remove fender and rear tire and wheel assembly. Drain transmission and disconnect upper and lower links from lift arm. Disconnect brake rod (44—Fig. MF123) by removing

yoke pin (47). Remove lever pin (45) and brake rod lock nut, then withdraw the brake rod lever (17). Remove cap screws retaining rear axle housing to differential (center) housing, then support and remove the final drive assembly.

Reinstall the unit in reverse order of removal, using only one standard gasket between differential (center) housing and rear axle housing to provide the correct differential carrier bearing adjustment.

153. AXLE SHAFT. To remove either rear axle shaft (15—Fig. MF119 or 10—Fig. MF120), proceed as follows: Jack up rear of tractor and remove rear fender and tire and wheel assembly. Drain the planetary drive housing. Lightly set and lock the brake on side from which axle shaft is being removed. CAUTION: If brake is released while axle is removed, brake discs will become misaligned and axle housing will need to be detached from center housing to reinstall shaft.

Scribe a line across the drive cover (22—Fig. MF119), planetary ring gear (30) and the rear axle housing (1). Remove the bolts retaining the drive cover to the rear axle housing; then remove the drive cover, wheel axle shaft and planet carrier assembly. Withdraw the rear axle shaft (15—Fig. MF119 or 10—Fig. MF120) from the axle housing.

Bearing cup (3—Fig. MF119), oil seal (4) and bearing cone (5) can be renewed at this time.

Install the shaft by reversing the removal procedure; be sure to check the wheel axle shaft bearing pre-load as outlined in paragraph 155 and align the previously affixed scribe marks.

154. R & R AND OVERHAUL WHEEL AXLE SHAFT AND PLANETARY GEAR ASSEMBLY. Proceed as follows: Drain planetary gear housing and jack up rear of tractor. Remove fender and rear tire and wheel assembly. Scribe a line across drive cover, ring gear and axle housing, then unbolt and remove the

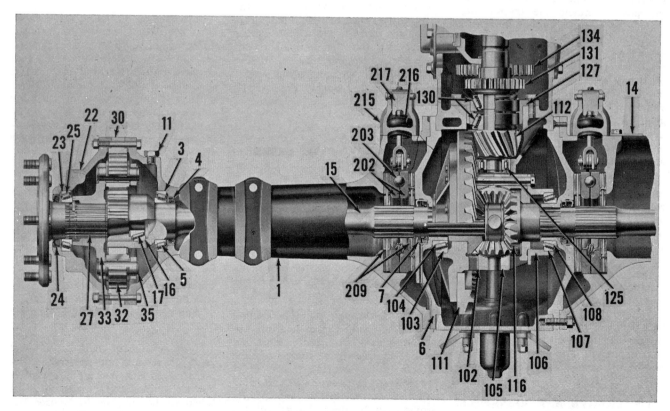

Fig. MF118—Cut-away view of the final drive assembly and differential housing used on the MF65 tractor.

1. Axle housing	17. Bearing cone	102. Case assembly	125. Pilot bearing
3. Bearing cup	22. Cover assy.	103. Bearing cup	127. Sleeve assembly
4. Oil seal	23. Bearing cup	104. Bearing cone	130. Bearing cone
5. Bearing cone	24. Oil seal	105. Differential case	131. Drive gear
6. Carrier plate	25. Bearing cone	106. Adapter assy	202. Actuating disc
7. Oil seal	27. Wheel axle	107. Bearing cup	203. Ball
11. Drain plug	30. Ring gear	108. Bearing cone	209. Lined disc
14. Cover plate	32. Pinion	111. Ring gear	215. Lever support
15. Axle shaft	33. Pinion carrier	112. Bevel pinion	216. Dust cover
16. Bearing cup	35. Pinion shaft	116. Differential gear	217. Brake lever

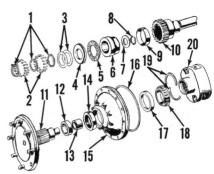

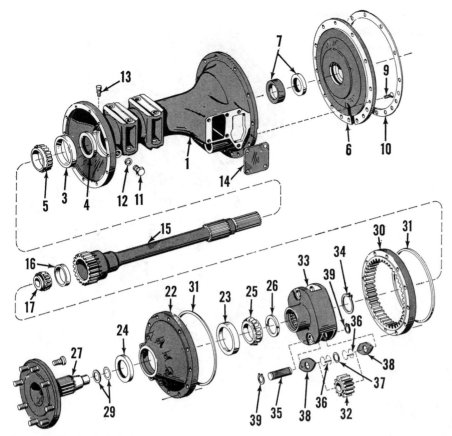

Fig. MF120—Late type final drive showing components.

1. Spacer	11. Wheel axle
2. Needle rollers	12. Axle sleeve
3. Shims	13. Axle sleeve
4. Thrust washer	14. Oil seal
5. Thrust bearing	15. Drive cover
6. Bearing cone	16. Gasket
7. Washer	17. Bearing cup
8. Snap ring	18. Bearing cone
9. Bearing cup	19. Ring halves
10. Long axle	20. Planet carrier

Fig. MF119—Exploded view of final drive, rear axle shaft, wheel axle shaft and components used on early models. Late models are the same except for planetary unit shown in Fig. MF120.

1. Housing, R. H.	11. Drain plug	23. Bearing cup	32. Pinion
3. Bearing cup	12. Gasket	24. Oil seal	33. Pinion carrier
4. Oil seal	13. Breather	25. Bearing cone	34. Snap ring
5. Bearing cone	14. Cover plate	26. Ring halves	35. Pinion shaft
6. Carrier plate	15. Rear axle shaft	27. Wheel axle	36. Roller needle
7. Oil seal assy.	16. Bearing cup	29. Shims	37. Washer
9. Allen screw	17. Bearing cone	30. Ring gear	38. Washer
10. Gasket	22. Drive cover	31. Gasket	39. Snap ring

wheel axle shaft, drive cover unit and planet carrier assembly.

On early models, press bearing cone (17—Fig. MF119) from end of wheel axle and remove shim pack (29). Extract snap ring (34) from inner end of wheel axle shaft (27) and slide the planet carrier assembly from axle shaft. Remove bearing support ring halves (26) and press the wheel axle shaft out of drive cover (22). Oil seal (24), bearing cup (23) and cone (25) can be renewed at this time.

On late models, remove snap ring (8—Fig. MF120) from end of wheel axle shaft and withdraw bearing assembly (1 through 7) and sleeves (12 and 13). Remove the retaining snap ring and withdraw planet carrier (20). Remove the bearing support ring halves (19) and press wheel axle shaft (11) out of bearing cone (18) and drive cover (15). Oil seal (14)

bearing cup (17) or cone (18) can be renewed at this time.

To remove the planet pinions (32—Fig. MF119) on all models, remove snap ring (39) from each side of pinion shaft (35); then remove the shaft, using a press. Carefully withdraw pinion (32) and thrust washers (38), being careful not to lose the 58 loose rollers (36) and spacer washer (37). Clean the pinion and bearings and renew any parts which are scored or worn. Reassemble by reversing the disassembly procedure, using heavy grease to hold loose rollers in position during assembly.

When assembling wheel axle, drive cover and planetary unit, use shim stock or installing sleeve on wheel axle splines to protect oil seal lip, and install drive cover and seal. Press the bearing cone (25—Fig. MF119 or 18

—Fig. MF120) on axle shaft only far enough to install the split bearing retainer. Install split retainers with chamfered edge away from bearing; then seat the bearing against split retainer, using the press. Install planet carrier assembly and retaining snap ring, then install and adjust wheel axle inner bearing as outlined in paragraph 155.

155. On early models, make a trial assembly by installing bearing cone (17—Fig. MF119) and all of the removed shims (29) except one 0.010 thick shim. Be sure to align the previously affixed scribe marks and, using a dial indicator, check and record the end play of the wheel axle shaft (27). Remove the wheel axle and drive cover assembly and bearing cone (17) and add shims (29) equal to the measured end play plus 0.001-0.005. This procedure will provide the wheel axle shaft bearings with the recommended preload of 0.001-0.005.

On late models, make a trial assembly by omitting one 0.010 thick shim (3—Fig. MF120). Use a dial indicator and check the end play of wheel axle shaft (11). Remove drive cover (15) and wheel axle shaft assembly, remove snap ring (8), bearing cone (6); then, add shims (3) equal to the measured thickness plus 0.001-0.005, to establish the recommended preload.

On all models, use new gaskets for final assembly, and tighten the assembly bolts to a torque of 45-50 Ft.-Lbs.

BRAKES

The MF65 tractor is equipped with seven-inch diameter double disc brakes (Fig. MF121) fitted internally in the inner end of each rear axle housing and separated from the differential housing by a plate (6—Fig. MF119). Machined surfaces on the plate and axle housing serve as stationary drums for the brake.

Fig. MF123 — Exploded view of the brake, actuating assembly and component parts.

2. Actuating disc
3. Actuating ball
4. Actuating spring
5. Plain link
6. Yoke link
7. Stud
9. Lined discs
13. Brake rod
15. Lever support
16. Dust cover
17. Brake lever
19. Lever return spring
20. Clip
21. Adjusting block
44. Connecting rod

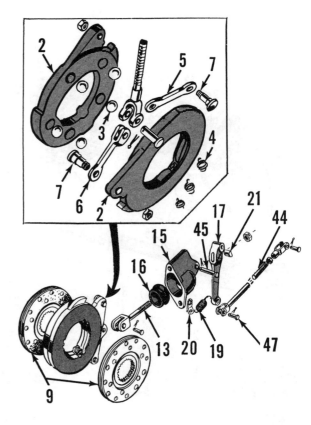

Fig. MF121—A partially exploded view of the brake assembly used on the MF65 tractor. Refer to Fig. MF123 for legend.

BELT PULLEY UNIT

Fig. MF122—Brake linkage is properly adjusted when pedal free play is 2½ to 3 inches. Refer to paragraph 156.

2½ TO 3"

156. ADJUSTMENT. To make a brake adjustment, proceed as follows: Turn the adjusting nut either way until a brake pedal free play of 2½-3 inches is obtained on both left and right brake pedals. Refer to Fig. MF122.

157. R&R AND OVERHAUL. Proceed as follows: Remove the final drive units as outlined in paragraph 152. Remove two Allen screws (9—Fig. MF119) retaining differential carrier plate (6) to axle housing and remove plate. Lift out both lined discs (9—Fig. MF121) and the actuating disc assembly. When removing right brake on tractors equipped with differential lock, the actuating fork and sliding coupler must be removed as outlined in paragraph 149. When tractor is assembled, check adjustment of differential lock as outlined in paragraph 148.

Separate the actuating discs and renew any parts which are questionable.

Assembly is the reverse of disassembly.

The belt pulley unit, Fig. MF124,, mounted at the rear of the rear axle center housing, is a self contained drive unit and is driven by the power take-off shaft. Belt pulley rotation in either direction can be obtained by mounting the unit so that the pulley is either to the right or left of the tractor center line. Pulley can also be located in the down position.

CAUTION: Never mount pulley on the right side unless a vertical muffler is installed. Never mount pulley in the "up" position, as the then top bearing will not receive proper lubrication.

158. OVERHAUL. Remove input shaft bearing housing (14—Fig. MF-124) and extract expansion plug (1). Remove nut (2) and bump the drive gear assembly (9) out of housing. Remove pulley, hub (21) and pinion shaft (7). The need and procedure for further disassembly is evident.

Both the driving pinion and gear are available as separate replacement parts.

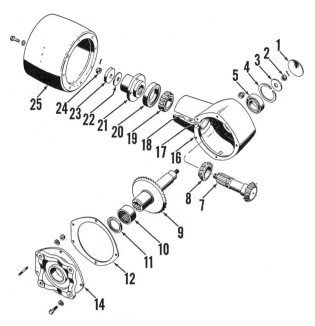

1. Expansion plug
2. Nut
3. Washer
4. Snap ring
5. Ball bearing
7. Pulley drive pinion
8. Bearing cone
9. Pulley drive gear
10. Needle bearing
11. Oil seal
12. Housing gasket
14. Bearing housing
16. Pulley housing
17. Bearing cup
18. Bearing cup
19. Bearing cone
20. Oil seal
21. Pulley hub
22. Oil seal
23. Washer
24. Nut
25. Pulley

LIVE PTO MAIN DRIVE GEARS

161. The pto main drive shaft pinion (35—Fig. MF107) is removed in conjunction with removing the transmission main drive shaft and gear (clutch shaft). Refer to paragraph 126 for tractors without "Multipower" transmission; or paragraph 134 for tractors equipped with "Multipower."

To remove the pto main drive shaft gear (48) after the drive pinion is out, first remove the transmission shifter rails and forks as outlined in paragraph 127 or 136. The gear (48) can then be removed through top opening in transmission housing.

LIVE PTO MAIN DRIVE SHAFT

162. To remove the live pto main drive shaft (52—Fig. MF107) on all models except "Multipower," first detach (split) transmission housing from engine as outlined in paragraph 122 and proceed as follows: Remove the front pto housing bearing cap (41) and extract snap ring (28) from forward end of shaft. Use two puller cap screws in the tapped holes provided in the pto shaft front bearing housing (44) and remove the housing and bearing assembly. Remove thrust washer (46). Remove the internal snap ring (47) from the pto main drive gear (48) and withdraw the main drive shaft.

Install the shaft by reversing the removal procedure and make certain that the small shoulder on thrust washer (46) is toward front.

When reassembling the unit, tighten nut (24) until a torque of 2-4 inch pounds is required to rotate shaft (7); then back the nut off until the cotter pin can be inserted. Tighten nut (2) until the bevel gears have a backlash of 0.004-0.006; then back the nut off until the cotter pin can be inserted.

POWER TAKE-OFF

OUTPUT SHAFT

159. To remove the pto output shaft, drain the system and unbolt rear bearing retainer (8—Fig. MF125) from center housing. Withdraw the shaft, seal and rear bearing assembly. Rear bearing (3) and seal (5) can be renewed at this time. The front needle bearing (12) can be removed after pulling its retainer (11) from the bore in center housing wall.

GROUND SPEED GEARS

160. To remove the ground speed drive gear, first remove the hydraulic lift cover as outlined in paragraph 168 and proceed as follows:

Working through the top opening in the rear axle center housing, remove the large cotter pin, collapse and remove the rear drive shaft assembly (XX—Fig. MF125). Remove snap ring (23) and withdraw the ground speed pto drive gear (24).

To remove the ground speed driven gear (15) after the drive gear has been removed, proceed as follows: Re-

move the pto shift cover assembly from left side of center housing and remove the pto output shaft as in paragraph 159. Remove the hydraulic pump locating dowel pins (one on each side of center housing) and withdraw the pump from the center housing. Remove the ground speed driven gear (15). Bushing (14) can be pulled from bore in sleeve (13).

When reassembling, install the hydraulic pump as in paragraph 166 and the hydraulic lift cover as in paragraph 168

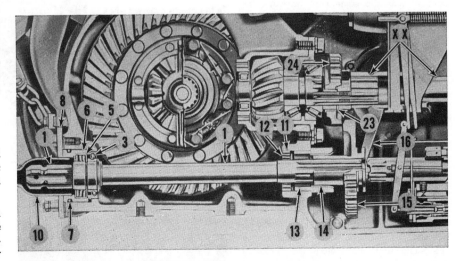

Fig. MF125—Sectional view of the differential housing showing the pto output shaft and the ground speed gears installation.

XX. Drive shaft
1. Pto shaft
3. Ball bearing
5. Seal
6. Seal retainer
7. "O" ring
8. Bearing retainer
10. Cap
11. Bearing retainer
12. Needle bearing
13. Sleeve
14. Bushing
15. Ground speed driven gear
16. Shift fork
24. Ground speed drive gear

To remove pto main drive shaft on "Multipower" models, remove the transmission as outlined in paragraph 135. Remove transmission input shaft as outlined in paragraph 139; the pto shaft front bearing cover plate; pto snap ring and spacer washer, then wtihdraw pto shaft rearward out of transmission. Install by reversing removal procedure.

HYDRAULIC LIFT

Most hydraulic system trouble is caused by dirt or gum deposits. The dirt may enter from the outside, or it may show up as the result of wear or partial failure of some part of the system. The presence of gummy deposits, however, usually results from inadequate fluids or from failure to drain and renew the fluid at the recommended intervals. These principles should be kept in mind when shooting trouble on the system and also when performing repair work on the pump, valves, and cylinders.

Thus, when disassembling the pump and valves unit, it is good practice generally to not remove any parts which can be thoroughly inspected while they are installed. Internal parts of the pump, valves, and cylinder, when removed, should be handled with the same care as would be accorded the parts of a Diesel pump or injector unit, and should be soaked or manually cleaned with an approved solvent to remove gum deposits. Unless you practice good housekeeping in your shop, do not undertake the repair of hydraulic equipment.

The transmission lubricant is the operating fluid for the hydraulic system. Recommended lubricant for summer is SAE 90W. For winter operation use SAE 80W.

The MF65 hydraulic system consists of a piston type pump unit which is submerged in the operating fluid and driven by the pto drive shaft; and a directly connected, single acting ram cylinder enclosed in the same housing. The control valve is located on the pump unit and meters the operating fluid at pump inlet. The rockshaft position can be automatically controlled by compression or tension on the upper implement attaching link, or by a cam on rockshaft ram arm. Refer to Fig. MF129 for a cross sectional view of rear axle center housing showing hydraulic system components.

ADJUSTMENTS

163. Whenever any malfunction of the hydraulic system is encountered and the cause is not readily apparent, it is recommended that the following adjustments be checked in the following sequence.

Detach implement from tractor and place lower lift links in their downmost position. Place the position control lever (Fig. MF126) in the "FAST" position as shown. Place the draft adjustment slide and the draft control lever together and between the punch marks on the draft control quadrant as shown. Remove the two special cap screws from the hydraulic lift lever link (Fig. MF127) and check the master control spring for end play by pushing and pulling on the clevis. If end play is present, loosen the Allen head set screw (S) in side of housing, unscrew retainer nut (N) and withdraw the master control spring assembly as shown in Fig. MF128. Remove pin (P) and turn the clevis on the control spring plunger until all spring end play is removed and the spring is snug but can still be rotated

by finger pressure. Insert Groove pin (P) and reinstall control spring assembly.

Turn the retainer nut (N—Fig. MF127) either way as required to eliminate all spring end play when checked by pushing in and pulling out on the clevis. Tighten the Allen head set screw (S). Note: Control spring end play will exist if retainer nut (N) is too tight as well as too loose.

Remove the inspection cover from right side of the rear axle center housing and check the control valve by moving the valve lever (Fig. MF-129) back and forth. The valve should work smoothly without any binding tendency. Use a punch or similar tool and wedge top of control valve lever rearward and away from vertical levers. Loosen nut (NN—Fig. MF130) and move the eccentric cam (C) out of contact with the cam arm.

Check the adjustment of each of the vertical levers individually. Desired adjustment is when a slight amount of finger pressure is required at lower end of lever to force it, against spring pressure, against the front of the slot

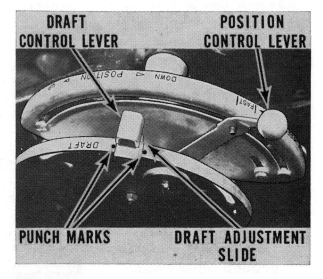

Fig. MF126 — Hydraulic system control quadrant. Control levers should be in the position shown when making any adjustment on the hydraulic system.

Fig. MF127—Master control spring can be checked for end play by disconnecting link from lift cover and pulling on the clevis.

Fig. MF128—Master control spring assembly removed from hydraulic lift cover. Spring end play can be adjusted by turning the clevis after removing pin (P).

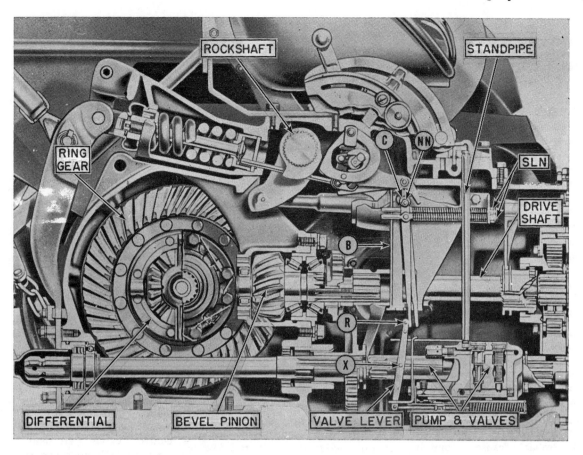

Fig. MF129—Sectional view of the differential housing showing the installation of the hydraulic lift system components.

in lower end of lever support bracket (B). If either of the self locking nuts (SLN) is too tight, the respective lever can be placed against the front of the slot without using pressure and the tight nut should be loosened. If, on the other hand, either of the nuts is too loose, excessive pressure will be required to push the respective lever against the front of the slot and the loose nut should be tightened. Adjustments can be made through side opening in center housing by using a short 1/2 inch box end wrench; however some mechanics prefer to remove the lift cover before making the adjustments.

Remove the previously installed wedge and release the control valve lever. While holding the vertical levers forward, tun the small eccentric cam (C) until it firmly contacts the cam arm without moving the vertical lever and tighten nut (NN), being careful not to rotate the cam farther than desired when tightening the nut.

To check the adjustment, move the position control lever down from the "FAST" position and into the response range. As the position control lever leaves the fast position, the vertical lever should begin to move the top of the control valve lever to the rear; thereby moving the control lever toward intake.

Move the position control lever back to the "FAST" position and attach an implement (two or three bottom plow) to tractor. Turn the self-locking nut (X–Fig. MF 129) either way as required so that there is just light contact between the ends of vertical levers and the roller (R) at top of the control valve lever.

This completes the internal adjustments and the inspection cover can be reinstalled on right side of center housing.

Remove the hitch pin from the differential housing and insert a long piece of 3/4 inch rod. Start engine and raise the lift with the position control lever until the distance between the center of the installed 3/4 inch rod and the center of the pin in the lift arm is 11 7/8 inches. Tighten the position control lever stop with the lever in this position.

Loosen the lower stop for the position control lever and carefully move the lever down until implement begins to raise. Allow the implement to raise part way; then, raise the position control lever until the implement just drops slowly. Tighten the lower stop at this point.

Adjust the lift arm retaining cap screws so that lift arms, when raised, will just drop of their own weight.

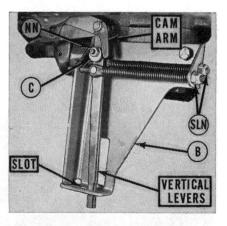

Fig. MF130 — Adjusting the control levers on the hydraulic lift cover. Adjustments can be made through inspection cover opening on right side of the center housing.

TROUBLE SHOOTING

164. With the hydraulic system properly adjusted as outlined in paragraph 163, the following procedure should facilitate locating any malfunction.

IMPLEMENT WILL NOT RAISE. Probable causes are:

a. Leak in system. Remove inspection cover and while system is operating, check for leaks at ram cylinder, stand pipe, control valve and pump side chambers.

b. Faulty relief valve. Refer to paragraph 165.

c. Broken or damaged internal pump parts. Usually indicated by a noisy pump.

d. Frozen ram cylinder. Remove side cover from center housing and see if relief valve is blowing.

IMPLEMENT LIFTS BUT WILL NOT LOWER. Probable cause is a damaged control valve spring. Remove inspection cover from center housing and determine if spring will put valve to discharge.

JERKY OR UNEVEN LIFT WHEN POSITION CONTROL LEVER IS RAISED. Probable cause is one or more faulty side chamber valves. Disassemble hydraulic pump and check for dirt or foreign material.

RELIEF VALVE BLOWS WHEN OPERATIONAL LEVER IS RAISED TO TRANSPORT POSITION. Probable causes are:

a. Check chains twisted.

b. Check chains installed in the lower holes of anchor brackets.

c. Lower links reversed.

ERRATIC ACTION OR POOR CONTROL WHEN OPERATING IN DRAFT CONTROL WITH LIGHT PRESSURE OR TENSION ON THE TOP LINK. Probable cause is damaged control linkage.

WHEN OPERATING IN POSITION CONTROL WITH KNURLED NUT SET ON QUADRANT, IMPLEMENT DOES NOT RETURN TO THE SAME POSITION WHEN THE OPERATIONAL LEVER IS RAISED AND LOWERED AGAIN TO STOP. Probable cause is eccentric rollers on position control link assembly.

SYSTEM OPERATING PRESSURE

165. The recommended operating pressure is 2300-2800 psi. To check the pressure, remove the pipe plug and install a suitable pressure gage in the position shown in Fig. MF131. Start engine, move the position control lever to the raise position and note the maximum operating pressure on the gage.

If the operating pressure is too high, the relief valve cartridge should be renewed. If the operating pressure is too low, remove the inspection plate from side of center housing, operate the system and check for visible fluid leaks. If there are no visible fluid leaks, and there is turbulence around the relief valve, renew the relief valve cartridge. If the installation of a new cartridge does not bring the system operating pressure within the specified range, a faulty hydraulic pump is indicated.

PUMP

NOTE: On models equipped with "Multipower" transmission, the transmission charging pump is mounted on top of the hydraulic system pump as shown in Fig. MF-132. Before removing the hydraulic pump on these models, the transmission pump must be removed as outlined in paragraph 143.

166. **REMOVE AND REINSTALL.** To remove the hydraulic pump assembly, first remove the lift cover assembly as outlined in paragraph 168. Working through the top opening in the differential housing, remove the large cotter pin, collapse and remove the rear drive shaft assembly (Fig. MF129).

Unbolt the pto output shaft rear bearing retainer from center housing and pull the shaft, seal and rear bearing assembly rearward. Disconnect the clutch release rod. Remove the hydraulic pump locating dowel pins (one on each side of center housing) and lift the pump from tractor. Notice that the dowel pin holes in pump are of different diameter.

When reassembling, install the pump and pto shaft, then tighten the right hand dowel first so that any required shifting will automatically be made in the left dowel when it is tightened.

167. **OVERHAUL.** To disassemble the removed pump, unscrew the adjusting nut on the control lever assembly (56—Fig. MF133). Remove guide (55), control valve plate (52), control valve (45) and associated washers, "O" rings and spacers. Remove locking wire and set screw (38). Remove snap rings (35 and 36) and withdraw oscillator assembly.

Remove cotter pin retaining front pump shaft coupling (16) to the camshaft (11) and slide coupling off. Remove cotter pin from oscillator link clevis pin (40). Remove the body bolts (9) and separate the pump body sections (2 & 3). Remove camshaft (11), pistons (13) and blocks (14 & 15).

Fig. MF131—Gage installation for checking the hydraulic system operating pressure.

Fig. MF132—The "Multipower" pump (M) is mounted on top of regular hydraulic pump as shown.

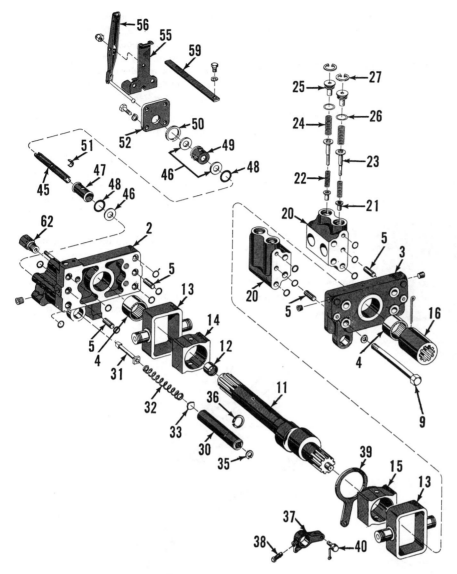

Fig. MF133—Exploded view of the hydraulic lift system pump. Pump assembly can be removed through top opening in center housing after removing lift cover. Note: Pump body may be equipped with bronze bushings instead of needle bearings (4). Refer to text.

2. Rear body assy.	23. Outlet valve	39. Cam follower
3. Front body assy.	24. Spring	40. Clevis pin
4. Bearing (or bushing)	25. Valve plug	45. Control valve
5. Locating pin	26. "O" ring	46. Washer
9. Body bolt	27. Snap ring	47. Retainer
11. Camshaft	30. Oscillator body	48. "O" ring
12. Needle bearing	31. Drive assembly	49. Spacer
13. Block piston	32. Control valve spring	50. Washer
14. Rear cam block	33. Retainer disc	51. Snap ring
15. Front cam block	35. Snap ring	52. Control valve plate
16. Front coupling	36. Snap ring	55. Guide bracket
20. Valve chamber	37. Oscillator link	56. Valve lever
21. Inlet valve	38. Link screw	59. Lever guide
22. Spring		62. Safety valve

Remove the snap rings (27); then thread a small cap screw into the valve plugs (25) and pull the plugs out. Remove the side valve assemblies (21 & 23) with their component parts.

After the pump is disassembled, thoroughly clean all parts and examine them for damage or excessive wear. Renew all "O" ring seals and any other parts which are questionable. Bearings (4 & 12) can be renewed at this time if necessary.

Note: Pump body needle bearing (4) have bronze bushings as an alternate for use in production and service.

To reassemble the pump, proceed as follows: Reinstall the side valve assemblies and the valve plugs. Install the oscillator cam follower (39) on the shoulder of the front block (15) and install the front block on the camshaft. Install the rear cam block (14) on the shaft. Follower (39) should now be between the cam blocks and

on shoulder of the front block. Install pistons (13) on cam blocks with machined edge of one piston facing the front body assembly and machined edge of other piston towards rear body assembly.

Slide pump body (2 & 3) over ends of shaft and install body bolts (9). Tighten the pump body bolts securely.

The remainder of the procedure is self-evident.

After the pump and lift cover are installed on tractor, check the system adjustments as outlined in paragraph 163.

LIFT COVER, WORK CYLINDER & ROCKSHAFT

168. **REMOVE AND REINSTALL ASSEMBLY.** To remove the lift cover, remove seat and disconnect upper lift links from lift arms. Remove transfer plate (Fig. MF134) from the hydraulic lift cover and withdraw the standpipe which is located directly under the transfer plate. Unbolt and remove the lift cover assembly from the rear axle center housing, but do not loosen the four nuts retaining work cylinder to lift cover.

Before installing the lift cover assembly, remove the inspection cover from right side of center housing and using a punch or similar tool, wedge top of control valve lever rearward so that when lift cover is installed, the vertical levers will be in front of control valve lever.

169. **WORK CYLINDER.** With the lift cover assembly removed from tractor, the procedure for removing

Fig. MF134—Before removing the hydraulic lift cover, remove transfer plate and the stand pipe under the plate.

the work cylinder, piston and/or rings is evident. When reassembling, be sure to renew the "O" ring between the lift cover and the cylinder. Refer to Fig. MF135.

170. **ROCKSHAFT.** The procedure for removing the rockshaft and renewing the bushings is evident. When reassembling, tighten the lift arm retaining cap screws to a point where

the arms, when held in a horizontal position, will just fall of their own weight.

AUXILIARY CONTROL VALVE

Two types of auxiliary valves are available for use with the MF65 hydraulic system, to control external remote cylinders. The dual control valve shown in Fig. MF138 can be used to operate either single or

double acting, remote cylinders. The control valve shown in Fig. MF139 is available in either a single spool or dual spool version for use with single acting cylinders only. Both valves are designed to mount directly on hydraulic lift cover in place of transfer plate (38—Fig. MF137).

171. **DOUBLE ACTING VALVE.** Refer to Fig. MF138. Valve body (6), valve spools (23) and shuttle valves (15) are matched parts and available as an assembly only. All other parts are available individually. Valve ports are machined for 3/8-inch, straight thread, tube fittings with "O" ring seal.

Switch valve knobs (1), located on back of shuttle valve body, must be turned out to operate single acting cylinders and turned in until they seat, to operate double acting cylinders. A body containing a third, iden-

Fig. MF135—Bottom side of the hydraulic lift cover with the work cylinder removed. Note the small "O" ring located between cylinder and cover.

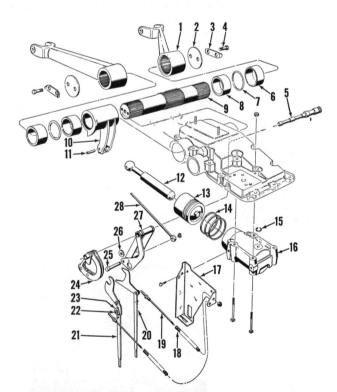

Fig. MF136 — Rockshaft, work cylinder and internal control levers and linkage.

1. Lift arm
2. Washer
3. Lock clip
4. Cap screws
5. Link shaft
6. Bushing
7. "O" ring
8. Bushing
9. Rock (lift) shaft
10. Ram arm
11. Roll pin
12. Connecting rod
13. Piston
14. Piston rings
15. "O" ring
16. Work cylinder
17. Lever support bracket
18. Spring
19. Spring guide
20. Draft control lever
21. Position control lever
22. Nut
23. Eccentric cam
24. Position control link assembly
25. Draft control pin
26. Roller cam
27. Draft control link assembly
28. Draft control rod

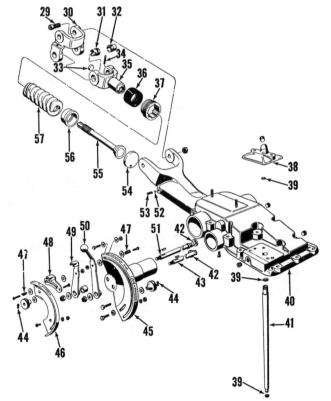

Fig. MF137 — Master control spring and external control linkage exploded from the hydraulic lift cover.

29. Special cap screw
30. Rocker link
31. Bushing (right)
32. Bushing (left)
33. Rubber plug
34. Groov pin
35. Clevis
36. Rubber boot
37. Retainer nut
38. Transfer plate
39. "O" ring
40. Lift cover
41. Stand pipe
42. Roller
43. Position control shaft
44. Roll pin
45. Position control quadrant
46. Draft control quadrant
47. Compression spring
48. Draft adjustment slide
49. Draft control lever
50. Position control lever
51. Draft control shaft
52. Lead ball
53. Set screw
54. Overload stop disc
55. Control spring plunger
56. Control spring seat
57. Master control spring

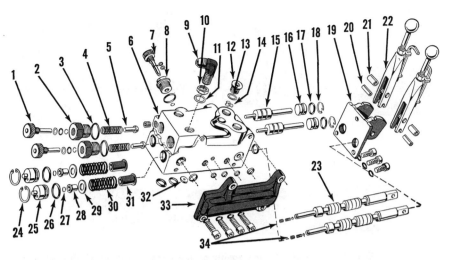

Fig. MF138—Exploded view of two-spool auxiliary control valve available as an accessory. A third spool attachment of identical design is available which is installed in place of cover (33).

1. Switch valve	13. "O" ring	24. Snap ring
2. Valve plug	14. Restrictor	25. Plug
3. "O" ring	15. Shuttle valve	26. "O" ring
4. Spring	16. Plug	27. Snap ring
5. Poppet valve	17. "O" ring	28. Collar
6. Valve body	18. Snap ring	29. Washer
7. Bypass valve	19. Front cover	30. Centering spring
8. Valve plug	20. Roll pin	31. Spring sleeve
9. Elbow	21. Roll pin	32. Restrictor
10. Lock nut	22. Lever asembly	33. Side cover
11. "O" ring	23. Valve spool	34. Detent release
12. Body plug		valve

A return detent is built into the control levers (22), to retain the valve in the raising position until the cylinder reaches the end of its stroke. To adjust the detent, loosen the locknut and turn the knob stem into lever to increase the release pressure; or out of lever to decrease pressure. Bypass valve (7) must be turned into valve housing until it seats, at all times.

172. **SINGLE ACTING VALVE.** Refer to Fig. MF139. A similar, single spool valve is available. All parts are available for service.

To use the valve, both levers on the internal system control quadrant must be moved to top of quadrant. To use the single spool valve; or to operate two remote cylinders with double spool valve, the tractor lower links must be locked in position to keep the system from neutralizing. To operate one single acting cylinder with the dual spool remote valve, move the left control lever (9) to raising (rear) position and lock in place with latch (10). The left outlet port must be capped or plugged, and the right control valve and lever used to operate remote cylinder.

tical valve assembly is available which may be installed instead of side cover (33).

To operate the remote control system, both levers on internal system control quadrant must be moved to top of quadrant. The tractor lower lift links must be locked in position or the passage to internal ram cylinder

blocked by moving the auxiliary valve left control lever (22) to the raising (rear) position and locking it in place.

When both a single acting and double acting cylinder are attached to valve at the same time, the single acting cylinder must be connected to the left of the double acting cylinder.

REMOTE CYLINDER

Two types of remote cylinders are used. The 2-inch cylinder shown in Fig. MF140 is used for single acting applications by installing breather (2) in rod end port; or for double acting applications by installing a second pressure hose. The 2½-inch cylinder shown in Fig. MF141 is used as a double acting cylinder only.

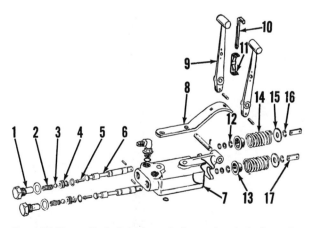

Fig. MF139 — Exploded view of the optional single acting acting remote control valve which is available in one-spool or two-spool model.

1. Valve plug	7. Valve body
2. Valve spring	8. Detent bracket
3. Variable valve ball	9. Control lever
4. Valve seat	10. Latch
5. Variable valve	11. Bracket
6. Valve spool	
12. Snap ring	
13. Spring seat	
14. Centering spring	
15. Spring seat	
16. Snap ring	
17. Link	

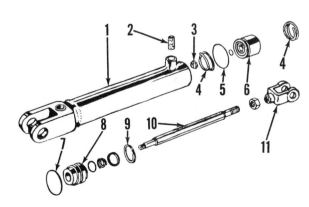

Fig. MF140—Exploded view of 2-inch remote cylinder which is used for single-acting or double-acting applications.

1. Cylinder	5. "O" ring	8. Cylinder bearing
2. Breather	6. Piston	9. Snap ring
3. Rod nut	7. "O" ring	10. Piston rod
4. Back-up ring		11. Clevis

173. To disassemble the 2-inch cylinder, remove the wrap-around snap ring (9—Fig. MF140) and withdraw the piston and rod assembly from cylinder.

To disassemble the 2½-inch cylinder, refer to Fig. MF141. First remove the hoses and fittings, cylinder stop (18) and end ring (19). Unbolt and remove piston rod stop sleeve (16). Extend the cylinder and remove bearing cover plate (2) from bearing. Slide the plate up out of way, bump the bearing (7) into cylinder far enough to remove snap ring (3); then withdraw piston rod, bearing and piston from cylinder.

On all models, renew "O" rings and seals whenever cylinder is disassembled. The need and procedure for further disassembly is evident.

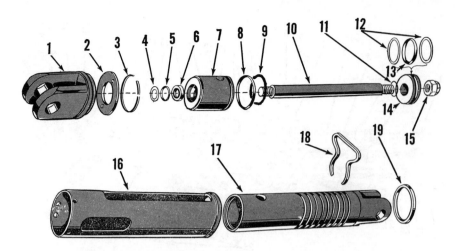

Fig. MF141—Exploded view of 2½-inch double acting remote cylinder which is optionally available.

1. Piston rod yoke
2. Bearing cover plate
3. Snap ring
4. Leather back-up ring
5. Shaft "O" ring
6. Oil seal
7. Bearing
8. Leather back-up ring
9. "O" ring
10. Piston rod
11. "O" ring
12. Leather back-up ring
13. "O" ring
14. Piston
15. Rod nut
16. Cylinder stop sleeve
17. Cylinder
18. Cylinder stop lock
19. Snap ring

NOTES

MASSEY FERGUSON

Models ■ MF85 ■ MF88 ■ MF Super 90 ■ MF Super 90 WR

Previously contained in I&T Shop Manual No. MF-20

SHOP MANUAL

MASSEY-FERGUSON

MODELS MF85-MF88-MF Super 90-MF Super 90 WR

Tractor serial number stamped on name plate on right side of battery box. Engine serial number stamped on engine name plate or on side of block.

INDEX (By Starting Paragraph)

CONDENSED SERVICE DATA

GENERAL	MF85-MF88 Non-Diesel	Diesel	MF Super 90 MF Super 90WR Non-Diesel	Diesel
Torque Recommendations	See End of Shop Manual			
Engine Make	Cont'l.	Cont'l.	Cont'l.	Perkins
Engine Model	E242	HD277	E242	4A 300
Number of Cylinders	4	4	4	4
Bore—Inches	3 7/8	4	3 7/8	4 1/2
Stroke—Inches	5 1/8	5 1/2	5 1/8	4 3/4
Displacement—Cu. In.	242	276.5	242	302.2
Compression Ratio				
Gasoline	7.31:1	......	7.31:1	......
LP-Gas	8.1:1	......	......	......
Diesel	......	15.2:1	......	16.1
Main Bearings (Number)	3	3	3	5
Cylinder Sleeves (Type)	Wet	Wet	Wet	Dry
Generator & Starter (Make)	—————————Delco-Remy———————			
Battery Terminal Grounded	Negative	Negative	Negative	Negative*

*MF Super 90 Diesel tractors use two 12V batteries connected in parallel.

TUNE-UP

	MF85-MF88 Non-Diesel	Diesel	MF Super 90 MF Super 90WR Non-Diesel	Diesel
Firing Order	1-3-4-2	1-3-4-2	1-3-4-2	1-3-4-2
Inlet Valve Tappet Gap (Hot)	0.016	0.014	0.016	0.010
Exhaust Valve Tappet Gap (Hot)	0.018	0.014	0.018	0.010
Inlet Valve Face Angle	30°	44°	30°	44°
Exhaust Valve Face Angle	45°	44°	45°	44°
Inlet Valve Seat Angle	30°	45°	30°	45°
Exhaust Valve Seat Angle	45°	45°	45°	45°
Ignition Distributor Make	AL or DR	......	DR	......
Injection Pump Make	......	Roosa	......	C.A.V.
Ignition Timing (Static)	**	......	**	......
Injection Timing	......	20°BTC	......	16°BTC
Distributor Point Gap	0.022	......	0.022	......
Distributor Cam Angle	See Text	......	25°-34°	......
Injector Nozzle Make	......	C.A.V.	......	C.A.V.
Nozzle Opening Pressure	1900 psi	......	2500 psi	......

**Gasoline, 5° BTC. LPG, 4° BTC.

TUNE-UP (Cont.)	MF85-MF88 Non-Diesel	Diesel	MF Super 90 MF Super 90WR Non-Diesel	Diesel
Carburetor Model				
Gasoline (M-S)	TSX 644	......	TSX 644	......
LP-Gas (Zenith)	A818-29	......	......	......
Float Setting	1/4 in.	......	1/4 in.	......
Engine Low Idle RPM	450	650	450	600
Engine High Idle RPM	2200	2275	2200	2175
Engine Loaded RPM	2000	2000	2000	2000
PTO No Load RPM	804	830	804	795
PTO Loaded RPM	731	731	731	731

SIZES—CAPACITIES—CLEARANCES

(Clearances in Thousandths)

	MF85-MF88 Non-Diesel	Diesel	MF Super 90 MF Super 90WR Non-Diesel	Diesel
Crankshaft Journal Diameter	2.624	2.873	2.624	2.999
Crankpin Diameter	2.248	2.498	2.248	2.749
Camshaft Journal Diameter				
Outboard Bearing	0.996	0.996	0.996	......
Front Bearing	1.996	1.996	1.996	2.057
Center Bearing	1.746	1.746	1.746	1.967
Rear Bearing	1.684	1.684	1.684	1.906
Piston Pin Diameter	1.109	1.250	1.109	1.4375
Main Bearing (diametral clearance)	0.5-2.7	1.5-3.5	0.5-2.7	3.0-5.5
Rod Bearing (diametral clearance)	1.4-2.4	1.3-3.8	1.4-2.4	2.8-4.5
Piston Skirt Clearance	2.5-3.5	3.5	2.5-3.5	(See Text)
Crankshaft End Play	6-8	5-8	6-8	8.5-15.5
Camshaft Bearing (diametral clearance)	2.5-4	See Text	2.5-4	See Text
Camshaft End Play	6-9	5-9	6-9	4-16
Cooling System—Qts.	15	22	15	15
Crankcase Oil—Qts.	7	7	7	10 1/2
Transmission, Differential Hydraulic Lift, Power Steering—Qts.	46	46	46	46
Final Drive Unit, Each—Pts.				
High Clearance	2	2	2	2
Standard Clearance	1	1	1	1

FRONT SYSTEM

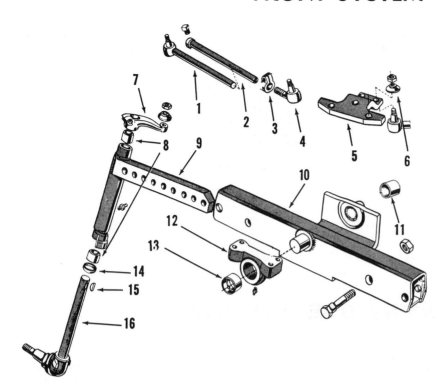

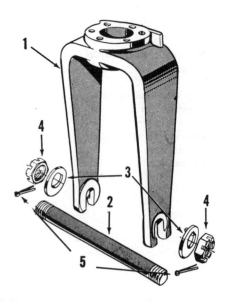

Fig. MF101 — View of single wheel fork used on High Clearance tractor. Fork bolts directly to vertical steering shaft flange.

1. Wheel fork
2. Axle shaft
3. Eccentric washer
4. Axle nut
5. Cotter pin

Fig. MF100—Standard and High Clearance front axle differ only in length of spindle and axle extension. Wheat & Rice models have one piece center axle and non-telescoping tie rod.

1. Tie rod	5. Center steering arm	9. Axle extension	13. Front pivot bushing
2. Right tube	6. Dust seal	10. Axle center member	14. Spindle thrust washer
3. Tie rod clamp	7. Outer steering arm	11. Rear pivot bushing	15. Woodruff key
4. Tie rod end	8. Spindle bushing	12. Front pivot bracket	16. Front spindle

MF-85 and MF Super 90 tractors are available in Standard Clearance and High-Clearance models. MF-88 and MF Super 90 WR (Wheat & Rice) is used to designate the non-adjustable axle, "Western" models. Many front end components are interchangeable for all models, and service procedures are identical except where obvious or otherwise noted.

Axle extension and spindle lengths are different for Standard and High-Clearance models; and tricycle type front end units are interchangeable with axle type on High-Clearance models only.

AXLE ASSEMBLY

All Axle Models

1. To remove the axle assembly (10 —Fig. MF100), support tractor under engine, remove lower grille panel and disconnect tie-rods from spindle steering arms. (On adjustable axle models, if axle center section is to be serviced, unbolt and remove both axle extension and wheel assemblies from center section.) Remove the four cap screws attaching axle front pivot bracket (12—Fig. MF100) to front

support and remove bracket. Pull center section forward and out of the rear pivot bushing which is located in the front support casting.

Inside diameter of new front and rear pivot bushings is 2.002-2.004. Pivot pin diameter for a new center member is 1.999-2.001. Renew center member and/or bushings if running clearance is excessive.

SPINDLE BUSHINGS

All Axle Models

2. Each spindle support contains two renewable bushings which require final sizing after installation to an inside diameter of 1.499-1.500. Recommended clearance between spindles and bushings is 0.0035-0.005.

TOE-IN AND TIE-RODS

All Axle Models

3. Tie-rod ends are of the automotive type and are not adjustable to compensate for wear.

Recommended toe-in of 0-⅛-inch is adjusted as follows:

On adjustable axle models, loosen tie-rod set screw at outer end and

clamp nut at inner end; then turn tube (2—Fig. MF-100) in or out as required. Make sure the set screw is in the bottom position and engaged in set screw hole.

On Wheat and Rice models, disconnect rod end assembly from steering arm and turn end in or out as required.

SINGLE WHEEL AND FORK

All Models So Equipped

4. The fork mounted single front wheel is carried in taper roller bearings which should be adjusted to provide a very slight rotational drag.

To remove the wheel fork, support tractor under engine and remove the grille lower panel. The upper portion of the yoke (1—Fig. MF101) is bolted to the vertical steering shaft flange by six cap screws.

DUAL WHEELS AND LOWER PEDESTAL

All Models So Equipped

5. Each of the dual wheels is mounted on taper roller bearings which should be adjusted to provide a very slight rotational drag.

To remove the lower pedestal and wheels assembly, support tractor under engine and remove the grille lower panel. The one-piece pedestal assembly is bolted to the vertical steering shaft flange by six cap screws.

POWER STEERING SYSTEM

NOTE: The maintenance of absolute cleanliness of all parts is of the utmost importance in the operation and servicing of the power steering system. Of equal importance is the avoidance of nicks or burrs on any of the working parts.

6. Fluid for the power steering system on all models is supplied by a PTO driven power steering pump (1—Fig. MF103) located on the right hand side of the transmission case. The oil reservoir is common to the transmission, differential and hydraulic system. The return oil from the power steering system operates the multiple disc PTO clutch (10). A review of the oil flow shown in Fig. MF103 will be beneficial in locating trouble.

TROUBLE-SHOOTING

All Models

7. The following paragraphs list the more common troubles encountered in the operation of the power steering system. Once the cause of trouble is determined, the procedure for correction is usually evident.

LOSS OF POWER ASSISTANCE. Most common cause is low system pressure or flow, which may be due to insufficient fluid in reservoir, defective pump or relief valve or plugged or restricted lines. Other possible causes are binding linkage or sticking or worn control valve.

POWER ASSISTANCE IN ONE DIRECTION ONLY. Could be caused by sticking or worn control valve and/or valve linkage, or faulty cylinder.

UNEQUAL TURNING RADIUS. Could be caused by binding or bent linkage, or improperly adjusted tie-rods.

ERRATIC STEERING CONTROL. Could be caused by plugged filter, low fluid level, sticking or worn control valve or faulty relief valve.

UNEQUAL TURNING EFFORT. Could be caused by sticking or worn control valve, faulty cylinder or binding mechanical linkage.

In checking for trouble, note the free back and forth movement of the worm shaft (7—Fig. MF109) when the steering wheel is turned with the engine not running. Movement should be approximately equal both directions from center. Listen for unusual noises

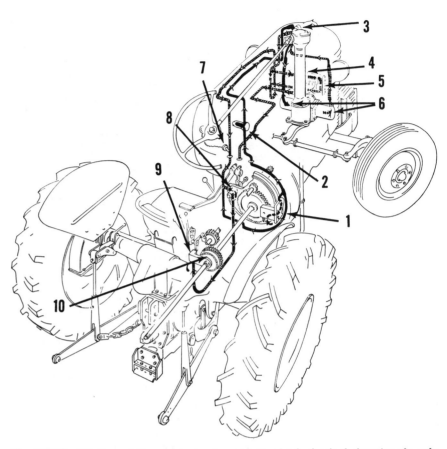

Fig. MF103—Oil flow of the power steering and pto clutch circuit designating the principal units of the system.

1. Power steering pump	3. Power steering control valve	5. Oil cooling radiator	7. Pto pressure-switch
2. Power steering filter	4. Vertical steering shaft	6. Power steering cylinder	8. Pto dump valve
			9. Pto selector valve
			10. Pto clutch

in pump and relief valve and check linkage carefully for wear or damage.

When cause of trouble has been determined, proceed as outlined in the appropriate subsequent paragraphs.

OPERATING PRESSURE AND RELIEF VALVE

All Models

8. A pressure test of the hydraulic circuit will disclose whether the pump, relief valve or some other unit in the system is malfunctioning. To make such a test, proceed as follows:

Connect a pressure test gage and shut-off valve into the pump pressure line at the control valve. Note that the pressure line is the steel line that connects to the top port of the control valve. Open the shut-off valve and

run engine until oil is warmed. Advance engine speed to 2000 rpm, close the shut-off valve and retain in the closed position only long enough to observe the gage reading. Pump may be seriously damaged if valve is left in the closed position for more than 10 seconds. If gage reading is 1500 psi with the shut-off valve closed, the pump and relief valve are O.K. and any trouble is located in the control valve, power cylinder and/or connections.

If the pump output pressure is more than 1500 psi, the relief valve is either improperly adjusted or stuck in the closed position. If the output pressure is less than 1500 psi, the relief valve is improperly adjusted, the filter or filters are plugged or the pump needs overhauling.

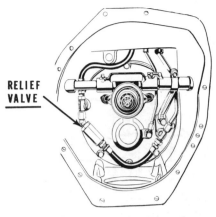

Fig. MF104—Inside of clutch housing showing location of power steering system relief valve.

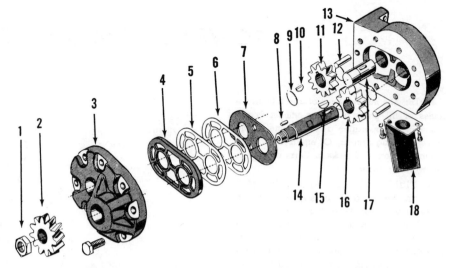

Fig. MF106—Exploded view of power steering pump showing components.

1. Gear retaining nut	7. Pump diaphragm	13. Pump body
2. Drive gear	8. Drive gear key	14. Drive gear shaft
3. Pump front plate	9. Snap ring	15. Lead gear key
4. "V" seal	10. Follow gear key	16. Lead gear
5. Leather gasket	11. Follow gear	17. Follow gear shaft
6. Fiber gasket	12. Dowel pin	18. Inlet filter

The power steering relief valve is located in the by-pass line between the pressure and return lines inside the clutch housing (refer to Fig. MF104). The pressure is adjusted by the addition of shims underneath the relief valve spring.

To service the pressure relief valve, first split the tractor between the engine and transmission case as outlined in paragraph 137.

POWER STEERING PUMP
All Models

9. **R&R AND OVERHAUL.** Remove pump from right side of transmission housing as follows:

Remove drain plug from left side of transmission case and drain out approximately five gallons of oil. Remove the six cap screws from pump mounting plate and remove plate and pump assembly as shown in Fig. MF-105.

Loosen the three Allen head cap screws securing pump to mounting plate until they protrude about $\frac{3}{16}$-inch.

Tap the screws lightly to free pump from mounting plate dowel pins and remove mounting plate. Note the installation of pump to plate and plate to transmission housing "O" ring seals.

Remove, inspect and clean the inlet screen (18—Fig. MF106).

Clamp pump drive gear (2) in a soft jawed vise and remove retaining nut (1), drive gear (2) and key (8). Inspect gear end of pump shaft and remove any burrs or roughness.

Remove the eight cap screws securing front plate (3) and remove same by tapping lightly. Do not tap on the shafts or pry with a screw driver. Remove pump gears and shafts.

Remove bronze faced diaphragm (7), fiber gasket (6), leather gasket

(5) and molded rubber "V" seal (4) from front plate. Inspect all parts for nicks and burrs and drive gear shaft (14) for broken keyway.

Renew gear shafts (14 and 17) if the diameter measures 0.6850 or less at any bearing area. Shaft bearings are pre-fitted in the housings and cannot be replaced individually.

Check pump body (13) for wear or scoring on gear face; renew body if wear exceeds 0.0015. Check gear pockets for wear or scoring and renew body if I.D. of gear pockets measure 1.715 or more. Check shaft bearings in body and front plate and renew body or plate if bearing I.D. measures more than 0.691.

Check gears for scoring or wear and renew same if gear width measures 0.4580 or less. Remove any sharp edges from gear teeth with emery cloth.

The bronze faced diaphragm, fiber gasket, leather gasket and molded rubber "V" seal should always be renewed when pump is reassembled. Using a blunt-pointed tool, install "V" seal in grooves of front plate with open side down. Make sure "V" seal is completely seated. Install leather gasket, fiber gasket and bronze faced diaphragm making sure they are completely seated.

Dip gear assemblies in oil and install in pump body. Note: Red painted end of idler shaft must be placed in body.

Fig. MF 105 — Removal of power steering pump and mounting plate from transmission housing.

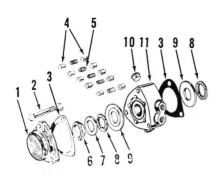

Fig. MF107 — Exploded view of steering valve. Valve body and spool are available only as an assembly.

1. Valve cover
2. Retaining cap screw
3. Gasket
4. Centering plunger
5. Centering spring
6. Retaining nut
7. Small thrust washer
8. Thrust bearing
9. Large thrust washer
10. Check valve
11. Valve assembly

Coat mating surfaces of body and front plate with light grease, install front plate and tighten the eight cap screws evenly to a torque of 15-20 Ft.-Lbs.

Reassemble remainder of pump by reversing disassembly procedure.

Whenever pump has been disassembled, it is necessary to run pump in for five minutes at low idle speed and five minutes at half throttle, no-load, before releasing tractor for service.

CONTROL VALVE

All Models

10. **R&R AND OVERHAUL.** Raise hood, remove left grille screen and remove left front grille closure from tractor. Disconnect hydraulic lines from control valve. Remove the three cap screws (2—Fig. MF107) and end cap (1). Remove the nut (6) from the worm shaft. Remove the small thrust washer (7), thrust bearing (8) and large thrust washer (9). Slide off the control valve and spool assembly (11) and remove the rear thrust washer and bearing.

Remove the valve centering plungers (4) and centering springs (5) from valve body and push spool from body bore. Thoroughly clean and examine all parts for damage or wear. The valve spool and body are mated parts and must be renewed as an assembly. The centering springs and plungers are available separately. The check valve (10) in the valve body can be removed and renewed as an assembly.

When reassembling, lubricate all parts in clean Automatic Transmission Fluid, Type "A" and place rear thrust bearing and thrust washer on shaft. Assemble centering plungers and springs in valve body and install spool so that identifying groove in I.D. of spool is toward same end of body as the port identification symbols "PR" and "RT". Reinstall the assembled valve body on the worm shaft with the symbols "PR" and "RT" to the outside and rear. Use two of the cap screws (2) to secure the valve body to the housing without the end cover and install the large thrust washer, thrust bearing and small thrust washer on shaft. Tighten the lock nut (6) to a torque of 20-30 Ft.-Lbs., back off ¼-turn and stake nut securely.

Remove the two temporarily installed cap screws, install the end cap and connect the power steering lines to the control valve.

Fig. MF108 — Vertical steering shaft assembly showing location of power cylinder rack adjusting plate and steering gear upper housing cover (1).

POWER STEERING CYLINDERS AND PISTONS

All Models

11. **R&R AND RENEW SEALS.** The power cylinders, pistons or seals can be renewed without removing any other steering units from the tractor. To remove the cylinders, first raise the hood and remove the left grille screen and the right hinged grille screen from the tractor. Loosen the two cap screws retaining the hood side panels to the front support and remove the lower sheet metal front panel by pulling it forward from its position on the front support. Unbolt and remove the lower grille closure panel, place a large pan under the tractor and drain the steering housing by removing the pipe plug at the front of the lower housing. Disconnect oil lines from the cylinder castings, then unbolt and withdraw cylinders from the pis-

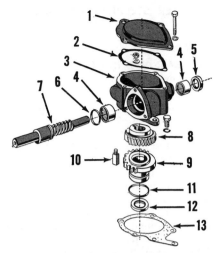

Fig. MF109 — Exploded view of steering gear upper housing. Backlash adjuster (9), "O" ring (11) and seal (12) are located in the vertical shaft housing.

1. Top cover	8. Worm gear
2. Gasket	9. Backlash adjuster
3. Upper housing	10. Tapered key
4. Worm shaft bearing	11. "O" ring
5. Seal	12. Shaft seal
6. "O" ring	13. Gasket
7. Steering worm	

1. Rack guide
2. Shim pack
3. Piston retaining bolt
4. Washer
5. "O" ring
6. Spacer
7. U cup packing
8. Piston
9. Rack
10. Gear
11. Bearing race
12. Thrust bearing
13. Cylinder housing
14. Gasket
15. Lower housing
16. Upper bearing
17. Lower bearing
18. Seal

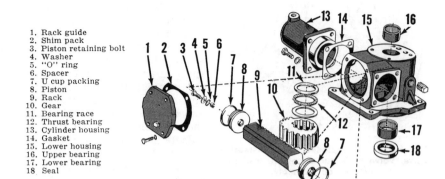

Fig. MF111—Exploded view of lower housing showing rack and pinion arrangement. Rack backlash of 0.001-0.008 can be adjusted by adding or removing shims (2). Cylinder housings (13) may be removed individually for service of piston seals.

tons. Refer to Fig. MF111. Piston seals (7) or pistons (8) can be renewed at this time. Cap screw securing piston to the steering rack contains a spacer to allow the piston to center in the cylinder. Piston rack may be removed after cylinders are off, by removing one piston and the rack guide (1).

WORM SHAFT AND UPPER GEAR

All Models

12. **ADJUSTMENT.** To adjust the backlash between the worm shaft and vertical shaft upper gear, proceed as follows: Remove top cover (1—Fig. MF109) from steering gear upper housing. With a suction pump, remove the

Fig. MF110—Using gage available to Massey-Ferguson dealers to locate worm gear 0.442 above shaft worm. Key nut (N) is tightened to hold gear in position.

housing lubricant, then remove Allen head locking screw from backlash adjuster (9—Fig. MF109). Rotate adjuster counter-clockwise until all backlash is removed; then back-off to nearest notch and reinstall locking screw. Fill housing until worm is barely covered, with a regular steering gear lubricant or SAE 80 or 90 Mild EP transmission gear oil. Reinstall cover.

12A. **R&R AND OVERHAUL.** Refer to Fig. MF109. To remove the worm shaft (7), vertical shaft upper gear (8) and upper gear housing, first raise the hood and disconnect oil lines from the control valve. If the worm shaft is to be removed for shaft, bearing or seal service, the control valve may be removed at this time as outlined in paragraph 10. Remove the four cap screws retaining the housing top cover, lift off the cover and remove the lubricant using a suction pump. Remove the two bolts securing the worm shaft rear yoke to the flexible coupling, loosen the yoke set screw and drive the yoke forward on the worm shaft far enough to clear the flexible coupling. The vertical shaft upper gear (8) is retained to and positioned on the shaft by a tapered key (10), secured with a nut and washer. Remove the nut and washer and loosen the gear on the vertical shaft by tapping the key down with a soft hammer. Remove the remaining cap screw retaining the upper housing to the vertical shaft housing and lift off the upper housing, worm shaft and vertical shaft upper gear as an assembly. Remove the yoke from the rear of the worm shaft, remove any nicks or burrs from the shaft and withdraw the shaft from the front of the housing. "O" ring (6), seal (5) and needle bearings (4) can be renewed at

this time. Vertical shaft eccentric bushing (9) can be removed by prying it up from its position in the vertical housing for the purposes of renewing seal (12) or "O" ring (11).

When reassembling the upper steering gear, rotate the eccentric bushing (9) clockwise to a position which will provide maximum clearance between worm and worm gear. Place the tapered key (10) in the gear and loosely install the retaining nut and washer. Using a serviceable gasket, install the gear, worm shaft and housing as an assembly on the vertical shaft housing and install the short cap screw to hold the housing in place. Using the special gage as shown in Fig. MF110 (or a straight-edge and ruler) position the gear on the vertical shaft so that 0.442 clearance is maintained between top line of gear hub and nearest edge of worm as shown. Tighten the key nut (N) to hold gear in position and adjust backlash as outlined in paragraph 12. Reinstall the control valve as outlined in paragraph 10.

VERTICAL SHAFT AND LOWER GEAR

All Models

13. **BACKLASH ADJUSTMENT.** The backlash between the power piston rack (9—Fig. MF111) and the vertical shaft pinion (10) can be adjusted without removing the unit from the tractor. To make the adjustment, raise the hood, remove the left grille screen and unbolt and remove the hinged right grille screen. Loosen the two cap screws retaining the hood side panels to the front support and remove the lower front panel by pulling it forward from the front support. Unbolt and remove the lower grille closure, place

a large drain pan under the tractor and drain the lower steering housing by removing the drain plug. Backlash between the piston rack and shaft pinion is controlled by shims (2) between the rack guide (1) and lower housing. With the wheels in a straight ahead position, remove the rack guide, remove the shim pack and reinstall the guide. While measuring with a feeler gage, tighten the guide retaining screws evenly until the gap between the housing and the guide flange is equal at all points and backlash between rack and gear is zero. Remember the feeler gage reading, remove the guide, and add shims equal to the measured clearance plus 0.001-0.008. Shims are available in thickness of 0.006, 0.010, 0.014 and 0.020.

13A. R&R AND OVERHAUL. To remove the vertical shaft and lower gear assembly, remove hood, side panels and upper grille closure plates. Remove steering gear upper housing as outlined in paragraph 12A, lower grille closure, and lower front panel. Drain steering housing. Disconnect and remove power steering pressure, return and bleed back lines, engine air cleaner and oil cooler radiator. Loosen the two lower radiator mounting cap screws and move the rear lower grille closure plate back away from steering housing. Remove center steering arm from vertical shaft flange on axle models or pedestal on tricycle models. Remove the four cap screws retaining vertical shaft housing to lower steering housing and lift housing from

shaft. While lower housing is still solidly bolted to tractor front support, loosen the large vertical shaft nut (N —Fig. MF112A). Unbolt and remove steering cylinders and rack guide and withdraw rack and pistons. Remove the cap screws retaining lower housing to front support and lift shaft and housing assembly from tractor. Remove shaft nut and lift housing, gear and thrust bearing from vertical steering shaft. Vertical shaft needle bearings (16 and 17—Fig. MF111) and seal (18) can be renewed at this time.

Note that one of the teeth on the sector gear is marked with an "O", the rack is marked with a punch mark between the two center teeth, and that the upper splines on the vertical shaft has a key flat milled in the shaft (see Fig. MF112). To reassemble the unit,

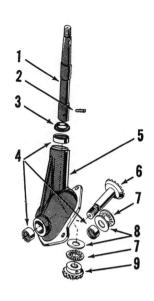

Fig. MF113—Exploded view of rear steering gear housing. Because of limited clearance in housing, roll pin (2) must be started from gear (9) with a punch and shaft rotated so that pin can be pulled with pliers. Shims (not shown) are available for gear backlash adjustment.

1. Steering shaft	6. Gear & shaft
2. Roll pin	7. Thrust bearing
3. Oil seal	8. Bearing race
4. Needle bearing	9. Gear
5. Housing	

Fig. MF112A — When assembling lower housing, tighten adjusting nut until 0.001-0.005 clearance is obtained between washer and lower housing as shown.

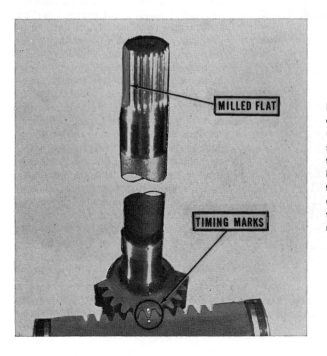

MILLED FLAT

TIMING MARKS

Fig. MF112 — When reassembling lower gear unit, shaft splines are inserted into sector gear so that milled flat is to the left as shown when "O" timing mark on sector gear is directly to the front. Rack tooth punch mark is indexed with timing mark on gear.

place the shaft in an upright position on a solid bench, with the milled flat to the left as shown. Place the shaft gear (10—Fig. MF111) in the housing so that the "O" mark on the gear tooth is up and toward the guide opening. Place thrust bearing (12) and washers (11) in the correct order on top of the gear and slide gear, thrust bearing and housing over upper end of the vertical shaft, engaging the gear with the shaft splines as the assembly is lowered. Check to be sure the "O" marked gear tooth and shaft milled flat are 90 degrees apart as shown in Fig. MF112 and install washer and shaft nut. With the assembly setting on the shaft flange, tighten the shaft nut until 0.001-0.005 clearance exists between the shaft washer and housing when measured with a feeler gage as shown in Fig. MF112A; then, stake nut to vertical shaft. Reinstall unit on tractor by reversing the removal procedure. When installing rack, make certain that rack and gear timing marks are in register as shown in Fig. MF112. Reinstall the upper housing as outlined in paragraph 12A, and adjust backlash as outlined in paragraphs 12 and 13.

Fig. MF114—Timing marks are provided on rear housing gears to properly locate steering wheel during assembly.

REAR STEERING GEAR
All Models

14. **R&R AND OVERHAUL.** Remove nut from steering wheel and remove wheel, Woodruff key, seal cap and seal from shaft. Remove ignition switch and starter button from lower instrument panel and remove panel. Drive out roll pin securing steering shaft to gear shaft. Remove the three cap screws securing rear gear housing to support and remove gear housing assembly.

Thoroughly clean the gear housing, place assembly in a vise and drive pin (2—Fig. MF113) securing pinion (9) to long shaft (1) a short distance into the gear. Note: Because of clearance in the housing, roll pin cannot be removed all the way with the punch. Rotate long shaft (1) until exposed portion of pin is visible, remove same with suitable pliers, and disassemble the remaining parts.

Examine bearings, shafts and gears for wear or damage and renew as necessary. When reassembling, vary number of 0.002 and 0.005 thick shims (not shown) to provide 0.000-0.001 gear backlash without binding. Mesh the punched timing marks on the two bevel gears to provide proper steering wheel spoke location (see Fig. MF114). Pack the housing with lubriplate prior to installation.

NON–DIESEL
ENGINE AND COMPONENTS

Fig. MF115—Side sectional view of 242 cubic inch engine showing oil flow. Bore and stroke are 3⅞ and 5⅛ respectively. Compression ratio is 7.3 to 1 for gasoline models and 8.1 to 1 for LP-Gas models.

R&R ENGINE WITH CLUTCH
All Non-Diesel Models

15. To remove the engine and clutch as an assembly, first drain the cooling system and if engine is to be disassembled, drain oil pan. Remove right and left side panels and power steering pressure, return and bleeder lines. Disconnect air cleaner hose, radiator hoses and radiator stay rod. Disconnect steering shaft at flexible coupling. Support tractor under transmission housing and front end assembly at top of steering housing. Unbolt front support casting from engine and roll front axle, support and radiator as an assembly away from tractor. Remove ignition switch and starter button from lower instrument panel and remove panel. Drive roll pin from steering shaft and remove shaft. Shut-off fuel and remove fuel tank. Disconnect heat indicator sending unit from water outlet elbow and cable from starting motor. Disconnect wires from coil and generator, tractor meter cable from the generator and oil gage line from right side of engine block. Disconnect choke rod from carburetor and governor control rod at rear. Remove exhaust pipe clamp brackets, loosen clamp at muffler and remove exhaust pipe. Support engine in a hoist and unbolt engine from transmission case.

CAUTION: On Multipower models, the master clutch must be detached from engine drive plate as outlined in paragraph 147.

CYLINDER HEAD
All Non-Diesel Models

16. **REMOVE AND REINSTALL.** To remove cylinder head, drain cooling system, shut-off fuel, disconnect fuel line and remove fuel tank front and rear mounting bolts. Remove ignition switch and starter button from lower instrument panel and remove panel. Disconnect steering shaft at flexible coupling, loosen set screw in front coupling and slide coupling forward as far as possible. Drive roll pin from rear of steering shaft, lift front of fuel tank and slide steering shaft forward out of tank. Remove fuel tank. Remove heat shield from manifold, disconnect and remove air cleaner pipe, disconnect and remove muffler (and upright pipe if used). Disconnect throttle and choke rods from carburetor, then unbolt and remove manifold. Remove upper radiator hose and disconnect heat indicator sending unit. Remove rocker arm cover and rocker arms assembly and push rods. Unbolt and remove cylinder head from tractor.

When reinstalling cylinder head, tighten the stud nuts from the center outward to a torque of 100-110 Ft.-Lbs. Manifold nut torque is 50-55 Ft.-Lbs. Valve tappet clearance is given in paragraph 17.

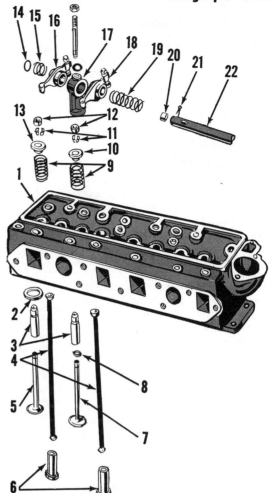

Fig. MF116—Valve train of the 242 cu. in. gasoline engine.

1. Cylinder head
2. Exhaust valve seat insert
3. Valve guide
4. Push rod
5. Exhaust valve
6. Valve tappet
7. Intake valve
8. Stem seal
9. Valve spring
10. Intake valve spring seat
11. Valve stem lock
12. Ball socket
13. Roto-cap assembly
14. Washer
15. Spring
16. LH rocker arm
17. Shaft support
18. RH rocker arm
19. Spring
20. Cork plug
21. Cotter pin
22. Rocker arm shaft

VALVES AND SEATS
All Non-Diesel Models

17. Valve tappet gap should be set hot to 0.016 for the intake and 0.018 for the exhaust. All valves are equipped with stem caps.

Intake valves seat directly in the cylinder head and the valve stems are sealed with neoprene oil guards. Exhaust valves are equipped with renewable type seat inserts and on gasoline models are fitted with positive type rotators. Replacement exhaust valve seat inserts should have an interference fit of approximately 0.003 in the cylinder head bore.

Intake valves have a face and seat angle of 30 degrees with a desired seat width of $\frac{1}{16}$-inch. Exhaust valves have a face and seat angle of 45 degrees with a desired seat width of $\frac{3}{32}$-inch. Seats can be narrowed using 15 and 75 degree stones.

Valve stem diameter is 0.3406-0.3414 for the intake, 0.3382-0.3390 for the exhaust.

VALVE GUIDES
All Non-Diesel Models

18. The pre-sized intake and exhaust valve guides are interchangeable

and can be driven from the cylinder head if renewal is required. Guides should be pressed into the cylinder head, using a piloted drift 0.002 smaller than the bore of the guide, until top end of guide is $1\frac{5}{32}$ inches from spring seat in cylinder head.

After installation ream the guides, if necessary, to provide the recommended stem to guide clearance of 0.0015 for the intake and 0.004 for the exhaust.

VALVE SPRINGS
All Non-Diesel Models

19. Intake and exhaust valve springs are interchangeable. Renew any spring that is rusted, discolored or does not meet the test specifications which follow:

Free length .... 2 5/64"
Lbs. test @ $1\frac{21}{32}$"............... 41-47
Lbs. test @ $1\frac{7}{32}$".............. 103-110

VALVE ROTATORS
All Non-Diesel Models

20. Normal servicing of the positive type exhaust valve rotators ("Roto-caps") consists of renewing the units. It is important however, to observe the valve action after the engine is started. The valve rotator action can be considered satisfactory if the valve rotates

a slight amount each time the valve opens.

VALVE TAPPETS
All Non-Diesel Models

21. Mushroom type valve tappets (cam followers) operate directly in machined bores of the cylinder block. The 0.6240-0.6245 diameter tappets are furnished in standard size only and should have a clearance of 0.0005-0.002 in the block bores. To remove the tappets, first remove the camshaft as outlined in paragraph 26.

ROCKER ARMS
All Non-Diesel Models

22. Rocker arms and shaft assembly can be removed after removing fuel tank and rocker arm cover. The rocker arms are right and left hand assemblies and not interchangeable. Desired clearance between new rocker arms and new rocker arm shaft is 0.0008-0.0014. Renew shaft and/or rocker arms if clearance exceeds 0.002. Oil holes in rocker arm shaft face toward valve spring.

When renewing rocker arm shaft, new cork plugs (20—Fig. MF116) should be installed in shaft.

Refer to paragraph 17 for valve tappet clearance.

VALVE TIMING
All Non-Diesel Models

23. Valves are properly timed when single punch marked tooth on crankshaft gear is meshed with double punch marked tooth space on camshaft gear as shown in Fig. MF117.

TIMING GEAR COVER
All Non-Diesel Models

24. To remove timing gear cover, remove front axle, front axle support and radiator from engine as outlined in paragraph 15. Remove fan blades and disconnect spring and throttle rod from governor. Remove starting jaw and crankshaft pulley. Remove the cap screws retaining oil pan to timing gear cover and loosen remaining oil pan cap screws. Unbolt and remove timing gear cover from engine.

Crankshaft front oil seal can be installed at this time and should be installed with lip facing the engine.

TIMING GEARS
All Non-Diesel Models

25. Timing gears can be renewed after removing timing gear cover as outlined in paragraph 24. Remove the governor assembly and using a suitable puller, remove the crankshaft timing gear. Remove retaining nut and remove camshaft gear.

Recommended timing gear backlash is 0.0015-0.003. Both crankshaft and camshaft gears are available in standard and oversizes or undersizes of 0.001 and 0.002. Both gears are stamped "S" (Standard), "U" (Undersize) or "O" (Oversize) and the amount to indicate the size. Any combination of standard, undersize or oversize gears which gives the desired backlash may be used.

Fig. MF117 — Camshaft and crankshaft gears showing location of timing marks.

During installation, mesh the single punch marked tooth on crankshaft gear with double mark on camshaft gear. Rearward movement of camshaft in operation is prevented by contact of camshaft gear with thrust plate. When camshaft gear is being installed with a drift, it is advisable to remove oil pan and buck-up camshaft with a heavy bar to prevent loosening of expansion plug at rear of camshaft.

CAMSHAFT
All Non-Diesel Models

26. To remove the camshaft, first remove the camshaft timing gear as outlined in paragraph 25. Remove fuel tank, rocker arm cover, rocker arms and shaft assembly and push rods. Push the tappets (cam followers) upward. Remove ignition distributor and camshaft thrust plate, then withdraw camshaft from front of engine.

Camshaft journal diameter is 0.996-0.9965 for outboard bearing, 1.9960-1.9965 for the front bearing, 1.7460-1.7465 for the center bearing and 1.6835-1.6840 for the rear bearing. Running clearance is 0.0025-0.004 in the bearings. Normal camshaft end play of 0.006-0.009 is controlled by the renewable thrust plate at the front end of the engine block.

ROD AND PISTON UNITS
All Non-Diesel Models

27. Piston and connecting rod assemblies are removed from above after removing cylinder head and oil pan. Pistons and rods are installed with the rod correlation marks facing the camshaft. Replacement rods are not marked and should be installed with the oil spray hole in the lower end of the rod facing away from the camshaft side of the engine. Tops of pistons are installed with arrow marked front facing toward front of engine.

Connecting rod nut torque is 70-75 Ft.-Lbs.

PISTONS, SLEEVES AND RINGS
All Non-Diesel Models

28. Aluminum alloy, cam ground pistons are supplied in standard size only and are available only in a kit consisting of pistons, pins, rings and sleeves. Recommended piston skirt clearance is 0.0025-0.0035. Piston skirt clearance is checked with a spring scale pull of 5-10 Lbs. using a 0.003 x ½-inch wide feeler gage 90 degrees from piston pin. Cylinder sleeves must be renewed when out-of-round or taper exceeds 0.008. With the piston and connecting rod assem-

bly removed from the cylinder block, use a suitable puller to remove the sleeves. Before installing the wet type sleeves, clean all cylinder block sealing surfaces. The top of the sleeve should extend 0.0015-0.004 above the top surface of the cylinder block. If the stand-out exceeds 0.004, check for foreign material under sleeve flange. Excessive standout will cause water leakage at the cylinder head gasket. To facilitate installation of sleeves, use a lubricant (Palm oil or vaseline) on the two neoprene sealing rings.

The top ring groove on some engines is fitted with a steel insert which is renewable. Top compression ring width is 0.093, second and third compression ring width is 0.125 and oil control ring width is 0.250. Ring sets are available in three types: (1) Original equipment rings having no expanders. (2) Re-Ring sets having flat spring expanders under oil control ring. (3) 2 in 1 sets having segmented oil control rings with flat steel expanders under second and third compression and oil control rings.

PISTON PINS
All Non-Diesel Models

29. The 1.1091-1.1093 diameter floating type piston pins are retained in the piston bosses by snap rings. Pin and retaining ring kits are available in standard size as well as oversizes of 0.003 and 0.005.

CONNECTING RODS AND BEARINGS
All Non-Diesel Models

30. Connecting rod bearings are of the shimless, non-adjustable, slip-in, precision type renewable from below after removing oil pan and connecting rod bearing caps. When installing new bearing shells, be sure that the projection engages milled slot in rod and cap and that rod and rod cap correlation marks are in register. Replacement rods are not marked and should be installed with the oil spray hole facing away from the camshaft side of the engine. Bearing inserts are available in 0.002, 0.010 and 0.020 undersize, as well as standard.

Crankpin diameter 2.248-2.249
Running clearance 0.0014-0.0024
Maximum allowable clearance . . . 0.003
Maximum allowable crankpin
  out-of-round 0.001
Maximum allowable crankpin
  taper 0.0015
Rod side play 0.006-0.009
Rod length C to C 8.373-8.377
Rod nut torque . . . . 70-75 Ft.-Lbs.

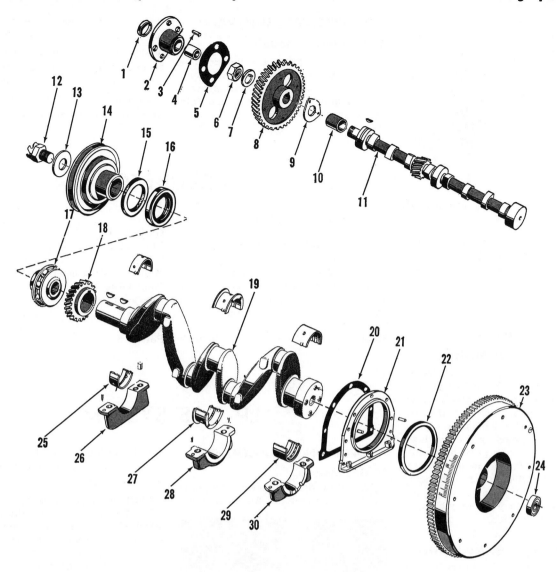

Fig. MF118—Engine camshaft, crankshaft and timing gears. Crankshaft front and rear seals are independent of main bearing caps.

1. Bearing support cap	11. Camshaft	21. Rear seal retainer
2. Camshaft bearing support	12. Starting jaw	22. Crankshaft rear seal
3. Dowel pin	13. Washer	23. Flywheel
4. Outer bushing	14. Fan drive pulley	24. Pilot bearing
5. Gasket	15. Front dust seal	25. Front bearing insert
6. Nut	16. Front oil seal	26. Front bearing cap
7. Washer	17. Governor assembly	27. Center bearing insert
8. Camshaft gear	18. Crankshaft gear	28. Center bearing cap
9. Thrust plate	19. Crankshaft	29. Rear bearing insert
10. Camshaft bushing	20. Rear seal gasket	30. Rear bearing cap

CRANKSHAFT AND MAIN BEARINGS

All Non-Diesel Models

31. Crankshaft is supported in three shimless, non-adjustable, slip-in, precision type main bearings renewable from below without removing the crankshaft. Main bearing caps are positively located to the engine block by dowels which are renewable.

Bearing inserts are available in 0.002, 0.010 and 0.020 undersize as well as standard.

Normal crankshaft end play of 0.006-0.008 is controlled by the flanged center main bearing inserts.

To remove crankshaft, it is necessary to remove engine, clutch, flywheel, rear oil seal retainer, timing gear cover, oil pan and bearing caps.

Check crankshaft journals for wear, scoring and out-of-round condition against the values listed below and in paragraph 30.

Main journal diameter.....2.624-2.625
Running clearance 0.0005-0.0027
Max. allowable out-of-round....0.001
Cap screw torque.......85-95 Ft.-Lbs.

CRANKSHAFT REAR OIL SEAL

All Non-Diesel Models

32. The crankshaft rear oil seal (22 —Fig. MF118) is contained in a one-piece retainer (21) and is renewable from the rear after removal of the clutch and flywheel.

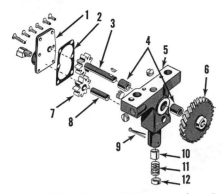

Fig. MF119—Exploded view of engine oil pump.

1. Cover	7. Follow gear
2. Gasket	8. Follow gear shaft
3. Lead gear & shaft	9. Cotter pin
4. Shaft bushings	10. Pressure relief valve
5. Pump body	11. Relief valve spring
6. Drive gear	12. Spring retainer

FLYWHEEL

Except Multipower Models

33. To remove the flywheel, separate the engine from the transmission case as outlined in paragraph 137, then remove the clutch unit from the flywheel. The starter ring gear can be renewed after removing the flywheel. To install a new ring gear, heat same to 500 degrees F. and install on flywheel with beveled end of teeth facing toward engine. Bolt holes are so arranged that flywheel can only be bolted on in one position.

Model MF Super 90 Multipower

33A. Because of the heavy clutch, multipower models use a thin engine drive plate instead of a flywheel as shown in Fig. MF191. The starter ring gear is attached to drive plate by screws which should be tightened to a torque of 23-25 ft.-lbs.

To service the drive plate or starter ring gear, refer to paragraphs 137 and 147.

OIL PUMP

All Non-Diesel Models

34. The gear type oil pump, shown exploded in Fig. MF119, is bolted to the bottom of number one main bearing cap, and is driven from the crankshaft timing gear. Pump is accessible after the oil pan is removed. Drive gear backlash of 0.010-0.012 is controlled by the addition of shims between the oil pump and main bearing cap.

Check the pump internal gears for wear or scoring and renew any damaged parts. Renew shaft and/or bushings if running clearance exceeds 0.001.

OIL PRESSURE RELIEF VALVE

All Non-Diesel Models

35. Plunger type relief valve (10—Fig. MF119) is located in oil pump body and can be adjusted with spacers inserted under the spring to maintain a pressure of 20-30 psi at 2200 engine rpm. At idle speed, pressure should be not lower than 15 psi. The piston of the relief valve should have a slide fit in its bore. The relief valve spring (11) should have a free length of 2 inches and should test 7¾-8¼ pounds at a working length of 1⅜ inches.

CARBURETOR

(Except LP-Gas)

36. All gasoline models are equipped with a Marvel-Schebler Carburetor Model TSX-644. This carburetor is balanced to operate with the dry element type secondary air filter installed on all models. Calibration data are as follows:

Float setting	¼-inch
Repair kit	286-1265
Gasket set	16-594
Float valve and seat	233-543
Idle jet	49-165
Main adjusting needle seat	36-294
Main adjusting needle	43-725
Idle adjusting needle	43-58
Nozzle	47-A23
Venturi	46-495
Power jet	49-179

LP-GAS SYSTEM

OPERATING ADJUSTMENTS

LP-Gas Models

37. Initial, dead engine adjustments on the carburetor are 2½ turns open

Fig. MF120—Pressure regulating carburetor showing location of adjustments.

1. Idle fuel adjustment
2. Main fuel adjustment
3. Throttle stop screw

for the idling screw (1—Fig. MF120) and 3½ turns open for the main fuel adjusting screw (2).

Start the engine and bring to operating temperature. Set throttle stop screw (3) to obtain an engine low idle speed of 450 rpm. Turn idle fuel adjustment (1) in or out until engine runs smoothly. Stop the engine and disconnect any three spark plug wires from the plugs. Restart engine, place throttle in wide open position and open main fuel adjustment (2) until highest engine rpm is obtained. After reinstalling plug wires, check idle fuel adjustment and readjust if necessary; then reset throttle stop screw to obtain a slow idle speed of 450 rpm.

Fig. MF121 — Left side view of LP-Gas tractor showing location of fuel filter and valves.

1. Vapor return valve
2. Liquid filler valve
3. Vapor withdrawal valve
4. Fuel filter
5. Liquid withdrawal valve

FUEL TANK AND LINES

LP-Gas Models

38. SERVICING. The pressure tank is fitted with fuel filler (2—Fig. MF-121), vapor return (1), pressure relief, bleeder, and liquid (5) and vapor (3) withdrawal valves which can only be serviced as complete assemblies. Before renewal is attempted on any of these units, drive the tractor to an open area and allow the engine to run until the fuel is exhausted; then open bleeder valve to allow any remaining pressure to escape. Fuel gage assembly consists of a dial face unit which can be renewed at any time, and a float unit which can only be renewed if the fuel tank is completely empty. Fuel tank is fitted with a ⅝ NC threaded boss to accommodate an eyebolt for removal. The safety relief valve is set to open at 312 psi pressure to protect the tank against excessive pressures. U.-L. regulations in most states prohibit any welding or repair on LP-Gas containers and the tank must be renewed rather than repaired in the event of damage. Fuel lines can be safely renewed at any time without emptying the tank if liquid and vapor withdrawal valves are closed and the engine is allowed to run until any fuel is exhausted.

39. FUEL FILTER. The cartridge type fuel filter (4—Fig. MF121) is located in the tee between the vapor and liquid withdrawal valves. To clean the disc-type filter, close both fuel tank withdrawal valves and run the engine until the fuel is exhausted from the system. Remove filter to vaporizer line at filter body, and filter body from filter head. Remove element and spring from filter body, clean element in a suitable solvent and dry thoroughly before reinstalling.

VAPORIZER

LP-Gas Models

40. OPERATION AND R&R. The LP-Gas vaporizer serves two purposes in the system. First, it vaporizes the liquid fuel as it is received from the fuel tank; and second, it acts as a primary regulator to reduce and control the pressure of the vapor prior to entering the carburetor.

The vaporizer is mounted in the water outlet elbow at the top of the cylinder head. Before removing vaporizer or disconnecting any fuel lines, close both tank withdrawal valves and allow engine to run until all fuel is exhausted from vaporizer, fuel lines and carburetor. Turn off the ignition switch after engine stops. Raise hood, drain radiator and completely remove the left hood side panel. Disconnect the vaporizer to carburetor connection and tank to vaporizer hose. Remove the vaporizer retaining cap screw and withdraw vaporizer from water outlet elbow. Refer to Fig. MF122 for an exploded view of vaporizer. Test and/or overhaul the unit outlined in paragraph 41.

41. TEST AND OVERHAUL. The removed vaporizer assembly can be tested for external or internal leaks without disassembly, and an almost complete diagnosis of vaporizer condition made. To test the unit, proceed as follows:

Connect the vaporizer inlet to a source of compressed air and completely immerse the unit in a water tank. External leaks will show up as air bubbles. Note especially the areas around vaporizer coil and mounting plate (1—Fig. MF122) and around

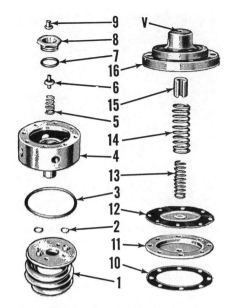

Fig. MF122 — Exploded view of vaporizer assembly.

1. Coil & plate	10. Gasket
2. "O" ring	11. Baffle
3. "O" ring	12. Diaphragm
4. Body	13. Inner spring
5. Valve spring	14. Outer spring
6. Valve	15. Spacer
7. "O" ring	16. Cover
8. Valve seat	V. Vent hole
9. Follower	

Fig. MF123—Typical LP-Gas engine fuel and air system.

Fig. MF127 — Assembled view of Zenith PC2 carburetor of the type used on most MF Super 90 LP-Gas tractors. The model PC1 carburetor used on earlier models is similar except that idle diaphragm assembly (3) is not used.

1. Idle speed adjusting screw
2. Idle needle
3. Idle diaphragm assy.
4. Main adjustment needle
5. Inlet diaphragm assy.
6. Inlet fitting

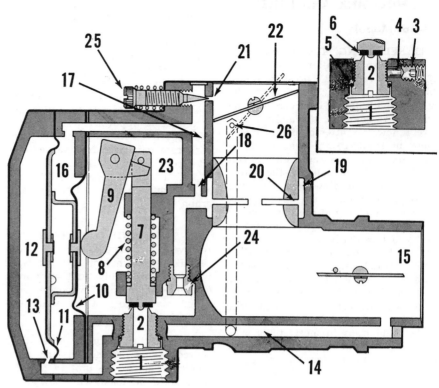

Fig. MF128—Cross-sectional view of pressure regulating carburetor showing main components. Inset shows fuel valve seat locking arrangement.

1. Fuel inlet
2. Fuel valve seat
3. Locking screw
4. Locking plug
5. "O" ring
6. Sealing disc
7. Fuel valve
8. Fuel valve spring
9. Diaphragm lever
10. Inner diaphragm
11. Outer diaphragm
12. Outer diaphragm chamber
13. Air passage orifice
14. Air passage
15. Air intake
16. Inner diaphragm chamber
17. Idle fuel passage
18. Idle orifice
19. Annulus
20. Venturi
21. Idle needle seat
22. Throttle fly
23. Pressure chamber
24. Main jet
25. Idle needle
26. Economizer orifice

diaphragm cover (16). Air bubbles emerging from vent hole (V) in top of diaphragm cover indicate a leaking diaphragm.

To check the vaporizer inlet valve and seat, install a low-pressure test gage in the vaporizer outlet port, and connect inlet to air pressure. The gage reading should be 9-11 psi and hold steady. If pressure continues to rise, a leaking fuel valve or valve seat "O" ring is indicated and vaporizer should be overhauled.

To disassemble the vaporizer, remove four alternate screws from diaphragm cover (16) and install aligning studs (Zenith Tool Part No. C161-195). Apply thumb pressure to top of diaphragm cover and remove the remaining diaphragm screws, diaphragm cover (16), spacer (15) and springs (13 and 14). To renew any part of the fuel valve assembly, remove the fuel valve seat (8) using a suitable socket wrench. To renew "O" rings (2 or 3), remove the four screws retaining coil and plate (1) to body, and withdraw the plate.

When reassembling, install fuel valve (6) with long stem toward diaphragm. Make sure all screw holes are aligned in gasket (10), baffle (11), diaphragm (12) and cover (16). Tighten the retaining screws evenly, leaving aligning studs installed until cover is tight. Recheck for leaks after assembly by immersing in water or using a soap solution.

CARBURETOR

LP-Gas Models

42. **OPERATION.** The Zenith pressure regulating carburetor serves both as a secondary regulator and as a carburetor. See Fig. MF128 for a cross sectional view of the early type, PC1 carburetor. NOTE: The late type PC2 carburetor is similar except for the idle diaphragm installed on carburetor as shown in Fig. MF127. Operation of both types of carburetors is similar.

The fuel valve seat (2—Fig. MF128) on most carburetors is adjustable so that position of diaphragm lever (9) can be varied with relation to the diaphragms. On these models the seat is locked in position by means of lock screw (3) and nylon plug (4) as shown in inset.

The two diaphragms (10 and 11) control the pressure and flow of incoming fuel to maintain the proper fuel-air mixture.

43. **R&R AND OVERHAUL.** To remove the carburetor, first close both withdrawal valves and allow engine to run until fuel is exhausted from regulator, lines and carburetor. Turn off ignition switch, disconnect choke and throttle linkage and the fuel inlet line. Unbolt and remove the carburetor assembly.

Remove the six screws securing the diaphragm cover, spacer and the two diaphragms, and remove the diaphragms. On late models, unbolt the idle diaphragm cover, remove cover and diaphragm; then, unbolt and remove the idle diaphragm housing from side of carburetor body.

Remove the fuel inlet fitting from bottom of carburetor and remove fuel valve seat (2—Fig. MF128). NOTE: The valve should be held off its seat by applying light pressure to diaphragm lever (9). On models so equipped, be sure locking screw (3) is loosened.

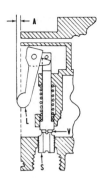

Fig. MF129—Schematic view of regulating valve assembly used on Zenith Pressure Regulating carburetors.

A. Adjustment
L. Lever
S. Valve seat
V. Valve

Remove the diaphragm lever shaft plug from side of carburetor body, and remove shaft, lever and valve assembly. Remove the main and idle needle valves and the main jet (24) using the special tool (Zenith C161-193). PC1 carburetors used on model MF85 and MF88 tractors are equipped with an economizer jet (not shown) which is located in pressure chamber (23). The economizer jet can be removed using the special main jet tool. The two jets are identical except for the size of the orifice. Make sure the yet having the smaller orifice is installed as the economizer jet.

Clean all metal parts in a suitable solvent and examine for wear or damage. Examine diaphragms for cracks, pinholes or deterioration and renew as necessary. Reassemble by reversing the disassembly procedure, using new gaskets and seals. When installing the fuel valve seat (2), hold valve open by applying pressure to valve lever (9), and tighten seat until it bottoms on models not equipped with locking screw (3). On all models equipped with locking screw and plug, use Step 2 of Zenith Tool C161-194; or adjust so that lever (9) is 1/16-inch to rear of rear surface of carburetor body as shown at (A—Fig. MF129). Tighen the locking screw (3—Fig. MF128) when adjustment has been obtained. When reinstalling the diaphragms, spacer, cover and gasket, make sure the parts are arranged so that air passages are open.

NON–DIESEL GOVERNOR

44. MINOR ADJUSTMENT. Warm up engine and adjust carburetor mixture. Push throttle lever up to the low idle position and adjust throttle stop screw to obtain an engine low idle speed of 450 to 500 rpm. Fully open throttle and check the high idle speed which should be 2100-2200 rpm.

If specified high idle speed is not obtained, open battery door and slide out battery. Using a $\frac{7}{16}$-inch end wrench, back-off nut on throttle link assembly (7—Fig. MF130). Lengthen link by means of turnbuckle to raise engine speed or shorten link to reduce engine speed.

With engine operating at 1000 rpm, check for surging or unsteady running. If surging exists, check for and remove any binding in the operating linkage.
Note: Engine speeds can be observed on tractormeter.

45. MAJOR ADJUSTMENT. With engine stopped and the carburetor throttle rod (12—Fig. MF130) disconnected at governor arm (17), pull throttle hand lever (1) down and check to see that governor arm moves to extreme rearward position without binding. If binding exists or if governor arm does not move fully rearward, proceed as in paragraph 46. With governor arm in rearward position, pull carburetor throttle rod rearward until throttle shaft is against stop in open position. Adjust the throttle control rod clevis (14) until it is necessary to move rod forward approximately $\frac{1}{32}$-inch to insert clevis pin through throttle rod and governor arm. Key the pin and tighten clevis lock nut.

Open battery door, remove battery and adjust throttle link (7) so that, with hand throttle lever in wide open position, the lower end of throttle control lower crank (8) clears front of battery box by $\frac{1}{8}$-inch. Tighten throttle link lock nut. Move throttle hand lever to fully closed position and adjust length of governor control rod until governor control spring plunger (16) lightly holds carburetor throttle shaft against idle stop. Tighten governor control rod lock nut, start engine and check for correct high idle speed of 2100-2200 rpm and low idle speed of 450-500 rpm; make minor adjustments as necessary.

If hand throttle tends to creep, remedy same by tightening star-spring retaining nut (5) on friction plate assembly until desired friction drag is obtained.

46. R&R AND OVERHAUL. First step in the removal of the governor is to remove the timing gear cover as outlined in paragraph 24. The governor fork and lever assembly, bearings and/or oil seal, Fig. MF130, located in the timing gear cover can be renewed at this time.

Remove the Woodruff key and withdraw the governor weight unit from the crankshaft.

To disassemble the governor weight unit remove the snap rings (1—Fig.

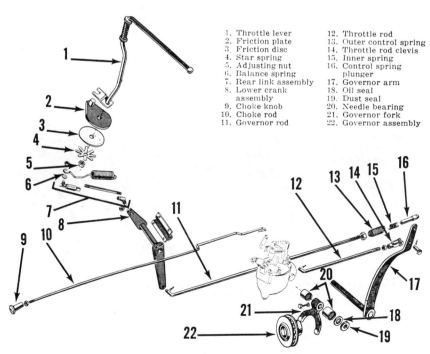

1. Throttle lever
2. Friction plate
3. Friction disc
4. Star spring
5. Adjusting nut
6. Balance spring
7. Rear link assembly
8. Lower crank assembly
9. Choke knob
10. Choke rod
11. Governor rod
12. Throttle rod
13. Outer control spring
14. Throttle rod clevis
15. Inner spring
16. Control spring plunger
17. Governor arm
18. Oil seal
19. Dust seal
20. Needle bearing
21. Governor fork
22. Governor assembly

Fig. MF130—Exploded view of throttle, choke and governor controls and linkage.

Fig. MF131—Exploded view of governor weight unit.

1. Snap ring
2. Fork base
3. Thrust bearing
4. Inner race
5. Ball
6. Governor driver
7. Outer race
8. Thrust washer

MF131). Renew the governor balls (5), the inner race (4) or outer race (7) if they are worn, scored or discolored.

Reassembly is the reverse of the disassembly procedure. When installing the timing gear cover, make certain the governor fork falls behind the governor weight unit.

NON-DIESEL COOLING SYSTEM

RADIATOR

All Non-Diesel Models

47. Radiator filler neck is fitted with a 7-lb. pressure cap which raises coolant boiling point approximately 19 degrees F. Check filler neck sealing surface for foreign material, dents or solder accumulations which would prevent proper sealing of cap. Check cap gasket for cracks or other signs of failure.

To remove radiator, first drain cooling system. Remove hood, side panels and grille assembly. Remove grille top closures, air cleaner pipe, oil coolant radiator and radiator hoses. Remove fan blades and mounting bolts and

withdraw radiator from right side of tractor.

THERMOSTAT
All Non-Diesel Models

48. The 180-degree thermostat is located in water outlet casting. To renew thermostat remove the fuel tank as outlined in paragraph 16, disconnect the upper radiator hose and unbolt and remove the upper coolant outlet elbow from the cylinder head.

WATER PUMP AND FAN
All Non-Diesel Models

49. To remove the water pump, re-

move the hood and side panels and drain the cooling system. Disconnect the lower radiator hose and unbolt the fan blades from the fan pulley. Unbolt the water pump body at the engine block and withdraw the pump.

To disassemble pump, remove the bearing housing to pump body bolts and remove the pump body (2—Fig. MF132) from the assembly. Remove fan pulley and hub assembly, pump impeller and seal. Remove the snap ring (9) and press the shaft and bearing assembly (8) forward out of the housing. Shaft and bearings are available as an assembled unit only.

NON-DIESEL IGNITION AND ELECTRICAL SYSTEM

SPARK PLUGS
All Non-Diesel Models

50. For normal service, all models come equipped with Champion D 14, AC 83S Comm or Auto-Lite BT-4 plugs. For heavy loads or severe service, Champion D-10, AC 82S Comm or Auto-Lite BT-3 may be used. For light loads or intermittent use in cold weather, Champion D-16, AC 85S Comm or Auto-Lite BT-8 may be needed. Electrode gap should be 0.025-0.027.

DISTRIBUTOR
All Non-Diesel Models

51. Gasoline tractors and late LP-Gas tractors are equipped with Delco-Remy 1112591 or Auto-Lite IAD-6004-2E distributors. LP-Gas models before serial number 804429 were equipped with Delco-Remy 1112598 distributor. Specification data follows:

DELCO-REMY 1112591
Breaker contact gap.............0.022
Breaker arm spring tension
  (measured at center of
  contact) ....17-21 oz.
Cam angle ....25°-34°

Advance data is in distributor degrees and distributor rpm.
Start advance ....0-2 @ 350
Intermediate advance...3.5-5.5 @ 500
Maximum advance........7-9 @ 1000

AUTO-LITE IAD-6004-2E
Breaker contact gap.............0.022
Breaker arm spring tension..17-20 oz.
Cam angle ....42°

Advance data is in distributor degrees and distributor rpm.
Start advance................0 @ 250
Intermediate advance.....4.75 @ 375
Maximum advance..........8 @ 650

DELCO-REMY 1112598
Breaker contact gap.............0.022
Breaker arm spring tension
  (measured at center of
  contact) ....17-21 oz.
Cam angle ....25°-34°

Advance data is in distributor degrees and distributor rpm.
Start advance..........0.5-2.5 @ 525
Maximum advance........3-5 @ 1000

51A. **IGNITION TIMING.** Crank engine until number one piston is coming up on compression stroke and continue cranking until flywheel marking of 5 degrees BTDC for gasoline models or 4 degrees BTDC for LP-Gas models aligns with timing pointer in inspec-

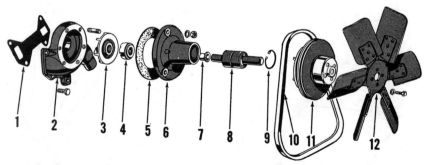

Fig. MF132—Exploded view of water pump and fan assembly. Wearing parts are renewable.

1. Gasket
2. Body
3. Impeller
4. Seal
5. Gasket
6. Bearing support
7. Slinger
8. Shaft & bearing assembly
9. Snap ring
10. Fan belt
11. Drive pulley
12. Fan

tion port (left side of engine block). See Fig. MF133. Set distributor contact gap at 0.022, loosen distributor clamping screw and rotate distributor until points just start to open. Tighten clamping screw to lock distributor in position.

This method sets static timing of distributor. If a timing light is available a running check should be made. Maximum advance at 450 engine rpm should be 5 degrees BTDC for gasoline tractors and 4 degrees BTDC for LP-Gas models. At 2000 engine rpm advance should be 19-23 degrees BTDC for gasoline tractors; 18-22 degrees BTDC for LP-Gas models with model 1112591 distributor; and 10-14 degrees BTDC for LP-Gas models with model 1112598 distributor. Firing order is 1-3-4-2.

GENERATOR & REGULATOR
All Non-Diesel Models

52. All models are equipped with a Delco-Remy 1100359 generator and Delco-Remy 1118981 regulator. Specification data are as follows:

DELCO-REMY 1100359 Generator
Brush spring tension...........28 oz.
Field draw
    Volts 12
    Amperes 1.58-1.67
Output (cold)
    Maximum amperes 20
    Volts 14.0
    RPM 2300

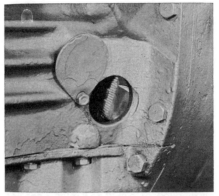

Fig. MF133—Left side of engine block with timing marks exposed. Static timing is 5° BTDC for gasoline models, 4° BTDC for LP-Gas models.

DELCO-REMY 1118981 Regulator
Cutout relay
    Air gap 0.020
    Point gap 0.020
    Closing voltage (range)....11.8-14.0
    Adjust to 12.8
Voltage regulator
    Air gap 0.075
    Voltage range 13.6-14.5
    Adjust to 14.0
Ground Polarity ....Negative

STARTING MOTOR
All Non-Diesel Models

53. All models are equipped with Delco-Remy 1107728. Specification data are as follows:

DELCO-REMY 1107728
Brush spring tension..35 oz. minimum
No-load test
    Volts 10.6
    Amperes (including solenoid)...112
    RPM 3240
Resistance test
    Volts 3.5
    Amperes (with solenoid)...320-385

MF85 & MF88 DIESEL ENGINE AND COMPONENTS
NOTE: REFER TO PARAGRAPHS 84 TO 131 FOR DIESEL ENGINE AND COMPONENTS USED IN SUPER 90 AND SUPER 90 WR

R&R ENGINE WITH CLUTCH
Models MF85 Diesel-MF88 Diesel

54. To remove the engine and clutch as an assembly, first drain the cooling system and if engine is to be disassembled, drain oil pan. Remove upright muffler, right and left side panels and power steering pressure, return and bleed lines. Disconnect air cleaner hose, radiator hoses and radiator stay rod. Disconnect steering shaft at flexible coupling. Support tractor under transmission housing and front end assembly at top of steering housing. Unbolt front support casting from engine and roll front axle, support and radiator, as an assembly, away from tractor. Remove starter switch from lower instrument panel and remove panel. Drive the roll pin from rear of steering shaft and withdraw shaft from front of fuel tank. Shut off the fuel and disconnect main fuel line from tank to sediment bowl, disconnect bleed back lines and lines leading to cold starting tank, remove cold starting tank then unbolt and remove the main fuel tank. Disconnect tractormeter cable from generator, wires from generator and cold starting unit, remove heat baffle from right side of engine and disconnect oil gage line. Disconnect cable from starting motor, wire from fuel pump and heat indicator sending unit from left side of engine. Remove governor and stop con-

trol rods from injection pump, support engine in a hoist and unbolt engine from transmission case.

CYLINDER HEAD
Models MF85 Diesel-MF88 Diesel

55. **REMOVE AND REINSTALL.** To remove cylinder head, drain cooling system, remove upright muffler and hood and side panels. Shut off the fuel, disconnect fuel and bleed-back lines and remove cold starting supply tank. Remove starter switch from lower instrument panel and remove panel. Remove the front and rear fuel tank mounting bolts, disconnect steering shaft at flexible coupling and drive out roll pin retaining rear of shaft. Slightly raise front end of fuel tank and withdraw the steering shaft, then lift off tank.

Disconnect the wire leading to the cold starting unit, remove air cleaner pipe and unbolt and remove the intake and exhaust manifolds.

Remove the injector pressure and bleed-back lines. Cap all exposed connections as lines are removed to prevent dirt entry, then unbolt and remove injectors. Remove upper radiator hose and disconnect heat indicator sending unit. Remove rocker arm cover, rocker arms assembly and push rods, then unbolt and remove cylinder head.

Install the cylinder head by reversing the removal procedure. Tighten the $\frac{9}{16}$-inch stud nuts to a torque of 130-140 Ft.-Lbs. in the sequence shown in Fig. MF136. Tighten the two $\frac{3}{8}$-inch stud nuts (18 and 19) to a torque of 35-40 Ft.-Lbs.

VALVES AND SEATS
Models MF85 Diesel-MF88 Diesel

56. Intake valves seat directly in cylinder head. Exhaust valves have renewable seat inserts. Replacement valve seat inserts are available only in 0.010 oversize, and should be installed with approximately a 0.003 interference fit in cylinder head counter bores.

Intake and exhaust valves have a face angle of 44 degrees, a seat angle of 45 degrees and a desired seat width of $\frac{3}{32}$-inch. Seats can be narrowed using 15 and 70 degree stones.

Intake valve stems are fitted with neoprene stem seals and have a stem to guide clearance of 0.0008-0.0021 with a suggested maximum wear limit of 0.004. Exhaust valves have a stem to guide clearance of 0.0035-0.005 with a suggested maximum wear limit of 0.007.

Intake valve stem diameter is 0.4344-0.4352. Exhaust valve stem diameter is 0.4315-0.4325. Adjust valve tappet gap to 0.014 hot, for both intake and exhaust.

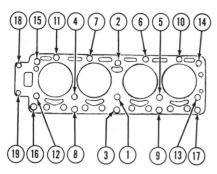

Fig. MF136—On diesel engine, tighten the 9/16-inch stud nuts to 130-140 Ft.-Lb. in sequence shown. Tighten nuts 18 and 19 to 35-40 Ft.-Lb.

VALVE GUIDES
Models MF85 Diesel-MF88 Diesel

57. Intake and exhaust valve guides are not interchangeable and can be driven from cylinder head if renewal is required. Press new guides into cylinder head with smaller O. D. of guide toward top of head until the distance from port end of guide to valve seat contact surface is 2 9/32-inches. Ream the guides after installation to an inside diameter of 0.4361-0.4365 to provide the desired stem to guide clearance of 0.0015 for the intake and 0.004 for the exhaust valves. Renew the guide if inside diameter exceeds 0.438.

VALVE SPRINGS
Models MF85 Diesel-MF88 Diesel

58. Intake and exhaust valve springs are interchangeable. Renew any spring which is rusted, discolored, or does not meet the test specifications which follow:

Free Length .... 2 ⅜-inches
Lbs. test @ 1⅞-inches..... 58-64
Lbs. test @ 1.521 inches..... 115-123
   Wear limit at open length of 1.521 inches is 105 lbs.

VALVE TAPPETS
Models MF85 Diesel-MF88 Diesel

59. The cylindrical type valve tappets operate directly in machined bores in the cylinder block. Tappets can be removed from above after cylinder head, rocker arms and push rods are removed or pushed up from below after the oil pan is off. Tappet diameter is 0.9975-0.9985, with a recommended diametral clearance of 0.0015-0.003 in block bores.
   Recommended valve tappet gap is 0.014 (hot) for both intake and exhaust.

ROCKER ARMS
Models MF85 Diesel-MF88 Diesel

60. The rocker shaft and rocker arms assembly can be removed after removing fuel tank and rocker arm cover. Desired clearance between rocker arms and shaft is 0.001. Renew rocker arms and/or shaft if clearance exceeds 0.005. The two ends of the rocker shaft are sealed with cork plugs retained by cotter pins. When reassembling, push the cork plugs into shaft only far enough to insert cotter pins, to avoid blocking the oil passages to end rocker arms. Refer to paragraph 56 for valve tappet clearance.

VALVE TIMING
Models MF85 Diesel-MF88 Diesel

61. Valves are properly timed when single punch-marked tooth on crankshaft gear is meshed with double punch-marked tooth space on camshaft gear as shown in Fig. MF137.

TIMING GEAR COVER
Models MF85 Diesel-MF88 Diesel

62. To remove the timing gear cover, first remove hood and side panels; then remove front axle, front axle support and radiator from tractor as a unit, as outlined in paragraph 54. Remove fan blades, starting jaw and crankshaft pulley. Remove the cap screws retaining oil pan to timing gear cover and loosen remaining oil pan cap screws. Unbolt and remove timing gear cover from engine.
   Crankshaft front oil seal can be installed at this time, and should be installed with lip facing engine.

TIMING GEARS
Models MF85 Diesel-MF88 Diesel

63. The timing gear train consists of the vibration damper drive and idler gears as well as camshaft and crankshaft gears, as shown in Fig. MF137. Removal of the vibration damper and service of the gears is covered in paragraphs 68 and 69. The camshaft and crankshaft timing gears are provided with threaded holes (A-Fig MF137) for attaching gear puller. To remove the camshaft timing gear after cover is off, remove nut (N) at forward end of camshaft and attach a gear puller at the two threaded holes (A). When reinstalling, remove oil pan and buck up camshaft with a heavy bar to prevent damage to injection pump drive gear or loosening of expansion plug at rear of camshaft. To remove the crankshaft gear, attach a suitable puller to threaded attaching holes (A).
   Recommended timing gear backlash is 0.000-0.002. Both crankshaft and camshaft gears are available in standard as well as oversizes and undersizes of 0.001 and 0.002. Gears are stamped "S" (Standard), "U" (Undersize) or "O" (Oversize) with a number to indicate the amount. Size number is lo-

Fig. MF137 — View of diesel engine timing gears with timing marks (TM) aligned. Camshaft and crankshaft gears are equipped with threaded attaching holes (A) for gear puller.

1. Camshaft gear
2. Crankshaft gear
3. Vibration damper idler
4. Vibration damper gear

cated inside the "O" or "U." In production, a number using the same code is stamped on the engine block at upper edge of timing gear cover gasket surface. This number indicates the basic total gear size required to establish the correct backlash in a new engine. When renewing the gears, install a crankshaft gear of the correct size to establish the recommended 0.004-0.007 backlash between crankshaft gear and vibration damper idler gear; then, using the block number as a guide, select a camshaft gear which will provide the recommended timing gear backlash. The condition of camshaft and bearing bores must be considered when using the guide number stamped on block.

When installing the gears, make sure the single punch-marked tooth on crankshaft gear is meshed with double punch-marked tooth space on camshaft gear, single punch-marked tooth on idler gear is meshed with double punch-marked tooth space on crankshaft gear, and single punch-marked tooth on damper drive gear is meshed with double punch-marked tooth space on idler gear as shown in Fig. MF137. Timing marks on vibration damper drive and idler gears will align only once in three revolutions of idler gear.

CAMSHAFT
Models MF85 Diesel-MF88 Diesel

64. The camshaft is supported by three bearing journals which ride in renewable bushings in the engine block, and by one outboard bearing located on timing gear cover. All bearing bores are fitted with renewable bushings which must be final sized after installation to provide recommended clearance. A helical cut gear at the center journal drives the vertically mounted injection pump. The recommended diametral clearance for camshaft journals is: Rear, 0.004-0.0057; center, 0.004-0.0057; front, 0.0035-0.0055, and outboard, 0.0025-0.0035.

To remove the camshaft, first drain cooling system and oil pan. Remove timing gear cover as outlined in paragraph 62, cylinder head as outlined in paragraph 55, and unbolt and remove the engine oil pan. Remove injection pump and drive gear as outlined in paragraph 78B. Remove camshaft timing gear, using a suitable puller, and unbolt and remove camshaft thrust plate. Push tappets upward in block bores and withdraw camshaft.

To renew the camshaft bushings, it is necessary to remove engine from

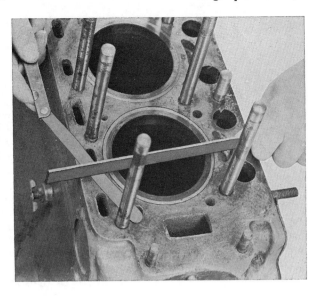

Fig. MF138 — Measure the standout of the cylinder sleeve above the block surface with straight edge and feeler gage as shown. Standout should measure 0.0015-0.004.

transmission case, remove clutch and flywheel and the expansion plug in camshaft bore at rear of engine block. Bushings should be removed and reinstalled with a piloted push-puller.

Camshaft journal diameter is 0.996-0.9965 for outboard bearing, 1.9955-1.9965 for front bearing, 1.7455-1.7465 for center bearing and 1.6835-1.6840 for rear bearing. Normal camshaft end play of 0.005-0.009 is controlled by the renewable thrust plate at front of engine block.

ROD AND PISTON UNITS
Models MF85 Diesel-MF88 Diesel

65. Piston and connecting rod assemblies are removed from above after removing cylinder head and oil pan. To gain access to the connecting rod bolts it is first necessary to remove the vibration damper frame as outlined in paragraph 68.

Connecting rods and caps are marked to correspond with the cylinders in which they are installed and should be installed with oil spray hole facing away from camshaft side of engine. The recommended connecting rod bolt torque is 85-95 Ft.-Lbs.

PISTONS, SLEEVES AND RINGS
Models MF85 Diesel-MF88 Diesel

66. Pistons are supplied in standard size only, and are available in kits consisting of piston and rings, or piston, pin, rings and sleeve. Each piston is fitted with three 1/8-inch wide compression rings, one 1/4-inch wide oil control ring and one 3/16-inch wide oil control ring. Top compression ring side clearance in groove should be 0.004-0.006, with a maximum wear limit of

0.008. All other rings have a recommended side clearance of 0.003-0.0045 with a maximum wear limit of 0.0065. Piston ring end gap should be 0.013-0.025 for all rings.

Piston skirt clearance can be considered satisfactory when a spring scale pull of 5-10 pounds is required to withdraw a 1/2-inch by 0.004 feeler gage from side of piston 90° from pin.

Cylinder sleeves must be renewed when out of round or taper exceeds 0.008. With piston and connecting rod assembly removed, use a suitable puller to remove the sleeve. Before installing the wet-type cylinder sleeve, thoroughly clean all cylinder block sealing surfaces. The top of the sleeve should extend 0.0015-0.004 above gasket surface of cylinder block. To measure the standout, insert the sleeve without seals into the cylinder block and use a straight edge and feeler gage as shown in Fig. MF138. Insufficient standout can be corrected by inserting the required amount of shims between cylinder block and sleeve flange. Shims are available in 0.0015 and 0.003. Too much standout is usually caused by the presence of dirt or scale on flange seating surface.

After sleeve standout is correctly adjusted, remove sleeves and install new sealing rings in the grooves at lower end of sleeves. Make certain sealing rings are not twisted, and coat outer surfaces with petroleum jelly or other suitable lubricant, align the connecting rod relief slots, and press sleeve in place with the palm of the hand.

After sleeves are in place, it is recommended that the cylinder block be filled with cold water to check for leaks at lower seals.

Fig. MF139—Vibration damper frame with timing marks (TM) aligned. Timing marks will align once in three revolutions of idler gear (3).

PISTON PINS
Models MF85 Diesel-MF88 Diesel

67. The 1.2498-1.2500 diameter full floating piston pins are available in 0.003 and 0.005 oversize as well as standard. Pin should have a light push fit in piston when piston is heated to 160 degrees F., and 0.0004-0.0006 clearance in connecting rod bushing.

VIBRATION DAMPER
Models MF85 Diesel-MF88 Diesel

68. **OPERATION AND R&R.** The Model HD277 diesel engine is equipped with a Lanchester type vibration damper mounted on a special frame underneath the engine block and driven by the crankshaft timing gear (See Fig. MF137). This frame also contains the engine oil pump. The damper consists of two unbalanced shafts which rotate in opposite directions at twice crankshaft speed, thus setting up a reaction to overcome the engine's natural tendency to vibrate. Correct timing is therefore extremely important. A vibration damper which is incorrectly installed may add to the vibration to cause actual engine damage.

To remove the vibration damper frame for overhaul or access to the rods or bearings, proceed as follows: Remove oil pan and the oil pump pressure lines from frame and engine block. Remove the retaining cap screws and pry the frame from the locating dowels.

To reinstall and time the vibration damper, first crank the engine until flywheel timing mark is at TDC when viewed through the timing window on left side of flywheel housing, then rotate the damper idler gear until the double punch-marked tooth space on idler gear is matched with single punch-marked tooth on the damper shaft gear (See Fig. MF139). The timing marks will align once in each three revolutions of the idler gear. While holding the idler gear in this position, carefully install the damper frame so that the single punch-marked tooth in idler gear meshes with the lower tooth space in crankshaft gear. Tighten the retaining cap screws to a torque of 42-47 Ft.-Lbs., and recheck the flywheel and damper timing marks. The crankshaft gear to idler gear timing marks may be checked with an in-

Fig. MF141 — When installing idler shaft, align correlation marks (CM) and bump shaft rearward until rear face is 11/32-inch below machined surface of frame.

spection mirror if desired. As the damper shaft turns at twice crankshaft speed, either No. 1 or No. 4 piston can be on the compression stroke.

69. **OVERHAUL.** To overhaul the removed damper unit, refer to Fig. MF140 and proceed as follows:

Remove the hexagon nut (N) on idler gear shaft and press shaft and gear from damper frame using an arbor press. The idler gear, shaft and bearing assembly is serviced only as a matched set. Remove oil pump strainer screen, inlet base, lead gasket and the shield from relief valve. Remove idler shaft pressure line (2) then unbolt and remove oil pump body (3) and body gears, from rear of frame. Overhaul data for the oil pump are given in paragraph 73. Drive the pin (P) from frame center web and counterweight idler shaft and, using a pointed tool scribe a correlation mark on shaft and frame as shown at (CM—Fig. MF141). Correlation mark is to insure that shaft be reinstalled in the same radial position. Using a soft drift to prevent damage to pressure line fitting threads, drive the idler shaft forward out of frame and counterweight. Shaft is a press fit in center web at pin location. Examine counterweight gear for chipped or damaged teeth and idler shaft and bushings for wear or scoring. The idler shaft should have 0.0025-0.004 clearance in idler counterweight bushings. If bushing renewal is indicated, press both bushings out gear end of counterweight. New bushing should be pressed into counterweight until rear edge is $\frac{1}{32}$-inch below rear surface of gear hub. If idler shaft must be renewed, install the shaft until rear end is $\frac{11}{32}$-inch below the rear frame surface as shown in Fig. MF141 and, using the existing $\frac{1}{16}$-inch hole as a guide, drill

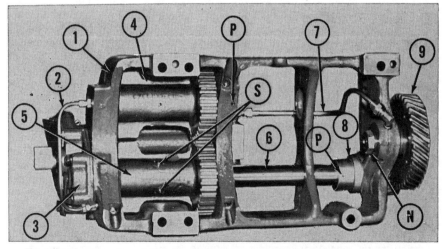

Fig. MF140—Top view of "Lanchester" vibration damper showing component parts.

1. Damper frame	5. Drive counterweight	9. Idler gear
2. Oil line	6. Drive shaft	N. Idler shaft nut
3. Oil pump	7. Oil line	P. Shaft pin
4. Idler counterweight	8. Collar	S. Set screw

a ¼-inch hole through frame web and shaft at (P—Fig. MF140) and install a tempered steel groove pin.

Before removing the drive shaft (6) from the counterweight and damper frame, measure the installed clearance between collar (8) and tab washer (TW—Fig. MF142) using a feeler gage. If the measured clearance is more than the recommended 0.003-0.005, renew the tab washers and/or thrust collar to obtain the specified clearance when the unit is reassembled. To remove the drive shaft, loosen the set screws (S—Fig. MF140) in counterweight (5), support shaft on under side and drive the pin (P) from collar (8) and shaft. Using a brass drift, drive the shaft (6) with gear attached, from counterweight and frame. Examine drive shaft for wear on oil pump drive splines or key seats, or wear or scoring at bushing areas. Examine gear for worn or damaged gear teeth. Do not remove drive gear from shaft unless renewal of one or the other is indicated. Examine counterweight for chipped or damaged gear teeth, worn key seat or looseness on drive shaft.

Replacement bushings for damper frame have pre-drilled holes which must be aligned with oil passages in frame casting. When installing new bushings, use air pressure after installation to check for proper alignment of bushing oil holes. When reinstalling drive shaft, heat counterweight to 200 degrees F. for easy installation of shaft. Make sure that tab washers and collar are properly located as shown in Fig. MF142 and insert shaft through frame and counterweight. It is recommended that a piloted sleeve be used over oil pump splines to protect rear bushing as shaft is installed. Tighten set screws (S—Fig. MF140) and stake in place, then reinstall collar pin (P) while

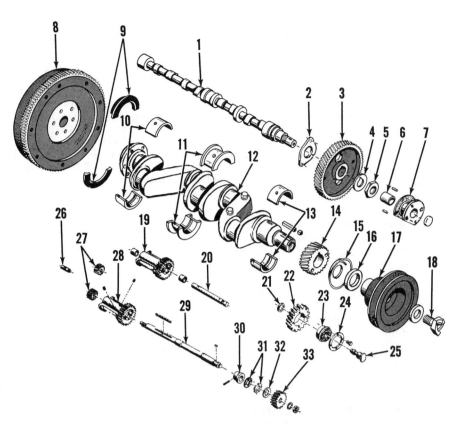

Fig. MF143—Exploded view of crankshaft, camshaft, timing gears and vibration damper, showing component parts.

1. Camshaft	12. Crankshaft	23. Idler bearing
2. Thrust plate	13. Front bearing	24. Retainer
3. Camshaft gear	14. Crankshaft gear	25. Idler shaft
4. Washer	15. Oil slinger	26. Idler shaft
5. Camshaft nut	16. Oil seal	27. Oil pump gears
6. Bushing	17. Crankshaft pulley	28. Counterweight
7. Housing	18. Starting jaw	29. Drive shaft
8. Flywheel	19. Idler counterweight	30. Collar
9. Rear oil seal	20. Idler shaft	31. Tab washer
10. Rear bearing	21. Spacer	32. Washer
11. Center bearing	22. Idler gear	33. Damper drive gear

supporting shaft on under side of collar. The counterweight gears must be meshed so that the counterweights are in the same plane, as shown in Fig. MF140.

Backlash between damper drive gear and idler gear should be 0.004-0.007. Measure the backlash in four places, turning the idler gear 90 degrees between each measurement. If a variation approaching or exceeding the given upper and lower limits is found, turn the idler gear 120 degrees from closest measurement and again measure backlash. If clearance is greater than previously recorded, examine damper drive shaft for straightness, and renew if necessary.

CONNECTING RODS AND BEARINGS
Models MF85 Diesel-MF88 Diesel

70. Connecting rod bearings are of the precision type, renewable from

below after removing oil pan, vibration damper frame and bearing caps. When installing new bearing shells, be sure that the projection engages milled slot in rod and cap and that correlation marks are in register. Replacement rods are not marked and should be installed with oil spray hole facing away from camshaft side of engine. Bearing inserts are available in 0.002, 0.010 and 0.020 undersize, as well as standard. Check the connecting rods, bearings and crankpin journals against the specifications which follow:

Crankpin diameter 2.498-2.500
Wear limit 2.497
Bearing diametral
  clearance 0.0013-0.0038
Wear limit 0.0048
Rod side play..........0.0065-0.010
Rod length C to C.....10.498-10.502
Rod bolt torque......85-95 Ft.-Lbs

Fig. MF142 — Install drive shaft thrust washers (TW) so that tab enters recess in frame boss.

CRANKSHAFT AND MAIN BEARINGS
Models MF85 Diesel-MF88 Diesel

71. Crankshaft is supported in three precision type main bearings renewable from below after removing oil pan, vibration damper frame and main bearing caps. Bearing inserts are available in 0.002, 0.010 and 0.020 undersize as well as in standard.

Normal crankshaft end play is 0.005-0.008 and is controlled by the flanged center main bearing inserts. Renew the inserts and/or crankshaft if end play exceeds 0.010.

To remove the crankshaft, first remove engine; then remove clutch, flywheel, timing gear cover and vibration damper frame.

Check crankshaft and main bearings against the specifications which follow and those listed in paragraph 70:

Main journal diameter....2.873-2.874
Wear limit 2.872
Diametral clearance.....0.0015-0.0035
Wear limit 0.0045
Main bearing bolt
   torque ....130-140 Ft.-Lbs.

CRANKSHAFT REAR OIL SEAL
Models MF85 Diesel-MF88 Diesel

72. An exploded view of the rear main bearing cap and crankshaft rear oil seal is shown in Fig. MF144. To remove the oil seal remove oil pan, vibration damper frame and rear main bearing cap. Loosen the other two main bearing caps. Remove lower seal pieces from main bearing cap and extract upper seal half from crankcase. Thoroughly clean all seal contacting surfaces and remove any sharp edges on bearing cap and crankcase which might cut seal during installation.

Using a thin coat of shellac, install side seals (A and D) in bearing cap. Install lower half of seal (B), making sure that flange (F) of main bearing cap fits into groove (F) of seal (B). Using light grease in groove (F) of upper seal half, slide seal into place around crankshaft and into engine

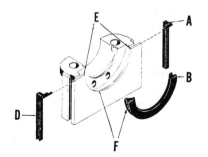

Fig. MF144—Rear main bearing cap showing component parts of rear oil seal. See text for installation.

block. Apply a light coating of rubber cement on mating surfaces of upper and lower seal (B) and allow cement to become tacky before installing bearing cap.

Reinstall the rear main bearing retaining cap screws and tighten all main bearing cap screws to the recommended torque of 130-140 Ft.-Lbs.

FLYWHEEL
Models MF85 Diesel-MF88 Diesel

72A. To remove the flywheel, first split tractor between engine and transmission case and unbolt and remove the clutch assembly. The starter ring gear can be renewed after removing the flywheel. To install a new ring gear, heat gear evenly to 500 degrees F. and install on flywheel with beveled end of teeth facing toward engine. Bolt holes are so arranged that flywheel can be installed in one position only.

OIL PUMP
Models MF85 Diesel-MF88 Diesel

73. The gear type oil pump is mounted on the rear face of the vibration damper frame and is driven by a spline connection on the rear end of the damper drive shaft. The vibration damper frame also serves as a mounting pad for the oil pump intake screen.

The oil pump can be removed after removal of the engine oil pan, without detaching the damper frame from engine block. Examine pump body and cover for wear or scoring and check body and gears against the specifications which follow:

Gear backlash 0.001-0.003
Radial clearance in body........0.003

Check damper drive shaft bushings at rear of frame and in pump body and renew if clearance is excessive. When reassembling, tighten the pump retaining cap screws to a torque of 12 Ft.-Lbs.

OIL PRESSURE RELIEF VALVE
Models MF85 Diesel-MF88 Diesel

74. The plunger type pressure relief valve is located in pump body and can be adjusted or renewed after removing the engine oil pan. Adjust the oil pressure to 40 psi at operating speed by means of shims.

MF85 & MF88 DIESEL FUEL SYSTEM

The diesel fuel system consists of three basic units; the fuel filters and lines, injection pump and injector nozzles. When servicing any unit associated with the fuel system, the maintenance of absolute cleanliness is of utmost importance. Of equal importance is the avoidance of nicks or burrs on any of the working parts.

Probably the most important precaution that service personnel can impart to owners of diesel powered tractors, is to urge them to use an approved fuel that is absolutely clean and free from foreign material. Because of the extreme pressures involved in the operation of pump and injectors, the working parts must be hand fitted with utmost care. While the filtering system will easily remove the larger particles of foreign material the greater danger exists in the presence of water or fine dust particles which might pass through an overloaded filter system. Proper care in fuel handling will pay big dividends in better service and performance.

TROUBLE SHOOTING
Models MF85 Diesel-MF88 Diesel

75. If the diesel engine does not start or does not run properly, and the diesel fuel system is suspected as the source of trouble, refer to the Diesel System Trouble-Shooting Chart

Figg. MF145—View of diesel engine showing fuel system.

1. Sediment bowl	B. Bleed screws
2. Fuel pump	D. Drain plug
3. Primary filter	R. Return line
4. Secondary filter	

and locate points which require further checking. Many of the chart items are self-explanatory, for other items refer to the appropriate following paragraphs.

FUEL FILTERS AND BLEEDING
Models MF85 Diesel-MF88 Diesel

76. **CIRCUIT DESCRIPTION AND MAINTENANCE.** Fuel from the fuel tank flows through the shut-off valve to the sediment bowl (1—Fig. MF145) which should be inspected daily and cleaned when necessary. Fuel then flows to the electric fuel pump (2) which contains a removable filter screen at inlet end. From the primary pump, fuel is forced through primary filter (3), through secondary filter (4) to the injection pump. The primary fuel filter contains a drain plug (D) which should be loosened periodically to remove any accumulation of water or sludge.

76A. **BLEEDING.** Due to the design of the units, the diesel fuel system is practically self-bleeding. When the fuel stop control (FS—Fig. MF146) is moved forward to the operating position, the electric switch (SW) is closed, thus activating the electric fuel pump. The Roosa-Master model DB injection pump is supplied with a far greater supply of diesel fuel than is actually burned in the engine. This excess fuel serves to lubricate and cool the pump parts and operate the advance mechanism and is then returned to the fuel supply tank through line (R—Fig. MF145). The return fuel also carries with it any air trapped in the system. When major units of the system have been

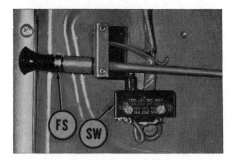

Fig. MF146—Fuel stop control with cover removed to show location of fuel pump switch (SW). Switch is closed when fuel stop (FS) is moved to operating position.

removed for service and are reinstalled, move the fuel stop to the operating position and loosen, in turn, the bleed screws (B—Fig. MF145) on primary and secondary fuel filters. Retighten screw when air-free fuel flows through vent opening. When the electric fuel pump has slowed to a uniform pulsation, loosen the four injector lines at injectors, move hand throttle to wide-open position and turn engine over with starter until air is exhausted from injector lines, then retighten connections. By the time injector lines have been bled, sufficient fuel should have entered injection pump body to provide initial pump lubrication. Any remaining air trapped in pump body will be bled back to fuel tank during normal operation.

INJECTOR NOZZLES
Models MF85 Diesel-MF88 Diesel

The HD277 engine used in models MF85 and MF88 diesel is equipped with C.A.V. pintle nozzles of the "delay", or "throttling" type. The delay nozzle limits the amount of fuel entering the combustion chamber during the start of the injection period, thus improving engine operation at idle speeds.

WARNING: Fuel leaves the injector nozzle with sufficient force to penetrate the skin. Keep exposed portions of your skin clear of nozzle spray when testing.

77. **TESTING AND LOCATING A FAULTY NOZZLE.** If rough or uneven engine operation, or misfiring, indicates a faulty injector, the defective unit can usually be located as follows:

With the engine operating at low idle speed, loosen the high pressure connection at each injector in turn. As in checking spark plugs, the faulty unit is the one which, when its line is loosened, least affects the running of the engine.

If a faulty nozzle is found and considerable time has elapsed since the injectors have been serviced, it is recommended that all injectors be removed and new or reconditioned units be installed, or the nozzles be serviced as outlined in the following paragraphs.

77A. **REMOVE AND REINSTALL.** Before loosening any fuel line connections, thoroughly clean the head

DIESEL SYSTEM TROUBLE-SHOOTING CHART

	Sudden Stopping of Engine	Lack of Power	Engine Hard to Start	Irregular Engine Operation	Engine Knocks	Excessive Fuel Consumption
Lack of fuel	★	★	★	★		
Water or dirt in fuel	★	★	★	★		
Clogged fuel lines	★	★	★	★		
Inferior fuel	★	★	★	★	★	
Faulty primary pump	★	★	★	★		
Faulty injection pump timing		★	★	★	★	★
Air traps in system	★	★	★	★		
Clogged fuel filters		★	★	★		
Deteriorated fuel lines	★	★	★	★		★
Faulty nozzle		★	★	★	★	★
Faulty injection pump		★	★	★	★	★

surface, lines and injectors with compressed air, if available, and by washing with diesel fuel or a suitable solvent. After disconnecting the pressure and leak-off lines, cap all connections to prevent entry of dirt or dust into fuel system. Loosen pressure line connections at injection pump to prevent bending the lines. Remove the retaining screws and carefully withdraw the injector assembly from cylinder head, being careful not to strike the tip end of nozzle against any hard surface.

Thoroughly clean the nozzle recess in cylinder head before reinserting the injector assembly. No hard or sharp tools should be used for cleaning. A piece of wood dowel or brass stock properly shaped, or an approved nozzle bore cleaner should be used. Install a new copper gasket, tighten injector retaining screws to a torque of 14-16 Ft.-Lbs., and bleed the injectors and lines as outlined in paragraph 76A.

77B. NOZZLE TESTER. A complete job of testing and adjusting the injector requires the use of a special tester such as that shown in Fig. MF147. Only clean approved testing oil should be used in tester tank.

The injector should be tested for spray pattern, seat leakage, back leakage and opening pressure as follows:

77C. SPRAY PATTERN. Operate tester handle until oil flows from injector connection, then attach the injector assembly. Close the valve to tester gage and operate tester handle a few quick strokes to purge air from injector and tester pump, and to make sure injector is not plugged or inoperative.

If a straight, solid core of oil flows from nozzle tip without undue pressure on tester handle, open valve to tester gage and remove cap-nut (1—Fig. MF148). Slowly depress the tester handle and observe the pressure at which core emerges. If opening pressure is not within the recommended range of 1850-1900 psi, loosen locknut (3) and turn adjusting cap (2) in or out until opening pressure is within the recommended range.

When opening pressure has been set, again close valve to tester gage and operate tester handle at approximately 100 strokes per minute while examining spray core. Fuel should emerge from nozzle opening in one solid core, in a straight line with injector body, with no branches, splits or atomization.

NOTE: The tester pump cannot duplicate the injection velocity necessary to obtain the operating spray-pattern of the delay type nozzles. Also absent will be the familiar popping sound associated with the nozzle opening of conventional nozzles. Under operating velocities, the observed solid core will cross the combustion chamber and enter the energy cell. In addition, a fine conical mist surrounding the core will ignite in the combustion chamber area above the piston. The solid core cannot vary more than 7½ degrees in any direction and still enter the energy cell. While the core is the only spray characteristic which can be observed on the tester, absence of core deviation is of utmost importance.

77D. SEAT LEAKAGE. The nozzle valve should not leak at pressures less than 1700 psi. To check for seat leakage, open the valve to tester gage and actuate tester handle slowly until gage pressure approaches 1700 psi. Maintain this pressure for at least 10 seconds, then observe the flat surface of nozzle body and the pintle tip for drops or undue wetness. If drops or wetness appear, the injector must be disassembled and overhauled as outlined in paragraph 77G.

77E. BACK LEAKAGE. A back leak test will indicate the condition of the internal sealing surfaces of the nozzle assembly. Before checking the back leakage, first check for seat leakage as outlined in paragraph 77D, then proceed as follows:

Turn the adjusting cap (2—Fig. MF148) inward until nozzle opening

pressure is set at 2350 psi. Release the tester handle and observe the length of time required for gage needle to drop from 2200 psi to 1500 psi. The time should be not less than 6 seconds. A faster drop would indicate wear or scoring between piston surface of nozzle valve or body (8), or improper sealing of pressure face surfaces (A, B & J—Fig. MF149). NOTE: Leakage at tester connections or tester check valve will show up as fast leak back in this test. If all injectors tested fail to pass this test, the tester, rather than the injector, should be suspected.

77F. OPENING PRESSURE. To assure peak engine performance, it is recommended that the four injectors installed in any engine be adjusted as nearly as possible, to equal opening pressures. The recommended

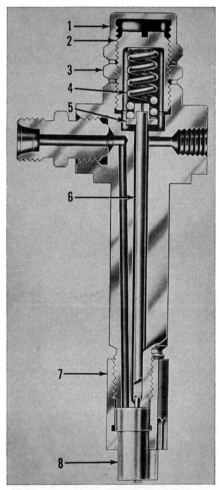

Fig. MF148—Cutaway view of injector assembly of the type used in MF85 diesel and MF88 diesel.

1. Cap nut	5. Spring seat
2. Adjusting cap	6. Valve spindle
3. Locknut	7. Nozzle holder nut
4. Spring	8. Nozzle assembly

Fig. MF147—To properly test and adjust an injector requires the use of a special tester such as the one shown.

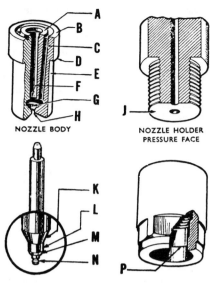

NOZZLE BODY   NOZZLE HOLDER PRESSURE FACE

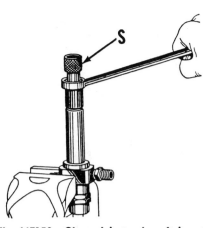

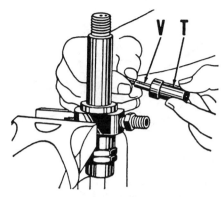

Fig. MF149—Inspect the disassembled injector at the points shown in the above views.

A. Nozzle body
   pressure face
B. Nozzle body
   pressure face
C. Fuel feed hole
D. Shoulder
E. Nozzle trunk
F. Fuel gallery
G. Valve seat
H. Pintle orifice
J. Holder pressure
   face
K. Valve cone
L. Stem
M. Valve seat
N. Pintle
P. Nozzle retaining
   shoulder

Fig. MF150—Clamp injector in soft jawed vise or use fixture when disassembling. When reassembling use centering sleeve (S) to center nozzle on injector assembly.

Fig. MF151—Withdraw nozzle valve (V) from body (T) with fingers as shown, or use special extractor shown in Fig. MF152.

opening pressure range is 1850-1900 psi. When a new spring (4—Fig. MF-148) is removed from parts stock and installed in an injector assembly, the injection pressure will drop quickly as the spring becomes seated under the constant compression. This rate of pressure drop is approximately 10 per cent. It is recommended that injectors containing new springs be initially set at 1980 psi opening pressure, and injectors with used springs at 1900 psi. After the opening pressure has been adjusted, tighten locknut (3) and reinstall cap nut (1), then recheck opening pressure to make sure adjusting cap has not moved.

77G. OVERHAUL. The maintenance of absolute cleanliness in the overhaul of injector assemblies is of utmost importance. Of equal importance is the avoidance of nicks or scratches on any of the lapped surfaces. To avoid damage to any of the highly machined parts, only the recommended cleaning kits and oil base carbon solvents should be used in the injector repair sections of the shop. The nozzle valve and body are individually fit and hand lapped, and these two parts should always be kept together as mated parts.

Before disassembling a set of injectors, cap the pressure line connec-

tions with a line nut with the hole soldered shut, or with a special metal cap, and immerse the units in a clean carbon solvent. While the injectors are soaking, clean the work area and remove any accumulation of discarded parts from previous service jobs. Remove the injectors one at a time from the solvent and thoroughly clean the outer surfaces with a brass wire brush. Be extremely careful not to damage the pintle end of the nozzle valve extending out of nozzle body. Rinse the injector in clean diesel fuel and test the injector as outlined in paragraphs 77B through 77F. Never disassemble an injector which can be adjusted and returned to service without disassembly.

If the injector unit must be disassembled, clamp the injector body in a soft jawed vise as shown in Fig. MF150, tightening only tight enough to keep injector from slipping, or use a holding fixture. Remove cap nut (1—Fig. MF148) and back off the adjusting cap (2) until all tension is removed from the spring, then remove the nozzle holder nut (7). Withdraw the nozzle valve (V—Fig. MF-151) from nozzle body (T) with the fingers as shown; or, if valve is stuck use the special extractor shown in Fig. MF152. NEVER loosen valve by tapping exposed pintle end of valve on a hard surface.

Examine the lapped pressure faces (A, B & J—Fig. MF149) of nozzle body and holder for nicks or scratches, and the piston (larger) portion of nozzle valve for scratches or scoring. Clean the fuel gallery (F—Fig. MF-149) with the special hooked scraper as shown in Fig. MF153, by applying

side pressure while the body is rotated. Clean the valve seat with the brass seat tool as shown in Fig. MF-154. Polish the seat with the pointed wooden polishing stick and a small amount of tallow as shown in Fig. MF155. Clean the pintle orifice from the inside, using the proper size probe. Polish the nozzle valve seat and pintle with a piece of felt and some tallow, loosening any particles of hardened carbon with a pointed piece of brass stock. Never use a hard or

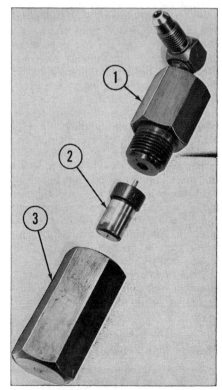

Fig. MF152—Hydraulic nozzle valve extractor for use with injector tester.

1. Adapter
2. Nozzle assembly
3. Nozzle cap

Fig. MF153—Use the special scraper to remove carbon from fuel gallery.

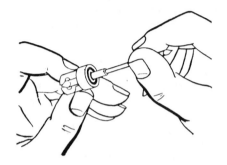

Fig. MF154—Use the brass seat tool to remove carbon from valve seat.

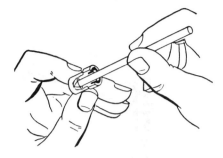

Fig. MF155—Use wooden stick and a small amount of tallow to polish valve seat.

sharp object such as a knife blade as any scratches will cause distortion of the injection core.

As the parts are cleaned, immerse in clean diesel fuel in a compartmented pan. Insert the nozzle valve into body underneath the fuel level and assemble valve body to nozzle holder while wet. Do not attempt to dry the parts with towels or compressed air because of the danger of dust particles remaining on the pressure faces of nozzle holder and nozzle body. Use the centering sleeve when reassembling, as shown in Fig. MF150. To use the sleeve, tighten the holder nut with the fingers while rotating centering sleeve in the opposite direction. When the nut is finger tight remove the sleeve and tighten the holder nut to a torque of approximately 60-65 Ft.-Lbs. Retest the injector as previously outlined. If injector fails to meet the tests and no leaks because of dust, were found upon disassembly, renew the nozzle assembly and any other parts suspected of being faulty.

INJECTION PUMP
Models MF85 Diesel-MF88 Diesel

The HD277 diesel engine used in models MF85 and MF88 is equipped with a Roosa-Master model DBG-VC435-1AJ injection pump. While service on the Roosa-Master model D pump can be accomplished with a relatively small investment in special equipment, the extreme care and specialized skills required as well as the availability of parts, put pump service beyond the scope of that which should be attempted in the average shop. This section therefore, will cover only the removal, installation and timing of the complete injection pump, including the governor.

78. OPERATION. The pump is of the single cylinder, opposed plunger, inlet metering, distributor type. The principal moving parts include the drive shaft (12—Fig. MF156), to which is attached the governor (10), pump rotor and transfer pump (2). These three components rotate as a unit at one half engine speed. Fuel enters the injection pump at the transfer pump (2) where the fuel pressure is increased to the amount required by the design of the engine and injection pump. Transfer pump pressure can vary from about 18 psi at engine idle speed to a maximum of about 130 psi. Top pressure is controlled by selection of the by-pass spring and is determined by engine design. A far greater volume of fuel is pumped by the transfer pump than is needed for engine operation. The excess, after passing through the regulating valve, fills the pump housing where it serves as a coolant and lubricant for the moving parts. This excess then passes from a fitting in governor cover through a return line, to the main fuel tank. At the same time, a supply of oil under transfer pump pressure is forced through a drilling (13) in the hydraulic head where it is fed into an annular groove connecting with the inner side of metering valve (3). When the rotor shaft turns so that one of the charging ports (4) is aligned with the port in the hydraulic head, a metered amount of fuel passes through the axial passage (5) to the cylinder area between the plungers (6). During this charging cycle the plunger rollers are passing through the relieved portion of the cam ring (7). The amount of fuel entering the cylinder area between the plungers is determined by the length of time the charging ports are open (engine speed), transfer pump pressure, and the position of the metering valve. As the rotor continues to turn, the pump plungers (6) are forced together by lobes, on cam

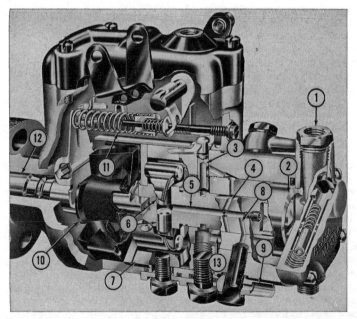

Fig. MF156—Cutaway view of typical Roosa-Master model D injection pump. Slight differences in construction from pump used, may be noted.

1. Inlet strainer	5. Rotor passage	10. Governor
2. Transfer pump	6. Cylinder plungers	11. Governor linkage
3. Metering valve	7. Cam ring	12. Drive shaft
4. Charging port	8. Discharge port	13. Fuel passage
	9. Injector line	

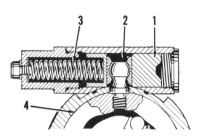

Fig. MF157—The automatic injection timing advance is hydraulically controlled by pressure from the transfer pump.

1. Servo piston   3. Advance spring
2. Cam advance screw   4. Pump cam ring

ring (7). At this time rotor position has closed the charging port (4) and opened one of the discharge ports (8), and the trapped fuel is forced out the connection (9) into the injector line.

As stated above, the amount of fuel entering the cylinder is determined by metering valve, engine speed and transfer pump pressure. The metering valve is connected by linkage (11) to the centrifugal governor (10); thus, when the engine speed is increased beyond the predetermined throttle setting, the delivery valve is rotated to cut down the fuel flow, maintaining a constant engine speed under varying loads. As the engine speed increases, the tendency for decreased fuel flow due to a shorter charging period is counteracted by increased transfer pump pressure at the higher speeds.

Under light loads when smaller amounts of fuel are injected, the

pump plungers (6) are moved out only a small amount. Under heavy loads when the maximum amount of fuel is injected, the plungers move fully out in the cylinder. As the rate of injection remains constant at a given engine speed and the end of injection occurs at the same point (top of cam lobes), the start of injection (timing point) must occur earlier under heavier loads. This tendency to advance injection timing as the load increases is countered by the automatic advance mechanism shown in Fig. MF157. This unit consists of a servo piston (1), cam advance screw (2) and return spring (3). The area behind the outer end of servo piston (1) is connected to a passage leading to the pressure side of the transfer pump. When the engine is started, transfer pump pressure forces the piston inward against spring (3). This movement rotates the cam ring (4) in the pump housing opposite the direction of rotor travel to advance the injection timing. The amount of fuel entering the advance piston area is controlled by the position of the metering valve. When the valve is moved to the full-load position, the flow of oil to the advance piston is completely cut off and spring pressure returns the cam ring (4) to the retard position. The initial injection point is thus maintained at the point desired for best engine operation.

78A. TIMING TO ENGINE. The injection pump should be timed at 20 degrees BTC. To check the timing proceed as follows:

Shut off the fuel, open the flywheel timing window (1—Fig. MF-158), and crank the engine until No. 1 piston is on the compression stroke and the 20 degree flywheel marking is centered on the timing pointer. Remove the pump timing window (2) and note the position of the pump timing marks. With the crankshaft in the 20 degree BTC position, the (lower) scribe line on the governor

cage should exactly align with (upper) scribe line on cam ring as shown in Fig. MF159. If it does not, loosen the two injection pump mounting nuts (A—Fig. MF158) and rotate the pump slightly until the marks are properly aligned. Tighten the mounting nuts securely before starting engine.

78B. REMOVE AND REINSTALL. To remove the injection pump, first thoroughly wash the pump and connections with clean diesel fuel, shut off the fuel and disconnect the throttle and stop linkage from injection pump. Remove the pump to injector lines and disconnect the fuel supply and bleed back lines. Cap all exposed fittings as the lines are removed, to prevent dirt entry. Remove the two injection pump retaining nuts (A—Fig. MF158) and lift the pump straight up until it is free of engine block.

Recommended backlash between injection pump drive gear (2—Fig. MF-160) and camshaft gear is 0.004-0.006. To adjust the backlash, lift the adapter (1) until it is free of locating pin (3) and rotate adapter until pin will enter the next serration in adapter flange. Turning adapter clockwise decreases backlash, turning counter clockwise increases clearance. Backlash is correct when adjusted to the closest point at which drive gear (2) can be fully inserted without binding.

To remove the drive gear, insert a small end wrench or wide bladed screw driver into slot and turn clockwise to thread the gear out of the camshaft worm. When installing gear, crank the engine until number one

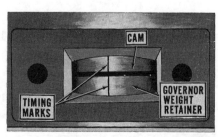

Fig. MF158—To time the injection pump open flywheel timing window (1) and remove pump timing window (2). Align timing marks by loosening retaining nuts (A) and rotating pump slightly.

Fig. MF159—Pump port is open for injection to number one cylinder when timing mark on governor case is aligned with timing mark on cam ring as shown.

Fig. MF160—Injection pump drive gear and adapter with injection pump removed. Backlash of drive gear (2) is adjusted by rotating adapter (1) on locating pin (3). See text.

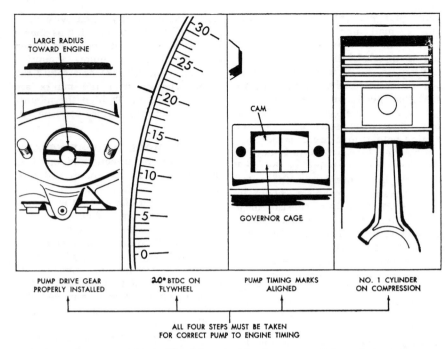

LARGE RADIUS TOWARD ENGINE

CAM

GOVERNOR CAGE

PUMP DRIVE GEAR PROPERLY INSTALLED | **20°** BTDC ON FLYWHEEL | PUMP TIMING MARKS ALIGNED | NO. 1 CYLINDER ON COMPRESSION

ALL FOUR STEPS MUST BE TAKEN FOR CORRECT PUMP TO ENGINE TIMING

Fig. MF161—To correctly install and time the injection pump, all four of the timing steps must be observed as shown.

Fig. MF162 shows the control linkage on injection pump. If the hand throttle lever tends to creep when operating at high governed speed, tighten the hand lever friction nut (5—Fig. MF163) slightly until hand lever will maintain its position. If either the recommended high or low idle speeds cannot be obtained, check for bent linkage or binding, then shift the position of throttle arm (19) on cross shaft (17).

COLD WEATHER STARTING AID
Models MF85 Diesel-MF88 Diesel

78D. The engine is equipped with a C. A. V. flame-type cold weather starting aid (Thermo-Start) located in the intake manifold. The unit consists of a valve, solenoid and heating element. Turning the starter switch to the "HEAT" position activates the heating element and the fuel valve solenoid. Fuel is then gravity fed to the heating coil where it vaporizes. As the starter switch is moved to the "HEAT-START" position the fuel is ignited thus heating the air entering the intake manifold.

The starting aid is a self contained unit and should be renewed if it fails to function properly. When installing the unit, make sure the arrow points in the direction of air travel in the manifold.

PRE-COMBUSTION CHAMBERS
(ENERGY CELLS)
Models MF85 Diesel-MF88 Diesel

79. The energy cells are located underneath the exhaust manifold on the right side of the cylinder head. The

piston is coming up on the compression stroke and continue cranking until the 20 degree BTC timing mark is aligned with the timing pointer. Turn the gear until the driving slot is almost crosswise to the engine block, large radius to the rear, then turn gear counterclockwise as it is fed downward into adapter. When gear is fully seated in adapter, driving slot should be lengthwise of block with large radius toward block as shown in Fig. MF161.

To install the injection pump, remove pump timing window and turn pump drive shaft until pump timing marks are in line and install pump on engine so that drive shaft tang enters

the slot in drive gear. Rotate pump slightly on mounting studs until pump timing marks are in line, then install and tighten retaining nuts. Recheck flywheel and pump timing marks before connecting the lines and linkage.

78C. **GOVERNOR ADJUSTMENT.** Both high-idle and low-idle governed speeds are factory pre-set on the injection pump and should not be changed. Governed speeds are as follows:

Low idle . . . . .650-690 RPM
High Idle . . . . .2250-2275 RPM

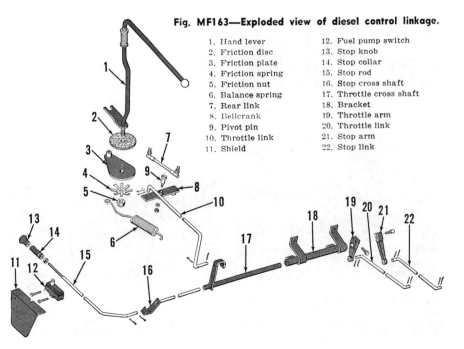

Fig. MF163—Exploded view of diesel control linkage.

1. Hand lever
2. Friction disc
3. Friction plate
4. Friction spring
5. Friction nut
6. Balance spring
7. Rear link
8. Bellcrank
9. Pivot pin
10. Throttle link
11. Shield

12. Fuel pump switch
13. Stop knob
14. Stop collar
15. Stop rod
16. Stop cross shaft
17. Throttle cross shaft
18. Bracket
19. Throttle arm
20. Throttle link
21. Stop arm
22. Stop link

Fig. MF162—Injection pump showing control linkage.

1. Fuel stop lever
2. Throttle link

3. Slow idle adjustment

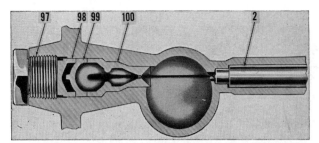

Fig. MF164—Combustion chamber showing action of energy cell. Injector (2) directs fuel spray across top of piston into energy cell (100) where vaporization and partial combustion takes place. Escaping gases return to area above piston where final combustion and expansion occurs.

2. Nozzle
97. Plug
98. Retainer
99. Cell cap
100. Energy cell

energy cell is composed of a cell body (100—Fig. MF164), cell cap (99) and retainer (98), held in the head by the threaded plug (97).

The necessity for cleaning the energy cell is usually indicated by excessive smoking, or a loss of power or fuel economy. To remove or clean the cells, first remove the exhaust manifold, then remove the threaded plug and retainer. Using a pair of thin nosed pliers, grasp the protruding tip of cell cap (99) and withdraw it from head. Clean the cell body with a hooked wire or special scraper.

To remove the cell body (100) for thorough cleaning or renewal, screw the end of a cell puller into the threaded portion of the body and withdraw the body; or, in the absence of a puller, remove the injectors and insert a brass drift through the injector nozzle opening and bump the energy cell body from the cylinder head.

After cleaning the removed parts, examine the mating surfaces of body and cap for roughness or pitting. Small imperfections can be removed by lapping with valve grinding compound. When reinstalling energy cells, make sure the seating surfaces in cylinder head are clean and free from carbon and tighten the retaining plug (97) enough to insure an air-tight seal.

MF85 & MF88 DIESEL COOLING SYSTEM

WATER PUMP
Models MF85 Diesel-MF88 Diesel

80. To remove the water pump, remove the hood and side panels and drain the cooling system. Disconnect the lower radiator hose and unbolt the fan blades from fan pulley. Unbolt water pump body at engine block and withdraw pump.

To disassemble pump, remove the six nuts securing shaft housing to pump body and lift off the housing, shaft, impeller and pulley as an assembly. Using an arbor press, remove fan pulley from shaft, remove snap ring at forward end of shaft housing and press shaft and bearings assembly from shaft housing and impeller. Shaft and bearings are a sealed unit and only available as an assembly or in a kit consisting of shaft, impeller, seal, snap rings and gasket.

RADIATOR
Models MF85 Diesel-MF88 Diesel

81. The radiator filler neck is fitted with a pressure cap which raises coolant boiling point. Check filler neck sealing surface for foreign material, dents or solder accumulations which would prevent proper sealing of cap. Check cap gasket for cracks or other signs of failure.

To remove radiator, first drain cooling system. Remove hood, side panels and grille assembly. Remove grille top closures, air cleaner pipe, oil coolant radiator and radiator hoses. Remove fan blades and radiator mounting bolts then withdraw radiator from right side of tractor.

MF85 & MF88 DIESEL ELECTRICAL SYSTEM

GENERATOR AND REGULATOR
Models MF85 Diesel-MF88 Diesel

82. Tractors are equipped with a Delco-Remy 1100359 generator and Delco-Remy 1118981 regulator. Specification data are as follows:

1100359 Generator
Brush Spring Tension ....28 oz.
Field Draw
  Volts 12
  Amperes 1.58-1.67
Output (cold)
  Maximum amperes 20
  Volts 14.0
  RPM 2300
1118981 Regulator
Cutout Relay
  Air gap 0.020
  Point gap 0.020
  Closing voltage (range) ...11.8-14.0
  Adjust to 12.3
Voltage regulator
  Air gap 0.075
  Voltage range 13.6-14.5
  Adjust to 14.0
Ground Polarity ....Negative

STARTING MOTOR
Models MF85 Diesel-MF88 Diesel

83. The model MF85 and MF88 diesel tractors are equipped with a Delco-Remy model 1113126 starter with a positive engaging, solenoid activated drive unit.

Starter service specifications are as follows:
Brush spring tension...48 oz. (min)
No-load test
  Volts 11.5
  Amperes 50
  RPM 6000
Lock test
  Volts 3.3
  Amperes 500
  Torque—Ft.-Lbs. 22

Fig. MF165 — Exploded view of diesel starter and drive assembly.

1. Drive housing
2. Snap ring
3. Spacer
4. Drive assembly
5. Thrust washer
6. Expansion plug
7. Expansion plug
8. Lever housing
9. Housing bolt
10. Dowel
11. Bushing
12. Armature
13. Washer
14. End plate
15. Pivot pin
16. Clip
17. Engaging arm
18. Pin
19. Snap ring
20. Spring retainer
21. Return spring
22. Retainer
23. Boot
24. Retainer
25. Solenoid plunger
26. Solenoid
27. Contact
28. Gasket
29. Terminal
30. Clip
31. Cover
32. Terminal

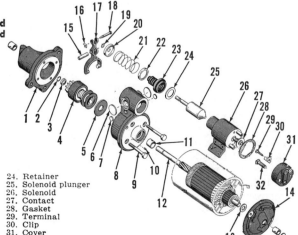

Fig. MF166—Left side view of HD277 diesel engine showing fuel system. Engine is equipped with Roosa-Master injection pump, C.A.V. injectors and Bendix electric fuel pump.

Fig. MF167—Right side view of engine. Full flow oil filter is horizontally mounted and contains check valve to prevent draining when engine is stopped.

PERKINS DIESEL ENGINE & COMPONENTS

NOTE: REFER TO PARAGRAPHS 54 TO 83 FOR DIESEL ENGINE AND COMPONENTS USED IN MF85 AND MF88.

Massey Ferguson Super 90 & Super 90 WR Diesel tractors are equipped with a four cylinder Perkins built engine having a bore of 4½ inches, a stroke of 4¾ inches, and a piston displacement of 302.2 inches.

R&R ENGINE ASSEMBLY

Models MF Super 90 Diesel-MF Super 90 WR Diesel

84. To remove the engine and clutch as an assembly, first drain cooling system, and if engine is to be disassembled, drain oil pan. Remove upright muffler, right and left side panels and power steering pressure, return and bleed lines. Disconnect air cleaner hose, radiator hoses and radiator stay rod. Disconnect steering shaft at flexible coupling. Support tractor under transmission and front end assembly from top of steering housing. Unbolt front support casting from engine and roll front axle, support and radiator, as an assembly, away from engine. Remove starter and light switch from lower instrument panel and remove panel. Drive the roll pin from rear of steering shaft and withdraw shaft from front of fuel tank. Shut off the fuel and

disconnect main fuel line from tank to fuel pump. Disconnect fuel and return lines leading to cold starting reserve tank; unbolt and remove reserve tank, then unbolt and remove the main fuel tank. Disconnect tractor-meter cable from generator, wires from generator and cold staring unit, and heat indicator bulb from right side of engine. Disconnect wires from starting motor and oil pressure gage line at union on left side of block. Disconnect throttle and stop control rods from injection pump, support engine in a hoist and unbolt engine from transmission case.

CAUTION: On Multipower models, the master clutch must be detached from engine drive plate as outlined in paragraph 147.

CYLINDER HEAD

Models MF Super 90 Diesel-MF Super 90 WR Diesel

85. **REMOVE AND REINSTALL.** To remove the cylinder head, it is necessary to remove hood, side panels, steering shaft, fuel tank; and either the front end unit or radiator, whichever is most convenient. Water

pump base and cylinder head are both secured by studs, making it impossible to remove head with radiator in place.

To remove radiator, remove shrouds and power steering line clamp, radiator hoses, air cleaner pipe and the two retaining cap screws from underneath; then slide radiator to right out of tractor frame.

Remove water pump, rocker arms, push rods and injector units; then, unbolt and remove head. Manifolds may be removed with head or separately.

Cylinder head bottom face may be planed to provide a smooth gasket surface, provided that no more than 0.020 is removed and that total thickness of head is not reduced to less than 3.22. Valves must be reseated after planing head, to provide the correct valve head clearance, as outlined in paragraph 86.

When installing head, coat both sides of gasket with a suitable gasket compound. Gasket is marked "TOP, FRONT" for proper assembly. Tighten cylinder head stud nuts to a torque of 105-110 ft.-lbs. in the sequence shown in Fig. MF170. Head should be re-torqued after engine is warm. Adjust valve tappet gap to 0.012, cold; or 0.010, hot. Install and bleed injector units as outlined in paragraph 115.

VALVES AND SEATS
Models MF Super 90 Diesel-
MF Super 90 WR Diesel

86. Intake and exhaust valves seat directly in cylinder head and are numbered in production. Seat inserts may be installed for exhaust valves, but must not be used for intake valves.

Fig. MF170 — When installing cylinder head on MF Super 90 and Super 90 WR diesel tractors, tighten the stud nuts to a torque of 105 - 110 ft.-lbs. in the sequence shown.

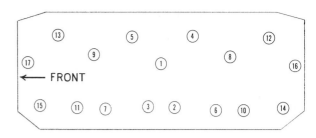

Fig. MF168—Right hand view of Perkins 4A 300 engine installed n Massey Ferguson MF Super 90 WR Diesel tractor. Engine is equipped with a Lanchester type engine balancer to minimize vibration.

Fig. FM169 — Left hand view of Perkins 4A 300 engine. A C.A.V., direct injection fuel system is used.

Both the intake and exhaust valves have a face angle of 44 degrees and a seat angle of 45 degrees. A narrow valve seat should be maintained.

Valves should be renewed if seated head depth exceeds 0.100 when measured with a straight edge and feeler gage as shown in Fig. MF171. In production engines, head depth is maintained within the limits of 0.061-0.076 for intake valves and 0.056-0.071 for exhaust valves. When installing exhaust valve seats or a new or reconditioned cylinder head, make sure the minimum recommended valve head depths are maintained.

Intake and exhaust valves have a stem diameter of 0.3735-0.3745, with a diametral clearance of 0.0015-0.004 in valve guide bores. Adjust valve tappet gap to 0.010 hot, for both intake and exhaust valves.

VALVE GUIDES
Models MF Super 90 Diesel-
MF Super 90 WR Diesel

87. Valve guides are shouldered for positive location, and are a press fit in cylinder head. Inside diameter of valve guide is pre-sized and should be 0.376-0.3775, with a diametral clearance of 0.0015-0.004 for valve stem.

Fig. MF171—Valve head depth in cylinder head should not exceed 0.100 when measured with straight edge and feeler gage as shown.

VALVE SPRINGS

Models MF Super 90 Diesel- MF Super 90 WR Diesel

88. Springs, retainers and locks are interchangeable for intake and exhaust valves. Prior to engine serial number 6900342, an inner and outer spring was used on each valve. Effective with the given serial number, the inner spring was omitted in production. No change was made in outer spring or retainer.

The one spring currently used has a free length of 2.10 inches, and should test 25-27 lbs. when compressed to the closed length of 1.78 inches. The inner spring, when used, should test 8-10 lbs. when compressed to the closed length of 1.53 inches.

Springs have no damper coils and may be installed either end up.

VALVE TAPPETS

Models MF Super 90 Diesel- MF Super 90 WR Diesel

89. The mushroom type valve tappets operate directly in machined bores in cylinder block and can be removed from below after camshaft has been removed. Tappet diameter is 0.7475-0.7485 with a recommended diametral clearance of 0.001-0.0033 in block bores.

ROCKER ARMS

Models MF Super 90 Diesel- MF Super 90 WR Diesel

90. The rocker shaft and rocker arms assembly can be removed after removing fuel tank and rocker arm

cover. Desired clearance between rocker arms and shaft is 0.0007-0.0035. Rocker arms are right hand and left hand units and are unbushed. If clearance is excessive, renew rocker arm and/or shaft.

VALVE TIMING

Models MF Super 90 Diesel- MF Super 90 WR Diesel

91. Valves are properly timed when timing marks are aligned as shown in Fig. MF172. If valve tappet gap is properly adjusted, inlet valve opens 9° BTDC and closes 41° ABDC. Exhaust valve opens 44° BBDC and closes 8° ATDC. To check the valve timing after rocker arm cover is removed, proceed as follows:

With the engine cold, make sure that tappet clearances for both valves on No. 4 cylinder are set exactly at the recommended 0.012. Insert 0.002 feeler gage between valve stem and rocker arm of No. 4 intake (rear) valve and have someone turn crankshaft in normal direction of rotation until feeler gage is pinched. This should occur at about 9° before TDC; or approximately 1⅛ inches ahead of "TDC 1" mark on flywheel when viewed through timing inspection hole on left side of engine block.

TIMING GEAR COVER

Models MF Super 90 Diesel- MF Super 90 WR Diesel

92. To remove the timing gear cover, first remove hood and side panels; then remove front axle, support and radiator from tractor as a unit as out-

lined in paragraph 84. Remove the starting jaw and crankshaft pulley and remove timing gear cover, being careful not to damage front oil seal.

Crankshaft front oil seal can be renewed at this time and should be installed with lip facing rear of cover. Front of seal should be recessed 3/32-inch below front of cover as shown in Fig. MF171A.

Timing gear front cover is not doweled. When installing cover, insert crankshaft pulley through seal and fit pulley over crankshaft splines. Center the seal, using crankshaft pulley as a pilot, and install and tighten cover retaining cap screws.

TIMING GEARS

Models MF Super 90 Diesel- MF Super 90 WR Diesel

93. The timing gear train consists of the engine balancer drive and idler gears as well as the camshaft, crankshaft and injection pump drive and idler gears, as shown in Fig. MF172.

Due to the odd number of teeth in idler gear (2), the timing marks (TM) will not align with every other revolution of the crankshaft. When renewing the gears or checking the timing, it is advisable to remove the idler gear (2) and reinstall with marks aligned as shown in Fig. MF172.

Recommended timing gear backlash is 0.003-0.006 for any two gears in timing gear train except those of the balancer unit. Suggested backlash for balancer idler and either of the mating gears is 0.006-0.009. Gears are

Fig. MF171A — Crankshaft front oil seal should be recessed 3/32-inch below front of cover as shown.

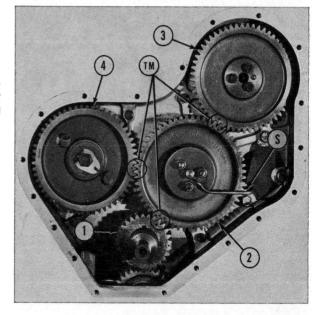

Fig. MF172—Engine timing gear train with cover removed, showing timing marks.

S. Oil spray hole
TM. Timing marks
1. Crankshaft gear
2. Idler gear
3. Injection pump drive gear
4. Camshaft gear

available in standard size only. If backlash is excessive, renew the gears concerned.

94. CAMSHAFT GEAR. The camshaft gear (4—Fig. MF172) is keyed to camshaft and retained by a cap screw, thrust plate and lock plate as shown. Camshaft gear may be removed with a suitable puller after removing retaining cap screw and plates.

95. CRANKSHAFT GEAR. The crankshaft gear (1—Fig. MF172) fits 0.0005 tight to 0.001 loose on crankshaft, and can usually be removed with a pry after timing gear cover is removed.

Crankshaft gear may be renewed without removing oil pan or balancer. Balancer counterweights are at the bottom when any piston is at TDC, and will automatically assume correct timing position with engine in upright position. When installing crankshaft gear, remove idler gear (2) and align visible timing marks as shown.

96. INJECTION PUMP DRIVE GEAR. The injection pump drive gear (3—Fig. MF172) attaches directly to injection pump rotor shaft and con-

tains a dowel for correct pump timing. Refer to Fig. MF172A. Pump may be removed and reinstalled without removing timing gear cover, by removing access cover and the three screws (S—Fig. MF172B).

97. IDLER GEAR. The idler gear (2—Fig. MF172) connects and times the other gears in timing gear train. Timing gears are pressure lubricated through a spray hole (S) in the oil spray pipe. To remove the idler gear, remove the oil spray pipe and the three self-locking nuts retaining idler thrust plate, then withdraw plate and idler gear. Idler gear should have 0.0028-0.0048 diametral clearance and 0.002-0.004 end play on idler hub. Renewable bushings in gear control both end play and running clearance. Idler hub fits 0.001 tight to 0.001 loose in a recess in engine block and is retained by the same three studs which retain the gear. Hub may be renewed if worn or scored.

98. BALANCER DRIVE GEARS. To renew the balancer drive gears, it is necessary to remove the engine oil pan, balancer unit, and the front main bearing cap which contains the balancer idler gear shaft. Refer to paragraphs 106, 107 and 108.

CAMSHAFT

Models MF Super 90 Diesel-MF Super 90 WR Diesel

99. The camshaft is supported by three bearings which are machined directly in the engine block. All bearings are pressure lubricated and the center camshaft bearing meters the oil supply for rocker shaft, rocker arms and valves.

To remove the camshaft, remove the timing gear cover, rocker arms assembly, push rods and oil pan. Remove the timing idler gear and turn camshaft until the two thrust plate retaining screws are accessible through holes in camshaft gear, then remove the retaining screws. Use a grease gun to deposit a small amount of viscous grease on each cam follower (tappet), push cam followers up to clear camshaft; then withdraw the camshaft assembly, being careful not to damage cam lobes, bearings, journals or cam followers.

Camshaft journal diameters are 2.0565-2.0575 for the front bearings; 1.966-1.967 for center bearing; and 1.905-1.906 for rear bearing. Normal diametral clearance is 0.002-0.005 for front and center bearings; and 0.004-0.007 for rear bearings. Normal camshaft end play of 0.004-0.016 is controlled by the thickness of the camshaft thrust plate, which should be 0.216-0.218. Renew thrust plate if end play is excessive.

Fig. MF172A — Injection pump drive gear is positively located on pump shaft by gear dowel and shaft slot (D).

Fig. MF172B—To remove the injection pump, first remove gear cover, then remove the three cap screws (S).

Fig. MF172C — Milled serrations positively locate connecting rod cap. Note also correlation numbers.

ROD AND PISTON UNITS

Models MF Super 90 Diesel-
MF Super 90 WR Diesel

100. Piston and connecting rod assemblies can be removed from above after removing cylinder head, oil pan, balancer assembly and rod caps.

Connecting rods and caps are marked to correspond with the cylinders in which they are installed, and should be installed with "F" marking on piston crown to front, and with correlation numbers on rod and cap to camshaft side of block. Connecting rod and cap are positively located by machined serrations as shown in Fig. MF172C.

When reinstalling, tighten the connecting rod cap screws to a torque of 100-105 ft.-lbs. and lock by bending the tab washer.

PISTONS, SLEEVES AND RINGS

Models MF Super 90 Diesel-
MF Super 90 WR Diesel

101. Pistons are supplied in standard size only, and are available in kits consisting of pistons, pins, rings and sleeves. Parts are also available individually.

Each piston is fitted with a plain faced compression ring in the top groove, which may be installed either

Fig. MF172D—Toroidal combustion chamber is offset in piston crown. "F" marks assure proper installation.

side up; two taper face compression rings which are marked "T" or "TOP" for correct installation; and two cast iron oil control rings which may be installed either side up. Suggested ring to groove clearance is 0.0019-0.0039 for the compression rings and 0.002-0.004 for oil control rings.

NOTE: Piston skirt clearance specifications are not supplied by the manufacturer. As a general rule, ring groove wear will determine the necessity for piston renewal; however, piston renewal due to skirt wear or collapse must be left to the discretion of the servicing mechanic.

The chrome plated, thin steel cylinder sleeves should be renewed when visible wear is present at top of ring travel, or if sleeve is damaged in any other way. The 0.085-0.087 thick cylinder sleeves fit 0.001 loose to 0.001 tight in the 4.5875-4.5885 block bores. When installing sleeves, make certain that block bore (including counterbore for sleeve flange) and the cylinder sleeve are absolutely clean and free of any particles of carbon or dirt. Also, examine sleeves and bores for nicks, burrs or damage. Chill the sleeves and press into block bore by hand. When correctly installed, top face of flange should be 0.004-0.011 below block top face. Allow a few minutes for sleeve to conform to shape of cylinder wall; then check for concentricity by measuring sleeve in two directions at top, center and bottom of sleeve. Measurement should be 4.5005-4.5025.

When installing pistons, make sure that "F" or "FRONT" marking is toward front of engine and that cavity in top of piston is offset toward in-

Fig. MF173—Filler block (B) and seal strip (S) assure positive oil seal at rear of engine, yet allow easy removal of rear main bearing cap.

jection pump side of block. New pistons have a skirt diameter of 4.4956-4.4966 at largest point of measurement.

PISTON PINS

Models MF Super 90 Diesel-
MF Super 90 WR Diesel

102. The 1.4375-1.4378 diameter floating type piston pins are retained in piston bosses by snap rings and are available in standard size only.

The connecting rod bushing is steel backed. To renew the bushing, carefully align oil hole in new bushing with the drilled rod; then press the old bushing out with new bushing. Ream bushing after installation to the recommended 1.4382-1.4392. Piston pin should have a thumb press fit in piston bosses after piston is heated to 120° F.

Fig. MF173A—When installing filler block, make sure pan gasket surface is flush by using straightedge as shown.

CONNECTING RODS AND BEARINGS

Models MF Super 90 Diesel-MF Super 90 WR Diesel

103. Connecting rod bearings are of the precision type, renewable from below after removing oil pan, balancer unit and bearing caps. When renewing bearing shells, be sure projection engages milled slot in rod and cap and that correlation marks are in register and face toward camshaft side. Upper and lower bearings are not interchangeable. Replacement rods should be marked with cylinder number in which they are installed. Bearings are available in 0.010, 0.020 and 0.030 undersize, as well as standard.

Crankpin diameter 2.7483-2.749
Bearing diametral
  clearance 0.0028-0.0045
Rod side clearance.....0.008 -0.017
Rod length C to C........... 8.500
Rod nut torque ....100-105 ft.-lbs.

CRANKSHAFT AND BEARINGS

Models MF Super 90 Diesel-MF Super 90 WR Diesel

104. The crankshaft is supported on five precision type main bearings renewable from below after removing oil pan, balancer unit and main bearing caps. To remove the rear main bearing cap, it is first necessary to remove the filler block, which is attached to rear oil seal retainer plate by three cap screws. Refer to Figs. MF173 and MF173A. When reinstalling the filler block, renew the seals (S) and make sure lower edge is flush with gasket surface of engine block, by using a straight edge as shown.

Main bearing caps are numbered from front to rear; and cap and engine flange are stamped with an identifying serial number as shown at (M—Fig. MF173B). Caps are aligned by ring dowels and cap and block serial numbers must be on same side as shown.

Normal crankshaft end play of 0.0085-0.0155 is controlled by renewable thrust washers at front and rear of center main bearing. Thrust washers in block and cap may be renewed, after cap is off, without removing crankshaft. Thrust washers are available in thicknesses of 0.007 oversize as well as standard. Bearing liners are available in standard size and undersizes of 0.010, 0.020 and 0.030. Front, center and rear bearing liners

Fig. MF173B — Block markings (M) indicate serial number of block and cap, also cap location. See text for further details. Crankshaft end play is controlled by renewable flange washers at center main bearing. End play may be measured using a feeler gage as shown at (C).

Fig. MF173C — Inverted engine block showing engine balancer installation. Engine oil pump is mounted on balancer frame.

1. Oil pump
2. Pressure relief valve
3. Oil inlet screen
4. Oil pressure line
5. Feed line

Fig. MF173D—To remove the balancer assembly, first disconnect pressure line at engine block.

Fig. MF174A—Oil pump may be removed from balancer frame after oil pan is removed.

Fig. MF174B — Balancer idler gear is attached to front main bearing cap as shown.

are interchangeable, as are the two intermediate bearings. Upper and lower bearing halves are not interchangeable.

When installing bearing caps, use new locking tabs and tighten the retaining cap screws to a torque of 145-150 ft.-lbs.

Crankshaft may be removed after removing bearing caps, timing gear cover and rear oil seal. Check the crankshaft against the values which follow:

Main journal diameter..2.9985-2.999
Crankpin diameter 2.7483-2.749
Bearing diametral clearance
　Main bearing ....0.003 -0.0055
　Crankpin 0.0028-0.0045
Crankshaft end play 0.0085-0.0155
Cap screw tightening torques
　Main bearings .....145-150 ft.-lbs.
　Connecting rods ....100-105 ft.-lbs.

ENGINE BALANCER
Models MF Super 90 Diesel- MF Super 90 WR Diesel

105. **OPERATION.** The Lanchester type engine balancer consists of two unbalanced shafts which rotate in opposite directions at twice crankshaft speed. The inertia of the shaft weights are timed to cancel out natural engine vibration, thus producing a smoother running engine. The balancer is correctly timed when the balance weights are at their lowest point as the pistons are at TDC or BDC of their stroke.

The balancer unit is driven by the crankshaft timing gear through an idler gear attached to the front main bearing cap. The engine oil pump is mounted at rear of balancer frame and driven by the balancer shaft.

Refer to Figs. MF173C through MF175B.

106. **REMOVE AND REINSTALL.** The balancer assembly can be removed after removing the oil pan and disconnecting the oil pressure line (4—Fig. MF173C) as shown in Fig. MF173D. Note the length of the securing cap screws. On some early tractors, two of the cap screws were 4 inches in length while the other four were 4¼ inches. On these tractors, install the two shorter cap screws at front of engine nearest the timing gears. On later tractors, the retaining cap screws are all of the same length.

When installing balancer with engine in tractor, timing marks will be difficult to observe. To install, turn the crankshaft until the TDC-1 flywheel timing mark is centered on timing pointer then install the balancer with counterweights hanging naturally. If carefully installed, timing will be correct. If engine is mounted on a stand, timing marks can be observed by removing timing gear cover.

107. **OVERHAUL.** Before disassembling the removed balancer unit, measure the end play of the driven

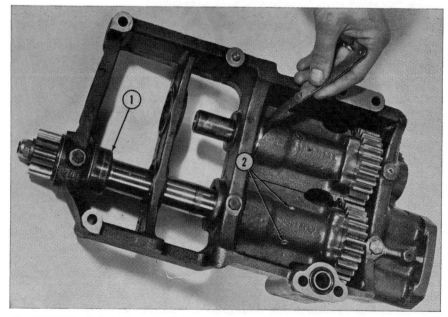

Fig. MF174C—Before disassembling the removed balancer assembly, first check end play of idler weight as shown. To remove the drive shaft, loosen the front collar (1) and the two weight retaining set screws (2). Refer to text.

weight assembly with a feeler gage as shown in Fig. MF174C. End play should be 0.006-0.022 and is controlled by thrust washers. Remove the oil pump (1—Fig. MF173D) and oil pressure relief valve (2).

To disassemble the drive shaft and weight, loosen the two Allen head set screws as shown in Fig. MF174D; unseat snap ring (3—Fig. MF175), drive the thrust collar (1) to the rear as shown, and remove Woodruff key (K). Turn the heavy part of counterweights (C) upward as shown, and

Fig. MF174D — Removing weight retaining set screws preparatory to disassembling drive shaft.

drift the shaft forward by working through oil pump mounting end. Refer to Fig. MF175A. (NOTE: With counterweights (CW) up as shown, the long flat key (K) will clear the center bushing as shaft is moved forward.) Withdraw the shaft while lifting out the contained counterweight, thrust washers and thrust collar.

To disassemble the driven counterweight, refer to Fig. MF175B. Loosen the locknut and remove the dog point set screw (2); then drift the shaft (1) rearward out of housing. When installing shaft (1), make sure shaft is turned so that set screw (2) can be fully inserted. The shaft has a stepped locating hole which will not allow it to be turned 180°, thus positively aligning the drilled oil passages.

The driven counterweight contains two bushings which are a press fit and must be installed with the oil groove toward the heavy mass of the counterweight. Align ream bushings to 1.002-1.0035 after installation.

The drive shaft bushings are retained in housing by the locating set screws (4). When installing new

bushings, align the pre-drilled oil holes in bushings with oil passages in housing and press the bushings into position. Drill holes for the locating set screws (4) after bushings are installed. The locating hole in rear (oil pump end) bushings must be drilled with a "Size I" (0.272) drill. The two front bushings require a 3/16-inch drill.

Align ream bushings after installation, to the following finished sizes:

Rear (oil pump end)
bushing ....0.876 -0.8772
Center bushing ....1.251 -1.2526
Front bushing 1.2823-1.2839

If gears must be removed from balance weights, be sure to mark them so that they can be reinstalled in their original position. Reassemble by reversing the disassembly procedure, making sure counterweights are timed so that flats are parallel as shown in Fig. MF174D.

108. BALANCER IDLER GEAR. The balancer idler gear is mounted on a shaft attached to front main bearing cap and secured by a snap

Fig. MF175—Unseat snap ring (3), drive the collar (1) rearward as shown, and remove Woodruff key (K). Counterweights (C) must be up as shown, as shaft is drifted forward out of housing.

Fig. MF175A — With counterweights (CW) in position shown, key (K) will clear center bushing as shaft is removed.

ring. Front face of gear contains a single punch-marked tooth which aligns with a double punch-marked tooth on crankshaft gear; and a double punch-marked tooth which aligns with a single punch-marked tooth on balancer drive gear. **These timing marks are assembly marks only.** Because of the odd number of teeth (37) in idler gear, timing marks will align only once in 259 revolutions of crankshaft. Timing marks being out of alignment does not necessarily mean that balancer is not correctly timed.

To remove the idler gear, it is necessary to remove oil pan, balancer unit and front main bearing cap; or, timing gear cover and timing gear housing.

Fig. MF175B—To remove idler weight shaft (1), first remove the stepped set screw (2). Drive shaft bushings (3) are retained by set screws (4). Refer to text.

CRANKSHAFT REAR OIL SEAL

Models MF Super 90 Diesel—MF Super 90 WR Diesel

109. The asbestos rope-type rear oil seal is contained in a two-piece seal retainer attached to rear face of engine block. Refer to Fig. MF173 or MF173A.

To renew the seal, remove the clutch and flywheel, then remove the seal retainer. Thoroughly clean the retainer and clamp each half in turn securely in a vise, being careful not to bend or mar the aluminum retainer. To install the rope seal start each end in its groove with about 0.010-0.020 protruding from end of retainer. After each end has been seated, work center of sealing strip into its groove with the fingers; then with a suitable round bar or hardwood dowel. Sealing strip must not be trimmed and must extend at each end not more than 0.020 or less than 0.010 when installed. Coat seal with engine oil when installing on engine.

Sealing surface of crankshaft contains a machined spiral groove 0.004-0.008 deep.

FLYWHEEL

Models MF Super 90 Diesel—MF Super 90 WR Diesel (Except Multipower)

CAUTION: Flywheel is only slightly piloted to crankshaft. Use caution when removing bolts, to prevent flywheel falling and causing damage or personal injury.

110. To remove flywheel, first separate engine from transmission case, then remove clutch. Flywheel ring gear can be renewed after flywheel is removed. Heat gear evenly to ap-

proximately 475° F. to install. Gear teeth are not beveled.

Flywheel is secured to crankshaft by six evenly spaced cap screws and is correctly timed when unused hole in crankshaft and flywheel are aligned. Tighten retaining cap screws to a torque of 75 ft.-lbs. Maximum allowable eccentricity is 0.012 and maximum allowable run-out is 0.001 per inch radius from flywheel center. If run-out or eccentricity exceeds the limits, remove flywheel and examine mating surfaces for burrs or dirt.

Models MF Super 90 Diesel—MF Super 90 WR Diesel (Multipower)

110A. Because of the heavy clutch assembly, multipower models use a thin engine drive plate instead of a flywheel, as shown in Fig. MF191. The starter ring gear is attached to drive plate by screws which should be tightened to a torque of 44-46 ft.-lbs.

To remove the drive plate or starter ring gear, refer to paragraph 147.

OIL PUMP

Models MF Super 90 Diesel—MF Super 90 WR Diesel

111. The gyrotor type oil pump is attached to rear of balancer housing and driven by the balancer drive shaft. Oil pump may be removed after removing oil pan, without removing balancer unit.

The oil pump is available for service as a complete assembly only. Inner to outer rotor clearance should not exceed 0.006 when measured with a feeler gage as shown in Fig. MF175D.

Fig. MF175C — Engine oil pump with end plate removed showing rotors.

Fig. MF175D—Measure clearance between rotors with feeler gage as shown. Clearance should not exceed 0.006.

Fig. MF176 — Rotor to body clearance should not exceed 0.010 when measured with feeler gage as shown.

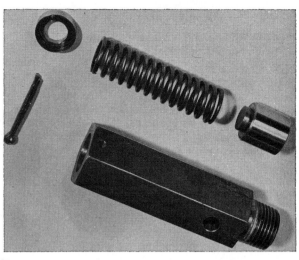

Fig. MF176A — Exploded view of oil pressure relief valve showing component parts.

Rotor to body clearance should not exceed 0.010 when measured as shown in Fig. MF176. Rotor end clearance in body should not exceed 0.003 when measured with straight edge and feeler gage. Renew pump if clearances are excessive.

RELIEF VALVE

**Models MF Super 90 Diesel-
MF Super 90 WR Diesel**

112. The plunger type oil pressure relief valve is mounted on the balancer frame as shown at (2—Fig. MF173D), and can be removed after removing oil pan.

Gage pressure at normal operating speed and temperature should be 25-35 psi or higher, and can be adjusted by renewing relief valve spring. Refer to Fig. MF176A for a disassembled view of pressure relief valve.

MF SUPER 90 & MF SUPER 90 WR DIESEL FUEL SYSTEM

The diesel fuel system consists of three basic units; the fuel filters and lines, injection pump and injector nozzles. When servicing any unit associated with the fuel system, the maintenance of absolute cleanliness is of utmost importance. Of equal importance is the avoidance of nicks or burrs on any of the working parts.

Probably the most important precaution that service personnel can impart to owners of diesel powered tractors, is to urge them to use an approved fuel that is absolutely clean and free from foreign material. Because of the extreme pressures involved in the operation of pump and injectors, the working parts must be hand fitted with utmost care. While the filtering system will easily remove the larger particles of foreign material, the greater danger exists in the presence of water or fine dust particles which might pass through an overloaded filter system. Proper care in fuel handling will pay big dividends in better service and performance.

TROUBLE SHOOTING
**Models MF Super 90 Diesel-
MF Super 90 WR Diesel**

113. If the engine does not start or does not run properly, and the diesel fuel system is suspected as the source of trouble, refer to the Diesel Trouble-Shooting Chart and locate points which require further checking. Many of the chart items are self-explanatory; for other items refer to the appropriate following paragraphs.

	Sudden Stopping of Engine	Lack of Power	Engine Hard to Start	Irregular Engine Operation	Engine Knocks	Excessive Fuel Consumption
Lack of fuel	★	★	★	★		
Water or dirt in fuel	★	★	★	★		
Clogged fuel lines	★	★	★	★		
Inferior fuel	★	★	★	★		
Faulty primary pump	★	★	★	★		
Faulty injection pump timing		★	★	★	★	★
Air traps in system	★	★	★	★		
Clogged fuel filters		★	★	★		
Deteriorated fuel lines	★					★
Faulty nozzle				★	★	★
Faulty injection pump		★	★	★	★	★

FUEL FILTERS AND BLEEDING

Models MF Super 90 Diesel—
MF Super 90 WR Diesel

114. **CIRCUIT DESCRIPTION AND MAINTENANCE.** Fuel from the tank flows to the sediment bowl on fuel pump at right rear of engine block. From the fuel pump (1—Fig. MF-176B) fuel passes to the primary filter (2), then to the final filter (3). From the final filter, fuel flows to the injection pump (1—Fig. MF176C). The sediment bowl should be inspected daily and cleaned when necessary. The element in primary filter (2—Fig. MF176B) should be renewed every 500 hours, and element in final filter (3) every 1,000 hours. Filter elements for both units are interchangeable.

115. **BLEEDING.** Air is bled from system from three bleed screws (B—Figs. MF176B and MF176C) with engine not running, while operating primer lever (P—Fig. MF176B) by hand. Order of bleeding is: (1) screw on primary filter; (2) upper screw on injection pump; then, (3) lower screw on injection pump. Loosen each bleed screw in turn until air-free fuel flows, then tighten screw before opening the next. CAUTION: Do not loosen bleed screws with engine running; design of system will allow air to be drawn into fuel.

If engine fails to start, loosen the four injector lines (C—Fig. MF176C) at injectors, move hand throttle to wide-open position and turn engine over with starter until fuel flows from loosened connections. Retighten connections (C); then start engine.

Fig. MF176B—Diesel fuel pump and filters showing method of cartridge renewal.

B. Bleed-screw
C. Filter cartridge
P. Priming lever
1. Fuel Pump
2. Primary filter
3. Final filter

INJECTOR NOZZLES

Models MF Super 90 Diesel—
MF Super 90 WR Diesel

The 4A 300 engine used in model MF Super 90 & Super 90 WR diesel is equipped with C. A. V. multiple hole nozzles which inject directly into combustion chamber in top of piston crown.

WARNING: Fuel leaves the injector nozzle with sufficient force to penetrate the skin. Keep exposed areas of body clear of nozzle spray when testing.

116. **TESTING AND LOCATING A FAULTY NOZZLE.** If rough or uneven engine operation, or misfiring, indicates a faulty injector, the defective unit can usually be located as follows:

With the engine operating at low idle speed, loosen the high pressure connection at each injector in turn. As in checking spark plugs, the faulty unit is the one which, when its line is loosened, least affects the running of the engine.

If a faulty nozzle is found and considerable time has elapsed since the injectors have been serviced, it is recommended that all injectors be removed and new or reconditioned units installed; or the nozzles serviced as outlined in the following paragraphs.

117. **REMOVE AND REINSTALL.** Before loosening any fuel line connections, thoroughly clean the head surface, lines and injectors with compressed air, if available, and by washing with diesel fuel or a suitable solvent. After disconnecting pressure and leak-off lines, cap all connections to prevent entry of dirt or dust into fuel system. Loosen pressure line connections at injection pump adapters to prevent bending the lines. Remove retaining stud nuts and clamp; then carefully withdraw injector assembly, being careful not to strike tip end of nozzle against any hard surface. Refer to Fig. MF176D.

Thoroughly clean nozzle recess in cylinder head before reinstalling injector. No hard or sharp tools should be used for cleaning. Install a new

Fig. MF176C — Left side of engine showing fuel system components.

1. Injection pump
2. Injector assemblies
B. Bleed screws
C. Pressure connections

Fig. MF176D — Injectors are retained in cylinder head by means of clamps as shown.

copper gasket and tighten the retaining stud nuts evenly. Bleed the injectors and lines as outlined in paragraph 115.

118. NOZZLE TESTER. A complete job of testing and adjusting the injector requires the use of a special tester such as that shown in Fig. MF177. Only clean, approved testing fuel should be used in tester tank.

The injector should be tested for spray pattern, seat leakage, back leakage and opening pressure as follows:

119. SPRAY PATTERN. Operate tester handle until oil flows from injector connection, then attach the injector assembly. Close the valve to tester gage and operate tester handle a few quick strokes to purge air from injector and tester pump, and to make sure injector is not plugged or inoperative.

If four, finely atomized sprays emerge from tip without undue pressure on tester handle, open valve to tester gage and check to see that opening pressure is somewhere near the recommended pressure of 2500 psi. If it is not, adjust pressure by removing nozzle cap (1—Fig. MF-177A), loosening locknut (3), and turning adjusting screw (5).

When opening pressure has been set, again close valve to tester gage and operate tester handle quickly several times while observing spray pattern. The four conical sprays should be symmetrical, finely atomized and equal in appearance, with no core or streaks. If spray pattern is not as described, or if one or more of the spray holes are plugged, overhaul nozzle as outlined in paragraph 123.

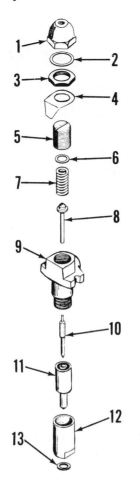

Fig. MF177A — Exploded view of C.A.V. injector assembly of the type used.

1. Cap nut	7. Pressure spring
2. Sealing washer	8. Spindle
3. Locknut	9. Body
4. Specification washer	10. Nozzle valve
5. Adjusting screw	11. Nozzle body
6. Washer	12. Nozzle holder nut
	13. Sealing washer

120. SEAT LEAKAGE. The nozzle valve should not leak at pressures below 2350 psi. To check for seat leakage, open valve to tester gage and actuate tester handle slowly until gage pressure approaches 2350 psi. Maintain this pressure for at least 10 seconds, then observe nozzle tip for drops of fuel or undue wetness. If evidence of leakage appears, injector must be overhauled as outlined in paragraph 123.

121. BACK LEAKAGE. A back leak test will indicate the condition of internal sealing surfaces of nozzle assembly. Before checking back leakage, first test for seat leakage as outlined in paragraph 120, then proceed as follows:

Actuate tester handle slowly until gage pressure approaches 2350 psi. Release tester handle and observe the

length of time required for gage needle to drop from 2200 psi to 1500 psi. The time should be not less than 6 seconds. A faster drop would indicate wear or scoring between piston surface of nozzle valve or body, improper sealing of pressure face surfaces between nozzle holder and valve body, or leakage at lines or connections. NOTE: Leakage at tester connections or tester check valve will show up as fast leak-back in this test. If all injectors tested fail to pass the leakage test, the tester rather than the injector should be suspected.

122. OPENING PRESSURE. To assure peak engine performance, it is recommended that the four injectors installed in any engine be adjusted as nearly as possible, to equal opening pressures. The recommended opening pressure for used injectors is 2500 psi. When a new spring is removed from parts stock and installed in an injector, it is recommended that the initial pressure be set at 2575 psi to allow for pressure drop as new spring seats in.

After injector has been overhauled, or has passed the other tests outlined in paragraphs 119 through 121, recheck the opening pressure of the set of injectors and adjust as outlined.

123. OVERHAUL. The maintenance of absolute cleanliness in the overhaul of injector assemblies is of utmost importance. Of equal importance is the avoidance of nicks, scratches or handling damage on any of the lapped or polished surfaces. To avoid damage, only the recommended cleaning kits and oil base carbon solvents should be used. The nozzle valve and body are individually fit and hand lapped, and these two parts should always be kept together and not intermixed.

Before disassembling a set of injectors, cap the pressure line connections with a line nut with hole soldered shut, or with a special metal cap, and immerse the units in a clean carbon solvent. While the injectors are soaking, clean the work area and remove any accumulation of discarded parts from previous service jobs. Remove injectors one at a time from the solvent and thoroughly clean the outer surfaces with a brass wire brush. Rinse the injector in clean diesel fuel and test as outlined in paragraphs 118 through 122. Never disassemble an injector which can be adjusted and returned to service without disassembly.

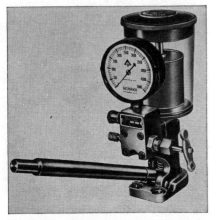

Fig. MF177 — Complete service on injector assemblies requires the use of a nozzle tester such as the one shown.

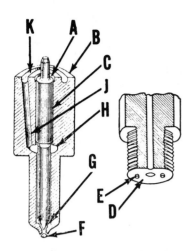

Fig. MF177B — Inspect the disassembled injector at the points indicated.

A. Nozzle body
   pressure face
B. Nozzle body
   pressure face
C. Nozzle valve
   piston surface
D. Nozzle holder
   pressure face

E. Dowel pins
F. Nozzle spray holes
G. Valve seat
H. Fuel gallery
J. Fuel passage
K. Fuel chamber

Fig. MF177C — Clean the fuel gallery using the special hooked scraper as shown.

Fig. MF177D — Clean nozzle valve seat using the brass seat scraper.

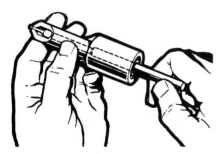

Fig. MF177E — Remove carbon from the dome cavity using the small hand tool furnished in kit.

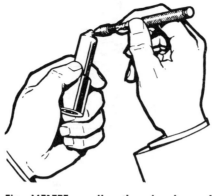

Fig. MF177F — Use the pin vise and cleaning needles to clean the nozzle spray holes.

Fig. MF177G — Clean the fuel feed hole of carbon and gum, using the small drill as shown.

If the injector must be disassembled, clamp the injector body in a holding fixture or soft-jawed vise, with only enough pressure to keep injector from slipping. Remove the protective cap, loosen locknut, then back off the pressure adjusting screw until all tension is removed from pressure spring. Remove the nozzle holder nut (12—Fig. MF177A) and withdraw the nozzle valve assembly. If valve cannot be removed from nozzle body with the fingers, return the assembly to the carbon solvent. A special nozzle valve extractor which uses hydraulic pressure from tester pump is available. Valve can sometimes be removed with a pair of pliers, but care should be exercised not to damage the parts. Examine the lapped pressure faces (A, B & D—Fig. MF177B) of nozzle body and holder for nicks or scratches, and the piston surfaces (C) of nozzle valve and body for scratches, wear or scoring. Clean the fuel gallery (H) with the special hooked scraper tool as shown in Fig. MF177C, by applying side pressure while body is rotated. Clean the nozzle valve seat with the brass seat scraper as shown in Fig. MF177D. Polish the seat with a pointed wooden polishing stick and a small amount of tallow. Use the special scraper to remove any carbon in dome cavity behind the spray holes as shown in Fig. MF177E, then clean the spray holes, using a 0.009 cleaning needle or broach, held in a pin vise as shown in Fig. MF177F. Leave only about 1/16-inch of cleaning wire protruding from vise, to resist bending. Clean the fuel passage with a small drill or similar tool as shown in Fig. MF177G.

As the parts are cleaned, immerse in clean diesel fuel in a compartmented pan. Insert the nozzle valve into body underneath the fuel level to prevent the inclusion of dust particles between piston surfaces of nozzle valve and body. If a reverse flushing attachment is available, flush the nozzle assembly after valve is inserted, then immediately assemble the injector. Make sure the sealing surfaces (A, B & D—Fig. MF177B) are absolutely clean, and that nozzle assembly is properly installed on dowels (E). Tighten the nozzle holder nut to a torque of 50 ft.-lbs., using the special nozzle nut socket if available. Retest the injector as previously outlined. If injector fails to meet the tests, and no dust or dirt is found on disassembly, renew the nozzle and any other parts suspected of being faulty.

Fig. MF178 — Injection pump with timing cover removed. Pump is timed with "E" mark aligned with timing mark on snap ring as shown.

INJECTION PUMP

Models MF Super 90 Diesel-
MF Super 90 WR Diesel

The fuel injection pump is a self-contained unit which includes the engine governor and components for delivering fuel in the properly metered amounts. The pump is horizontally mounted on left side of engine and

driven by a gear in the timing gear train. This section will cover only the removal, installation and timing of the complete injection pump, including the governor.

124. TIMING. The injection pump should be timed at 16° BTDC, with the "E" timing mark on rotor aligned with scribe line on snap ring as shown in Fig. MF178.

To check the timing, shut off the fuel, remove timing hole plug from left side of flywheel housing, and crank engine until No. 1 piston is coming up on compression stroke and the "STAT/16" timing mark on flywheel is centered on timing pointer as shown in Fig. MF178A. Remove the pump timing window (5—Fig. MF-178B) and note the position of pump timing marks, which should appear as shown in Fig. MF178. If pump timing marks are not exactly aligned, loosen the two injection pump mounting stud nuts and rotate pump slightly on timing gear housing until marks are aligned.

125. REMOVE AND REINSTALL. When pump is originally installed and timed, scribe lines are made on injection pump body and timing gear cover as shown at (M—Fig. MF178B). If pump is removed for routine service, pump may be reinstalled without regard to crankshaft position by aligning the scribe lines.

To remove the injection pump, first thoroughly wash pump and connections with clean diesel fuel or solvent.

Shut off the fuel and disconnect throttle and stop linkage. Disconnect injector, fuel supply and bleed back lines and cap the exposed fittings. Remove access cover from front of timing gear cover and remove the three cap screws (S—Fig. MF178C). Remove the two stud nuts securing pump to timing gear housing and withdraw the pump.

Install by reversing the removal procedure, making sure the scribe lines (M—Fig. MF178B) are aligned. A dowel pin located in pump timing gear fits a milled slot in pump shaft to assure correct timing if the original pump is installed and timing gears are not disturbed. Timing may be checked as outlined in paragraph 124. if advisable. Complete the installation by connecting the lines and bleeding the system as outlined in paragraph 115.

GOVERNOR

Models MF Super 90 Diesel-
MF Super 90 WR Diesel

126. The diesel engine governor is integral with the injection pump and only the speed and linkage adjustments will be covered in this section.

With the engine at operating temperature, turn the idle speed adjusting screw (2—Fig. MF178B) to obtain an engine slow idle speed of 575-625 rpm. The high speed adjusting screw (4) is factory sealed and should maintain a high idle, no-load engine speed of 2175 rpm. Governor lever (3) should contact both the slow idle (2)

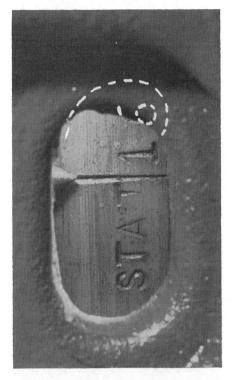

Fig. MF178A — "STATIC 16" flywheel timing mark aligned with timing pointer for pump timing.

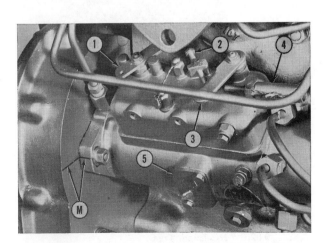

Fig. MF178B — Left side view of engine showing timing marks (M) on timing gear cover and pump housing. If the same pump is installed, pump can be correctly timed without checking crankshaft location by aligning the marks.

M. Scribe lines
1. Stop lever
2. Slow speed
   adjusting screw
3. Governor lever
4. High speed
   adjusting screw
5. Pump timing window

Fig. MF178C — To remove the injection pump, remove the gear cover and the three gear retaining cap screws (S).

and high speed (4) stop screws as throttle is moved fully to high and low speed positions. If it does not, loosen the throttle arm clamping screw on cross shaft at left front of battery box, and shift throttle arm on cross shaft.

COLD WEATHER STARTING AID

Models MF Super 90 Diesel- MF Super 90 WR Diesel

127. The engine is equipped with a C. A. V. flame-type cold weather starting aid (Thermo-Start) located in the intake manifold. The unit consists of a valve, solenoid and heating element. Turning the starter switch to the "HEAT" position activates the heating element and the fuel valve solenoid. Fuel is then gravity fed into the heating coil where it vaporizes. As the starter switch is moved to the "HEAT-START" position, the fuel is ignited, thus heating the air entering the intake manifold.

The starting aid is a self contained unit and should be renewed if it fails to function properly. When installing the unit, make sure the arrow points in the direction of air travel in the manifold. (Toward the rear.)

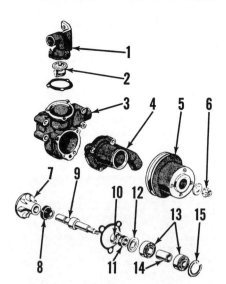

Fig. MF179 — Exploded view of water pump and associated parts.

1. Outlet elbow	9. Shaft
2. Thermostat	10. Retainer
3. Rear housing	11. Felt seal
4. Front housing	12. Flange
5. Pulley	13. Bearing
6. Nut	14. Spacer
7. Impeller	15. Snap ring
8. Seal	

MF SUPER 90 & MF SUPER 90 WR DIESEL COOLING SYSTEM

WATER PUMP

Models MF Super 90 Diesel- MF Super 90 WR Diesel

128. To remove the water pump it is first necessary to remove the radiator as outlined in paragraph 129; or the front end unit as outlined in paragraph 15.

Remove the fan blades, then unbolt and remove the front body assembly (4—Fig. MF179). Remove nut (6) and fan pulley (5); remove snap ring (15), then press shaft and bearings out of housing (4) and impeller.

When assembling the water pump, press the rear bearing (13) on shaft until it bottoms, with shielded end toward impeller end of shaft. Install spacer (14), then install front bearing (13) with open end toward spacer.

Fill the cavity between bearings half full of water pump grease; then assemble by reversing the disassembly procedure. Press the impeller on shaft until end clearance of blades is 0.012-0.032 when measured with feeler gage as shown in Fig. MF179A.

RADIATOR

Models MF Super 90 Diesel- MF Super 90 WR Diesel

129. To remove the radiator, first drain cooling system and remove hood, side panels and grille assembly. Remove grille top closures, air cleaner pipe and fan shrouds. Remove radiator hoses and radiator mounting bolts and withdraw radiator from right side of tractor.

Fig. MF179A — Press the impeller on shaft until clearance is 0.012-0.032 when measured as shown.

MF SUPER 90 & MF SUPER 90 WR DIESEL ELECTRICAL SYSTEM

GENERATOR AND REGULATOR

Models MF Super 90 Diesel- MF Super 90 WR Diesel

130. Delco-Remy 1100362 generator and Delco-Remy 1118981 regulator are used. Specification data are as follows:

Generator

Brush spring tension . . . . 28 oz.
Field Draw
  Volts 12
  Amperes 1.58-1.67

Output (cold)
  Maximum amperes 20
  Volts 14.0
  RPM 2300
Regulator
Cutout Relay
  Air gap 0.020
  Point gap 0.020
  Closing voltage (range) . . . 11.8-14.0
  Adjust to 12.3
Voltage Regulator
  Air gap 0.075
  Voltage range 13.6-14.5
  Adjust to 14.0
Ground Polarity . . . . Negative

STARTING MOTOR

Models MF Super 90 Diesel- MF Super 90 WR Diesel

131. Delco-Remy model 1113139 starter with positive engaging, solenoid actuated drive unit is used. Starter service specifications are as follows:
Brush spring tension . . . . 80 oz. (min.)
No-Load Test
  Volts 11.5
  Amperes (Min., incl. solenoid) 57
  RPM 5000
Lock Test
  Volts 3.4
  Amperes 500
  Torque—Ft.-Lbs. 22

ENGINE CLUTCH
All Models Except Multipower

PEDAL ADJUSTMENT

All Models Except Multipower

132. Adjustment to compensate for wear is made by rotation of clutch pedal on the clutch release shaft as shown in Fig. MF184. Adjustment is correct when a clearance of 3/8-inch exists between pedal and foot rest as release bearing contacts clutch fingers.

To make the adjustment, loosen clamp bolt (1) and rotate clutch release shaft (2) clockwise with a wrench until contact is felt between release bearing and clutch fingers. Position pedal with the proper clearance and tighten clamp bolt (1).

CLUTCH ASSEMBLY

All Models Except Multipower

133. **REMOVE AND REINSTALL.** To remove the clutch assembly, first split tractor as outlined in paragraph 137. Loosen cover screws evenly to avoid damaging cover, and remove cover assembly (25—Fig. MF185), lined plate (28) and cover spacers (26 & 27) if used. NOTE: Gasoline and LP-Gas tractors use six 7/16-inch spacers between flywheel and

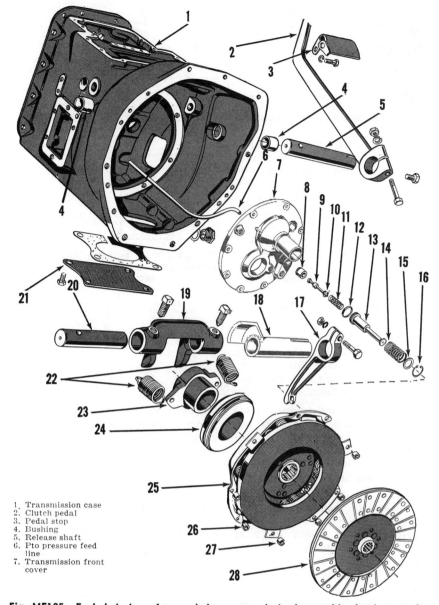

1. Transmission case
2. Clutch pedal
3. Pedal stop
4. Bushing
5. Release shaft
6. Pto pressure feed line
7. Transmission front cover

Fig. MF185—Exploded view of transmission case and clutch assembly showing pto dump valve and fork.

8. Pto relief valve	16. Snap ring	22. Clutch pedal return springs
9. Relief valve shaft	17. Dump valve fork	23. Release bearing carrier
10. Snap ring	18. Transmission brake cam	24. Release bearing
11. Relief valve spring	19. Clutch release fork	25. Cover & pressure plate assembly
12. "O" ring	20. Clutch fork pivot shaft	26. Short spacer
13. Dump valve	21. Inspection cover	27. Long spacer
14. Dump valve spring		28. Clutch disc
15. Retainer washer		

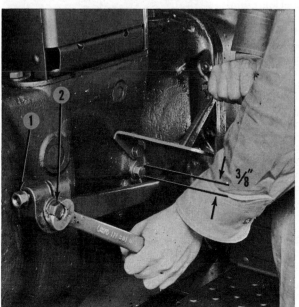

Fig. MF184 — Adjusting the clutch pedal free play.

1. Clamp bolt
2. Clutch release shaft

clutch cover and three 5/8-inch spacers between pto drive hub and flywheel. Because of additional flywheel rim thickness, diesel tractors use three 3/16-inch spacers between pto drive hub and flywheel only. When installing the clutch assembly, make sure that the correct thickness spacers are used in the proper location.

Clutch pilot and throw-out bearings are sealed units and should be renewed if damaged or worn.

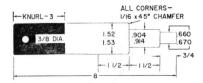

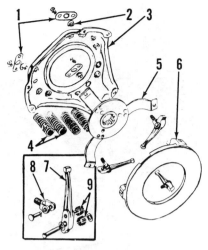

Fig. MF185A—Dimensions for special pilot tool necessary for realigning clutch disc and pressure plate assembly to flywheel.

A special clutch aligning tool is necessary to properly align the clutch during installation. Dimensions for the tool are shown in Fig. MF185A. Install clutch disc (28—Fig. MF185) with short hub toward flywheel and make sure the proper spacers are used when installing cover. Before reconnecting transmission case to engine, check the adjustment of pto dump valve as outlined in paragraph 135, and transmission input shaft brake as outlined in paragraph 136.

134. OVERHAUL. Examine clutch disc (28—Fig. MF185) for glazing, wear or other damage. Disc must be renewed if facings are worn to within 3/32-inch of rivet heads.

To disassemble the pressure plate and cover assembly, remove the cap screws retaining thrust plates (1—Fig. MF185B). Place the assembly in a press and apply pressure while removing release lever yoke nuts (2).

Pressure springs (4) should have a free length of approximately 3½ inches, and should test 165 lbs. when compressed to a height of 2.121 inches. Pressure plate (6) may be resurfaced providing not more than 1/16-inch of metal is removed.

Release lever pivots each contain 19 uncaged needle rollers (9). To remove pivot pins, grind off riveted end and be careful not to lose any of the loose rollers. When reassembling, make sure the longer rollers are used in adjusting yoke pivots.

To adjust the release levers, use the clutch disc to be installed, and assemble clutch cover to flywheel. Adjust each lever in turn to a height of 7/16-inch below the countersunk inner rim of cover, using clutch adjusting gage (Nuday—MFN750) or a suitable depth gage and straight edge.

PTO PUMP VALVE

All Models So Equipped

135. To adjust the pto dump valve, loosen the clamp screw (B—Fig. MF185C). Make sure clutch pedal is against the stop on upper foot rest, then measure the distance (A) between front of fingers on dump valve

Fig. MF185B — Exploded view of clutch pressure plate and cover assembly, showing component parts.

1. Thrust plate	6. Pressure plate
2. Adjusting nut	7. Release lever
3. Cover	8. Adjusting yoke
4. Springs	9. Needle rollers
5. PTO drive hub	

fork and inside edge of valve front land. This distance should be 1.570 for Models MF85 and MF88; and 1.12 for Models MF Super 90 and MF Super 90 WR. Position the fork on clutch release shaft to obtain the proper distance then tighten clamp bolt (B).

TRANSMISSION SHAFT BRAKE

All Models Except Multipower

136. To measure the clearance of the transmission shaft brake plunger, remove the transmission drain plug and withdraw approximately five gallons of transmission fluid. Unbolt and remove the power steering pump assembly from right side of transmission housing. Working through pump mounting hole, measure clearance between brake plunger and disc with a feeler gage as shown in Fig. MF186. Clearance should be 0.030-0.050, with clutch pedal against stop on upper foot rest. Adjust by turning the adjusting screw (7—Fig. MF187).

NOTE: Adjustment and measurement can be made through top cover opening without removing power steering pump, if top cover is removed for other work.

SPLITTING TRACTOR

Note: While the following procedures apply specifically to non-diesel models, the procedure for diesel models is identical except for obvious differences in disconnecting fuel lines, wiring, etc. Because the switch

Fig. MF185C — With clutch pedal against stop, loosen clamp bolt (B) and adjust dump valve release fork until clearance (A) is 1.570 for model MF85 or MF88; or 1.12 for model MF Super 90 or MF Super 90WR.

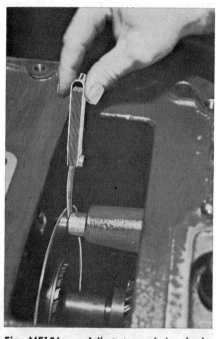

Fig. MF186 — Adjust transmission brake plunger to 0.040 clearance with clutch fully engaged. Adjustment can be made through pto pump mounting opening if transmission top cover is not removed.

for the electric fuel pump on MF85 and MF88 diesel models is automatically connected and mechanically disconnected, be sure wiring at switch is disconnected before removing fuel lines or pump connections.

On Multipower models, the same general procedures apply except that multipower clutch assembly must be detached from engine drive plates as outlined in paragraph 147.

137. To detach engine from transmission, proceed as follows: Drain cooling system, remove hood side panels and hood. Shut-off fuel and remove fuel line. Remove power steering oil lines. Remove ignition switch and starter button from lower instrument panel and remove panel. Drive roll-pin from rear end of steering shaft. Open battery door, remove fuel tank rear mounting bolts and block up between fuel tank and engine assembly. Disconnect heat indicator sending unit from head, and cable from starting motor. Disconnect wires from coil and generator, tractormeter cable from drive, choke rod from carburetor and oil gage line from cylinder block. Disconnect governor control rod at rear and remove exhaust pipe. Support engine in a hoist and place a rolling floor jack under transmission case. Unbolt engine from transmission and separate the tractor halves.

TRANSMISSION AND CONNECTIONS
(All Models Except Multipower)

SEE PARAGRAPHS BEGINNING WITH 147 FOR MULTIPOWER TRANSMISSION

138. **REMOVE AND REINSTALL.** To remove the complete transmission housing assembly, first detach engine from transmission as outlined in paragraph 137 and proceed as follows: Disconnect tail light wire at connector, open battery door and remove battery. Disconnect pto warning light wire at pressure switch and starter switch wires at safety switch. Unbolt battery box from transmission housing and remove battery box, steering gear assembly and instrument panel from transmission housing as a unit. Remove both step plates, support transmission housing and rear axle center housing separately, then unbolt and separate the units.

139. **R&R TOP COVER.** Shifter housing can be removed by unbolting and lifting off. Top cover can be removed after battery box and shifter housing are removed.

140. **OVERHAUL.** Data on overhauling the various transmission components are as follows:

141. PTO INPUT SHAFT & GEAR, FRONT COVER AND PTO COUNTERSHAFT AND GEAR. To remove the pto input shaft and gear (13—Fig. MF187), and front cover, first split the tractor between engine and transmission case as outlined in paragraph 137 and proceed as follows: Disconnect the clutch throwout springs and remove clutch throwout bearing and hub. Remove throwout fork set screws, clutch pedal clamping bolt and clutch pedal. Loosen clamping bolt on pto dump valve fork. Drive right and left throwout shafts outward to free throwout fork and remove fork and shafts. Remove transmission brake cam (8—Fig. MF187) and pto dump valve fork by withdrawing from inside of case. Slide transmission brake plunger (4) forward out of cover as an assembly. Unbolt and remove power steering and pto pump from right side of case. Disconnect power steering pressure, return and bleed lines from front cover and reposition same to top of case to clear front cover. Unbolt and remove front cover (19), with pto input gear (13), pto countershaft gear (48) and countershaft (53) attached.

To remove pto countershaft and countershaft gear, first disengage snap ring (51) from shaft groove at rear of pto countershaft gear. Tap shaft (53) forward to force expansion plug (20) out of cover. Remove the other snap ring (51) from front end of shaft and remove shaft from gear. Remove snap ring (52) from gear hub and press gear out of bearing (50), then remove snap ring (49) at rear of bearing and press bearing from cover. Reinstall bearing (50) in cover with shielded side to the rear. The pto input shaft and gear (13) can be removed after the pto countershaft gear is removed by extracting the snap ring (14) from cover and withdrawing bearing and shaft as an assembly.

Oil seals (12 & 18) in cover and pto input shaft can be renewed at this time. If bearing (15) is removed from shaft, it must be reinstalled with shielded side toward the gear end of shaft.

The transmission brake plunger parts (1, 2, 3, 4, 5, 6 & 7) are serviced individually.

Before reinstalling the assembled transmission front cover, first remove and check the pto clutch oil line (6—Fig. MF185). Both ends must be parallel to each other and line length must be $13\frac{15}{16}$ inches when measured on this parallel line. When reinstalling cover, make certain oil line engages in the proper hole at each end of line.

142. SHIFTER RAILS AND FORKS. To remove the transmission shifter rails and forks, transmission top cover (7—Fig. MF188) must first be removed and the transmission housing separated from the rear axle center housing. Remove the four detent springs and balls (21) from the top of the transmission case. Remove the set screw from the planetary shifter fork (18) and remove the fork, shifter hub and stub shaft. Remove the set screws from the remainder of the shifter forks and selector blocks, remove stop plate (12) and shifter rails and forks from transmission case. Note: The second and third speed shifter rail (15) has an interlocking pin (22) in its forward end, with in-

Fig. MF187—Exploded view of transmission shafts and gears.

1. Transmission brake stop
2. Outer spring
3. "O" ring
4. Transmission brake plunger
5. Inner spring
6. Plunger
7. Adjusting screw
8. Cam
9. Input gear & shaft
10. Transmission brake disc
11. Snap ring
12. Inner oil seal
13. Pto input gear
14. Snap ring
15. Input bearing
16. Snap ring
17. Gasket
18. Outer seal
19. Transmission front cover
19A. "O" ring
20. Expansion plug
21. Planetary coupler
22. Planetary rear cover
23. Thrust washer
24. Planetary carrier assembly
25. Ring gear
26. Planetary front cover
27. Shim
28. Main shaft rear bearing
29. Snap ring
30. Main shaft
31. Needle bearings
32. Main shaft 1st gear
33. Main shaft 4th gear
34. Main shaft 2nd & 3rd gear
35. Snap ring
36. Snap ring
37. Main shaft front bearing
38. Snap ring
39. Countershaft rear bearing
40. Countershaft reverse gear
41. Spacers
42. Countershaft 1st gear
43. Countershaft 4th gear
44. Countershaft 2nd gear
45. Needle bearing
46. Countershaft front bearing
47. Countershaft
48. Pto countershaft gear
49. Snap ring
50. Pto countershaft bearing
51. Snap ring
52. Snap ring
53. Pto countershaft
54. Reverse idler shaft
55. Retaining screw
56. Thrust washers
57. Needle bearings
58. Retaining washers
59. Reverse idler gear

terlock balls (23) in the housing. All rails must be in neutral before one can be withdrawn. When reassembling, reverse the disassembly procedure making certain interlock pin and balls are properly placed. Rails should be installed with milled flats to top and rear and interlock grooves to center. Forks and rails are not interchangeable.

143. MAIN (SLIDING GEAR) SHAFT AND INPUT SHAFT. To remove the transmission main shaft (30—Fig. MF187), first remove the transmission assembly (paragraph 138) and the shifter rails and forks (paragraph 142).

Remove the four cap screws from the planetary gear housing. Remove plate (22), thrust washer (23), plane-

tary gear assembly (24 & 25), inner plate (26) and shim (27) from the rear of the transmission case.

Withdraw main shaft (30) and rear bearing (28) from rear of case. After main shaft has been withdrawn, input shaft (9) can be removed as follows: First remove transmission front cover (paragraph 141) and remove snap ring (11) and brake disc (10) from front of input shaft. Remove snap ring (35) securing bearing (37) to rear of shaft and using a suitable adapter, drive input shaft forward out of mainshaft front bearing, being careful not to lose or damage the thirteen uncaged pilot bearing rollers (31).

When reassembling, first reinstall the input shaft and secure with lock ring (35). Using cup grease, install the thirteen rollers of the main shaft pilot bearing in the bore of the input shaft. Slide the main shaft and bearing through the rear of the case and position the sliding gears as shaft is moved forward. Note: All sliding gears are installed with the shifter fork groove toward the front of the transmission. The thirty-one tooth first and reverse gear (32) is installed on the shaft first, followed by the sixteen tooth fourth gear (33) and the twenty-seven tooth second and third gear (34). When installing the front end of the main shaft in the uncaged pilot bearing (31), be careful not to pull the shaft back out once it has entered. If the rollers are pulled out of position, they may jam the main shaft, causing the tractor to creep with the transmission in neutral. After completion of assembly, check pilot bearing location by rotating input shaft with transmission in neutral. Input shaft should rotate freely without binding the main shaft.

144. COUNTERSHAFT. To remove the countershaft (47—Fig. MF187), first remove the main shaft, front cover (19) and brake disc (10), as outlined in paragraph 143 and proceed as follows: Remove the snap ring (38) from the rear of the countershaft, and using a suitable adapter, drive the countershaft forward and out of the case. The needle bearing (45) inside the countershaft and the needle bearing (46) in the front of the transmission case can be renewed at this time.

When reassembling, install the gears and spacers in the following order: Spacer (41); twenty-seven tooth second gear (44) with hub forward; thirty-six tooth fourth gear (43) with hub rearward; spacer (41); twenty-two tooth first gear (42) with hub rearward; spacer (41); and eighteen tooth reverse gear (40). With countershaft fully in place, wedge a wooden block in the front compartment of the transmission housing to prevent countershaft from moving forward while rear bearing is being drifted on. Using a 2-inch ID pipe, tap the rear bearing into position; install snap ring (38) and remove the wedge. On MF Super 90 and Super 90 WR, the snap ring is available in selective thicknesses of 0.093, 0.103, 0.113 and 0.123. Select and install the snap ring which provides the tightest fit in groove.

Note: The three spacers (41) on the countershaft are interchangeable.

145. REVERSE IDLER. The reverse idler can be removed after the transmission main shaft has been removed (paragraph 143). To remove the reverse idler gear (59—Fig. MF187), first remove the locking bolt (55) then remove the reverse idler shaft (54) through the rear of the case. The reverse idler gear rotates on two rows of twenty-eight uncaged needle bearings (57), separated by three identical needle retaining washers (58). The thrust washers (56) are also interchangeable. The reverse idler gear is installed hub forward, with a bronze thrust washer on each side of the gear. Make certain that the locking bolt indexes with the retaining hole in the shaft.

146. PLANETARY UNIT. The planetary unit, composed of parts (21 through 27—Fig. MF187), can be removed as follows after detaching the transmission from the rear axle housing: Remove shifter fork set screw, planetary shifter fork, and hub. Remove the four cap screws retaining the planetary unit to the transmission case and remove the plate (22), thrust washers (23), planetary gear assembly (24 & 25) and inner plate.

The planetary gears rotate on two rows of twenty-seven uncaged needle bearings separated by three interchangeable spacer washers. To renew the bearings, press out the shafts using a suitable arbor. Bearing needles, spacer washers and thrust washers are available individually as are the shafts and gears. When pressing the shafts into place during assembly,

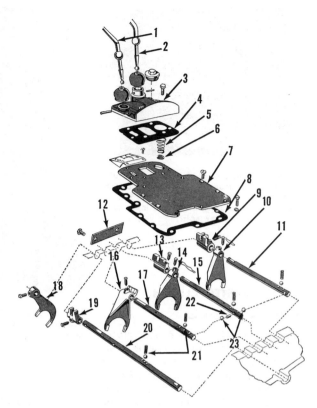

Fig. MF188 — Exploded view of shifter rails and forks. Interlocking plunger (22) and balls (23) prevent movement of other rails when transmission is in any gear.

1. Planetary shift lever
2. Shift lever
3. Housing
4. Gasket
5. Spring
6. Spring seat
7. Transmission cover
8. Gasket
9. Selector block 4th
10. Fork 4th
11. Rail 4th
12. Stop plate
13. Selector block 2nd & 3rd
14. Fork 2nd & 3rd
15. Rail 2nd & 3rd
16. Fork 1st & reverse
17. Rail 1st & reverse
18. Planetary fork
19. Planetary shift block
20. Planetary rail
21. Detent ball & spring
22. Interlock pin
23. Interlock ball

a short piece of material slightly smaller than shaft size will help keep the parts in alignment and prevent damage to the bearing needles and thrust washers.

To install the planetary unit, assemble the ring gear (25—Fig. MF187), thrust washer (23), inner plate (26) and shim (27) and install on transmission case with flat edge of inner plate and shim to the bottom. Install planet carrier (24) with gears, thrust washer (23) and outer plate (22) and secure with the four cap screws. Reinstall shifter fork and coupler.

MULTIPOWER TRANSMISSION

Massey-Ferguson Super 90 & Super 90 WR tractors are optionally equipped with a "Multipower" transmission, which is a modification of the standard, dual range, four speed transmission. The "Multipower" unit provides an additional, hydraulically operated high-low range which can be shifted while the tractor is in motion, thus providing a total of sixteen forward and four reverse speeds. Power for the "Multipower" hydraulic clutch is supplied by the return oil from the power steering system, which also supplies the operating fluid for the PTO clutch.

SPLITTING TRACTOR

Multipower Models

147. To detach (split) engine from transmission, follow the general procedure outlined in paragraph 137, with the following important exceptions:

The Master Clutch Must Be Detached From Engine Drive Plate before the units are separated, by removing the starter unit; then, working through starter hole, removing the six cap screws which retain clutch assembly to plate.

As soon as the dowels are free, the transmission case may be independently lowered, as there are no shafts or other alignment problems. CAUTION: Do not allow the master clutch unit to slip from end of transmission input shaft splines. Clutch unit is extremely heavy and is not retained in any way after the units are separated. If master clutch is not to be removed, a retaining clip should be attached to front of transmission case to prevent accidental removal.

Fig. MF189—Removal of multipower clutch assembly from transmission housing can be faciliated by the use of a lifting tool as shown. Clutch slips from splined input shaft after split.

MASTER CLUTCH

Multipower Models

148. Removal of the master clutch assembly after splitting engine from transmission requires the use of a special tool as shown in Fig. MF189. The weight and position of clutch prevents efficient removal without use of such a tool.

The master clutch unit consists of a master disc (10—Fig. MF190) and a high-range disc (12) which are hydraulically engaged and mechanically released through the action of springs (14). The master clutch is controlled by the regular clutch pedal, and the high-range clutch by a small lever on dash panel.

To disassemble the removed clutch assembly, refer to Fig. MF191. Place clutch unit, forward end down, on a bench and remove the securing cap screws. The two piston assemblies (9) are interchangeable as are the piston rings (7 and 8). Clutch discs (10 and 12) are identical except for the hubs. The center drive plate (11) may be installed either side forward, while remainder of parts must be installed as shown in Figs. MF190 and MF191. Note that hub of high-range clutch disc (12) pilots into machined portion of front cover (16) and that hub of master clutch (10), pilots into high range clutch; thus eliminating the necessity for alignment during assembly or installation.

Clutch piston (9) may be refaced provided not more than 0.020 of metal is removed. The center drive plate (11) may be refaced provided not more than 0.025 is removed from either face.

Mig. MF190 — Cross sectional view of the hydraulically operated, dual plate multipower clutch, showing location of components. Refer to Fig. MF191 for legend.

CONTROL VALVE

Multipower Models

149. **PRESSURE CHECK.** The "Multipower" and Power Take-Off operating pressure should be 225 psi, and can be tested by removing the access plate (A—Fig. MF192), and installing a test gage in one of the plug holes (T); or easier, by installing a gage in place of the PTO pressure gage sender unit in power steering return line on right side of tractor. A pressure lower than specified could be caused by a leak in PTO or multipower system, a weak spring (35—Fig. MF193) or a sticking valve (34).

150. **REMOVE AND REINSTALL.** To remove either the master control valve housing (9—Fig. MF193) or the high-range valve housing (31), it is necessary to remove the master clutch as outlined in paragraph 148, and remove the transmission front cover (20). The master control valve (11) operates the transmission brake plunger (22) by metering pressurized fluid to plunger bore when clutch pedal is depressed. NOTE: The area on both sides of brake plunger is pressurized; however, the plunger is moved rearward due to the larger area on front side of plunger.

When reassembling, place two drops of LOCTITE on threads of cap screws retaining front cover (20). Adjust the multipower control lever

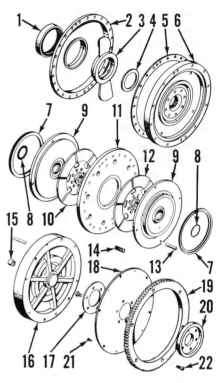

Fig. MF191—Exploded view of multipower clutch and associated parts.

1. Oil seal	12. Clutch disc
2. Cover	(high range)
3. Oil slinger	13. Dowel
4. Washer	14. Release spring
5. Gasket	15. Pilot bushing
6. PTO drive cover	16. Front cover
7. Piston seal	17. Retainer
8. Piston seal	18. Engine drive plate
9. Piston	19. Starter ring gear
10. Main clutch	20. Retainer
disc	21. Screw (ring gear)
11. Center drive	22. Screw (clutch
plate	mounting)

until valve (32) moves fully in either direction when dash lever is moved. Adjust clutch pedal until the total travel measures 4½ inches when measured from the pedal pad to the stop bolted to transmission housing.

TRANSMISSION R&R

Multipower Models

151. To remove the complete transmission, first drain transmission and rear axle center housing; then split transmission from engine as outlined in paragraph 147. Remove the master clutch as outlined in paragraph 148. Disconnect tail light wire at connector, open battery door and remove battery or batteries. Disconnect other interfering wiring, unbolt battery box from transmission housing and remove battery box, rear steering gear unit and instrument panel assembly, as a unit, from transmission. Remove both step plates, support transmission housing and rear axle center housing separately, then unbolt and separate the units.

OVERHAUL

Multipower Models

152. **PTO INPUT SHAFT, PTO COUNTERSHAFT AND TRANSMISSION FRONT COVER.** To remove the transmission front cover and the contained PTO driving gears and shafts, first split transmission from engine assembly as outlined in paragraph 147 and remove master clutch as outlined in paragraph 148. Refer to Fig. MF192. Disconnect and remove the multipower linkage (ML) and the clutch linkage (CL). Remove the power steering pump from right side of transmission housing and disconnect and remove all interfering hydraulic lines and tubes. Remove the cap screws retaining the transmission front cover (20—Fig. MF193) to transmission housing, then remove the cover with PTO driving mechanism attached. NOTE: The cover retaining cap screws are installed with LOCTITE. If difficulty is encountered in removal, heat the screws to approximately 500° F., and remove while hot.

To remove the PTO countershaft (18—Fig. MF195), gear (20), and bearing (22) from the front cover, first release the snap ring (19) at rear of gear and bump the shaft forward to dislodge expansion plug (24). Remove snap ring (23) and bump gear (20) rearward out of bearing (22). Remove snap ring (21) from

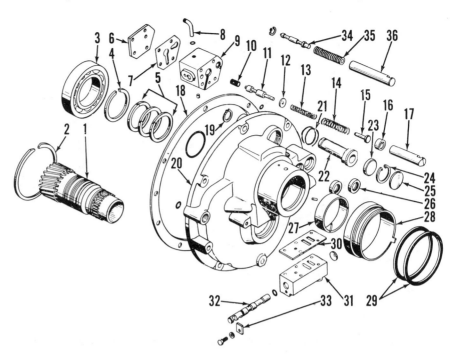

Fig. MF193 — Exploded view of transmission front cover showing PTO input shaft, control valves and associated parts.

1. PTO input shaft	13. Spring	25. Expansion plug
2. Snap ring	14. Spring	26. Oil seal
3. Bearing	15. Push rod	27. Bushing
4. Snap ring	16. Bushing	28. Slinger drive
5. Sealing rings	17. Plunger	29. Seals
6. Plate	18. Gasket	30. Gasket
7. Gasket	19. Snap ring	31. Valve body
8. Oil tube	20. Front cover	32. Multipower valve
9. Valve body	21. Brake return	33. Retainer
10. Valve spring	spring	34. Regulating valve
11. Valve (master clutch)	22. Brake plunger	35. Spring
12. Washer	23. Plug	36. Plunger
	24. Snap ring	

Fig. MF192 — Front view of multipower transmission housing with clutch removed.

A. Access cover
R. Relief valve
CL. Clutch linkage
ML. Multipower shift linkage

cover and bump bearing from housing. Assemble by reversing the disassembly procedure.

To remove the PTO input shaft (1—Fig. MF193), first remove the PTO countershaft gear as previously outlined. Remove snap ring (2) and bump shaft (1) and bearing (3) rearward out of cover, being careful not to damage the sealing rings (5).

When reinstalling the front cover, make sure the "O" rings which seal hydraulic passages are in place, and that PTO oil feed line enters holes in front cover and rear wall of transmission housing. Install cover retaining cap screws using two drops of LOCTITE on threads of each screw.

153. **INPUT SHAFTS AND MULTIPOWER OVERRUNNING CLUTCH.** To remove the transmission master input shaft (9—Fig. MF194), remove the front cover as outlined in paragraph 152; then, withdraw the shaft and overrunning clutch assembly (Items 1, 2, 4, 5, 6, 10, 13, 14, 15 & 16).

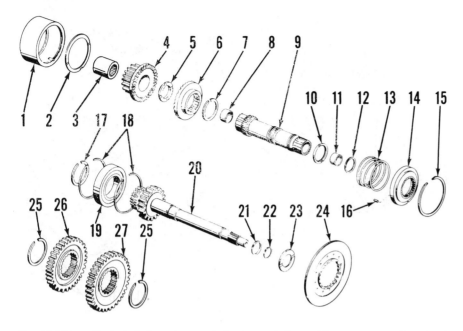

Fig. MF194 — Exploded view of transmission input shafts, multipower overrunning clutch and countershaft drive gears.

1. Clutch housing
2. Thrust washer
3. Needle bearing
4. Master drive gear
5. Thrust washer
6. Clutch jaw
7. Snap ring
8. Bushing
9. Main input shaft
10. Thrust washer
11. Bushing
12. Sealing ring
13. Clutch spring
14. Clutch cover
15. Snap ring
16. Dowel
17. Snap ring
18. Snap ring
19. Bearing
20. Input shaft (high speed)
21. Sealing ring
22. Sealing ring
23. Thrust washer
24. Brake plate
25. Snap ring
26. Countershaft drive gear (high speed)
27. Countershaft drive gear (low speed)

Oil sealing rings on both transmission input shafts can be renewed at this time, as can needle bearing (3) and bushings (8 & 11). Overrunning clutch can be disassembled by removing snap ring (15) and withdrawing the parts. When reassembling, make sure that dowel (18) is in place and enters groove in housing (1).

To remove the high range input shaft (20) it is first necessary to remove the complete transmission assembly as outlined in paragraph 151, and transmission mainshaft as in paragraph 155. Remove snap ring (25) from forward end of transmission countershaft and withdraw the mult-power driven gears (26 and 27). Remove snap ring (18) from rear of bearing (19) and work input shaft (20) and bearing assembly rearward out of transmission case. Assemble by reversing the disassembly procedure.

154. SHIFTER RAILS AND FORKS. The transmission shifter rails and forks are identical to those used on tractors without multipower, and can be serviced as outlined in paragraph 142.

155. TRANSMISSION MAIN SHAFT. To remove the transmission main shaft, first drain the fluid from transmission and rear axle center

Fig. MF195 — Exploded view of gears and shafts used in multipower transmission.

1. Main shaft
2. Snap ring
3. Bearing
4. Bearing roller
5. 1st speed gear
6. 4th speed gear
7. 2nd & 3rd speed gear
8. Snap ring
9. Bearing
10. Countershaft gear
11. Spacer
12. Countershaft gear
13. Countershaft gear
14. Countershaft gear
15. Needle bearing
16. Needle bearing
17. Countershaft
18. PTO countershaft
19. Snap ring
20. PTO countershaft gear
21. Snap ring
22. Bearing
23. Snap ring
24. Expansion plug
25. Reverse idler shaft
26. Thrust washer
27. Spacer
28. Needle roller
29. Reverse idler gear
30. Retaining screw

housing and detach (split) tractor at rear of transmission case. NOTE: If service is required on transmission units other than planetary unit, shifter mechanism, or main shaft and sliding gears; transmission must be removed as outlined in paragraph 151. Remove planetary unit as outlined in paragraph 146 and shifter rails and forks as in paragraph 142.

Withdraw the main shaft (1—Fig. MF195) and bearing (3) rearward out of transmission case while lifting the sliding gears (5, 6 & 7) out through top opening of case. NOTE: The front sliding gear (7) should be moved to the rear before main shaft is withdrawn, and care taken that the thirteen loose bearing rollers (4) remain in pilot bore of high-range input shaft and not lost in transmission case.

When installing the main shaft, note that all sliding gears are assembled with shifter fork groove to the front and in the order shown in Fig. MF195.

Make sure that the loose bearing rollers (4) are properly positioned and do not bind the shaft.

156. TRANSMISSION COUNTER-SHAFT. To remove the transmission countershaft (17—Fig. MF195), first remove transmission assembly as outlined in paragraph 151, front cover as in paragraph 152 and main shaft as in paragraph 155. Remove the snap ring (8) from rear of countershaft and, using a suitable step plate in end of shaft, bump countershaft (17) forward out of rear bearing (9) and the shaft gears and spacers.

When reassembling, note that the three spacers (11) are interchangeable; gear (14) is installed with longest hub forward and gears (12 & 13) with longest hub rearward. Gear (10) may be installed with ether side forward. Snap ring (8) is available in thicknesses of 0.093, 0.103, 0.113 and 0.123. Select and install the snap ring which provides the tightest fit in snap ring groove.

157. REVERSE IDLER. The reverse idler (29—Fig. MF195) can be removed after the transmission main shaft has been removed as outlined in paragraph 155. Remove the locking bolt (30) and withdraw the reverse idler shaft (25) rearward out of case. The reverse idler gear rotates in two rows of twenty-eight uncaged rollers (28) separated by three identical needle retaining washers (27). The thrust washers (26) are also interchangeable. The reverse idler gear is installed hub forward, with a bronze thrust washer on each side of gear. Make certain that the locking bolt indexes with the retaining hole in shaft.

158. PLANETARY UNIT. The planetary range unit is identical to that used on tractors not equipped with multipower transmission, and may be serviced as outlined in paragraph 146.

DIFFERENTIAL AND MAIN DRIVE BEVEL GEAR

DIFFERENTIAL AND BEVEL GEAR

All Models

166. REMOVE AND REINSTALL. To remove the differential unit, first drain the transmission and rear axle center housing. Remove the hydraulic system control frame on models so equipped. Support rear of tractor and remove step plates or operators platform; then remove both final drive units. Remove the center housing rear cover (C—Fig. MF199) and support differential assembly by blocking up between ring gear and floor of center housing.

The socket head cap screws which retain the differential carrier plates (P) are secured with "LOCTITE" and are staked in place. To remove the differential carriers, remove the staking and heat bolt heads to a temperature of approximately 500° F. to free the LOCTITE bond. Remove bolts while they are hot.

Preload of carrier bearings is not adjustable. When reinstalling differential, the ring gear is assembled to

left of bevel pinion on standard clearance models; and to right of bevel pinion on high clearance models.

Reinstall by reversing the removal procedure. If main drive bevel gears are renewed, check the backlash as outlined in paragraph 168. Install the carrier plates with oil drain holes down, using new cap screws. After checking the backlash adjustment, place two drops of LOCTITE on the threads of each screw, and tighten. Using a small punch and hammer, strike the head of each screw sharply to be sure screws are seated, then retighten securely. Stake into groove in carrier plate, using a center punch.

167. OVERHAUL. Before disassembling the removed differential, place correlation marks on the two case halves for correct assembly in the same relative position. Refer to Fig. MF200 for an exploded view of the differential unit.

Backlash between axle gears and spider gears should be 0.003-0.011. If backlash is excessive, renew thrust washers and/or the gears concerned.

Tighten the cap screws retaining the case halves to a torque of 50-55 ft.-lbs. and secure with safety wire.

The main drive bevel gear may be bolted or riveted to the differential case. The bevel ring gear is available only as a matched set with the main drive bevel pinion, and new bolts and nuts for attaching ring gear are included in the kit. To renew the bevel ring gear on riveted assemblies, drill out rivets carefully, using a ⅜-inch drill, and assemble using the bolts provided. Apply two drops of LOCTITE to threads of each bolt and tighten the nuts to a torque of 110-120 ft.-lbs. When renewing bevel gears, differential case or bearings, check and adjust the backlash as outlined in paragraph 168.

168. BACKLASH ADJUSTMENT. The main drive bevel gear backlash should be 0.008-0.012, and is adjusted by means of shims installed between bevel ring gear and differential case. Shims are available in thicknesses of 0.003 and 0.007. With the unit assembled, measure the backlash using a dial indicator. Remove differential assembly from center housing and bevel ring gear from differential; then, add or remove shims as required. Assemble as outlined in paragraphs 166 and 167.

MAIN DRIVE BEVEL PINION

All Models

169. **REMOVE AND REINSTALL.** To remove the bevel pinion, first remove the differential assembly as outlined in paragraph 166. Remove the hydraulic lift cover as outlined in paragraph 195; then unbolt and remove rear axle center housing from transmission case. Remove the hydraulic pump support pins from each side of center housing and slide pump forward out of housing. Unstake the locknut (13—Fig. MF208) and remove the nut and the two tab washers (11 and 12). Remove the four cap screws retaining front bearing carrier (8) and remove carrier and front bearing cone (10) by using forcing screws in tapped holes of carrier. Bevel pinion, rear bearing cone (2) and pto ground drive gear (3) can now be withdrawn rearward out of housing.

When reassembling, heat the front bearing cone (10) in oil or an oven to a temperature of approximately 180° F. Keep bearing heated until bevel pinion, pto drive gear (3) and front bearing carrier (8) are in place, then install front bearing, tab washers and locknut. Tighten the locknut until a rolling torque of 18-20 inch pounds is required to keep the bevel pinion shaft turning. Be sure to stake locknut in position after adjustment is complete.

REAR AXLE AND FINAL DRIVE

REMOVE AND REINSTALL

All Models

170. To remove one or both final drive assemblies, first remove fenders and step plates or operators' platform. Remove hydraulic system control frame rockers on tractors so equipped, as outlined in paragraph 198. Remove brake linkage on side to be overhauled, and support tractor under rear axle center housing. Remove wheel and wheel disc assembly. Support final drive assembly with a chain hoist by attaching to outer fender boss for standard clearance models, or at drop housing for high clearance models. Remove cap screws retaining final drive assembly to center housing, and withdraw the unit being careful not to damage the rear axle inner oil seal.

OVERHAUL

High Clearance Models

171. To remove the drop housing assembly, first support the tractor under rear axle center housing and remove the wheel assembly. Remove

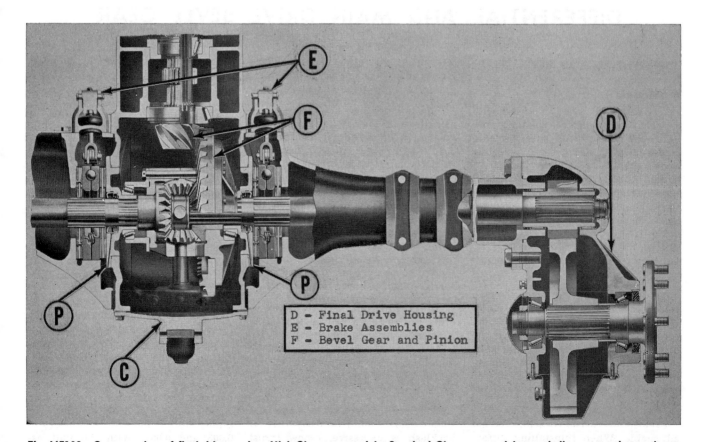

D - Final Drive Housing
E - Brake Assemblies
F - Bevel Gear and Pinion

Fig. MF199—Cutaway view of final drive used on High Clearance models. Standard Clearance models are similar except that a planetary final drive is used, and ring gear is located on opposite side of drive pinion.

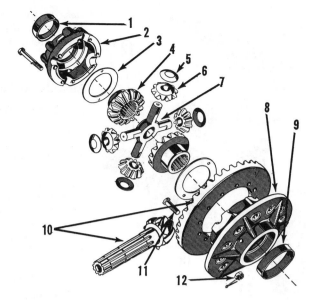

Fig. MF200 — Exploded view of differential assembly. Differential carrier bearings and bearing cups are not interchangeable.

1. Outer cup
2. Outer differential case
3. Thrust washer
4. Differential side gear
5. Thrust washer
6. Differential pinion
7. Differential spider
8. Differential case
9. Bearing cup
10. Bevel gear & pinion (matched)
11. Bevel gear bolt
12. Nut & cotter pin

Fig. MF201 — Differential carrier plate and seal showing location of oil drain hole. Carrier plate is to be installed so that relief hole is down.

The long axle shaft is supported on either side of the pinion by needle bearings (8) in drop housing and axle housing. End thrust is taken on a roller thrust bearing (5) and two interchangeable thrust washers (4) which are retained in drop housing bore by snap rings (3). Thrust bearing is packed with Lubriplate during assembly, and protected on shaft end by a lip type seal (6). To remove the seal, drive seal toward inside after thrust bearing has been removed. Install outer support bearing (8) in drop housing with a suitable adapter until flush with bore. Do not bottom the bearing.

To remove the wheel axle shaft (11), bearings (13 & 18), seal (12) and bull gear (22), first remove bearing cap (21), groove pin (19) and adjusting nut (20). Unseat snap ring (16) from its groove in wheel axle, and suitably support drop housing in the bed of a press as near as possible to axle flange while pressing axle from gear and housing.

The bull gear is installed with long hub toward outside of tractor, and spacer (15) between hub and outer bearing. During reassembly, lubricate axle shaft outer seal (12) and install over axle shaft, lip facing away from flange; then, press bearing cone (13) on shaft until it bottoms. Install spacer (14) on axle shaft. Place axle shaft assembly flange down on a bench and lower the drop housing and bull gear assembly over wheel axle shaft while guiding axle splines through bull gear. Install and seat snap ring (16), then work outer seal into place using screw drivers or similar tools. Install inner bearing and adjusting nut and tighten nut until a rolling torque of 45-55 inch pounds is required to maintain axle shaft in motion. Reinstall groove pin and cap after adjustment is complete.

oil pan from underside of drop housing with the oil remaining in pan. Drill a hole in expansion plug (1—Fig. MF202), remove plug and small snap ring (2) on end of axle shaft. Unbolt and withdraw drop housing from axle housing.

The long axle shaft (9) can be removed from axle housing without further disassembly; however, before axle shaft is withdrawn, apply a light pressure to brake pedal and lock securely to prevent movement of brake discs and subsequent difficulty in re-inserting axle shaft.

Fig. MF202 — Exploded view of final drive used on MF High Clearance tractors.

1. Expansion plug
2. Snap ring
3. Snap ring
4. Thrust washer
5. Thrust bearing
6. Oil seal
7. Drop housing
8. Needle bearing
9. Shaft & gear assembly
10. Gasket
11. Wheel axle
12. Oil seal
13. Cone & roller assembly
14. Bearing cup
15. Spacer
16. Snap ring
17. Bearing cup
18. Cone & roller assembly
19. Groove pin
20. Bearing nut
21. Dust cap
22. Bull gear
23. Gasket
24. Oil pan

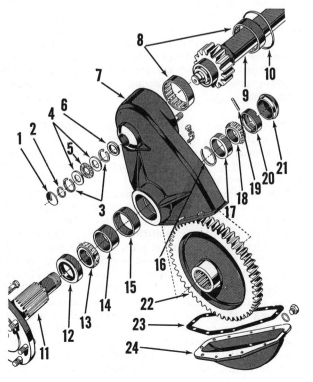

Models MF85-MF88, Standard Clearance

172. To remove the rear axle and planetary drive assembly, support tractor under rear axle center housing and remove tire and rim assembly and wheel disc. Drain the planetary assembly. Scribe a line across drive cover (25—Fig. MF203), ring gear (23) and axle housing so that units can be reassembled in the same relative position; then unbolt and remove the drive cover unit, wheel axle and

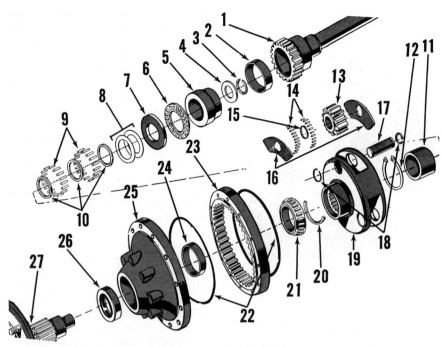

Fig. MF203—Planetary final drive as used on MF85 standard and MF88 tractors.

1. Axle shaft	8. Shim pack	15. Retainer washer
2. Bearing cup	9. Needle bearings	16. Thrust washer
3. Snap ring	10. Retainer washers	17. Pinion shaft
4. Washer	11. Sleeve	18. Snap ring
5. Bearing cone	12. Snap ring	19. Planetary carrier
6. Thrust bearing	13. Planetary pinion	20. Half-moon retainer
7. Thrust washer	14. Needle bearing	ring

21. Cone & roller assembly	25. Cover assembly
22. Gasket	26 Oil seal
23. Ring gear	27. Wheel axle
24. Bearing cup	

planetary carrier as an assembly. The wheel axle shaft rotates on a tapered roller outer bearing (21) and a needle roller and thrust roller cone inner bearing (4 through 10).

The long axle shaft (1) can be removed at this time; however, before shaft is withdrawn apply light pressure to brake pedal and lock securely to prevent movement of brake discs and subsequent difficulty in reinserting axle shaft. After removal of long axle shaft, the outer bearing, bearing cup and seal can be renewed.

Disassamble wheel axle shaft and planetary unit by first removing snap ring (3) from end of axle and removing bearing cone (5), thrust roller (6), thrust washer (7) and shim pack (8) from axle. The wheel axle rotates on two rows of uncaged needle bearings (9) contained in the bearing cone and held in place by three interchangeable retainer washers (10). End thrust is taken by the thrust roller bearing (6) and adjusted by means of shims (8) to provide a preload of 0.001-0.005. Procedure for preload adjustment is outlined paragraph 173.

Remove collar (11) and snap ring (12) from wheel axle shaft and slide planet carrier assembly (19) from

shaft. Remove the bearing support ring halves (20) and press the wheel axle shaft out of drive cover and outer bearing.

Planetary gears are held in position by a shaft (17) secured with snap rings (18) on either end. The gears (13) contain two rows of uncaged needle rollers (14) separated by retaining washers (15). If renewal of planet pinions or bearings is indicated, press shafts (17) from carrier with a suitable press after removing snap rings (18).

173. When reassembling the final drive unit, install outer bearing cone (21) with a drift only far enough to install bearing support ring halves (20). Any excess clearance will give a false reading on final drive preload adjustment. Reassemble remainder of final drive unit by reversing the disassembly procedure, omitting one 0.010 thick shim from shim pack (8). Align the previously affixed scribe marks and tighten three of the retaining bolts. Using a dial indicator, check and record the end play of the wheel axle shaft. Remove wheel axle and drive cover assembly; remove snap ring (3), bearing cone (5), and the thrust bearing (6 & 7). Add to shim pack (8), shims equal in thick-

ness to the previously measured end play plus 0.001-0.005 to obtain the recommended preload.

Models MF Super 90-MF Super 90 WR, Standard Clearance

174. To remove the rear axle and planetary drive assembly, support tractor under rear axle center housing and remove tire and rim assembly and wheel disc. Drain the planetary unit. Scribe a line across drive cover (14—Fig. MF204), ring gear (17) and axle housing (1) so that units can be reassembled in same relative position. Remove the retaining bolts and withdraw wheel axle, drive cover and planetary carrier as a unit.

The long axle shaft (6) can be removed at this time. Before shaft is withdrawn, apply slight pressure to brake pedal and lock securely to prevent movement of brake disc and subsequent difficulty in installing shaft.

Disassemble the wheel axle shaft and planetary assembly by removing the shaft locks (27), withdrawing planet pinion shafts (26) and lifting out planet pinion assemblies (25) and sun gear (8). Be careful not to lose any of the loose needle rollers (24) located in planet pinion bores.

Loosen retainer (9) by cutting it with a sharp chisel, and remove retainer and split ring (10); then remove wheel axle from planet carrier and drive cover using a press.

Bearing cone (18), cup (15) and outer oil seal (13) can be renewed at this time. When reassembling, lubricate lip of seal (13). Make sure splines of wheel axle (11) are absolutely clean and free of grease and that sealing ring (12) is properly positioned in recess of planetary carrier hub. Coat the splines of wheel axle (11) with "LOCTITE" and press planetary carrier on axle until split ring (10) can be installed. Retainer (9) is a shrink fit over end of axle and split ring, and should be heated to approximately 500° F. for installation.

The taper roller bearings should be pre-loaded 0.001-0.005, and are adjusted by shim packs (5). The split shims (5) are available in thicknesses of 0.002, 0.005 and 0.010, and must be installed in sets of four to provide a tight oil seal.

Make a trial assembly of the planetary unit using an additional 0.010 shim (5). (More, if bearings are renewed.) Install and tighten at least

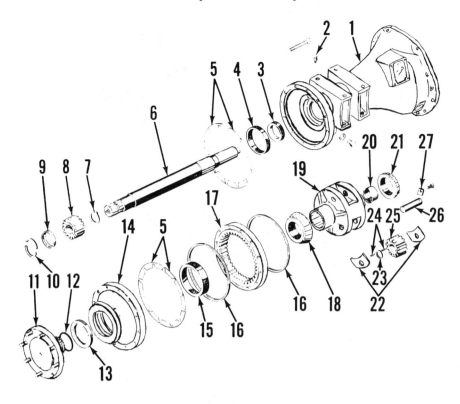

three of the retaining bolts; then measure the end play of wheel axle using a dial indicator. When end play has been measured and recorded, loosen the retaining bolts and remove shims equal to the measured end play plus 0.001-0.005. Shims may be removed from either side of ring gear (17), but must be removed in sets of four. NOTE: A shim pack of approximately equal thickness should be used on each side of planetary ring gear.

Tighten the retaining bolts and nuts to a torque of 55 - 60 ft.-lbs. on final assembly.

BRAKES

175. ADJUSTMENT. To make a brake adjustment, turn the adjusting nut either way as required until a brake pedal free play of 2½-3 inches is obtained on both left and right brake pedals. See Fig. MF205.

176. R&R AND OVERHAUL. First remove the final drive units as outlined in paragraph 170 and remove brake assemblies from axle. Separate the actuating discs (2—Fig. MF206) and renew any parts which are questionable.

Fig. MF204 — Exploded view of rear axle and planetary final drive assembly used on MF Super 90 and MF Super 90WR.

1. Axle housing	10. Split ring	19. Planet carrier
2. Breather	11. Wheel axle	20. Bushing
3. Oil seal	12. "O" ring	21. Bearing cone
4. Bearing cup	13. Oil seal	22. Thrust washer
5. Shims	14. Drive cover	23. Spacer washer
6. Long axle shaft	15. Bearing cup	24. Needle rollers
7. Snap ring	16. Gasket	25. Planet pinion
8. Sun gear	17. Ring gear	26. Pinion shaft
9. Retainer	18. Bearing cone	27. Retainer

Fig. MF205 — Procedure for adjusting right brake linkage. Left linkage is adjusted until brake action is the same with the pedals locked together.

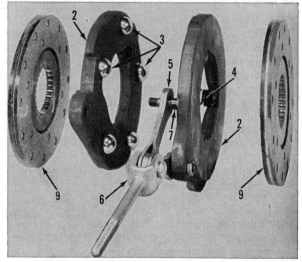

Fig. MF206 — A partially exploded view of the brake assembly on all models.

2. Actuating disc	6. Brake rod
3. Actuating ball	7. Stud
4. Spring	9. Lined discs
5. Yoke link	

BELT PULLEY UNIT

The belt pulley is a self contained accessory unit driven by the tractor pto shaft. Pulley rotation in either direction is obtained by mounting the unit so that the pulley is either to the right or left of the tractor center line.

177. OVERHAUL. Remove input shaft bearing housing (7—Fig. MF207) and extract expansion plug (27). Remove nut (26) and bump the drive gear assembly (11) out of housing. Remove pulley, hub (19) and pinion shaft (13). The need and procedure for further disassembly is evident.

Both the driving pinion and gear are available as separate replacement parts.

When reassembling the unit, tighten nut (22) until a torque of 2-4 inch pounds is required to rotate shaft (13); then, back the nut off until the cotter pin can be inserted. Tighten nut (26) until the bevel gears have a backlash of 0.004-0.006; then, back the nut off until the cotter pin can be inserted.

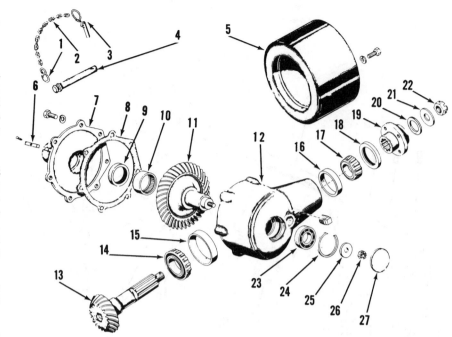

Fig. MF207—Exploded view of belt pulley attachment. Key (6) locates pulley on pto housing.

1. Chain link	11. Drive gear	19. Pulley hub
2. Chain	12. Gear housing	20. Sealing washer
3. Pin	13. Pulley drive pinion	21. Washer
4. Retaining pin	14. Cone & roller	22. Nut
5. Pulley	assembly	23. Ball bearing
6. Locating key	15. Bearing cup	24. Snap ring
7. Bearing housing	16. Bearing cup	25. Washer
8. Gasket	17. Cone & roller	26. Nut
9. Oil seal	assembly	27.. Expansion plug
10. Needle bearing	18. Oil seal	

POWER TAKE-OFF

OPERATION

All Models

178. The power take-off shaft is selectively coupled to the transmission pto countershaft by a hydraulically operated, wet type multiple disc clutch located in the rear axle center housing; or, to the main bevel pinion shaft by a sliding gear arrangement. Hydraulic power for engagement of the multiple disc clutch is obtained from the return oil from the power steering system. The clutch can be disengaged by moving the pto selecter lever to "Neutral" position or, on most models, by pushing the engine clutch pedal through to the second stage disengaged position. The clutch pedal relieves the hydraulic pressure by means of a dump valve located in the transmission front cover plate. Refer to Fig. MF103 for a schematic view of the working parts of the system. The ground speed pto is obtained by moving the pto selector lever to

"Ground Drive," which disengages the pto clutch and couples the pto output shaft to a gear located on the tractor main drive bevel pinion.

A pressure switch is located in the power steering return line and connected to a pto warning light on the instrument panel. Whenever the hydraulic pressure drops below 120 psi in the power steering return line, the switch closes and the warning light lights up. A pressure drop is normal when the foot clutch is depressed through the second stage, and the light should come on at that time.

TROUBLE SHOOTING

All Models

A basic understanding of the inter-relationship of the pto with allied tractor components will allow the cause of pto malfunction to be pin-pointed with a fair degree of accuracy and very little trouble.

The hydraulic pump can be checked by starting the engine and operating the power

steering. If power steering operates satisfactorily, the pump can be eliminated as the cause of trouble.

Damage to the pto gear train and shafts ahead of the pto multiple disc clutch, can be checked by operating the hydraulic lift system. As the hydraulic lift pump is driven by the pto shaft, satisfactory operation of the lift system would indicate that trouble is located in pto multiple disc clutch or output shaft.

The output shaft can be checked by means of the ground drive pto. If pto functions properly in ground drive, the output shaft and multiple disc clutch splines can be considered satisfactory.

Methods of procedure in eliminating some of the other problems are as follows:

179. WARNING LIGHT STAYS ON. Start the tractor and test power steering system operation. If system does not operate properly, check for trouble in pump or lines. If power steering system operates properly, stop engine, remove pressure switch from power steering return line and install a pressure gage of at least 300 psi capacity. Start engine and run at operating

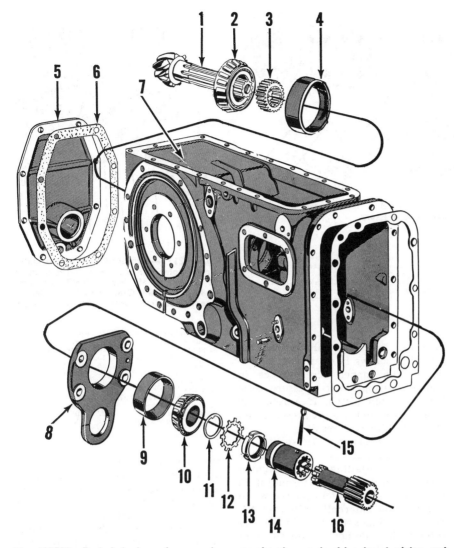

Fig. MF208—Exploded view of rear axle center housing, main drive bevel pinion and associated parts.

1. Bevel pinion	5. Rear end plate	9. Bearing cup	13. Adjusting nut
2. Cone & roller	6. Gasket	10. Cone & roller	14. Main drive coupler
3. Pto ground drive gear	7. Rear axle center housing	11. Tab washer	15. Cotter pin
4. Bearing cup	8. Front bearing carrier	12. Tab lock washer	16. Main drive shaft

speed. Gage pressure should read 150-250 psi (225 psi on multipower models); if it does not, trouble is located in pto relief and dump valve assembly, pto feed line, pto selector valve or pto clutch. If pressure reading is satisfactory, renew the pressure switch.

180. **WARNING LIGHT WILL NOT LIGHT.** Check electrical wiring, bulb and pressure switch.

181. **PTO SHAFT WILL NOT OPERATE.** Check power steering to determine pump condition. Check hydraulic system operation to eliminate possibility of gear train failure. Depress clutch pedal through second stage and note warning light. If all operate properly, trouble is due to plugged pto feed line, faulty pto selector valve, faulty pto clutch or faulty power train from pto clutch rearward. Remove and service the units as outlined in the following paragraphs.

PTO DUMP VALVE

All Models Except Multipower

182. **R&R AND OVERHAUL.** The pto dump valve is located on transmission front cover plate, and serves the dual purpose of pto relief valve and control valve. The entire return flow from the power steering system except that necessary to operate the pto clutch piston is returned to the reservoir through the valve. Movement of the dump valve forward by the operation of the clutch pedal, re-

lieves the pressure on relief valve and allows the fluid to flow without pressure.

To remove the valve, first split the tractor between engine and transmission housing as outlined in paragraph 137. Loosen the clamp screw on dump valve fork and remove the snap ring which retains valve to front cover. Valve parts are serviced separately. When reinstalling valve, refer to paragraph 135 for adjustment. NOTE: Dump valve is not used on Multipower models.

Multipower Models

182A. The multipower regulating valve serves as regulating valve for pto clutch pressure. Refer to paragraph 149.

PTO CLUTCH

All Models

183. **REMOVE AND REINSTALL.** To remove the pto clutch, first split tractor between transmission and rear axle center housing and remove hydraulic pump as outlined in paragraph 191. Remove drive shaft assembly (14 and 16—Fig. MF208); then unstake and remove drive pinion adjusting nut (13). Unbolt and remove drive pinion front bearing carrier (8) by using puller screws. Remove center housing rear end plate (5) and pto output shaft assembly, and working through rear opening in center housing, remove clutch hub snap ring (8—Fig. MF213) from rear of clutch. Bump clutch assembly forward out of housing.

Rear oil feed line and pto control valve can be removed for service at this time as outlined in paragraph 186.

Reinstall the pto clutch by reversing the disassembly procedure. It will be impossible to rotate the drive pinion to check bearing adjustment without removing final drive units and differential assembly. A close approximation of the proper rolling torque can be obtained as follows:

Heat the bevel pinion front bearing cone (10—Fig. MF208) in hot oil to a temperature of 180° F. and immediately reinstall carrier plate, pinion bearing, washers and adjusting nut. Tighten the adjusting nut until pinion is locked tight. Tap front end of pinion shaft sharply with a soft hammer, back off the adjusting nut and retighten with a torque wrench to a torque of 25 - 30 ft.-lbs. and lock in place. Fig. MF210 shows the special socket used by Massey-Ferguson dealers for pinion bearing adjustment.

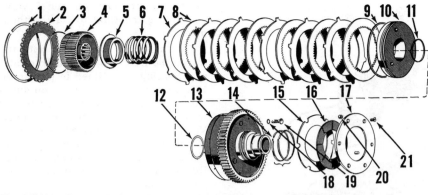

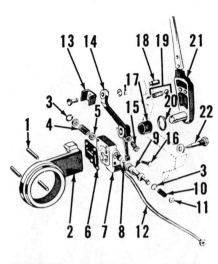

Fig. MF209—Exploded view of pto clutch, which is located in the rear axle center housing and engaged by hydraulic pressure.

1. Snap ring	7. Clutch plate	12. Inner piston ring	17. Brake retaining
2. Retainer plate	8. Clutch disc	13. Clutch housing	plate
3. Thrust washer	9. Outer piston ring	14. Sealing rings	18. "O" ring
4. Clutch hub	10. Clutch piston	15. Brake pressure plate	19. Brake plunger
5. Spring retainer nut	11. Thrust washer	16. Brake disc	20. Spacer
6. Return spring			21. Retainer screw

Fig. MF211 — Exploded view of pto clutch control valve and associated parts used on MF Super 90, MF Super 90WR and some MF85 and MF88. Other models are similar, except modulator valve (8) is not used.

1. Brake locating pin	12. Feed tube
2. Support assembly	13. Gear fork
3. Snap ring	14. Shifter finger
4. Modulating valve	15. Set screw
spring	16. Adjusting screw
5. Washer	17. Spacer
6. Gasket	18. Stop pin
7. Valve body	19. Detent assembly
8. Modulating valve ball	20. "O" ring
9. Control valve	21. Selector lever
10. Return spring	22. Control valve
11. Retaining washer	locking pin

184. OVERHAUL. Inlet passage sealing rings (14—Fig. MF209) may be removed from rear hub by unlocking with a pointed tool and expanding with snap ring pliers.

Remove the large snap ring (1) from front of clutch housing and remove retainer plate (2), brass thrust washer (3) and clutch hub (4). Remove the clutch friction discs (8) and plates (7).

Place the clutch housing assembly in a suitable press and remove retainer nut (5) and spring (6), applying sufficient pressure to keep parts from flying when threads are disengaged. Remove piston (10) by bumping housing to jar piston loose. Steel sealing rings (9 and 12) on housing inner hub and piston, are of the interlocking type. Renew piston, rings, and/or housing if they are scored, worn or otherwise damaged.

The six screws (21) which retain the pto clutch brake assembly are installed with LOCTITE and staked in place. To disassemble the brake assembly, remove the staking and heat the screw heads to a temperature of 480° F. Screws must be discarded when removed. Clutch brake parts are renewable individually if badly worn, scored or otherwise damaged. When reassembling, use new "O" rings (18) and new screws (21). Apply two drops of LOCTITE to each of the screws (21) and use the valve support (2—Fig. MF211) as a pilot to properly position brake disc (16—Fig. MF209) while screws (21) are being tightened. Restake the screws (21) after brake is assembled.

NOTE: Proper installation of the retaining screws is of utmost importance. The tapped screw holes in clutch housing enter pressure area behind clutch piston. If a retaining screw works out, loss of pressure and clutch failure will result.

Install the piston (10), spring (6) and retainer nut (5). Use a suitable press while installing the nut. Tighten spring retainer nut to a torque of 75 ft.-lbs. and stake in four places. When installing the externally splined clutch plates (7), index the tabs in housing splines which do not contain oil relief holes. Install the assembled clutch as outlined in paragraph 183.

PTO CONTROL VALVE
All Models

185. OPERATION. The pto control valve is a closed center valve. A fluid pressure of 150-250 psi (225 psi on multipower models) is maintained at the valve inlet when engine is running. When the selector lever is moved to "Engine PTO" position, this pressure is directed through passages in

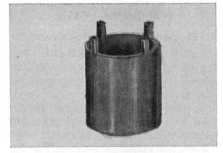

Fig. MF210 — Special four-point socket for adjusting drive pinion bearing preload with a torque wrench without removing differential assembly. See text.

pto support housing (2—Fig. MF211) to the pto clutch assembly. Fluid pressure thus engages the clutch and locks the pto output shaft to the hydraulic lift pump camshaft to complete the power circuit. When the selector lever is moved to the "Neutral" position, pressure from the pump is blocked at valve (9) and the area behind clutch piston is opened to the sump. The clutch piston return spring, releases the clutch and at the same time applies the clutch brake, stopping the rotation of the pto output shaft. Moving the selector lever to "Ground PTO" position mechanically connects the sliding pto drive gear on main drive bevel pinion with the gear teeth on pto housing, turning the pto output shaft through the main drive gears in direct relation to tractor ground speed. The clutch brake will be applied during ground pto operation, but pressure is not enough to cause damage to the parts.

186. R&R AND OVERHAUL. The pto control valve can be removed after first removing the pto clutch as outlined in paragraph 183 and withdrawing rear feed line (12—Fig. MF-211). Remove the locating screw (22) from outside center housing directly below the selector lever and slide the valve forward out of the housing.

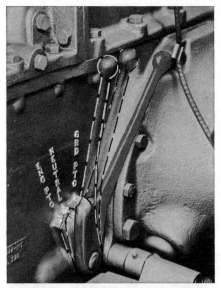

Fig. MF212 — View of pto selector lever showing shift positions. Pto may be shifted into engine drive independently of engine clutch.

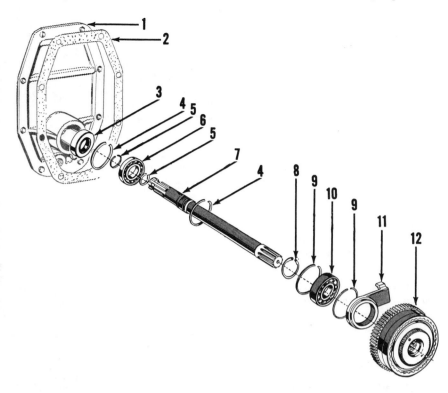

Fig. MF213—Exploded view of pto output shaft and associated parts.

1. Rear end plate	4. Snap ring	7. Pto output shaft	10. Front bearing
2. Gasket	5. Snap ring	8. Snap ring	11. Pto support
3. Oil seal	6. Rear bearing	9. Snap ring	12. Pto clutch assy

To remove the selector lever, remove set screw (15) from pto shifter finger (14) and slide the lever out of finger and center housing. Detent ball and spring (19), "O" ring (20) and sleeve (17) may be serviced at this time.

All MF Super 90 and Super 90 WR models and some MF85 and MF88 models use a control valve containing a modulator valve, such as that shown in Fig. MF211. Other MF85 and MF88 models are equipped with a valve similar in appearance and operation, but without the modulating feature which permits feathering of the pto clutch engagement. Component parts of both valves are serviced individually. The modulating type valve can be identified by the adjusting screw (16) on valve contact end of shifter finger (14). When installing the modulating type valve, shift the selector lever to "Neutral" position, and turn the adjusting screw (16) to provide 0.045-0.055 clearance between adjusting screw (16) and valve (9). NOTE: If this clearance cannot be obtained, remove shifter finger (14) and grind lower arm of finger until specified clearance is obtainable.

PTO OUTPUT SHAFT AND SEAL.
All Models

187. To remove the pto output shaft (7—Fig. MF213), unbolt and remove center housing rear plate (1) from tractor. Remove front snap ring (4) from end plate and remove shaft and bearing as an assembly. Remove rear snap ring (4) and seal (3). Install seal with lip toward front (inside) of plate.

HYDRAULIC SYSTEM

All models are equipped with a scotch yoke type piston pump which is mounted in the rear axle center housing and driven by the power take-off shaft. On models equipped with three-point hitch, the hydraulic pump contains a control valve which meters the operating fluid entering the pump. The pump outlet is channeled directly to the lift cylinder with a connection to lowering side of control valve. In models equipped with three-point lift, therefore, fluid is pumped only during the raising cycle of the system.

On models without three-point lift, the control valve is omitted, the pump inlet being open to the fluid supply in the sump. Except for the control valve, the pumps on all models are identical and parts are interchangeable.

The remote control valve and cylinder are standard equipment on models MF88 and MF Super 90 WR, and are available as accessory equipment on other models.

Most hydraulic system trouble is caused by dirt or gum deposits. The dirt may enter from the outside or it may show up as the result of wear or partial failure of some part of the system. The presence of gummy deposits however, usually results from inadequate fluids or from failure to drain and renew the fluid at the recommended intervals. These principles should be kept in mind when shooting trouble on the system and also when performing repair work on the pump, valves and cylinders.

When disassembling the pump and valves unit, it is good practice generally to not remove any parts which can be thoroughly inspected while they are installed. Internal parts of the pump, valves and cylinder, when removed, should be handled with the same care as would be accorded the parts of a diesel pump or injector unit, and should be soaked or cleaned with an approved solvent to remove gum deposits. Unless you practice good housekeeping in your shop, do not undertake the repair of hydraulic equipment.

The transmission lubricant is the operating fluid for the hydraulic system. Automatic Transmission Fluid, Type "A" is recommended for year around use.

ADJUSTMENTS

Models With Three-Point Lift

188. Whenever malfunctions of the hydraulic system exist and the cause is not readily apparent, it is recommended that a check of the adjustments be made in the following sequence:

Detach the implement from tractor and place tractor in a relatively level position. The area should be as clean as possible and free from blowing dirt and trash. Examine control spring in control frame to make certain that front spring seat (5—Fig. MF214) contacts rear edge of control frame front support plates and rear spring seat contacts front edge of control spring adjusting nut (3). If they do not, readjust control spring as follows: Remove swinging drawbar, then remove the rear four bolts in swinging drawbar pivot bracket and loosen the remainder of the bolts. Slide rear support bracket with rear drawbar bracket attached out to the rear, remove the remainder of the bolts and remove drawbar pivot bracket, control frame roller and upper plate. Loosen Allen head set screw (6) in control frame and control spring support. Back off control spring adjusting nut (3) and control spring adjusting screw (1) approximately one-half turn. Place a prybar through one control frame rocker and pry control frame back to make certain front spring seat is seated in frame. Tighten control spring adjusting screw until all end play has just been eliminated, then tighten one-eighth turn further and lock with set

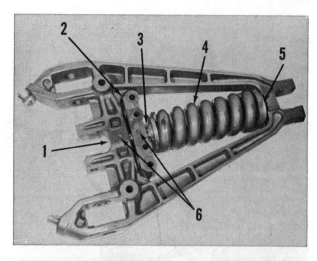

Fig. MF214 — Hydraulic system draft control frame viewed from underneath indicating points of control spring adjustment.

1. Adjusting screw
2. Control spring support
3. Adjusting nut
4. Control spring
5. Front spring seat
6. Set screws

Fig. MF215 — System hydraulic controls.

1. Position control lever
2. Draft control lever
3. Implement lowering speed dial
4. Ram cylinder shut off valve
5. Draft response dial

screw. Release pressure on control frame and check to see that front spring seat contacts rear edge of front support plate. Screw out control spring adjusting nut until forward edge contacts rear spring seat, then screw nut out an additional one-eighth turn and lock with set screw. Reinstall control frame roller, plates and support bracket.

Check oil level to see that it is within operating range and remove center housing side cover plate with parking brake attached. Start engine and adjust control valve inner linkage as follows: Move draft (2—Fig. MF215) and implement position (1) control levers to their lower-most position; loosen the locking screw on the implement lowering dial plate (3) and rotate the plate and pointer fully in a counter-clockwise direction. Working through the side plate opening in the housing, loosen the lock nut (1—Fig. MF216) on the clevis end of the valve control rod (2) and thread the rod into the

Fig. MF216 — Control valve and draft response dashpot adjusting points for adjustments made through rear axle center housing access plate.

1. Control valve locknut
2. Control valve rod
3. Dashpot needle valve set screw
4. Dashpot piston locking screw

clevis until the lower links begin to rise. When the links are in approximately a horizontal position, back the rod out until links are stationary, and tighten locking nut.

Adjust draft control linkage as follows: Move the implement lowering pointer (3—Fig. MF215) clockwise to fast position. Move draft control lever (2) up until it is centered between the punch marks on the quadrant. Loosen the lock nut (4—Fig. MF217) on the external draft control rod and thread adjusting nut down against the tube until the lower links begin to rise. When links reach horizontal position, back nut off until links are held stationary and lock in place.

Move draft control lever to lower-most position and position control lever (1—Fig. MF215) to the transport stop at the top of the quadrant. When system has returned to neutral and links are in transport position, place scribe marks across lift arm hub and top cover, and move position control lever rearward past the stop until relief valve blows. The difference in the two scribe marks should be $\frac{1}{16}$-$\frac{3}{16}$ inches. If minor adjustment of transport stop will not obtain this measurement, other adjustments are incorrect or internal linkage is damaged or out of adjustment.

Attach a heavy implement or weight to lower links and adjust implement

Fig. MF217 — Draft control external linkage showing main components.

1. Actuating pin
2. Draft control bellcrank
3. Draft control link
4. Adjusting nut
5. Draft control upper crank

lowering dial as follows: Move adjusting pointer to fully counter-clockwise position, move draft control lever to lower-most part of quadrant, and raise weight to transport position with position control lever. Move position control lever to lower position and rotate implement lowering pointer slowly clockwise until lowering speed of weight is such that 5-10 seconds are required for lower links to move from transport to fully lowered position. When this position has been found, move dial stop against pointer and lock in position.

Adjust draft response as follows: Loosen locking screw on draft response dial (5—Fig. MF215), move dial to fully clockwise position and lock in place. Move draft response pointer to fully counter-clockwise (slow) position. Loosen locking screw (3—Fig. MF216) on dash pot needle valve and relock. Check to see that rod is free and that no end play exists. Loosen locking screw (4) on dash pot spring collar and retighten so that dash pot piston is in upper-most position and collar is resting on guide at top of rod. With draft response pointer in slow position, apply pressure down on dash pot rod at collar. Rod should move slowly down as dash pot oil is exhausted.

TROUBLE-SHOOTING
Models With Three-Point Lift

189. Items to check in eliminating some of the problems encountered are as follows:

IMPLEMENT WILL NOT RAISE.

a. Leak in system. Remove inspection plate and check for leaks at ram cylinder, stand pipe, control valve, relief valve and pump side chambers.

b. Broken or damaged parts. Check for oil turbulence to indicate pump pto shaft is turning. Damaged pump parts are usually indicated by noise.

c. Plugged intake strainer. Remove and clean strainer.

d. Frozen ram cylinder or shut-off valve closed. Indicated by relief valve blowing.

IMPLEMENT LIFTS BUT WILL NOT LOWER. Probable cause is damaged control valve spring or sticking control valve.

JERKY OR UNEVEN LIFT. One or more faulty side chamber valves. Remove and disassemble pump.

RELIEF VALVE BLOWS WHEN LEVER IS RAISED TO TRANSPORT POSITION.

a. Check chains installed wrong.

b. Shut-off valve closed.

c. Broken or bent stop pin or internal parts.

ERRATIC CONTROL.

a. Damaged control linkage.

b. Implement or top link improperly adjusted. See operator's manual.

c. Sticking control valve.

SYSTEM OPERATING PRESSURE
All Models

190. The recommended operating pressure is 2500 psi. To check pressure on models equipped with three-point lift, remove hex plug (P—Fig. MF215) in the right front corner of the front lift cover and install gage. On Western models remove oil line from rear port of control valve. (Note: Threads used are SAE straight tube threads with "O" ring seals. Do not use pipe thread adapters.) Start engine, move the draft control lever to the top of the quadrant or hold valve lever back and observe gage reading.

If the operating pressure is too high, the relief valve assembly should be renewed. If the operating pressure is too low, remove the inspection plate from the side of the center housing, operate the system and check for visible leaks. If there are no visible leaks and there is turbulence around the relief valve (24—Fig. MF218) renew the valve assembly. If the installation of a new valve does not bring the pressure up to the specified limits, a faulty hydraulic pump is indicated.

PUMP
Models With Three-Point Lift

191. **REMOVE AND REINSTALL.** To remove the hydraulic pump, first drain the fluid reservoir at transmission, center housing and differential drain plugs and disconnect and remove the tail-light wire. Support the tractor underneath the transmission housing, place a rolling floor jack underneath

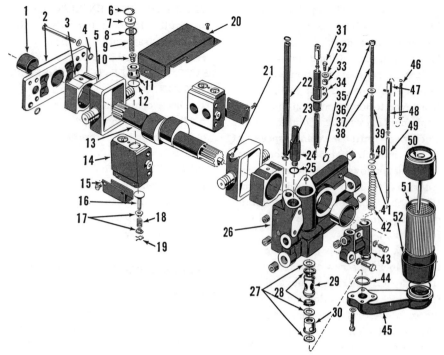

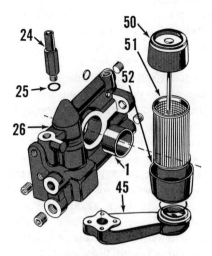

Fig. MF219—Front body assembly used on Western model hydraulic pump. See Fig. MF218 for legend.

Fig. MF218—Exploded view of the hydraulic lift system pump used on models with three-point lift. Pump used on models without three-point lift is identical except for pump front body which does not contain control valve or dashpot assembly. See Fig. MF219 for front body assembly without valves.

1. Camshaft bushing	19. Hair pin	36. Collar
2. Rear body	20. Pump shield	37. Spring
3. Cam block	21. Oscillator pin	38. Washer
4. "O" ring	22. Stand pipe	39. Dashpot plunger
5. Piston	23. "O" ring	40. Snap ring
6. Snap ring	24. Relief valve	41. Dashpot piston
7. Valve plug	25. "O" ring	42. Piston spring
8. "O" ring	26. Front body	43. Dashpot body
9. Spring	27. Sealing washer	44. Washer
10. Outlet valve	28. "O" ring	45. Strainer housing
11. Outlet valve seat	29. Body	46. Push rod
12. "O" ring	30. Spacer	47. Coupler
13. Camshaft	31. Retaining screw	48. Spring
14. Valve chamber	32. Washer	49. Needle valve
15. Inlet valve guard	33. Spacer	50. Strainer cover
16. Inlet valve	34. Control valve	51. Element
17. Spring seat	assembly	52. Strainer bowl
18. Spring	35. "O" ring	

the rear axle center housing, remove the bolts securing center housing to the transmission housing and separate the tractor. Remove the draft response needle valve (49—Fig. MF218) by loosening the coupler set screw and telescoping the two halves together. Remove the dash pot rod (39) by loosening the set screw in the collar (36) and pulling rod up out of the dash pot; then down out of the retaining bracket. Disconnect the control valve by unscrewing the link out of the clevis. Remove transfer plate from the hydraulic lift cover and withdraw the stand pipe (22) which is located directly under the transfer plate. Remove the hydraulic pump support pins from each side of the center housing and slide the pump forward out of the housing.

192. **OVERHAUL.** Remove pump filter (51) and sheet metal pump cover (20). Remove control valve assembly (34) by removing oscillating arm cap screw (31) and spacer (33) and lifting out valve assembly. Remove control valve body (29) and sealing washers (27) by unbolting and removing intake filter inlet (45) from pump body. Remove the two snap rings (6) from each of the two pump side chambers (14) and remove the four plugs (7), springs (9) and outlet valves (10). Turn pump over and remove inlet valve guards (15). Remove inlet valve retaining hairpins (19), spring seats (17) and spring (18). Make certain that each piston in turn is withdrawn from side chamber completely on its intake stroke, then push the intake valve (16) through its bore forcing the discharge

valve seat (11) and intake valve (16) out the top of the pump. To remove the control valve from the control rod and oscillator, remove the machine screw and cone-shaped spring from bottom end of valve. In reassembly, tighten machine screw until it bottoms then back-off two turns. Procedure for further disassembly of the pump is evident.

After the pump is disassembled, thoroughly clean and examine all parts for damage or excessive wear. Renew all "O" ring seals and any other parts which are questionable.

After pump is reassembled and reinstalled in tractor, adjust the system as outlined in paragraph 188.

Models Without Three-Point Lift

193. **REMOVE AND REINSTALL.** Removal procedure is identical to that given in paragraph 191 with the following exceptions: No control valve, draft response needle valve or dash pot rod are used so procedure for removing them can be disregarded. External two spool control valve must be removed to withdraw the stand-pipe.

With these exceptions proceed as outlined in paragraph 191.

194. **OVERHAUL.** The hydraulic pump used on Western models is identical to that used on the other models with the exception that no control valve or dash pot is used. Overhaul procedure given in paragraph 192 can be used, disregarding instructions pertaining to control valve, control valve body and adjustment.

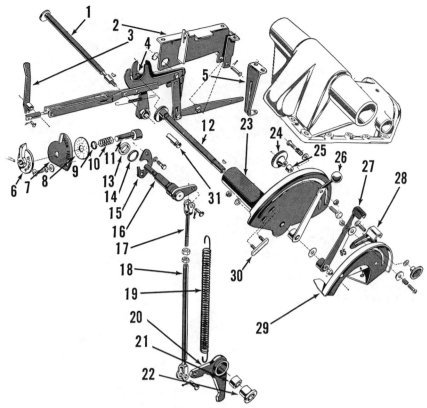

Fig. MF220—Exploded view of hydraulic control linkage.

1. Draft control shaft	9. "O" ring	17. Adjusting yoke	25. Spacer
2. Bracket	10. Spring	18. Draft control yoke	26. Position control lever
3. Follower arm	11. Implement lowering cam	19. Actuating spring	27. Draft control lever
4. Control linkage	12. Position control shaft	20. Draft control bell crank	28. Draft control slide
5. Lever guide	13. Bearing	21. Bearing	29. Draft control quadrant
6. Implement lowering speed pointer	14. Oil seal	22. Trunnion	30. Transport stop
7. Implement lowering dial	15. Retaining plate	23. Position control quadrant	31. Anchor pin
8. Friction disc	16. Draft control upper crank	24. Adjusting knob	

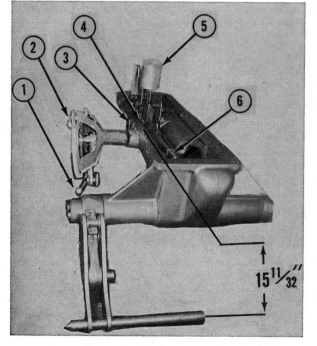

Fig. MF221 — Lift cover adjustment. Position rockshaft as indicated and lock in place.

1. Position control lever against transport stop
2. Draft control lever at opposite end of quadrant
3. Implement lowering dial and pointer fully counter-clockwise
4. 0.002 clearance between implement lowering cam and link arm
5. Six pound weight applying pressure to arm
6. Point of adjustment

LIFT COVER, WORK CYLINDER AND ROCKSHAFT

Models With Three-Point Lift

195. **REMOVE AND REINSTALL.** To remove lift cover, remove seat and disconnect upper lift links from lift arms. Remove transfer plate (8—Fig. MF222) from front lift cover and withdraw standpipe. Remove plate (15—Fig. MF220) retaining draft control upper crank (16) to center housing and remove crank by rotating it counter-clockwise while sliding it out, to clear connecting link (17) from tube (18). Remove center housing inspection side plate with parking brake attached and disconnect control valve linkage by screwing control rod out of yoke. Unbolt and remove front and rear cover as an assembly by using a chain fall. Approximate balance of lift cover can be obtained by passing the

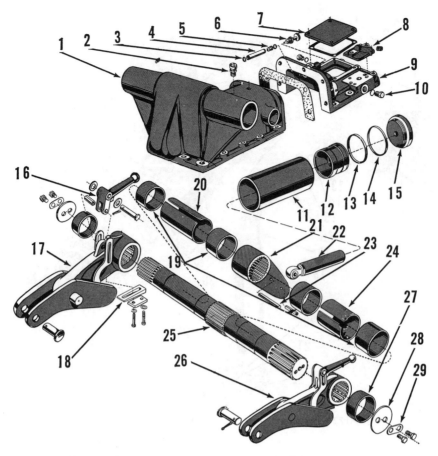

Fig. MF222—Rockshaft, work cylinder and cover assembly.

1. Rear cover
2. Breather
3. "O" ring
4. Connector pipe
5. Valve seat
6. Shut-off valve
7. Inspection cover
8. Transfer plate
9. Front cover
10. Pressure plug
11. Cylinder sleeve
12. Piston
13. Piston seal
14. "O" ring
15. Cylinder end
16. Locking handle
17. Lift arm
18. Adjusting wedge
19. Shaft bearing
20. Spacer
21. Ram arm
22. Piston rod
23. Follower pin
24. Spacer
25. Rockshaft
26. Lift arm
    assembly
27. Spacer
28. Retainer plate
29. Retainer clip

two ends of a chain in front of each rockshaft boss and fastening chain at rear of cover.

196. OVERHAUL. Shut-off valve (6—Fig. MF222) may be removed by unscrewing packing nut. Valve is renewed as an assembly. Valve seat (5) is tapped with ¼-20 screw threads for removal. Seat is sealed with an "O" ring. To remove control linkage, disconnect follower arm clevis pin at rear of linkage and unbolt and remove bracket from cover. To remove work cylinder, unbolt and remove front section of cover (9), remove Allen screw at front of lift cylinder, and with a suitable spanner, unscrew and remove cylinder end (15). Piston (12) may now be removed from cylinder (11) by bumping it forward out of cylinder. If cylinder is to be removed and reinstalled, mark cylinder so same end will be installed to the front. Install piston seal (13) with flared end to head of piston. When reinstalling front section

of cover, make certain that machined gasket surface of front and rear covers are in the same plane by measuring with a straight edge.

197. ADJUSTMENTS. Adjustment of the position control linkage can only be accomplished with the lift cover removed. To make the adjustment proceed as follows:

a. Invert lift cover on bench and support with blocks so that rockshaft and quadrant levers are free to move.

b. Loosen the locking screw on the implement lowering speed dial (7—Fig. MF220) and rotate dial and pointer (6) counter-clockwise as far as possible.

c. Locate transport stop (30) at top of position control quadrant in center of slots and lock in place.

d. Move position control lever (1—Fig. MF221) against transport stop at top of quadrant and draft control lever (2) in opposite direction to bottom of quadrant.

e. Unstake cap screws in end of rockshaft and tighten until sufficient preload is applied to hold rockshaft in position. Position lift links 15$\frac{11}{32}$ inches below the surface of the lift cover by laying a straight edge on the gasket surface and measuring to center of lift link hole (See Fig. MF221). Note: Lift link used for this measurement must not be bent.

f. Duplicate control valve spring tension by applying approximately a 6-lb. pressure to end of valve lever (5). At this time, position control follower arm (3—Fig. MF220) should be in contact with the follower pin (23—Fig. MF222) projecting from the ram arm, and position control crank should clear implement lowering cam (11—Fig. MF220) by 0.002-0.004. Adjustment is made by screwing follower rod in or out of clevis at rear end.

g. Readjust cap screws in end of rockshaft until all end play is removed and lift arms will fall of their own weight.

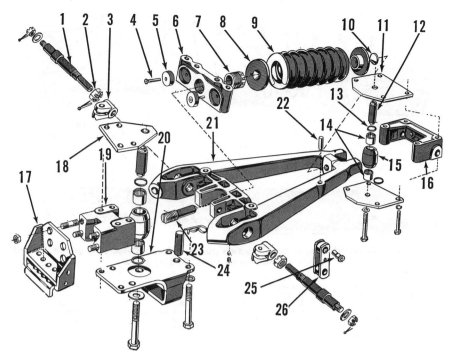

Fig. MF223—Exploded view of draft control frame and hangers used.

1. Lower link
   support shaft
	7. Adjusting nut	14. Needle bearing	19. Rear support
			    bracket

2. Shaft lock nut	8. Control spring
	   seat
		15. Control frame
		    roller
			20. Rear support
			    plate

3. Stabilizer
   connector
	9. Control spring	16. Front support
		    bracket
			21. Control frame

4. Bumper retainer
   pin
	10. Front abutment	17. Rear drawbar
		    support
			22. Draft control pin

5. Support bumper	11. Front support
	    plate
		18. Upper support
		    plate
			23. Adjusting screw

6. Control spring support	12. Roller shaft		24. Drawbar pin
	13. Oil seal		25. Pivot pin
			26. Rocker assembly

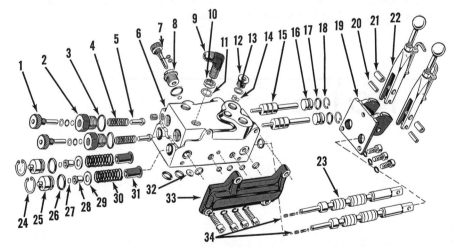

Fig. MF224—Exploded view of two-spool control valve available as an accessory on the models with three-point lift and used as standard equipment on Western models. A third spool attachment of identical design is available to replace the side cover.

1. Switch valve	10. Lock nut	19. Front cover	28. Collar
2. Valve plug	11. "O" ring	20. Roll pin	29. Washer
3. "O" ring	12. Body plug	21. Roll pin	30. Centering spring
4. Spring	13. "O" ring	22. Lever assembly	31. Spring sleeve
5. Poppet valve	14. Restrictor	23. Valve spool	32. Restrictor
6. Valve body	15. Shuttle valve	24. Snap ring	33. Side cover
7. Bypass valve	16. Plug	25. Plug	34. Detent release valve
8. Valve plug	17. "O" ring	26. "O" ring	
9. Elbow	18. Snap ring	27. Snap ring	

CONTROL FRAME

Models With Three-Point Lift

198. **R&R AND OVERHAUL.** To remove the hydraulic system control frame (21—Fig. MF223), first remove the swinging drawbar, then disconnect and remove the tractor lower links. Disconnect the rocker arms from the rear axle housings by removing the cap screws and using a ¾-N.C. bolt, pipe bushing and heavy washer as a puller for the upper pivot pins (25). Pivot pins are threaded for removal. Place a floor jack underneath the frame from the rear and support the frame at the rear of the control spring. Remove the rear drawbar support (19) by removing the cap screws securing it to the center housing and sliding it to the rear out of assembly. Remove remainder of cap screws in rear support plate (20) and remove plate, rear roller (15) and upper plate (18). Remove the two cap screws securing control spring rear support (6) to center housing, lower jack until control frame is free and balance frame on jack while rolling out from under rear of tractor. To remove rocker assembly (26) and lower link support shaft (1), remove cotter pin from lock nut (2) and use lock nut to force shaft from frame by screwing castellated portion of nut against stabilizer connector (3).

Control spring (9), spring seats (8) and rear support (6) can be removed by loosening set screw which locks control spring adjusting screw (23) to frame, and backing out pin. In reassembly, tighten adjusting screw (23) until all end play has been removed from spring then tighten one-eighth turn to provide recommended preload. Nylon plugs to protect the threads are used underneath the locking screws of the control spring adjusting nut and adjusting screw.

When reinstalling control frame and rockers, leave rocker pivot pins (25) loose until end play has been removed from control spring as outlined in paragraph 188.

EXTERNAL CONTROL VALVE

All Models So Equipped

199. The dual spool control valve is designed to operate either single or

double acting remote cylinders. Valve is an open center type self centering spool valve. Valve body (6—Fig. MF-224), valve spools (23) and shuttle valves (15) are matched parts and available as an assembly only. All other parts are available individually. Valve ports are machined for ⅜-tube, straight thread fittings with an "O" ring seal.

Disassembly and service procedures are evident and should present no problem. Switch valve knobs (1), located at the back of shuttle valve body must be screwed out to operate single acting cylinders and screwed in to operate double acting cylinders. When both a single and double acting cylinder is connected to the same valve, the single acting cylinder must be connected to the left of the double acting cylinder. The by-pass valve (7) on left side of housing is not used, and must be screwed in at all times. A return detent is made available in the control levers to retain the valve in the raised position until the piston reaches the end of its stroke. To adjust, loosen locking nut on lever and turn handle in to increase release detent pressure or out to decrease pressure.

Valve body is ported to receive ⅜-inch SAE straight tube threads with "O" ring seals. Do not attempt to use NPT pipe fittings in the ports.

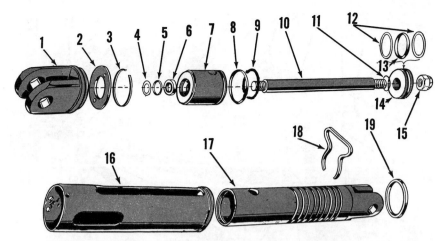

Fig. MF225—Exploded view of double acting power cylinder. Cylinder is ported for SAE straight tubing threads with "O" ring seals.

1. Piston rod yoke	6. Oil seal
2. Bearing cover plate	7. Bearing
3. Snap ring	8. Leather back-up ring
4. Leather back-up ring	9. "O" ring
5. Shaft "O" ring	10. Piston rod

11. "O" ring	16. Cylinder stop sleeve
12. Leather back-up ring	17. Cylinder
13. "O" ring	18. Cylinder stop lock
14. Piston	19. Stop ring
15. Rod nut	

DOUBLE ACTING CYLINDER

All Models So Equipped

200. The double acting cylinder is an ASAE 2½-inch cylinder with an 8-inch stroke. Cylinder is ported to receive SAE straight tube fittings with an "O" ring seal.

To disassemble the cylinder, first remove the hoses and fittings and slip off the cylinder stop (18—Fig. MF225) and end ring (19). Unbolt and remove the piston rod stop sleeve (16), extend the cylinder and remove the bearing cover plate (2) from the bearing and slide plate up piston rod out of the way. Bump bearing (7) into cylinder barrel to free snap ring (3) and remove snap ring. Piston rod (10), bearing and piston (14) can then be removed from barrel as a unit. Procedure and need for further disassembly is evident. Renew "O" rings and seals whenever cylinder is disassembled. Inspect and renew other parts as necessary.

NOTES

MASSEY-FERGUSON

Models ■ MF1080   ■ MF1085

Previously contained in I&T Shop Manual No. MF-32

SHOP MANUAL

MASSEY-FERGUSON

MODELS MF1080 - MF1085

Tractor serial number stamped on instrument panel name plate. Engine serial number stamped on right hand side of engine block.

INDEX (By Starting Paragraph)

CONDENSED SERVICE DATA

GENERAL	MF1080	MF1085
Torque Recommendations	See End of Shop Manual	
Engine Make	Perkins	Perkins
Engine Model	A4.318	A4.318
Number of Cylinders	4	4
Bore, Inches (mm)	4.5(114.3)	4.5(114.3)
Stroke, Inches (mm)	5.0(127)	5.0(127)
Displacement, Cu. In.	318	318
Compression Ratio	17.5:1	17.5:1
Main Bearings, No. of	5	5
Cylinder Sleeves	Dry	Dry
Forward Speeds	6 or 12	8 or 12
Reverse Speeds	2 or 4	2 or 4
TUNE-UP		
Firing Order	1-3-4-2	1-3-4-2
Valve Tappet Gap—		
Cold—Inch (mm)	0.012 (0.30)	0.012 (0.30)
Holt—Inch (mm)	0.010 (0.25)	0.010 (0.25)
Valve Face Angle	45°	45°
Valve Seat Angle	45°	45°
Injection Timing BTDC	28° BTDC	20° STATIC
Timing Mark Location	Crankshaft Pulley*	

*Some Early 1080 Models Used Flywheel Timing Marks.

CONDENSED SERVICE DATA CONT.

TUNE-UP (Cont.'d)	MF1080	MF1085
Battery—		
Volts	12	12
Capacity Amp/Hr.	95	95
Ground Polarity	Negative	Negative
Injectors—		
Opening Pressure, psi	2575	2575
Atmospheres	175.2	175.2
Kg/cm^2	181	181
Spray Hole Diameter	0.0118-0.0126 in.	0.0118-0.0126 in.
	0.30-0.32 mm	0.30-0.32 mm
Governed Speeds—		
Engine		
Low Idle rpm	800-850	800-855
High Idle rpm	2170-2200	2170-2200
Loaded rpm	2000	2000
1000 rpm PTO		
High Idle	1080	1080
Loaded	1000	1000
540 rpm PTO		
High Idle	690	690
Loaded	627	627
Horsepower** at PTO Shaft	81.23	81.58
Hydraulic System—		
Maximum Pressure, psi	2900-3100	2900-3100
Atmospheres	197.3-210.9	197.3-210.9
Kg/cm^2	203.9-218.0	203.9-218.0

**According to Nebraska Test.

SIZES-CAPACITIES-CLEARANCES

	MF1080	MF1085
Crankshaft Journal Diameter		
Inches (mm)	2.999(76.175)	2.999(76.175)
Crankpin Diameter, Inches (mm)	2.749(69.825)	2.749(69.825)
Camshaft Journal Diameter		
Front, Inches (mm)	2.057 (52.248)	2.057(52.248)
Center, Inches (mm)	1.966(49.936)	1.966(49.936)
Rear, Inches (mm)	1.905(48.387)	1.905(48.387)
Crankshaft Bearing Clearance		
Main Bearings, Inch	0.0025-0.0045	0.0025-0.0045
(mm)	0.064-0.114	0.064-0.114
Crankpin, Inch	0.002-0.0042	0.002-0.0042
mm	0.051-0.107	0.051-0.107
Crankshaft End Play, Inch	0.005-0.016	0.005-0.016
mm	0.127-0.406	0.127-0.406
Piston to Cylinder Clearance, Inch	0.007-0.011	0.007-0.011
mm	0.178-0.278	0.178-0.278
Camshaft Bearing Clearance		
Front & Center, Inch	0.002-0.005	0.002-0.005
mm	0.051-0.127	0.051-0.127
Rear Bearing, Inch	0.004-0.007	0.004-0.007
mm	0.102-0.178	0.102-0.178
Cooling System Capacity, Quarts	16.5	16.5
Liters	15.6	15.6
Crankcase Capacity, Quarts +	10	10
Liters +	9.5	9.5
Transmission, Differential and Hydraulic Lift		
Capacity, Gallons	8.5	8.5
Liters	32.2	32.2
Planetary Final Drive Capacity (Each),		
Quarts	1¾	3
Liters	1.65	2.8
Power Steering System Capacity,		
Quarts	3	3
Liters	2.8	2.8

+ Add 1 quart (0.95 liter) if filter is changed.

FRONT SYSTEM

All models are equipped with either a standard or adjustable, wide front axle and hydrostatic steering system. Refer to Figs. 1 and 2 for exploded views of axle assemblies.

AXLE ASSEMBLY

All Models

1. Refer to Fig. 1 for an exploded view of standard axle unit used on Western models and to Fig. 2 for adjustable axle unit. Pivot pins are a part of the welded axle or center section and are not available separately. The pre-sized bushings are renewable in axle brackets. Hole in bushing should be at top when bushing is installed.

Spindle bushings must be align reamed to 1.5 inch after installation. Clean out lubricant pocket between bearing area after reaming.

TIE RODS AND TOE-IN

All Models

2. The automotive type tie rods should be adjusted to provide 0-1/4 inch toe-in when wheels are in straight-ahead position. Adjust both tie rods an equal amount to obtain the setting.

POWER STEERING SYSTEM

LUBRICATION AND BLEEDING

All Models

3. The hydrostatic steering hand pump and steering cylinder assembly (steering motor) are lubricated by the operating fluid. Massey Ferguson M-1129A Transmission & Hydraulic oil is

recommended. Type A, Suffix A Automatic Transmission Fluid can be used; do not use other automatic transmission fluids.

The system is self-bleeding, but steering should be cycled and reservoir refilled as often as necessary until level stops dropping. Proceed as follows:

Stop engine and remove filler plug (See Fig. 4). If steering motor has been disassembled or drained, fill reservoir then start and idle engine, adding fluid as level lowers until system is stabilized. Install filler plug loosely, cycle the system then recheck, adding fluid as necessary to maintain full reservoir. Tighten plug securely when fluid level ceases to drop.

ADJUSTMENT

All Models

4. **OPERATING PRESSURE.** The power steering pump pressure relief valve should be set to open at 1500-1600 psi. The ball type pressure relief valve is located on bottom side of pump rear body as shown in Fig. 5. Turn screw in or out, using an Allen wrench, until specified pressure is obtained. If correct relief pressure cannot be obtained, or if pressure drops rapidly, overhaul power steering pump as outlined in paragraph 12 or hydrostatic hand pump as in paragraph 9.

TROUBLE SHOOTING

All Models

5. Internal fluid leakage is objection-

able only to the extent that performance is affected. External leakage will be obvious, the only possible place where external leakage will not be visible would be the pump shaft seal which is ported to pump intake side.

Normal internal leakage of steering system will permit approximately 2 revolutions per minute of the steering wheel with moderate steering pressure applied an wheels fully turned to right or left.

NOTE: DO NOT attempt to operate pump at relief pressure for one minute; the figure is given as a guide, not as a test.

If leakage is faster when turning in one direction than the other, the trouble is probably in steering cylinder; if slippage is approximately equal, the hand pump may be at fault.

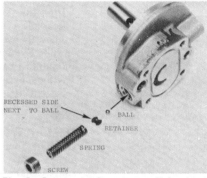

Fig. 5—Power steering pump relief valve is located in pump rear body as shown.

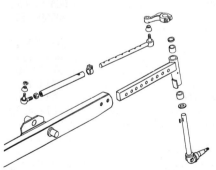

Fig. 2—Exploded view of adjustable axle used on Row Crop Models. Front support and associated parts are as shown in Fig. 1.

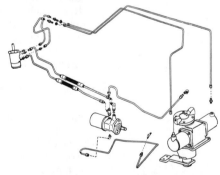

Fig. 6—Schematic view of hydrostatic steering system showing lines and components.

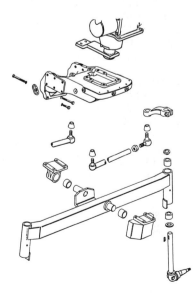

Fig. 1—Exploded view of front axle, front support and associated parts used on Western Models.

Fig. 4—Left hand view of engine showing power steering pump filler plug.

STEERING CYLINDER (MOTOR)

Two types of steering motor units have been used. The type shown in Fig. 8 is equipped with a rack guide (RG) which is bolted to front of housing as shown. Steering motor without rack guide is shown in Fig. 7. Refer to the appropriate following paragraphs depending upon type of steering motor.

Models Without Rack Guide

Because of the welded center steering arm (14—Fig. 11) the steering motor cannot be removed without some disassembly; and most overhaul can be accomplished without removal. Paragraph

Fig. 7—Grille screens removed showing installed power steering motor. Unit shown is type without rack guide.

Fig. 8—Partially installed view of steering motor which has rack guide (RG). Refer to Fig. 13 for exploded view of this type.

Fig. 9—Right cylinder must be removed as shown, before unit can be removed from tractor.

6 outlines the procedure for removal and installation of the unit for other work or for renewal; if overhaul is planned, proceed as outlined in paragraph 6A and disassemble the unit only as far as required to accomplish the necessary repairs.

6. REMOVE AND REINSTALL. To remove the steering cylinder assembly (motor) for other work or for renewal, first turn wheels to a straight ahead position and shut off engine. Remove radiator grille screen as shown in Fig. 7 and disconnect cylinder lines, lube line and bleed line from motor unit.

Disconnect tie rods from center steering arm. Unbolt and remove right cylinder assembly as shown in Fig. 9.

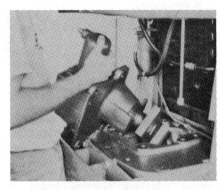

Fig. 10—Steering motor without rack guide is removed as shown. The right side cylinder must be removed first.

Reaching underneath the tractor, pull rear end of center steering arm to the right until it contacts limit stop; remove the four retaining cap screws and lift off steering motor as shown in Fig. 10.

Install by reversing the removal procedure. Tighten the cap screws retaining cylinder sleeve to a torque of 50-55 ft.-lbs. Screws attaching steering motor to front housing should be tightened to 190-285 ft.-lbs. torque. Fill and bleed steering system as outlined in paragraph 3.

6A. OVERHAUL. Refer to Fig. 11 for an exploded view of steering motor unit. To overhaul the installed unit, first remove radiator grille screen as shown in Fig. 7 and proceed as follows:

Disconnect cylinder lines, lube line and bleed line. Remove both cylinders (1—Fig. 11). Pry out welch plug (2), remove snap ring (3) and thrust washer (4); then unbolt and remove top cover (5).

Lift out pinion gear (10), then withdraw rack (18) with pistons attached, through either side opening. Reach through top opening and remove snap ring (11), then unbolt and lift off motor housing (8). Center steering arm and shaft unit can be lifted out after disconnecting steering tie rods.

If Teflon piston rings (19) are to be renewed, heat rings in water to ap-

Fig. 11—Exploded view of hydrostatic steering motor showing component parts of type without rack guide.

1. Cylinder
2. Expansion plug
3. Snap ring
4. Thrust washer
5. Top cover
6. Upper bushing
7. Gasket
8. Housing
9. Gasket
10. Pinion gear
11. Snap ring
12. Lower bushing
13. Oil seal
14. Arm & Shaft Assy
15. Spacer
16. Piston
17. O-ring
18. Rack
19. Piston ring

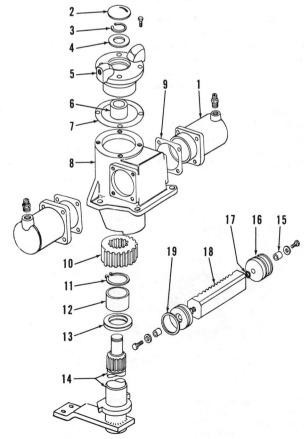
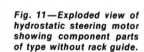

proximately 180°F. to soften and expand the rings. Install in piston grooves then clamp both pistons in a ring compressor to allow seals to return to original size.

Bushings (6 & 12) are pre-sized and will not require reaming after installation. Install lower bushing (12) with open end of oil groove to inside, with outer edge 1/16-inch below chamfer in bore. Install seal (13) with lower edge flush to 1/64-inch below edge of housing bore.

Tighten cap screws retaining pistons (16) to rack (18) to a torque of 30-35 ft.-lbs. Tighten top cover retaining cap screws to 33-38 ft.-lbs. and cylinder retaining cap screws to 50-55 ft.-lbs.

After installing lower shaft snap ring (11) turn steering shaft until steering arm points straight to rear as shown in Fig. 12 and install rack and pinion with timing punch marks aligned as shown.

Install a new welch plug (2—Fig. 11) using Hydraulic Sealant (Brown) LOCTITE. Fill and bleed steering system as outlined in paragraph 3 after assembly is completed and lines connected.

Models With Rack Guide

Steering cylinders, pistons or piston rings can be renewed or backlash adjusted without removal of unit from the tractor. If complete overhaul is indicated, steering motor should be removed as outlined in paragraph 7.

7. REMOVE AND REINSTALL. To remove the complete steering motor as a unit, first disconnect center steering arm from spindle. Remove front grille, disconnect steering cylinder lines, then unbolt and lift out steering motor assembly.

Install by reversing the removal procedure. Make sure that steering spindle (9—Fig. 13) and rack (16) are centered before installing center steering arm. Use Grade "C" (Blue) LOCTITE on cap

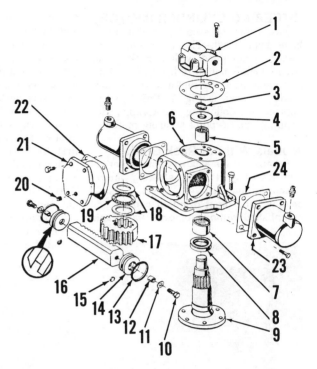

Fig. 13—Exploded view of hydrostatic steering motor which has rack guide and needle bearings.

1. Cover
2. Gasket
3. Snap ring
4. Thrust washer
5. Needle bearing
6. Spindle housing
7. Needle bearing
8. Oil seal
9. Steering spindle
10. Cap screw
11. Washer
12. Spacer
13. Piston ring
14. Piston
15. Piston ring
16. Rack
17. Pinion
18. Bearing races
19. Thrust bearing
20. Drain plug
21. Rack guide
22. Shim
23. Cylinder
24. Gasket

screws attaching center steering arm or pedestal to spindle (2) and tighten cap screws to a torque of 175-200 ft.-lbs. Steering motor is self-bleeding. Screws attaching steering motor to front housing should be tightened to 190-285 ft.-lbs. torque.

7A. OVERHAUL. When disassembling the steering motor either on or off the tractor, unbolt and remove steering cylinders (23—Fig. 13) before removing the rack guide (21).

Piston can be removed by removing cap screw (10) after cylinder is off. Piston uses spacer (12) to free the installed piston on rack for alignment purposes. Piston should be free to turn with bolt (10) tightened. "O" ring (15) is fitted between piston (14) and rack (16) and seals inner hole in piston when hydraulic pressure is applied. Tighten bolt (10) to a torque of 30-35 ft.-lbs. when unit is reassembled.

Steering gear backlash should be 0.001-0.008. Measure the backlash as follows: Assemble rack (16) and rack guide (21) without any shims (22). Tighten screws retaining rack guide to housing evenly to a torque of 30-35 ft.-lbs. Measure and record clearance between flange of rack guide and housing. Remove rack guide and install shims (22) equal in thickness to measured clearance plus 0.001-0.008. Shims (22) are available in four thicknesses from 0.006 to 0.023. Make sure rack is centered on gear (17) when unit is assembled.

To remove steering spindle (9), first remove cylinders, rack guide (21) and rack (16), then unbolt and remove housing cover. (1). Unseat and remove

snap ring (3), then bump spindle (9) downward out of splines of gear (17).

Top edge of bearing (5) should be approximately 1/16-inch below flush with top surface of housing. Lower bearing (7) should be approximately 1/4-inch from seal counterbore. Seal (8) should be approximately 1/16-inch above lower surface of housing. Large chamfer of washer (4) should be toward top.

An "X" mark is located on steering spindle (9) which is not visible when correctly assembled. Mark shaft across (180°) from the stamped "X", then align this mark with timing mark on gear as shown in Fig. 14. Washers (18—Fig. 13) are not identical. The thin washer should be at top and thick washer below the thrust bearing (19) as shown in Fig. 14. Install rack with timing marks on gear and rack aligned as shown in Fig. 15. Check and adjust rack backlash as outlined in previous paragraph.

Fig. 12—Steering motor is correctly assembled when timing marks on rack and pinion are aligned with steering arm pointing straight to rear.

Fig. 14—The steering spindle shaft must be correctly timed to the gear, however "X" mark provided on shaft is not visible. Mark shaft opposite "X" mark and align this new mark with timing mark on gear.

HYDROSTATIC HAND PUMP

All Models

8. REMOVE AND REINSTALL. To remove the hydrostatic hand pump, first remove left accessory panel, disconnect the four hydraulic lines; then unbolt and remove hand pump unit. Refer to Fig. 16.

Install by reversing the removal procedure. Fill and bleed steering system as outlined in paragraph 3 after unit is installed.

9. OVERHAUL. Refer to Fig. 17 for a partially exploded view of the Char-Lynn ORBITROL hydrostatic hand pump. Rotary valve spool (13), sleeve (14) and housing (12) are available only as a matched set which also includes reaction springs (9), cross pin (10) and component parts of check valve (5, 6, 7 & 8).

To disassemble the removed Orbitrol unit, first remove the cap screws retaining metering pump to bottom of housing and remove items (15) through (19). Remove the four screws retaining top cover (2) and lift off cover. Insert a suitably bent wire through port nearest check valve plug (5) and push out the plug as shown in Fig. 18. Remove check valve seat using an Allen wrench and remove check valve ball and spring (Fig. 19). Carefully push control valve

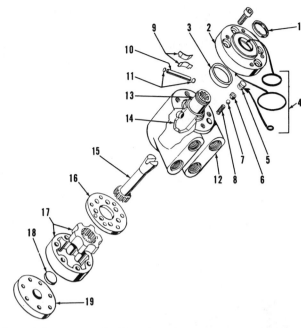

Fig. 17—Exploded view of ORBITOL hydrostatic hand pump showing component parts.

1. Oil seal
2. Cover
3. Bushing
4. O-rings
5. Valve plug
6. Valve seat
7. Valve ball
8. Valve spring
9. Centering springs
10. Drive pin
11. Teflon disc
12. Valve body
13. Valve spool
14. Valve sleeve
15. Drive link
16. Metering plate
17. Rotor assembly
18. Disc
19. End plate

spool and sleeve assembly out bottom of housing as shown in Fig. 20.

NOTE: Be careful valve unit does not bind. Parts are fit to extremely close tolerance and a twisting motion may be required for withdrawal.

Remove nylon cross pin discs (11—Fig. 17) and cross pin (10), then separate spool and sleeve as shown in Fig. 21. Thoroughly clean all parts in a

suitable mineral solvent and blow dry with air. Inspect lapped surfaces for scoring or other damage. Mating surfaces of cover (19—Fig. 17), rotor assembly (17) and metering plate (16) can be hand lapped to remove burrs, slight scratches or other imperfections. Use 600 grit abrasive paper or lapping compound. Rinse in clean solvent and blow dry after lapping is complete. Rotor

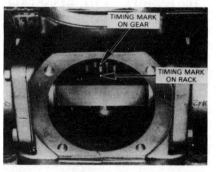

Fig. 15—Timing mark on gear and mark on rack must be aligned when assembling the steering rack.

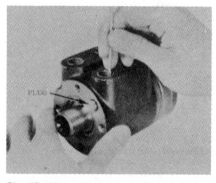

Fig. 18—Use a bent wire and work through valve port to push out check valve plug as shown.

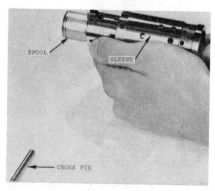

Fig. 20—Push spool and sleeve assembly out bottom of housing as shown.

Fig. 16—Left accessory panel removed from 1080 model showing installed ORBITOL unit. Model 1085 is similar.

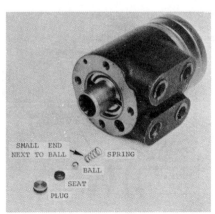

Fig. 19—Check valve components removed from valve body.

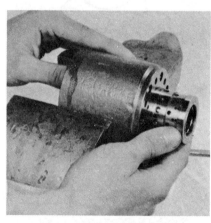

Fig. 21—Spool can be pushed from sleeve after removing cross pin.

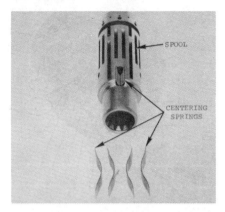

Fig. 22—View showing correct installation of centering springs (arch to arch in sets of three, notched edge down).

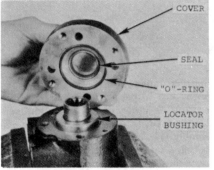

Fig. 23—Be sure O-rings, seal and bushing are properly installed when reassembling.

Fig. 24—When properly assembled, slot in drive link will align with valley of inner rotor as shown.

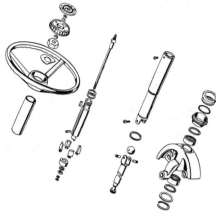

Fig. 26—Exploded view of adjustable steering column.

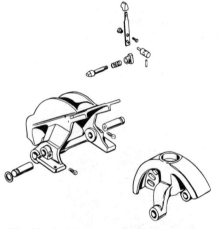

Fig. 27—Steering column pivot housing and associated parts showing release latch mechanism.

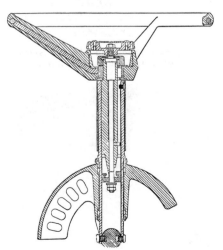

Fig. 25—Cross sectional view of adjustable steering column. Refer to Fig. 26 for exploded view.

units are only available as a matched set.

Inspect centering springs (9) for fractures or distortion. Springs should have a minimum arch of 7/32-inch when measured at center. Inspect spool, sleeve and body for nicks, scoring or wear. If any part is damaged, renew the housing assembly which includes spool sleeve and check valve. Renew O-rings (4) when reassembling the unit.

Tighten check valve seat (6) to a torque of 150 inch-pounds. Install springs (9) with notched edge down, in sets of three as shown in Fig. 22.

Insert the assembled spool & sleeve unit from bottom of valve body using a twisting motion. Do not allow sleeve to move up beyond flush with upper machined surface of body. Be sure pin slot in drive shaft (15—Fig. 17) is aligned with valley of inner rotor as shown in Fig. 24. Rotor has six gear teeth and 12 splines; if drive slot is improperly aligned, steering unit will operate in reverse of hand pressure. Tighten the seven lower cap screws evenly to a torque of 250 inch-pounds and the four cap screws in top cover to a torque of 220 inch-pounds.

Install the assembled unit as outlined in paragraph 8 and bleed as in paragraph 3.

10. CONTROL COLUMN. The adjustable steering control column is shown in cross section in Fig. 25 and exploded in Figs. 26 and 27. Bearings, column shaft and slide should be repacked when disassembled, with Lubriplate or equivalent.

POWER STEERING PUMP

All Models

11. REMOVE AND REINSTALL. The gear driven, gear-type power steering pump is mounted on left hand side of engine block and driven by the timing idler gear. Mounting cap screws enter pump housing from front and heads are located on timing gear front cover as shown in Fig. 35. Fill and bleed the system as in paragraph 3 after pump is installed.

12. OVERHAUL. Refer to Fig. 28 or Fig. 29 for an exploded view of power steering pump and associated parts.

Fig. 28—Exploded view of one type of power steering pump showing component parts. Refer to Fig. 29 for other pump.

1. Drive gear
2. Oil seal
3. Cover
4. Seal
5. Brass seal
6. Rubber spacer
7. Bearing
8. Pumping gears
9. Pump body
10. Rear cover
11. Relief valve
12. Retainer
13. Filter
14. Reservoir

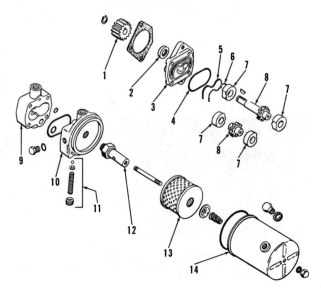

On pump shown in Fig. 28, pumping gears (8) should be renewed in pairs. Bearings (7) are interchangeable and provide end seal surface for pumping gears as well a shaft support. End clearance is controlled by hydraulic pressure on back side of bearings. Renew bearings in sets of four if wear is a factor in replacement, and install with chamfered side next to gears. Brass seal (5) and rubber spacer (6) takes the place of a one-piece rubber seal used on some early models. When assembling the pump, tighten through-bolts to a torque of 30 ft.-lbs. and install drive gear (1) with paint mark forward.

For both types of pumps, observe the following if relief valve (11—Fig. 28 or Fig. 29) is to be disassembled: Count the number of turns necessary to remove the plug and reinstall at the same distance; then check relief pressure as outlined in paragraph 4 after pump is reinstalled.

On pump shown in Fig. 29, scribe alignment marks across front cover (3), body (9), rear cover (10) and reservoir (14) before disassembling. Affix timing marks to teeth of gears (8) if original gears are to be reinstalled. Lips of seal (2) should be toward pump. The "V" groove of seal (15) should be toward front cover (3). Small (1/16-inch) holes in gasket (16), shield (17) and wear plate (18) should be up toward pressure side of pump. Bronze side of wear plate (18) must also be toward gears (8). Notch (N) on flat surface of pump body must be over holes (H). Bronze side of

thrust plate (19) must be toward gears and large cut-out section must be down toward inlet side of pump. Tighten

pump assembly screws (20) to 225-250 in.-lbs. torque. Counterbore of gear (1) should be away from pump.

ENGINE AND COMPONENTS

All models are equipped with a Perkins Model A4.318, four cylinder diesel engine having a bore of 4.5-inches, a stroke of 5.0-inches and a displacement of 318 cu. in.

R&R ENGINE WITH CLUTCH

All Models

13. To remove the engine and clutch as a unit, first drain cooling system and if engine is to be disassembled, drain oil pan. Remove hood, batteries, battery carrier and both side rails.

Disconnect both radiator hoses at engine, air inlet tube at manifold and four hydraulic lines at instrument panel support. Remove power steering lubricant and bleed lines.

Support tractor under transmission housing and front end assembly from a hoist, then unbolt and remove front axle, front support and radiator assembly as a unit.

Shut off fuel and disconnect fuel lines, throttle controls, and temperature indicator sending units. Disconnect power steering pressure and return lines at pump and oil gage line at engine block. Disconnect wiring. Support engine at lifting brackets, remove attaching cap screws and lift engine away from clutch housing.

Install by reversing the removal procedure. Bleed fuel system as outlined in paragraph 44 and adjust throttle controls as in paragraph 54 after engine in installed.

CYLINDER HEAD

All Models

14. To remove the cylinder head, first drain cooling system and remove hood. Remove batteries and battery carrier. Disconnect fuel lines and air inlet pipe, then unbolt and remove intake manifold with fuel filters attached. Disconnect the two steering lines (and Multipower oil cooler lines, if so equipped) at front and rear and lift off the lines.

Remove fan belt, fan blades and both coolant hoses leading to water pump.

Unbolt and remove exhaust manifold. Disconnect heat indicator sending unit. Remove leak-off pipe, injection pressure lines and injectors. Remove rocker arm covers and rocker arms assembly. Loosen, then remove cylinder head nuts and lift off cylinder head with water pump attached.

Cylinder head may be resurfaced provided not more than 0.020-inch is removed; nozzle protrusion does not exceed 0.23-inch; or cylinder head thickness is not reduced to less than 3.230-inches. If cylinder head is re-machined, test for leaks at 30 psi using warm water.

Cylinder head gasket is marked "TOP FRONT" for proper installation. Install gasket dry and tighten cylinder head stud nuts to correct torque using the sequence shown in Fig. 30. Engines before serial number 318 UA1148 were equipped with regular cylinder head nuts which should be tightened to 105-110 ft.-lbs. torque. Engines serial number 318UA1148 and above were originally equipped with phosphated cylinder head nuts. The late phosphated cylinder head nuts may be installed on any engine and the correct torque for these nuts is 85-90 ft.-lbs. On all models, bleed fuel system as outlined in paragraph 44 and adjust valve clearance as in paragraph 20 after head is installed. Retorque cylinder head and readjust valve clearance (to 0.010-inch Hot) after engine reaches normal operating temperature. Cylinder head nuts should be retorqued and valve clearance should be readjusted after about 50 to 100 hours of operation.

VALVES AND SEATS

All Models

15. Both the intake and exhaust valves seat directly in cylinder head, with seats recessed a specified amount for correct valve head positioning. Reseat at 45°, using a 1-5/8-inch stone for exhaust valves and a 1-7/8-inch stone for intake valves, to prevent cutting

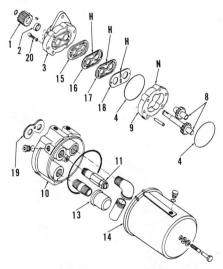

Fig. 29—Exploded view of Parker-Hannifin power steering pump. Refer to text for assembly notes.

1. Drive gear	13. Filter
2. Seal	14. Reservoir
3. Front cover	15. "V" seal
4. Seals	16. Gasket
8. Pumping gears	17. Shield
9. Pump body	18. Wear plate
10. Rear cover	19. Thrust plate
11. Relief valve	20. Screws

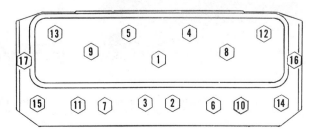

Fig. 30—Cylinder head tightening sequence. Recommended tightening torque depends upon type of nuts used. Refer to text.

outer portion of head around the seat. Specified valve seat width is 1/16-3/32-inch for intake valves and 3/32-7/64-inch for exhaust valves, remove only enough metal to clean and true valve seat face.

Recommended valve face angle is 45°. In production assembly, intake valve heads are recessed 0.057-0.071-inch below machined surface of cylinder head when measured with a straight edge and feeler gage as shown in Fig. 31. Exhaust valves are recessed 0.052-0.065-inch. Renew valve and/or remachine-cylinder head if measurement exceeds 0.140 for any valve. The manufacturer provides seat inserts for exhaust valves, only, as a repair item.

VALVE GUIDES

All Models

16. Intake and exhaust valve guides are interchangeable. Guides contain a shouldered step which acts as a seat for inner and outer valve springs and properly positions guide in cylinder head.

Inside diameter of new guide is 0.375-0.376-inch and specified stem-to-guide clearance is 0.0015-0.0035-inch for intake valves or 0.002-0.004-inch for exhaust valves. Replacement guides are presized and do not require reaming after installation.

VALVE SPRINGS

All Models

17. Springs, retainers and locks are interchangeable for intake and exhaust valves. Inner and outer springs are equipped with close coils at one end, which should be installed next to valve guide.

Renew the springs if they are distorted, heat discolored, or fail to meet the test specifications which follow. Renew other parts if they show wear or damage.

Fig. 31—Using a straight edge and feeler gage to check valve head height. Refer to paragraph 15.

INNER SPRING
Lbs. test at
  1-9/16-inches 14.63-16.17
Lbs. test at
  1-9/64-inches 31.83-35.17

OUTER SPRING
Lbs. test at
  1-25/32-inches 38-42
Lbs. test at
  1-23/64-inches 68.5-75.7

CAM FOLLOWERS

All Models

18. The mushroom type tappets (Cam Followers) operate directly in machined bores in cylinder block and can be renewed after removing camshaft as outlined in paragraph 29. The 0.7475-0.7485-inch diameter tappets should have a diametral clearance of 0.001-0.0033-inch in block bores.

ROCKER ARMS

All Models

19. To remove the rocker arm cover for access to rocker arms or tappets, first disconnect air intake hose at manifold and reposition air intake pipe. Remove breather pipe then unbolt and remove rocker arm cover.

Rocker arm shaft can be removed at this time. The cast iron rocker arms are interchangeable and are equipped with renewable bushings. Press bushings into rocker arm bores with oil holes aligned and finish ream to the correct inside diameter of 0.6245-0.6255-inch for 1080 models, 0.7505-0.7520-inch for 1085 models. Recommended bushing to shaft maximum clearance is 0.0035-inch. Minimum clearance should

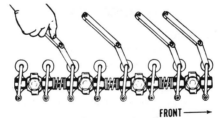

Fig. 32—With "TDC" timing marks aligned and No. 1 piston on compression stroke, adjust the indicated valves to 0.012 cold.

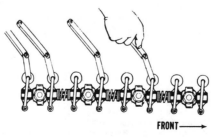

Fig. 33—With "TDC" marks aligned and No. 4 piston on compression stroke, adjust the indicated valves to 0.012 cold.

be 0.0008-inch for 1080 models 0.001-inch for 1085 models. Renew shaft if grooved or scored, and bushings if diametral clearance exceeds 0.006-inch.

Install rocker arm shaft with oil feed holes facing toward valve stem side of cylinder block. Tighten rocker shaft bracket stud nuts to a torque of 21-24 ft.-lbs. and adjust tappet gap as outlined in paragraph 20.

VALVE CLEARANCE (TAPPET GAP)

All Models

20. The recommended valve clearance with engine cold is 0.012-inch for both the intake and exhaust valves. Cold (static) setting of all valves can be made from just two crankshaft positions, using the procedure shown in Figs. 32 and 33. Proceed as follows:

Remove the rocker arm cover as outlined in paragraph 19 and turn crankshaft until TDC timing mark is aligned with timing pointer.

NOTE: On models prior to Engine Serial Number 318 UA 373, flywheel timing marks were used and timing port was located on engine adapter plate just below left side rail and closed by a pipe plug. On later engines, timing marks are located on crankshaft pulley as shown in Fig. 34.

Check the rocker arms for front and rear cylinders. If rear rocker arms are tight and front rocker arms loose, No. 1 piston is on compression stroke; adjust the four tappets shown in Fig. 32. If front rocker arms are tight and rear rocker arms loose, No. 4 piston is on compression stroke; adjust the four tappets shown in Fig. 33. Turn crankshaft one complete turn until TDC timing mark is again aligned, then adjust the remaining tappets.

Recheck the adjustment, if desired, with engine running at slow idle speed.

Fig. 34—View of engine showing timing pointer and timing marks on crankshaft pulley. On some early engines, timing marks are on flywheel.

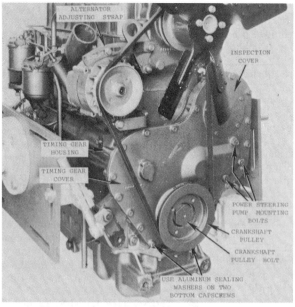

Fig. 35—Front view of engine showing parts which must be removed before removal of timing gear cover.

Recommended tappet clearance with engine at operating temperature is 0.010-inch. Clearance may be adjusted with engine stopped and crankshaft positioned as described previously for Cold (static) setting.

Fig. 36—A special tool is provided to center crankshaft seal bore when timing gear cover is installed.

Fig. 37—The double-stepped centering tool can also be used to install and locate the oil seal.

TIMING GEAR COVER

All Models

21. To remove the timing gear cover, drain cooling system and remove hood, batteries, battery carrier and both side rails. Disconnect radiator hoses at engine, air inlet tube at manifold and four hydraulic lines at instrument panel support. Remove power steering lubricant and bleed lines.

Support tractor under transmission housing and front end assembly from a hoist; then unbolt and remove front axle, front support and radiator assembly as a unit.

Refer to Fig. 35 and remove fan blades, fan belt and power steering pump. Remove crankshaft pulley bolt and washer from front of crankshaft and check to be sure pulley locating marks are present and visible, before removing pulley.

NOTE: Some early engines had timing marks located on flywheel, and crankshaft pulley was not timed.

Remove the cover retaining cap screws, noting that aluminum sealing washers are used on the two lower screws, then lift off cover.

Crankshaft front oil seal can be renewed at this time. If special tools are available, install cover with seal removed and use the single-stepped side of Tool MFN 306A to center cover seal bore over crankshaft as shown in Fig. 36. Install and tighten the retaining cap screws, then use double-stepped side of Special Tool MFN 306A to install and position oil seal as shown in Fig. 37.

If special tools are not available, install new seal with lip to rear with front edge recessed 0.360-0.370-inch into seal bore when measured from front of cover; then use crankshaft pulley to center seal bore when installing cover.

On all models, tighten crankshaft pulley retaining cap screw to a torque of 280-300 ft.-lbs. and complete the assembly by reversing disassembly procedure.

TIMING GEARS

All Models

22. Fig. 38 shows a view of timing gear train with cover removed. Before

Fig. 38—Front view of engine with timing gear cover and oil pan removed, showing timing marks aligned. Because of the odd number of teeth in idler gears, all marks will not align with each camshaft revolution. The fact that marks are not aligned does not necessarily indicate indorrect engine timing.

attempting to remove any of the gears, first remove rocker arm cover and rocker arms to avoid the possibility of damage to pistons or valve train if camshaft or crankshaft should either one be turned independently of the other.

Timing gear backlash should be 0.003-0.006-inch between any of the gears in main timing gear train; or 0.006-0.009-inch between any of the gears in balancer gear train. Replacement gears are available in standard size only. If backlash is not within limits, renew gears, idler shafts, bushings or other items concerned.

NOTE: Because of the odd number of teeth in idler gear, all timing marks will not align with each revolution of the camshaft. Failure of marks to align does not necessarily indicate engine is mistimed.

To remove the timing gears or time the engine, refer to the appropriate following paragraphs:

23. IDLER GEAR AND HUB. The timing idler gear (Fig. 38) should have a diametral clearance of 0.001-0.003-inch on the 1.998-1.999-inches diameter hub. Suggested end play is 0.002-0.005-inch. Hub is a light press fit in timing gear housing bore and can be loosened with a soft hammer if renewal is indicated. Due to uneven spacing of hub studs, hub can only be installed in one posi-

tion. The semi-finished idler gear bushing should be pressed in place if renewal is indicated and reamed to an inside diameter of 2.000-2.002-inches after installation.

Timing gears are pressure lubricated through idler gear hub and spray tube (Fig. 39) which directs a continuous stream of oil against tooth face of rotating idler gear. Be sure spray hole is open and clean and that end of tube is properly inserted in positioning hole in timing gear housing. Tighten retaining stud nuts to a torque of 21-24 ft.-lbs. after aligning timing marks as outlined in paragraph 27.

24. CAMSHAFT GEAR. The camshaft gear (Fig. 41) is pressed and keyed to shaft and retained by a special cap screw (1), tab washer (2) and retaining plate (3). Use a suitable puller to remove the gear. Use the retaining plate and cap screw to draw gear into position when reinstalling. Tighten cap screw to a torque of 41-45 ft.-lbs. and lock in place by bending tab washer (2). Time the gears as outlined in paragraph 27 when reinstalling.

25. INJECTION PUMP DRIVE GEAR. The injection pump drive gear (Fig. 43) is retained to pump adapter by three cap screws. When installing

the gear, align dowel pin (D) in gear with slot in adapter hub, then install retaining cap screws. The injection pump drive gear and adapter are supported by injection pump rotor bearings. Tighten retaining cap screws to a torque of 19-21 ft.-lbs. and time the gears as outlined in paragraph 27.

26. CRANKSHAFT GEAR. The crankshaft timing gear is a transition fit on shaft, and can usually be removed with a pry after timing gear cover is off.

Crankshaft gear may be renewed without removing oil pan or engine balancer. Balancer counterweights are at the bottom when any piston is at TDC, and will automatically assume correct timing position when engine is upright. When installing crankshaft gear, remove idler gear and retime engine as outlined in paragraph 27.

27. TIMING THE GEARS. To install and time the gears, first install crankshaft, camshaft and injection pump drive gears as outlined in paragraphs 24, 25 and 26, with timing marks to front. Turn the shafts until timing marks are pointing toward idler gear hub as shown in Fig. 38, then install idler gear with all marks aligned.

NOTE: Mark on engine balancer idler gear may not be aligned, because of the odd number of teeth in gear. Balancer will be correctly timed, however, if crankshaft gear has not been removed; or if crankshaft key is pointing straight up (See Fig. 42) when crankshaft gear is installed.

Refer to paragraph 23 for installation and securing of idler gear.

TIMING GEAR HOUSING

All Models

28. To remove the timing gear housing, first remove timing gears as outlined in paragraphs 23 through 26 and injection pump as in paragraph 55. If

Fig. 39—Spray tube (arrow) is ported to engine oil gallery, and hole in tube provides oil for timing gear lubrication.

Fig. 41—Camshaft gear is pressed and keyed to shaft, and retained by cap screw (1), tab washer (2) and retaining plate (3).

Fig. 43—Align dowel pin and slot (D) when installing injection pump gear.

Fig. 40—Idler gear hub is a light push fit in recess in engine block.

Fig. 42—Removing camshaft gear.

oil pan has not been removed, take out the five front oil pan cap screws. Remove the cap screws securing timing gear housing to block front face and lift off housing.

Install by reversing the removal procedure. Tighten cap screws retaining timing gear housing to cylinder block to a torque of 20 ft.-lbs. and the five cap screws securing oil pan to 12-15 ft.-lbs. The four oil pan cap screws should be tightened first after all screws are installed.

CAMSHAFT

All Models

29. To remove the camshaft, first remove timing gear cover as outlined in paragraph 21. Remove fuel lift pump, rocker arm cover, rocker arms assembly and push rods. Remove timing idler gear as outlined in paragraph 23. Secure cam followers (tappets) in their uppermost position and, working through holes provided in camshaft gear remove thrust plate retaining capscrews as shown in Fig. 44. Camshaft, gear and thrust plate can now be withdrawn as a unit.

The camshaft runs in three bearings. The front bearing bore contains a bushing which is not available as a service item, while the two rear journals runs directly in engine block. Recommended diametral clearance is 0.002-0.005-inch for the two front journals or 0.004-0.007-inch for rear journal. End play of 0.007-0.016-inch is controlled by the camshaft thrust plate which rides on a machined shoulder of front bearing journal and is retained to front face of block by two cap screws. Thrust plate clearance (camshaft end play) can be measured with a feeler gage when assembly is out. If mea-

sured clearance exceeds 0.018-inch, renew the plate.

NOTE: When measuring clearance with feeler gage, be sure a shoulder worn into thrust plate does not give a false reading.

Camshaft journal diameters are as follows:

Front journal 2.0565-2.0575
Center journal 1.966-1.967
Rear journal. 1.905-1.906

ROD AND PISTON UNITS

All Models

30. Connecting rod and piston units are removed from above after removing cylinder head, oil pan, engine balancer and rod bearing caps. Cylinder numbers are stamped on the connecting rod and cap. When reinstalling, make sure correlation numbers are in register and face toward camshaft side of engine. When installing connecting rod caps, use new rod locks and tighten cap screws to a torque of 100-105 ft.-lbs.

PISTONS, SLEEVES AND RINGS

All Models

31. The aluminum alloy, cam ground pistons are supplied in standard size only and available only in a kit consisting of piston, pin and rings for one cylinder. The toroidal combustion chamber is offset in piston crown and piston is marked "FRONT" or "F" for proper installation.

Pistons should be assembled to connecting rods with number on side of rod and cap toward (camshaft) side of engine.

Each piston is fitted with three compression rigns and one oil control ring, all located above the piston pin. The plain faced chrome top ring may be installed either side up in groove. The internally stepped cast iron 2nd & 3rd

compression rings are marked "BOTTOM" for proper installation. The cast iron oil control ring is equipped with an integral spring expander and may be installed either side up. Specifications for fitting piston rings are as follows:

Eng Gap	Inch
Top ring	.0.018-0.025
2nd & 3rd	
compression rings	.0.014-0.021
Oil ring	.0.018-0.025
Side Clearance	
Top ring	.0.0025-0.0045
2nd & 3rd	
compression ring	.0.0015-0.0035
Oil ring	.0.0025-0.0045

The production cylinder sleeves are 0.001-0.003-inch press fit in cylinder block bores and are unfinished. Service sleeves are in transition fit (0.001-inch tight to 0.001-inch clearance) in block bores and cylinder walls are finished. Both types have been provided for field installation, refer to the following paragraphs for procedure:

Use a heavy duty sleeve puller capable of high pulling force, and Sleeve Puller Plate MFN326 for removal. Thoroughly clean cylinder bore, giving particular attention to counterbore.

Production sleeves can be identified by the rough, unfinished inner bore. Lightly coat outside with clean oil and press into position with suitable equipment. When correctly installed, top of sleeve should be within the limits of 0.025-0.031-inch above block face as shown in Fig. 46. After installation, carefully bore and finish hone cylinder wall to an inside diameter of 4.5025-4.5035-inches.

NOTE: Inside diameter of the semi-finished sleeve bore is 4.467-4.471-inches. Thoroughly clean the finished bore with soap and water then with

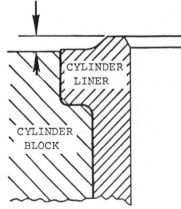

Fig. 44—Working through holes provided in camshaft gear, remove cap screws retaining camshaft thrust plate.

Fig. 45—View of camshaft thrust plate with gear removed.

Fig. 46—Installed cylinder sleeve (liner) should extend 0.025-0.031 above gasket face of block as shown.

clean oil, until rag shows no discoloration from contact with cylinder wall.

The finished service sleeves may be chilled with dry ice or in a deep-freeze, then fully installed by hand. Sleeve may also be coated with oil and installed with a press (without chilling) if preferred. Do not lubricate chilled sleeve. When correctly installed, top of sleeve should be within the limits of 0.025-0.031-inch above block face as shown in Fig. 46. Maximum allowable distortion of the installed, finished sleeve is 0.0025-inch.

Cylinder sleeves should be renewed if scored, or if wear or taper exceeds 0.006.

NOTE: Finished inside diameter of early service sleeves was 4.5005-4.5025-inches. If sleeves are early type, hone to currently recommended ID of 4.5025-4.5035-inches after installation.

PISTON PINS

All Models

32. The 1.4375-1.4378-inches diameter, floating type piston pins are retained in piston bosses by snap rings and are available in standard size only. The renewable connecting rod bushing must be final sized after installation to provide the recommended diametral clearance of 0.0004-0.0017-inch for the pin. Be sure the pre-drilled oil hole in bushing is properly aligned with hole in top of connecting rod and install bushing from chamfered side of bore. Piston pin should be a thumb press fit in piston after piston is heated to 100-120°F.

CONNECTING RODS AND BEARINGS

All Models

33. Connecting rod bearings are precision type, renewable from below after removing oil pan, balancer unit and rod bearing caps. When renewing bearing shells, be sure that the projection

Fig. 47—Crankshaft end play is controlled by thrust washers installed at front and rear of center main bearing as shown. Tab on lower half and slot in cap prevents thrust washer rotation in block.

engages milled slot in rod and cap and that the correlation marks are in register and face toward camshaft side of engine.

Connecting rod bearings should have a diametral clearance of 0.002-0.0042-inch on the 2.7483-2.749-inches diameter crankpin. Recommended connecting rod side clearance is 0.008-0.017-inch. Bearings are available in undersizes 0.010, 0.020 and 0.030 as well as standard; however, crankshaft is originally "Tufftrided" and must be retreated if reground to smaller diameter. Upper and lower connecting rod inserts are not interchangeable. Be sure that insert with locking tang closest to side is in rod (top). Renew connecting rod bolt locks and tighten cap screws to a torque of 100-105 ft.-lbs. when units are reinstalled.

CRANKSHAFT AND BEARINGS

All Models

34. The crankshaft is supported in five precision type main bearings. To remove the rear main bearing cap, it is first necessary to remove the engine clutch and flywheel. All other main bearing caps can be removed after removing oil pan and engine balancer.

Upper and lower inserts are not interchangeable, the upper (block) half being slotted to provide pressure lubrication to crankshaft and connecting rods. Inserts are interchangeable in pairs for the 1st, 3rd & 5th main bearings; and for the 2nd and 4th. Crankshaft end play is controlled by thrust washers installed on front and rear of center main bearing as shown in Fig. 47. Be sure that steel back is positioned next to block and cap and the grooved, bearing surfaces next to crankshaft thrust faces. Lower half is positioned in cap by a tab and upper half can be "rolled" out after cap is removed, without removing crankshaft.

Front bearing cap contains mounting hub for balancer idler gear (Fig. 48). All main bearing caps are equipped with ring dowels around cap screws for positive cap location. Be sure ring dowels are in position when caps are reinstalled. Main bearing caps are numbered from front to rear, and cap and engine flange are stamped with an identifying serial number. Cap and block serial numbers should be on same side when caps are installed.

Bearing inserts are available in undersizes of 0.010, 0.020 and 0.030-inch as well as standard; however, crankshaft is originally "Tufftrided" and must be retreated if reground to smaller diameter. Thrust washers which limit crankshaft end play are available in standard thickness and oversizes of 0.0075-inch. Recommended

main bearing diametral clearance is 0.0025-0.0045-inch and recommended crankshaft end play is 0.005-0.016-inch. Tighten main bearing retaining cap screws to a torque of 140-150 ft.-lbs. and secure by bending tabs on locking washers. When renewing rear main bearing, refer to paragraph 38 for installation procedure of rear seal and oil pan bridge piece. Check the crankshaft journals against the values in inches which follow:

Main journal diameter 2.9985-2.999
Crankpin diameter 2.7483-2.749

ENGINE BALANCER

All Models

35. **OPERATION.** The Lanchester type engine balancer consists of two unbalanced shafts which rotate in opposite directions at twice crankshaft speed. The inertia of the shaft weights is timed to cancel out natural engine vibration. The balancer is correctly timed when the balance weights are at their lowest point when pistons are at TDC and BDC of their stroke.

The balancer unit is driven by the crankshaft timing gear train through an idler gear carried on front main bearing cap. The engine oil pump is mounted at rear of balancer frame and driven by balancer shaft. See Fig. 49.

36. **REMOVE AND REINSTALL.** The engine balancer assembly can be removed after removing oil pan and mounting cap screws; and disconnecting oil delivery pipe at engine block. Refer to Fig. 49 for installed view (engine inverted).

When installing balancer with engine in tractor, engine timing marks cannot

Fig. 48—Front main bearing cap contains mounting hub for balancer idler gear as shown.

Refer to paragraph 41 for overhaul of engine oil pump and to paragraph 36 for installation of balancer assembly.

Fig. 49—Installed view of balancer showing engine oil pump and associated parts.

CRANKSHAFT REAR OIL SEAL

All Models

38. The asbestos rope type rear oil seal is contained in a two-piece retainer attached to rear face of engine block of early engines. Later engines use lip seal which is pressed into a one piece seal retainer. On all models, seal retainer can be removed after removing the flywheel. The early flywheel cannot be used with later lip type rear seal. The late flywheel can be used with either type of seal.

The rope type crankshaft seal is precision cut to length, and must be installed in retainer halves with 0.010-0.020-inch of seal ends protruding from each end of retainer. Do not trim the seal. To install the seal, clamp each half of retainer in a vise as shown in Fig. 52. Make sure seal groove is clean. Start each end of seal in groove with the specified amount of seal protruding. Allow seal rope to buckle in the center until about an inch of each end is bedded in groove, work center of seal

be seen. Because of the odd number of teeth in balancer idler gear, timing marks will align only occasionally and cannot be used as an installation guide. To install the balancer without removing timing gear cover for complete alignment of timing marks, proceed as follows:

Turn crankshaft until No. 1 and No. 4 pistons are at the top of their stroke and "TDC" timing mark on crankshaft pulley is aligned with timing pointer. Install balancer frame with balance weights hanging normally. If carefully installed, timing will be correct. If engine is mounted on a stand, timing marks can be observed by removing timing gear cover. Refer to Fig. 38. Tighten balancer retaining cap screws evenly to a torque of 32-36 ft.-lbs. and reconnect oil delivery pipe, then complete tractor assembly by reversing disassembly procedure.

37. OVERHAUL. Refer to Fig. 49 for an installed view of engine balancer and to Fig. 50 for an exploded view. To disassemble the removed balancer unit, unbolt and remove oil pump and housing rear cover assembly (8).

The set screws retaining balance weights (7) to shafts (15 & 16) are installed using Grade "C" (Blue) LOCTITE. Loosen the set screws, then push balance shafts (15 & 16) forward out of balance weights and frame.

NOTE: If Allen head set screws in balance weights cannot be loosened, heat screws slightly to break the LOCTITE bond. Also be careful not to damage bushings in balancer frame with the flat keys as shafts (15 & 16) are removed. Bushings are not available as service items.

Recommended diametral clearance of balance shafts in frame bushings is 0.003-0.005. New shaft diameters are 1.311-1.3115 for front journals and 1.0925-1.093 for rear. Renew shafts, frame or balancer unit if clearances are excessive.

When assembling the balancer, use Grade "C" (Blue) LOCTITE for installing the screws retaining gears (6) to balance weights (7) and the set screws retaining balance weights to shafts (15 & 16). Also make sure flat surfaces of weights are aligned when installed, as shown in Fig. 51.

Idler gear shaft can be pressed out of front main bearing cap if renewal is indicated. Bushing in balancer idler gear is also renewable. Ream bushing after installation to a finished ID of 1.500-1.5015.

Fig. 50—Exploded view of engine balancer, engine oil pump and associated parts.

1. Snap ring
2. Thrust washer
3. Idler gear
4. Bushing
5. Shaft
6. Gear
7. Weight
8. Plate
9. Pump body
10. Valve plunger
11. Spring
12. Cap

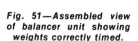

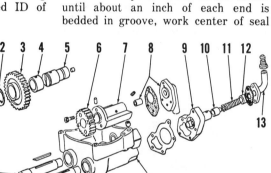

13. Pipe
14. Frame
15. Shaft
16. Shaft

Fig. 51—Assembled view of balancer unit showing weights correctly timed.

into position then roll with a round bar as shown. Repeat the process with other half of seal.

With lip type seal, lip should be toward engine. Be sure seal is square in retainer.

On all models, coat mating surfaces of cylinder block, bridge piece and seal retainer with a suitable non-hardening sealing compound and install using new end seals. Align gasket surfaces of bridge piece and cylinder block using a straight edge, and tighten retaining cap screws. Screws for bridge piece and seal retainers should be torqued to 12 ft.-lbs.

FLYWHEEL

All Models

39. To remove the flywheel, first separate engine from transmission housing and remove the clutch. Flywheel is secured to crankshaft flange by six evenly spaced capscrews. A seventh, unused hole in flywheel and crankshaft is aligned during installation.

The starter ring gear can be renewed after flywheel is removed. Heat ring gear evenly to approximately 475°F. and install on flywheel with beveled end of teeth facing front of engine.

Install flywheel using new locks, and tighten retaining cap screws to a torque of 74-80 ft.-lbs. Maximum allowable face runout is 0.001-inch for each inch from flywheel centerline to point of measurement.

OIL PAN

All Models

40. The heavy cast iron oil pan serves as the tractor frame and attaching point for tractor front support. To remove the oil pan, support tractor underneath transmission housing and drain oil pan. Loosen side rail rear cap screws and remove cap screws securing oil pan to cylinder block, front support and transmission housing. Lower oil pan carefully when bolts are removed.

Install by reversing the removal procedure. Tighten the cap screws securing oil pan to cylinder block to a torque of 19-21 ft.-lbs., front cap screws entering timing gear cover to 12-15 ft.-lbs. and rear screws entering bridge piece to 10 ft.-lbs. torque.

OIL PUMP

All Models

41. The rotary type oil pump is mounted on engine balancer frame as shown in Fig. 51 and Fig. 54. Oil pump is driven by balance weight shaft by means of a driving slot (Fig. 54).

Fig. 54—Oil pump is driven from balance weight shaft by means of slot as shown.

Fig. 52—Use a round bar to bed the asbestos rope seal in retainer half. Refer to text for details.

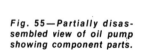

Fig. 55—Partially disassembled view of oil pump showing component parts.

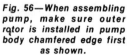

Fig. 53—Use a straight edge to align the cylinder block bridge piece (B) when seal is installed. Use new end seals (S).

Fig. 56—When assembling pump, make sure outer rotor is installed in pump body chamfered edge first as shown.

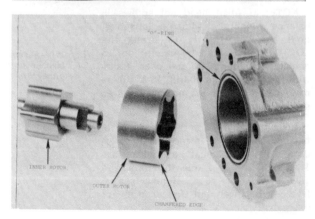

Examine oil pump parts for wear or scoring against the inch values which follow:

Outer Rotor to
  Body Clearance 0.010-0.012
Inner to Outer
  Rotor Clearance 0.0025-0.0045
Rotor End
  Clearance 0.001-0.0035

Refer to Figs. 54, 55 and 56 for disassembled views. When assembling pump, make sure outer rotor is installed chamfered edge first in body as shown. With the exception of relief valve parts, gaskets and O-rings, oil pump parts are not available separately. If oil pump is damaged or excessively worn, renew pump unit. Tighten retaining cap screws to a torque of 15-20 ft.-lbs. when installing.

RELIEF VALVE

All Models

42. The plunger type oil pressure relief valve is located in oil pump body as shown in disassembled view, Fig. 57. Free length of relief valve spring is 1-9/16 inches and spring should test approximately 16 lbs. when compressed to a height of 1 inch. Relief pressure is 50-65 psi.

DIESEL FUEL SYSTEM

The diesel fuel system consists of three basic units; the fuel tank and filters, injection pump and injector nozzles. When servicing any unit associated with the diesel fuel system, the maintenance of absolute cleanliness is of utmost importance. Of equal importance is the avoidance of nicks or burrs on any of the working parts.

Probably the most important precaution that service personnel can impart

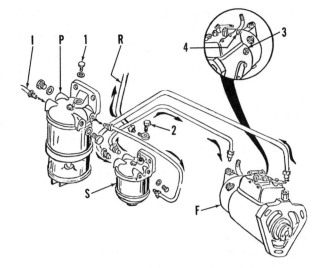

Fig. 58—Schematic view of diesel injection pump, fuel filters and lines of the type used on early models. Late tractors use parallel type twin filters as shown in Fig. 59. Numbers refer to bleed screws which should be loosened in sequence.

F. Fuel injection pump
I. Inlet line
P. Primary filter
R. Return line
S. Secondary filter

to owners of diesel powered tractors is to urge them to use an approved fuel that is absolutely clean and free from foreign materials. Extra precaution should be taken to make certain that no water enters the fuel storage tanks. Because of the high pressures and degree of control required of injection equipment, extremely high precision standards are necessary in the manufacture and servicing of diesel components. Extra care in daily maintenance will pay big dividends in long service life and the avoidance of costly repairs.

FUEL FILTERS AND LINES

All Models

43. OPERATION AND MAINTENANCE. Refer to Fig. 58 for a schematic view of fuel flow through filters and injection pump on early models. Late models are similar except that parallel type fuel filters are used as shown in Fig. 59. A much greater volume of fuel is circulated within the system than is burned in the engine, the excess serving as a coolant and lubricant for

the injection pump. Fuel enters the primary filter through inlet connection (I—Fig. 58 or 59) where it passes through the water trap and first stage filter elements. Both lines leading to injection pump (F) and the system return line (R), enter a common passage in secondary filter of filter head. The greater volume of filtered fuel is thus recirculated between the filter unit and injection pump, with the inlet line supplying little more fuel to the system than is actually burned in the engine.

Inspect the glass bowl at bottom of primary filter (P) daily (both bowls on parallel filter unit) and drain off any water or dirt accumulation. Renew primary filter element at 500 hour intervals and secondary filter element at 1000 hour intervals for normal tractor operation. Renew both elements and clean the tank and lines if evidence of substantial water contamination exists.

44. BLEEDING. To bleed the system, make sure tank shut-off is open, have an assistant actuate the manual lever on fuel lift pump and proceed as follows:

Refer to Fig. 58. Numbers indicate

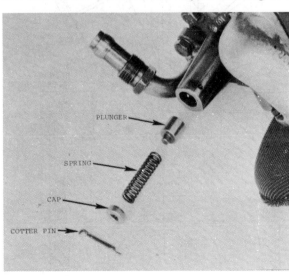

Fig. 57—Exploded view of oil pump relief valve showing component parts.

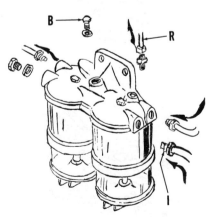

Fig. 59—Late, parallel type twin filters showing direction of fuel flow.

B. Bleed screw
I. Inlet line
R. Return line

17

bleed screws in correct bleeding order; on late models with parallel filter unit, the single bleed screw (B—Fig. 59) takes the place of the two screws (1 & 2—Fig. 58). Loosen each bleed screw in turn and continue to operate fuel lift pump until air-free fuel flows from bleed screw hole; then tighten screw and repeat the procedure with the next bleed screw. Operate the manual lever on fuel lift pump for approximately ten extra strokes after tightening bleed screw (4) on injection pump, to expel any air remaining in bleed-back lines to filter.

NOTE: Air in governor housing removed by bleed screw (4) will not prevent tractor from starting and running properly; however, condensation in the trapped air can cause rusting of governor components and eventual pump malfunction. Do not fail to bleed governor housing even though tractor starts and runs properly.

With the fuel supply system bled, push in the fuel shut-off control, partially open throttle lever and attempt to start the tractor. If tractor fails to fire, loosen compression nut at all injector nozzles and turn engine over with starter until fuel escapes from all loosened connections. Tighten compression nuts and start engine.

FUEL LIFT PUMP

All Models

45. The fuel lift pump (Fig. 60) is mounted on right side of engine block and driven by the camshaft. Pump parts are available separately or in a rebuilding kit. Output delivery pressure should be 5-8 psi.

INJECTOR NOZZLES

All Models

All models are equipped with C.A.V. multihole nozzles which extend through the cylinder head to inject the fuel charge into combustion chamber machined in crown of piston. Refer to Fig. 61.

WARNING: Fuel leaves the injector nozzle with sufficient force to penetrate the skin. Keep exposed portions of your body clear of nozzle spray when testing.

46. TESTING AND LOCATING A FAULTY NOZZLE. If rough or uneven engine operation or misfiring indicates a faulty injector, the defective unit can usually be located as follows:

With engine running at the speed where malfunction is most noticeable (usually slow idle speed), loosen the compression nut on high pressure line for each injector nozzle in turn, and listen for a change in engine performance. As in checking spark plugs, the faulty unit is the one which, when its line is loosened, least affects the running of the engine.

If a faulty nozzle is found and considerable time has elapsed since the injectors have been serviced, it is recommended that all nozzles be removed and checked, or that new or reconditioned units be installed. Refer to the following paragraphs for removal and test procedure.

47. REMOVE AND REINSTALL. Before loosening any fuel lines, thoroughly clean the lines, connections, injectors and engine area surrounding the injector with air pressure and solvent spray. Disconnect and remove the leak-off line, disconnect pressure

line and cap all connections as they are loosened, to prevent dirt entry into the system. Remove the two stud nuts and withdraw injector unit from cylinder head. If injector cannot be withdrawn, NUDAY Puller Tool MFN330 should be used.

Thoroughly clean the nozzle recess in cylinder head before reinstalling injector unit. It is important that surface be free of even the smallest particle of carbon or dirt which could cause the injector to be cocked and result in blow-by. No hard or sharp tools should be used in cleaning. Do not re-use the copper sealing washer located between injector nozzle and cylinder head, always install a new washer. Each injector should slide freely into place in cylinder head without binding. Make sure that dust seal is reinstalled and tighten the retaining stud nuts evenly to a torque of 10-12 ft.-lbs. After engine is started, examine injectors for blow-by, making the necessary corrections before releasing tractor for service.

48. TESTING. A complete job of testing and adjusting the injector requires the use of special test equipment. Only clean, approved testing oil should be used in tester tank. The nozzle should be tested for opening pressure, seat leakage, back leakage and spray pattern. When tested, the nozzle should open with a sharp popping or buzzing sound, and cut off quickly at end of injection with a minimum of seat leakage and a controlled amount of back leakage.

Before conducting the test, operate tester lever until fuel flows, then attach the injector. Close the valve to tester gage and pump tester lever a

Fig. 60—Exploded view of fuel lift pump showing component parts.

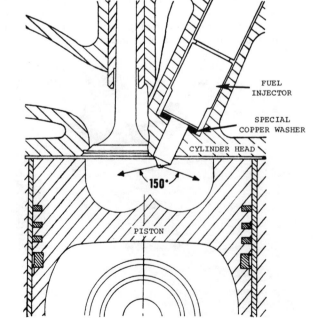

Fig. 61—Schematic view of combustion chamber showing injector spray angle relative to engine.

few quick strokes to be sure nozzle valve is not plugged, that four sprays emerge from nozzle tip, and that possibilities are good that injector can be returned to service without overhaul.

NOTE: Spray pattern is not symmetrical with centerline of nozzle tip.

The apparently irregular location of nozzle holes (See Figs. 61 & 63) is designed to provide the correct spray pattern in combustion chamber as shown.

If adjustment is indicated by the preliminary tests, proceed as follows:

49. OPENING PRESSURE. Open the valve to tester gage and operate tester lever slowly while observing gage reading. Opening pressure should be 2575 psi. If opening pressure is not as specified, remove the injector cap nut (1—Fig. 64), loosen locknut (3) and turn adjusting sleeve (5) as required to obtain the recommended pressure.

NOTE: When adjusting a new injector or an overhauled injector with a new pressure spring (6), set the pressure at 2650 psi to allow for initial pressure loss as spring settles in.

50. SEAT LEAKAGE. The nozzle tip should not leak at a pressure less than 2425 psi. To check for leakage, actuate tester lever slowly as the gage needle approaches 2425 psi and observe the nozzle tip. Hold the pressure for ten seconds; if drops appear or if nozzle tip is wet, the valve is not seating properly and injector must be disassembled and overhauled as outlined in paragraph 53.

51. BACK LEAKAGE. If nozzle seat as tested in paragraph 50 was satisfactory, check the injector and connections for wetness which would indicate leakage. If no visible external leaks are noted, bring gage pressure to 2200 psi, release the lever and observe the time required for gage pressure to drop from 2200 psi to 1470 psi. For a nozzle in good condition, this time should not

be less than six seconds. A faster pressure drop would indicate a worn or scored nozzle valve piston or body, and the nozzle assembly should be renewed.

NOTE: Leakage of the tester check valve or connections will cause a false reading, showing up in this test as excessively fast leakback. If all injectors tested fail to pass the test, the tester rather than the units should be suspected as faulty.

52. SPRAY PATTERN. If leakage and pressure are as specified, operate the tester handle several times while observing spray pattern. Four finely atomized, equally spaced, conical sprays should emerge from nozzle, tip, with equal penetration into the surrounding atmosphere.

If pattern is uneven, ragged or not finely atomized, overhaul the nozzle as outlined in paragraph 53.

NOTE: Spray pattern is not symmetrical with nozzle tip; refer to Fig. 61.

53. OVERHAUL. Hard or sharp tools, emery cloth, grinding compound, or other than approved solvents or lapping compound should not be used. An approved nozzle cleaning kit is available through an C.A.V. service agency and other sources.

Wipe all dirt and loose carbon from exterior of nozzle and holder assembly. Refer to Fig. 64 and proceed as follows:

Secure the nozzle in a soft jawed vise or holding fixture and remove the cap nut (1). Loosen the jam nut (3) and back off adjusting sleeve (5) enough to completely unload pressure spring (6). Remove the nozzle cap nut (12) and nozzle body (11). Nozzle valve (10) and body (11) are matched assemblies and should never be inter-mixed with other assemblies.

Place all parts in clean calibrating oil or diesel fuel as they are removed. Clean the exterior surfaces with a soft wire brush, soaking in an approved carbon solvent if necessary, to loosen hard carbon deposits. Rinse the parts in clean neutral solvent as they are removed from carbon solvent solution, to

prevent etching of polished surfaces.

Polished surfaces should be relatively bright without scratches or dull patches. Mating faces (F—Fig. 70) of nozzle body and holder must be absolutely clean and free from nicks, scratches or foreign materials as these surfaces must form a high pressure joint.

Clean out the small fuel feed channels using a suitable wire or drill bit as shown in Fig. 65. Insert the special groove scraper (Fig. 66) into nozzle body until nose of scraper enters fuel

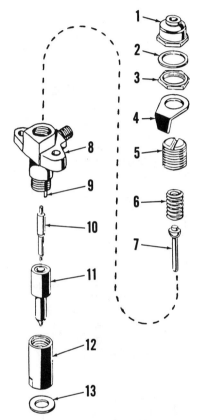

Fig. 64—Exploded view of C. A. V. injector nozzle and holder assembly. Correct opening pressure is indicated on tab (4).

1. Cap nut
2. Gasket
3. Locknut
4. Tab
5. Adjusting sleeve
6. Spring
7. Valve spindle
8. Nozzle holder
9. Dowel
10. Nozzle valve
11. Nozzle body
12. Nozzle nut
13. Seat washer

Fig. 62—A suitable injector tester is required to completely test and adjust the injector nozzles.

Fig. 63—Nozzle holes (arrows) are not located an equal distance from nozzle tip.

Fig. 65—Clean the small fuel feed hole using a suitable tool.

gallery as shown. Press nose of scraper hard against side of cavity and rotate body to clean all deposits from gallery. Using seat scraper in a rotating motion, clean carbon from valve seat as shown in Fig. 67. Clean dome cavity in nozzle tip with the 1.2 mm Dome Cavity Scraper as shown in Fig. 68. Use pin vise and 0.011 inch diameter cleaning wire to remove all carbon from the four spray holes in nozzle tip as shown in Fig. 69.

Examine the stem and seat end of nozzle valve (10—Fig. 64) and remove any carbon deposit using a clean, lint free cloth. Use extreme care, however, as any burr or small scratch may cause valve leakage or spray pattern distortion. If valve seat has a dull circum-

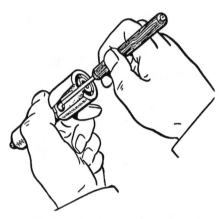

Fig. 66—Use the hooked scraper to clean fuel gallery.

Fig. 67—Clean valve seat using brass scraper as shown.

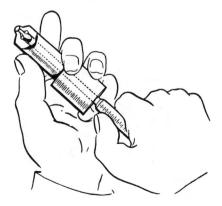

Fig. 68—Clean dome cavity in nozzle tip using the special reamer as shown.

ferential ring indicating wear or pitting, or if valve is blued, the valve and body assembly should be turned over to an authorized diesel service station for possible reclaiming.

Before reassembling, thoroughly rinse all parts in clean diesel fuel and make sure that all carbon is removed from nozzle holder nut. Install nozzle assembly and cap nut making sure that valve stem is located in hole of holder body and that the two dowels (9) enter proper holes in nozzle body. Tighten holder nut (12) to a torque of 50 ft.-lbs.

Reinstall spindle (7), spring (6) and adjusting sleeve (5), if these items were removed. Connect the injector to a nozzle tester and readjust opening pressure as outlined in paragraph 49. Recheck the injector as outlined in paragraphs 50, 51 and 52. If injector fails to pass these tests, renew needle and nozzle assembly.

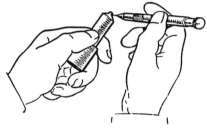

Fig. 69—Clean spray holes in nozzle tip using a pin vise and 0.011 wire probe.

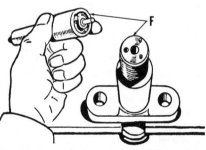

Fig. 70—Lapped pressure faces (F) must be clean when nozzle is installed.

Fig. 71—Installed view of injection pump showing drive gear retaining cap screws and linkage adjustments.

INJECTION PUMP

All Models

The injection pump is a completely sealed unit. No service work of any kind should be attempted on the pump or governor unit without the use of special pump testing equipment and special training. Inexperienced or unequipped personnel should never attempt to overhaul a diesel injection pump.

54. ADJUSTMENT. The slow idle stop screw (Fig. 71) should be adjusted with engine warm and running, to provide the recommended slow idle speed of 800-850 rpm. Check to make sure that throttle arm contacts the slow idle screw and high speed screw when throttle lever is moved to slow and fast positions. Also check to be sure that the stop lever arm moves fully to operating position when stop button is pushed in, and shuts off the fuel to injectors when stop button is pulled. The maximum engine speed adjusting screw is factory set and the adjustment is sealed. Governed speed under load should be 2000 rpm, with a high idle (no load) speed of 2170-2220 rpm. Refer to paragraph 56 for pump timing adjustment.

55. REMOVE AND REINSTALL. Before attempting to remove the injection pump, thoroughly wash the pump and connections with clean diesel fuel or an approved solvent. Disconnect throttle control rod (T—Fig. 73) from governor arm by removing nut (N) and withdrawing stud (S) from arm. Disconnect stop control cable from stop lever and cable housing from bracket. Remove the inspection cover from timing gear cover as shown in Fig. 71 and remove the three cap screws retaining injection pump drive to pump shaft. Disconnect fuel inlet, outlet and high pressure lines from pump, capping

Fig. 72—Check to see that timing scribe marks are present and aligned when pump is installed.

all connections to prevent dirt entry. Check to see that timing scribe marks (Fig. 72) are present and aligned, remove the three flange stud nuts, then withdraw the pump as shown.

Normal installation of injection pump can be accomplished without reference to crankshaft timing marks or internal timing marks on injection pump. Be

Fig. 73—Injection pump showing linkage and adjustments.

H. High speed stop screw	N. Nut
I. Idle speed stop screw	S. Stud
	T. Throttle link

Fig. 74—Injection pump with timing cover removed, showing "E" timing mark properly aligned with square end of snap ring. Refer to text.

sure timing scribe marks (Fig. 72) are aligned and reverse the removal procedure. Bleed fuel system as outlined in paragraph 44. Check injection pump timing if necessary, as outlined in paragraph 56.

56. PUMP TIMING TO ENGINE. The injection pump drive shaft contains a milled slot (Fig. 72) which engages a dowel pin (D—Fig. 43) in injection pump drive gear. Thus, injection pump can be removed and reinstalled without

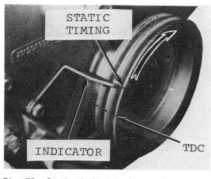

Fig. 75—Static timing marks are located on crankshaft pulley of must models as shown. Early 1080 models may have timing marks on flywheel which are visible after removing plug from left side.

regard to timing position.

NOTE: Injection pump gear cannot become unmeshed from idler gear without removal of timing gear cover, therefore timing is not disturbed by removal and installation of pump.

To check the pump timing, shut off the fuel and remove pump timing window. Refer to Fig. 74. Turn the crankshaft until the static timing mark (On flywheel, engines prior to Ser. No. 318 UA 373; crankshaft pulley on later models) is properly aligned with timing pointer. Refer to Fig. 75. With timing marks aligned, "E" scribe mark on injection pump rotor should align with square end of snap ring as shown in Fig. 74. The mounting holes in pump mounting flange are elongated to permit minor timing variations. Correct static timing is 28° BTDC for 1080 models, 20° BTDC for 1085 models.

If timing marks cannot be properly aligned by shifting pump on mounting studs, timing gear cover must be removed as outlined in paragraph 21 and the gears retimed.

THROTTLE LINKAGE

All Models

57. Refer to Fig. 76 for an exploded view of throttle linkage and associated parts. Nominal length of hand throttle rod (6) should be 8 15/16-9 inches and center distance of clamp bolt of spring anchor (9) should be 7 1/8 inches from center of ball joint (7). With hand throttle lever (1) 1/16-inch from top of slot in instrument panel, governor arm on injection pump should be in contact with maximum speed stop screw and foot throttle pedal (17) should be 1/16-inch from surface of operator's platform. Adjust hand throttle by loosening clamp screw in throttle lever arm (12) and turning arm on cross shaft. Adjust foot throttle by loosening clamp screw

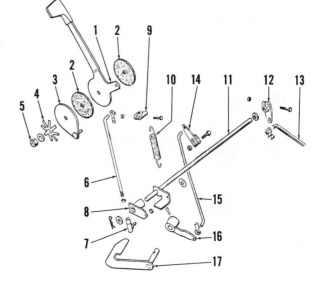

Fig. 76—Exploded view of throttle linkage and associated parts.

1. Hand lever
2. Disc
3. Friction plate
4. Spring
5. Adjusting nut
6. Rod
7. End
8. Arm
9. Spring anchor
10. Compensating spring
11. Cross shaft
12. Lever arm
13. Rod
14. Lever
15. Rod
16. Pedal arm
17. Throttle pedal

in foot throttle arm (14) and turning arm on shaft. Adjust friction nut (5) if necessary until a breakaway pressure of 9 lbs. is required to move the lever.

FUEL TANKS

REMOVE AND REINSTALL

All Models

58. The dual, saddle type fuel tanks are located at each side of tractor underneath the operator's platform. Refer to Figs. 77 and 78. Tanks are interconnected by a crossover hose at lower front and a vent line at top center. Tanks can be lowered on a wheel dolly or floor jack and rolled out to rear as shown in Fig. 78.

COOLING SYSTEM

RADIATOR

All Models

59. A 7 psi pressure type radiator cap is used. Cooling system capacity is 16.5 U.S. quarts.

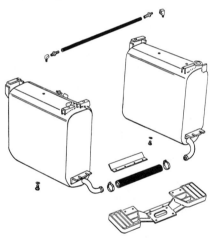

Fig. 77—Exploded view of saddle type dual fuel tanks showing crossover hose and vent line.

Fig. 78—Tanks can be lowered on wheel dolly or floor jack and rolled out to rear. Refer to paragraph 58.

To remove the radiator, first drain cooling system and remove hood and grille. Remove air inlet tube and oil hydraulic lines leading to front of tractor. Remove the eight cap screws attaching fan shroud to radiator and move shroud back toward engine. Disconnect upper and lower radiator hoses and remove cap screws securing radiator to support frame then lift radiator straight up out of frame assembly. Install by reversing the removal procedure.

THERMOSTAT

All Models

60. The by-pass type thermostat is located in the water pump outlet elbow on top of water pump housing. Thermostat starts to open at a temperature of 177°-183°F. and is completely open at

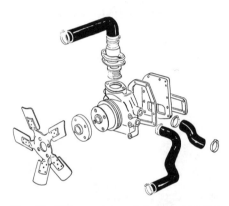

Fig. 79—Water pump and hoses, showing proper assembly.

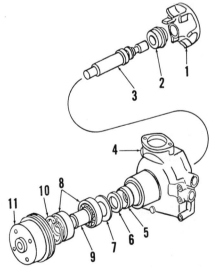

Fig. 80—Exploded view of water pump showing component parts.

1. Impeller	7. Flange
2. Seal	8. Bearings
3. Shaft	9. Spacer
4. Housing	10. Snap ring
5. Retainer	11. Pulley
6. Front seal	

208° F. Refer to Fig. 79 for an exploded view of water pump and associated parts including thermostat.

WATER PUMP

All Models

61. Refer to Fig. 80 for an exploded view of water pump and associated parts. To disassemble the removed pump, use a press or puller and remove pulley (11), then press pump shaft (3) and impeller (1) out rear of housing. Remove pump seal (2) and snap ring (10) and; applying pressure to flange (7), press the two bearings (8) and spacer (9) forward out of housing bore.

With the exception of impeller (1) and housing (4), all parts are available individually. Impeller is available in a repair kit which includes all gaskets and items (2, 3 and 5 through 10). If housing (4) is damaged, renew the pump. Assemble as follows:

Install rear bearing (8) on shaft (3), shielded side first. Install spacer (9) then front bearing (8) open side first. Pack both bearings and the cavity between bearings ½ full of High Temperature Lithium Grease such as Shell Alvania #2. Position flange (7) on shaft with cupped side toward bearing. Install front seal (6) and retainer (5) on shaft against the flange (7), then press shaft, bearing, flange and seal assembly into pump housing (4) until it bottoms. Install snap ring (10). Support rear end of shaft and press pulley on shaft until it bottoms against front bearing. Apply a light coating of sealing compound around outside brass portion of rear seal (2) and install the seal, applying pressure only to outer flange. Rotate shaft and pulley to check for binding or misassembly, then press impeller on rear of shaft until a clearance of 0.027-0.035-inch exists between impeller blades and housing when measured through outlet hose connection as shown in Fig. 81.

Fig. 81—Measure impeller clearance using a feeler gage as shown.

ELECTRICAL SYSTEM

ALTERNATOR & REGULATOR

Models With Separate Regulator

62. ALTERNATOR. A "DELCO-TRON" generator (alternator) is used. Units are negative ground.

The only test which can be made without removal and disassembly of alternator is output test. Output should be approximately 32 amperes at 5000 alternator rpm.

IMPORTANT: Outlet terminal post of alternator is equipped with a fuse/washer as shown in Fig. 82. If fuse burns out, charging output is cut off from white (output) wire. Current still flows to the blue wire leading to voltage regulator terminal. With charging flow to the battery cut off, voltage rises in the blue (control) wire causing the regulator to cut back the charging current. If generator shows no output, check the fuse before proceeding with further disassembly. Do not attempt to bypass the fuse or connect white wire directly to output terminal. Renew the fuse if damaged, and reassemble as shown in Fig. 82.

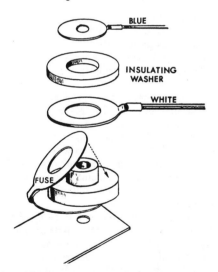

Fig. 82—Exploded view of alternator output terminal components showing output fuse/washer which wraps around terminal insulator. Check the fuse if output circuit is dead.

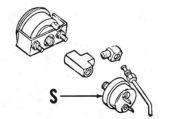

Fig. 83—Alternator field coils are energized by closing of oil pressure switch (S), thus breaking charging circuit if key switch is not turned. Refer to text.

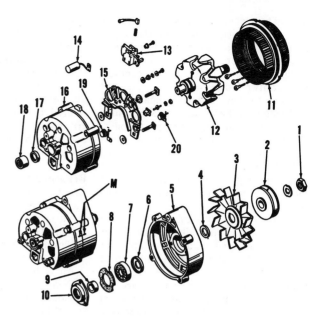

Fig. 84—Exploded view of alternator of the type used with external regulator.

1. Pulley nut
2. Drive pulley
3. Fan
4. Collar
5. Drive end frame
6. Slinger
7. Bearing
8. Gasket
9. Collar
10. Bearing retainer
11. Stator assembly
12. Rotor assembly
13. Brush holder
14. Capacitor
15. Heat sink
16. Slip ring end frame
17. Felt seal & retainer
18. Needle bearing
19. Negative diode (3 used)
20. Positive diode (3 used)

Also, alternator field coils are energized by battery current supplied through the oil pressure switch (S—Fig. 83) which is attached by a tee fitting to underneath side of oil pressure gage. Check switch or wiring if proper charging current cannot be obtained. NOTE: Fuel gage will also show EMPTY if oil pressure switch or wiring is defective.

To disassemble the alternator, first place match marks (M—Fig. 84) on the two frame halves (5 & 16), then remove the four through bolts. Pry frame apart with a screwdriver between stator frame (11) and drive end frame (5). Stator assembly (11) must remain with slip ring end frame (16) when unit is separated.

NOTE: When frames are separated, brushes will contact rotor shaft at bearing area. Brushes MUST be degreased if they are to be re-used.

Clamp the iron rotor (12) in a protected vise only tight enough to permit loosening pulley nut (1). Rotor and end frame can be separated after pulley is removed. Check bearing surfaces of rotor shaft for wear or scoring. Examine slip ring surfaces for scoring or wear and windings for overheating or other damage. Check rotor for grounded, shorted or open circuits using an ohmmeter as follows:

Refer to Fig. 85 and touch the ohmmeter probes to points (1-2) and (1-3); a reading near zero will indicate a ground. Touch ohmmeter probes to the two slip rings (2-3); reading should be 4.6-5.5 ohms. A higher reading will indicate an open circuit and a lower reading will indicate a short. If windings are satisfactory, mount rotor in a lathe and check runout at slip rings using a dial indicator. Run-out should not exceed 0.002. Slip ring surfaces can

be trued if runout is excessive or if surfaces are scored. Finish with 400 grit or finer polishing cloth until scratches or machine marks are removed.

Disconnect the three stator leads and separate stator assembly (11—Fig. 84) from slip ring end frame assembly. Check stator windings for grounded or open circuits as follows: Connect ohmmeter leads successively between each pair of stator leads. Readings should be equal and relatively low. A high reading would indicate an open lead. Connect ohmmeter leads to any stator lead and to stator frame. The three stator leads have a common connection in the center of the windings, a reading other than infinity would indicate a grounded winding circuit. A short circuit within the stator windings cannot be readily determined by test because of the low resistance of the windings.

Three negative diodes (19) are located in slip ring end frame (16) and three positive diodes in heat sink (15). Diode should test at or near infinity in one direction when tested with an ohmmeter, and at or near zero when

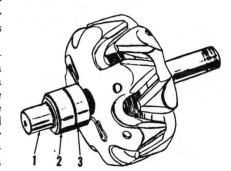

Fig. 85—Removed rotor assembly showing test points to be used when checking for grounds, shorts and opens.

ohmmeter leads are reversed. Renew any diode with approximately equal meter readings in both directions. Diodes must be removed and installed using an arbor press which contacts only outer edge of diode. Do not attempt to drive a faulty diode out of end frame or heat sink, as shock may cause damage to the other good diodes. If all diodes are being renewed, make certain the positive diodes (marked with red printing) are installed in heat sink and negative diodes (marked with black printing) are installed in end frame.

Brushes are available only in an assembly which includes brush holder (13). Brush springs are available for service and should be renewed if heat damage or corrosion is evident. If brushes are re-used, make sure all grease is removed from surface of brushes before unit is reassembled. When reassembling, install brush springs and brushes in holder, push brushes up against spring pressure and insert a short piece of straight wire through hole (W—Fig. 86) and through end frame (16—Fig. 84) to outside. Withdraw wire only after alternator is assembled.

Capacitor (14) connects to the heat sink and is grounded to end frame. Capacitor protects the diodes from voltage surges.

Remove and inspect ball bearings (7). If bearing is in satisfactory condition, fill bearing ½-full with Delco-Remy Lubricant No. 1960373 and reinstall. Inspect needle bearing (18) in slip ring end frame. This bearing should be renewed if its lubricant is exhausted; no attempt to re-lubricate the bearing should be made. Press old bearing out toward the inside and new bearing in from outside until bearing is flush with outside of end frame. Saturate felt seal with SAE 20 oil and install seal and retainer assembly.

Reassemble alternator by reversing the disassembly procedure. Tighten pulley nut to a torque of 45 ft.-lbs.

NOTE: A battery powered test light can be used instead of an ohmmeter for all electrical checks except shorts in rotor winding; however when checking diodes, test light must not be more than 12 volts.

62A. REGULATOR. A Delco-Remy standard two-unit regulator is used. Except for output (fuse) terminal (Fig. 82) on alternator, quick disconnect plugs are used at regulator and alternator. Production regulator is riveted to shock mount; service units are shipped less mount and are attached with screws. Test specifications are as follows:

Regulator Model 1119513

Ground Polarity	Negative
Field Relay	
Air Gap	0.015
Point Opening	0.030
Closing Voltage Range	3.8-7.2
Voltage Regulator	
Air Gap	0.067*
Point Opening	0.014
Voltage setting:	
at 65°F.	13.9-15.0
at 85°F.	13.8-14.8
at 105°F.	13.7-14.6
at 125°F.	13.5-14.4
at 145°F.	13.4-14.2
at 165°F.	13.2-14.0
at 185°F.	13.1-13.9

*The specified air gap setting is for bench repair only; make final adjustments to obtain specified voltage, with lower contacts opening at not more than 0.4 volt less than upper contacts. Temperature (ambient) is measured ½-inch away from regulator cover and adjustment should be made only when regulator is at normal operating temperataure.

Models With SI Series Alternator

63. ALTERNATOR WITH INTEGRAL REGULATOR. A Delco-Remy 10-SI type 106 alternator with integral, solid state regulator is used on 1085 models. Several different alternators have been used. Be sure of model when testing. Tractors without cabs use 37 or 42 amp units and models with cabs use alternators rated at 61 or 63 amps. Cold output RPM is alternator RPM (not engine speed).

1100578

Field Current (80°F.)	
Amperes	4.0-4.5
Volts	12
Cold Output	
Amperes at 2000 RPM	22
Amperes at 5000 RPM	33
Rated Hot Output (Amperes) at	
Maximum Operating Speed	37

1100579 & 1100583

Field Current (80°F.)	
Amperes	4.0-4.5
Volts	12
Cold Output	
Amperes at 2000 RPM	25
Amperes at 5000 RPM	37
Rated Hot Output (Amperes) at	
Maximum Operating Speed	42

1100581

Field Current (80°F.)	
Amperes	4.0-4.5
Volts	12
Cold Output	
Amperes at 2000 RPM	30
Amperes at 5000 RPM	55
Rated Hot Output (Amperes) at	
Maximum Operating Speed	61

1100589

Field Current (80°F.)	
Amperes	4.0-4.5
Volts	12
Cold Output	
Amperes at 2000 RPM	30
Amperes at 5000 RPM	57
Rated Hot Output (Amperes) at	
Maximum Operating Speed	63

To disassemble the alternator, first scribe matching marks (M—Fig. 87) on the two frame halves (4 and 16), then remove the four through-bolts. Pry frame apart with a screwdriver between stator frame (12) and drive end frame (4). Stator assembly (12) must remain with slip ring end frame (16) when unit is separated.

NOTE: When frames are separated, brushes will contact rotor shaft at bearing area. Brushes MUST be cleaned of lubricant if they are to be reused.

Clamp the iron rotor (13) in a pro-

Fig. 86—Exploded view of brush holder assembly. Insert wire in hole (W) to hold brushes up. Refer to text.

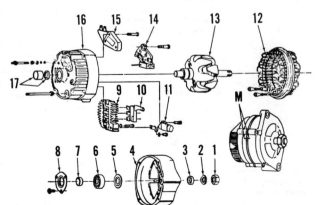

Fig. 87—Exploded view of alternator with internal mounted solid state regulator. Note match marks (M) on end frames.

1. Pulley nut
2. Washer
3. Spacer (outside drive end)
4. Drive end frame
5. Grease slinger
6. Ball bearing
7. Spacer (inside drive end)
8. Bearing retainer
9. Bridge rectifier
10. Diode trio
11. Capacitor
12. Stator
13. Rotor
14. Brush holder
15. Solid state regulator
16. Slip ring end frame
17. Bearing & seal assembly

tected vise, only tight enough to permit loosening of pulley nut (1). Rotor end frame can be separated after pulley and fan are removed. Check bearing surface of rotor shaft for visible wear or scoring. Examine slip ring surface for scoring or wear, and rotor winding for overheating or other damage. Check rotor for grounded, shorted or open circuits using an ohmmeter as follows:

Refer to Fig. 85 and touch the ohmmeter probes to points (1-2) and (1-3); a reading near zero will indicate a short circuit to ground. Touch ohmmeter probes to the slip rings (2-3); reading should be 5.3-5.9 ohms. A higher reading will indicate an open circuit and a lower reading will indicate an internal short. If windings are satisfactory, mount rotor in a lathe and check runout at slip rings using a dial indicator. Runout should not exceed 0.002. Slip ring surfaces can be trued if runout is excessive or if surfaces are scored. Finish with 400 grit or finer polishing cloth until scratches or machine marks are removed.

Before removing stator, brushes or diode trio, refer to Fig. 88 and check for gounds between points A and C and B to C with an ohmmeter, using the lowest range scale. Then reverse the lead connections. If both A to C readings or both B to C readings are the same, the brushes may be grounded because of defective insulating washer and sleeve at the two screws. If the screw assembly is not damaged or grounded, the regulator is defective.

To test the diode trio, first remove the stator. Then remove the diode trio, noting the insulator positions. With an ohmmeter, check between points A and D (Fig. 89) and then reverse the ohmmeter lead connections. If diode trio is good it will give one high and one low reading. If both readings are the same, the diode trio is defective. Repeat this test at points B and D and at C and D.

The rectifier bridge (Fig. 90) has a grounded heat sink (A) and an insulated heat sink (E) that is connected to the output terminal. Connect ohmmeter to the grounded heat sink (A) and to the flat metal strip (B). Then reverse the ohmmeter lead connections. If both readings are the same, the rectifier bridge is defective. Repeat this test between points A and C, A

and D, B and E, C and E, and D and E. Capacitor (11—Fig. 87) connects to the rectifier bridge and grounds to end frame, and protects the diodes from voltage surges.

Test the stator windings for grounded or open circuits as follows: Connect ohmmeter leads successively between each pair of leads. A high reading would indicate an open circuit.

NOTE: The three stator leads have a common connection in the center of the windings. Connect ohmmeter leads between each stator lead and stator frame. A very low reading would indicate a shorted circuit. A short circuit within the stator windings cannot be readily determined by test because of the low resistance of the windings.

Brushes and springs are available only as an assembly which includes brush holder (14—Fig. 87). If brushes are reused, make sure all grease is removed from surface of brushes before unit is reassembled. When reassembling, first install regulator and then brush holder, springs and brushes. Push brushes up against spring pressure and insert a short piece of straight wire through hole and through end frame to outside. Be sure that the two screws at points A and B (Fig. 88) have insulating washers and sleeves.

NOTE: A ground at these points will cause no output, or controlled output. Withdraw the wire under brushes only after alternator is assembled.

Remove and inspect ball bearing

Fig. 88—Test points for brush holder. Refer to text.

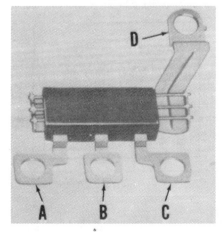

Fig. 89—Diode trio test points. Refer to text.

Fig. 90—Bridge rectifier test points. Refer to text.

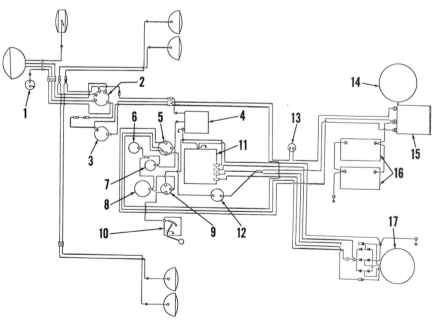

Fig. 91—Wiring diagram typical of that used for 1080 models.

1. Auxiliary socket	6. Oil pressure gage	10. Fuel sending unit	14. Starter motor
2. Light switch	7. Temperature gage	11. Regulator	15. Solenoid
3. Starter switch	8. Tachometer	12. Oil pressure switch	16. Batteries
4. Inverter	9. Fuel gage	13. Neutral switch	17. Alternator
5. Ammeter			

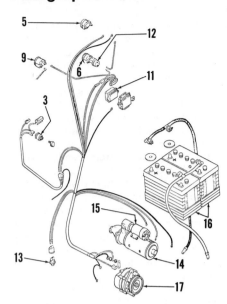

Fig. 92—Schematic view of wiring. Refer to Fig. 91 for parts identification.

(6—Fig. 87). If bearing is in satisfactory condition, fill bearing ¼-full with Delco-Remy lubricant No. 1948791 and reinstall. Inspect needle bearing (17) in slip ring end frame. This bearing should be renewed if its lubricant supply is exhausted; no attempt should be made to relubricate and reuse the bearing. Press old bearing out toward inside and press new bearing in from outside until flush with outside of end frame. Saturate felt seal with SAE 20 oil and install seal.

Reassemble alternator by reversing the disassembly procedure. Tighten pulley nut to a torque of 50 ft.-lbs.

STARTING MOTOR

All Models

64. A Delco-Remy starting motor is used. Specifications are as follows:

Delco-Remy 1113676
Brush spring tension......35 oz. (min.)
No-Load test
Volts.........................9
Amperes (w/solenoid).......125-185
RPM....................4700-7600

Starter drive pinion clearance is not adjustable, however, some clearance must be maintained between end of pinion and starter drive frame, to assure solid contact of the heavy-duty magnetic switch. Normal pinion clearance should be within the limits of 0.010-0.140 inch. Connect a 6-volt battery to solenoid terminals when measuring pinion clearance, to keep armature from turning.

CIRCUIT DESCRIPTION

All Models

65. Refer to Fig. 91 or Fig. 93 for a schematic wiring diagram. The two 95

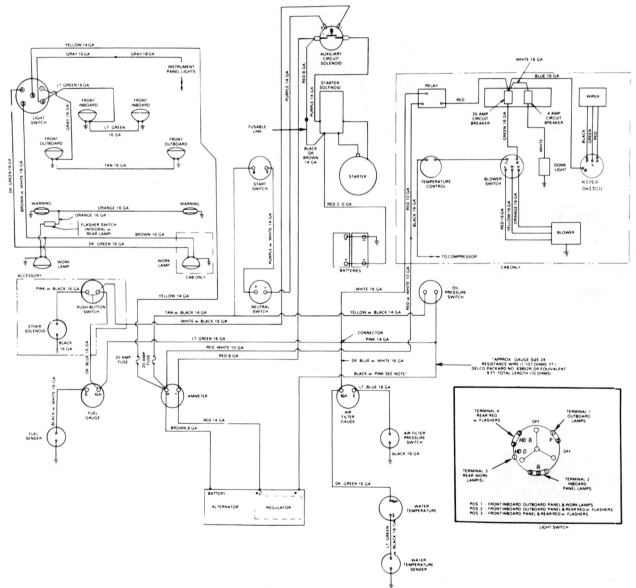

Fig. 93—Wiring diagram typical of all 1085 models.

Ampere Hour, 12 Volt batteries are connected in parallel and negative posts are grounded. Make sure both batteries are of equal capacity and condition. Quick disconnect couplings are used at all main connections.

ENGINE CLUTCH

All Models are equipped with a single disc, split torque clutch, with power to the drive wheels being interrupted by depressing the clutch pedal and continuous power being supplied by splined hub in clutch cover to the shaft which drives the hydraulic pumps and power take-off shaft.

ADJUSTMENT

All Models

66. FREE PLAY. Refer to Fig. 94 for an external view of clutch release shaft. To adjust the free play, insert a punch or small bar in hole in release shaft and rotate shaft clockwise until throwout bearing contacts clutch fingers. At this time, clearance between release arm stop and housing boss should be 1/8-inch when measured as shown. If it is not, loosen clamp bolt and rotate the arm on shaft until clearance is correct. Tighten clamp bolt

Fig. 94—Clutch free play adjustment is correct when clearance between release arm stop and housing boss measures 1/8 inch as shown.

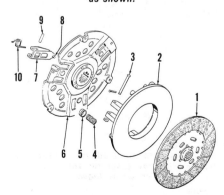

Fig. 95—Exploded view of split torque clutch of the type used.

1. Clutch disc	6. Cover
2. Pressure plate	7. Release lever
3. Pin	8. Adjusting screw
4. Spring	9. Pin
5. Cup	10. Spring

securely and recheck the clearance.

67. LINKAGE ADJUSTMENT. With free play adjusted as outlined in paragraph 66, adjust length of connecting link until linkage pin can be inserted with release arm against housing stop and pedal raised by hand as high as it will go.

TRACTOR SPLIT

All Models

68. To detach engine from transmission assembly, first drain cooling system and remove hood and side panels. Remove batteries and battery box, side rails and instrument panel side sheets. Disconnect hydraulic steering lines, fuel line, oil gage line and temperature indicator sending unit. Disconnect electric wiring leading to starter and alternator. Disconnect throttle and stop controls.

Support both halves of tractor separately, remove the attaching cap screws and separate the tractor. Rejoin the tractor by reversing the split procedure. Tighten the 7/16-inch diameter flange cap screws to a torque of 55 ft.-lbs. and the 5/8-inch diameter screws to 155 ft.-lbs.

OVERHAUL

All Models

69. Refer to Fig. 95 for an exploded view of clutch disc and pressure plate assembly. To remove the clutch after tractor is split, remove the six retaining cap screws evenly and lift off the unit.

Before disassembling, mark pressure plate and cover for correct assembly. Position the clutch assembly in a press as shown in Fig. 97 and apply only enough pressure to relieve tension on pins. Remove pins (3—Fig. 95) and lift off cover.

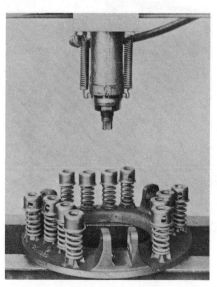

Fig. 96—Springs and spring cups correctly installed on pressure plate.

Inspect release levers and pins for wear or damage and pressure plate for scoring, heat checks or wear at actuating pin holes. Pressure plate may be refaced if facilities are available. Inspect pto drive hub in cover for spline wear or looseness. Springs color coded light blue for 1080 models should have free length of approximately 2¾ inches, and should test 160 lbs. when compressed to a height of 1-13/16 inches. Springs for 1085 models are not color coded but should have free length of 3.19 inches and should test 195 lbs. when compressed to 2.06 inches. On all models, inspect springs for heat discoloration or other damage.

Assemble by reversing the disassembly procedure, making certain that previously affixed marks on cover and pressure plate are aligned. Install pins with heads leading in normal rotation as shown in Fig. 97. Clutch fingers should be adjusted to equal height after installation using a new

Fig. 97—Release lever pins should be installed with heads leading in normal direction of rotation, as shown.

1. Floating pin
2. Anti-Rattle spring

Fig. 98—Clutch assembly mounted on flywheel. Tighten cover cap screws (1) to a torque of 75-85 ft.-lbs. and check finger adjustment (2).

clutch disc and adjusting gage MFN 202E.

Install the clutch assembly using a suitable pilot such as Massey Ferguson Special Tool, MFN 753. Tighten clutch cover retaining cap screws (1—Fig. 98) alternately and evenly to a torque of 75-85 ft.-lbs.

TRANSMISSION

Model MF1080 tractors may be equipped with a gear change transmission having six forward and two reverse speeds; or a modified. "Multipower" version which is additionally equipped with a hydraulically operated high-low range unit which may be shifted while tractor is moving under load. This "Multipower" transmission provides twelve forward speeds and four reverse speeds.

Model MF1085 tractors may be equipped with a "Multipower" transmission which is similar to type used on MF1080 models; or a gear change transmission which has eight speeds forward and two reverse speeds.

Many service procedures and parts are similar for the three different types of transmissions. Refer to the appropriate following section for service of a specific type: Six speed, Eight speed or "Multipower" (12 Speed).

TRANSMISSION REMOVAL

All 1080 Models

70. To remove the complete transmission unit from tractor, first drain transmission and hydraulic system fluid. Remove the operator's platform and disconnect and remove fuel tanks as outlined in paragraph 58. Detach (split) engine from clutch housing as

outlined in paragraph 68. Unbolt and remove transmission top cover, steering support and instrument panel as a unit as shown in Fig. 99.

Reaching through transmission top opening, disconnect auxiliary hydraulic pump flexible hose at front end. Disconnect brake control rods. Support transmission and rear axle center housing separately, remove the attaching bolts and lift off the transmission assembly.

Install by reversing the removal procedure. Tighten transmission to center housing flange cap screws to a torque of 55 ft.-lbs.

All 1085 Models

70A. To remove the complete transmission unit from tractor, first drain transmission and hydraulic system fluid. Remove the operator's platform, then detach (split) engine from clutch housing as outlined in paragraph 68. Remove instrument console (Fig. 100) then unbolt and remove shift cover from top of transmission. Disconnect hydraulic line inside transmission case and remove fuel tanks. Support transmission and rear axle center housing separately, remove attaching bolts and lift off the transmission assembly.

Install by reversing removal procedure. Tighten transmission to center housing screws and shift cover retaining screws to 55 ft.-lbs. torque.

TRANSMISSION TOP COVER

All 1080 Models

71. **REMOVE AND REINSTALL.** To remove the transmission top cover, first drain cooling system and remove hood and side covers. Remove batteries and battery box. Disconnect hydraulic steering lines, oil gage line and tem-

perature indicator sending unit. Disconnect wires from starter, alternator and starter safety switch. Disconnect brake pedal and clutch control rods and "Multipower" shift linkage if tractor is so equipped. Disconnect "Multipower/IPTO" oil cooler and filter lines from clutch housing connections. Remove flange mounting bolts and lift lines from clutch housing connections. Remove flange mounting bolts and lift off instrument panel, steering support and transmission top cover as a unit.

Overhaul the cover controls as outlined in paragraph 71A. When installing the cover, make sure levers (15, 16 & 17—Fig. 101) properly engage shift forks and planetary range selector in transmission case and that all plungers shift through the three detent positions, then install and tighten the cover cap screws.

If steering support is detached from cover or shifting linkage disconnected, make sure 2nd & 3rd speed arm (5) connects to left plunger (12) and that 1st & Reverse speed link (6) connects to center plunger (11). Shorten or lengthen the three links (7, 8 & 9) if necessary, until transmission shifts easily and fully to all gear positions. Complete the installation by reversing the removal procedure.

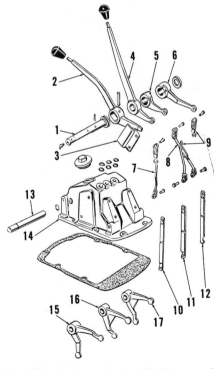

Fig. 101—Dash mounted shift linkage and transmission top cover for 1080 models is shown exploded.

1. Shaft	10. Plunger
2. Gear shift lever	11. Plunger
3. Pivot bracket	12. Plunger
4. Range shift lever	13. Pivot shaft
5. Shift arm	14. Top cover
6. Shift arm	15. Lever
7. Link	16. Lever
8. Link	17. Lever
9. Link	

Fig. 99—On 1080 models, transmission top cover, instrument panel and steering support can be removed as a unit after clutch split as shown.

Fig. 100—View showing model 1085 instrument console being removed.

71A. OVERHAUL. To overhaul the removed transmission cover, refer to Fig. 101. Remove the retaining snap ring and withdraw pivot shaft (13) and levers (15, 16 and 17). Remove links (7, 8 & 9) and withdraw plungers (10, 11 &

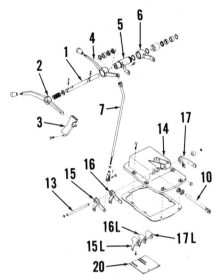

Fig. 101A—Exploded view of 1085 shift controls for models with "Multipower".

1. Shaft	13. Pivot shaft
2. Lever	14. Cover
3. Guide	15. Belcrank (Planetary)
4. Lever (Planetary)	15L. Lever
5. Shift arm	16. Belcrank (2nd & 3rd)
(2nd & 3rd)	16L. Lever
6. Shift arm	17. Belcrank (1st
(1st & Rev.)	& Rev.)
7. Link rod	17L. Lever
10. Plunger (3 used)	20. Guide plate

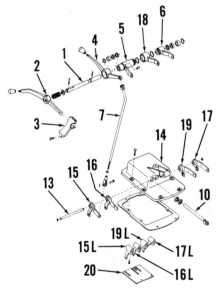

Fig. 102—Exploded view of shift controls for models with 8-speed transmission.

1. Shaft	15. Planetary belcrank
2. Lever	15L. Lever
3. Guide	16. Belcrank (2nd & 3rd)
4. Lever	16L. Lever
5. Shift arm	17. Belcrank (1st
(2nd & 3rd)	& Rev.)
6. Shift arm	17L. Lever
(1st & Rev.)	18. Shift arm (4th)
7. Link rod	19. Belcrank (4th)
10. Plunger (4 used)	19L. Lever
13. Pivot shaft	20. Guide plate
14. Cover	

12) downward out of housing (14). O-ring seals are used for plunger bores and bore for pivot shaft (13).

To disassemble the shift linkage, drive out the pin securing lever (2) to shaft (1) and withdraw shaft from bore in steering support. Assemble by reversing the disassembly procedure and install and adjust as outlined in paragraph 71.

All 1085 Models

72. REMOVE AND REINSTALL. To remove the shift cover from top of transmission, remove center plate from operator's platform and lower plate from instrument console. Disconnect shift rods from bellcranks, remove screws attaching shift cover to top of transmission, then carefully lift the cover off.

Refer to paragraph 71A for overhaul of "Multipower" models; paragraph 72A for eight speed models. When installing, be sure that levers properly engage shift forks and planetary range selector in transmission case and that plungers shift through the detent positions, then install and tighten screws to 55 ft.-lbs. torque. Complete installation by reversing removal procedure.

72A. OVERHAUL. For eight speed models, refer to Fig. 102 and Fig. 103. Be sure that forks are correctly installed as shown and safety wire the forks to the rails. Use "Loctite" or equivalent on screws retaining guide plate. Position all shift rails and forks in neutral, and be sure that all detent pins and springs are located in transmission case. Position gasket as shown in Fig. 105 and carefully lower the shift cover into position making sure that forks in cover correctly engage gear selector forks in transmission. Tighten the cover screws to 55 ft.-lbs. torque.

CONTROL VALVE

A valve body (2—Fig. 106) is used on all models. On "Multipower" models, the body contains the Multipower Shift Valve and the Multipower/IPTO regulating valve. On models without Multipower (6 or 8 speed), only the regulating valve is used and the bore for the Multipower shift valve is not drilled.

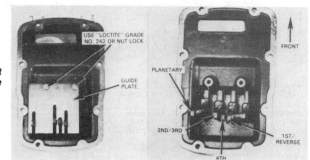

Fig. 103—Views of shift cover and rails in the cover of 8-speed models.

All Models

73. REMOVE AND REINSTALL. To remove the "Multipower" control valve or IPTO regulating valve, first detach engine from transmission housing as outlined in paragraph 68. Remove clutch release bearing, release fork and pivot shafts. Remove brake

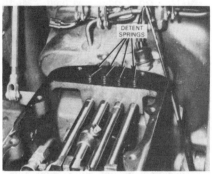

Fig. 104—Detent springs protrude through holes as shown.

Fig. 105—Gasket should cover detent springs when correctly installed.

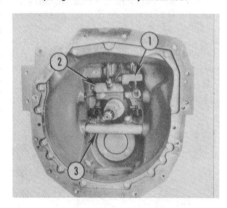

Fig. 106—Front view of transmission case after clutch split, showing multipower shift bracket (1), shift control valve (2) and brake cross shaft (3).

cross shaft. Disconnect shift linkage (models so equipped) and oil inlet tube. Unbolt and remove shift lever bracket and linkage assembly from Multipower models, remove retaining cap screws; then withdraw shaft retainer, IPTO input shaft and valve housing as a unit from transmission housing.

NOTE: On models without "Multipower" the transmission input shaft will be removed with housing and outer shaft. On "Multipower" models, the Multipower clutch unit prevents removal of transmission shaft until transmission is disassembled.

When detaching control valve from retainer, note that left, front mounting screw is sealed with a copper washer. Make sure this washer is reinstalled in the correct location when unit is reassembled.

Use a new gasket when reinstalling the valve. Make sure the long cap screw with copper sealing washer is installed in the correct hole and tighten all screws evenly to a torque of 36-48 In.-Lbs.

74. OVERHAUL. To overhaul the removed "Multipower" control valve, refer to Fig. 107 and proceed as follows:

Remove fitting (4) and withdraw shift valve (7). Remove plug (12), washer (11) and spring (10) then bump valve body (8) against a wooden block if necessary, to dislodge regulating valve (9) from its bore.

Clean all parts in a suitable solvent. Discard O-rings (5 & 6) and gaskets (2 & 11) and use new parts when reassembling. Carefully examine all parts for wear, scoring or other damage. All parts are available individually.

When reassembling the valve unit, make sure regulating valve spool (9) is

installed wide land first. Tighten plug (12) to a torque of 17-22 ft.-lbs.

NOTE: Shift valve (7) will not be used on models without "Multipower". Where shift valve is used, be sure shift valve is inserted before fitting (4) is installed.

Install the assembled valve on tractor as outlined in paragraph 73 and adjust shift linkage as in paragraph 75.

MULTIPOWER SHIFT LINKAGE
All Multipower Models

75. Shift valve spool (7—Fig. 107) must move fully to high and low positions as Multipower control lever (10—Fig. 108) is moved top and bottom of control slot.

To make the adjustment, move hand lever (10) to "HIGH" position as shown in inset. Loosen clamp bolt securing link (6) to lower rod (4) and push lower control rod (4) firmly downward as far as it will go. Retighten clamping bolt to lock the adjustment.

SHIFTER RAILS AND FORKS
All 1080 Models

76. To remove the shifter rails and forks, first remove transmission top cover as outlined in paragraph 71 and detach transmission from rear axle center housing.

Unwire and remove the set screws retaining selector and shifter forks to

rails. Remove detent spring and plunger assemblies (1—Fig. 109) and stop plate (5); then withdraw shifter rails and forks from transmission case.

Forks (4) are interchangeable but rails (2 & 3) are not. Rails should be installed with milled flat to top rear and selector lock grooves to center. Interlock pin (6) prevents the movement of the other rail when either selector rail is moved from the neutral position.

The neutral safety switch should be removed from transmission case before planetary shift rail (9) is withdrawn. Neutral safety switch plunger fits in detent notch in planetary shift rail.

All 1085 Models

76A. For models with "Multipower", remove the transmission top cover as outlined in paragraph 71 and detach transmission from rear axle center housing. Refer to paragraph 76 for remainder of removal and assembly procedure.

On models with eight speed transmission, remove the transmission shift cover as outlined in paragraph 72 and detach transmission from rear axle center housing. Carefully remove de-

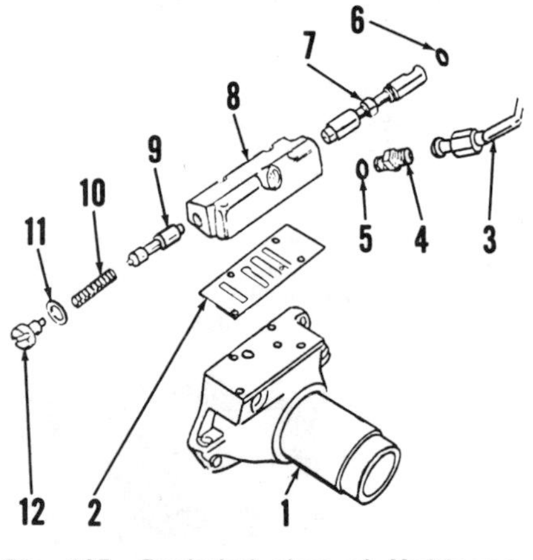

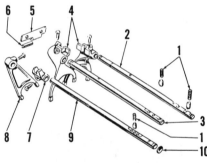

Fig. 109—Exploded view of transmission shifter rails, forks and associated parts typical of 6 and 12 speed models.

1. Shift detents
2. Second/high rail
3. Low/reverse rail
4. Shift forks
5. Stop plate
6. Interlock pin
7. Selector
8. Planetary shift fork
9. Planetary shift rail

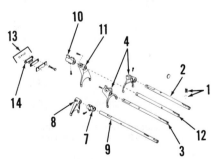

Fig. 109A—Exploded view of 8-speed transmission shifter rails, forks and associated parts. Refer to Fig. 109 for parts identification except the following.

10. Fourth gear selector
11. Fourth gear shift fork
12. Fourth gear shift rail
13. Interlock balls and pin
14. Interlock plate

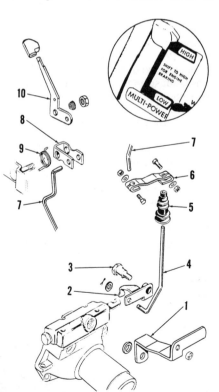

Fig. 107—Exploded view of Multipower Shift control valve and associated parts.

1. Housing
2. Gasket
3. Pressure line
4. Fitting
5. O-ring
6. O-ring
7. Shift valve
8. Valve body
9. Regulating valve
10. Spring
11. Gasket
12. Plug

Fig. 108—Exploded view of Multipower clutch control linkage and associated parts.

1. Bracket
2. Shift lever
3. Pivot bolt
4. Lower link rod
5. Boot
6. Link
7. Upper link rod
8. Spacer
9. Spring
10. Lever

tent springs and plungers (Fig. 110). Remove safety wire and all set screws (S) except for the center rail selector. Slide the planetary shift fork (8) from rail and coupler from planetary. Remove the shift rail lock mechanism from rear of case and be careful not to lose the steel balls. Slide planetary shift rail from case while removing selector. Withdraw right hand shift rail from case while removing fork. Slide left shift rail from case and shift fork. Loosen set screw from center shift selector and slide rail from case. Identify shift rails and forks as they are removed to facilitate assembly.

When assembling, slide center shift rail into case and through the selector and shift fork. Tighten selector set screw and install safety wire. Rotate rail until shift fork set screw can be tightened fully, then install safety wire. Install the left shift rail and fork, then right shift rail and fork into respective bores, tighten set screws and safety wire. Slide the planetary shift rail through rear of case and locate fork over end with set screw to the outside. Slide rail forward into case and into selector. Tighten set screws and install safety wire. Be sure that interlocking pin is through the center rail and that all rails are in "NEUTRAL". Install steel balls in lock mechanism and locate at rear of case. The steel balls should be at both ends of the center rail interlock pin and the widest flat of lock plate should be over planetary rail. Tighten screws attaching the interlock

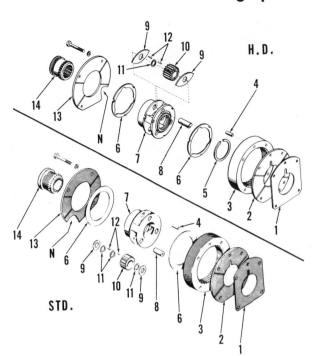

Fig. 111—Exploded view of rear (range) planetary unit and associated parts. Some difference will be noted between standard (STD.) and heavy duty (H.D.) units.

1. Front shim
2. Front plate
3. Ring gear
4. Dowel
5. Snap ring
6. Thrust washer
7. Planet carrier
8. Pinion shaft
9. Side washer
10. Pinion
11. Spacer washer
12. Needle rollers
13. Rear plate
14. Shift coupler

Fig. 110—View showing shift rails, forks, detents and springs. Set screws (S) should be safetied with wire after assembly.

mechanism to 30-35 ft.-lbs. torque. Locate the coupler around shift fork, them slide coupler into planetary while piloting shift fork over the rail. Tighten and safety wire the set screws.

REAR PLANETARY UNIT

All Models

77. To remove the rear planetary unit, first detach transmission from axle center housing as outlined in paragraph 106. Remove the planetary shift fork (8—Fig. 110) and coupling from rear of transmission case. Remove the four retaining screws and withdraw rear cover (13—Fig. 111), rear thrust washer (6) and planet carrier (7). Work planetary ring gear (3) and dowels (4) from case using screwdrivers. Remove planetary front cover and shim (1).

Planet pinion shafts (8) are a tight press fit in planet carrier (7) and heavy duty models are equipped with a snap ring (5) which holds pinion shaft in position. Use a suitable press for removing and installing pinion shafts.

Assemble and install unit as follows: Apply a light coat of grease to one side of side washers (9), bearing needles (12) and spacer washers (11). Position one side washer (9) on bench with greased side up and locate pinion (10) over the washer. On standard duty planetary unit, install one of the nine spacer washers (11), one row of bearing needles (12), a second spacer (11), the second row of bearing needles (12), a third spacer (11) and the side washer (9). On heavy duty units, the side washers (9) are oval with a flat along one side and only three spacers (11) are used, one in each pinion. Install one row of needles (12), a spacer (11), the

second row of needles and the side washer (9). Assemble the two remaining pinions for all models. Position the carrier (7) in a press with the hub side down, install the pinion, bearing and side washers assembly then press pinion shaft into carrier until flush. On heavy duty planetary, flat side of side washers should be toward inside of carrier and the snap ring (5) should be installed after all pinion shafts are in place. Gap in snap ring should be located between pinion shafts. On all models, position front plate (2) on ring gear (3) with oil grooves toward inside. Install shim (1) with cutaway on shim and front plate aligned and toward top when assembled to rear of case. Tap into position on rear case to be sure that dowels are fully into case. Use a light coat of grease to hold thrust washer (6) to the front of carrier (7) making sure that tangs on washer engage notches of carrier. Insert carrier assembly into ring gear carefully so that thrust washer doesn't fall. Make sure that carrier is fully into ring gear and over main shaft. Install the remaining thrust washer, making sure that tangs engage notches of carrier. Install the rear cover plate (13) with oil grooves inward. If unit is heavy duty, the cutout of cover plate should be over the bottom left mounting bolt hole. If unit is standard duty, the cutout of cover plate should be over bottom right mounting bolt hole. Install the bottom left mounting bolt on all models without lock washer. Install the three remaining bolts with lock washers, then torque all four mounting bolts to 30-35 ft.-lbs. Complete assembly by reversing removal procedure.

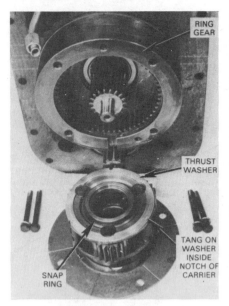

Fig. 112—View of heavy duty rear planetary carrier assembly removed from ring gear.

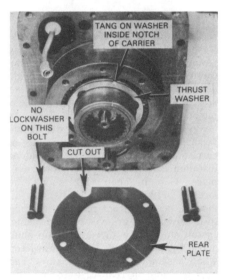

Fig. 113—On heavy duty rear planetary assemblies, cut-out of rear plate should be over the lower left mounting screw. On standard planetary the cut-out should be over lower right mounting screw.

PTO DRIVE SHAFT AND INPUT SHAFT SEALS

All 1080 Models

78. Remove the pto input shaft and housing assembly as outlined in paragraph 73.

NOTE: On models without "Multipower" transmission, the transmission input shaft will be removed with the input shaft housing and pto input shaft unit.

To disassemble the removed unit, withdraw the transmission input shaft on models without "Multipower". Unseat and remove the large snap ring (9—Fig. 118) retaining bearing (5) to shaft housing (2) and bump shaft (6)

and bearing (5) rearward out of housing (2). Bearing can be removed from shaft after removing snap ring (4). Seals (1 & 8) are both installed with sealing lips to rear. Inner seal (8) seats agsinst a shoulder in bore of shaft (6) for positive positioning. Special tools (MFN 849 and MFN 849A) are available from Massey Ferguson, Inc., for service on the seal.

Assemble by reversing the disassembly procedure.

All 1085 Models

78A. Remove the pto input shaft and housing assembly as outlined in paragraph 73. Refer to paragraph 74 for service of "Multipower" models. To disassemble the removed unit on models with eight speed transmission, proceed as follows: Unseat and remove the large snap ring (Fig. 114) retaining bearing to the housing, then bump shaft and bearing rearward out of housing. Bearing can be removed from shaft after removing smaller snap ring. Both seals should be installed with lips toward rear. Inner seal should seat against shoulder in bore of IPTO input shaft for positive positioning. Special tool (MFN 742) is available for installing the inner seal. Special tools (MFN 850 and MFN 850A) are recommended for installing the bearing and seal to correct depth in retainer (Fig. 115). Fill cavity between seal and bearing in retainer with Lithium grease (M-1105 or equivalent) before assembling. Seal protectors should be used when sliding the IPTO shaft into the retainer (Fig. 116) and when installing the shaft and retainer assembly over the main input shaft (Fig. 117). Remainder of assembly is reverse of disassembly.

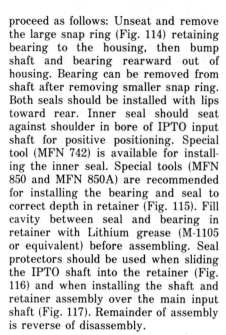

Fig. 114—View of IPTO input shaft and retainer used on 8-speed models.

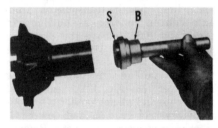

Fig. 115—Special tool (MFN 850 and MFN 850A) can be used to install bearing to correct depth using side (B) and seal to correct depth using side (S).

Fig. 116—Seal protector (MFN 0741) should be used when inserting shaft through seal in retainer.

Fig. 117—Seal protector (FT 357) should be used to prevent splines on shaft from cutting seal inside IPTO shaft. Be sure that special washer is installed.

MAIN (OUTPUT) SHAFT

All 1080 Models

79. To remove the Main (Output) shaft (19—Fig. 118), first remove rear planetary unit as outlined in paragraph 77 and shifter rails and forks as in paragraph 76.

Move low speed sliding gear (20) forward until interlock groove in main shaft splines is exposed. Insert a large blade screwdriver or similar tool in groove at rear of gear, then insert a large pry bar between front of gear (20) and center wall of transmission housing as shown in Fig. 119. Pry shaft assembly rearward until front bearing (18—Fig. 118) clears the bore in transmission case. Remove snap ring (17) and, using low speed gear (20) as a slide hammer, bump bearing (18) from front of shaft. Withdraw shaft (19) and rear bearing (23) out from rear while lifting gears (20 & 21) out top opening.

When reinstalling, place cluster gear (21) on shaft with smaller gear to front, and low gear (20) on shaft with shift fork groove toward cluster gear. Use Special Tool MFN 458 (Fig. 120) or similar thin spacer plate as shown, and bump shaft forward into bearing bore. Install snap ring (17—Fig. 118) then complete the assembly by reversing disassembly procedure.

All 1085 Models

79A. To remove the Main (Output) shaft from models with "Multipower", first remove the rear planetary unit as outlined in paragraph 77 and the shifter rails and forks as outlined in paragraph 76A. Refer to paragraph 79 for remainder of service for models with "Multipower".

For models with eight speed transmission proceed as follows: Remove all shift rails and forks as outlined in paragraph 76A and the rear planetary unit as outlined in paragraph 77. Insert a heavy screwdriver into groove of main shaft, place a hardwood block between the center rib of case and first gear, then pry the shaft rearward (Fig. 121). Remove snap ring from groove in front of main bearing (Fig. 122), then slide snap ring from rear of bearing back against the first gear (Fig. 123). Position a 2-inch open end wrench or similar tool between bearing and rear snap ring, then pull main shaft rearward sharply to remove front bearing (Fig. 124). Remove the snap ring from shaft and withdraw shaft while lifting gears out of case. The inner bearing in front end of shaft should be pressed into bore until flush with counterbore.

Fig. 119—Removing main (output) shaft on 6-speed or 12-speed models.

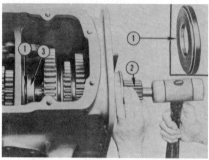

Fig. 120—A special tool (MFN 458) or equivalent is required for installing mainshaft front bearing as shown.

1. Special tool
2. Mainshaft
3. Front bearing

The rear bearing should be installed with outer snap ring toward rear. Smaller gear of cluster should be to front. Low gear should be installed with shift fork groove toward cluster gear.

TRANSMISSION INPUT SHAFT

Models With 6-Speed Transmission

80. Transmission input shaft (13—Fig. 118) can be removed as outlined in paragraph 73. Examine caged bearing (15) for wear and bearing race in shaft bore for pitting or ridging. Be sure the

Fig. 121—Pry the main (output) shaft back as shown using a block (hammer) and large screwdriver. An 8-speed model is shown.

Fig. 122—Remove snap ring from groove and slide snap ring back against first gear (Fig. 123).

Fig. 123—View showing snap ring back against first gear for removal purposes.

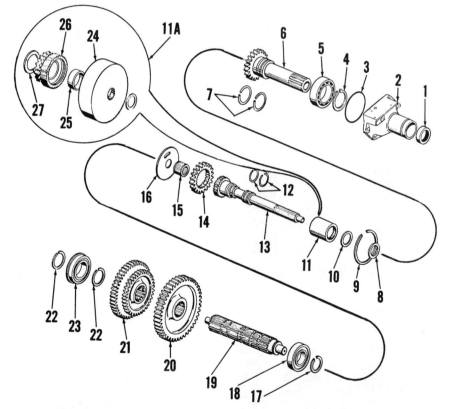

Fig. 118—Exploded view of transmission upper shafts and associated parts for 6-speed and 12-speed models. Circled parts (11A) replace spacer (11) in Multipower models.

1. Oil seal	8. Seal	13. Input shaft	20. Gear
2. Retainer	9. Snap ring	14. Input pinion	21. Cluster gear
3. O-ring	10. Thrust washer	15. Pilot bearing	22. Snap ring
4. Snap ring	11. Spacer (6-speed models)	16. Spacer	23. Bearing
5. Bearing	11A. Multipower clutch	17. Snap ring	24. Multipower clutch
6. IPTO input shaft	12. Seal rings	18. Bearing	25. Bushing
7. Seal rings		19. Main shaft	26. Overdrive pinion

Fig. 124—A 2-inch open end wrench or similar tool can be used to remove front bearing.

bearing and spacer collar (11) are installed when unit is reassembled.

Models With 8-Speed Transmission

80A. Remove the pto input shaft and retainer as outlined in paragraph 73 and the shift rails and forks as outlined in paragraph 76A. Slide the spacer washer from front of transmission input shaft and remove the front sliding gear (Fig. 126). Remove the rear thrust washer and withdraw the input shaft as shown in Fig. 127. Reassemble in reverse order of disassembly, making sure that grooves of rear thrust washer are toward rear as shown in Fig. 128. The shift groove of front main shaft gear should be toward rear (Fig. 126). Refer to paragraph 73 and use appropriate seal protectors when assembling.

"Multipower" Models

81. REMOVE AND REINSTALL. To remove the transmission input shaft and "Multipower" clutch, first remove transmission unit as outlined in paragraph 70. Remove pto input shaft and housing assembly as outlined in paragraph 73 and main output shaft as in paragraph 79.

Move input shaft forward slightly and remove thrust spacer (16—Fig. 118), then withdraw input shaft (13) rearward out of transmission while lifting clutch (24) and associated parts out top opening.

Examine the caged bearing (15) and sealing rings (12) for wear. Inspect polished bearing and sealing surfaces of shaft for wear or other damage.

Overhaul the removed "Multipower" clutch as outlined in paragraph 82 and assemble by reversing the removal procedure.

82. "MULTIPOWER" CLUTCH. To disassemble the removed "Multipower" clutch unit (24—Fig. 118), place unit on a clean bench with overdrive pinion (26) up. Apply slight pressure to clutch retainer plate (2—Fig. 129) and remove snap ring (1) with a narrow blade screwdriver. Completely disassemble the clutch and examine component parts for wear or scoring. Renew piston sealing rings (7 & 8) whenever clutch is disassembled. When installing the piston, carefully compress the outer sealing ring (7) using a narrow blade screwdriver or similar tool, and work the piston into its bore. The inner ring (8) will normally compress because of chamfer in inner bore of piston, if care is used in assembly.

When assembling clutch plates, note that clutch drum (9) contains six bleed holes which are evenly spaced, and that driving plates (5) have six external driving lugs. With piston installed, refer to Fig. 130 and install the first drive plate on top of piston with lugs (L) one spline clockwise from bleed holes (B). Install an internally splined clutch disc (D) then the second plate with drive lug (L) one spline clockwise from lug on first plate as shown. Repeat the procedure for remainder of discs and plates. Place the piston return springs (3—Fig. 129) on the driving lugs of first plate installed, then install retainer plate (2) and snap ring (1).

IPTO COUNTERSHAFT AND DRIVE GEAR

All 6 and 12 Speed Models

83. To remove the IPTO countershaft (11—Fig. 131) and drive gear (10), first remove the transmission assembly as outlined in paragraph 70 or 70A and disassemble upper shaft as in the appropriate preceding paragraph 80 or 81.

Remove housing cap (1), snap ring (3) and thrust washer (4). Thread two 3/8-inch NC puller bolts into threaded holes of bearing housing (7) and remove housing and bearing assembly. Remove shaft (11) rearward out of transmission countershaft while lifting drive gear (10) out top opening.

When assembling, make sure snap ring (9) is installed in inner bore of drive gear (10) and assemble by reversing the disassembly procedure. Front end of ipto countershaft (11)

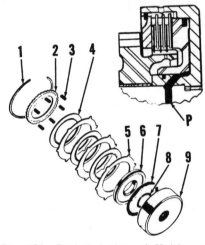

Fig. 129—Exploded view of Multipower clutch and associated parts. Inset shows cross-sectional view and pressure passage (P) which enters clutch through the drilled input shafts.

1. Snap ring	6. Piston
2. Retaining plate	7. Piston ring
3. Release springs	8. Piston ring
4. Clutch discs	9. Clutch housing
5. Clutch plates	

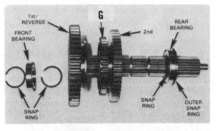

Fig. 125—View of main shaft and gears used in 8-speed models. Gear marked with (G) may be third or fourth speed.

Fig. 127—On 8-speed models, the main input shaft should be withdrawn from top as shown.

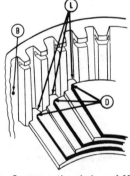

Fig. 130—Cross-sectional view of Multipower clutch showing recommended method of assembly. Refer to text.

B. Bleed hole	L. Drive lugs on
D. Clutch discs	clutch plates

Fig. 126—Groove for shift fork should be toward rear of front gear on 8-speed models.

Fig. 128—Grooved side of thrust washer should be toward rear.

contains a threaded hole which can be used to pull shaft into front bearing and bearing housing into transmission case as shown in Fig. 132. Apply non-hardening gasket sealer to threads of bolts retaining housing cap (1—Fig. 131) before installing cap.

All 8-Speed Models

83A. To remove the IPTO countershaft, first remove the transmission assembly as outlined in paragraph 70A and transmission input shaft as outlined in paragraph 80A. Remove the front cover, then remove "O" ring from around the retainer, snap ring from end of shaft and flat washer. Refer to Fig. 133. Use jack screws in the two threaded holes of bearing retainer to pull the bearing from front of shaft. Pull pto shaft to rear and lift gear out

top of case (Fig. 134). Be careful not to damage rear bearing of pto shaft when removing.

When assembling, make sure that snap ring is installed in inner bore of drive gear and hub is toward front as shown in Fig. 134. Front of pto countershaft has a threaded hole which can be used to pull shaft into front bearing and bearing housing into transmission case. Be sure to install flat washer, snap ring and "O" ring as shown in Fig. 133. Apply a non-hardening sealer to the cap screws and install cover. Tighten the four cap screws attaching cover and bearing retainer to 40-45 ft.-lbs. torque. Complete assembly by reversing disassembly.

COUNTERSHAFT

All 6 and 12 Speed Models

84. To remove the countershaft, first

remove transmission assembly as outlined in paragraph 70 or 70A and IPTO countershaft and drive gear as in paragraph 83. Unseat and remove rear snap ring (21—Fig. 131).

On models without "Multipower", remove front snap ring (12) and lift off countershaft drive gear (13).

On models with "Multipower", insert a 7/8 x 36 inch puller bolt through hollow countershaft. Install a heavy, 2½ inch OD flat washer and nut on rear of puller bolt and a crossbar, washer and nut on front end as shown in Fig. 135. Pull countershaft forward a slight amount until snap ring (17—Fig. 131) is exposed, then unseat and move the snap ring forward on shaft. Remove the through-bolt and slide countershaft rearward slightly. Install

Fig. 132—IPTO countershaft is drilled and tapped at front end for installation as shown.

Fig. 133—View showing IPTO front shaft. Refer to text for removal procedure.

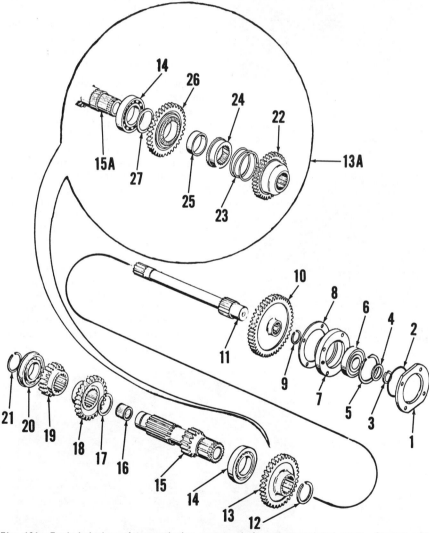

Fig. 131—Exploded view of transmission countershaft and associated parts. Circled parts (13A) replace gear (13) in Multipower models.

1. Housing cap	9. Snap ring	15. Countershaft (6-speed)
2. O-ring	10. IPTO drive gear	15A. Countershaft (Multipower)
3. Snap ring	11. IPTO countershaft	16. Pilot bearing
4. Thrust washer	12. Snap ring	17. Snap ring
5. Snap ring	13. Gear (6-speed)	18. Gear
6. Bearing	13A. Multipower gears	19. Gear
7. Bearing housing	14. Bearing	
8. Gasket		

20. Bearing	
21. Snap ring	
22. High range gear	
23. Spring	
24. Jaw coupler	
25. Bushing	
26. Low range gear	
27. Thrust washer	

Fig. 134—The pto front gear must have internal snap ring installed before assembling to shaft. Hub side of gear is toward front.

Massey Ferguson Special Tool MFN 830 around standard drive gear (26) and overdrive gear (22) as shown in

Fig. 135—Using a 7/8 x 36 inch puller bolt and welded crossbar to pull countershaft; refer to text.

Fig. 136—Massey Ferguson Special Tool MFN 830 (1) being used to remove countershaft Multipower gears.

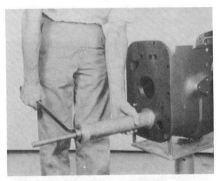

Fig. 137—Using 7/8 x 36 inch puller bolt and pipe spacer to install countershaft.

Fig. 138—View showing removal/installation of snap ring which retains constant mesh gear to front of countershaft on 8-speed models.

Fig. 136. Tighten the two clamp bolts on tool by hand until spring (23—Fig. 131) is compressed, remove snap ring (12) and lift standard drive gear (26) overdrive gear (22) and associated parts out top opening as a unit.

NOTE: If special tool MFN 830 is not available, snap ring (12) must be removed and items (22, 23, 24 and 26) must be removed individually, working against spring pressure while snap ring and gear (22) are removed.

On "Multipower" models, countershaft (15A) can be moved forward out of case while bearing (20) is removed from rear and gears (18 & 19) lifted out top opening.

Fig. 139—The snap ring at rear of countershaft must be removed before countershaft can be bumped forward for removal of snap ring shown in Fig. 140.

Fig. 140—Snap ring shown should be removed from groove and positioned in small section of shaft between groove and first gear.

Fig. 141—View of countershaft and gears correctly assembled. Constant mesh gear is shown at (1), first gear at (3) and second speed gear at (5). Gears (2 & 4) are either third or fourth speed depending upon specific transmission.

On models without "Multipower", install puller bolt as shown in Fig. 135 as previously outlined, and pull shaft forward out of case while lifting gears (18 & 19—Fig. 131) out top opening.

Install by reversing the removal procedure. A pipe spacer and the 36-inch puller bolt can be used for installing the countershaft rear bearing as shown in Fig. 137.

All 8-Speed Models

84A. To remove the countershaft, remove the IPTO countershaft and drive gear as outlined in paragraph 83A and the main shaft as outlined in paragraph 79A. Remove snap rings from front (Fig. 138) and rear of countershaft (Fig. 139), then bump shaft forward and remove bearing from rear. Withdraw snap ring from groove in shaft just forward of countershaft fourth gear (Fig. 140). Slide shaft to rear and remove gears from front of shaft. Slide shaft toward front and remove gears from rear of shaft. Shaft can be withdrawn from front of case.

Inspect all parts for wear or damage. The IPTO shaft bearing should be installed in bore at rear of countershaft, snap ring should be around shaft ahead of groove and the front bearing should be on shaft before installing shaft in case. Insert shaft from front of case and install gears (4 & 5—Fig. 141). Slide countershaft toward rear and install gears (1 & 2). Slide countershaft forward and position snap ring in groove just in front of gear (4). Position bearing with snap ring over rear of shaft. Block shaft in position (Fig. 142) and drive bearing onto shaft just far enough to install snap ring (Fig. 139). Install snap ring (Fig. 138) at front of shaft, then complete assembly by installing main shaft and IPTO countershaft as described in appropriate paragraphs.

REVERSE IDLER ASSEMBLY

All Models

85. The reverse idler shaft and gear assembly can be removed from transmission housing after removing the main (output) shaft as outlined in paragraph 79 or 79A.

The reverse idler gear shaft (6—Fig. 144) is retained in housing bore by clip (5) and cap screw (3). The cluster gear (5—Fig. 145) contains two rows (of 28 each) of loose needle rollers which use the shaft and gear as inner and outer races. Removal and installation is facilitated by using Special Tool MFN 800K or a piece of bar stock of shaft diameter and a length equal to the inside distance between housing bosses,

to retain the loose needle rollers inside the gear during removal and installation. Remove the retaining clip and push the special tool through front boss and into gear, displacing the reverse idler shaft. Lift out gear (5), gear washers (2) and spacer (3) as an assembly along with special tool and loose needle rollers. Assemble by reversing the removal procedure.

PUMP

"Multipower" Models

86. Power for the "Multipower" disc clutch is supplied by the gear type auxiliary pump which mounts on top of the regular hydraulic pump in rear axle center housing. Refer to paragraph 113

for testing procedure and to paragraph 128 for overhaul.

OIL COOLER AND FILTER

All Models

87. Refer to Fig. 146 for an exploded view of oil cooler, filter and associated parts. Oil filter and oil cooler radiator operate continuously at "Multipower" or IPTO system pressure and flow. Make sure units are in good condition and that connections are tight.

DIFFERENTIAL, BEVEL GEARS AND FINAL DRIVE

All models are equipped with a planetary final drive unit located at outer end of rear axle housings. A mechanically actuated, jaw-type differential lock is standard equipment.

DIFFERENTIAL

All Models

88. **REMOVE AND REINSTALL.** The ring gear and differential unit can be removed after removing the complete left final drive unit as outlined in paragraph 98.

Fig. 144—Rear view of transmission case showing countershaft rear bearing and reverse idler shaft.

1. Snap ring	4. Tab washer
2. Needle bearing	5. Shaft clip
3. Cap screw	6. Shaft

Fig. 145—Inside view of transmission case showing reverse idler gears installed.

1. Special Tool	3. Spacer
(MFN 800K)	4. Shaft
2. Gear washers	5. Gear

Fig. 142—Block shaft as shown when installing the rear bearing on countershaft.

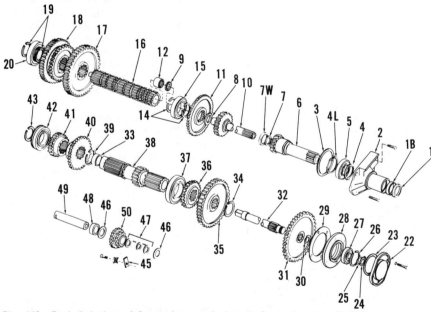

Fig. 143—Exploded view of 8-speed transmission shafts and gears. Planetary is installed on rear of shaft (16) which doubles the four speeds provided by the gears shown.

1. Oil seal	11. Gear	26. Snap ring	
1B. Bearing	12. Bearing	27. Bearing	39. Snap ring
2. Retainer	14. Snap rings	28. Bearing	40. Gear
3. O-ring	15. Bearing	29. Gasket	41. Gear
4. Snap ring	16. Main (output) shaft	30. Ring	42. Bearing
4L. Snap ring	17. Gear	31. PTO gear	43. Snap ring
5. Bearing	18. Cluster gear	32. PTO front shaft	45. Retainer plate
6. Pinion assembly	19. Snap rings	33. PTO bearing	46. Washers
7. Seal	20. Bearing	34. Snap ring	47. Bearing rollers and
7W. Washer	22. Cap	35. Constant mesh gear	spacers
8. Spacer	23. O-ring	36. Gear	48. Spacer
9. Spacer	24. Snap ring	37. Bearing	49. Shaft
10. Pinion shaft	25. Washer	38. Countershaft	50. Reverse idler

Carrier bearing pre-load should be 0.002 loose to 0.010 tight. Pre-load is adjusted by installing a selective thickness shield (3—Fig. 147) behind bearing cup (4) in carrier plate (2). Refer to paragraph 89 for pre-load adjustment.

Bevel gear backlash should be 0.004-0.020. Backlash is adjusted by shims interposed between ring gear (15) and differential case (17). Refer to paragraph 90 for adjustment procedure.

89. BEARING PRE-LOAD. Carrier bearing preload will need to be checked only when major parts are renewed or when trouble exists. To check the adjustment, differential assembly and left axle housing should be installed, and hydraulic lift (top) cover and right axle housing removed. Proceed as follows:

Unbolt and remove right carrier plate (2—Fig. 147) from axle housing and remove bearing cup (4) and shield (3). Reinstall bearing cup using the thickest available shield. Install carrier

plate (2) on rear axle center housing using three evenly spaced cap screws and omitting the gasket. Tighten cap screws evenly and carefully until carrier bearing end play is removed. Measure the clearance between carrier plate and center housing at all three cap screw locations, average the clearance and install appropriate feeler gages of equal thickness next to each cap screw. Tighten cap screws snugly and recheck for carrier bearing end play using a pry bar.

The thickness of installed shield is correct if 0.005-0.020 thickness feeler gages are required. If clearance is greater than 0.020, remove bearing cup (4) and install a different shield (3) of appropriate less thickness. If clearance is less than 0.005, check for the presence of two gaskets behind left carrier plate (20) and make sure shim (7) is installed. If unit is correctly assembled, install an additional 0.010

shim (7) and recheck.

Check backlash as outlined in paragraph 90 after pre-load is adjusted and before tractor is assembled.

Shields (shims) (3) are available in four thicknesses from 0.025 to 0.045 inch, shims (7) are 0.010 inch thick and one is used. The paper gaskets for carrier plates (2 & 20) have a compressed thickness of approximately 0.007 inch.

90. BACKLASH ADJUSTMENT. The manufacturer recommends that backlash not be adjusted or changed on gears that have been run and a wear pattern established. Backlash should only be checked in conjunction with bearing pre-load as outlined in paragraph 89, and after pre-load adjustment has been completed. Backlash can be checked through transmission top opening using a dial indicator. Recommended backlash is 0.004-0.020 inch.

Adjustment is made by adding or removing shims between ring gear and differential case as shown in Fig. 148. Shims are available in thicknesses of 0.003 and 0.007 inch and adding or removing shims will change backlash by an amount approximately equal to the change in shim pack thickness. Install ring gear flange nuts using Grade AV (Red) LOCTITE and tighten to a torque of 110-120 ft.-lbs.

91. OVERHAUL. To disassemble the removed differential unit, refer to Fig. 147. Remove the eight through-bolts and lift off differential lock coupler (6), shim (7) and differential case half (8). Note that axle gears (10 & 14) have different length hubs as shown in inset views. Backlash of differential gears is controlled by thrust washers (9 & 13). Renew axle gears (10 & 14) in pairs and differential pinions (12) in sets of four. When assembling the differential unit, make sure axle gears are correctly installed and that match marks on case halves (8 & 17) are aligned. If ring gear was removed, reinstall using Grade AV (Red) LOCTITE and tighten flange nuts to a torque of 110-120 ft.-lbs. Tighten differential case through-bolts to a torque of 75-85 ft.-lbs.

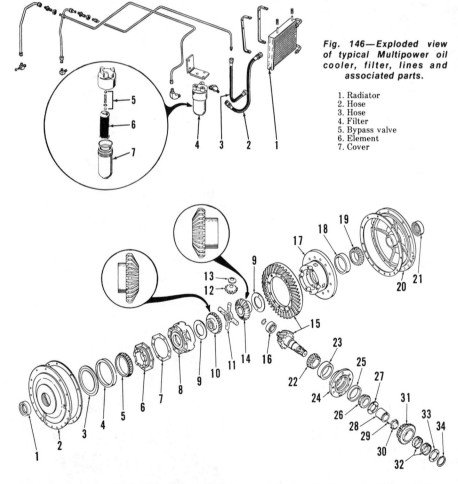

Fig. 146—Exploded view of typical Multipower oil cooler, filter, lines and associated parts.

1. Radiator
2. Hose
3. Hose
4. Filter
5. Bypass valve
6. Element
7. Cover

Fig. 147—Exploded view of differential, bevel gears and associated parts. Axle side gears are shown in profile to illustrate difference in hub length.

1. Oil seal	9. Thrust washers	half
2. Carrier plate	10. Right axle gear	18. Bearing cup
3. Spacer shield	11. Spider	19. Bearing cone
4. Bearing cup	12. Differential pinions	20. Carrier plate
5. Bearing cone	13. Thrust washer	21. Oil seal
6. Coupling half	14. Left axle gear	22. Bearing cone
7. Shim	15. Bevel gears	23. Bearing cup
8. Differential case	16. Pilot bearing	24. Bearing carrier
half	17. Differential case	25. Bearing cup

26. Bearing cone	
27. Thrust washer	
28. Bearing sleeve	
29. Drive key	
30. Adjusting nut	
31. IPTO reduction gear	
32. Bearings	
33. Thrust washer	
34. Snap ring	

Fig. 148—Shims may be used to adjust bevel gear backlash.

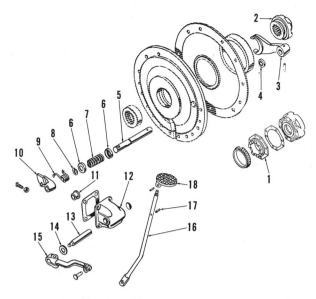

DIFFERENTIAL LOCK

All Models

92. OPERATION. The mechanically actuated differential lock assembly is standard equipment. When differential lock foot pedal is depressed, the axle half of coupler is forced inward to contact the differential case half of coupler. If slippage is occurring at one wheel depressing the pedal will cause the coupler dogs to lock the differential case to the right axle. The differential and both drive wheels then rotate together as a unit. As soon as contact is made by the coupler dogs, the pressure will keep the differential lock engaged and foot pedal may be released. When ground traction on both wheels again becomes equal, coupler dog contact pressure will be relieved and the coupler will automatically disengage. Refer to Fig. 149 for an exploded view of differential lock actuating mechanism.

93. ADJUSTMENT. The differential lock coupler should be fully engaged when differential lock pedal clears platform by ¼ -½ inch. If adjustment is required, loosen the clamp bolt which

secures lever arm (15—Fig. 149) to camshaft (13). Insert a punch in hole in camshaft and fully engage differential lock, reposition pedal and tighten clamp bolt.

94. REMOVE AND REINSTALL. To remove the differential lock coupler halves (1 & 2—Fig. 149) first drain transmission and remove right final drive unit as outlined in paragraph 98. To remove differential case coupling (1), first remove bearing cone using a

pry bar as shown in Fig. 150. Remove the cap screws securing coupling to differential case and lift off the coupling and accompanying shim pack.

NOTE: These same cap screws hold differential case together; be careful not to dislodge differential case half while cap screws are removed. When installing, tighten differential case through-bolts to a torque of 75-85 ft.-lbs.

To remove axle half of coupler (2—Fig. 149), remove the spring pin securing shift fork (3) to actuating shaft (5) and the two countersunk screws securing carrier plate to axle housing, then slide the plate, fork and coupler carefully off end of drive axle. Install by reversing the removal procedure.

To overhaul the differential lock cap (12) and associated parts, refer to Fig. 151 and unhook the return spring using a blade screwdriver as shown. Loosen locknut and set screw in actuating cam and withdraw camshaft (13—Fig. 149). When assembling, reverse the disassembly procedure and make sure approximately 0.010 inch end play exists in camshaft before tightening lock screw in cam (10).

Special tools (MFN 684A and MFN 648B) are almost essential in dis-

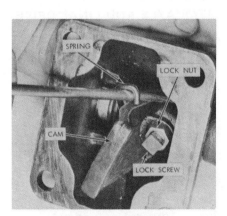

Fig. 151—Unhooking pedal return spring from differential lock cam.

Fig. 153—Installing holding clamp (MFN 684A) in preparation for removing actuating shaft.

Fig. 150—Right carrier bearing must be removed before differential case coupler half bolts can be removed.

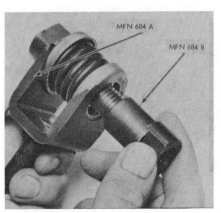

Fig. 152—Special tools are provided for compressing actuating shaft spring.

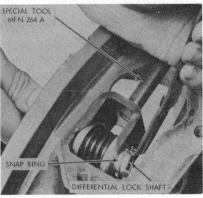

Fig. 154—With special tool installed as shown in Fig. 153, remove snap ring and withdraw shaft.

assembling and reassembling the differential lock actuating shaft. Refer to Fig. 152. To disassemble the shaft, first turn camshaft to compress the spring (Fig. 153) and install holding clamp (MFN 648A) over spring and guides as shown. Remove snap ring as shown in Fig. 154 and carefully withdraw actuating shaft, spring, tool and associated parts. Use compressor tool (MFN 684B) to remove and install spring and guides in holding clamp as shown in Fig. 152.

Adjust differential lock as outlined in paragraph 93 after tractor is assembled.

BEVEL GEARS

All Models

95. BEVEL PINION. The main drive bevel pinion is available only in a matched set which also includes the main drive bevel ring gear and attaching bolts and nuts. To remove the bevel pinion, first split tractor as outlined in paragraph 106, remove left final drive unit as in paragraph 98. Disconnect IPTO pressure lines and unbolt and remove center housing left side cover. Remove differential. Remove the hydraulic pump retaining dowels and slide hydraulic pump forward (Fig. 155) and lift out IPTO clutch and associated parts as shown. Remove snap ring (Arrow—Fig. 156) and lift off IPTO reduction gear as shown in Fig. 157. Remove the six cap screws retaining bearing carrier to center housing and use two of the removed cap screws as forcing screws to remove pinion and bearing unit.

To disassemble the unit, clamp pinion in a protected vise and unstake and remove pinion adjusting nut (30—Fig. 147). Withdraw bearing sleeve (28), key (29), washer (27) and front bearing cone (26); then lift off bearing carrier (24) containing both bearing cups (23 & 25). Pilot bearing (16) can be removed after removing the retaining snap ring. Bearing cups (23 and 25) in carrier (24) are available individually but new bearing carrier contains both cups and locating dowel (alignment pin—Fig. 158).

When assembling bevel pinion, tighten the nut (30—Fig. 147) to obtain a rolling torque of 18-22 In.-Lbs. for the pinion shaft bearings. Stake nut into two splines when adjustment is correct. Install by reversing the removal procedure, making sure dowel (alignment) pin (Fig. 158) is properly aligned. Tighten carrier flange cap screws to a torque of 75-80 ft.-lbs.

96. BEVEL RING GEAR. The main drive bevel ring gear is available only in a matched set which includes the pinion and the ring gear attaching bolts and nuts.

To remove the main drive bevel ring gear, first remove differential assembly as outlined in paragraph 88. Ring gear retaining nuts are installed with LOCTITE, Grade AV (red) and heat may be required for removal. In some factory installations, ring gear may be riveted and rivets should be carefully drilled to keep from damaging the holes in differential carrier flange.

When installing the ring gear, make sure that mating flanges of ring gear and differential case are absolutely clean and free from nicks and burrs. Use two drops of LOCTITE, Grade AV (red) on each attaching bolt and tighten the nuts to a torque of 100-120 ft.-lbs. Runout of ring gear should not exceed 0.002 inch.

REAR AXLE AND FINAL DRIVE

All Models

97. R&R WHEEL AXLE & PLANET

Fig. 156—Remove snap ring (arrow) and withdraw IPTO reduction gear.

Fig. 155—Hydraulic pump package moved forward for removing IPTO clutch and valve unit.

Fig. 157—IPTO reduction gear rides on bevel pinion shaft as shown.

CARRIER ASSEMBLY. To remove the wheel axle and planet carrier as a unit, suitably support tractor and drain final drive planetary housing. Remove wheel and tire unit and fender assembly. Remove top cap screw from drive cover and install a longer screw with suitable lifting eye as shown in Fig. 159; remove the retaining cap screws and lift off the unit with a hoist as shown.

If unit is disassembled, make sure oil drain plug in drive cover is to the bottom when reassembled. Install by reversing the removal procedure. Tighten axle housing to planetary ring gear cap screws to a torque of 65-70 ft.-lbs. and fill to level of fill plug with correct amount of Massey Ferguson M-1129A EP Gear Oil or equivalent.

98. R&R COMPLETE FINAL DRIVE. To remove either final drive assembly as a unit, first drain transmission and hydraulic system, suitably support rear of tractor and remove the fender and rear wheel & tire unit. Disconnect hydraulic lift arm knuckle, raise the rockshaft and block in raised position. Disconnect brake rod. If right axle is being removed, disconnect dif-

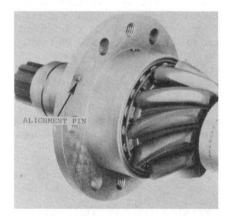

Fig. 158—Make sure alignment pin enters drilled hole in center housing when assembly is installed.

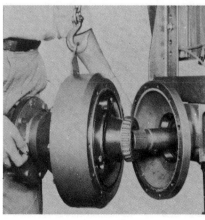

Fig. 159—Attach a lifting eye as shown, to lift off wheel axle and planet carrier assembly.

ferential lock lever. Remove the castellated nut, two hex head cap screws and socket head screw securing lift link adapter to axle housing and drawbar support and remove the adapter. Support axle unit from a hoist as shown in Fig. 160, remove the retaining stud nuts, then remove the housing as shown.

Install by reversing the removal procedure. Tighten the retaining stud nuts to a torque of 75-85 ft.-lbs.

99. OVERHAUL FINAL DRIVE UNIT. Remove the unit as outlined in paragraph 97. Remove the retaining cap screws and lift off the planetary ring gear. Refer to Fig. 161.

Cap screws securing retainer plates

Fig. 160—Wrap a chain around outer fender boss to remove complete final drive unit.

(5) are installed with LOCTITE; remove the screws, retainers and pinion shafts (10). Carefully withdraw the three pinions (8) along with their thrust washers (6) and loose needle bearings.

NOTE: Each planetary pinion contains two rows of (22 each) loose needle rollers (7) separated by a washer (9).

Lift out sun gear (3) through side opening of planet carrier (4). Using the Special Nut (MFN 265) and Forcing Screw (MFN 764-3), remove planet carrier from wheel axle splines as shown in Fig. 162. Remove snap ring (11—Fig. 161) from splines of wheel axle (18) and using a suitable press, remove axle shaft from drive cover (15) and bearing cone (12).

Dust shield (17) is retained to wheel axle (18) by wheel studs which are swaged in place. Do not disassemble unless renewal is necessary. Outer seal (16) should be installed in drive cover (15) with felt dust seal to outside and with outside edge flush with cover bore. Coat outside of seal with LOCTITE Hydraulic Sealer or equivalent before installing. Using a suitable length of 3-inch pipe, assemble wheel axle, drive cover (15) and bearing cone (12); pressing bearing cone on shaft splines only far enough to allow instal-

lation of snap ring (11). After snap ring is installed, press on end of axle shaft until bearing is firmly seated against snap ring.

Press planet carrier (4) on axle shaft until it bottoms. Make sure sun gear (3) is installed before installing planet gears. Install cap screws for retainer plates (5) using LOCTITE, Grade C (blue). Adjust final drive bearings as outlined in paragraph 100 and reinstall as in paragraph 97.

100. BEARING ADJUSTMENT. Planet carrier bearing pre-load of 0.002-0.012 inch is adjusted by means of shims (21—Fig. 161) installed between bearing cup (22) and shoulder of axle housing (19). Shims are available in thicknesses of 0.005, 0.010 and 0.015 inch.

To check the adjustment after unit is overhauled, first lock the brake and withdraw main axle shaft (24—Fig. 161). Using a suitable slide hammer, remove bearing cup (22) and add approximately 0.030 inch thickness of shims to existing shim pack (21). Reinstall bearing cup and install the assembled wheel axle and planet carrier assembly, omitting main axle shaft (24) and gasket seal (1). Install three equally spaced flange cap screws and tighten finger tight. Bump outer end of wheel axle (18) sharply to make sure bearings are seated. Equalize the gap between axle housing (19) and ring gear (2) by tightening the three installed cap screws as required; then measure gap in three places next to installed cap screws and record the measurement. Remove wheel axle and planet carrier assembly and bearing cup (22), then remove shims equal in thickness to the measured gap plus 0.010 inch, to establish the desired pre-load. Install the parts removed for trial assembly and install planetary unit as outlined in paragraph 97.

101. AXLE SHAFT & HOUSING. Axle shaft (24—Fig. 161) can be

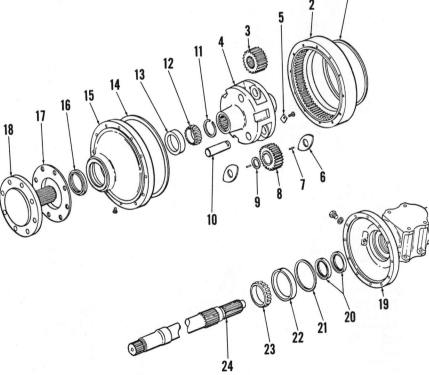

Fig. 161—Exploded view of planetary final drive unit showing component parts.

1. Inner gasket	7. Needle roller	13. Bearing cup	19. Axle housing
2. Ring gear	8. Planet pinion	14. Outer gasket	20. Oil seals
3. Sun gear	9. Spacer washer	15. Drive cover	21. Shim
4. Planet carrier	10. Pinion shaft	16. Oil seal	22. Bearing cup
5. Clip	11. Snap ring	17. Dust shield	23. Bearing cone
6. Side washer	12. Bearing cone	18. Wheel axle	24. Axle shaft

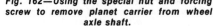

Fig. 162—Using the special nut and forcing screw to remove planet carrier from wheel axle shaft.

withdrawn after removing wheel axle shaft and planetary unit as in paragraph 97 and locking brake to keep brake discs properly aligned. Axle housing outer oil seals (20) can be removed after shaft is out. The two seals are identical and both should be installed with lip toward outside, toward planetary unit. Inner seal should bottom against shoulder in housing bore and outer seal installed flush with shoulder in bearing cavity of housing. An oil drain hole is drilled into the space between the seals leading to outside of flange as shown at (W—Fig. 164). Make sure drilling is open and clean.

Main axle inner oil seal is located in differential carrier plate as shown in Fig. 165 for models with dry type brakes. When installing the seal, make sure both seal lips face away from brake contact surface and that oil drain holes are aligned. Right carrier plate contains the differential lock shift fork; refer also to paragraph 94. Refer to paragraph 103 for brake overhaul.

BRAKES

All models are equipped with disc type brakes. Brake assemblies on some

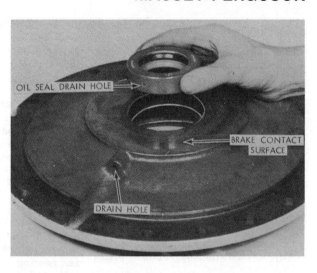

Fig. 165—On models with dry brakes, drain holes in oil seal and carrier plate must be aligned when seal is installed. Both seal lips face away from brake contact surface.

Fig. 163—Shims shown are used to adjust planet carrier bearing preload.

Fig. 164—Wire (W) being used to make sure oil drain hole is open and clean on models with dry brakes.

tractors are designed to be operated in oil (Fig. 167A) while other assemblies must be operated in a completely dry compartment (Fig. 167). Principles of operation and much service is the same or similar; however, parts and suggested lubricants are different. On dry type brakes, be careful to assure dry operation, being especially careful not to damage the seals which prevent oil

Fig. 166—Turn brake adjusting nut until pedal free play is 2½-3 inches. Make sure both sides are adjusted equally.

Fig. 167—Exploded view of dry, disc type individual wheel brakes, control linkage and associated parts. Parts shown are for 1080 model, but similar parts are used on some 1085 models.

1. Pedal shaft
2. Brake pedal
3. Parking pawl
4. Brake pedal
5. Brake rods
6. Brake lever
7. Lever support
8. Return spring
9. Lever
10. Cross shaft
12. Lever
13. Lever
14. Adjusting nut
15. Adjusting block
16. Boot
17. Brake rod
18. Lined discs
19. Actuating disc

from differential compartment or final drive planetary compartment from entering brake compartment. Lubricate the brake reaction pin with molybdenum disulphide lubricant and do not lubricate any other part of the brake actuating disc assembly.

ADJUSTMENT

All Models

102. To adjust the disc-type brakes, turn each adjusting nut (Fig. 166) as required until brake pedal free play is 2½-3 inches, measured at pedal pad. Adjust both pedals equally. Pedal height can be equalized by adjusting actuating rod (5—Fig. 167 or 167A).

OVERHAUL

All Models

103. To remove the brake assemblies, first remove complete final drive units as outlined in paragraph 98 and on right side remove differential lock fork and coupling (paragraph 94). Remove adjusting nut (14—Fig. 167 or 167A), block (15) and lever support (7).

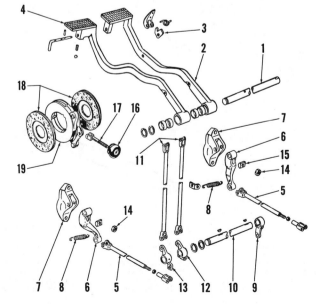
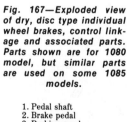

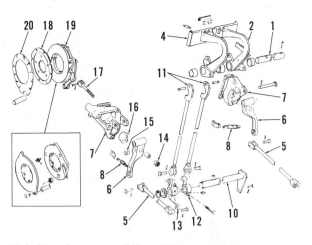

Fig. 167A—Exploded view of wet type brake used on some models. Six friction discs (18) and four intermediate plates (20) are used in each assembly. Refer to Fig. 167 for parts identification.

arms and control beam at draft control spring yoke. Disconnect IPTO shift rod and hydraulic lines at left side cover and response linkage on right side. Remove transfer cap, standpipe and hydraulic lift cover flange cap screws. Install guide studs in two holes for flange bolts as shown in Fig. 171; attach a hoist to console as shown in Figs. 170 and 171; then carefully lift off the console and lift cover as a unit.

Disconnect hydraulic pressure hoses leading from pump and valve unit to left side cover and remove the cover. Disconnect pressure hose leading to Multipower control valve. Support

Unbolt and remove differential carrier plate.

Brake discs (18) and actuator (19) can be removed at this time. Brake discs must not bind on axle splines. If evidence of oil is present on dry type brakes, find and correct the cause before proceeding with brake overhaul. On all models, unhook the springs and separate the actuator unit. Examine actuating balls and ramps for wear, pitting or other damage and renew as required.

Brake discs for dry type are equipped with one metallic and one organic lining; make sure metallic lining contacts the actuator for both discs. The types of lining are used because of the difference in heat dissipation between actuator and brake contact surfaces of axle housing and carrier plate.

On wet type brakes, six friction discs (18—Fig. 167A) and four intermediate plates (20) are used in each brake assembly.

On all models, adjust brakes after assembly as outlined in paragraph 102.

POWER TAKE-OFF

All models are equipped with an Independent Power Take-Off which is driven by a flywheel mounted "Split-Torque" clutch and controlled by a hydraulically actuated multiple disc clutch contained in the rear center housing. A dual speed output shaft is used (Fig. 177), with a reduction idler gear carrier on main drive bevel pinion. Outer shaft rotates at ASAE 540 rpm at crankshaft speed of 1720. The inner shaft rotates at 1000 rpm at 2000 rpm crankshaft speed.

OPERATION

All Models

104. The IPTO and hydraulic pump drive shaft is splined into a hub contained in the Split Torque Clutch

Cover (paragraph 69) and turns continuously when engine is running.

The IPTO control lever is mounted on left side of seat console and actuating lever on center housing left side cover. Pushing lever down disengages the hydraulically actuated multiple disc clutch and engages the brake. Moving lever to upper position releases the hydraulic brake and engages the multiple disc IPTO clutch.

Standby hydraulic pressure to actuate the clutch and brake is provided by the auxiliary gear type pump which also provides power for the Multipower clutch and/or auxiliary hydraulic system if tractor is so equipped.

REMOVE AND REINSTALL

All Models

105. Before removing any of the IPTO components, including the output drive shaft, it is first necessary to detach (split) tractor between transmission and rear axle center housing as follows:

106. **TRACTOR SPLIT.** To detach (split) tractor between transmission and rear axle center housing for service on IPTO or hydraulic system components, first drain transmission and hydraulic system fluid. Remove operator's platform, seat and fuel tanks. Disconnect lift links at rockshaft

Fig. 168—Carrier plate removed from left axle housing showing installed brake assembly.

Fig. 169—Actuating disc assembly showing housing stops.

Fig. 170—Console and lift cover can be removed as a unit as shown. Refer also to Fig. 171.

Fig. 171—Rear view of tractor showing lift cover removal. Refer also to Fig. 170.

transmission and rear axle center housing separately. Disconnect brake rods. Remove bolts securing center housing to transmission assembly and roll the units apart.

Attach and assemble by reversing the disassembly procedure, making sure vertical control lever on cover is to rear of lever on hydraulic pump when cover is installed. Tighten flange bolts and hydraulic lift cover retaining cap screws to a torque of 50-55 ft.-lbs.

NOTE: Some models may be equipped with a center housing bridge piece as shown in Fig. 171A. The bridge piece provides additional structural rigidity to tractors subjected to severe operating conditions. Install as follows; tighten all flange bolts and top cover cap screws except those securing bridge piece. Tighten bolts (4) finger tight. With nuts (2) loose, hold nuts (3) and tighten bolts (1) to 25-30 ft.-lbs. Tighten bolts (4) to 80-85 ft.-lbs. then nuts (2) to 35-55 ft.-lbs.

107. IPTO CLUTCH AND VALVE UNIT. To remove the IPTO clutch and valve unit, first split tractor as outlined in paragraph 106. Remove hydraulic pump retaining dowels as shown in Fig. 172, slide hydraulic pump forward and lift IPTO clutch and valve unit out top opening as shown. Install by reversing

Fig. 171A—Some models are equipped with a center housing bridge piece as shown. Refer to note following paragraph 106 for installation procedure.

1. Bolts
2. Flange nuts
3. Bridge nuts
4. Cap screws

Fig. 172—Hydraulic pump package moved forward for IPTO clutch removal.

the removal procedure.

108. OUTPUT SHAFTS AND GEARS. To remove the IPTO output shafts and gears, first split tractor as outlined in paragraph 106 and remove clutch unit as in paragraph 107.

Remove snap ring (Arrow—Fig. 173) and withdraw IPTO reduction gear and bearing from bevel pinion shaft. Remove snap ring (Arrow—Fig. 174) from 1000 rpm output shaft and lift off the gear. Remove front snap ring (Arrow—175) from 540 rpm output shaft, lift off gear and remove rear snap ring (Fig. 176).

Remove output stub shaft and pto shield, if installed, and output shaft

Fig. 173—Remove snap ring (arrow) and withdraw IPTO reduction gear.

Fig. 174—Remove lower snap ring (arrow) and gear from 1000 rpm shaft.

Fig. 175—Remove front snap ring (arrow) and gear from 540 rpm shaft.

retainer from rear of tractor; then withdraw IPTO output shaft assembly rearward out of rear axle center housing. Install by reversing the removal procedure.

OVERHAUL

All Models

109. OUTPUT SHAFT. Refer to Fig. 178 for an exploded view of output shafts, gears and associated parts. To disassemble the removed shaft unit, use a suitable tool (MFN 248A or equivalent) and remove seal (4) from rear of 540 rpm shaft (9). Unseat and remove snap ring (6), then slide 1000 rpm shaft (8) and bearing (7) rearward out of outer shaft (9). Bearing (7) can be removed after removing snap ring (5). Front needle bearing (10) in hollow (540 rpm) shaft (9) should be removed only if renewal is indicated. Install flush with shoulder of step in shaft bore. Needle bearing (16) in center

Fig. 176—Rear snap ring must be removed from 540 rpm shaft before shafts can be withdrawn.

Fig. 177—Rear view of tractor with cover removed (below), 540 rpm shaft installed (upper left) and 1000 rpm shaft installed (upper right).

housing wall should be recessed 11/32 inch from front face of bore if renewed. Seals (4 & 13) should both be installed lip forward when unit is assembled.

110. IPTO CLUTCH. To disassemble the removed IPTO clutch and valve unit, place the assembly on a bench, valve (front) side down. Remove snap ring (1—Fig. 179), retainer plate (2), drive plates (3), separator springs (4) and driven plates (5). Lift out drive hub (6). Using two pairs of pliers, grasp strengthening ribs of piston (7) and lift out piston.

Remove snap ring (16) and thrust washer (15). Unbolt and remove brake cylinder (20) and piston (19); then carefully withdraw valve housing (14) and associated parts. Brake disc (12) is retained to clutch housing by two groove pins (10).

To disassemble the modulating valve (parts 1 through 6—Fig. 180) apply slight pressure while unseating the internal expanding snap ring (5).

Spacer ball (4) is available in alternate diameters of ¼, 9/32, 5/16 and 11/32 inch. Ball is used to establish collapsed length of modulating valve within the recommended 4.030-4.060

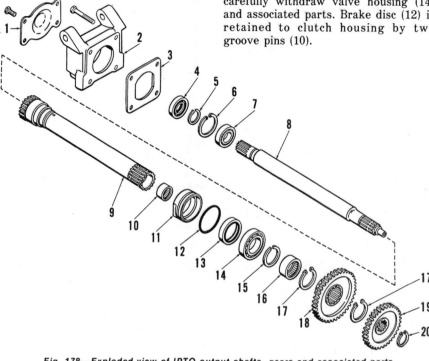

Fig. 178—Exploded view of IPTO output shafts, gears and associated parts.

1. Cover		17. Snap ring
2. Housing	7. Bearing	12. O-ring	18. 540 rpm (driven)
3. Retainer	8. 1000 rpm shaft	13. Oil seal	gear
4. Oil seal	9. 540 rpm shaft	14. Bearing	19. 1000 rpm (drive)
5. Snap ring	10. Needle bearing	15. Snap ring	gear
6. Snap ring	11. Retainer	16. Needle bearing	20. Snap ring

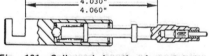

Fig. 180—Exploded view of modulating valve, valve housing and associated parts.

1. Retainer	6. Valve spool
2. Plunger	7. Connector
3. Spring	8. Housing
4. Spacer ball	9. Plug
5. Retaining ring	

Fig. 181—Collapsed length of modulating valve should be 4.030-4.060 inches when measured as shown. Length is adjusted by installing a different size spacer ball (4—Fig. 180).

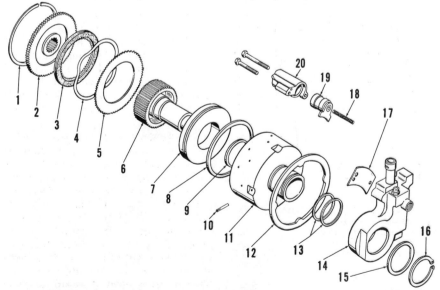

Fig. 179—Exploded view of IPTO multiple disc clutch and hydraulic brake unit.

1. Snap ring	6. Drive hub	11. Clutch housing	16. Snap ring
2. Pressure plate	7. Piston	12. Brake disc	17. Wear plate
3. Drive plates	8. Piston ring	13. Sealing rings	18. Return spring
4. Separator springs	9. Piston ring	14. Valve housing	19. Piston & shoe assy.
5. Driven plates	10. Groove pins	15. Thrust washer	20. Brake cylinder

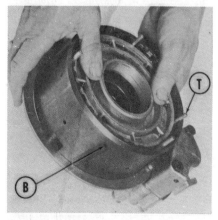

Fig. 182—Use assembly tool kit (T) to install piston as shown. Oil bleed holes (B) are used in assembly as outlined in text and shown in Fig. 183.

inches as shown in Fig. 181.

When assembling the clutch unit, use tool kit MFN 768 for installing the piston as shown at (T—Fig. 182). If assembly tool kit is not available, cut five or six 2-inch pieces of 3/16-inch rod to serve as guides for piston installation. With piston in place, install center hub (6—Fig. 179) and, beginning with an externally splined separator plate (5), alternately install seven separator plates (5), six wave springs (4) and six friction discs (3). Push down on the last separator plate to compress the wave springs and insert two Allen wrenches (or rod ends) in bleed holes (B—Fig. 182) as shown in Fig. 183, to hold the plate in place. Install the remaining friction disc and wave spring, then install pressure plate (2—Fig. 179) and snap ring (1). The restraining Allen wrenches can be removed at this time.

NOTE: If new friction discs are installed, they should be soaked for 30 minutes in transmission and hydraulic fluid before installation.

Install the assembled clutch unit as outlined in paragraph 107.

111. HYDRAULIC PUMP. Power for the IPTO clutch and brake is supplied by the gear type auxiliary pump which also supplies the Multipower transmission unit and/or auxiliary hydraulic system if tractor is so equipped. Refer to paragraph 128 for overhaul procedure.

Fig. 183—Insert Allen wrenches or rod ends in bleed holes (B—Fig. 182) to hold separator springs compressed during clutch assembly. Refer to text.

Fig. 184—Front view of assembled IPTO clutch unit.

1. Modulating valve
2. Snap ring
3. Thrust washer
4. Clutch housing
5. Brake disc

HYDRAULIC SYSTEM

The hydraulic system consists of a pto driven piston type pump which is submerged in operating fluid; and a single acting ram cylinder enclosed in the same housing. A control valve is located in the pump unit which meters the operating fluid at pump inlet. The rockshaft position can be automatically controlled by compression or tension on the upper implement attaching link, by a cam on the ram arm, or by pressure in the ram cylinder; to control rockshaft height and/or to transfer implement weight to rear tires for additional traction.

A gear-type auxiliary pump supplies pressure for the Multipower clutch and IPTO clutch; and pressure and flow for remote hydraulic cylinder applications.

The transmission lubricant is the operating fluid for the hydraulic system. Massey-Ferguson M-1129A Fluid is recommended.

TROUBLE SHOOTING

All Models

112. SYSTEM CHECKS. Before attaching an implement to tractor, start the engine, move response control lever to "FAST" position and inner quadrant lever to "TRANSPORT" po-

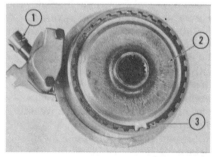

Fig. 185—Rear view of assembled IPTO clutch unit.

1. Cotter pin
2. Pressure plate
3. Snap ring

Fig. 186—View of operator's compartment showing hydraulic control quadrant and levers.

sition. With engine running at slow idle speed, check to make sure that rockshaft moves through full range of travel as draft control lever is moved to "DOWN" and "UP" positions. Using the draft control lever, stop and hold the movement with lower links in an approximately horizontal position. Lever should be centered between sector marks on quadrant.

Attach an overhanging implement such as a fully mounted plow, to the links. With draft control lever in "UP" position and engine running at slow idle speed, raise and lower the implement a little at a time using inner quadrant lever in rear sector. Implement should move in response to the lever and hold steady after completion of movement, through full range of rockshaft travel. Move lever to "TRANSPORT" position and scribe a line across lift arm hub and lift cover. Move lever forward to "PRESSURE" sector. The scribed lines should be separated 1/8-3/16 inch.

Hold the implement clear of the ground using draft control lever. If draft control spring is properly adjusted, system should respond to the application of pressure or lifting force to rear of implement, lowering when pressure is applied and raising when rear of implement is lifted.

Check pressure and flow as outlined in paragraph 113. If system fails to perform as indicated, adjust as outlined in the appropriate following paragraphs and/or overhaul the system.

113. PRESSURE AND FLOW. The internal hydraulic system pump should deliver a flow of 7.0 gpm at 2000 engine rpm and 2500 psi pressure. Maximum pressure setting should be 2900-3100 psi. Auxiliary hydraulic system pump should deliver 8 gpm flow at 2000 engine rpm. Maximum auxiliary pressure setting should be 2300-2600 psi. The low pressure (accessory) section of auxiliary pump has a relief

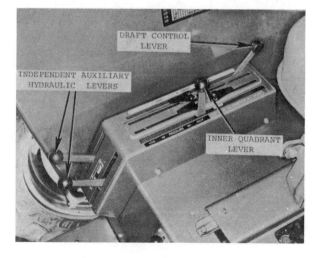

pressure setting of 650-800 psi in pump body and a regulated pressure of 240-350 psi at the IPTO—Multipower Regulating Valve located on transmission input shaft housing.

Internal hydraulic system pressure and flow can be tested at transfer cap at right, front corner of hydraulic lift cover, or at socket head plugs on either side of lift cover at front.

Auxiliary pump pressure and flow can be tested at breakaway couplings for remote cylinders.

IPTO and Multipower pressure can be checked at test plug located in rear axle center housing left side cover (IPTO Shift Cover). Regulated pressure will vary considerably as oil temperature rises, and tests should be made at operating temperature.

ADJUSTMENTS

All Models

114. CONTROL SPRING ADJUSTMENT. To check the master control spring, disconnect control beam at upper end and pivot beam down to expose the spring. Check for end play by pulling and pushing on spring clevis. If end play is present, loosen Allen head set screw (3—Fig. 187) in side of housing and pull back rubber boot (9) to expose adjusting nut (8). Unscrew nut (8) using the special spanner wrench (FT-358) and withdraw master control spring assembly from top cover.

Grasp spring (7) while holding clevis (10) and attempt to rotate spring on plunger (5). Spring should fit snugly with no end play but should still turn with moderate effort. If adjustment is incorrect, drive groove pin (11) out of clevis and thread clevis on or off plunger (5) until end play is just eliminated. Tighten plunger (5) if necessary until slot is aligned and reinstall pin (11). Reinstall control spring assembly in housing and turn adjusting nut (8) until end play is just eliminated, then retighten set screw (3) to a torque of 40-65 In.-Lbs.

NOTE: End play will be present if adjusting nut (8) is either too loose or too tight.

115. VALVE SYNCHRONIZATION. The control valve, located in main pump body, must be synchronized with control linkage located on top cover whenever unit has been disassembled or as a check when trouble exists. To synchronize the linkage, first remove operator's platform and fuel tanks (paragraph 58), and attach a temporary auxiliary fuel tank to supply line. Drain the system down until right side cover containing response control lever and dipstick can be removed and remove the cover.

Insert locating pin MFN 1080C through holes (L—Fig. 188) in outer quadrant plate (1), draft control lever (2), position control lever (3) and inner quadrant plate (4); then loosen linkage screws (A) if necessary and reposition draft control lever (6) between index marks on quadrant. Tighten adjusting screws and remove locating pin.

Using a piece of soft wire as shown in Fig. 189, tie the response plunger in the fully down position as shown. Install a suitable pressure gage in transfer cap port, attach a heavy weight to lower links and start and run engine at approximately 1000 rpm.

Place draft control lever in "TRANSPORT" position and inner quadrant lever in "HIGH PRESSURE" position. Lift arms should raise until ram arm strikes center housing; place a chisel mark across lift arm and housing. Move inner quadrant lever to "TRANSPORT" position and using chisel marks as a guide, check rockshaft movement. Chisel marks should separate approximately 5/32-inch, if adjustment is incorrect, turn control valve adjusting screw (Fig. 189) clockwise to increase the amount of drop or counter-clockwise to decrease drop.

Move inner quadrant lever to "PRESSURE" range and continue to move lever forward until weight

lowers, then move lever until weight is held steady just clear of ground. With engine running, turn Pressure Control Adjusting Screw (Fig. 189) clockwise until vertical lever begins to pulsate; then counter-clockwise until lever steadies and pressure gage does not fluctuate more than 100 psi.

If it is not possible to make the adjustments or if system fails to respond, lift cover must be removed and internal adjustments made as outlined in paragraph 117.

116. RESPONSE ADJUSTMENT. To adjust the response control move control lever to approximately 3/16-inch from "SLOW" end of slot. Remove the plug immediately below lever pivot on right side cover. Using a socket and extension, reach through plug port and loosen clamp screw in response plunger, thus allowing response needle to

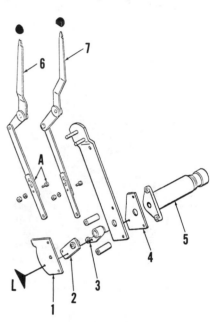

Fig. 188—Exploded view of control quadrant and lever assembly showing component parts.

1. Outer plate	
2. Draft control lever	6. Draft control lever
3. Position control lever	7. Inner quadrant lever
4. Inner quadrant plate	A. Adjusting bolts
5. Support	L. Locating pin holes

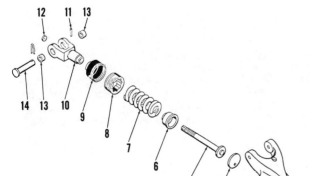

Fig. 187—Exploded view of hydraulic lift control spring assembly and associated parts.

1. Lift cover
2. Nylon plug
3. Lock screw
4. Overload stop
5. Plunger
6. Spring seat
7. Control spring
8. Adjusting nut
9. Rubber boot
10. Clevis
11. Groove pin
12. Plug
13. Bushing
14. Pin

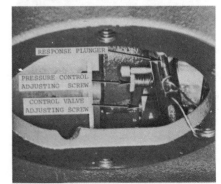

Fig. 189—Right side cover removed for valve synchronization. Response plunger should be wired down in SLOW position as shown.

center by spring pressure. Retighten screw to 25-35 In.-Lbs. and reinstall plug.

117. INTERNAL LINKAGE ADJUSTMENT. To adjust the internal lift linkage, first check and adjust master control spring as outlined in paragraph 114 and remove hydraulic lift cover as in paragraph 118.

NOTE: Three special tools are required to properly make the internal hydraulic adjustments. Tools are: MFN 1080B, Gage—Pressure Control Valve; MFN 1080C—Locating Pin & Gage; MFN 1080D—Gage, Draft, Position & Pressure Control.

Invert lift cover on a bench and block so that rockshaft is free to move through its full range of travel. Turn shaft until lift arms are in their mid position and lock in place by tightening cap screw in one end of shaft. With draft control rod in contact with control spring plunger, use locating pin as a gage and adjust self locking screw in draft control link as shown in Fig. 190.

Remove pressure control lever and insert locating pin through both levers and brackets (Fig. 191), and install special tool MFN 1080D as shown. Apply 3-lbs. rearward pressure to end of vertical control lever and turn draft control adjusting bolt (Fig. 192) if necessary, until a clearance of 0.002 inch exists between end of vertical lever and gage pin of tool.

Withdraw locating pin (MFN 1080C) and move draft control lever to transport position as shown in Fig. 193, then re-insert pin through position control lever only. Rotate rockshaft until lift arms are in transport position and lock in place. Apply 3-lbs. rearward pressure to vertical control lever and turn position control adjusting screw until clearance between special tool MFN 1080D and vertical lever is 0.002 inch.

Reinstall pressure control lever and, with inner quadrant lever still fixed by locating pin (MFN 1080C) as shown in Fig. 193, turn pressure control adjusting screw (Fig. 194) until it just touches gage block in special tool MFN 1080D.

Refer to Fig. 195 and, while cover is off check and adjust height of pressure control valve adjusting screw using special tool MFN 1080B. Adjustment is correct when 0.100 inch clearance exists between head of screw and gage block on special tool. Adjust, if necessary, by holding valve plunger and turning adjusting screw after loosening locknut.

HYDRAULIC LIFT COVER

All Models

118. REMOVE AND REINSTALL. If lift cover is being removed for other service, cover and console can be removed as a unit as outlined in paragraph 106 and illustrated in Figs. 170 and 171. If lift cover is to be disassembled, remove seat, platform and interfering sheet metal, then unbolt and remove the cover. Use guide studs in two holes (G—Fig. 196) from which cap screws were removed, remove transfer cap and withdraw standpipe and use a suitable hoist.

When reinstalling cover, make sure vertical control lever on cover enters to rear of lever of pump body, tighten cover retaining cap screws to a torque of 50-55 ft.-lbs. and reassemble by reversing the disassembly procedure.

Fig. 190—Locating pin is also used to adjust limit stop screw as shown.

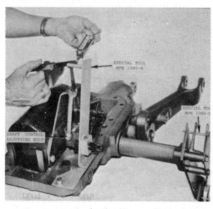

Fig. 192—Adjusting draft control linkage.

Fig. 194—Adjusting pressure control linkage.

Fig. 191—Special tools are required to satisfactorily adjust lift cover as shown.

Fig. 193—Adjusting position control linkage.

Fig. 195—Adjust pressure control valve while cover is off.

119. OVERHAUL. To disassemble the removed lift cover, invert the cover on blocks high enough to clear lift arms. Loosen set screw in hole (3—Fig. 198), pull back rubber boot (4) and remove control spring assembly as shown. Remove locking screw from hole (L—Fig. 200) then rotate and remove quadrant control levers. Remove the pin securing pressure control lever to bracket and lift off the lever. Unscrew adjusting bolt (Fig. 199) from vertical lever, unhook position control cam spring then lift off the lever. Insert cotter pins through hole in draft and position control spring rods as shown in Fig. 200, to retain the springs; then unbolt and remove lever bracket from side of ram cylinder.

Remove plug from outside of lift cover which aligns with shaft (4—Fig. 201). Loosen set screw (3) and slide shaft (4) through plug hole out of cams (1 & 2), then lift out the cams. Remove the four stud nuts securing ram cylinder to lift cover and lift off cylinder and piston assembly. Bump open end of cylinder on a block of wood if necessary, to remove the piston.

Unlock and remove cap screws from ends of rockshaft, remove lift arms then withdraw the shaft.

Assemble by reversing the disassembly procedure. Leave set screw (3—Fig. 201) loose until quadrant is installed and cams aligned; then push shaft (4)

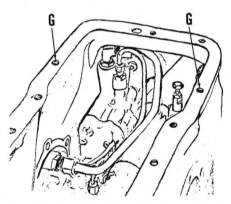

Fig. 196—Install guide studs in indicated cap screw holes to properly position the cover.

Fig. 197—Assembled view of hydraulic cover linkage.

into housing until end play is eliminated but cams do not bind before tightening set screw (3). Rockshaft and lift arms are provided with master splines for correct alignment. Tighten cap screws in end of shaft to eliminate all end play but allow free movement, then lock in place by bending locking clip. Adjust lift cover as outlined in paragraph 117, control spring as in paragraph 114 and valve synchronization as in paragraph 115.

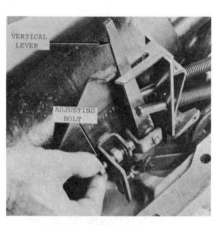

Fig. 198—Removing control spring assembly.

1. Special wrench 　　　　3. Lock screw hole
2. Adjusting nut 　　　　　4. Rubber boot

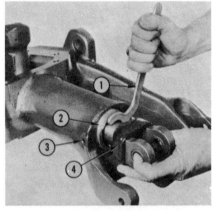

Fig. 199—Disassembling control linkage.

Fig. 200—Quadrant support is retained by lock screw in hole (L).

MAIN HYDRAULIC PUMP

All Models

120. REMOVE AND REINSTALL. To remove the main and auxiliary pump package, first split tractor as outlined in paragraph 106 and remove IPTO clutch and valve unit as in paragraph 107. Rotate the pump package until bottom is forward and up as shown in Fig. 202, then lift pump package out front opening of center housing. Install by reversing the removal procedure.

121. DISASSEMBLE AND REASSEMBLE. To disassemble the removed hydraulic pump package, first unbolt and remove pressure control valve (1—Fig. 203). Disconnect and remove filter assembly (2); then remove auxiliary pump and drive as a unit by slipping it forward off of main hydraulic pump splines.

Remove control valve lever on early models by using a 1/8-inch Allen wrench to thread the adjusting screw out of pivot block as shown in Fig. 204. Lift the lever up and away from the

Fig. 201—Disassembling control cams.

1. Position control link 　　3. Locking screw
2. Draft control link 　　　4. Pivot shaft

Fig. 202—After IPTO clutch is removed, turn pump package with bottom side up and forward, then withdraw through front opening as shown.

Fig. 203—Removed hydraulic pump package before disassembly.

1. Pressure control valve
2. Inlet filter
3. Auxiliary pump & drive

Fig. 204—Removing control valve lever from early pump.

Fig. 205—Removing rear cover plate.

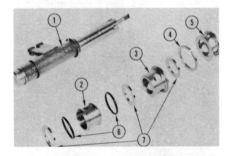

Fig. 206—Partially disassembled view of control valve and associated parts.

1. Valve & oscillator
2. Front spacer
3. Center spacer
4. Washer
5. Rear spacer
6. O-rings
7. Sealing washers

control valve and remove the cross pin and two rollers. Disconnect control valve oscillator drive pin, remove retaining nuts and front pump body; then withdraw control valve and oscillator assembly. Remove rear cover plate and response dashpot spring (Fig. 205) then, using a hooked brass rod, withdraw control valve washers and spacer spools (2 through 5—Fig. 206)

Fig. 207—Rear view of pump housing with cover removed showing control valve spacers and response dashpot piston installed.

from pump body bore. Remove camshaft, valve bodies, pistons and associated parts as a unit from pump body (10—Fig. 208).

Overhaul the removed assemblies as outlined in the appropriate following paragraphs and assemble by reversing the disassembly procedure. Tighten housing stud nuts to a torque of 30-35 ft.-lbs. Turn adjusting screw into valve control lever (Fig. 204) one complete turn as an initial adjustment, then adjust as in paragraph 115 after tractor is assembled and running.

122. CONTROL VALVE. The control valve is serviced as a matched assembly which consists of valve unit (4—Fig. 209) and three sealing washers (7—Fig. 206). Washers are self aligning and sealed by spacers; and provide the "lands" machined on the conventional valve spool.

To disassemble the valve and oscillator assembly (1—Fig. 206), refer to Fig. 209, unseat and remove retaining ring (1) from internal groove in oscillator housing (10) and withdraw valve (4) and parts (2 through 8) from housing as a unit. Remove retaining

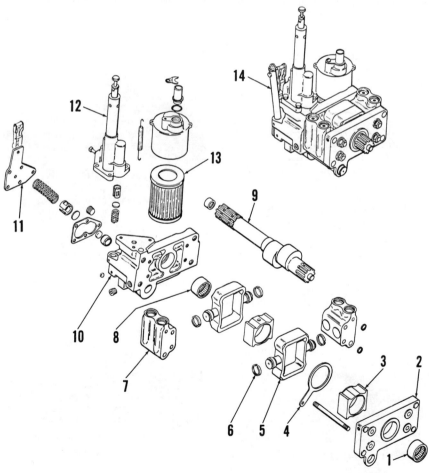

Fig. 208—Exploded view of main hydraulic pump showing component parts.

1. Needle bearing
2. Front body
3. Cam block
4. Oscillator lever
5. Piston
6. Piston ring
7. Valve chamber
8. Needle bearing
9. Camshaft
10. Pump body
11. Rear cover
12. Pressure control
13. Inlet filter
14. Response control tube

ring (7) from end of collar (6) and while holding against spring pressure, push out pin (8) and remove spring and associated parts. Assemble by reversing the disassembly procedure.

123. VALVE CHAMBERS. Both valve chambers (7—Fig. 208) are identical and interchangeable, and each contains two sets of valve components (5 through 12—Fig. 210). To disassemble the chamber, remove snap ring (12) and thread a ¼-inch bolt into plug (11) to provide leverage. Plug can be pulled by hand. A reseating reamer (Tool No. N-6007) is available which dresses both seats in one bore at the same time. If tool is used, remove only enough metal to smooth the seats. All parts are available individually.

124. **PRESSURE CONTROL VALVE.** The pressure control valve

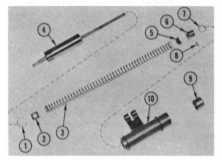

Fig. 209—Exploded view of control valve and oscillator unit.

1. Snap ring	6. Collar
2. Disc	7. Snap ring
3. Spring	8. Pin
4. Valve	9. Plug
5. Guide	10. Oscillator

shown exploded in Fig. 211 serves the dual purpose of providing pressure relief for the main hydraulic pump during normal operation; and providing weight transfer for added traction when using some types of pulled or mounted implements.

The valve is internally ported to ram cylinder passage and cylinder pressure acts against servo piston in valve (4) which is held seated by the variable rate spring (8). The spring is compressed to the maximum when inner quadrant lever is in Position Control or Constant Pumping sectors of quadrant and is released at a uniform rate as lever is moved toward "LOW" end of "PRESSURE" range.

All parts are available individually. Refer to Fig. 211 for disassembly and assembly sequence and to paragraph 117 for adjustment procedure.

125. RESPONSE CONTROL MECHANISM. Draft response dashpot is located in pump body (10—Fig. 208), and dashpot can be serviced after removing rear cover plate (11). Response control needle is located in tube (14) and can be removed as shown in Fig. 212 after removing snap ring (8—Fig. 213). Adjust as outlined in paragraph 116 after tractor is assembled. Note that when clamp screw (6) is loosened, needle is centered by equal pressure of the two springs (2).

With response lever in slow position and a heavy weight on lower links, rockshaft should start lowering at a slow rate then increase in speed as rockshaft lowers, with draft control lever at bottom of quadrant. Approxi-

mate lowering time should be 2 seconds with hot oil.

AUXILIARY PUMP DRIVE

All Models

126. The auxiliary pump gear train is shown exploded in Fig. 214 and installed in Fig. 216. Gear train and auxiliary pump may be removed as a unit after removing pump package as outlined in paragraph 120, then removing inlet filter unit.

Gear train mounting plates serve as pump support and gear train must be disassembled to remove the pump. Idler gear (9—Fig. 214) contains 22 loose needle rollers (10) which may fall out as unit is disassembled. Gear backlash is not adjustable; renew parts if backlash exceeds 0.015 inch between

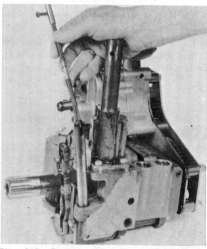

Fig. 212—Response needle can be withdrawn after removing retaining snap ring.

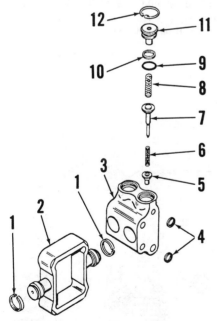

Fig. 210—Exploded view of valve chamber and associated parts.

1. Piston ring	7. Outlet valve
2. Piston	8. Spring
3. Chamber	9. O-ring
4. O-rings	10. Backup ring
5. Inlet valve	11. Plug
6. Spring	12. Snap ring

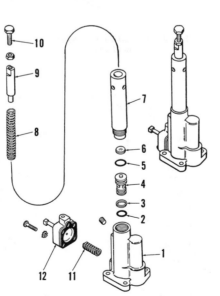

Fig. 211—Exploded view of pressure control valve.

1. Housing	6. Support
2. O-ring	7. Tube
3. Backup ring	8. Spring
4. Valve	9. Plunger
5. O-ring	10. Adjusting screw

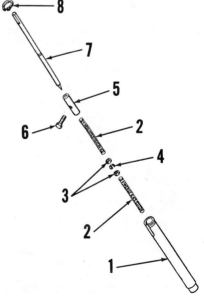

Fig. 213—Exploded view of response needle and associated parts.

1. Tube	5. Barrel
2. Spring	6. Clamp screw
3. Washer	7. Needle
4. Retainer	8. Snap ring

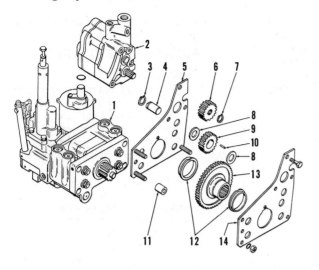

being careful not to lose the 22 loose needle bearings (10—Fig. 214); then remove pump from rear mounting plate (5).

Install by reversing the removal procedure. Tighten drive train retaining bolts to a torque of 30-35 ft.-lbs. when unit is assembled. Fig. 217 shows an assembled view of pump package and Figs. 215 and 216 show installed views.

128. OVERHAUL. To disassemble the removed pump, remove cap screws and stud nut securing end cover (23—Fig. 218); cover will be forced off by pressure of spring (22). Do not remove high pressure adjusting screw (24); screw is staked in place and setting should not be changed. Pry rear body (16) from front body (6) using a suitable tool in the notches provided. If pump housings show evidence of wear or scoring in gear pocket area, renew the pump. If housings are usable, gears, relief valves and seals can be renewed.

When assembling the pump, omit high pressure relief valve spring (22) and its washer, and use the body cap screws to draw bodies together over dowel pins. Remove screws and cover, and install relief valve spring. Tighten body cap screws to a torque of 18-20 ft.-lbs. in final assembly. When installing low pressure relief valve (parts 2 through 5), tighten adjusting plug (1) until it bottoms, back out four full turns and stake in place. This procedure should establish relief valve pressure within the recommended range of 650-800 psi.

Fig. 214—Exploded view of auxiliary pump drive gear train.

1. Main pump
2. Auxiliary pump
3. Snap ring
4. Idler shaft
5. Rear mounting plate
6. Pump drive gear
7. Snap ring
8. Side washers
9. Idler gear
10. Needle rollers
11. Spacer
12. Bushings
13. Drive gear
14. Front plate

any two gears. Tighten retaining bolts to a torque of 30-35 ft.-lbs.

AUXILIARY PUMP

All Models

127. REMOVE AND REINSTALL. To remove the auxiliary pump, first re-move pump package as outlined in paragraph 120, remove inlet filter unit; and slide pump and drive unit forward off of main hydraulic pump camshaft splines. Disassemble the drive train,

Fig. 215—Installed view of hydraulic pump package showing hydraulic lines attached.

Fig. 217—Assembled view of hydraulic pump package.

Fig. 216—Front view of rear axle center housing with pump package installed.

Fig. 218—Exploded view of dual section auxiliary pump showing component parts.

1. Adjusting plug
2. Seat
3. Spring
4. Poppet
5. Ball
6. Front body
7. Seal
8. Seal
9. Pressure plate
10. Dowel stud
11. Follow gear
12. Driven gear
13. O-ring
14. Seat
15. Dowel
16. Rear body
17. Follow gear
18. Driven gear
19. O-ring
20. Piston
21. Plate
22. Spring
23. End cover
24. Adjusting screw

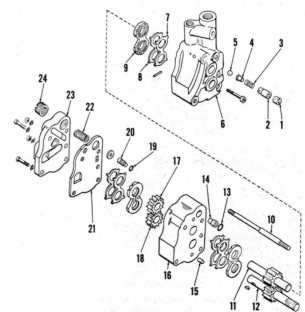

AUXILIARY VALVE

All Models

129. Refer to Fig. 219 for an exploded view of the 2 spool auxiliary valve and to Fig. 220 for single spool valve. To disassemble either valve, remove snap ring (16—Fig. 219 or 220) and plug (15). Back out spool detent screw (or screws) (2) and unbolt and remove valve end cap (1); then carefully withdraw spool and spring assembly from spring end of valve body. Valve spools (9) and bodies (7) are available only as a part of the complete valve unit. Items (10 through 13) are available as an assembly which also includes items (17 through 24). Back flow check valve assembly (8) comes in a kit containing a spacer which is only used in single valve; discard the spacer when kit is used in 2 spool valve. When unit is assembled, tighten detent (2) until lever will hold in raising or lowering position but can be easily released by hand pressure on lever.

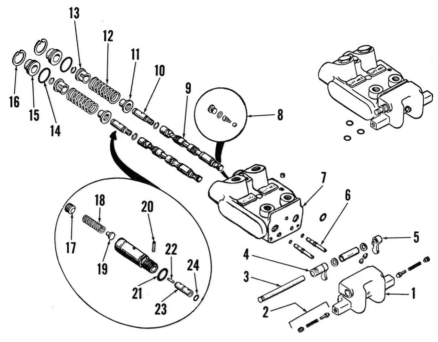

Fig. 219—Exploded view of 2-spool auxiliary valve showing component parts.

1. End cap	7. Body	13. Sleeve	19. Guide
2. Detent assy.	8. Check valve	14. O-ring	20. Pin
3. Inner rod	9. Spool	15. Plug	21. O-ring
4. Outer lever	10. Spool plug	16. Snap ring	22. Pin
5. Inner lever	11. Sleeve	17. Nylon screw	23. Guide
6. Switch valve	12. Spring	18. Spring	24. O-ring

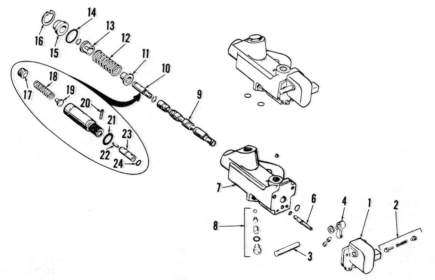

Fig. 220—Exploded view of single spool control valve. Refer to Fig. 219 for parts identification.

NOTES

MASSEY FERGUSON

Models ■ MF1100   ■ MF1130

Previously contained in I&T Shop Manual No. MF-24

SHOP MANUAL

MASSEY FERGUSON

MODELS
MF1100 - MF1130

**Tractor serial number stamped on instrument panel name plate.
Engine serial number stamped on side of engine.**

I N D E X (By Starting Paragraph)

CONDENSED SERVICE DATA

GENERAL

	MF1100 Gasoline & LP Gas	MF1100 Diesel	MF1130 Diesel
Torque Recommendations	—— See End of Manual ——		
Engine Make	Waukesha	Perkins	Perkins
Engine Model	F320-G	A6.354	AT6.354
Number of Cylinders	6	6	6
Bore—Inches	4-1/8	3-7/8	3-7/8
Stroke—Inches	4	5	5
Displacement—Cu. In.	320	354	354
Compression Ratio	8:1	16:1	16:1
Main Bearings, No. of	4	7	7
Cylinder Sleeves	None	Dry	Dry
Forward Speeds	—— 6 or 12 ——		
Reverse Speeds	—— 2 or 4 ——		

TUNE-UP

	MF1100 Gasoline & LP Gas	MF1100 Diesel	MF1130 Diesel
Firing Order	—— 1-5-3-6-2-4 ——		
Valve Tappet Gap (Cold)			
Intake	0.012	0.012	0.012
Exhaust	0.022	0.012	0.012
Compression Pressure @ Cranking Speed	......	......	......
Valve Face Angle	45°	44°	44°
Valve Seat Angle	45°	45°	45°
Timing (Ign or Inj)			
Static	See Par. 87	22° BTC	28° BTC
High Idle	See Par. 87	22° BTC	......
Timing Location	—— Flywheel ——		
Battery			
Volts	12	12	12
Capacity Amp/Hr	95	95*	95*
Ground Polarity	Negative	Negative	Negative
*Each Battery—Two Batteries Used, Connected in Parallel.			
Distributor Contact Gap	0.016	......	......
Spark Plug Size	18 mm	......	......
Electrode Gap	0.025	......	......
Injectors			
Opening Pressure	......	2500 psi	2795 psi
Spray Hole Dia.	......	0.011	0.011

TUNE-UP (Cont'd.)

	MF1100 Gasoline & LP Gas	MF1100 Diesel	MF1130 Diesel
Governed Speeds			
Engine			
Low Idle	750-800 rpm	——800-850 rpm ——	
High Idle	2450 rpm	—— 2400 rpm ——	
Loaded	2200 rpm	—— 2200 rpm ——	
1000 rpm PTO			
High Idle	1225 rpm	—— 1200 rpm ——	
Loaded	1100 rpm	—— 1100 rpm ——	
540 rpm PTO			
High Idle	661 rpm	—— 654 rpm ——	
Loaded	594 rpm	—— 594 rpm ——	
Horsepower @ PTO Shaft	89*	93.5	120.5
*Gasoline Tractor, Mfr's Rating—	Diesel Models, According to Nebraska Test		
Hydraulic System			
Maximum Pressure	—— 2550 psi ——		
Rated Delivery	—— 20 gpm ——		

SIZES—CAPACITIES—CLEARANCES

(Clearances in Thousandths)	MF1100 Gasoline & LP Gas	MF1100 Diesel	MF1130 Diesel
Crankshaft Journal Diameter	2.625	2.999	2.999
Crankpin Diameter	2.625	2.499	2.499
Camshaft Journal Diameter			
Front	1.750	1.997	1.997
2nd	1.749	1.987	1.987
3rd	1.749	1.977	1.977
Rear	1.749	1.967	1.967
Crankshaft Bearing Clearance			
Main Bearings	1.5-4.5	2.5-4.5	2.5-4.5
Crankpin	0.5-3.0	1.5-3.0	1.5-3.0
Crankshaft End Play	4.5-8.5	2-14	2-14
Piston to Cylinder Clearance	3-4	7-10	7-10
Camshaft Bearing Clearance			
Front	1.5-3.0	2.5-4.5	2.5-4.5
Other Journals	2.0-3.5	2.5-5.5	2.5-5.5
Cooling System—Quarts	21	22	22
Crankcase—Quarts	8	17	17
Transmission, Differential and Hydraulic Lift—Gal.	—— 22 ——		

FRONT SYSTEM

Refer to Figs. 1 through 4 for exploded views of the different front end types available, and to the following paragraphs for pertinent disassembly and overhaul data.

AXLE ASSEMBLY

All Models

1. Refer to Fig. 2 for an exploded view of Western Type axle assembly and to Fig. 3 for adjustable axle unit. Front support, steering motor and associated parts are shown in Fig. 1.

To remove either axle as an assembly, disconnect center steering arm (1—Fig. 2) from steering spindle, appropriately support the tractor and unbolt and remove axle pivot brackets (2 & 9—Fig. 2 or 3). When reinstalling, align the center bolt hole of steering arm (1—Fig. 2) with "X"—marked bolt hole on steering spindle flange. Use Grade "C" (Blue) LOC-TITE on cap screws securing steering arm (1) and tighten the screws to a torque of 175-200 ft.-lbs.

TIE RODS AND TOE-IN

All Models

2. Automotive type tie rod ends are used. Recommended toe-in is 0-⅜ inch. Adjust both tie rods an equal amount to obtain the required setting.

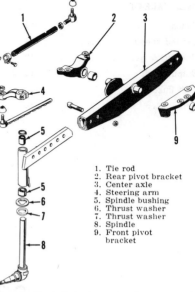

1. Tie rod
2. Rear pivot bracket
3. Center axle
4. Steering arm
5. Spindle bushing
6. Thrust washer
7. Thrust washer
8. Spindle
9. Front pivot bracket

Fig. 3—Exploded view of adjustable axle.

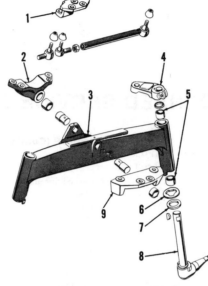

Fig. 2—Exploded view of Western Type front axle and associated parts.

1. Center steering arm
2. Rear pivot bracket
3. Axle
4. Steering arm
5. Spindle bushing
6. Thrust washer
7. Thrust washer
8. Spindle
9. Front pivot bracket

Fig. 4—Exploded view of single wheel tricycle pedestal and associated parts.

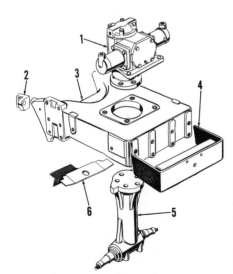

Fig. 1—Front support, steering motor and dual wheel tricycle pedestal shown exploded. Front support and steering motor are interchangeable on all tractors.

1. Steering motor
2. Engine support
3. Front support
5. Pedestal
6. Shield

POWER STEERING SYSTEM

All models are equipped with a closed center hydraulic system which provides standby power for operation of the hydrostatic steering system as well as the other hydraulic functions of operation and control of the tractor. This section covers only the Hydrostatic Power Steering Hand Pump (steering control valve), Steering Cylinder Assembly (steering motor), connecting lines and attaching parts. Refer also to HYDRAULIC SYSTEM Section for information on main hydraulic pump, priority valve, relief valve and operating fluid.

LUBRICATION AND BLEEDING
All Models

3. The hydrostatic steering hand pump and steering cylinder assembly (steering motor) are lubricated by the operating fluid; refer to paragraph 126 for fluid type and checking procedure. The hand pump (P—Fig. 5) is lubricated by the high-pressure oil as the system is operated. The shaft, bearings and gears of the steering motor (C) are lubricated by return oil from the lubrication passages of the main hydraulic pump. The power steering system is self-bleeding.

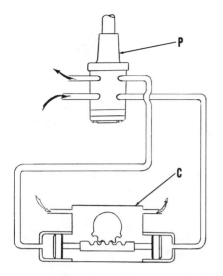

Fig. 5 — Schematic view of hydrostatic power steering hand pump (P) and steering motor (C) showing direction of oil flow.

Fig. 6—Exploded view of hydraulic lines and associated parts running to front of tractor.

1. Tube filter to oil cooler
2. Tube oil cooler to regulating valve
3. Pump pressure tube
4. Steering motor bleed tube
5. Steering cylinder tube
6. Steering cylinder tube
7. Filter inlet tube
8. Pump inlet tube
9. Steering lubrication tube
10. Main hydraulic pump
11. Oil cooler
12. Bracket
13. Hydraulic filter
14. Steering motor

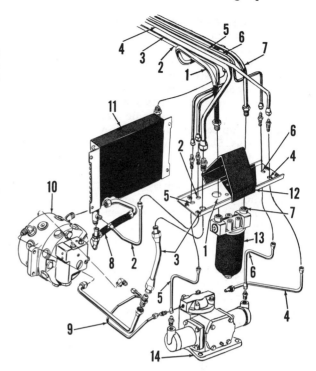

ADJUSTMENT

All Models

4. The only adjustment required is adjustment of cylinder rack clearance (backlash). If clearance is excessive, wear is usually indicated and cylinder unit should be removed as outlined in paragraph 6 and overhauled as in paragraph 7.

TROUBLE SHOOTING

All Models

5. If malfunction of the steering system exists, first make an operational check of some other hydraulic unit to eliminate the hydraulic system as the cause of trouble. The procedure for completely checking the hydraulic system is given beginning with paragraph 127.

External fluid leakage is not generally objectional from an operational standpoint and should be corrected only to prevent visible loss of fluid.

Internal leakage is objectionable only to the extent that performance is affected. Normal leakage of the steering system will permit approximately 2 revolutions per minute of the steering wheel with normal steering pressure applied and wheels fully turned to right or left steering stop. If considerable difference exists when turning in one direction only, the trouble is probably in steering cylinder. If slippage is approximately equal in either direction, the hand pump may be at fault.

STEERING CYLINDER (MOTOR)

All Models

Steering cylinders, pistons or piston rings can be renewed or backlash adjusted without removal of unit from the tractor. If complete overhaul is indicated, steering motor should be removed as outlined in paragraph 6.

6. **REMOVE AND REINSTALL.** To remove the complete steering motor as a unit, first disconnect center steering arm from spindle on axle models or support front of tractor and remove

lower pedestal on tricycle tractors. On all models remove front grille and the complete left side grille and grille casting.

Disconnect filter lines (1 & 7—Fig. 6) and unbolt and remove hydraulic filter unit (13) from support bracket (12). Disconnect lower cylinder lines (5 & 6) and lubrication lines (4 & 9), then unbolt and lift out steering motor (14).

Install by reversing the removal procedure. When attaching center steering arm or tricycle lower pedes-

Fig. 7—Partially installed view of steering motor.

1. Steering motor
2. Front support
3. Oil cooler mounting lug

tal to spindle (2—Fig. 8), make sure "X" marked cap screw hole (3) points directly to the rear and install center steering arm or pedestal with front wheels in a straight-ahead position. Use Grade "C" (Blue) LOCTITE on cap screws attaching center steering arm or pedestal to spindle (2) and tighten cap screws to a torque of 175-200 ft.-lbs. Steering motor is self-bleeding.

7. **OVERHAUL.** When disassembling the steering motor either on or off the tractor, unbolt and remove steering cylinders (23—Fig. 9) before removing the rack guide (21).

NOTE: ID of cylinder is relieved at outer end. If piston and rack assembly is allowed to bottom in cylinder, ring will expand into relief making it impossible to withdraw piston without damaging piston ring.

Piston can be removed by removing cap screw (10) after cylinder is off. Piston uses spacer (12) to free the installed piston on rack for alignment purposes. Piston should be free to turn with bolt (10) tightened. "O" ring (15) is fitted between piston (14) and rack (16) and seals inner hole in piston when hydraulic pressure is applied. Tighten bolt (10) to a torque of 30-35 ft.-lbs. when unit is reassembled.

Steering gear backlash should be 0.001-0.008. Measure the clearance between rack (16) and guide (21) using a feeler gage, after one cylinder has been removed. Backlash MUST be measured with rack centered on gear, backlash increases slightly when turned from center position to allow for wear adjustment. Shims (22) are

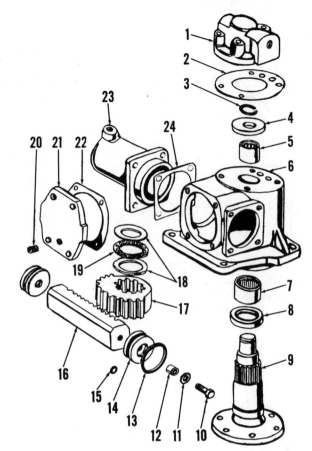

Fig. 9—Exploded view of hydrostatic steering motor.

1. Cover
2. Gasket
3. Snap ring
4. Thrust washer
5. Needle bearing
6. Spindle housing
7. Needle bearing
8. Oil seal
9. Steering spindle
10. Cap screw
11. Washer
12. Spacer
13. Piston ring
14. Piston
15. Piston ring
16. Rack
17. Pinion
18. Bearing races
19. Thrust bearing
20. Drain plug
21. Rack guide
22. Shim
23. Cylinder
24. Gasket

available in four thicknesses from 0.006 to 0.023. Make sure rack is centered on gear (17) when unit is assembled.

To remove steering spindle (9), first remove cylinders, rack guide (21) and rack (16), then unbolt and remove housing cover (1). Unseat and remove snap ring (3), then bump spindle (9) downward out of splines of gear (17).

Needle bearings (5 & 7) and seal (8) can be renewed at this time; install bearings and seal with outer edges approximately flush with their respective bores.

If steering motor has been removed, reinstall as outlined in paragraph 6. Refer to Fig. 6 for an exploded view of hydraulic lines and associated parts.

HYDROSTATIC HAND PUMP

All Models

The hydrostatic hand pump (Char-Lynn ORBITROL) is a remote hydraulic control valve and metering valve which supplies pressurized fluid to the power steering motor. The unit also serves as a manual control valve and pump to provide steering control when hydraulic power is absent. No mechanical linkage exists between the steering wheel and tractor front wheels.

8. **REMOVE AND REINSTALL.** To remove the hydrostatic power steering hand pump, first remove hood, hood side panels and battery platform. Clean the area surrounding pump ports, disconnect hydraulic lines; then unbolt and remove the hydrostatic hand pump.

Fig. 8—Installed steering spindle viewed from below.

1. Right steering stop
2. Steering spindle flange
3. "X" marked capscrew hole
4. Left steering stop

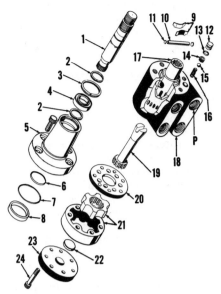

Fig. 10—Exploded view of ORBITROL hydrostatic hand pump showing component parts.

1. Input shaft	13. O-ring
2. Snap ring	14. Ball seat
3. Snap ring	15. Check valve
4. Bearing	16. Spring
5. Shaft housing	17. Sleeve
6. Quad ring	18. Valve housing
7. O-ring	19. Drive link
8. Bushing	20. Spacer plate
9. Centering springs	21. Gear set
10. Drive pin	22. Disc
11. Teflon disc	23. End plate
12. Valve plug	24. Cap screw

Install by reversing the removal procedure. Cycle the system several times to bleed air from motor and lines. The system is self-bleeding.

9. **OVERHAUL.** Refer to Fig. 10 for a partially exploded view of the Char-Lynn Model ZC2-51 Orbitrol used on models with adjustable steering wheel. The unit used on models with fixed wheel is similar except shaft and bearing (items 1 through 4) are omitted and upper flange (5) is modified. Rotary valve sleeves (17) and housing (18) are available as a matched set only.

To disassemble the removed Orbitrol unit, first remove cap screws (24) and withdraw items (19) through (23). Unbolt and remove upper control flange (5) and shaft as a unit. Withdraw the double control sleeve (17) as a unit from botttom after metering unit is removed. Examine the sleeve unit to determine method of assembly, then push out centering springs (9) and drive pin (10) and separate the sleeves. Insert a bent wire through port (P) in valve body and push plug (12) from body; then remove check valve seat (14) using an Allen wrench. Remove check ball (15) and spring (16).

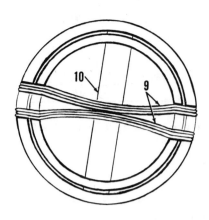

Fig. 11—Schematic end view of steering valve sleeves showing correct assembly of centering springs (9) and drive pin (10). Refer also to Fig. 10 and paragraph 9.

Wash all parts in a suitable solvent. Inspect centering springs (9) for fractures or distortion. Springs should have a minimum arch of $\frac{7}{32}$-inch when measured at center. Inspect plates (23 & 20) for wear or scoring. Plates should be discarded if measurable wear exists, however a visible polish pattern should not be confused with wear. Inspect control valve sleeves and body for nicks, scoring or wear. If any part is damaged, renew the housing assembly which includes the sleeves and check valve.

When assembling the unit, tighten check valve seat (14) to a torque of 150 inch-pounds. Install springs (9) with notched edge down in sets of three as shown in Fig. 11. Insert the assembled sleeve valve unit from bottom of valve body using a twisting motion. Do not allow sleeve to move up beyond flush with upper machined surface of body. Be sure pin slot in drive shaft (19—Fig. 10) is aligned with a valley of inner rotor as shown in Fig. 12. Rotor has six gear teeth and 12 splines; if drive slot is im-

Fig. 12—When properly assembled, slot in drive link must align with valley of inner gear as shown.

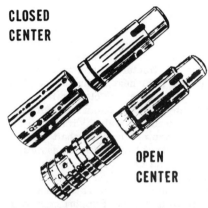

Fig. 13—Comparative views of Open Center and Closed Center ORBITROL Steering valve sleeves. Closed center sleeves must be used on Models MF 1100 and MF 1130.

properly aligned, steering unit will operate in reverse when hydraulic pressure is applied. Tighten the seven lower cap screws (24—Fig. 10) evenly to a torque of 250 inch-pounds and the four cap screws in upper plate (5) to a torque of 220 inch-pounds. Install the assembled unit as outlined in paragraph 8.

NON-DIESEL ENGINE

AND COMPONENTS

All Non-Diesel Models are equipped with a six cylinder Waukesha engine having a bore of 4⅛ inches, a stroke of 4 inches and a piston displacement of 320 cubic inches.

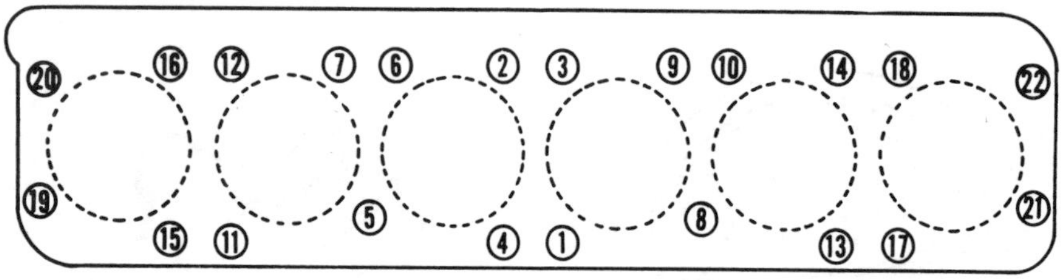

Fig. 14—On non-diesel engines, cylinder head cap screws must be torqued to 90-100 ft.-lbs. in the sequence shown.

R&R ENGINE WITH CLUTCH

All Models

10. To remove the engine and clutch as a unit, first drain cooling system and if engine is to be disassembled, drain oil pan. Remove hood, grille and hood rear side panels. Remove battery and battery box; then disconnect and remove the power steering and main hydraulic lines running to front of tractor. Disconnect wiring harness and heat indicator sending unit. Disconnect air cleaner hose and radiator hoses. Back out the four capscrews securing main hydraulic pump drive adapter to crankshaft pulley. Support tractor under transmission housing and unbolt front support casting from engine, then roll front axle, support and radiator as an assembly away from engine. Support engine in a hoist and unbolt engine from transmission case.

Install by reversing the removal procedure. Adjust the governed speed as outlined in paragraph 82 after tractor is reassembled.

CYLINDER HEAD

All Models

11. **REMOVE AND RINSTALL.** To remove the cylinder head, first remove hood and side panels. On LP Gas Models, fuel tank must be removed. On all models, disconnect and remove interfering main hydraulic system and power steering lines. Disconnect carburetor linkage and fuel line, and unbolt and remove carburetor and manifolds. Remove rocker arm cover, rocker arms, push rods and thermostat housing; then, unbolt and remove the cylinder head.

When installing the cylinder head, reverse the removal procedure. Tighten the cylinder head cap screws to a torque of 90-100 ft.-lbs. using the sequence shown in Fig. 14. Adjust valve tappet gap using the procedure outlined in paragraph 13.

VALVES AND SEATS

All Models

12. Intake valves seat directly in the cylinder head while exhaust valves have renewable valve seat inserts. Replacement exhaust valve seat inserts are provided in oversizes of 0.031 and 0.062 as well as standard. Valve seats should be installed with an interference fit of 0.003-0.005.

Valve face and seat angle is 45° for all valves, with a recommended seat width of $\frac{3}{32}$ inch. Seats can be narrowed using 15 and 70 degree stones.

13. **VALVE TAPPET GAP.** The recommended cold tappet gap setting is 0.012 for intake valves and 0.022 for exhaust valves. Cold (static) setting of all valves can be made from just two crankshaft positions using the procedure shown in Figs. 15 and 16, as follows:

Remove timing port plug from right side of engine adapter plate and turn crankshaft until TDC timing mark is aligned. Check the rocker arms for front and rear cylinders. If rocker arms on rear cylinder are tight, No. 1 cylinder is on the compression stroke and the six tappets indicated in Fig. 15 can be adjusted. If rocker arms on front cylinder are tight, No. 6 cylinder is on the compression stroke and the six tappets indicated in Fig. 16 can be adjusted. After adjusting the indicated tappets, turn crankshaft one complete revolution until TDC

timing mark is again aligned and adjust the clearance of the remaining tappets. Hot tappet adjustment is impractical because of tractor design.

VALVE GUIDES

All Models

14. The pre-sized valve guides are interchangeable for intake and exhaust valves. Valve guides are shouldered for correct location. The manufacturer suggests that distortion is possible if excessive pressure is applied after guide is seated. To avoid the possibility, it is recommended that guide be pressed into head until 0.002-0.010 clearance exists between guide shoulder and machined surface of head.

Inside diameter of new valve guide is 0.3755-0.3765. Recommended stem-to-guide diametral clearance is 0.0025-0.0045 for intake valves and 0.0035-0.0055 for exhaust valves. Renew guides if clearance exceeds the recommended maximum by 0.003.

VALVE SPRINGS

All Models

15. Intake and exhaust valve springs are not interchangeable. Renew any spring which is rusted, discolored, distorted, or fails to meet the test specifications which follow:

Intake Valve Springs:
    Free length (approx)....2 49/64 In.
    Lbs. test @ 2 $\frac{11}{32}$ In. 52-60
    Lbs. test @ 1 $\frac{15}{16}$ In. 114-134

Exhaust Valve Springs:
    Free length (approx.) ....2 $\frac{11}{16}$ In.
    Lbs. test @ 2 15/64 In. 50-56
    Lbs. test @ 1 53/64 In. 122-132

VALVE ROTATORS

All Models

16. Normal servicing of the positive type exhaust valve rotators consists of renewing the units. It is important, however, to observe the operation af-

Fig. 15—With TDC timing mark aligned and No. 1 piston on compression stroke, adjust the indicated valves as outlined in paragraph 13.

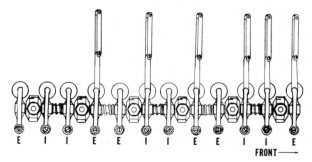

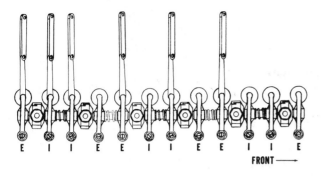

Fig. 16—With TDC timing mark aligned and No. 6 piston on compression stroke, adjust the indicated valves as outlined in paragraph 13.

ter valves are assembled. The valve rotator can be considered satisfactory if the valve turns a slight amount each time valve opens.

VALVE TAPPETS

All Models

17. The mushroom type tappets (cam followers) operate directly in machined bores in cylinder block. The 0.623-0.6235 diameter tappets should have a diametral clearance of 0.0015-0.003 in block bores. Tappets can be removed after removing camshaft as outlined in paragraph 21.

Refer to paragraph 13 for tappet gap adjustment procedure.

ROCKER ARMS

All Models

18. The rocker arm shaft is positively positioned by the support bolts which pass through drilled holes in the shaft and support brackets. Oil holes for rocker arms must be located on top of shaft and must be properly positioned before oil feed line can be installed.

The rocker arms should have 0.0015-0.0035 clearance on the 0.742-0.743 shaft. Bushings for the rocker arms are not available; if clearance is excessive, renew the rocker arms. Refer to paragraph 13 for valve tappet gap adjustment procedure.

TIMING GEAR COVER
All Models

19. **REMOVE AND REINSTALL.** To remove the timing gear cover, first remove front axle, support assembly and radiator as a unit, following the general procedure outlined in paragraph 10. Remove the fan blades and fan belt. Remove the cap screw and washer retaining the crankshaft pulley (4—Fig. 18) on crankshaft (10). Using the tapped holes provided and a suitable center to protect the threads for cap screw (1), attach a puller to crankshaft pulley (4); tighten the puller to remove the slack, then strike forcing screw sharply with a hammer to bounce out the tapered retainer (3). Pulley can now be easily removed. Disconnect and remove the governor, then unbolt and remove timing gear cover.

Crankshaft front oil seal can be renewed at this time. Install seal from front of cover with cork side of seal facing forward.

Install timing gear cover by reversing the removal procedure. Adjust the governor after tractor is assembled, as outlined in paragraph 82.

TIMING GEARS
All Models

20. Timing gear backlash should be 0.004-0.006. Camshaft end play is controlled by a thrust button and spring located in front end of camshaft, and

end thrust is against a brass expansion plug located in rear end of camshaft bore in block.

Gears are properly timed when "C" timing marks align as shown in Fig. 19. The crankshaft timing gear can be removed after removing timing gear cover as outlined in paragraph 19. To remove camshaft gear, camshaft should first be removed as in paragraph 21.

CAMSHAFT
All Models

21. To remove the camshaft, first remove timing gear cover as outlined in paragraph 19 and rocker arms as in paragraph 18. Remove the oil pan, oil pump and distributor. On gasoline models, remove fuel pump. Block up or support the cam followers and carefully withdraw camshaft forward out of engine block.

Camshaft gear can be pressed from camshaft after shaft is removed. When installing the gear, make sure "C" timing mark is toward the front.

The front camshaft bore contains a renewable bushing. All other camshaft journals ride in unbushed bores in engine block. Front bushing is pre-sized, and must be installed with oil hole aligned. Camshaft diameter is 1.7485-1.7495 for all journals. Examine the brass expansion plug at rear of cylinder block whenever camshaft is removed, and renew if indicated. The brass expansion plug carries camshaft thrust which is rearward when engine is running. To renew the expansion plug, it is first necessary to remove engine from tractor and remove clutch, flywheel and flywheel adapter plate.

ROD AND PISTON UNITS
All Models

22. Connecting rod and piston units are removed from above after removing cylinder head and oil pan.

Cylinder numbers are stamped on front side of pistons on new engines; however connecting rods are not normally marked. Connecting rods are properly installed when lower connecting rod bolt is on camshaft side of engine block.

Tighten connecting rod cap screws to a torque of 85-90 ft.-lbs. when rod and piston units are installed.

PISTONS, RINGS AND CYLINDERS
All Models

23. The cam ground, aluminum alloy pistons operate directly in cylinder block bores and are available in standard size and oversizes of 0.020 and 0.040.

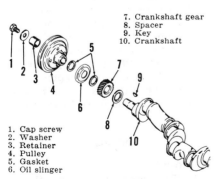

7. Crankshaft gear
8. Spacer
9. Key
10. Crankshaft

1. Cap screw
2. Washer
3. Retainer
4. Pulley
5. Gasket
6. Oil slinger

Fig. 18—Crankshaft pulley uses taper and retainer (3) to tighten pulley on shaft. Refer to paragraph 19 for removal procedure.

Fig. 19—View of timing gears with cover removed, showing timing marks properly aligned.

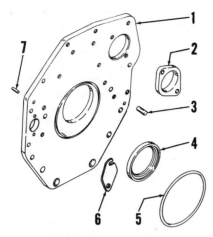

Fig. 20—Engine adapter plate and associated parts.

1. Adapter plate
2. Starter adapter
3. Dowel
4. Oil seal
5. O-ring
6. Timing hole cover
7. Dowel pin

Standard cylinder bore is 4.125-4.126 and maximum allowable taper is 0.006. Production pistons are letter sized for selective fit, the identifying marking "A, B, C or D" being stamped on head of piston. Desired piston skirt clearance is 0.003-0.004, measured at right angle to piston pin at bottom of skirt and bottom of piston bore below ring travel area.

Each piston is fitted with two compression rings and one segmented oil control ring. The top compression ring is chrome faced. Desired piston ring end gap is 0.0013-0.0023 for compression rings. Desired side clearance is 0.002-0.0035 for top compression ring or 0.002-0.004 for second ring. End gap and side clearance of segmented oil control ring is not measured. Stagger the end gaps when pistons are installed.

PISTON PINS

All Models

24. The 1.2494-1.2497 diameter full floating type piston pins are retained in piston bosses by snap rings and are available in standard size and 0.005 oversize.

Two bushings are fitted in piston pin end of connecting rod and are installed with outer edges flush with outer edges of rod bore and cut-away portions of bushings aligned with oil hole in rod. Hone bushings after installation to provide a thumb push fit (0.0004-0.0009 clearance) for piston pin. If oversize piston pins are being used, piston bosses must be honed to provide a tight push fit (0.0002-0.0004

clearance) for pin. Maximum allowable clearance for piston pin is 0.0019 in rod and 0.0008 in piston.

CONNECTING RODS AND BEARINGS

All Models

25. Connecting rod bearings are precision type, renewable from below after removing oil pan and bearing caps. When renewing bearing shells, be sure that the projections engage milled slot in rod and cap and are installed together on side away from camshaft side of engine. Upper and lower bearing halves are interchangeable.

Bearings are available in standard size and undersizes of 0.020 and 0.040. Connecting rod bearings should have a diametral clearance of 0.0005-0.003 on the 2.624-2.625 diameter crankpin. Recommended connecting rod side clearance is 0.0075-0.0135. Tighten the self-locking connecting rod cap screws to a torque of 85-90 ft.-lbs.

CRANKSHAFT AND BEARINGS

All Models

26. The crankshaft is supported in four precision type main bearings. All bearing caps can be removed from below after removing oil pan.

Upper and lower bearing liners are not interchangeable, the difference being in the positioning of oil holes. Liners are interchangeable for all locations except rear-intermediate, which contains the flanged liners to control crankshaft end play.

Recommended main bearing diametral clearance is 0.0015-0.0045 on the 2.624-2.625 main journals. Recommended crankshaft end play is 0.004-0.0085. Tighten the retaining stud nuts to a torque of 105-115 ft.-lbs. when installing main bearing caps.

To remove the crankshaft, first remove engine as outlined in paragraph 10; then remove crankshaft pulley, timing gear cover, clutch, flywheel, engine adapter plate, oil pan, oil pump and main and connecting rod bearing caps.

REAR OIL SEAL

All Models

27. The crankshaft rear oil seal is housed in the engine adapter plate as shown in Fig. 20. Oil seal can be removed after detaching engine from transmission as outlined in paragraph 93 and removing clutch and flywheel. Before removing seal, however, be sure that leak is at the seal. If doubt exists, remove adapter plate (1) and

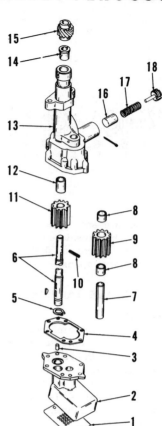

Fig. 21—Exploded view of gear type engine oil pump.

1. Screen
2. Cover
3. Dowel
4. Gasket
5. Snap ring
6. Drive shaft
7. Idler gear shaft
8. Bushing
9. Follow gear
10. Spring pin
11. Driven gear
12. Bushing
13. Body
14. Bushing
15. Drive gear
16. Relief valve plunger
17. Spring
18. Adjusting screw

examine or renew the sealing "O" ring (5). Adapter plate is positively located on engine block by dowel pins (3).

FLYWHEEL

All Models

28. To remove the flywheel, first separate engine from transmission housing and remove the clutch. Flywheel is secured to crankshaft mounting flange by six evenly spaced cap screws which also retain the pto shaft drive hub. When installing the flywheel, make sure No. 1 and No. 6 pistons are at TDC, then align timing mark with window in adapter plate. Tighten the flywheel retaining cap screws to a torque of 65-70 ft.-lbs.

OIL PAN

All Models

29. The heavy cast iron oil pan serves as an attaching point for front support. To remove the oil pan, first

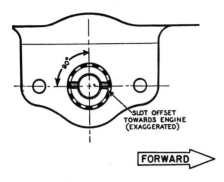

Fig. 22—Oil pump is properly timed when distributor drive slot is positioned as shown, with No. 1 piston at TDC on compression stroke.

Upper bushing (14)
    Inside diameter 0.4955-0.4965
    Diametral clearance 0.001-0.0025
Lower bushing (12)
    Inside diameter 0.626-0.6265
    Diametral clearance 0.0015-0.0025
Idler gear bushings (8)
    Inside diameter 0.626-0.6265
    Diametral clearance 0.0035-0.0045

31. **RELIEF VALVE.** Normal operating pressure should be 35-40 psi at operating speed and 12 psi at slow idle speed. The setting of relief valve (16—Fig. 21) in oil pump can be adjusted by removing the cotter pin and turning adjusting screw (18). Oil pan must be removed for access to adjusting screw.

DIESEL ENGINE AND COMPONENTS

All Diesel Models are equipped with a Perkins six cylinder engine having a bore of 3⅞ inches, a stroke of 5 inches and a displacement of 354 cubic inches. The engine used in Model MF1100 is naturally aspirated while Model MF1130 is turbocharged. Compression ratio of both engines is 16.0:1.

R&R ENGINE WITH CLUTCH

All Diesel Models

33. To remove the engine and clutch as a unit, first drain cooling system and if engine is to be disassembled, drain oil pan. Remove hood, grille and hood rear side panels. Remove batteries and battery box; then disconnect and remove the power steering and main hydraulic lines running to front of tractor. Disconnect wiring harness, heat indicator sending unit, fuel lines and throttle controls. Disconnect air cleaner and radiator hoses and on Model MF1130, disconnect engine oil cooler lines running to radiator. Back out the four capscrews securing main hydraulic pump drive adapter to crankshaft pulley. Support tractor under transmission housing and unbolt front support casting from engine. Support engine from hoist, remove the attaching cap screws and swing engine from tractor frame.

Install by reversing the removal procedure. Bleed the fuel system as outlined in paragraph 66 and adjust throttle controls as in paragraph 76.

CYLINDER HEAD

All Diesel Models

34. To remove the cylinder head, first drain cooling system and remove hood and side panels. Remove grille, batteries and battery carrier, then disconnect and remove the power steering and main hydraulic system lines running to front of tractor. Disconnect front throttle rod at injection pump end and horizontal rod at bellcrank, then unbolt and remove channel support and throttle rods as an assembly.

On Model MF1130, disconnect turbocharger oil lines, air cleaner and inlet manifold hoses, unbolt and remove inlet manifold; then unbolt and remove exhaust manifold and turbocharger as a unit. On Model 1100, remove the manifolds. On all models, remove rocker arm cover breather pipe and injector pressure lines as an assembly.

Disconnect upper and lower coolant hoses and temperature indicator sending unit from water pump body. Un-

OIL PUMP

All Models

30. The gear type oil pump (Fig. 21) is driven by the camshaft and can be removed after removing oil pan as outlined in paragraph 29. When installing the pump, it will be necessary to remove the ignition distributor, turn crankshaft until No. 1 piston is at TDC on compression stroke; then align distributor drive slot in top of oil pump drive shaft as shown in Fig. 22. Complete the installation by reversing the removal procedure. Time the ignition as in paragraph 87.

To overhaul the removed oil pump, refer to Fig. 21 and proceed as follows: Unbolt and remove pump cover (2) and withdraw pump idler gear (9). Drive out the roll pin (10); then press drive shaft (6) and gear (11) downward out of drive gear (15) and housing (13). Remove snap ring (5) and press gear from shaft if renewal is indicated. Oil pump gears (9 & 11) should measure 1.747-1.748 in thickness and have 0.002-0.003 radial clearance in body bores. Gear bores in pump body should measure 1.742-1.745 in depth, allowing the installed gears to extend 0.002-0.006 above gasket face of body. Refer to the following table for installed diameter and clearance of bushings:

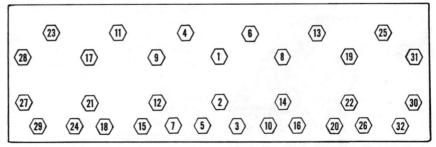

Fig. 25—On diesel engines, tighten cylinder head cap screws to a torque of 85 ft.-lbs. using the sequence shown.

bolt and remove the fan blades. Disconnect belt tightener arm from water pump body and remove fan belt from water pump pulley. Remove rocker arm top cover, rocker arms and push rods; then remove rocker arm lower cover. Remove the cylinder head stud nuts, lift the head until clear of top of studs, then move head rearward until fan pulley is clear of radiator shroud.

Cylinder head gasket is marked "TOP-FRONT" for proper installation. Head gasket should be installed dry. Tighten cylinder head stud nuts to a torque of 85 ft.-lbs. using the sequence shown in Fig. 25. Adjust valve tappet gap after head is installed as outlined in paragraph 40.

VALVES AND SEATS

All Diesel Models

35. On Model MF1100, both the intake and exhaust valves seat directly in the cylinder head. Renewable inserts are used in all valve seats on Model MF1130. On all models valve heads should be recessed a specified amount into the cylinder head and inserts should be installed or renewed when specifications are exceeded. Clearance can be checked using a straight edge and feeler gage as shown in Fig. 26. Production clearances are held within the limits of 0.035-0.045 for intake valves or 0.029-0.039 for exhaust valves. A maximum clearance of 0.060 is allowed for either valve.

On most engines, the exhaust valve seat is lowered by machining seat as shown in Fig. 27, rather than by using a 15° stone after seat is installed. When reseating the valves, use a 1½-inch OD stone, cut down if necessary, in order to reach the seat. On intake

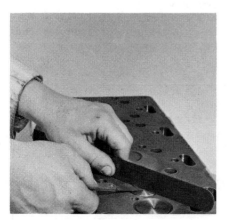

Fig. 26—Using a straight edge and feeler gage to check valve head height. Refer to paragraph 35.

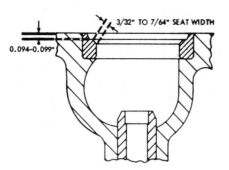

Fig. 27—Cross sectional view of exhaust valve seat profile used on most engines. Refer to paragraph 35.

valves, recess valve head by using a 15° stone above valve face contact area.

Valve face angle is 44° and seat angle is 45° for all valves, with a recommended seat width of 3/32 to 7/64-inch.

VALVE GUIDES

All Diesel Models

36. The cast iron valve guides are not interchangeable for intake and exhaust valves. Refer to Fig. 28. Exhaust valve guide bores are counterbored at valve-head end as shown. Press all guides into head until distance (A) between top end of guide and spring seat of head measures ⅝-inch.

Inside diameter of a new guide is 0.375-0.376 and desired diametral clearance is 0.0015-0.0035 for intake valves or 0.002-0.004 for exhaust valves.

VALVE SPRINGS

All Models

37. Springs, retainers and locks are interchangeable for intake and exhaust valves; however, an umbrella type oil deflector is installed on intake valve stem. Model MF1130 uses an inner and outer spring and seating washer differs from that used with single spring. The outer spring used with turbocharged models is identical with that used on models with single spring. Springs may be installed either end up. Renew the springs if they are distorted, discolored, or fail to meet the test specifications which follow, or the other parts if they show signs of wear or damage.

INNER SPRING:
Lbs. Test @ 1 9/16 inches 13-17
Lbs. Test @ 1⅛ inches 32-36
OUTER SPRING:
Lbs. Test @ 1 25/32 inches 38-42
Lbs. Test @ 1 23/64 inches 69-75

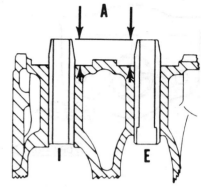

Fig. 28—Distance (A) should measure ⅝-inch for all guides. Difference in design of intake guide (I) and exhaust guide (E) is also shown.

CAM FOLLOWERS

All Models

38. The mushroom type tappets (cam followers) operate in machined bores in the cylinder block and can be renewed after removing camshaft as outlined in paragraph 46. The 0.7475-0.7485 diameter tappets should have a diametral clearance of 0.0015-0.0038 in crankcase bores.

ROCKER ARMS

All Models

39. The rocker arm shaft assembly can be removed from left side on turbocharged engine or right side on other models; after removing the respective hood side panel and rocker arm cover.

NOTE: Push rods can drop into oil pan if dislocated at lower end. Be careful not to lose push rods into engine when removing or installing rocker arms.

Fig. 29 shows an assembled view of the rocker arm shaft. Right and left hand rocker arm units, mounting brackets and the oil feed tube must be positioned as shown. Oil feed holes in shaft must be installed toward valve stem side. Make sure the sealing "O" ring does not roll out of retaining grooves in feed tube as shaft unit is installed. Some mechanics prefer to position the "O" ring on tube below the positioning grooves and allow it to roll into proper position as shaft is installed.

Desired diametral clearance between rocker arms and shaft is 0.001-0.0035. Renew shaft and/or rocker arms if clearance is excessive. Tighten rocker arm support cap screws to a torque of 50-55 ft.-lbs.

Fig. 29—Assembled view of rocker arm shaft showing parts properly installed.

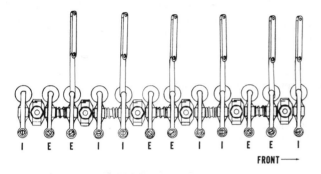

Fig. 30—With TDC timing marks aligned and No. 1 piston on compression stroke, adjust the indicated valves to 0.012 cold.

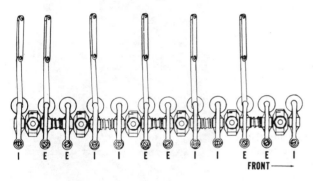

Fig. 31—With TDC timing marks aligned and No. 6 piston on compression stroke, adjust the indicated valves to 0.012 cold.

VALVE TIMING

All Models

41. Timing gears are keyed and valve timing will be correct if timing marks are properly aligned as outlined in paragraph 44.

TIMING GEAR COVER

All Models

42. The timing gear cover (housing) (12—Fig. 32) can only be removed after removing the camshaft gear (7) and auxiliary drive shaft gear (10). The crankshaft front oil seal (11) seals against crankshaft pulley hub, and seal and seating surface can be inspected and renewed if necessary, after removing crankshaft pulley.

To remove the timing gear cover or renew the oil seal, first support tractor and remove front end as a unit as outlined in paragraph 33.

43. **CRANKSHAFT PULLEY AND FRONT OIL SEAL.** If front oil seal only is to be renewed, remove crankshaft pulley retaining cap screw (1—Fig. 33) and washer (2). Using a center, attach a suitable puller to crankshaft pulley (4) and apply slight pressure with puller forcing screw then strike forcing screw sharply with a hammer to bounce out the tapered locking ring (3). Pulley can now be easily withdrawn.

VALVE TAPPET GAP

All Models

40. The recommended cold tappet gap setting is 0.012 for both the intake and exhaust valves. Cold (static) setting of all valves can be made from just two crankshaft positions, using the procedure outlined in this paragraph and illustrated in Figs. 30 and 31.

Remove timing plug from right front side of flywheel adapter housing and turn crankshaft until "TDC" timing mark on flywheel is aligned with timing pointer. Check the rocker arms for front and rear cylinders. If rear rocker arms are tight and front rocker ams have clearance, No. 1 piston is on compression stroke; adjust the six tappets shown in Fig. 31. After adjusting the indicated tappets, turn crankshaft one complete turn until the "TDC" mark is again aligned and adjust the remaining tappets.

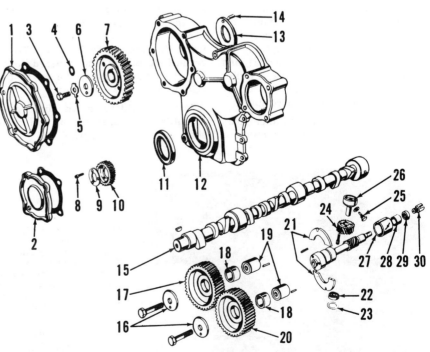

Fig. 32—Plates (1 & 2) and gears (7 & 10) must be removed before timing gear housing (12) can be removed for access to remainder of gears. Refer to text for details.

1. Cap screw
2. Washer
3. Locking ring
4. Pulley
5. Oil slinger
6. Crankshaft timing gear
7. Woodruff key
8. Crankshaft

Fig. 33—Crankshaft front pulley (4) is retained by tapered lock ring (3). Refer to paragraph 43 for recommended removal procedure.

Coat outer edge of new seal with a hardening type gasket cement and install seal with lip toward crankshaft gear. Front face of seal must be recessed approximately ¼-inch in seal bore of cover.

Thoroughly clean and degrease the splines on crankshaft and locking ring (3). Coat the splines with Grade AV (Red) LOCTITE and reinstall pulley and locking ring. Install washer (2) on capscrew (1) with inner chamfer next to capscrew head and install and tighten cap screw to a torque of 275 ft.-lbs.

44. COVER (HOUSING) AND GEARS. If the cover (12—Fig. 32) is to be removed, turn crankshaft until TDC timing mark on flywheel is aligned and No. 1 cylinder is on compression stroke. Remove crankshaft pulley as outlined in paragraph 43. Remove the two covers (1 and 2) and the gears (7 and 10), using a puller to remove camshaft gear (7). Cover can now be unbolted and removed.

The crankshaft gear and idler gears are now accessible for inspection or service. Allowable backlash between any two gears in timing gear train is 0.003-0.006. Diametral clearance of idler gears to hubs is 0.001-0.003 and installed end play of idler gears is 0.0023-0.0123. Bushings in idler gears must be reamed after installation to an inside diameter of 1.375-1.377. Oil holes are provided in gear hub and gear to supply lubrication to gear teeth.

Camshaft and auxiliary drive shaft can be withdrawn from cylinder block after timing gear housing is removed. Refer to paragraph 45 for details on auxiliary drive unit overhaul and to paragraph 46 for camshaft removal.

Before timing gear cover can be reinstalled, it is first necessary to remove the injection pump for proper retiming of auxiliary shaft. Remove the right hand idler gear (17) and reinstall with timing punch marks on idler gears and crankshaft gear aligned as shown in Fig. 34.

NOTE: Left Hand Idler Gear may not have timing marks and marks may be ignored if present. Marks are not required for timing injection pump drive gear.

Because of the odd number of teeth in idler gears, complete timing marks will only occasionally align and cannot therefore be used as a timing check.

Make sure the camshaft, auxiliary drive shaft and their thrust washers are properly positioned, then reinstall the timing gear cover and loosely install the retaining cap screws. Center the crankshaft oil seal (or oil seal bore) in cover over the crankshaft, using the Special Tool MFN 747B or crankshaft pulley and tapered locking ring; then tighten the cover retaining cap screws securely.

Install auxiliary drive shaft gear as outlined in paragraph 45 and camshaft gear as in paragraph 46.

45. AUXILIARY DRIVE SHAFTS AND GEARS. The auxiliary drive shaft is driven by the timing gear train and drives the engine oil pump and injection pump. The auxiliary drive unit is shown exploded in Fig. 37 and in cross section in Fig. 38.

The auxiliary drive shafts and gears are pressure lubricated by the engine oiling system as shown at (P—Fig. 38). Renewable fitting (11) provides spray lubrication for drive worm.

To check the auxiliary shaft timing, remove injection pump as outlined in paragraph 77 and turn engine crankshaft until No. 1 piston is at TDC on compression stroke. Auxiliary shaft timing is correct if notches (N—Fig. 39) in pump drive shaft and adapter plate align as shown.

To remove the pump drive gear, unbolt and remove adapter (A). On

Fig. 35—Timing idler gear withdrawn to show pressure oil passages (O) which provide lubrication to timing gear train. Refer also to Fig. 36.

Fig. 34 — Front view of engine block with timing marks properly aligned. Timing gear train will never appear as shown, because shaded gears (AS & CS) must be removed for access to timing gear housing.

AS. Accessory shaft gear
CS. Camshaft gear
CR. Crankshaft gear

Fig. 36—Pressure oil passage (O) in idler gear hub and block. Refer also to Fig. 35.

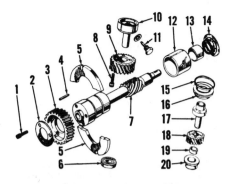

Fig. 37—Exploded view of accessory shaft and gears and injection pump drive shaft. Items (15 through 20) are used on turbocharged models instead of items (6, 9 & 10) which are used only on Model MF 1100.

1. Screw	11. Oil spray tube
2. Retainer plate	12. Bushing
3. Drive gear	13. Bushing
4. Dowel pin	14. Oil seal
5. Thrust washer	15. O-ring
6. Bearing	16. Thrust sleeve
(Model MF1100)	17. Pump drive
7. Accessory shaft	(Model MF 1130)
8. Screw	18. Worm gear
9. Worm gear	(Model MF 1130)
(Model MF 1100)	19. Bushing
10. Pump drive	20. Lower collar
(Model MF 1100)	

models without turbocharger, the lower ball bearing (6—Fig. 37), gear (9) and shaft (10) must be removed as a unit. If a suitable puller and slide hammer is available, the complete assembly can sometimes be withdrawn from above without further disassembly. The alternative method is to remove the oil pan and oil pump, then tap the assembly upward using a suitable drift and hammer. On models with turbocharger, remove thrust sleeve (16) from top opening, then lift out shaft (17) and gear (18) as a unit. Thrust collar (20) and bushing (19) are a press fit in housing, and can only be removed after removing oil pan and oil pump.

NOTE: Some early tractors without turbocharger require the use of shims underneath the ball bearing (6). Where shims are required, shim pack thickness is stamped on housing next to engine Serial Number. If housing is not stamped, shims are not required.

The Allen head capscrews (8) which retain the bronze worm gear (9 or 18) are secured with LOCTITE and gears are a press fit on shaft. When renewing gear or shaft, make certain that oil slots and cap screw holes are aligned and that countersunk ends of cap screw holes in gear are to the bottom, away from shaft hub. On turbocharged models only, coat inner bore of gear with Grade

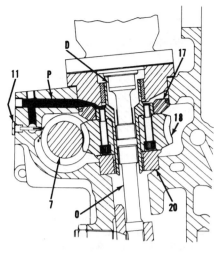

Fig. 38—Cross sectional view of accessory drive housing and injection pump drive used on turbocharged models. Other models are similar. Note drilled lubrication pressure passage (P) with pressurized oil delivery shown in black. Oil spray tube (11) directs lubricant to worm on accessory shaft (7). Refer to Fig. 37 for parts identification except for injection pump drive (D) and oil pump drive (O).

"C" (Blue) LOCTITE. On all models, press gear tightly against shaft shoulder, coat cap screw threads with Grade "C" LOCTITE, then install and tighten cap screws.

If suitable equipment is available for finish honing concentric with housing, the bushing may be renewed in injection pump adapter housing on all models, and lower thrust collar (20) on turbocharged models. Recommended shaft-to-bushing clearance for models without turbocharger is 0.001-0.0048. On models with turbocharger, adapter housing (upper) bushing should have 0.001-0.0042 clearance and thrust collar (lower) bushing 0.0016-0.004 clearance.

Auxiliary shaft thrust washers (5) fit in machined slots in shaft (7). Outer portion of washers fit a machined recess in front face of engine block and are retained by timing gear housing. Normal shaft end play of 0.0025-0.009 is adjusted by renewing the thrust washers and/or shaft.

Auxiliary drive shaft should have a diametral clearance of 0.001-0.0042 in front bushing (12) and 0.001-0.0036 in rear bushing (13). Bushings are precision type and will not require reaming after installation. Make sure oil holes are aligned when bushings are installed.

Auxiliary shaft can be withdrawn from front after removing timing gear housing as outlined in paragraph 44.

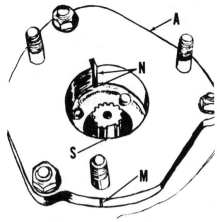

Fig. 39—Injection pump drive adapter with pump removed showing timing marks. Refer to paragraph 45 for shaft timing procedure.

A. Adapter
M. Pump timing mark
S. Master spline
N. Worm gear timing notches

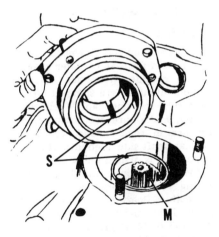

Fig. 40—Drive adapter must be removed to retime injection pump drive gear. Refer also to Fig. 39.

When reinstalling the auxiliary shaft, align engine timing marks as shown in Fig. 34 and pump drive shaft timing notch as shown in Fig. 39. The holes in auxiliary shaft drive gear (3—Fig. 37) are slightly elongated for proper alignment. Tighten cap screws (1) to a torque of 20-22 ft.-lbs. when alignment is correct.

46. CAMSHAFT. To remove the camshaft, first remove the rocker arms unit as outlined in paragraph 39 and timing gear housing as in paragraph 44. Unbolt and remove the fuel lift pump from right side of engine block. Remove the valve side covers and secure all cam followers in their uppermost position; withdraw camshaft forward out of engine block.

Normal camshaft bearing clearance is 0.0025-0.0045 for front bearing or 0.0025-0.0055 for other journals. Cylinder block bores are unbushed. Camshaft end play of 0.004-0.016 is controlled by a thrust washer located between front journal and camshaft gear. The thrust washer fits in a recess in front of engine block and is retained by timing gear housing and located by a dowel pin.

When reassembling the engine, reposition idler gear with timing marks aligned as outlined in paragraph 44 and shown in Fig. 34. With timing gear housing reinstalled, turn camshaft and gear until timing punch mark on idler gear is located between the two marks on camshaft gear and draw the gear into position using a suitable puller bolt. Tighten the camshaft gear retaining cap screw to a torque of 40-50 ft.-lbs. and lock in place by bending the tab washer.

ROD AND PISTON UNITS

All Models

47. Connecting rod and piston units are removed from above after removing the cylinder head and oil pan. Cylinder numbers are stamped on the connecting rod and cap. When reinstalling, make sure correlation numbers are in register and face away from camshaft side of engine. The connecting rod is drilled on turbocharged engines, and upper and lower connecting rod bearings are not interchangeable. Bearings are available in standard size and undersizes of 0.010-0.020 and 0.040. Regrinding is recommended ONLY for the induction hardened crankshaft used in MF1100 production engines. DO NOT regrind the tuftrided shaft used in Model MF1130

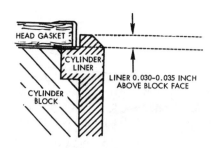

Fig. 42—Cross sectional view of cylinder block showing cylinder liner properly installed.

HEAD GASKET
CYLINDER LINER
LINER 0.030-0.035 INCH ABOVE BLOCK FACE
CYLINDER BLOCK

production and as service item for all engines. When installing connecting rod caps, use new self-locking nuts and tighten to a torque of 65-70 ft.-lbs.

PISTONS, SLEEVES AND RINGS

All Models

48. The aluminum alloy, cam ground pistons are supplied in a standard size only and are available in a kit consisting of piston, pin and rings for one cylinder. The toroidal combustion chamber is offset in piston crown as shown in Fig. 41 and piston is marked "FRONT" for proper assembly.

Each piston is fitted with a plain faced chrome top ring which may be installed either side up. The internally stepped cast iron 2nd & 3rd compression rings must be installed with the marking "BTM" toward the bottom, away from piston crown. Second and third compression rings are chrome faced on turbocharged engines. Engines without turbocharger use identical cast iron oil control rings in fourth and fifth grooves above and

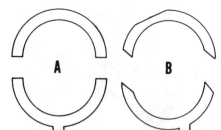

Fig. 44—Old type (A) and new type (B) crankshaft thrust washers are not interchangeable. Refer to paragraph 51.

below the piston pin. Rings may be installed either side up. Turbocharged engines use a segmented chrome oil control ring in fourth groove and piston does not contain a fifth groove. Specifications for fitting piston rings are as follows:

Model 1100
End Gap
    Top ring 0.018-0.022
    Other rings 0.014-0.019

Side Clearance
    Compression rings 0.0019-0.0039
    Oil control rings 0.0025-0.0045

Model 1130
End Gap
    Top ring 0.021-0.026
    Other compression rings 0.015-0.020

Side Clearance
    Top ring 0.003-0.005
    Other compression
    rings 0.0019-0.0039

The production cylinder sleeves are 0.001-0.003 press fit in cylinder block bores. Service sleeves are a transition

Fig. 41—Top view of piston showing combustion chamber recess and piston markings.

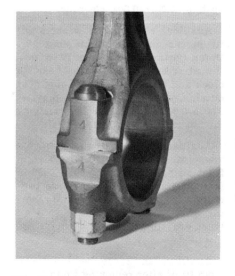

Fig. 43—Assembled view of serrated connecting rod and cap.

Fig. 45—Rear view of cylinder block and crankshaft showing rear oil seal partially removed.

Fig. 46—Use a round bar to bed the asbestos rope seal in retainer half. Refer to text for details.

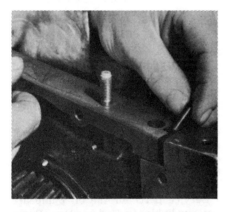

Fig. 47—Installing end seals in cylinder block bridge piece. Refer to paragraph 52.

Fig. 48—Use a straight edge to align gasket faces of cylinder block and bridge piece.

fit (0.001 tight to 0.001 clearance). When installing new sleeves, make sure sleeves and bores are absolutely clean and dry, then chill the sleeves and press fully into place by hand. When properly installed, liner should extend 0.030-0.035 above gasket surface of cylinder block as shown in Fig. 42. Recommended piston skirt clearance is 0.007-0.010.

PISTON PINS

All Models

49. The full floating piston pins are retained in piston bosses by snap rings. Piston pin diameter is 1.3748-1.375 for models without turbocharger or 1.4998-1.500 for models with turbocharger. Piston pins are available in standard size only. On all models, the renewable connecting rod bushing must be final sized to provide a diametral clearance of 0.00075-0.0017 for the pin. Be sure the predrilled oil hole (or holes) in bushing are aligned with holes in connecting rod and install bushing from chamfered side of rod bore. Piston pin should be a thumb press fit in piston after piston is heated to 160° F.

CONNECTING RODS AND BEARINGS

All Models

50. Connecting rod bearings are precision type, renewable from below after removing oil pan and connecting rod bearing caps. When renewing bearing shells, be sure that the projection engages milled slot in rod and cap and that the correlation marks are in register and face away from camshaft side of engine. On models with turbocharger, the upper shell is equipped with an oil hole to provide pressure lubrication to the drilled rod, and upper and lower shells are not interchangeable.

Connecting rods are graded as to weight and a code number is etched on big end of rod. Three code numbers are used for each model; 11, 12 and 13 for Model 1100; and 13, 14 and 15 for Model 1130. The largest number indicates the heaviest rod. When reenwing a connecting rod, use the same weight code as the removed rod. Replacement rods should be marked with cylinder number corresponding to that used on removed rod.

Bearings are available in standard size and undersizes of 0.010, 0.020 and 0.040. Regrinding is recommended ONLY for the induction hardened crankshaft used in MF1100 production engines. DO NOT regrind the tuftrided shaft used in Model MF1130 pro-

duction and as service shaft for all models.

Connecting rod bearings should have a diametral clearance of 0.0015-0.003 on the 2.499-2.4995 diameter crankpin. Recommended connecting rod side clearance is 0.0095-0.0145. Renew the self-locking connecting rod nuts and tighten to a torque of 65-70 ft.-lbs.

CRANKSHAFT AND BEARINGS

All Models

51. The crankshaft is supported in seven precision type main bearings. To remove the rear main bearing cap, it is first necessary to remove the engine, clutch, flywheel and rear oil seal. All other main bearing caps can be removed after removing the oil pan.

Upper and lower bearing inserts are not interchangeable, the upper (block) half being slotted to provide pressure lubrication to crankshaft and connecting rods. Inserts are interchangeable in pairs for all journals except the center main bearing. Two types of thrust washers have been used, which are not interchangeable; refer to Fig. 44. When installing the early type (A), make sure the steel back is positioned next to block and cap and the grooved, bronzed surface next to crankshaft thrust faces. The new (B) type cannot be installed backward.

Bearing inserts are available in undersizes of 0.010, 0.020 and 0.030 as well as standard; and thrust washers are available in standard thickness of 0.089-0.093, and oversizes of 0.007. Regrinding is recommended ONLY for the induction hardened crankshaft used in MF1100 production engines. DO NOT regrind the tuftrided shaft used in Model MF1130 production and as service shaft for all models.

Recommended main bearing diametral clearance is 0.0025-0.0045 and recommended crankshaft end play is 0.002-0.014. Tighten the main bearing retaining cap screws to a torque of 145-150 ft.-lbs. and secure by bending the tabs on locking washers. When renewing rear main bearing, refer to paragraph 52 for installation of rear seal and oil pan bridge pieces. Check the crankshaft against the values which follow:

Main journal diameter 2.9985-2.999
Crankpin diameter 2.499-2.4995

CRANKSHAFT REAR OIL SEAL

All Models

52. The asbestos rope type rear oil seal is contained in a two-piece seal retainer attached to rear face of en-

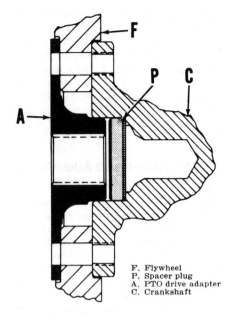

F. Flywheel
P. Spacer plug
A. PTO drive adapter
C. Crankshaft

Fig. 49—Cross sectional view of crankshaft and flywheel showing construction details.

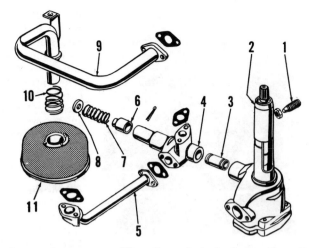

Fig. 51 — Partially exploded view of oil pump and associated parts.

1. Set screw
2. Oil pump
3. Delivery pipe
4. Delivery housing
5. Pressure pipe
6. Relief valve piston
7. Valve spring
8. Valve cap
9. Intake pipe
10. Spring
11. Intake screen

gine block as shown in Fig. 45. The seal retainer can be removed after removing flywheel and engine adapter plate.

The rope type crankshaft seal is precision cut to length, and must be installed in retainer halves with 0.010-0.020 of seal ends projecting from each end of retainer. Do not trim the seal. To install the seal, clamp each half of retainer in a vise as shown in Fig. 46. Make sure seal groove is clean. Start each end in groove with the specified amount of seal protruding. Allow seal rope to buckle in the center until about an inch of each end is bedded in groove, work center of seal into position, then roll with a round bar as shown. Repeat the process with the other half of seal.

When installing the cylinder block bridge piece at front or rear, lightly coat metal-to-metal contact faces with a non-hardening gasket cement and insert the end seals as shown in Fig.

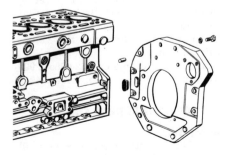

Fig. 50—View of flywheel adapter used on diesel models.

47; then use a straight edge as shown in Fig. 48 to make sure gasket faces are flush.

FLYWHEEL
All Models

53. To remove the flywheel, first separate engine from transmission housing and remove the clutch. Flywheel is secured to crankshaft mounting flange by six evenly spaced cap screws which also retain the pto shaft drive hub as shown in Fig. 49. The flywheel is not doweled. A seventh, unthreaded hole is drilled in crankshaft flange and a matching hole in flywheel rim; when these two holes are aligned, flywheel timing marks will be properly positioned.

The starter ring gear can be renewed after flywheel is removed. Heat ring gear evenly to approximately 475° F. and install on flywheel with beveled end of teeth facing front of engine.

Check flywheel runout with a dial indicator after flywheel is installed. Maximum allowable flywheel runout is 0.001 for each inch measurement is taken from flywheel centerline. Tighten the flywheel retaining cap screws to a torque of 75-80 ft.-lbs. and secure with lock wire.

ADAPTER PLATE
All Models

54. The engine adapter plate (Fig. 50) can be removed after removing flywheel as outlined in paragraph 53. The adapter plate is positively located on engine block by dowel pins and secured by six cap screws.

OIL PAN
All Models

55. The heavy cast-iron oil pan serves as part of the tractor frame and attaching point for the tractor front support. To remove the oil pan, support tractor underneath the transmission housing, drain the oil pan and remove the retaining screws. Use

a rolling floor jack to lower the oil pan away from engine block. Install by reversing the removal procedure.

OIL PUMP
All Models

56. The rotary type oil pump is mounted on lower side of engine block and driven by the injection pump drive shaft. Refer to Fig. 51 for an exploded view of oil pump and associated parts.

To remove the oil pump after oil pan is off, first unbolt and remove intake tube (9) and screen (11). Remove the cap screws retaining delivery tube (5) and housing (4) to cylinder block. Loosen the locknut and back out the set screw (1) on outside of engine block; then withdraw oil pump (2), delivery housing (4) and associated parts as a unit.

Service parts for the oil pump are not available individually. Pump cover and rotors may be removed for inspection or cleaning. Inner rotor is retained to shaft by pins and removal from shaft is not advised. Be sure to install outer rotor in pump body with chamfered outer edge to bottom of gear pocket, away from pump cover. Install oil pump by reversing the removal procedure. Make sure tapered end of set screw (1) fully locates the pump.

NOTE: On some models, a cap nut may be used on set screw (1) instead of the locknut shown.

RELIEF VALVE
All Models

57. The piston type relief valve (6—Fig. 51) is contained in oil delivery housing (4). The relief valve is set to open at 50-60 psi and oil pressure should be maintained at a minimum of 25 psi on engines without turbocharger or 30 psi on engines with turbocharger, with engine at working temperature and normal operating speed. Oil delivery housing and relief valve parts are available individually.

GASOLINE FUEL SYSTEM

CARBURETOR

All Models

58. **ADJUSTMENT.** All gasoline models are equipped with a Zenith Model 69J carburetor with a fixed main jet. Idle mixture adjustment screw (17—Fig. 52) is an air-bleed screw and clockwise rotation enriches the mixture. Initial setting is ⅞-1 turn open from closed position. Final adjustments must be made under operating conditions with engine at normal operating temperature. Adjust slow idle speed to 750-800 rpm by turning stop screw (16).

59. **OVERHAUL.** Refer to Fig. 52 for an exploded view of Zenith carburetor of the type used. When dis-

assembling the carburetor, accelerator pump piston must be withdrawn from bore in fuel bowl (2). Separate the housings carefully to prevent damage to accelerator pump or float.

Use the vacuum piston as a slide hammer to jar the piston retainer from throttle body bore, then withdraw pump piston assembly (11). The disc type inlet check valve (10) can be removed from bottom of pump piston bore if renewal is indicated. Valve will be damaged in removal and must be renewed. Remove by threading the special Zenith Tool (C161-15) into valve body and tapping on hooked end of tool to loosen valve. Install using Zenith Tool (C161-197) and a light hammer. Valve is properly posi-

tioned when tool bottoms. Accelerating jet (5) contains the outlet check valve and unit is renewed as an assembly. Examine piston assembly (11) and bores in throttle body and fuel bowl for scoring, scratches or excessive wear. Renew piston assembly or complete carburetor as required, fuel bowl (2) or throttle body (19) are not available separately.

Float setting is 1 $\frac{19}{32}$-inch, measured from farthest edge of float to gasket surface of throttle body with body inverted and inlet needle valve closed. Adjust by carefully bending levers close to float body. Use new gaskets and packing when assembling. Check mixture adjustments after installation as outlined in paragraph 58.

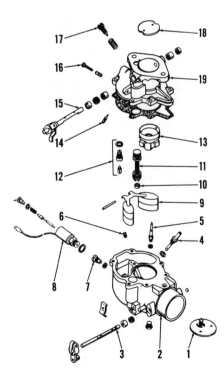

Fig. 52—Exploded view of Zenith Model 69J carburetor of the type used.

1. Choke plate	11. Accelerator pump
2. Fuel bowl	12. Inlet needle
3. Choke shaft	13. Venturi
4. Discharge jet	14. Idle jet
5. Accelerator jet	15. Throttle shaft
6. Well vent	16. Idle speed stop
7. Main jet	screw
8. Shut-off solenoid	17. Idle mixture needle
9. Float	18. Throttle plate
10. Check valve	19. Throttle body

LP-GAS SYSTEM

OPERATING ADJUSTMENTS

All LP-Gas Models

61. Initial setting of power adjustment screw (1—Fig. 54) is 3¾-4¼ turns open from the closed position. Slow idle mixture is adjusted by shortening or lengthening the adjustable drag link (2). Final adjustments must be made under operating conditions of temperature and load.

The recommended high-idle speed of 2425-2475 rpm is adjusted by

changing the tension on governor spring at nuts (2—Fig. 55). Recommended slow idle speed of 725-775 rpm is adjusted by governor arm stop screw (1). A "Choke Stop Screw" (13—Fig. 56) is located on throttle shaft arm and should be adjusted to provide a slow engine idle speed of approximately 500 rpm when screw contacts stop. To check the choke stop screw adjustment, disconnect the governor-to-carburetor link or pull out the choke cable.

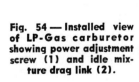

Fig. 54 — Installed view of LP-Gas carburetor showing power adjustment screw (1) and idle mixture drag link (2).

FUEL TANK AND LINES

All LP-Gas Models

62. The pressure tank is fitted with fuel filler, vapor return, pressure relief, and liquid and vapor withdrawal valves which can only be renewed when fuel tank is completely empty. The fuel gage unit consists of a dial face unit which can be renewed at any time, and a float unit which can only be renewed when tank is empty of fuel. The safety relief valve is set to open at 312 psi pressure to protect the tank against excessive pressures and adjustment should not be attempted. Fuel lines, filters, regulator or carburetor can be renewed without emptying the fuel tank if liquid and vapor withdrawal valves are closed and engine run until fuel is exhausted. UL Regulations in most states prohibit any welding or repair on LP-Gas containers, and the tank must be renewed rather than repaired, in case of damage. The fuel filter cartridge is of the throw-away type, renewable as a unit.

CONVERTER

All LP-Gas Models

63. Refer to Figs. 57 and 58 for an exploded view of the Century Model H converter. Back plate (1—Fig. 57) and gasket (2) seals the heat ex-

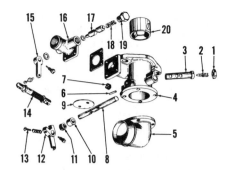

Fig. 56—Exploded view of LP-Gas carburetor of the type used.

1. Locknut	11. Seal
2. Power adjustment screw	12. Lever
3. Spray bar	13. Idle speed stop screw
4. Throttle body	14. Drag link
5. Air horn	15. Lever
6. Stop pin	16. Metering valve body
7. Plug	17. Metering valve
8. Throttle shaft	18. Spring
9. Throttle plate	19. Plug
10. Bushing	20. Venturi

changer passages for both the coolant liquid and the fuel; it is imperative that gasket surfaces of plate (1) and body (3) be perfectly flat to prevent coolant leaks into fuel system. Disassemble primary and secondary regulator as follows:

Remove the screws retaining primer cover (17) and lift off cover and primer plunger (19) as a unit. Remove secondary regulator cover (15) and diaphragm assembly. Remove the

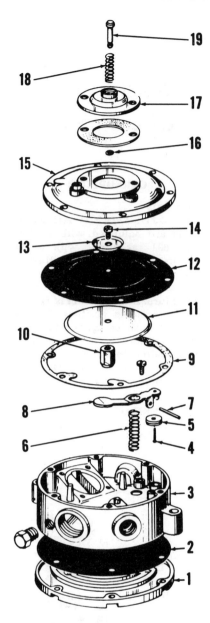

Fig. 57—Exploded view of LP-Gas regulator and associated parts.

1. Back plate	11. Plate
2. Gasket	12. Diaphragm
3. Body	13. Plate
4. Pin	14. Screw
5. Valve seat	15. Cover
6. Spring	16. O-ring
7. Shaft	17. Cover
8. Lever	18. Spring
9. Gasket	19. Primer
10. Button	

two screws which retain valve lever shaft (7) and lift out low pressure valve lever (8) as an assembly. Remove inlet cover (1—Fig. 58) and lever (5); then remove high pressure cover (12) and lift out spring (11) and diaphragm assembly.

Wash metal parts in solvent and air dry. Renew high pressure diaphragm (9) if hardened or cracked and high pressure valve (4) if sealing surface of seat is damaged. If high pressure diaphragm must be disassembled,

Fig. 55—Left side view of LP-Gas engine showing points of adjustment and fuel system components.

1. Idle speed stop screw	3. LP-Gas regulator
2. High speed governor adjustment	4. Slow idle adjustment
	5. Carburetor

make sure that legs of damper spring (7) align with flats of link (6) and are parallel with edges of diaphragm. The damper spring legs apply pressure to sides of diaphragm slot in heat exchanger body (13) and correct positioning is essential. Proper installation is also important. Four aligning pins (Century Part No. M-501) should be used to position the parts during assembly. Proceed as follows; install the aligning pins in screw holes in converter body (13). Install a new gasket (8) over aligning pins; then install the assembled diaphragm, making sure legs of damper spring (7) properly enter slot in heat exchanger body. Position spring (11), small end down as shown, then install cover (12) over aligning pins making sure spring is properly located. Insert a screwdriver or similar tool through opening for inlet cover (1) and hold up on diaphragm link (6); carefully push down on diaphragm cover (12), compressing spring (11). Remove aligning pins one at a time and immediately install and tighten cover screw before removing next aligning pin. Release the pressure on diaphragm link (6) only after all screws are tightened.

Examine low pressure valve seat (5 —Fig. 57) and renew if damaged; then reinstall low pressure lever assembly. Lay a straight edge across gasket surface of heat exchanger body

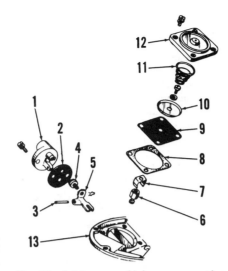

Fig. 58—Inlet cover, high pressure valve and associated parts.

1. Inlet cover	8. Gasket
2. Gasket	9. Diaphragm
3. Pivot pin	10. Plate
4. Valve seat	11. Spring
5. Valve lever	12. High Pressure
6. Diaphragm link	cover
7. Damper spring	13. Converter body

(3) and measure distance from straight edge to nearest point of valve lever (8). Adjust, if necessary, to $\frac{5}{16}$-inch by bending lever. Renew low pressure diaphragm (12) if its condition is questionable; then reinstall gasket (9), diaphragm assembly and cover (15), leaving cover retaining

screws loose. Using a pair of needle-nosed pliers and working through opening for primer cover (17), grasp head of diaphragm screw (14) and pull diaphragm plate (13) firmly into contact with cover (15); then tighten cover retaining screws evenly. Complete the assembly by reversing the disassembly procedure.

CARBURETOR

All LP-Gas Models

64. Refer to Fig. 56. The carburetor is simply constructed and overhaul procedures are self evident. Use a thread sealant when installing reducer bushing in metering valve housing (16) and do not overtighten. Overtightening may cause valve to bind. If metering valve (17) does not turn freely, remove lever (15) and plug (19) and push metering valve from housing. Clean the valve and housing bore with an oil base solvent. Coat valve with Lubriplate and reinsert in housing bore. If valve is not now free, chuck lever end of valve in a slow-speed drill and lap the valve using tallow or penetrating oil as a lubricant. DO NOT use a lapping compound or allow valve and housing to become excessively warm. If condition is questionable, renew the parts.

DIESEL FUEL SYSTEM

The diesel fuel system consists of three basic units; the fuel tank and filters, injection pump and injector nozzles. When servicing any unit associated with the diesel fuel system, the maintenance of absolute cleanliness is of utmost importance. Of equal importance is the avoidance of nicks and burrs on any of the working parts.

Probably the most important precaution that service personnel can impart to owners of diesel powered tractors is to urge them to use an approved fuel that is absolutely clean and free from foreign material. Extra precaution should be taken to be sure that no water enters the fuel storage tanks. Because of the high pressures and degree of control required of injection equipment, extremely high precision standards are necessary in the manufacture and servicing of diesel components. Extra care in daily maintenance will pay big dividends in extended service life and the avoidance of costly repairs.

FUEL FILTERS AND LINES

All Diesel Models

65. **OPERATION AND MAINTENANCE.** Refer to Fig. 59 for a schematic view of fuel flow through filters and injection pump.

NOTE: Actual location of filters may differ somewhat from that shown. The camshaft actuated, diaphragm type fuel lift pump is not illustrated.

A much greater volume of fuel is circulated within the system than is burned in the engine, the excess serving as a coolant and lubricant for the injection pump. Fuel enters the primary filter (P) through inlet connection (I) where it passes through the water trap and first stage filter element. Both lines leading to injection pump (F) and the system return line (R) enter a common passage in

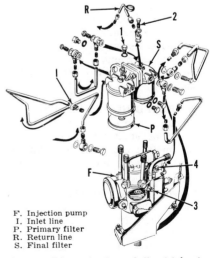

F.	Injection pump
I.	Inlet line
P.	Primary filter
R.	Return line
S.	Final filter

Fig. 59—Schematic view of diesel injection pump and filters. Direction of fuel flow is indicated by arrows. Bleed screws and proper bleeding order are indicated by numbers (1, 2, 3 & 4).

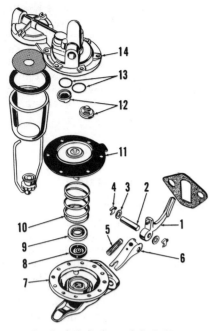

Fig. 60—Exploded view of fuel lift pump.

1. Actuating lever	8. Seal
2. Shaft	9. Retainer
3. Washer	10. Spring
4. Retainer	11. Diaphragm
5. Spring	12. Valve
6. Link	13. Gasket
7. Body	14. Cover

secondary filter (S) and are separated from the filter inlet only by the filtering element. The greater volume of filtered fuel is thus recirculated between the secondary filter (S) and injection pump (F), with the inlet line supplying little more fuel to the system than is actually burned in the engine.

Inspect the glass bowl at bottom of primary filter (P) daily and drain off any water or dirt accumulation. Completely drain the primary filter at 100 hour intervals and renew the element each 500 hours. Renew element in secondary filter (S) every 1000 hours. Renew both elements and clean the tank and lines if evidence of substantial water contamination exists.

66. BLEEDING. To bleed the system, make sure tank shut-off is open, have an assistant actuate the manual lever on fuel lift pump on right side of engine block, and proceed as follows:

Loosen the air vent (1—Fig. 59) on primary filter (P) and continue to operate the lift pump until air-free fuel flows from vent plug hole. Tighten plug (1); then loosen vent plugs (2, 3 and 4) in order, while continuing to operate the fuel lift pump. Tighten each plug as air is expelled, and proceed to the next.

NOTE: Air in governor housing relieved by bleed screw (4) will not prevent tractor from starting and running properly; however, condensation in the trapped air can cause rusting of governor components and eventual pump malfunction. Do not fail to bleed governor housing even though the tractor starts and runs properly.

Operate the manual lever on fuel lift pump for approximately ten extra strokes after tightening vent plug (4), to expel any air remaining in bleed-back lines to filter.

With the fuel supply system bled, push in the stop button, partially open throttle lever and attempt to start the tractor. If tractor fails to fire, loosen compression nut at all injector nozzles and turn engine over with starter until fuel escapes from all loosened connections. Tighten compression nuts and start engine.

FUEL LIFT PUMP
All Diesel Models

67. The fuel lift pump (Fig. 60) is mounted on right side of engine block and driven by the camshaft. Pump parts are available separately or in a rebuilding kit. Output delivery pressure should be 5-8 psi.

INJECTOR NOZZLES
All Diesel Models

All models are equipped with C. A. V. multi-hole nozzles which extend through the cylinder head to inject the fuel charge into combustion chamber machined in crown of piston.

WARNING: Fuel leaves the injector nozzle with sufficient force to penetrate the skin. Keep exposed portions of your body clear of nozzle spray when testing.

Fig. 61—A suitable injector tester is required to completely test and adjust the injector nozzles.

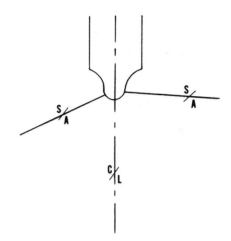

Fig. 63—Nozzle spray pattern is not symmetrical with centerline of nozzle tip.

68. TESTING AND LOCATING A FAULTY NOZZLE. If rough or uneven engine operation or misfiring indicates a faulty injector, the defective unit can usually be located as follows:

With engine running at the speed where malfunction is most noticeable (usually slow idle speed), loosen the compression nut on high pressure line for each injector nozzle in turn, and listen for a change in engine performance. As in checking spark plugs, the faulty unit is the one which, when its line is loosened, least affects the running of the engine.

If a faulty nozzle is found and considerable time has elapsed since the injectors have been serviced, it is recommended that all nozzles be removed and serviced or that new or reconditioned units be installed. Refer to the following paragraphs for removal and test procedure.

69. REMOVE AND REINSTALL. Before loosening any fuel lines, thoroughly clean the lines, connections, injectors and surrounding engine area with air pressure and solvent spray. Disconnect and remove the leak-off line, disconnect pressure line and cap all connections as they are loosened, to prevent dirt entry into the system. Remove the two stud nuts and withdraw injector unit from cylinder head.

Thoroughly clean the nozzle recess in cylinder head before reinstalling the injector unit. It is important that seating surface be free of even the smallest particle of carbon or dirt which could cause the injector unit to be cocked and result in blow-by. No hard or sharp tools should be used in cleaning. Do not re-use the copper

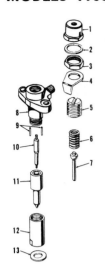

Fig. 64—Exploded view of C. A. V. injector nozzle and holder assembly. Correct opening pressure is indicated on tab (4).

1. Cap nut
2. Gasket
3. Locknut
4. Tab
5. Adjusting screw
6. Spring
7. Valve spindle
8. Nozzle holder
9. Dowels
10. Nozzle valve
11. Nozzle body
12. Nozzle nut
13. Seat washer

sealing washer located between injector nozzle and cylinder head, always install a new washer. Each injector should slide freely into place in cylinder head without binding. Make sure that dust seal is reinstalled and tighten the retaining stud nuts to a torque of 10-12 ft.-lbs. After engine is started, examine injectors for blow-by, making the necessary corrections before releasing the tractor for service.

70. **TESTING.** A complete job of testing and adjusting the injector requires the use of special test equipment. Only clean, approved testing oil should be used in the tester tank. The nozzle should be tested for opening pressure, seat leakage, back leakage and spray pattern. When tested, the nozzle should open with a sharp popping or buzzing sound, and cut off quickly at end of injection with a minimum of seat leakage and a controlled amount of back leakage.

Before conducting the test, operate tester lever until fuel flows, then attach the injector. Close the valve to tester gage and pump tester lever a few quick strokes to be sure nozzle valve is not plugged, that four sprays emerge from nozzle tip, and that possibilities are good that injector can be returned to service without overhaul.

NOTE: Spray pattern is not symmetrical with centerline of nozzle tip, but forms a "tilted" pattern as shown in Fig. 63. Hole location is engineered to provide the correct spray pattern in the combustion chamber.

If adjustment is indicated in the preliminary tests, proceed as follows:

71. OPENING PRESSURE. Open the valve to tester gage and operate tester lever slowly while observing gage reading. Opening pressure should be 2500 psi on Model 1100; or 2795 psi on Model 1130 (Turbocharged engine). If opening pressure is not as specified, remove the injector cap nut (1—Fig. 64), loosen locknut (3) and turn adjusting sleeve (5) as required to obtain the recommended pressure.

NOTE: When adjusting a new injector or an overhauled injector with a new pressure spring (6), set the pressure at 2775 psi for Model 1100 or 3015 psi for turbocharged engine. The additional pressure is to allow for initial pressure loss as the spring takes a set.

Early engines in Model 1130 used an injector with an operating pressure of 2575 psi. Pressure should be raised to the presently recommended 2795 psi whenever engine is serviced.

72. SEAT LEAKAGE. The nozzle tip should not leak at a pressure less than 200 psi under the recommended operating pressure; or 2300 psi for Model 1100 or 2600 psi for Model 1130. To check for leakage, actuate tester lever slowly and as the gage needle approaches the suggested test pressure, observe the nozzle tip. Hold the pressure for 10 seconds; if drops appear or nozzle tip is wet, the valve is not seating and injector must be disassembled and overhauled as outlined in paragraph 75.

73. BACK LEAKAGE. If nozzle seat as tested in paragraph 72 was satisfactory, check the injector and connections for wetness which would indicate external leakage. If no leaks are noted, bring gage pressure to 2250 psi, release the lever and observe the time for gage pressure to drop from 2250 psi to 1500 psi. For a nozzle in good condition, this time should not be less than six seconds. A faster drop would indicate a worn or scored nozzle valve piston or body, and the nozzle assembly should be renewed.

NOTE: Leakage of the tester check valve or connections will cause a false reading, showing up in this test as excessively fast leakback. If all injectors tested fail to pass the test, the tester rather than the units should be suspected as faulty.

74. SPRAY PATTERN. If leakage and pressure are as specified when tested as outlined in paragraphs 71 through 73, operate the tester handle several times while observing the spray pattern. Four finely atomized, equally spaced, conical sprays should emerge from nozzle tip, with equal penetration into the surrounding atmosphere.

If pattern is uneven, ragged or not finely atomized, overhaul the nozzle as outlined in paragraph 75. NOTE: Spray pattern is not symmetrical with centerline of nozzle tip, but tilted as shown in Fig. 63.

75. **OVERHAUL.** Hard or sharp tools, emery cloth, grinding compound, or other than approved solvents or lapping compounds must never be used. An approved nozzle

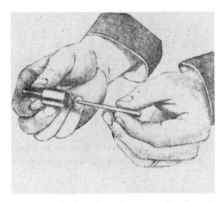

Fig. 65—Clean the pressure chamber in nozzle tip using the special reamer as shown.

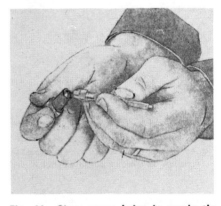

Fig. 66—Clean spray holes in nozzle tip using a pin vise and 0.010 wire probe.

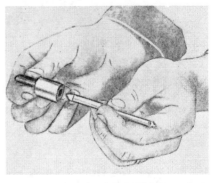

Fig. 67—Clean the valve seat using brass scraper as shown.

cleaning kit is available through any C. A. V. Service Agency and other sources.

Wipe all dirt and loose carbon from exterior of nozzle and holder assembly. Refer to Fig. 64 and proceed as follows:

Secure the nozzle in a soft-jawed vise or holding fixture and remove the cap nut (1). Loosen jam nut (3)

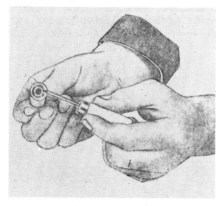

Fig. 68—Use the hooked scraper to clean annular groove in top of nozzle body.

Fig. 69—Polish the nozzle seat using mutton tallow on a wood polishing stick.

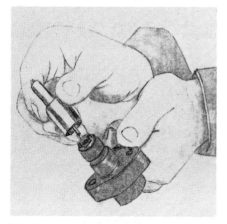

Fig. 70—Make sure locating dowels are carefully aligned when nozzle body is reinstalled.

and back off the adjusting sleeve (5) to completely unload the pressure spring (6). Remove the nozzle cap nut (12) and nozzle body (11). Nozzle valve (10) and body (11) are matched assemblies and must never be intermixed. Place all parts in clean calibrating oil or diesel fuel as they are removed. Clean the exterior surfaces with a soft wire brush, soaking in an approved carbon solvent if necessary, to loosen hard carbon deposits. Rinse the parts in clean diesel fuel or calibating oil immediately after cleaning, to neutralize the carbon solvent and prevent etching of polished surfaces. Clean the pressure chamber of nozzle tip using the special reamer as shown in Fig. 65. Clean the spray holes in nozzle with an 0.010 (0.24 mm) wire probe held in a pin vise as shown in Fig. 66. Wire probe should protrude from pin vise only far enough to pass through spray holes (Approximately $\frac{1}{16}$-inch), to prevent bending and breakage. Rotate pin vise without applying undue pressure.

Clean valve seats by inserting small end of brass valve seat scraper into nozzle and rotating tool. Reverse the tool and clean upper chamfer using large end. Refer to Fig. 67. Use the hooked scraper to clean annular groove in top of nozzle body (Fig. 68). Use the same hooked tool to clean the internal fuel gallery.

With the above cleaning accomplished, back flush the nozzle by installing the reverse flusher adapter on injector tester and nozzle body in adapter, tip end first. Secure with the knurled adapter nut and insert and rotate the nozzle valve while flushing. After nozzle is back flushed, seat can be polished using a small amount of tallow on end of polishing stick, rotating the stick as shown in Fig. 69.

Light scratches on valve piston and bore can be polished out by careful use of special injector lapping compound only, DO NOT use valve grinding compound or regular commercial polishing agents. DO NOT attempt to reseat a leaking valve using polishing compound. Clean thoroughly and back flush if lapping compound is used.

Reclean all parts by rinsing thoroughly in clean diesel fuel or calibrating oil and assemble valve to body while immersed in the cleaning fluid. Reassemble the injector while still wet. With adjusting sleeve (5—Fig. 64) loose, reinstall nozzle body (11) to holder (9), making sure valve (10) is installed and locating dowels aligned as shown in Fig. 70. Tighten nozzle cap nut (12—Fig. 64) to a torque of 50 ft-lbs. Do not overtighten; distortion may cause valve to stick and overtightening cannot stop a leak caused by scratches or dirt on the lapped metal surfaces of valve body and nozzle holder.

Retest and adjust the assembled injector assembly as outlined in paragraphs 71 through 74.

Fig. 71—Installed view of injection pump showing slow idle stop screw (1). High speed stop screw is sealed when pump is adjusted on test stand.

NOTE: If overhauled injector units are to be stored, it is recommended that a calibrating or preservative oil rather than diesel fuel be used for the pre-storage testing. Storage of injectors containing diesel fuel for more than thirty days may result in the necessity of recleaning prior to use.

INJECTION PUMP

All Models

The injection pump is a completely sealed unit. No service work of any kind should be attempted on the pump or governor unit without the use of special pump testing equipment and training. Inexperienced or unequipped service personnel should never attempt to overhaul a diesel injection pump.

76. ADJUSTMENT. The slow idle stop screw (1—Fig. 71) should be adjusted with engine warm and running, to provide the recommended slow idle speed of 800-850 rpm. Also check to make sure that governor arm contacts both stop screws when hand lever is moved through full travel. Also check to make sure that stop lever arm moves fully to operating position when stop button is pushed in, and shuts off the fuel when stop button is pulled. The high speed stop screw is set at the factory and the adjustment is sealed. Governed speed under load should be 2200 rpm, with a high idle (no-load) speed of 2380-2420 rpm. Refer to paragraph 78 for pump timing adjustment.

77. REMOVE AND REINSTALL. Before attempting to remove the injection pump, thoroughly wash the pump and connections with clean diesel fuel or an approved solvent. Disconnect throttle control rod from governor arm and disconnect and remove stop cable and bracket. Disconnect high pressure fuel lines at both ends, unbolt bracket from cylinder head, then lift off bracket and lines as a unit. Cap all connections as lines are removed. Disconnect and cap fuel inlet and return lines at pump, remove the remaining attaching stud nut and lift off the pump.

Normal installation of injection pump can be accomplished without reference to crankshaft timing marks or internal timing marks on injection pump. Be sure timing scribe lines (T—Fig. 72) are aligned when pump is installed. The master spline on pump drive shaft aligns with splines in pump to assure proper timing if auxiliary shafts are correctly installed as outlined in paragraph 45. Bleed the fuel system after installation as outlined in paragraph 66. If necessary to check injection pump timing, refer to paragraph 78.

78. PUMP TIMING TO ENGINE. Because of the master splines on pump drive shaft, pump timing will be correct if gears are properly installed and timing marks (T—Fig. 72) are aligned. To check gear installation, it is necessary to remove the injection pump as outlined in paragraph 77, then proceed as in paragraph 45.

Static timing (end of injection) should occur at 22° BTDC on Model 1100 and 28° BTDC on Model 1130. A static timing mark for No. 1 cylinder is stamped on front face of flywheel as shown in Fig. 73 and is visible through timing window in adapter housing on left side of engine block. To check the injection pump timing without removing the pump, proceed as follows:

Turn engine over by hand until No. 1 piston is on compression stroke and the indicated "Static Timing" mark is aligned in timing window. Remove the timing cover (C—Fig. 72) from injection pump body; the "F" mark on pump rotor should be aligned with the square end of injection pump snap ring. Minor adjustment can be made by loosening the injection pump mounting stud nuts and shifting the pump mounting flange. To make a major timing adjustment, retime the auxiliary drive shaft as outlined in paragraph 45.

DIESEL TURBOCHARGER

OPERATION

Model MF1130

79. The exhaust driven turbocharger supplies air to the intake manifold at above normal atmospheric pressure. The additional air entering the combustion chamber permits an increase in the amount of fuel burned, and increased power output over an engine of comparable size not so equipped.

The use of the engine exhaust to power the compressor increases engine flexibility, enabling it to perform with the economy of a small engine on light loads yet permitting a substantial horsepower increase at full load. Horsepower loss because of altitude or atmospheric pressure changes is also largely eliminated.

The turbocharger contains one moving part. A rotating shaft which carries an exhaust turbine wheel on one end and a centrifugal air compressor on the other. The rotating member is precisely balanced and capable of rotative speeds up to 100,000 rpm. Bearings are of the floating sleeve type and the unit is lubricated and cooled by a flow of engine oil under pressure.

Fig. 72—Pump should be properly timed when timing scribe lines (T) are aligned as shown. Timing window (C) can be removed and timing checked as outlined in paragraph 78 if conditions warrant.

Fig. 73—A static timing mark is stamped on front face of flywheel as shown. The 28 degree timing mark is used on Model MF 1130. Static timing for Model MF 1100 is 22° BTDC.

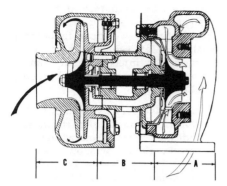

Fig. 74—Cross sectional view of diesel turbocharger showing (A) turbine housing, (B) shaft section and (C) compressor section.

Refer to Fig. 74 for a cross sectional schematic view of a HOLSET turbocharger of the type used. The turbine housing (A) attaches to the exhaust manifold opening and exhaust gases are directed into the turbine wheel by a nozzle ring which allows entry throughout the circumference of the turbine.

The shaft section (B) contains the sleeve bearings and oil reservoir. The intake section (C) houses the centrifugal compressor.

SERVICE

Model MF1130

80. In a naturally aspirated diesel engine (without turbocharger) an approximately equal amount of air enters the cylinders at all loads, and only the amount of fuel is varied to compensate for power requirements. Turbocharging may supply up to 3 times the normal amount of air under full load.

All diesel engines operate with an excess of air under light loads. In a naturally aspirated engine, most of the air is used at full load, and increasing the amount of fuel results in a higher smoke level with little increase in power output. Turbocharging provides a variation of air delivery, and a turbocharged engine operates with an excess of air up to and beyond the design capacity of the engine. When more fuel is provided, the turbocharger speed and air delivery increase, resulting in additional horsepower and heat, with little change in smoke level. Smoke cannot, therefore, be used as a guide to safe maximum fuel setting in a turbocharged engine. DO NOT increase horsepower output above that given in CONDENSED SERVICE DATA tables at the front of this manual.

The turbocharger is serviced only as a new or rebuilt unit, and disassembly is not recommended. Turbine and compressor wheel blades can be inspected through exhaust pipe and

air cleaner hose openings after unit is removed.

CAUTION: Do not operate the turbocharger without adequate lubrication. When turbocharger is first installed, turn engine over with starter with fuel cut-off knob pulled until oil gage starts to register pressure; then run engine at slow idle speed for at least two minutes before opening the throttle or putting engine under load.

Some of the precautions to be observed in operating and servicing a turbocharged engine are as follows:

Do not operate at wide-open throttle immediately after starting. Allow engine to idle until turbocharger slows down before stopping engine.

Because of increased air flow, care of air cleaner and connections is of added importance. Check the system and condition of restriction indicator whenever tractor is serviced, by manually restricting air intake pipe opening while engine is running.

FUEL TANK

REMOVE AND REINSTALL

Gasoline and Diesel Models

81. Gasoline and diesel models are equipped with dual, saddle type fuel tanks located at each side of tractor underneath the operator's platform. Refer to Fig. 75. Tanks are interconnected by a transfer pipe or hose (4) at lower front and vent hose (8) at upper rear.

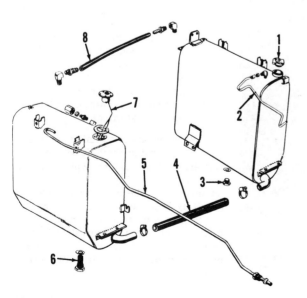

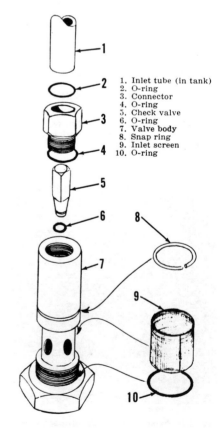

1. Inlet tube (in tank)
2. O-ring
3. Connector
4. O-ring
5. Check valve
6. O-ring
7. Valve body
8. Snap ring
9. Inlet screen
10. O-ring

Fig. 75 — Exploded view of mid-mounted saddle tanks of the type used on gasoline and diesel models.

1. Filler cap
2. Return line
3. Drain plug
4. Crossover hose
5. Fuel line
6. Inlet screen
7. Fuel gage sender unit
8. Vent line

Fig. 76—On some early models an inlet check valve is used instead of inlet screen (6—Fig. 75). Refer to paragraph 81 for removal and installation information.

Some early diesel models are equipped with an inlet check valve and strainer which can be removed from outside. The unit is shown exploded in Fig. 76. The unit must be assembled before installation, then slipped up over inlet line (1) which is part of right-hand fuel tank. Use care to keep from damaging O-ring (2) located in connector nut (3). Later models use inlet strainer (6—Fig. 75) instead of check valve and strainer unit.

NON-DIESEL GOVERNOR

ADJUSTMENT

All Models

82. Recommended governed speeds are as follows:

Low Idle ....750-800 rpm
High Idle ....2425-2475 rpm
Loaded Speed ....2200 rpm

To adjust the governor, disconnect carburetor to governor rod (6—Fig. 79) at either end. Pull hand throttle down to be sure governor is fully open, then adjust the length of carburetor rod until governor arm must be moved forward about $\frac{1}{32}$-inch be-

fore rod can be reconnected. Start and warm the engine. Normal high speed adjustment is made at the two nuts (H) to change the tension of governor spring, and normal slow speed adjustment is made at stop screw (9) in bracket on water pump housing. Refer to Fig. 80 for an exploded view of control linkage. Adjust linkage as necessary to obtain full range of governor travel using either the hand or foot throttle. If hand throttle lever creeps, tighten friction nut (20).

OVERHAUL

All Models

83. Governor housing (5—Fig. 79) is positively located on timing gear cover (1) by two dowels. When governor housing is removed, shaft (4) and gear (3) are free to come with housing or to remain with engine. Governor can be removed and overhauled without additional engine disassembly. Only those parts shown in Fig. 79 are available as service items.

COOLING SYSTEM

RADIATOR

All Models

84. A 7 psi pressurized cooling system is used on all models. Cooling system capacity is approximately 21 U. S. Quarts on non-diesel models and 22 U. S. Quarts on diesel models.

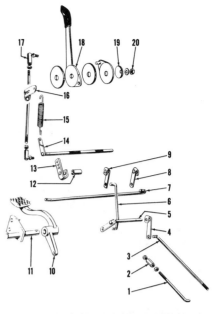

Fig. 80—Exploded view of throttle controls and associated parts.

1. Front governor rod	11. Pivot
2. Turnbuckle	12. Spacer
3. Rear governor rod	13. Arm
4. Throttle arm	14. Cross shaft
5. Cross shaft	15. Spring
6. Link rod	16. Bracket
7. Pedal rod	17. Ball joint
8. Throttle arm	18. Lever
9. Throttle arm	19. Cup washer
10. Pedal	20. Nut

On turbocharged models, an engine oil cooler is built into radiator core.

To remove the radiator, first drain cooling system and remove hood, grille and interfering power steering and main hydraulic lines. Remove air cleaner and oil cooler radiator, disconnect coolant hoses, then unbolt and remove radiator and shroud as a unit.

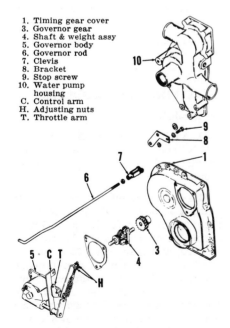

1. Timing gear cover
3. Governor gear
4. Shaft & weight assy
5. Governor body
6. Governor rod
7. Clevis
8. Bracket
9. Stop screw
10. Water pump housing
C. Control arm
H. Adjusting nuts
T. Throttle arm

Fig. 79—Exploded view of non-diesel governor, timing gear cover and associated parts. Low idle stop screw (9) is installed in bracket (8) which mounts on water pump housing (10).

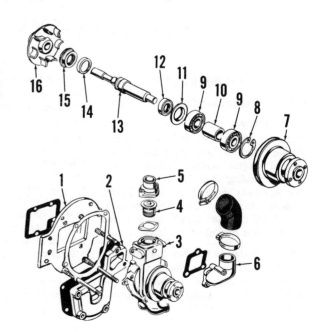

Fig. 81 — Exploded view of water pump, thermostat and associated parts used on diesel models.

1. Mounting plate
2. Water connection
3. Water pump housing
4. Thermostat
5. Water outlet
6. Water inlet
7. Fan pulley
8. Snap ring
9. Bearing
10. Spacer
11. Seal flange
12. Seal
13. Shaft
14. Retainer
15. Water seal
16. Impeller

1. Fan pulley
2. Snap ring
3. Shaft assembly
4. Slinger
5. Water pump housing
6. Seal
7. Seat
8. Impeller
9. Thermostat
10. Elbow
11. Water pump cover

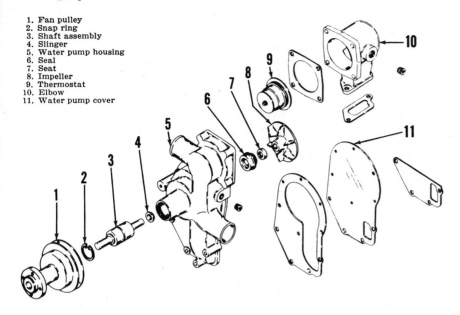

Fig. 82—Exploded view of water pump and associated parts used on non-diesel models.

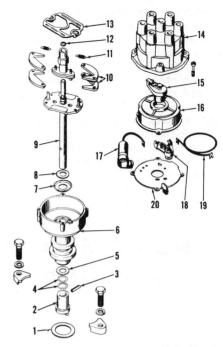

Fig. 83—Exploded view of ignition distributor and associated parts.

1. Gasket
2. Drive coupling
3. Spring pin
4. Shims
5. Spacer washer
6. Housing
7. Seal
8. Washer
9. Shaft
10. Governor weight
11. Spring
12. Oil wick
13. Plate
14. Cap
15. Rotor
16. Dust cover
17. Condenser
18. Point set
19. Primary lead
20. Breaker plate

THERMOSTAT

All Models

85. On diesel models, the thermostat is contained in the water pump housing as shown in Fig. 81 and retained by outlet elbow (5). On non-diesel models, refer to Fig. 82. Thermostat (9) is contained in water pump housing (5) and retained by water pump elbow (10). Opening temperature is 160° F. on non-diesel models, 167°-176° F. on diesel models.

WATER PUMP

All Models

86. Refer to Fig. 81 for an exploded view of diesel water pump and associated parts and to Fig. 82 for view of non-diesel pump.

On non-diesel models, water pump bearing and shaft is a sealed unit. On diesel models, pack the area between bearings (9) half full of high melting point grease when reassembling. Water pump can be removed without removing radiator.

Advance data is in distributor degrees and distributor rpm.

Start advance	0-2 @ 275
Intermediate advance..	5-11 @ 400
Intermediate advance....	9-11 @ 800
Maximum advance	14-16 @ 1300

IGNITION AND ELECTRICAL SYSTEM

DISTRIBUTOR

All Non-Diesel Models

87. **TIMING.** Timing marks are located on flywheel and are visible through timing hole in right-hand side of engine adapter plate. Initial (static) timing is 2° BTDC for gasoline models and TDC for LP-Gas Models. Maximum advance timing is 28° BTDC and 26° BTDC for gasoline and LP-Gas models respectively.

Firing order is 1-5-3-6-2-4 and distributor shaft rotates clockwise when viewed from cap end. Advance mechanism starts to operate at approximately 800 crankshaft rpm.

88. **OVERHAUL.** Refer to Fig. 83 for an exploded view of the distributor. Centrifugal advance mechanism can be checked for binding or broken springs by turning rotor (15) clockwise after removing cap (14). Bushings are not available for housing (6); renew housing and/or shaft (9) if clearance is excessive. Shims (4) are available in thicknesses of 0.005 and 0.010 for adjusting shaft end play which should be 0.002-0.010. Test specifications are as follows:

Breaker contact gap 0.016
Breaker arm spring tension
   (measured at center of
   contact) ....17-21 oz.
Cam angle (degrees) 31-34

ALTERNATOR & REGULATOR

All Models

89. **ALTERNATOR.** A "DELCO-TRON" generator (alternator) is used on all models. Units are negative ground.

The only test which can be made without removal and disassembly of alternator is output test. Output should be approximately 30 Amperes @ 5000 alternator rpm. To disassemble the alternator, first put match marks (M—Fig. 84) on the two frame halves (5 and 16), then remove the four through-bolts. Pry frame apart with a screwdriver between stator frame (11) and drive end frame (5). Stator (11) must remain with slip ring end frame (16) when unit is disassembled.

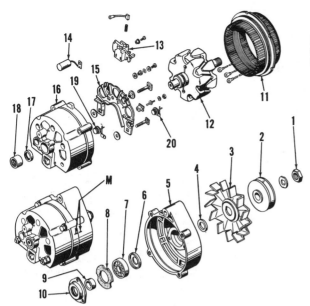

Fig. 84 — Exploded view of DELCOTRON alternator of the type used on all models.

1. Pulley nut
2. Drive pulley
3. Fan
4. Collar
5. Drive end frame
6. Slinger
7. Bearing
8. Gasket
9. Collar
10. Bearing retainer
11. Stator assembly
12. Rotor assembly
13. Brush holder
14. Capacitor
15. Heat sink
16. Slip ring end frame
17. Bearing retainer
18. Bearing
19. Negative diode
20. Positive diode
M. Match marks

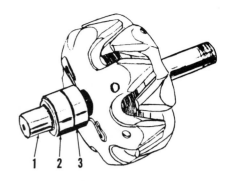

Fig. 85—Removed rotor assembly showing probe points for testing rotor.

NOTE: When frames are separated, brushes will fall out on shaft at bearing area. Brushes MUST be cleaned of lubricant if they are to be re-used.

Clamp the iron rotor lightly and carefully in a protected vise only tight enough to permit loosening of pulley nut (1). Rotor and end frame can be separated after pulley is removed. Check the bearing surfaces of rotor shaft for visible wear or scoring. Examine slip ring for scoring or wear, and windings for overheating or other damage. Check rotor for grounded, shorted or open circuits using an ohmmeter as follows:

Refer to Fig. 85 and touch the ohmmeter probes to points (1-2) and (1-3); a reading near zero will indicate a ground. Touch ohmmeter probes to the two slip ring segments (2 & 3); reading should be 4.6-5.5 ohms, a higher reading will indicate an open circuit, a lower reading will indicate a short. If windings are satisfactory, mount the rotor between lathe centers and check runout at slip ring using a dial indicator. Runout should not exceed 0.002. Surface can be trued if runout is excessive or if surface is scored. Finish with 400 grit polishing cloth until scratches or machine marks are removed.

Disconnect the three stator leads and separate stator from slip ring end frame. Check for continuity and for grounds to stator frame. The three leads have a common connection in the center of the windings. Short circuits within the windings cannot be readily determined by test, because of the low resistance.

Three diodes (19—Fig. 84) are located in slip ring end frame (16) and three diodes (20) in heat sink (15). Diodes should test at or near infinity in one direction when tested with an ohmmeter, and at or near zero when meter leads are reversed. Renew any diode with approximately equal meter readings in both directions. Diodes must be removed and installed using an arbor press and suitable tool which contacts only the outer edge of the diode.

Brushes are available only in an assembly which includes brush holder (13). If brushes are re-used, make sure all grease is removed from surface of brushes before unit is reassembled. When reassembling alternator, install both brushes and their springs in holder, push brushes up against spring pressure and insert a short piece of straight wire through hole (W—Fig. 86) and through end frame (16—Fig. 84) to outside. Withdraw the wire only after alternator is reassembled.

Capacitor (14) connects to "BATTERY" terminal and is grounded in alternator frame. Capacitor protects the diodes from voltage surges. A shorted capacitor will cause a dead short in battery wiring and burn out the twin fuses located near the starter solenoid. Press the old capacitor out and new capacitor in, working from outside of slip ring end frame (16), and check new unit for shorts after capacitor is connected.

Ball bearing (7) and needle bearing (18) should be filled ¼-full with Delco-Remy Bearing Lubricant when alternator is reassembled. Over-filling

may cause lubricant to be thrown into alternator resulting in malfunction. Assemble by reversing the disassembly procedure. Tighten pulley nut (1) to a torque of 50-60 ft.-lbs.

NOTE: A battery powered test light can be used instead of ohmeter for all electrical checks except shorts in rotor winding.

90. REGULATOR. A DELCO-REMY standard two-unit regulator is used. Quick disconnect plugs are used at regulator and alternator. Production regulator is riveted to shock mount; service units are shipped less mount and are attached with screws. Test specifications are as follows:

Regulator Model 1119513

Ground polarity	Negative
Field Relay	
Air Gap	0.015
Point Opening	0.030
Closing Voltage Range	3.8-7.2
Voltage Regulator	
Air Gap	0.067*
Point Opening	0.014
Voltage Setting	13.9-14.8*

* The specified air gap setting is for bench repair only; make final adjustment to obtain specified voltage, with lower contacts opening at not more than 0.4 volt less than upper contacts. The given voltage settings are for am-

Fig. 86 — Removed brush holder showing hole (W) for wire to position brushes for installation. Refer to text.

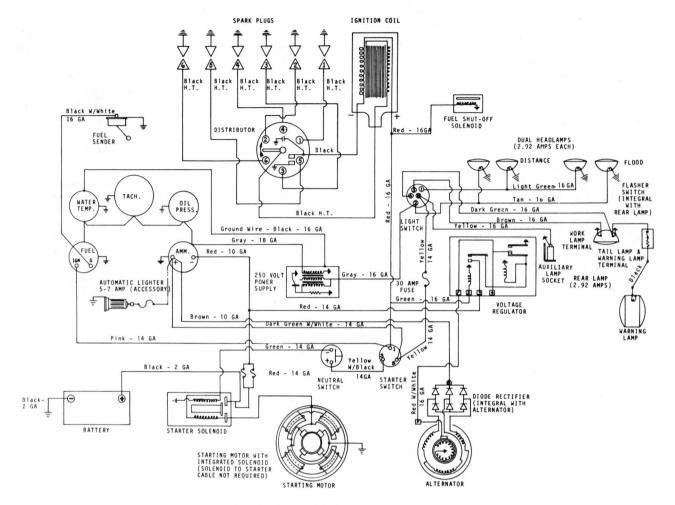

Fig. 87—Wiring diagram typical of that used on non-diesel models.

bient temperature of 100° F. or less. Regulator is temperature compensated.

STARTING MOTOR

All Models

91. Delco-Remy starting motors are used. Specifications are as follows:

Model 1108324 (Non-Diesel)

Brush spring tension ....35 oz.

No-Load Test

Volts	....10.6
Amperes	49-87
Minimum rpm	6200

Resistance Test

Volts	....4.3
Minimum amperes (w/solenoid)	290
Maximum amperes (w/solenoid)	425

Model 1113139 (Diesel)

Brush spring tension ....35 oz.

No-Load Test

Volts	....11.5
Amperes	57-70
Minimum rpm	5000

Lock Test

Volts	....3.4
Amperes	500
Torque, ft.-lbs.	22

NOTE: Resistance test is conducted with armature securely locked, but torque is not measured. Vary the resistance until voltmeter registers the value shown, then note ammeter reading which should be within the specified range.

Starter drive pinion clearance is not adjustable, however, some clearance must be maintained between end of pinion and starter drive frame to as sure solid contact of the heavy duty magnetic switch. Normal pinion clearance should be within the limits or 0.010-0.140. Connect a 6-volt battery to solenoid terminals when checking pinion clearance to keep armature from turning.

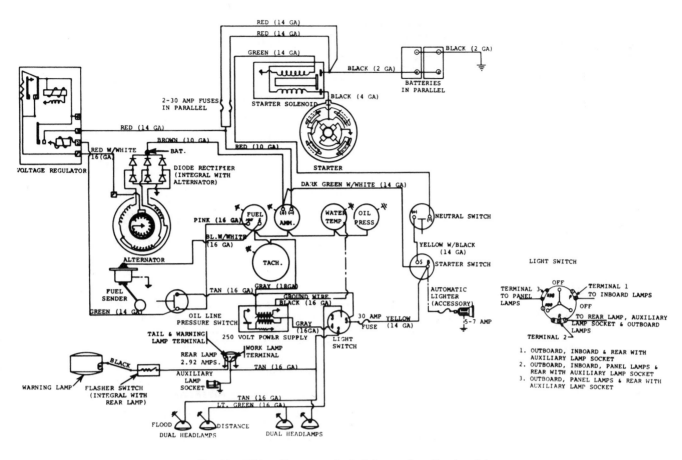

Fig. 88—Wiring diagram typical of that used on diesel models.

ENGINE CLUTCH

ADJUSTMENT

All Models

92. Clutch pedal free play should be 1½ inches when measured between pedal and platform as shown in Fig. 89. If adjustment is required, put a suitable wrench on adjusting flats (A—Fig. 90) of throwout shaft (6). Loosen clamp bolt (C). Turn shaft (6) until throwout bearing contacts clutch release fingers, reposition clutch pedal with specified clearance; then re-tighten clamp bolt (C).

TRACTOR SPLIT

All Models

93. To detach engine from transmission assembly for clutch service, first remove operator's platform and detach hydraulic lines and wires leading to transmission and center hous-

Fig. 89 — Clutch pedal free play would be 1½ inches when measured from stop on step plate as shown.

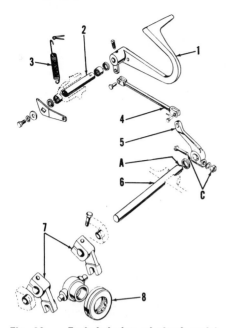

Fig. 90 — Exploded view of clutch pedal linkage.

1. Pedal
2. Pivot shaft
3. Return spring
4. Pedal link
5. Release arm
6. Cross shaft
7. Fork
8. Release bearing
A. Adjusting flats
C. Clamp bolt

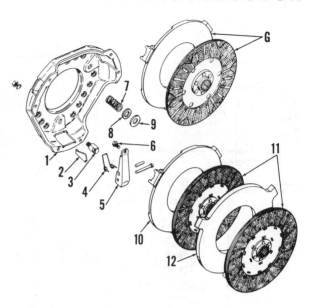

Fig. 92 — Exploded view of clutch used on late models. Clutch disc and pressure plate (G) used on non-diesel models are interchangeable with components used on diesel tractors.

1. Cover
2. Wear plate
3. Pivot
4. Spring
5. Release lever
6. Adjusting screw
7. Spring
8. Fiber washer
9. Washer
10. Pressure plate
11. Clutch discs
12. Separator plate

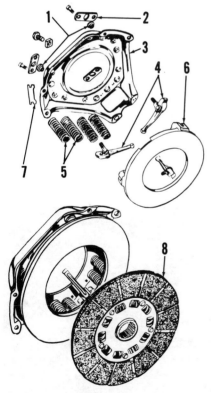

Fig. 91—Exploded view of clutch used on early diesel models.

1. Adjusting nuts
2. Locks
3. Cover
4. Release levers
5. Springs
6. Pressure plate
7. Wear plate
8. Clutch disc

ing. Remove shield, brake pedals and brake valve. Disconnect clutch rod, park linkage rod and Hi-Lo shift rod. Remove rear, lower cap screw from each grille side panel. Remove rear screw attaching channel support to thermostat housing and loosen front screw. Remove the two cap screws attaching battery support to cylinder head and all cap screws securing transmission top cover to transmission housing.

Using a hoist, raise steering support and transmission top cover approximately two inches until shift levers clear transmission housing, then support the transmission top cover and associated parts by blocking up between valve cover and hood assembly.

Support front and rear sections of tractor separately, remove clutch housing attaching cap screws and sep-

arate the tractor, lowering transmission as necessary as parts are separated. Assemble by reversing the disassembly procedure.

OVERHAUL
All Models

94. Refer to Fig. 91 for exploded view of early clutch and associated parts and to Fig. 92 for late units. Late diesel models use two identical clutch discs (11) and separator plate (12). The clutch pressure plate and clutch disc used on gasoline and LP-Gas models is interchangeable with diesel, except only one clutch disc (11) is used and separator plate (12) is omitted.

On all models, install clutch disc (or discs) with long end of hub to rear. Clutch levers on all models

Fig. 93—Adjusting clutch finger height using the special tool MFN 766.

should be checked and adjusted if necessary, using new clutch discs and Massey-Ferguson Special Tool MFN 766 as shown in Fig. 93. Adjustment is made using nuts (1—Fig. 91) on early models, or adjusting screws (6—Fig. 92) on late tractors. Re-usable worn clutch discs can be reinstalled after adjustment is made. All fingers must be adjusted to within 0.005 of equal height. If pressure plate assembly is to be dissembled, use a suitable press and Fig. 91 or 92 as a guide. All parts are available individually.

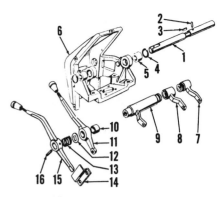

TRANSMISSION AND CONNECTIONS

All models are equipped with a basic three-speed, sliding gear transmission coupled to a manually selected, constant mesh dual range unit located in rear axle center housing at output end of main transmission unit.

Available as a factory installed option is a hydraulically actuated dual range unit located on the transmission input shaft which can be manually shifted while tractor is moving under load.

Transmissions with the hydraulically actuated dual range option are referred to as "Multi-Power Transmissions" and are capable of 12 forward and 4 reverse speeds. Transmissions without the option are referred to as "Six-Speed Transmissions" and are capable of 6 forward and 2 reverse speeds.

Service procedures on the two transmissions are similar except where otherwise indicated.

REMOVE AND REINSTALL

All Models

95. To remove the transmission assembly from tractor, first drain transmission and rear axle center housing. On all except LP-Gas Models, drain fuel tank. Split the tractor between engine and clutch housing as outlined in paragraph 93.

Remove lower step plates and tool box. On early models, loosen the clamps on fuel tank connecting tube, slide tube to either side and remove the tube. On late models, remove fuel tank connecting hose. On all models, remove fuel tank lower front support bracket and vent hose or tube connecting the tanks at top rear, then unbolt and remove the tanks.

Fig. 95—Exploded view of instrument panel support showing shift levers and linkage.

1. Shaft	9. 2nd & 3rd hub
2. Pin	10. Bushing
3. Interlock key	11. Lever
4. Snap ring	12. Washer
5. Fiber bushing	13. Spring
6. Support	14. Bracket
7. Park hub	15. Lever
8. 1st & reverse hub	16. Pin

Suitably block the rear axle center housing to prevent it from tipping to front or rear. Swing transmission housing from the top and remove the attaching flange bolts, then lift off the transmission assembly.

The Power Take-Off spline coupling (Fig. 94) may remain with transmission or with rear axle center housing when units are separated. When reinstalling the transmission unit, place coupling on rear unit as shown, with chamfered end of splines forward. Complete the installation by reversing the removal procedure.

Fig. 94 — Front view of rear axle center housing showing pto spline coupling properly installed. Chamfered end of splines should be forward.

ADJUSTMENT

All Models

96. **SHIFT LINKAGE.** Except for the clutch release linkage, the only adjustment required is adjustment of link rods running from instrument panel mounted shift levers to the actuating levers on transmission housing. Refer to Figs. 95, 96, 97 and 98 for exploded views of the complete shift linkage.

A suggested method of making the adjustment is to complete the assembly except for connecting the shift rod yokes to levers (7, 8, 9 & 11—Fig. 95). Shift the main transmission levers (19 & 21—Fig. 96) to neutral (center) detent, Hi-Lo shift lever (Fig. 97) to "LOW" position and park pawl (Fig. 98) to "DRIVE" position. Adjust the connecting rods (17 & 18—Fig. 96, 29—Fig. 97 and 39—Fig. 98) until connecting pins can be easily inserted with main shift lever in neutral position and Hi-Lo shift lever in down position. With pins installed, check to be sure that shifting can be accom-

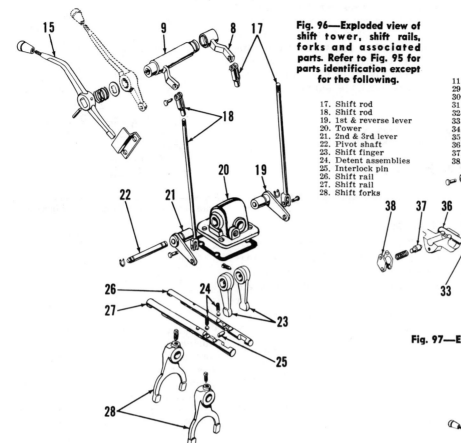

Fig. 96—Exploded view of shift tower, shift rails, forks and associated parts. Refer to Fig. 95 for parts identification except for the following.

17. Shift rod
18. Shift rod
19. 1st & reverse lever
20. Tower
21. 2nd & 3rd lever
22. Pivot shaft
23. Shift finger
24. Detent assemblies
25. Interlock pin
26. Shift rail
27. Shift rail
28. Shift forks

11. Lever
29. Shift rod
30. Inner arm
31. Outer arm
32. Link
33. Lever
34. Finger
35. Shift fork
36. Shift rail
37. Detent plunger
38. Plate

Fig. 97—Exploded view of Hi-Lo shift linkage.

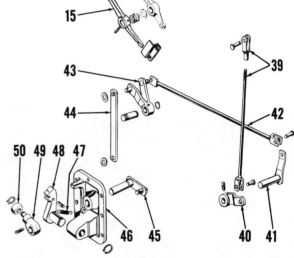

Fig. 98 — Exploded view of park lock and linkage.

7. Park hub
15. Shift lever
39. Shift rod
40. Inner arm
41. Outer arm
42. Link
43. Bellcrank
44. Link
45. Lever
46. Park cover
47. Spring
48. Pawl
49. Cam arm
50. Roller

plished to all detent positions without the levers bottoming or main shift lever binding through the neutral slot. If linkage cannot be properly adjusted, check for wear or damage and correct the trouble before proceeding further.

OVERHAUL

Multi-Power Models

97. INPUT SHAFT AND FRONT COVER. Refer to Fig. 99 for an exploded view of input shaft, front cover and associated parts used on Multi-Power models. Fig. 100 shows main transmission shafts and gears used on Six-Speed models. Except for input shaft, front cover and countershaft, parts shown in Fig. 100 are interchangeable between the two transmission types.

If only the input shaft, front cover, Multi-Power clutch and associated parts are to be removed, detach clutch housing from engine as outlined in paragraph 93. If transmission is to be disassembled, remove the unit as in paragraph 95.

Remove the clutch release shaft, forks and bearing. Remove the Multi-Power inlet hydraulic tube and withdraw front pto shaft (11—Fig. 100).

Remove the cap screws securing transmission front cover (4—Fig. 99) to housing, noting that top, center cap screw is machined to serve as a locating dowel. This special screw must be reinstalled in this same location. Lift out the input shaft, front cover and Multi-Power clutch as an assembly.

The Multi-Power output gear (56) and its bearing will remain in housing when input shaft is removed. If output gear must be removed, bump the gear and bearing rearward out of housing after removing mainshaft as outlined in paragraph 99 and rear snap ring (10—Fig. 100).

To disassemble the removed input shaft unit, refer to Fig. 99 and proceed as follows: Lift off the jaw clutch (54) and spring (52). Unseat and remove snap ring (53), spacer washer (51) and overdrive pinion (50). Remove snap ring (47), pressure plate (46), friction discs (45) and separator plates (42). Lift out the nineteen separator springs (44) and roller guides (43). Using two pairs of pliers and grasping the strengthening rib, work the piston (41) from its bore in clutch drum (37). Thrust washer (39) may or may not use the locating dowels shown.

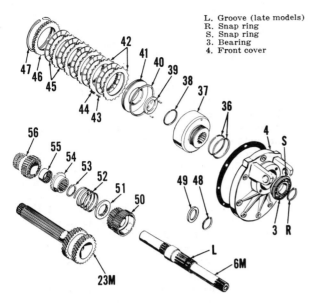

L. Groove (late models)
R. Snap ring
S. Snap ring
3. Bearing
4. Front cover

Fig. 99 — Exploded view of transmission front cover, input gears, Multi-Power clutch and associated parts. Remainder of unit is similar to that shown in Fig. 100.

6M. Input shaft
23M. Countershaft
36. Sealing rings
37. Clutch hub
38. Piston ring
39. Thrust washer
40. Piston ring
41. Piston
42. Separator plates
43. Roller guides
44. Separator springs
45. Friction discs
46. Pressure plate
47. Snap ring
48. Snap ring
49. Spacer washer
50. Overdrive pinion
51. Spacer washer
52. Spring
53. Snap ring
54. Jaw clutch
55. Needle bearing
56. Output gear

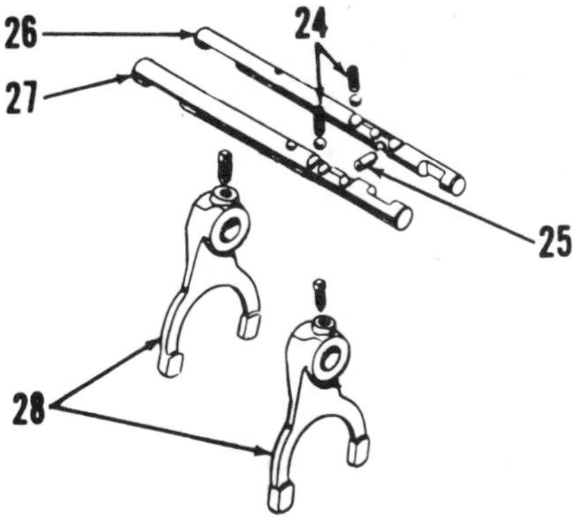

Fig. 101—Exploded view of main transmission shifter rails and forks. Refer to text and to Fig. 96.

Unbolt and remove the input shaft bearing retainer (1—Fig. 100). Remove snap ring (R—Fig. 99) from front groove of input shaft (6M), then bump the shaft and clutch drum rearward out of cover (4).

Tractors after Transmission Serial Number DU 501 have a second snap ring groove (L) in input shaft (6M), and clutch drum (37) is shortened to provide room for snap ring (48). Only the late (short) drum is available for service, and spacer washer (49) is provided for installing late drum on early shaft without snap ring groove (L).

Input shaft (6M) has an inner tube at rear of shaft which is sealed at each end by an O-ring. The tube provides passage for lubricant to the Multi-Power jaw clutch, and removal is necessary only if leakage is suspected.

Examine clutch plates (42) and discs (45) for heat discoloration, wear or other damage. Discs should be renewed if they measure less than 0.092 in thickness.

When reassembling the input shaft and Multi-Power clutch, lubricate piston rings and install piston (41) in clutch drum then insert input shaft through drum. On late model units using snap ring (48), select the thickest snap ring which will fit in groove (L). Snap ring (48) is available in four thicknesses of 0.109, 0.115, 0.120 and 0.125. With snap ring (48) or spacer (49) installed if required, remove snap ring (S) from outer race of bearing (3) and press bearing on front of shaft, snap ring groove forward; then select the thickest snap ring (R) which can be installed in front groove of shaft (6M). Snap ring (R) is identical to snap ring (48) and

available in the four thicknesses indicated above. With bearing and snap ring installed, clutch drum end play should not exceed 0.004.

Position input shaft and clutch drum assembly with front end down and install thrust washer (39) and one separator plate (42), then install the nineteen separator springs and guide rollers in alternate splines of clutch drum with lower ends resting on tabs of the installed separator plate. Install overdrive pinion (50) then alternately install the five clutch discs (45) and remaining four separator plates (42), beginning and ending with a clutch disc (45). Install pressure plate (46) and snap ring (47).

Install new bronze-teflon sealing rings (36) in grooves on clutch hub (37), then carefully install the assembled shaft and clutch unit in front cover (4). Secure with snap ring (S) and install bearing retainer (1—Fig. 100).

Make sure clutch spring (52—Fig. 99) and jaw clutch (54) are properly positioned on input shaft before reinstalling front cover. Tighten cover retaining cap screws to a torque of 33-38 ft.-lbs. and complete the assembly by reversing the disassembly procedure.

NOTE: Make sure the machined special (doweled) cap screw is installed in top center position in cover (4).

Six Speed Models

98. INPUT SHAFT AND FRONT COVER. Refer to Fig. 100 for an exploded view of main transmission shafts, gears and associated parts.

Front cover (4), bearing (3) and associated parts can be removed after splitting engine from clutch housing

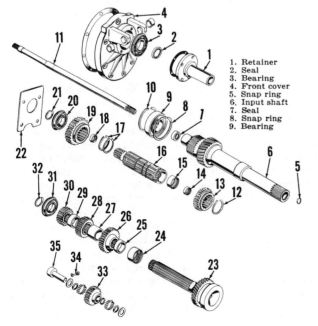

Fig. 100—Exploded view of main transmission shafts, gears and associated parts used on six speed models.

1. Retainer
2. Seal
3. Bearing
4. Front cover
5. Snap ring
6. Input shaft
7. Seal
8. Snap ring
9. Bearing
10. Snap ring
11. PTO input shaft
12. Snap ring
13. 2nd & 3rd gear
14. Needle bearing
15. Needle bearing
16. Main shaft
17. Snap rings
18. Needle bearings
19. Low-reverse gear
20. Bearing
21. Snap ring
22. Rear plate
23. Countershaft
24. Needle bearing
25. Spacer
26. Gear
27. Spacer
28. Gear
29. Spacer
30. Gear
31. Bearing
32. Snap ring
33. Reverse idler
34. Lock screw
35. Shaft

as outlined in paragraph 93. To remove the input shaft (6), it is first necessary to remove the transmission as in paragraph 95 and transmission main shaft as in paragraph 99.

To remove the front cover after clutch split or transmission removal, first remove clutch release shaft, forks and bearings. Unbolt and remove bearing retainer (1) to protect the seal (2). Remove the cap screws retaining front cover (4) to transmission housing, noting that top, center cap screw is machined to serve as a locating dowel. Remove front cover and bearing (3) as an assembly, with input shaft (6) and rear bearing (9) remaining in transmission housing.

To remove input shaft (6) and bearing (9) with mainshaft (16) and front cover removed, unseat rear snap ring (10) and bump input shaft and bearing rearward out of bearing bore. Install by reversing the removal procedure.

When reinstalling front cover (4), note that special (doweled) cap screw is installed in upper, center position and tighten all screws to a torque of 33-38 ft.-lbs.

All Models

99. MAIN SHAFT AND SLIDING GEARS. To remove the transmission main shaft and sliding gears, first remove transmission assembly as outlined in paragraph 95. Remove retainer plate (22—Fig. 100) from rear of transmission housing and slide pto front shaft (11) forward out of upper shafts. Remove the locking set screws in shift forks (28—Fig. 101) and with the other rail in neutral, slide either of the shift rails (26 or 27) rearward out of forks and housing. Retrieve detent balls and springs (24) and interlock pin (25) as rails are removed.

Slide main shaft (16—Fig. 100) to the rear until 2nd-3rd sliding gear (13) can be lifted from shaft, unseat and remove snap rings (17), then slide shaft (16) rearward out of low-reverse sliding gear (19) and housing unit.

Rear bearing (20) can be pressed from shaft after removing snap ring (21). Inner bore contains three caged needle bearings (14, 15 and 18) which can be renewed using Massey-Ferguson Special Tools MFN 800A, MFN 800P and MFN 800Q, or by carefully measuring the location of rear bearing before removal and installing new bearing in same position. The two front bearings should be installed just beyond flush with edge of bearing bores. The new shaft is shipped with bearings installed.

Reinstall the main shaft by reversing the removal procedure. Note that the two shift forks (28—Fig. 101) are interchangeable and both installed with long end of hub to rear.

100. COUNTERSHAFT. To remove the transmission countershaft (23M—Fig. 99 or 23—Fig. 100), first remove the mainshaft as outlined in paragraph 99 and the transmission front cover as in paragraph 97 or 98.

Remove snap ring (32—Fig. 100) from rear of countershaft. Place suitable wood blocks (about 2 inches thick) between second gear (26) and housing wall and, using a suitable puller, remove countershaft (23—Fig. 100 or 23M—Fig. 99) forward out of transmission case, lifting off gears and spacers as shaft is removed.

When installing the countershaft, snap ring (32—Fig. 100) is supplied in selective thicknesses of 0.093, 0.103, 0.113 and 0.125. Insert shaft from front, assembling spacers and gears as shown. Block front of countershaft inside of front cover flange and, using a short section of 2 inch ID pipe, tap bearing (31) onto rear of shaft until

it bottoms on shaft shoulder. Select the thickest snap ring (32) which can be installed. Countershaft end play should not exceed 0.010.

101. REVERSE IDLER SHAFT. To remove the reverse idler gear (33—Fig. 100) and shaft (35), first remove the main shaft as outlined in paragraph 99. Remove the locking bolt (34) from left side of transmission housing and tap the shaft (35) rearward, lifting gear and thrust washers out as shaft is withdrawn. Be careful not to tip the gear and spill the 56 loose needle rollers in transmission case as gear is withdrawn.

Assembly can be facilitated by using the Massey Ferguson Special Tool MFN 800K or a piece of shaft about 0.010 shorter than reverse idler slot in housing. Use the shaft to retain the loose needle rollers and thrust washers and install the gear with long hub forward. Shaft (35) will push the special tool out as it is inserted. Align the counterbore and install locking bolt (34), then complete the assembly by reversing the disassembly procedure.

REAR TRANSMISSION, CENTER HOUSING, MAIN DRIVE BEVEL GEARS, AND DIFFERENTIAL

The rear axle center housing contains the Hi-Lo (Range) transmission, main drive bevel gears and differential, as well as power take-off gears, shafts and clutch on models so equipped. The hydraulic system charging pump is driven by gear teeth machined on outer edge of pto clutch drum, and all tractors contain the crankshaft driven pto drive shafts back to and including the clutch drum.

TRACTOR SPLIT
All Models

102. To detach (split) the tractor between transmission and rear axle center housing, first drain transmission and center housing. On all ex-

cept LP-Gas Models, drain fuel tank. Remove seat console lower front panel and operator's platform.

Remove lower step plates and tool box. On early models, loosen the clamps on fuel tank connecting tube, slide tube to either side and remove the tube. On late models, remove fuel tank lower front support bracket and vent hose or tube connecting the tanks at top rear, then unbolt and remove the tanks.

Disconnect park linkage rod and Hi-Lo shift rod at rear. Disconnect hydraulic lines and wiring running to front of tractor. Suitably support both sections and remove attaching flange

Fig. 102—Front view of rear axle center housing. Spline coupling for pto shaft must be installed with chamfered end of splines forward.

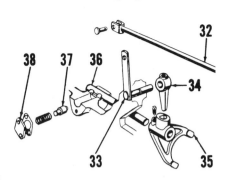

Fig. 103 — Exploded view of Hi-Lo shift mechanism. Refer to text and to Fig. 97.

bolts, then roll front of tractor forward away from rear axle center housing.

The Power Take-Off spline coupling (Fig. 102) may remain with transmission or with rear axle center housing when units are separated. When reconnecting the tractor, place coupling in rear unit as shown, with chamfered end of splines forward. Complete the assembly by reversing the disassembly procedure.

RANGE TRANSMISSION

All Models

103. **SHIFTER RAIL & FORK.** Refer to Fig. 103 for an exploded view of range transmission shift rail and fork located in rear axle center housing. Actuating lever mounted on instrument panel is shown in Fig. 95 and adjustment is covered in paragraph 96.

To remove the rails and forks located in center housing, first detach (split) transmission from center housing as outlined in paragraph 102. Disconnect remainder of hydraulic lines from center housing front, top cover; remove the cap screws and lift off the cover with hydraulic system charging pump attached.

Remove detent plate (38—Fig. 103), spring and plunger (37). Loosen the

set screw in fork (35) and slide shift rail (36) forward out of housing bore. Install by reversing the removal procedure. Actuating lever (33) and arm (34) can be removed if desired, by loosening set screw in arm.

104. **INPUT CLUSTER GEAR.** To remove the input cluster gear shaft (2—Fig. 104), first detach (split) transmission from rear axle center

housing as outlined in paragraph 102. Disconnect remainder of hydraulic lines from center housing front, top cover and remove the cover and attached hydraulic system charging pump.

Remove pto output housing or center housing rear cover as outlined in paragraph 119. Withdraw rear pto shaft or stub shaft (11—Fig. 105). Move pto drive shaft (9) rearward to expose snap ring (10), unseat and remove the snap ring; then slide pto drive shaft (9) rearward out of tractor.

With pto drive shaft removed, unseat front snap ring (2) and move cluster gear (6) forward until front bearing (3) is out of housing wall. Remove snap ring (1), then bearing (3) using suitable pulling equipment. Lift cluster gear (6) out top opening of housing.

Fig. 104 — Top view of rear axle center housing with front hydraulic cover removed.

1. Intake tube
2. Cluster gear

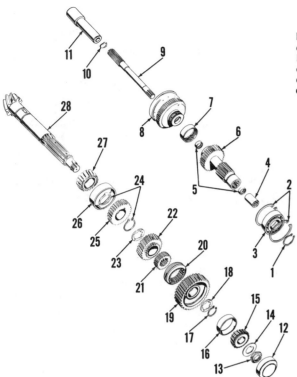

Fig. 105—Exploded view of bevel pinion shaft and Hi-Lo shift gears on models without pto. Clutch drum (8) drives the hydraulic system charging pump.

1. Snap ring
2. Snap rings
3. Bearing
4. Coupling
5. Needle bearings
6. Cluster gear
7. Needle bearing
8. PTO clutch drum
9. PTO drive shaft
10. Snap ring
11. Stub shaft
12. Bearing cap
13. Shaft nut
14. Thrust washer
15. Bearing cone
16. Bearing cup
17. Snap ring
18. Thrust washer
19. Low gear
20. Shift collar
21. Hub
22. High gear
23. Thrust washer
24. Snap ring
25. Gear
26. Bearing cup
27. Bearing cone
28. Bevel pinion

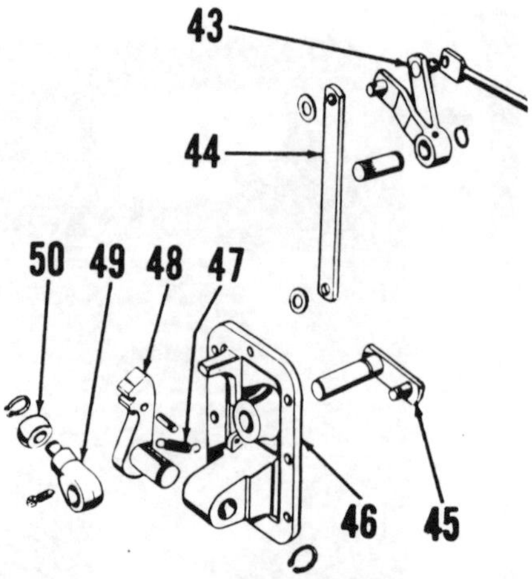

Fig. 106—Exploded view of park lock cover and associated parts. Park pawl (48) meshes with gear (25—Fig. 105) on bevel pinion shaft. Refer also to Fig. 98.

To remove the park lock cover, it is first necessary to drain rear axle center housing and drain and remove fuel tanks.

PARK LOCK.

All Models

106. The park lock cover mounts on left side of rear axle center housing. An exploded view is shown in Fig. 106.

Park pawl (48) engages the gear (25—Fig. 105) which is keyed to bevel pinion shaft and which also serves as the ground speed pto drive gear on models so equipped.

105. **BEVEL PINION SHAFT.** To remove the main drive bevel pinion shaft (28—Fig. 105), first detach (split) rear axle center housing from transmission as outlined in paragraph 102, remove input cluster gear as in paragraph 104 and differential assembly as in paragraph 107 or 110.

Remove pinion shaft bearing cap (12). Shaft nut (13) is staked and secured with Grade C LOCTITE. Block pinion shaft from turning and remove the nut.

Unseat front snap ring (24) and bump pinion shaft rearward to remove front bearing cone (15). Unseat and remove snap ring (17) at front of pinion shaft; then slide shaft rearward out of housing, lifting gears and associated parts out top opening as shaft is withdrawn.

Gear (25) is installed with long hub to rear. Only one snap ring (24) is used on Model MF1130 and snap ring is located ahead of gear (25). Install shift collar hub (21) with blind spline over cross drilling in shaft. Install nut (13) using Grade C (blue) LOCTITE and tighten nut to provide a rolling torque of 12-20 inch-pounds to shaft. Complete the assembly by reversing the disassembly procedure.

NOTE: Bearing cap (12) plays an important part in the centrifugal lubrication of gears (19 and 22). Make sure cap is installed when shaft is reassembled.

DIFFEERENTIAL AND BEVEL RING GEAR

Models Without Differential Lock

107. **REMOVE AND REINSTALL.** To remove the differential assembly, first drain transmission and center housing. Remove both final drive units as outlined in paragraph 112 and power take-off housing as in paragraph 119.

Working through rear opening in center housing, block up between rim

Fig. 107 — Use a wood block as protection and roll differential assembly rearward out of center housing as shown.

of main drive bevel gear and bottom of housing. Unbolt and remove both carrier bearing retainer plates and brake pressure plates as assemblies, allowing differential unit to rest on previously installed block. Roll differential and ring gear rearward from center housing as shown in Fig. 107. Overhaul the removed unit as outlined in paragraph 108.

Mesh position and backlash of main drive bevel gears are not adjustable. When installing differential, make sure the stamped word "TOP" on carrier bearing retainer plate is upward. Install cap screws using Grade C (blue) LOCTITE and tighten screws securely. Complete the assembly by reversing removal procedure and bleed brakes as outlined in paragraph 115.

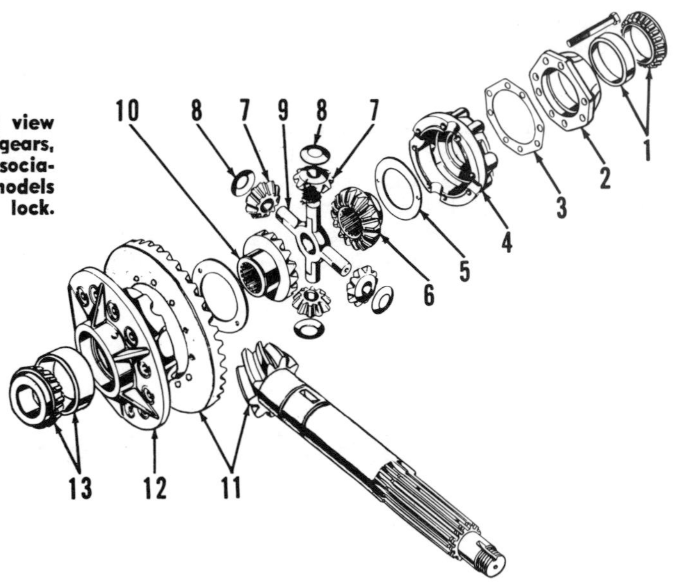

108. OVERHAUL. Refer to Fig. 108 for an exploded view of differential assembly. Differential case bolts are installed with LOCTITE and are not safety wired.

Main drive bevel ring gear and pinion (11) are available only as a matched assembly. Ring gear is riveted to differential case (12) at original assembly, but service parts are provided with bolts and nuts. Install with Grade C (blue) LOCTITE and tighten to a torque of 110-120 ft.-lbs.

The differential case is available only as a matched assembly which includes adapter (2), shim pack (3), case halves (4 & 12) and the two carrier bearing cups. The 0.004 thick shims (3) are used to obtain proper bearing adjustment at factory assembly and thickness of shim pack should not be changed.

The two side gears (6 & 10) are not interchangeable, the one with the longer (2½ inch) hub being installed in gear half of case (12).

When assembling the differential unit, use Grade C (blue) LOCTITE on differential case bolts and tighten to a torque of 80-85 ft.-lbs.

Models With Differential Lock

109. OPERATION. The differential lock assembly consists of a hydraulically actuated multiple disc clutch which selectively locks the left axle gear to differential cage, causing both rear wheels to rotate together as a unit. The differential lock clutch is automatically released if either or both brakes are applied, and can be released by pulling up on actuating valve knob or momentarily applying brake.

Fig. 108—Exploded view of main drive bevel gears, differential and associated parts used on models without differential lock.

1. Bearing
2. Adapter
3. Shim
4. Case half
5. Thrust washer
6. Side gear
7. Differential pinion
8. Thrust washer
9. Spider
10. Side gear
11. Ring gear & pinion
12. Case half
13. Bearing

Power to operate the differential lock clutch is supplied by the low pressure, low volume circuit which also supplies working fluid to the hydraulic seat circuit and Multi-Power and Power Take-Off hydraulic clutches.

Refer to Fig. 109 for an exploded view of differential assembly and to paragraph 166 for service procedures on hydraulic system components other than clutch.

110. REMOVE AND REINSTALL. To remove the differential assembly, first drain transmission and center housing. Remove both final drive units as outlined in paragraph 112 and power take-off housing as in paragraph 119.

Working through rear opening in center housing, block up between rim of main drive bevel gear and bottom of housing. Unbolt and remove both carrier bearing retainer plates and brake pressure plates as assemblies, allowing differential unit to rest on previously installed block. Roll differential unit rearward from center housing as shown in Fig. 107. Overhaul the removed unit as outlined in paragraph 111.

Mesh position and backlash of main drive bevel gears is not adjustable. When installing differential, make sure the stamped word "TOP" on carrier bearing retainer plate is upward. Install cap screws using Grade C (blue) LOCTITE and tighten screws securely. Complete the assembly by

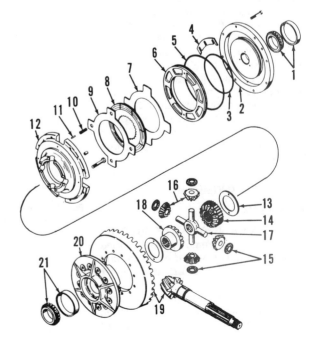

Fig. 109—Exploded view of differential lock clutch, main drive bevel gears, differential and associated parts used on models with differential lock.

1. Bearing
2. Clutch cover
3. O-ring
4. Shims
5. O-ring
6. Piston
7. Separator plate (end)
8. Clutch disc
9. Separator plates
10. Separator springs
11. Roller guides
12. Case half
13. Thrust washer
14. Side gear
15. Thrust washers
16. Differential pinions
17. Spider
18. Side gear
19. Ring gear & pinion
20. Case half
21. Bearing

reversing the removal procedure and bleed brakes as outlined in paragraph 115.

111. OVERHAUL. Refer to Fig. 109 for an exploded view of differential assembly. Differential case bolts are installed with LOCTITE and are not safety wired. Before differential case bolts can be removed, it is first nec-essary to unbolt and remove differential lock clutch cover (2) and clutch plates (7 through 9). Be sure and save shims (4) for reinstallation. Shims are used to obtain proper bearing adjust-ment at factory assembly.

Main drive bevel ring gear and pin-ion (19) are available only as a matched assembly. Ring gear is riv-eted to differential case (20) at orig-inal installation, but service parts are provided with bolts and nuts. Install with Grade C (blue) LOCTITE and tighten to a torque of 110-120 ft.-lbs.

The differential case is available only as a matched assembly which in-cludes the two differential case halves (12 & 20), adapter (2), shims (4) and associated parts. Thickness of shim pack (4) must not be changed.

REAR AXLE AND FINAL DRIVE

All models use a planetary final drive unit located in inner end of axle housing next to center housing. Final drive lubri-cant reservoir is common with transmis-sion, differential and hydraulic reservoir. Axle stub shafts are free floating, with in-ner ends supported by differential side gears and outer ends serving as planetary sun gears. Service on any part of axle or final drive requires removal of unit as outlined in paragraph 112.

REMOVE AND REINSTALL

All Models

112. To remove either final drive unit as an assembly, first drain trans-mission and hydraulic system fluid, suitably support rear of tractor and remove rear wheel and fender. Re-move drawbar and drawbar frame.

Attach a hoist to final drive hous-ing as shown in Fig. 110. Remove the attaching stud nuts and cap screws and slide the complete final drive as-sembly away from rear axle center housing. Stub axle and brake disc may remain with center housing or with final drive unit. When reinstalling final drive, place stub axle and brake in final drive unit.

Use a new gasket and install by re-versing the disassembly procedure. Tighten retaining stud nuts to a torque of 155 ft.-lbs. and cap screws to 170 ft.-lbs.

OVERHAUL

All Models

113. **AXLE & BEARINGS.** With final drive unit removed as outlined in paragraph 112, withdraw stub axle (4—Fig. 111) and lift off brake cover plate (2).

Hold axle from turning and remove the two cap screws (3—Fig. 112), then withdraw planet carrier (2). Recover and save the shim pack (5—Fig. 113) interposed between end of axle shaft (1) and retainer plate contained in planet carrier (4). Shims adjust axle bearing preload which should be 0.001-0.005.

Using a sharp chisel, carefully cut outer seal (13—Fig. 114) from axle housing (22); then bump axle and outer bearing cone from housing.

Outer bearing cone and both bear-ing cups can be removed with suit-able pulling equipment. When assem-bling, pack outer bearing with lithium base multipurpose grease. Insert axle carefully through inner seal (11), then install inner bearing cone and thrust washer (9). Install planet carrier us-ing the removed shim pack plus one 0.010 shim. Install and securely tighten the two retaining cap screws, then using a dial indicator, measure and record the axle end play. Remove cap screws and planet carrier, then remove shims (2) in thickness equal to measured end play plus 0.001-0.005 to obtain the specified preload for axle bearings. On final assembly, in-stall the two retaining cap screws us-ing Grade C (blue) LOCTITE and tighten to a torque of 80-85 ft.-lbs. In-stall outer seal using a suitable seal-ant, after planet carrier is installed. Complete the assembly by reversing the disassembly procedure.

114. **PLANET CARRIER.** The final drive planet carrier (4—Fig. 113) is shown exploded in Fig. 114. After re-moving the unit as outlined in para-graph 113, disassemble as follows:

Place the carrier, inner (flat) side down on a bench. Cut the safety wires and remove cap screws and shaft re-tainers (8).

NOTE: On some early models, cap screws may not be drilled and may be retained in carrier with LOCTITE. On these units, it is recommended that new type retainers (8), Part No. 517 790 M1 and new drilled head cap screws 365 475 X1 be installed when unit is reassembled.

Fig. 110—Removing right final drive unit.

Fig. 111—Inside view of final drive housing showing brake cover plate (2), planet carrier (3) and stub shaft (4).

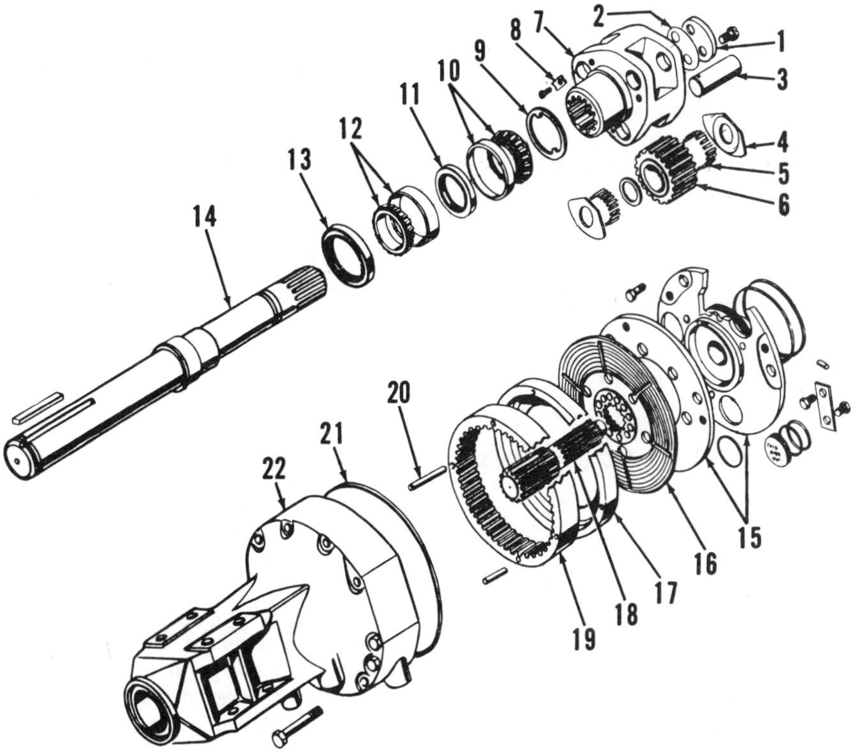

Fig. 114—Exploded view of final drive and brakes showing component parts.

Fig. 112—Planet carrier (2) can be removed after removing cap screws (3). Planetary ring gear (1) is not serviced separately from axle housing.

Lift out the planetary pinion shafts (3) and slide the pinions (6), thrust washers (4) and loose needle rollers as assemblies from the carrier. Retainer plate (1) can be withdrawn through pinion opening after removing one pinion.

Each planet pinion (6) contains two rows of 25 each loose needle rollers (5) separated by a spacer washer and retained by thrust washers (4). The manufacturer supplies a spring-loaded tool (MFN 800DD) which can be used

for assembly and installation of planet pinions and bearings. The tool can be compressed for insertion of the assembled pinion and thrust washers. Springs will expand tool into shaft bore for perfect alignment for shaft insertion as shown in Fig. 115.

Note: Be sure retainer plate (1) is properly positioned in carrier before the last pinion is installed.

Tighten shaft retainer cap screws to a torque of 20 ft.-lbs. and secure with safety wire. Install the axle and planet carrier as outlined in paragraph 113.

NOTE: Planetary ring gear (19) is available only as an assembly with axle housing (22). Do not attempt to remove ring gear from housing, if either is damaged, renew the assembly.

Fig. 113—Shims (5) interposed between end of axle shaft (1) and retainer plate for planet carrier (4) control preload for axle bearings.

Fig. 115—Using the special tool to reassemble planet carrier. Refer to text.

SPECIAL TOOL
MFN 800 DD

BRAKES

All models are equipped with individual hydraulic power brakes for each rear wheel, the power being supplied by the tractor central hydraulic system. Brakes can be manually operated by depressing the brake pedals, when main hydraulic system is inoperative.

The wet type disc brakes are located on stub axle shafts between differential and final drive planetary gears. Brakes can be serviced after removing final drive housings as outlined in paragraph 112.

ADJUSTMENT AND BLEEDING

All Models

The only adjustment required of the brakes is adjustment and equalization of pedal height. Adjust and bleed the brakes as outlined in the appropriate following paragraphs. If brakes cannot be satisfactorily adjusted, remove and overhaul control valve as outlined in paragraph 156 or brake pistons as in paragraph 117.

115. **BLEEDING.** To bleed the hydraulic brakes, refer to Fig. 116 and proceed as follows: Loosen jam nuts (1) on both brakes and back out bleed screws (2) two full turns. Start tractor engine and operate at 1000 rpm. Slowly depress both brake pedals until resistance is felt and hold pedals depressed for 30 seconds.

Release the pedals, tighten both bleed screws to a torque of 6-10 ft.-lbs. and secure with locknuts. Adjust pedal height as outlined in paragraph 116.

116. **PEDAL ADJUSTMENT.** Pedal height adjustment is correct when, with engine running, solid resistance is felt when pedals clear stop on platform by ⅝-inch and both pedals are equally adjusted.

To make the adjustment, disconnect brake pedal interlock, refer to Fig. 117 and proceed as follows: Loosen locknuts (1) and with engine running and pedal depressed, turn brake rod (2) until adjustment is correct.

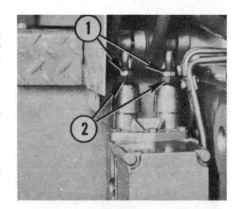

Fig. 117—To adjust brake pedal linkage, loosen locknuts (1) and turn brake rods (2). Refer to paragraph 116 for details.

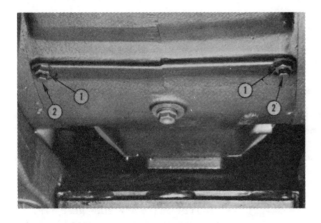

Fig. 116 — To bleed the brakes, loosen locknut (1) and back out bleed screw (2) two turns. Hold pedals depressed for 30 seconds with engine running to expel air from system.

Fig. 119—Brake disc removed from stub axle (4), showing cover plate (2) and planet carrier (3).

Fig. 118—View of final drive housing showing brake disc (1) installed on stub axle (2).

Fig. 120—Differential carrier plate (2) removed from center housing showing O-rings (1 & 3).

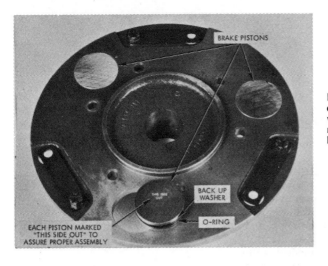

Fig. 121—Outer face of differential carrier plate with brake pressure plate removed. The leaf type brake return springs are shown.

Fig. 123—The stamped word "TOP" on differential carrier plate must be up as shown.

Test drive the tractor with pedal interlock engaged. If tractor pulls to left or right when brakes are applied, repeat bleeding and adjustment procedure. If trouble cannot be corrected, overhaul brake valve as outlined in paragraph 156 or brake pistons as in paragraph 117. Test procedure for hydraulic brake valve is given in paragraph 140.

OVERHAUL

All Models

117. **BRAKE DISCS AND ASSOCIATED PARTS.** Before any service can be performed on brake discs or actuating pistons, final drive unit must be removed as outlined in paragraph 112.

With final drive off, brake disc (1—Fig. 118) can be removed as can the brake cover plate (2—Fig. 119). To remove the brake pressure plate, pistons or O-rings, first unbolt and remove differential carrier plate as shown in Fig. 120. Place unit on a bench, brake pressure plate down, and remove the three cap screws securing return springs to plate. Lift off retainer plate, brake pistons and return springs as shown in Fig. 121. Brake pistons are

marked "THIS SIDE OUT" for proper installation. Back-up washer for brake piston O-ring goes to outside as shown. Fig. 122 is a cross section view of brake actuating unit. O-rings (1 & 3—Fig. 120) seal brake pressure passages.

When installing differential carrier plate and brake assembly, make sure word "TOP" stamped on differential carrier plate hub is upward as shown in Fig. 123. Install retaining cap screws using Grade C (blue) LOCTITE and tighten screws securely. Complete the assembly by reversing the removal procedure and bleed brakes as outlined in paragraph 115.

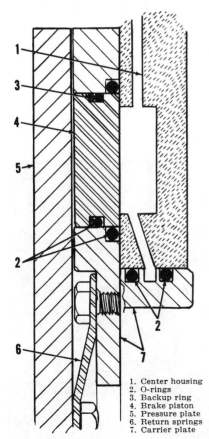

1. Center housing
2. O-rings
3. Backup ring
4. Brake piston
5. Pressure plate
6. Return springs
7. Carrier plate

Fig. 122—Cross sectional view of brake actuating unit showing component parts.

Fig. 124—Rear view of engine flywheel with clutch removed, showing splined pto drive adapter.

POWER TAKE-OFF

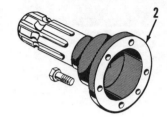

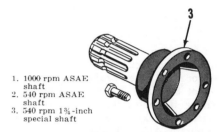

1. 1000 rpm ASAE shaft
2. 540 rpm ASAE shaft
3. 540 rpm 1¾-inch special shaft

Fig. 126 — PTO output accessory shafts available for installation on tractor.

The power take-off input shaft is splined to an adapter attached to engine crankshaft along with flywheel as shown in Fig. 124. The front shaft runs inside the hollow clutch shaft and transmission mainshaft, and is connected by a coupling to the pto drive shaft located in rear axle center housing.

The hydraulic pump drive gear is an integral part of pto clutch drum, and all tractors are equipped with flywheel adapter, front pto shaft, coupling, pto drive shaft and pto clutch drum, whether or not a pto output shaft and controls are used.

The multiple disc pto clutch is hydraulically actuated by the low volume, low pressure circuit of the main hydraulic system. Tests and service of pump and valves is covered in the HYDRAULIC SYSTEM section of this manual.

OUTPUT SHAFT

All Models

118. The 540 and 1000 rpm pto output shaft drives consist of outer and inner flanges as shown in Fig. 125. Optional output shafts are shown in Fig. 126.

Any service on output shaft drive units requires removal of pto output housing as outlined in paragraph 119 and overhaul of output shaft gear train as in paragraph 120.

PTO OUTPUT HOUSING AND GEARS

119. **REMOVE AND REINSTALL.** To remove the pto output housing on models equipped with power take-off (or center housing rear cover on other models), first drain transmission and center housing and remove interfering drawbar frame. On models with draft control hitch, remove rear dress panel cover (1—Fig. 129) and draft control springs (2). Remove bolts (1—Fig. 130) securing draft control linkage (2) to output housing.

Move pto shift lever to "Engaged" (up) position on later models without ground speed pto, or to "Engine" (down) position on models with ground speed pto. Remove attaching flange bolts and lift off pto housing or rear cover.

PTO rear drive shaft (1—Fig. 131) can be withdrawn after cover is removed, but be sure pto coupling is engaged to prevent coupler or coupler gear from dropping out of position. Also be sure shim pack with thrust washer (2) is not lost or damaged when rear cover is removed. Shims control pto rear drive shaft end play which should be 0.005-0.026. Shims are available in thicknesses of 0.015-0.030 and 0.062. Clearance can be checked with a feeler gage after re-

moving center housing front, top cover as outlined in paragrap 147 or park lock cover as in paragraph 106.

Use Grade C (blue) LOCTITE on flange retaining cap screws when reinstalling pto housing or rear cover, and tighten screws securely. Complete the assembly by reversing the disassembly procedure. Refer to paragraph 172 for draft control linkage adjustment.

Fig. 125—Rear view of tractor with 1000 rpm pto output shaft installed. The 540 rpm shaft attaches to outer flange.

Fig. 127—Assembled view of tractor with standard swinging drawbar and 3-point, draft control linkage.

Fig. 128—Assembled view of tractor with heavy duty swinging drawbar without 3-point lift.

Fig. 131—Rear axle center housing with pto output housing removed, showing rear pto drive shaft (1) and thrust washer (2).

Fig. 129—To remove the pto housing, first remove rear panel cover (1) and draft control springs (2), then refer to Fig. 130.

Fig. 132—To disassemble the pto output housing, remove the cap screws (1) and lift off gear cover (2) as shown in Fig. 133.

Fig. 130 — Remove the bolts (1) securing draft control linkage (2) to pto output housing, then lift off linkage with dress panel.

Fig. 133—Disassembling the pto output gears.

Fig. 134—Removing cluster gear.

1. Cover	3. Shaft
2. Shaft	4. Input gear

Fig. 136 — Assembled view of 540 rpm and 1000 rpm output shafts. Refer to Fig. 137 for cross sectional view.

120. OVERHAUL. To disassemble the removed output housing, remove the three cap screws (1—Fig. 132) and lift off gear cover (2) as shown in Fig. 133. Note the location and position of thrust washers as housing is disassembled.

When disassembling the 540 and 1000 rpm output drive flanges, unseat and remove the snap rings, gears and thrust bearings from inside of housing, then withdraw both flange shafts as a unit as shown in Fig. 136. The shafts contain 66 loose needle rollers assembled in two rows of 33 each, separated by a machined pipe spacer. Disassemble carefully to prevent loss or damage of any of the loose needle rollers. Fig. 137 shows a cross sectional view of output flange shafts properly assembled. Install shafts carefully from rear of housing. The three thrust roller bearings have one thrust race each,

the other race being machined on the 540 and 1000 rpm driven gears (3—Figs. 138 and 139). Renew shaft seals whenever unit is disassembled, and other parts if their condition is questionable. Assemble by reversing the disassembly procedure, using Figures 132 through 139 as a guide, then reinstall as outlined in paragraph 119.

PTO CLUTCH

All Models

121. REMOVE AND REINSTALL. To remove the pto drive shaft, pto hydraulic clutch and/or hydraulic pump drive gear, first remove pto output housing or center housing rear cover

as outlined in paragraph 119 and center housing front, top cover as in paragraph 147.

Withdraw oscillator crank (3—Fig. 140) by pulling forward firmly. On models with ground speed pto, withdraw pto rear drive shaft to the rear and move coupler gear (2—Fig. 141) to right, rear corner of compartment as

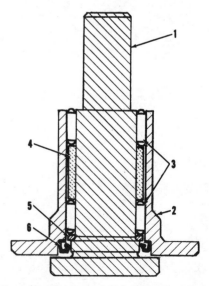

Fig. 135—Power take-off housing cover with input gear and idler installed.

Fig. 137—Cross sectional view of pto output flange shafts showing bearing and seal installation.

1. 1000 rpm shaft	4. Spacer
2. 540 rpm shaft	5. Thrust washer
3. Needle rollers	6. Oil seal

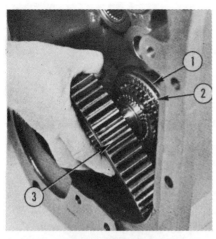

Fig. 138—Installing 540 rpm output gear (3). Note proper positioning of thrust bearing race (1) and thrust roller bearing (2) Front race is machined on gear (3).

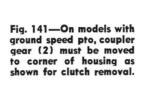

Fig. 141—On models with ground speed pto, coupler gear (2) must be moved to corner of housing as shown for clutch removal.

1. Clutch drum
2. Coupler gear
3. Snap ring
4. PTO drive shaft

Fig. 139—The 540 rpm output flange shaft is retained by snap ring (1). Thrust bearing (2) is backed up by a thrust race (not shown) and machined race on 1000 rpm gear (3).

shown. On models without ground speed drive, lift out coupler after removing rear drive shaft. On models without power take-off, remove stub shaft.

On all models, move pto drive shaft (4) rearward and unseat and remove snap ring (3); then withdraw shaft (4) rearward out of clutch drum (1). With pto drive shaft removed, move clutch or clutch drum rearward as

shown in Fig. 142 and lift out through top opening.

NOTE: Production changes have been made in pto clutch and associated parts, most of which do not materially affect service procedures. Early tractors were equipped with ground speed drive gears for pto shaft which is not used in late models. On early tractors, hydraulic pump drive used spur gears and oscillator crank was plain as shown in Fig. 140. On late tractors, pump

Fig. 140 — Top view of rear axle center housing with front hydraulic cover removed.

1. PTO clutch drum
2. Coupler gear
3. Oscillator crank

Fig. 142 — Move clutch rearward against housing wall, then lift out top opening.

1. Coupler gear
2. Clutch drum

Fig. 143—On late models, oscillator crank is equipped with renewable shoe as shown.

Fig. 146 — Push piston fully in place by hand as shown.

drive gear teeth were helical cut and oscillator crank was equipped with a renewable shoe as shown in Fig. 143. Parts are not individually interchangeable. Early tractors were equipped with cast iron sealing rings on hub of clutch drum. Late models use teflon bronze rings, and late type should be installed whenever early tractors are disassembled.

Lubricate teflon bronze sealing rings with petroleum jelly and install clutch unit by reversing removal procedure.

122. OVERHAUL. With pto clutch removed as outlined in paragraph 121, disassemble as follows: Place clutch drum hub-end down on a bench and applying slight pressure to plate (9—

Fig. 145—Using special dowels and ice pick to compress piston ring.

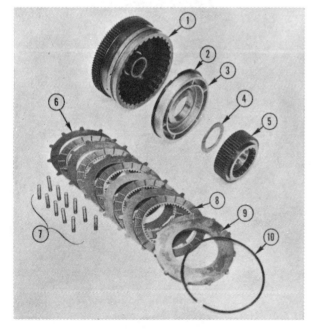

Fig. 144 — Disassembled view of pto clutch showing component parts.

1. Clutch drum
2. Piston ring
3. Piston
4. Thrust washer
5. Clutch hub
6. Separator plates
7. Separator springs
8. Clutch discs
9. Pressure plate
10. Snap ring

Fig. 144), unseat and remove snap ring (10). Remove items (4 through 9). Five clutch discs (8), five separator plates (6) and eleven springs (7) are used.

The manufacturer recommends a special tool set consisting of eleven dowels, to be used as a ring compressor when removing and installing the clutch piston. Grasp piston (3) by two opposite stiffening webs, using suitable pliers, and work the piston upward in drum. Evenly space about half the dowels in splines of drum and work lower end behind ring (2); then withdraw the piston and ring assembly.

Examine the parts and renew any which are worn or damaged. When assembling the clutch, work piston ring into drum as shown in Fig. 145; then, using an ice pick or similar tool, install the eleven dowels evenly spaced as shown in Fig. 146. Place one separator plate (6—Fig. 144) on top of piston, then alternate clutch discs and separator plates, with lugs of separator plates one spline either way from lug of bottom plate as shown in Fig. 147. Position the eleven separator springs on lugs of first separator plate as shown, then complete the assembly by reversing the disassembly procedure. Install the clutch as outlined in paragraph 121.

PTO SHIFT LINKAGE
All Models

123. Fig. 148 shows an exploded view of pto shift linkage and associated parts. Items (1), (2), and (3E) are used on early models with ground speed pto. Item (3L) is two-position lever for late models. Other items are similar on all models. Sealing O-ring (9) fits in channeled groove of housing bore for shaft (8).

Fig. 147 — Separator springs are installed on lugs of first separator plate as shown.

HYDRAULIC

SYSTEM

All models have a closed center, constant pressure hydraulic system which can logically be divided into three circuits as follows:

LOW PRESSURE/LOW VOLUME Circuit. Operates at a volume of 6 gpm and pressure of 210-275 psi to provide the working fluid for Multi-Power, PTO and Differential Lock clutches and Hydraulic Seat. All valves are closed-center type and fluid discharges through combining valve into the Low Pressure/High Volume Circuit.

LOW PRESSURE/HIGH VOLUME Circuit, also referred to as "Charging Circuit." Supplies working fluid on demand to high pressure, piston pump and provides constant circulation to system filter and oil cooler. Circuit is supplied by 18 gpm gear type pump and unused flow from low volume circuit. Charging circuit pressure is regulated at 90-100 psi.

HIGH PRESSURE (MAIN) Circuit. Supplies working fluid for power steering and power brakes on a priority basis, with sufficient reserve capacity for hydraulic working cylinders. Pressure is maintained at 2100-2400 psi, with available volume to 20 gpm.

The transmission lubricant is the operating fluid for the hydraulic system. Massey-Ferguson M-1127 Fluid is recommended. System capacity is 22 U. S. Gallons. Refill capacity is 17 U. S. Gallons, as 5 gallons will remain when reservoir is drained.

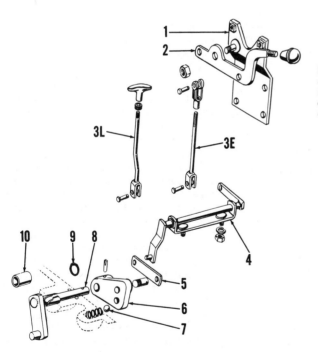

Fig. 148—Exploded view of pto shift linkage. Early model parts (1, 2 & 3E) are for tractors with ground speed pto drive gears.

1. Bracket
2. Lever
3E. Link (early
3L. Link (late)
4. Bellcrank
5. Link
6. Lever
7. Detent ball
8. Shift crank
9. O-ring
10. Bushing

OPERATION

All Models

125. The hydraulic system operating fluid is drawn from reservoir in rear axle center housing through a screen type inlet filter. The gear type dual charging pump is engine driven by the power take-off input shaft. Fluid from the low-volume side of dual pump flows under pressure to the hydraulic seat and differential lock valve, and to the Multi-Power and PTO control valve. All units are equipped with closed-center valves; if no demand for fluid exists, the fluid passes through the combining valve where it is mixed with the flow from the high-volume side of charging pump.

GROUND SPEED GEARS

Models So Equipped

124. The ground speed drive gear is keyed to main drive bevel pinion. The gear is installed on all tractors and serves as engaging point for park lock pawl. Procedure for removing the drive gear is given in paragraph 105.

The ground speed driven gear also serves as pto shift coupling. Gear is shown at (2—Fig. 140). Procedure for removal of ground speed driven gear is given in paragraph 121.

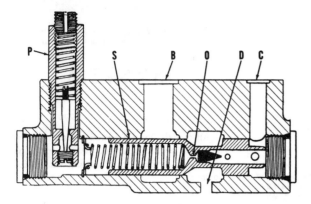

Fig. 149—Cross sectional view of hydraulic system regulating valve which controls charging pump pressure and flow.

B. Bypass port
C. Sensing control port
D. Dump port
O. Control orifice
P. Regulating pilot valve
S. Valve spool

The combined flow from the charging pump is open to two passages. One leads through the system filter and oil cooler to the inlet side of the main hydraulic pump, then to control port (C—Fig. 149) of regulating valve. Four to six gpm passes through the regulating valve spool orifice (O) to the sensing pressure regulating valve (P), where sensing pressure is maintained at 90-100 psi. If there are no other demands, the remainder of charging pump flow passes from bypass port (B) to dump port (D) through the passage opened by movement of regulating valve spool (S).

The radial piston type main hydraulic pump is front mounted and driven by engine crankshaft. The main hydraulic pump attempts to maintain a pressure of 2100-2400 psi at all valves of the high pressure main hydraulic system. When this pressure is reached, flow from the main hydraulic pump ceases, the eight radial pistons are held away from the actuating cam and the pump drive

shaft continues to rotate without performing work. The main pump fluid delivery is maintained at the flow necessary to satisfy system requirements up to maximum delivery of 20 gpm.

LUBRICATION, TESTING & ADJUSTMENT

All Models

126. LUBRICATION. The hydraulic system fluid serves as a lubricant for the power steering motor and hand pump, transmission, differential, final drive and pto gear train. The transmission and rear axle center housing serves as system reservoir.

The only recommended fluid is MASSEY-FERGUSON M-1127 FLUID. System capacity is 22 U.S. Gallons; when system is not disassembled, however, only about 17 gallons will be drained from the system and only this amount is required for normal refill. To check the fluid level, be sure tractor is on level ground and re-

Fig. 150 — Hydraulic oil level dipstick is on left side of transmission housing as shown.

move dipstick from left side of transmission housing as shown in Fig. 150. Fill through dipstick opening if fluid is needed.

System filter (Fig. 151) behind radiator grille should be drained each 500 hours and cartridge renewed or cleaned.

NOTE: On early models, a renewable paper filter cartridge was used and a bypass valve was built into filter body as shown in Fig. 152. Late models use a permanent filter which can be cleaned and reused; and the bypass passage is blocked. The late type filter is recommended when service is required. When late type filter is installed, bypass must be blocked by discarding the poppet spring and using the seating washer, spacer and bolt to make the poppet inoperative.

Transmission should be drained and refilled each 1000 hours, and intake screen should be removed and cleaned when unit is drained. Refer to Fig. 153 for view of the two drain plugs and intake screen cover plate. Do not attempt to flush the system. If oil is badly contaminated, system must be disassembled for cleaning.

127. QUICK TEST. A quick test of the high pressure system can be performed using a pressure gage as follows:

Connect pressure gage to remote coupler and pressurize that coupler, with engine running first at 1,000 rpm, then at 2000 rpm. Gage reading should be 2100-2400 psi on both tests.

Flow can be tested at breakaway coupling on early models equipped with poppet type auxiliary valves. Late (spool type) auxiliary valves have an adjustable flow restrictor and flow test is inconclusive. To make the test, connect flow meter inlet to breakaway coupling and direct the flow meter outlet back to reservoir. A plug opening is provided at rear left side of rear axle center housing as shown in Fig. 154. Make sure all hydraulic functions are in neutral. Start and operate engine at 1000 rpm. Move auxiliary lever to pressurize the coupling containing flow meter and lock the lever in open position. Operate flow meter on high scale and adjust load valve to 1000-2000 psi. Allow fluid to reach an operating temperature of 90°-140° F., then check the measured flow which should be 9 gpm MINIMUM.

Slow the engine to 850 rpm and readjust load valve to 1200 psi. Measure the flow which should be at least 5 gpm.

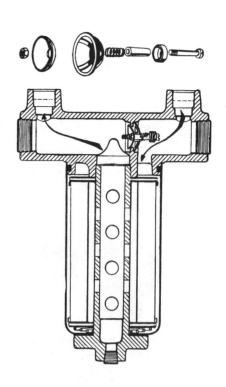

Fig. 151—Front view of tractor showing hydraulic parts.

1. End plug
2. Filter body
3. Cover
4. Center port
5. Drain plug

Fig. 152—Cross sectional view of hydraulic system main filter assembly. Bypass valve is shown exploded at top.

If flow and/or pressure are not as specified, refer to pump tests and adjustments outlined in paragraph 130 to pinpoint the possible causes of trouble. For specific tests on system circuits and components, refer to the applicable paragraphs 128 through 140.

128. CHARGING CIRCUIT PRESSURE & FLOW. Fig. 155 shows an exploded view of rear axle center housing front, top cover and associated parts. To obtain access to front cover for service or testing, first remove the ring from manual unloading valve, then remove operator's platform center panel and padding.

All tests should be conducted with tractor hydraulic system fluid at operating temperature of 90°-140° F.

129. PRESSURE TESTS. Remove plug (C—Fig. 155) from sensing port and install a suitable 0-400 psi pressure gage. With engine running at any speed above slow idle, sensing pressure should be 90-100 psi. Pressure should not drop below 80 psi at slow idle speed. If pressure is incorrect, remove port plug (E) and, working through plug opening, turn the exposed slotted head plug clockwise to raise the pressure or counter-clockwise to lower pressure. Adjust the pressure to 95 psi @ 1000 engine rpm.

Remove plug from low volume port (B) and install 0-400 psi pressure gage. Low volume system pressure should not be less than 210 psi @ 1000

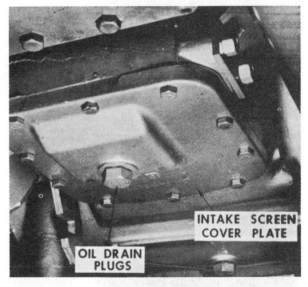

Fig. 153—Bottom view of rear axle center housing showing hydraulic system fluid plugs and intake screen cover.

INTAKE SCREEN COVER PLATE

OIL DRAIN PLUGS

rpm with gear pump or 225 psi @ 1000 rpm with rotor pump. With either pump, pressure should not be more than 275 psi @ 2000 engine rpm.

Check the combined system pressure by installing gage in high volume port plug (D). Pressures should be approximately the same as those given for low volume circuit.

Substantially higher pressures at low volume pressure port may indicate a sticking combining valve in pump body. If pressure is normal at 1000 rpm but higher than normal at 2000 rpm, check for plugged filter (models without by-pass) or restricted

Fig. 154—Left, rear corner of rear axle center housing with fitting (1) and flow meter return line (2) attached.

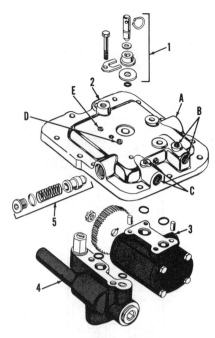

Fig. 155—View of front hydraulic cover showing main parts exploded.

1. Unloading valve
2. Cover
3. Dual charging pump
4. Regulating valve
5. Safety relief valve
A. Charging supply port
B. Low volume port & plug
C. Sensing control port & plug
D. High volume port plug
E. Adjusting port plug

2000 rpm
Gear pump ....22 gpm
Rotor pump ....20 gpm

Low pump output can be caused by clogged intake screen, air leaks or restrictions in intake tube, faulty manual unloading valve, faulty safety relief valve or faulty pump.

To check the safety relief valve (5—Fig. 155) leave flow meter connected and install a master gage in plug port (D). With engine running at 850 rpm and load valve on flow meter open, start closing load valve and note pressure at which a noticeable drop in flow occurs. Pressure should be 180 psi. Close flow meter load valve to block the flow. Pressure at plug port should be a minimum of 210 psi @ 1000 engine rpm and maximum of 350 psi @ 2000 rpm.

Low volume flow can be checked separately by connecting flow meter inlet to low volume outlet port (B). Adjust flow meter load valve to main-

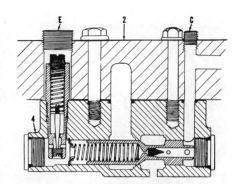

Fig. 156—Cross sectional view of hydraulic cover with regulating valve installed. Refer to Fig. 155 for parts identification.

tain a pressure of 175 psi and run engine at indicated speeds. Minimum flow should be as specified for gear (early) or rotor (late) pump.

1000 rpm
Gear pump ....2 gpm
Rotor pump ....3 gpm
2000 gpm
Gear pump ....5 gpm
Rotor pump ....6.5 gpm

lines. Low pressure at high rpm may indicate a sticking safety relief valve (5).

130. FLOW. To check the combined flow from charging pump, disconnect charging supply tube from elbow at port (A—Fig. 155) and connect flow meter inlet line to elbow. Direct flow meter outlet to sump (Fig. 154).

CAUTION: Operate tractor only long enough with charging supply tube disconnected to make the tests.

Make sure PTO valve lever is in "OFF" position, Multi-Power valve lever is in "Low" position, seat valve lever is in neutral and differential lock lever is in "Off" position.

Start and run engine at the indicated speeds, maintaining a pressure of 150 psi by adjusting the flow meter load valve. Minimum flow should be as specified for gear (early) or rotor (late) pump.

800 rpm (Idle)
All models ....8 gpm
1000 rpm
Gear pump ....11 gpm
Rotor pump ....10 gpm

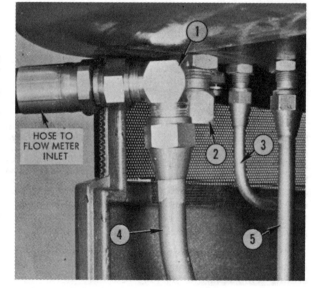

Fig. 157 — Recommended method of attaching flow meter inlet line for testing main pump pressure and flow. Flow meter return line is attached as shown in Fig. 154.

1. Elbow
2. Capped fitting
3. Power steering pressure tube
4. Main supply tube
5. Sensing line

HOSE TO FLOW METER INLET

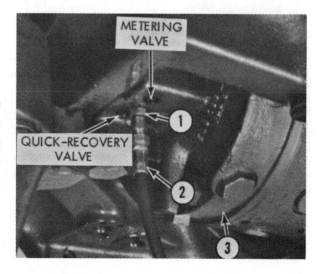

Fig. 158—Bottom view of main hydraulic pump (3) showing installation of fitting (1) and pressure gage tube (2) in control (intermediate) pressure port. Also shown are adjusting plugs for metering valve and quick recovery valve.

METERING VALVE

QUICK-RECOVERY VALVE

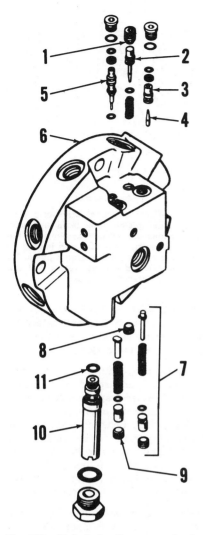

Fig. 159—Main hydraulic pump valve body showing valves exploded.

1. Adjusting plug
2. Dampening valve
3. Metering valve spool
4. Metering valve piston
5. Quick recovery valve
6. Valve body
7. Metering valve
8. Port plug
9. Adjusting plug
10. Relief plug
11. O-ring

at front end of valve housing. Install a 0-500 psi pressure gage to tee, using suitable tubing connections.

With fluid at operating temperature and engine running at 1000 rpm, slowly move pto control lever toward "On" position while watching gage pressure. Pressure should rise smoothly with movement of lever and hold at any selected pressure, until gage reading is 100 psi; then rise quickly to system pressure of 210-225 psi as lever is moved further toward "On".

If pto pressure cannot be properly controlled, remove valve as outlined in paragraph 158 and overhaul as in paragraph 159 or 160.

132. MAIN PUMP PRESSURE & FLOW. To check the main hydraulic pump pressure and flow, first remove front grille section. Disconnect main pump outlet tube (4—Fig. 157) at top end and loosen bottom connection. Install cap (2) to prevent fluid loss. Turn outlet tube (4) away from fitting and retighten lower connection. Attach flow meter inlet hose to tube (4) and connect flow meter outlet to sump as shown in Fig. 154. Remove the center plug from lower face of main pump front cover and attach a hose (2—Fig. 158) leading to 0-5000 psi pressure gage.

Start and run engine at 1000 rpm and adjust flow meter load valve to obtain a pressure of 1900 psi. Allow fluid temperature to reach 140°-160° F. test temperature.

133. RELIEF VALVE CHECK. With engine running and fluid at test temperature, adjust flow meter load valve until a flow of 3-5 gpm is obtained. Turn the metering valve adjustable plug (Fig. 158) clockwise until pres-

sure ceases to rise on flow meter pressure gage. Reading should be 2600-2800 psi.

CAUTION: On models with early pump, be careful not to bottom the screw, or metering valve needle may be damaged.

If maximum obtainable pressure is below 2600 psi, remove and renew the relief valve cartridge (10—Fig. 159) or O-ring (11). Relief valve cartridge (10) may be accurately checked after removal, using a hand test pump. Pressure must be within the range of 2600-2800 psi and cartridge must not leak at 2000 psi.

134. QUICK RECOVERY VALVE. After checking relief valve as outlined in paragraph 133, back out metering valve adjusting plug (Fig. 158) until flow meter pressure gage registers 1400 psi with engine running at 1000 rpm and flow adjusted to 3 gpm.

Flow meter pressure gage should be unstable, if it is not, quickly close the flow meter load valve then open valve to permit 0.5 gpm flow. If pressure holds steady, turn quick recovery valve adjusting plug clockwise until pressure fluctuates. Increase flow to 3 gpm, pressure should remain unstable. Pressure should stabilize at 1600 psi and be unstable at 1400 psi. Turning quick recovery plug clockwise decreases stability, counterclockwise increases stability.

135. DAMPENING VALVE. With quick recovery valve adjusted as outlined in paragraph 134, engine running at 1000 rpm and flow adjusted to 3 gpm, turn metering valve adjusting screw to obtain a pressure of 2050 psi on flow meter pressure gage, which is the correct setting.

PTO valve lever must be in "Off" position, Multi-Power valve lever in "Low" position, seat valve lever in neutral and differential lock valve lever in "Off" position when making the test. Measured flow should not drop more than 1.2 gpm when Multi-Power valve lever is moved to "High" position, nor more than 0.5 gpm when either the PTO valve lever or Differential Lock valve lever is moved to "On" position. A greater drop would indicate excessive leakage in the respective clutches.

131. PTO MODULATING VALVE. To check the pto modulating valve, remove left instrument panel cover and the cap plug from pto outlet tee

Fig. 160 — Flow meter connected in series with main pump supply line to check condition of system.

1. Elbow
2. Power steering tube
3. Elbow
4. Main supply tube
5. Sensing line

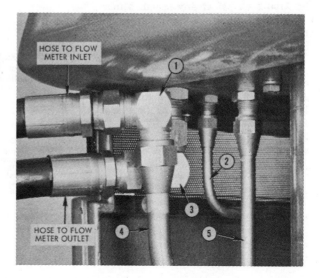

HOSE TO FLOW METER INLET

HOSE TO FLOW METER OUTLET

"DIVERTER" HOSE – PART OF MFN-1100 "HYDRA-ANALYZER" KIT

Fig. 161 — Diverter hose installed for checking priority valve. Refer to paragraph 137.

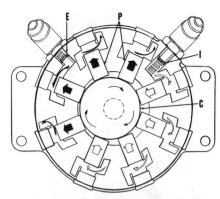

Fig. 162—Cross sectional schematic view of the radial piston type main hydraulic pump. Pistons are pushed outward by the rotating eccentric cam (C) and inward by hydraulic pressure from charging pump. Inlet valves (I) and exhaust valves (E) are interchangeable.

Check the intermediate pump pressure registered on gage installed in pump cover (2—Fig. 158). Pressure should be 375-425 psi. On earlier pumps, dampening valve (2—Fig. 159) may be maladjusted if pressure is incorrect. On late models, dampening valve may be staked, blocked or eliminated, and a fixed orifice drilled in distributor valve piston.

NOTE: If proper adjustment cannot be obtained of dampening valve, quick recovery valve, or standby pressure; if pump is unstable and cannot be corrected; or, if main relief valve seems to be damaged due to overheating, refer to paragraph 145 for possible causes.

136. **OUTPUT.** Increase engine speed to 2000 rpm and adjust flow meter load valve to obtain an output pressure of 1900 psi on flow meter pressure gage. Main pump output flow should be 16 gpm.

Slowly close the flow meter load valve and note the gage pressure when flow stops. Shut-down pressure should be 2100-2400 psi at any speed. Minor adjustments of shutdown pressure can be made by turning metering valve adjusting plug.

Slow engine speed to 1000 rpm and open flow meter load valve to obtain a flow of 5 gpm; then quickly close load valve while observing pressure. Pressure should momentarily rise to 3200-3400 psi, then quickly drop to shutdown pressure and hold steady.

137. **HIGH PRESSURE SYSTEM.** To check the high pressure system, shut off the engine, disconnect flow meter outlet line from fitting in sump and connect to high pressure outlet tube as shown in **Fig. 160.**

Open the load valve on flow meter and start and run engine with all hydraulic units inactive. The pressure registered on flow meter indicates system leakage. Pressure should be 2100-2400 psi and flow should be 0-½ gpm.

138. **PRIORITY VALVE.** Connect both breakaway couplings of one valve bank with a hose as shown in Fig. 161 and start the engine with valve in neutral. With engine at slow idle speed, move valve lever to operating position while watching flow meter pressure gage. Pressure should drop from standby pressure and hold steady at about 1300 psi.

A low or substantially higher reading indicates a malfunctioning priority valve. The valve may be disassembled and cleaned as outlined in paragraph 154, but parts are not available for overhaul of the unit.

139. **POWER STEERING.** Minor malfunction which does not materially affect steering operation will have no effect on hydraulic system.

As a quick hydraulic test, start and run engine at any speed. Turn steering wheel either way against steering stop. Continue to apply pressure at steering wheel. Leakage may be considered excessive if steering wheel can be turned one revolution in less than 30 seconds or if pump flow increases more than 0.25 gpm over reading with steering wheel released.

Refer to paragraphs 3 through 9 for service on steering components.

140. **BRAKE VALVE.** After bleeding brakes as outlined in paragraph 115 and adjusting linkage as in paragraph 116, check for leakage and valve malfunction as follows:

Fig. 163 — Right hand view of tractor showing main hydraulic pump.

1. Lower radiator hose
2. Pump drain line
3. Pump inlet line
4. Radiator drain
5. Drive coupler bolt
6. Pump pressure line

Fig. 164—Left hand view of tractor showing main hydraulic pump partially prepared for removal.

1. Mounting bolt
2. Inlet fitting
3. Drain line
4. Drive coupler flange

Fig. 165—Using the special tool for removing main hydraulic pump. Refer also to Fig. 166. Adjusting screw (1) is used to align pump for installation, and locking bar (2) is used to hold pump when removing mounting bolts.

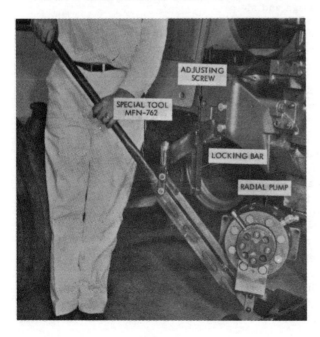

Fig. 166 — The special tool is designed to pivot pump from tractor as shown.

Firmly depress brake pedals one at a time with engine running at 1000 rpm, while watching flow meter gage. Flow should not increase more than 0.25 gpm when either pedal is depressed. If leakage is excessive, isolate the trouble by disconnecting the tube leading from valve to center housing and capping the connection. If leak continues, valve is at fault. If leak stops, brake unit is at fault. Refer to paragraph 117 for service on brake pistons, or paragraph 157 for service on brake valve.

Brake valve operation is based on a principal of proportional pressure, with 30 psi output pressure being transmitted for each 1 lb. pressure applied to brake pedal. Pressure should increase smoothly with increased pedal pressure until standby pressure of hydraulic system is reached.

The brake valve is equipped with equalizer piston assemblies which should open and equalize the output pressure when both brake pedals are applied at an unequal rate. Test the equalizer action by installing a pressure gage in each brake line. With engine running, depress either pedal until about 1000 psi is obtained on that brake line. While holding that pedal steady slowly depress the other pedal while watching pressure gages. Pressure should rise smoothly as pedal is depressed until about 400 psi is developed; then equalizer piston should open and both gages show equal pressure. Repeat the test, depressing pedals in reverse order. An equalizing point significantly higher or lower than 400 psi could incidate a malfunctioning equalizing valve.

With engine not running, either or both pedals must develop and hold pressure on third pedal stroke.

MAIN PUMP

All Models

141. **OPERATION.** The radial piston type main hydraulic pump mounts on front support casting and is driven by engine crankshaft.

Fluid from the charging pump is provided at main pump inlet at moderate pressure. Unless flow is blocked by movement of the distributing valve, fluid passes through the inlet valves and pushes the radial pump pistons (P—Fig. 162) down into contact with pump camshaft (C). When demands for pressurized fluid are met, the distributing valve closes inlet valves off from charging pressure and pump pistons fail to follow cam. Pumping action thus stops until a drop in standby pressure signals a demand for additional fluid. The metering action of

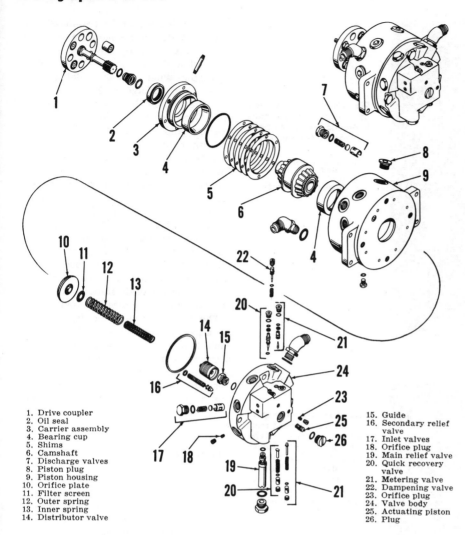

1. Drive coupler
2. Oil seal
3. Carrier assembly
4. Bearing cup
5. Shims
6. Camshaft
7. Discharge valves
8. Piston plug
9. Piston housing
10. Orifice plate
11. Filter screen
12. Outer spring
13. Inner spring
14. Distributor valve
15. Guide
16. Secondary relief valve
17. Inlet valves
18. Orifice plug
19. Main relief valve
20. Quick recovery valve
21. Metering valve
22. Dampening valve
23. Orifice plug
24. Valve body
25. Actuating piston
26. Plug

Fig. 167—Exploded view of crankshaft driven radial piston pump.

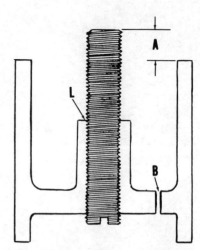

Fig. 168—Cross sectional view of distributing valve piston. Refer to paragraph 144 for details.

A. Adjustment
B. Orifice
L. Locknut location

the distributing valve determines volume delivered, up to the maximum capacity of the system.

Pressurized fluid enters the pump cover at a charging pressure of up to 100 psi. The distributing valve piston is spring loaded in the open position. Orifices lead from the charging pressure passages into bearing and camshaft area of pump and provide lubrication. Fluid from the main pump high pressure gallery passes through a metering valve to the area behind the distributing valve piston, then through fixed orifices to the charging passage, to balance the high pressure system requirements.

The control cover also contains a "Quick Recovery" valve which is spring loaded to the open position and closed by hydraulic pressure. The quick recovery valve connects the intermediate (control) pressure passage to charging passage. A sudden drop in standby pressure causes the quick re-

covery valve to open, dumping pressure from the control passage and allowing the distributing valve to completely open. A full supply of high pressure fluid is thus provided to meet systems demands.

142. REMOVE AND REINSTALL. To remove the main hydraulic pump, first drain radiator and remove lower radiator hose.

On axle models, suitably support front of tractor, unbolt axle pivot brackets and move front axle forward to provide rom for pump removal. On tricycle models, remove left front wheel and turn front spindle fully to left to provide clearance. On all models, disconnect inlet, outlet and lubrication lines. Unscrew the cap screws (5—Fig. 163) retaining drive coupler to crankshaft pulley and slide the coupler forward against pump body.

NOTE: The main hydraulic pump weighs more than 100 pounds and means must be provided for lifting the pump down out of front support. The manufacturer provides a tool for lifting out the pump as shown in Figs. 165 and 166.

Attach the tool or other suitable tool to pump body and remove the mounting bolts, then swing the pump away from tractor frame.

Install the pump by reversing the removal procedure, making sure that drive coupler properly aligns with crankshaft pulley before tightening the pump mounting bolts securely.

143. OVERHAUL. With pump removed as outlined in paragraph 142, refer to Fig. 167 and proceed as follows:

Carefully withdraw pump coupling (1). Remove two of the ⅝-11 cap screws securing valve body (24) to piston housing (9) and install 5 inch aligning studs. Loosen and remove the other two cap screws evenly, then carefully remove the cover and distributing valve springs (12 & 13).

144. DISTRIBUTING VALVE. The distributing valve piston (14—Fig. 167) can be removed from valve body bore after removing valve body. Three different distributing valves have been used. The latest valve contains a built-in dampening orifice (B—Fig. 168) in web of valve. This valve must

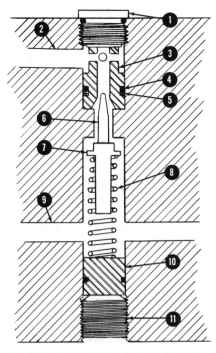

Fig. 169—Cross sectional view of late type metering valve showing method of assembly.

1. Upper plug
2. High pressure passage
3. Valve sleeve
4. O-ring
5. Backup ring
6. Valve needle
7. Valve guide
8. Valve spring
9. Control passage
10. Spacer
11. Adjusting plug

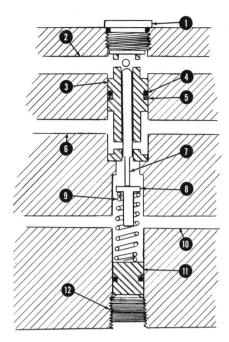

Fig. 170—Cross sectional view of quick recovery valve, which is spring loaded in the open position. When pressure in passage (2) drops below standby pressure, spring (9) moves valve (7) up, connecting the moderate pressure "Control" passage (6) and inlet passage (10) through the "Step" in valve (7). The rapid drop in control pressure allows the distributing valve to fully open, permitting the radial pump to operate at full capacity until pressure balance is restored.

1. Plug
2. High pressure passage
3. Valve sleeve
4. O-ring
5. Backup ring
6. Control passage
7. Valve plunger
8. Valve guide
9. Spring
10. Low pressure passage
11. Spacer
12. Adjusting plug

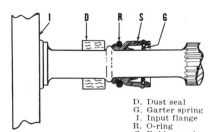

Fig. 171—Cross sectional view of pump coupler showing proper installation of shaft seal. The felt dust seal (D) is cemented in place as outlined in paragraph 146.

D. Dust seal
G. Garter spring
I. Input flange
R. O-ring
S. Rubber seal

pumps by readjustment of the metering valve and dampening valve.

Several attempts may be required on some early models to obtain the proper balance. On early models where trouble exists, it is often less expensive to install the latest type, pre-adjusted valve body and dampening valve assembly from parts stock.

NOTE: The main relief valve can be damaged by an improperly adjusted distributor valve stop screw. It is recommended that relief valve be removed and checked as outlined in paragraph 133, whenever trouble has been encountered with stop setting.

145. VALVE BODY. With valve body off as in paragraph 142 and distributing valve removed as in paragraph 144, disassemble as follows:

Remove the snap ring from secondary relief valve (16—Fig. 167) and fish out the valve if necessary, using a small wire. Unscrew plug and remove main relief valve (19). Remove quick recovery valve (20) and metering valve (21), working from both ends. Remove plug (26) and actuating piston (25).

only be used with valve body (24—Fig. 167) containing the "Vee-Slotted" distributor valve sleeve, and adjustable dampening valve (22) MUST be blocked or eliminated. When installing a distributor valve containing the fixed orifice in place of one without the orifice, completely turn down and stake the dampening valve (22).

On late models with dampening orifice, distance (A—Fig. 168) is factory adjusted and locked by installing a jam nut at (L). Adjustment should not be changed nor stop screw removed. On other models, install a suitable jam nut and adjust distance (A) to 0.150-0.155, only if pumping action is unstable or standby pressure cannot be properly adjusted.

The adjustment (A) of stop screw plays an important part in the operation of the pump. When stop screw contacts orifice plate (10—Fig. 167), the amount of fluid metered to main pump pistons must balance the flow through metering valve (21). If distance (A—Fig. 168) is too great, flow to the pistons will exceed the amount that can be bypassed through the metering valve and standby pressure will be too high. If distance (A) is too

small, flow to pistons can be completely shut off causing instability of the standby pressure, especially at slower engine speeds. One third ($\frac{1}{3}$) turn of stop screw will change standby pressure about 350 psi. This can be partially compensated for on earlier

Fig. 172 — Top view of front hydraulic cover with platform removed.

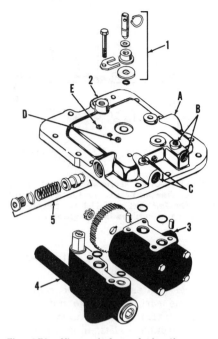

Fig. 173—View of front hydraulic cover showing main parts exploded.

1. Unloading valve	B. Low volume port
2. Cover	& plug
3. Dual charging	C. Sensing control port
pump	& plug
4. Regulating valve	D. High volume port
5. Safety relief valve	plug
A. Charging supply	E. Adjusting port plug
port	

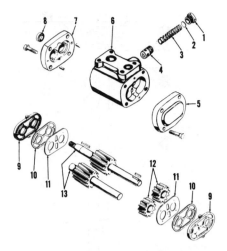

Fig. 174—Exploded view of gear type dual charging pump showing component parts.

1. Plug	8. Shaft seal
2. Shim	9. Diaphragm seal
3. Valve spring	10. Gasket
4. Combining valve	11. Diaphragm
5. End cover	12. Low volume gears
6. Pump body	13. High volume gears
7. End cover	

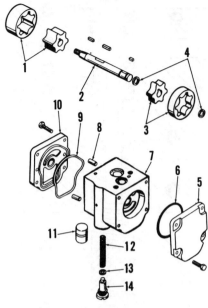

Fig. 175—Exploded view of rotor type dual charging pump.

1. High volume gears	8. Dowel
2. Drive shaft	9. Seal
3. Low volume gears	10. End cover
4. Snap rings	11. Combining valve
5. End cover	12. Valve spring
6. O-ring	13. Shim
7. Pump body	14. Plug

Remove the eight inlet valves (17), being careful the valve discs do not fall into inlet manifold and become lodged. Check the valve seats in body for damage.

The two orifice plugs (18 and 23) will not need to be removed unless their condition is questionable. The plugs are interchangeable. Piston guide (15) is installed with LOCTITE and will not need to be removed unless O-ring or guide are to be renewed or trouble is suspected. Clean and inspect all parts and renew any which are damaged or questionable.

Two types of metering valve have been used, with only the later type being available for service. The parts are available only in a kit which includes the sleeve, valve, guide and spring. When assembling the new type valve, refer to Fig. 169 and proceed as follows:

Working from bottom side of valve body, install valve guide (7), spring (8), spacer plug (10) and adjusting plug (11) in the order given. Make sure the O-ring is installed on spacer plug (10). Install O-ring (4) and backup ring (5) as shown on valve sleeve (3), and install sleeve in top of bore; then drop valve needle (6) into sleeve (3), chamfered end up. Install and tighten plug (1).

When installing the quick recovery valve, refer to Fig. 170. Install valve guide (8), spring (9), spacer plug (11) and adjusting plug (12) in bottom side of body. Install O-ring (4) and backup ring (5) as shown on valve sleeve (3), then install sleeve with smaller OD to top. Drop in the piston (7) with small end down. Install and tighten plug (1).

If valve guide (15—Fig. 167) has been removed, install using Type A (Red) LOCTITE and tighten to a torque of 60-90 inch pounds. Install actuating piston (25), then install and tighten plug (26) to a torque of 25-30 ft.-lbs.

The eight intake valves, guides and springs (17) are interchangeable with each other and with outlet valves (7). The individual parts can be installed in valve body either side out. Tighten the plugs to a torque of 45 ft.-lbs.

146. PISTON HOUSING. To disassemble the piston housing (9—Fig. 167), first remove the eight chamber plugs (8) and pump pistons, keeping pistons in their proper order so they can be installed in their mated bores. Remove the cap screws securing carrier (3) to pump body and remove the carrier by carefully prying it straight out. Remove and save shim pack (5). Shims control preload of the tapered camshaft bearings.

With pistons and carrier removed, shaft and bearing unit can be lifted out. Camshaft and bearings are available only as an assembly, renew the unit if any part is damaged.

Remove plugs and outlet valves (7), being careful that valves do not fall into discharge manifold of pump. Examine valve seats in piston housing for damage.

Pump pistons should have a clearance of 0.0003-0.0015 in their bores. Pistons and pump housing are factory matched and not available separately.

When reassembling the pump, first install discharge valves (7). Parts are interchangeable and can be turned either side out. Tighten plugs to a torque of 45 ft.-lbs.

Place an additional 0.007 shim in shim pack (5) and reinstall camshaft (6) and carrier (3); leaving out the seal (2) and sealing O-ring. Install and tighten three alternate carrier cap screws; then using a dial indicator, measure camshaft end play. Remove the carrier, then remove shims equal to the measured end play plus 0.000-0.003 to establish the desired preload. Shims (5) are available in thicknesses of 0.003, 0.005, 0.006, 0.007 and 0.020. Install O-ring on carrier (3) and seal (2), then reinstall the carrier, tightening retaining cap screws to a torque of 30-35 ft.-lbs.

Install pump pistons in bores from which removed, with flat side next to camshaft. Install and tighten plugs to a torque of 140-150 ft.-lbs.

Install valve body on pump housing, making sure pump inlet and out-

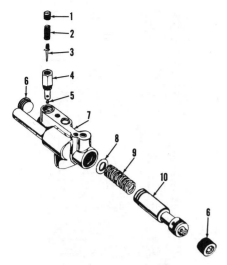

Fig. 176 — Exploded view of regulating valve assembly. Component parts are not serviced.

1. Adjusting plug   6. End plug
2. Spring   7. Valve housing
3. Pilot valve   8. Spring seat
4. Valve body   9. Spring
5. Valve stop   10. Valve spool

let ports are approximately 90 degrees apart and pointing up as shown in assembled view of Fig. 167. Tighten the four retaining cap screws to a torque of 155-170 ft.-lbs.

The hydraulic pump coupling seal has been changed by the addition of a felt dust seal (D—Fig. 171). Whenever pump is disassembled, renew the seal as follows: Clean the shaft and cement the felt seal (D) in place as shown. using Scotch Grip Industrial Adhesive No. 847 or equivalent. Position boot seal (S) and secure to shaft with O-ring (R), then install garter spring spreader (G). Lubricate outer portions of felt seal and boot seal, then reinstall shaft in pump. Reinstall pump as outlined in paragraph 142.

CHARGING PUMP

All Models

147. REMOVE AND REINSTALL. The hydraulic system charging pump is mounted on front hydraulic cover as shown in Fig. 173. To remove the cover, first remove the ring from manual unloading valve, then remove operator's platform center panel and padding.

Remove the pin securing seat cylinder to bracket mounted on cover and move the cylinder out of the way. Disconnect hydraulic tube assemblies and remove flange bolts. Pry the cover from locating dowels, then lift off cover with charging pump and regulating valve attached.

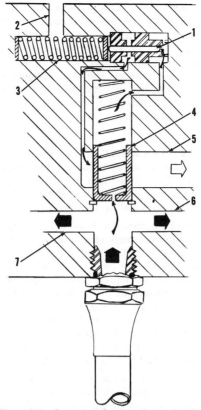

Fig. 177—Cross sectional schematic view of priority valve.

1. Pilot piston   5. Hydraulic lift port
2. Bleed port   6. Brake pressure port
3. Spring   7. Steering pressure
4. Priority piston   port

Remove the two cap screws securing pump to underneath side of top cover and lift off the pump. Renew O-rings and gaskets, and install pump and cover by reversing the removal procedure.

148. OVERHAUL GEAR PUMP. An exploded view of the gear-type dual charging pump is shown in Fig. 174. To disassemble the pump, remove pump drive gear and the two end plates (5 & 7). Remove low volume pump gears (12) and the drive gear key, then withdraw high volume gears and shafts as a unit. Remove plug (1) and shims (2), then withdraw spring (3) and combining valve (4).

When assembling the pump, work the diaphragm seal (9) into cover (5 or 7), open side first. Gasket (10) and diaphragm (11) must fit into molded portion of seal. Bronze side of diaphragm (11) must be toward the gears. End plates (5 & 7) must be installed with shaft relief holes away from output (mounting) side of pump body.

Tighten the end plate retaining cap screws to a torque of 27-30 ft.-lbs. and plug (1) to a torque of 30-35 ft.-lbs.

149. OVERHAUL ROTOR PUMP. Refer to Fig. 175 for an exploded view of rotor type pump. To disassemble the removed pump unit, first remove pump drive gear and key, then remove rear end plate (10). Remove high volume rotor assemblies (1) and inner rotor drive key. Remove front end plate (5), then withdraw shaft (2) and rotor assembly (3) from front of pump. Remove plug (14), shim (13), spring (12) and combining valve (11).

Assemble the pump by reversing the disassembly procedure.

REGULATING VALVE

All Models

150. REMOVE AND REINSTALL. The regulating valve assembly is mounted on front hydraulic cover as shown at (4—Fig. 173). Remove front cover as outlined in paragraph 147. Remove the attaching cap screws and lift off the regulating valve.

Reinstall by reversing the disassembly procedure. After tractor is reassembled, adjust sensing pressure as outlined in paragraph 129.

151. OVERHAUL. The regulating valve is available only as an assembly and individual parts are not renewed. The regulating valve may, however, be disassembled for inspection and cleaning. Refer to Fig. 176 for exploded view.

Spool (10) must slide freely in its bore. Center drilling of spool contains a 0.014 metering orifice and most units contain a screen filter ahead of orifice.

Pilot valve (3) is prevented from seating by valve stop (5). If sensing pressure cannot be properly adjusted and no other trouble is encountered. check the setting of pilot valve stop as follows: Clamp pilot valve body (4) lightly in a vise, right side up with both ends accessible. Note the setting of adjusting plug (1) and remove the plug and spring (2). Mount a dial indicator with button bearing on top end of needle valve (3). Back out the threaded stop (5) until dial indicator needle ceases to move; indicator reading should be 0.004-0.007. Adjust to those limits by turning stop (5) until needle is raised the indicated amount. When setting is correct, the nylock insert in threads of stop (5) should not be visible; if it is,

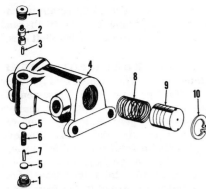

Fig. 178—Exploded view of priority valve showing component parts. Valve is serviced only as an assembly.

1. Plug
2. Valve sleeve
3. Pilot piston
4. Valve housing
5. Spring disc
6. Pilot spring
7. Pilot pin
8. Spring
9. Priority piston
10. Snap ring

grind a small amount off bottom end of valve needle (3) and readjust.

Assemble the regulating valve by reversing the disassembly procedure, using Fig. 176 as a guide.

PRIORITY VALVE

All Models

152. **OPERATION.** The priority valve is shown schematically in Fig. 177 and exploded in Fig. 178. Pressure supply ports (6 & 7—Fig. 177) leading to brake valve and power steering are located ahead of priority piston (4), while piston must unseat to supply fluid flow to hydraulic lift system port (5). When there is no flow, pressures are equal throughout the valve.

One end of the pilot valve piston (1) is open to high pressure while the other end is open to atmospheric pressure of bleed port (2). Before priority piston (4) can move to open lift system port (5), two things must happen in the priority valve. System pressure must overcome the pressure of spring (3), allowing pilot piston to move down and open the pilot passage to lift port (5), and lift port pressure must drop below system pressure. In normal operation, pilot flow (small arrows) through piston orifice and pilot piston, unbalances the system to allow pilot piston to open whenever movement of one of the control valves signals a fluid demand.

153. **REMOVE AND REINSTALL.** To remove the priority valve, first remove the hood, batteries, battery

platform and voltage regulator. Remove right and left side covers from steering column support. Disconnect input line, brake line, steering line, bleed line and output line from valve housing. Remove the two mounting bolts and lift off the priority valve assembly. Install by reversing the removal procedure.

154. **OVERHAUL.** The priority valve is available as an assembly only. To disassemble the valve for cleaning or inspection, refer to Fig. 178 and proceed as follows:

Working through inlet port, unseat and remove snap ring (10), priority piston (9) and spring (8). Remove lower plug (1), spring discs (5), spring (6) and pilot pin (7). NOTE: Some valves may not have upper plug (1) and renewable sleeve (2). On these models, jar valve body (4) downward on a bench or wood block to dislodge the piston (3). If piston is stuck, valve must be renewed.

On models with upper plug, remove the plug (1) and sleeve (2). Piston (3) may be pushed from sleeve if necessary.

Assemble the valve by reversing the disassembly procedure.

BRAKE VALVE

All Models

155. **OPERATION.** A schematic view of brake valve is shown in Fig. 179. Except for inlet pressure fitting (1) and inlet check valve (2), the parts are duplicated for right and left brakes. Initial movement of brake pedal depresses power piston (5) and closes off escape passage (4). Further movement of the pedal opens the modulating needle valve (3) allowing fluid pressure to enter through inlet check valve (2) and pass through connecting passage (8) to brake passage fitting (12). At the same time, fluid pressure reacts on the bottom side of power piston (5) and equalizing piston (9).

When pedal pressure is exceeded by pressure from the fluid, the pedal moves upward closing the modulating valve and opening escape passage (4). The upward movement of equalizing valve (9) in response to actuating pressure opens the connecting passage (10) which leads to the opposite equalizing valve. If both equalizing valves are opened simultaneously by depressing both pedals, actuating

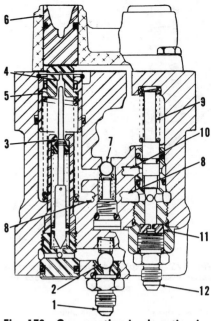

Fig. 179—Cross sectional schematic view of brake valve. Except for inlet pressure fitting (1) and check valve (2), parts are duplicated for right and left sides.

1. Inlet fitting
2. Check valve
3. Modulating valve
4. Escape passage
5. Power piston
6. Plunger
7. Manual check ball
8. Actuating passage
9. Equalizing piston
10. Equalizing passage
11. Orifice plate
12. Outlet fitting

pressures are equalized through connecting passage (10).

One pound of pressure at brake pedal will generate 30 psi of actuating pressure in the power braking system.

In manual brake operation, a greater pedal travel depresses the power piston (5) and manual piston (33—Fig. 180), and trapped fluid is pumped through outlet port (12—Fig. 179) to actuating pistons. The reverse flow restrictors (11) slow the return flow of fluid allowing piston area to refill through reservoir check valve when pedal is released. Three strokes of pedal should build up and hold pressure for manual operation.

156. **REMOVE AND REINSTALL.** To remove the brake valve, first remove cover panel and right foot rest. Disconnect lines from brake valve and remove snap ring from pedal pivot shaft. Remove the mounting cap screws securing valve to clutch housing slide the pedals off pivot shaft while piloting valve housing from sump return line.

Install by reversing the removal procedure. Bleed the brakes as outlined in paragraph 115 and adjust pedal linkage as in paragraph 116.

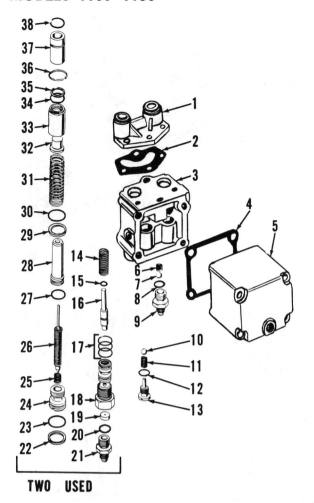

TWO USED

Fig. 180—Exploded view of brake valve. Refer also to Fig. 179.

1. Cap
2. Gasket
3. Valve body
4. Gasket
5. Reservoir
6. Spring
7. Check ball
8. O-ring
9. Inlet pressure fitting
10. Check ball
11. Spring
12. O-ring
13. Plug
14. Spring
15. O-ring
16. Equalizing piston
17. O-rings
18. Valve spool
19. Orifice plate
20. O-ring
21. Outlet fitting
22. Backup washer
23. O-ring
24. Snubber plug
25. Spring
26. Snubber needle
27. O-ring
28. Cartridge
29. Backup washer
30. O-ring
31. Spring
32. Power piston
33. Manual piston
34. O-ring
35. Backup washer
36. Snap ring
37. Plunger
38. O-ring

Gaskets and O-rings are available in a seal kit which also includes snap rings (36). Renew any other parts which are questionable, and reassemble using Figs. 179 and 180 as a guide.

MULTI-POWER & PTO CONTROL VALVE

The Multi-Power and PTO control valves are combined in a common housing. Valves are supplied by Cessna or Wooster and all valves are closed center type. Refer to Figs. 181 through 185 for exploded views.

All Models

158. **REMOVE AND REINSTALL.** Remove left instrument panel cover and disconnect all lines leading to valve. Disconnect lever yokes from valve spools. Remove the three bolts securing valve housing to instrument panel support and lift off the valve assembly.

When installing the valve, make sure lever links are adjusted to permit full movement of valve spools to both detent positions and that lines are reconnected in proper order.

159. **OVERHAUL CESSNA VALVE.** To remove Multi-Power valve spool (9—Fig. 181 or 182), first remove detent plug, shim (3), spring (1) and ball (2). Remove snap ring from each end of spool (9) and pull spool rearward until O-ring in front end of valve body (4) is exposed. Extract the O-ring, then push spool forward out of body bore. Install by reversing the removal procedure, moving spool only far enough in either direction to install the O-ring.

To remove the pto valve spool (5 through 8—Fig. 181 or 183), push valve forward into "ON" (detent) po-

157. **OVERHAUL.** Refer to Fig. 180 for an exploded view of brake valve. To disassemble the removed valve, first unbolt and remove reservoir cover (5) and gasket (4). Remove valve cap (1) and gasket (2).

Using a C-clamp, press, or other suitable tool, depress manual piston (33) and remove snap ring (36). Slowly release the pressure and re-

move manual piston (33), power piston (32), spring (31) and associated parts.

Remove snubber plug (24), snubber needle (26), cartridge (28) and associated parts working from bottom of housing. Unscrew and remove equalizer valve spool (18), valve (16) and spring (14) as a unit.

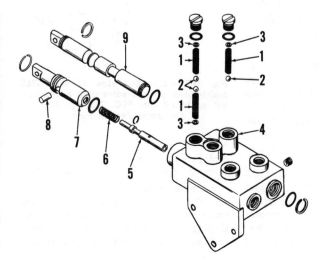

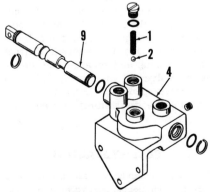

Fig. 181—Exploded view of CESSNA Multi-Power and PTO control valve showing component parts.

1. Detent spring
2. Detent ball
3. Shim
4. Valve body
5. PTO modulating valve
6. Spring
7. PTO valve spool
8. Nylon plug
9. Multi-Power valve spool

Fig. 182—Exploded view of CESSNA Multi-Power control valve used on models not equipped with pto. Refer to Fig. 181 for parts identification.

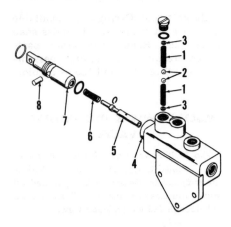

Fig. 183—Exploded view of CESSNA PTO control valve used on models not equipped with Multi-Power. Refer to Fig. 181 for parts identification.

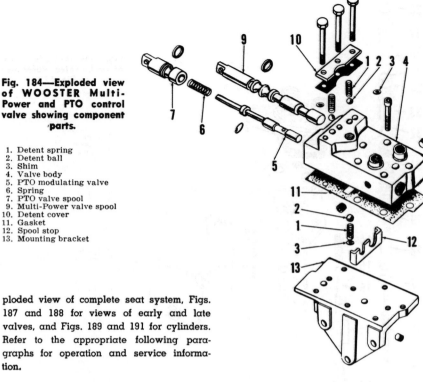

Fig. 184—Exploded view of WOOSTER Multi-Power and PTO control valve showing component parts.

1. Detent spring
2. Detent ball
3. Shim
4. Valve body
5. PTO modulating valve
6. Spring
7. PTO valve spool
9. Multi-Power valve spool
10. Detent cover
11. Gasket
12. Spool stop
13. Mounting bracket

sition and extract retaining snap ring. Remove detent plug and upper shim (3), spring (1) and ball (2). Turn valve spool (7) 90° in either direction until lever eye is aligned with detent plug bore, and allow lower detent ball (2) and spring (1) to push nylon plug (8) out of spool cross drilling. Extract lower detent ball and spring through cross drilling and out detent plug opening. The pto valve spool can now be withdrawn. Remove the snap ring and withdraw modulating valve (5) and spring (6) from valve spool. Install by reversing the removal procedure.

160. OVERHAUL WOOSTER VALVE. To disassemble the removed Wooster valve, refer to Figs. 184 or 185. First remove the three hex head cap screws, detent cover (10) and the two upper detent springs (1), balls (2) and shims (3).

Remove the six socket head cap screws and lift valve body (4) off of bracket (13). The pto valve lower detent ball, spring and shim will be free to fall out when parts are separated, as will spool stop (12). Do not lose any of the parts. The valve spools can be withdrawn from their bores after stop (12) is removed. Extract the snap ring and separate the modulating valve (5) and spring (6) from pto valve spool (7). Assemble by reversing the disassembly procedure.

SEAT HYDRAULICS

The seat hydraulic system consists of a closed center, poppet type valve, a hydraulic positioning cylinder and an air-filled accumulator cylinder teed to the hydraulic cylinder line. Refer to Fig. 186 for an ex-

ploded view of complete seat system, Figs. 187 and 188 for views of early and late valves, and Figs. 189 and 191 for cylinders. Refer to the appropriate following paragraphs for operation and service information.

All Models

161. OPERATION. The seat valve is connected to the hydraulic system Low Pressure/Low Volume circuit. Seat is raised or lowered by directing fluid to the single acting cylinder (11—Fig. 186), and air cushioning obtained by transfer of fluid between the positioned cylinder and accumulator (10). Fluid leaking by the cylinder piston is returned to reservoir through vent tube (12).

Settling of the seat where no fluid loss is visible would indicate a leaking cylinder piston or leaking return sealing washer in control valve. An increase in seat height would indicate a leaking inlet sealing washer in valve. Loss of cushioning action would indicate failure of accumulator (10).

Remove and overhaul the system components as outlined in the appropriate following paragraphs.

162. CONTROL VALVE. Refer to Fig. 187 for an exploded view of early type seat valve and Fig. 188 for view of late valve. Valves are similar in appearance but fluid lines are reversed, the front elbow fitting being the pressure fitting in early valve and return fitting in late valve.

To remove the valve from all models, first remove knobs from seat valve, differential lock valve spool and/or pto shift lever, then unbolt and remove dress cover panel. Disconnect hydraulic fitting connections

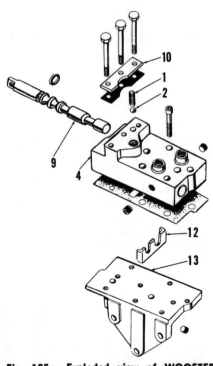

Fig. 185 — Exploded view of WOOSTER Multi-Power control valve used on tractors not equipped with PTO. Refer to Fig. 184 for parts identification.

and remove the valve through-bolts, then lift off the valve assembly.

Remove wrap-around snap ring (7—Fig. 187 or 188) and withdraw valve cam (11). Remove pressure fitting (1) and return fitting (8) or plug (14) and withdraw valve springs (3),

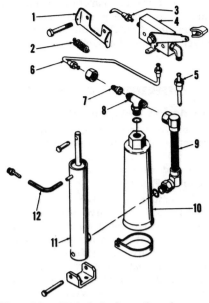

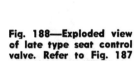

Fig. 186 — Exploded view of main component parts of seat hydraulic system. Late type is shown.

1. Bracket	7. Adapter
2. Centering spring	8. Tee
3. Inlet pressure fitting	9. Hose
4. Control valve	10. Accumulator
5. Return fitting	11. Cylinder
6. Control line	12. Vent tube

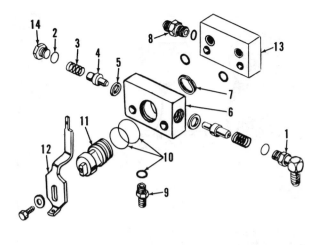

Fig. 187—Exploded view of early type seat control valve showing component parts.

1. Inlet pressure fitting
2. O-ring
3. Spring
4. Poppet
5. Sealing washer
6. Valve body
7. Snap ring
8. Return fitting
9. Seat cylinder fitting
10. O-rings
11. Valve cam
12. Lever
13. Adapter

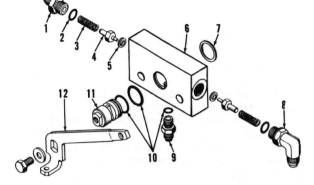

Fig. 188—Exploded view of late type seat control valve. Refer to Fig. 187 for parts identification.

poppet valves (4) and sealing washers (5). Renew sealing washers and O-rings, and any other parts which are damaged or questionable. Reassemble by reversing the disassembly procedure.

163. ACCUMULATOR. A bladder type accumulator is used on all models. Refer to (10—Fig. 186). The accumulator bag is air-filled at 80-85 psi and the compressibility of the air cushion serves as the seat spring. The line and fitting resistance to the transfer of trapped fluid between seat cylinder and accumulator serves as a damper and shock absorber for seat.

The accumulator is sealed and cannot be repaired or re-charged. If the accumulator bladder fails or loses air, the accumulator canister will fill with hydraulic fluid as seat is repositioned and the springiness of the air-ride seat will be lost. Correct the condition by renewing the accumulator assembly.

164. SEAT CYLINDER (CESSNA). The seat cylinder can be removed after removing the lower front panel from seat console. Two types of seat cylinder have been used. The CESSNA cylinder can be identified by welded port end construction of cylinder barrel and by the wrap-around snap ring at rod end seating in bearing rather than cylinder barrel.

To disassemble the removed cylinder, refer to Fig. 190. Remove outer snap ring (7) and push bearing (11) into barrel until inner snap ring (8) is exposed. Remove inner snap ring, then withdraw piston rod, piston and bearing as an assembly.

NOTE: Four different cylinder bearings have been used as shown in Fig. 189, only the last of which is serviced. If the removed bearing contains one of the O-rings shown in cross section, a new bearing and late seal kit must be obtained.

If piston is removed from piston rod (6), tighten nut (2) to a torque of 30-35 ft.-lbs. when reinstalling. A new locknut must be used whenever piston is removed. The piston contains a Teflon piston ring (3) which uses an O-ring (4) expander as shown in cross section in Fig. 190. The Teflon ring is not elastic but can be stretched when warmed in oil or water. The stretched ring will gradually return to its original size. A suggested method of installation is as follows:

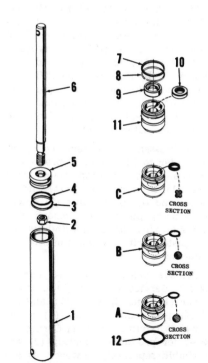

Fig. 189—Exploded view of CESSNA seat cylinder used optionally with the WOOSTER cylinder shown in Fig. 191. Old type bearings (A, B & C) are no longer serviced.

1. Barrel	7. Snap ring
2. Rod nut	8. Snap ring
3. Teflon ring	9. Wiper
4. O-ring	10. Seal
5. Piston	11. Bearing
6. Piston rod	12. O-ring

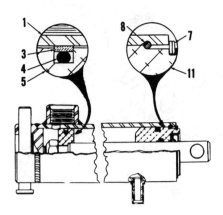

Fig. 190—Cross sectional view of CESSNA seat cylinder showing method of assembly. Refer also to Fig. 189 for exploded view and parts identification.

Install O-ring in bottom of piston ring groove. Warm Teflon ring (90°-120° F.) in water or oil and stretch only enough for installation. Compress with a suitable ring compressor for a few minutes until ring returns to original size. Remove the compressor, lubricate the ring and install piston rod assembly carefully to prevent damage to piston ring. Complete the assembly by reversing the disassembly procedure.

165. SEAT CYLINDER (WOO-STER). The seat cylinder can be removed after removing lower front panel from seat console. Two types of seat cylinder have been used. The Wooster cylinder can be identified by the removable end plug in port end of cylinder and by wrap-around snap ring at rod end engaging an internal groove in piston barrel.

To disassemble the removed cylinder, refer to Fig. 191 and proceed as follows: Remove wrap-around snap ring (1) from bottom (port) end of barrel and bump end plug (7) downward using piston rod. Remove plug and snap ring (5), then withdraw rod and piston unit downward out of cylinder barrel.

Remove bearing from rod end of barrel. Service bearing (9) contains wiper (10) and lip seal (11), and replaces production bearing (2) if renewal is required. Sealing O-rings (3 & 4) for production bearing (2) are available separately however, as are wiper (10) and lip seal (11).

If piston is removed from piston rod, use a new lock nut (17) and tighten to a torque of 30-35 ft.-lbs.

The piston contains a Teflon piston ring (16) which uses O-ring (15) as an expander. The Teflon ring is not

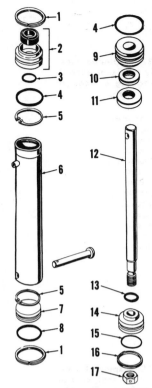

Fig. 191 — Exploded view of WOOSTER seat cylinder of the type optionally used.

1. Snap ring	9. Bearing (service)
2. Bearing (production)	10. Wiper seal
3. O-ring	11. Oil seal
4. O-ring	12. Piston rod
5. Snap ring	13. O-ring
6. Barrel	14. Piston
7. Plug	15. O-ring
8. O-ring	16. Teflon ring
	17. Rod nut

elastic, but can be stretched when warmed in oil or water. The stretched ring will gradually return to its original size. A suggested method of installation is as follows:

Install O-ring in bottom of piston ring groove. Warm Teflon ring (90°-120° F.) in water or oil and stretch only enough for installation. Compress with a suitable ring compressor for a few minutes until ring returns to original size. Remove the compressor, lubricate the ring and install pis-

ton rod assembly carefully from rod end of barrel, to prevent damage to piston ring. Complete the assembly by reversing the disassembly procedure.

DIFFERENTIAL LOCK VALVE

All Models So Equipped

166. On models equipped with differential lock, the valve attaches to seat valve and uses seat valve pressure and return lines for fluid flow. Refer to Fig. 192.

To remove the valve, first remove knobs from seat valve and differential lock valve spool, then unbolt and remove front console cover panel. Disconnect hydraulic fitting connections from seat valve and connections (4, 5 and 6) from differential lock valve. Remove the valve through-bolts (10), then remove the differential lock valve (3) and seat valve (9). Install by reversing the removal procedure.

To disassemble the removed valve, loosen locknut securing brake plunger housing (10—Fig. 193) to valve housing (2), then remove plunger housing (10). Spool valve (1) and brake plungers (6 & 7) can now be removed. Use Fig. 193 as a guide when reassembling the valve.

HYDRAULIC LIFT SYSTEM (3-POINT LINKAGE)

All Models So Equipped

167. OPERATION. The three-point linkage hydraulic lift system offers three types of control for the lift arms, selected by turning the selector knob as shown in Fig. 194.

DRAFT CONTROL. In "Draft Control", lift arm height is determined by the selected setting and controlled by top link pressure or tension on the

Fig. 192—Exploded view of differential lock valve and attaching parts.

1. Bracket
2. Knob
3. Valve
4. Differential lock supply line
5. Right brake line
6. Left brake line
7. Gasket
8. Transfer plate
9. Seat valve
10. Through-bolts

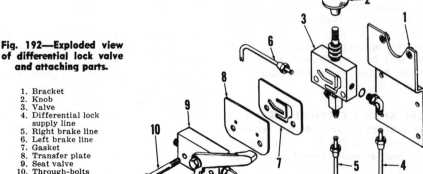

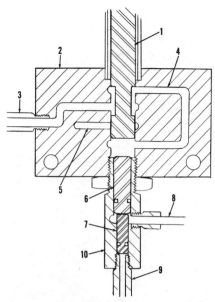

Fig. 193—Cross sectional schematic view of differential lock valve showing component parts.

1. Valve spool
2. Valve body
3. Supply line
4. Return passage
5. Pressure passage
6. Left brake plunger
7. Right brake plunger
8. Left brake line
9. Right brake line
10. Plunger housing

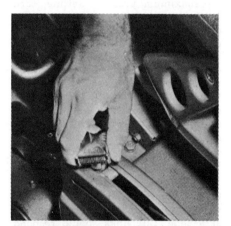

Fig. 194—Selector knob can be turned to three control positions after moving lever to top of quadrant as shown.

Fig. 195—Lift arm lock is located on left side of lift cover housing as shown.

leaf-type draft control springs. Draft control is the middle detent position on selector knob.

POSITION CONTROL. In "Position Control", lift arm height is maintained at a selected setting by an actuating lobe on rockshaft. Position control is obtained by turning selector knob (Fig. 194) counter-clockwise as far as possible.

PRESSURE CONTROL. In "Pressure Control", a selected pressure is maintained in rockshaft lift cylinders, the pressure being controlled by a servo piston. Pressure control is obtained by turning selector knob clockwise as far as possible.

The master control valve is spring loaded in the "LIFT" position and is moved to neutral or lowering position by action of the control linkage.

168. QUICK CHECKS. For a quick check of hydraulic system, first be sure lift arm lock is "OPEN" as shown in Fig. 195 and flotation latch is locked as shown in Fig. 196.

Attach a heavy weight to lower links. Weight should allow rockshaft to travel through complete arc and should not be attached to upper link.

Start and run engine at 1000-1600 rpm, then move control lever to top of quadrant. Weight should raise. With lever at top of quadrant, check to see that selector knob turns easily to all three detent positions.

Turn selector knob to "Position Control" (counter-clockwise detent) and slowly move control lever downward on quadrant. When lever is ½-⅝ inch from top of slot, lift links

should start to lower. Move lever down and up quadrant a little at a time. Lift links should respond to movement of control lever, completely lowering when lever is at bottom of quadrant slot and stopping at any selected intermediate point.

Lower the weight to ground and insert a ⅝ inch gage block between draft linkage ratio lever and lever guide, at point shown at (A—Fig. 198), then turn draft control handwheel (B) until gage block is lightly supported. Move control lever to top of quadrant and turn selector knob to "DRAFT" (center detent) position; then move control lever about ⅔ the distance to bottom of quadrant midway between the two sector marks (1—Fig. 199). Lower links should remain suspended in any selected position when control lever is within ¼-inch of midway between marks, raising when lever is moved upward slightly, or lowering when lever is moved downward.

169. TROUBLE SHOOTING. Relief pressure for the three-point hitch is controlled by maximum setting for pressure control. Instead of a relief valve opening when maximum setting is exceeded, control valve is returned to neutral by a pilot valve. If relief pressure malfunctions, check and/or adjust the setting of pressure control linkage.

Draft control linkage is returned to neutral in transport position by the "POSITION CONTROL" linkage. If "Position Control" linkage fails to return the main valve to neutral, valve will be neutralized by "Pressure Control" linkage when lift pistons reach the end of their stroke.

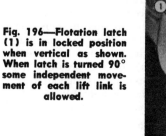

Fig. 196—Flotation latch (1) is in locked position when vertical as shown. When latch is turned 90° some independent movement of each lift link is allowed.

Fig. 197—Response control knob (1) is located on seat side of dress panel.

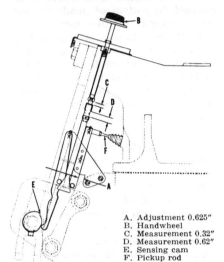

A. Adjustment 0.625"
B. Handwheel
C. Measurement 0.32"
D. Measurement 0.62"
E. Sensing cam
F. Pickup rod

Fig. 198—Schematic view of draft control sensing linkage used on models so equipped.

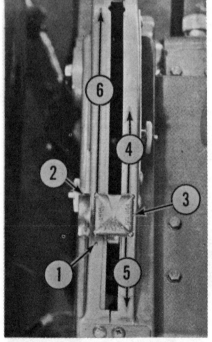

Fig. 199—View of control quadrant showing draft control positions.

1. Sector marks
2. Adjustable locator
3. Control lever
4. Decreasing depth
5. Increasing depth
6. Transport

Fig. 200 — View of lift cover with front access plate removed.

1. Position control adj. screw
2. Draft control bellcrank
3. Pressure control adjusting rod

If lever must be moved ½-⅝ inch from top of quadrant before lift links start to lower, the position control linkage adjustment is probably satisfactory.

If lift hunts, or searches, in neutral position; or is unduly noisy; leakage or improperly adjusted pressure control linkage is to be suspected.

170. CHECK AND ADJUST. First be sure lift arm lock is "OPEN" as shown in Fig. 195 and flotation latch is locked as shown in Fig. 196.

Attach a heavy weight to lower links. Weight should allow links to travel through complete arc and should not be attached to upper link.

Remove upper port plug from either rockshaft cylinder and install a suitable pressure gage in port opening. Remove lower front dress panel from seat console and front access plate from hydraulic lift cover. Remove access plate from right side of cover housing.

Start and run engine at 1000-1600 rpm, move control lever to top of quadrant and note gage pressure. The gage will either show the maximum pressure setting (which should be 1800-2000 psi), or the pressure necessary to support the attached weight (which will be considerably less than maximum pressure setting).

Working through the opening in front of lift cover, back off the locknut on position control adjusting screw (1—Fig. 200). If gage reading is considerably less than maximum pressure setting, back out the adjusting screw (1) counter-clockwise until

gage pressure rises sharply; then turn screw in until pressure drops. Tighten locknut while holding the adjusting screw. If gage reading was originally the maximum pressure setting, turn screw in (clockwise) until pressure drops then tighten locknut. After completing the initial adjustment, move control lever slowly forward on quadrant and measure lever position when lift links first start to lower. Distance should be ½-⅝ inch. Readjust position control adjusting screw (1) if necessary. Turning screw clockwise will increase the distance, counter-clockwise will decrease distance.

With selector knob in "Position Control", move lever forward (downward) on quadrant a short distance at a time. Lift links should respond by moving a corresponding amount then stopping. Lift cylinders should be fully retracted when control lever reaches bottom of quadrant.

With lift arms lowered, turn lift arm lock knob to "LOCK" position, then move control lever back to top of quadrant. Gage pressure should now be 2000 psi. If it is not, loosen the locknut and adjust servo relief valve rod (3). Lengthen rod to increase pressure or shorten rod to decrease pressure.

Turn selector knob on control lever to "PRESSURE" (clockwise detent) position, then move lever slowly down the quadrant while watching gage reading. Pressure should drop steadily at a uniform rate as lever is moved; and should read 50-250 psi when lever reaches bottom of quadrant.

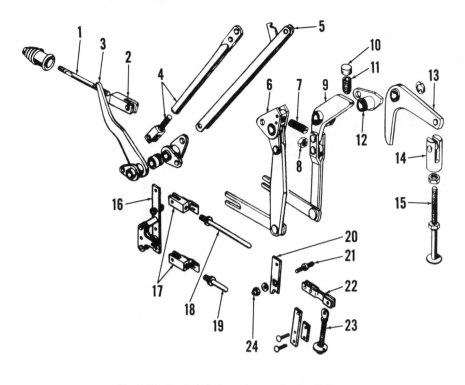

Fig. 201—Exploded view of control valve linkage.

1. Pickup rod	8. Locknut	16. Valve lever
2. Clevis	9. Draft control	17. Clevis
3. Position control	bellcrank	18. Pressure control
pickup	10. Rubber plug	rod
4. Position control	11. Stop screw	19. Main valve rod
link	12. Pressure control	20. Drive link
5. Draft control link	bellcrank	21. Pivot stud
6. Position control	13. Pressure control	22. Dashpot beam
bellcrank	lever	23. Dashpot piston rod
7. Adjusting screw	14. Yoke	24. Nut
	15. Adjusting rod	

Move control lever to top of quadrant. Working through access plate opening on right side of lift cover, lengthen pressure control piston rod (18—Fig. 201) until pressure gage reading begins to fluctuate or cycle; then shorten rod until cycling stops.

NOTE: If piston rod (18) is too short, main control valve will not return to neutral and fluid entering through the main valve must escape through the servo relief valve, overworking and overheating the valve. Make sure adjustment of piston rod (18) is correct.

Turn lift arm lock knob to "OPEN" position and check to see that lift arms raise and lower normally.

Move control lever to top of quadrant and turn selector knob to "DRAFT" (center detent) position; then move control lever down the quadrant until aligned with "dimple" approximately ¼ the distance from top. Lower the lift links all the way by turning hand wheel (Fig. 202), then check the clearance between draft link (9—Fig. 201) and stop screw (11). Clearance should be 0.002-0.004; if it is not, remove rubber plug (10) from top of lift cover and readjust the stop screw (11).

171. MAIN VALVE ADJUSTMENT. If proper adjustment cannot be obtained as outlined in paragraph 170, check the main spool valve adjustment as follows:

Prepare the tractor as outlined in paragraph 170. Turn selector knob to "POSITION" control and move control lever to bottom of quadrant. With engine running and working through right access plate opening, press down on vertical dashpot link (20—Fig. 203) as shown by heavy arrow. Be certain lower notch engages spacer on actuating lever (16). Temporarily tighten locknut (24) to hold the link (20) in lowermost position.

Shorten control valve rod (19) until lower links begin to raise. Lengthen rod (19) when lower links are horizontal, until weight settles one inch in 5 seconds, measured at outer end of links.

Fig. 202 — Draft control handwheel.

1. Hand wheel
2. Increasing depth
3. Decreasing depth

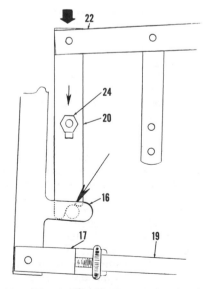

Fig. 203—Schematic view of main valve linkage showing adjustment procedure. Refer to Fig. 201 for parts identification and to paragraph 171 for details.

Fig. 204—With no tension or pressure on upper link, clevis pin must clear bracket stop by the amount shown.

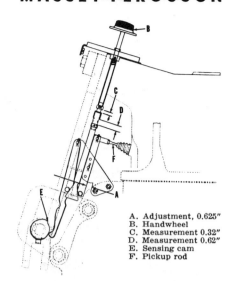

A. Adjustment, 0.625"
B. Handwheel
C. Measurement 0.32"
D. Measurement 0.62"
E. Sensing cam
F. Pickup rod

Fig. 205—Schematic view of draft control sensing linkage used on models so equipped.

Loosen locknut (24) until dashpot linkage is free to move; then readjust all the linkage as outlined in paragraph 170.

172. SENSING LINKAGE ADJUSTMENT. To adjust the draft control sensing linkage, first check to make sure that $\frac{3}{8}$-$\frac{7}{16}$ inch clearance exists between top link clevis pin and top link bracket as shown in Fig. 204. There must be no pressure or tension on upper link when measurement is taken, and clevis pin must be attached to lower holes in sensing springs.

If measurement is incorrect, remove the two bolts securing pin to sensing springs and turn the pin 180° until the other flats contact springs. Reinstall the bolts and re-measure. If measurement is still incorrect, check for bent or damaged springs or attaching parts.

Move control lever to top of quadrant and turn selector knob to "DRAFT" (center detent) position, then move the lever ⅔ the way down the quadrant until centered midway between the two sector marks. Upper link clevis must be hanging down and point of cam (E—Fig. 205) must be turned away from contact with pickup finger as shown. Turn draft control handwheel (B) counter-clockwise if necessary, until a ⅝-inch gage block can be inserted between ratio lever and bracket as indicated at (A); then turn handwheel (B) clockwise until gage block is just supported. With linkage thus positioned, distance (C) between clevis and upper stop nut should be approximately $\frac{5}{16}$ inch and distance (D) between clevis and lower stop nut should be approxi-

mately ⅝ inch. Reposition the stop nuts if necessary, by loosening setscrew and turning nut.

With engine running and linkage positioned as outlined, rockshaft should hold position without raising or lowering. Turning draft control handwheel (B) ¼-turn should cause rockshaft to raise or lower. If adjustment is incorrect, loosen the locknut and turn pickup rod (F) in or out as required.

173. RESPONSE ADJUSTMENT. With hydraulic fluid at operating temperature, quadrant lever in draft control and approximately 1000 lbs. of weight attached to lower links, lowering time should not be more than 3 seconds with response control knob in "FAST" position, nor more than 6 seconds with control knob in "SLOW" position.

To adjust the response control, refer to Fig. 206 and proceed as follows: Slightly loosen set screws (2). Move response control knob to "SLOW" position and screw dashpot needle valve (1) in by hand until seated. Tighten setscrews (2) to a torque of 25-35 ft.-lbs. and recheck.

174. OVERHAUL. The draft control sensing linkage, rockshaft arm and cylinders, or attaching links can be removed and installed without removing hydraulic lift cover. To obtain access to control valve, rockshaft or internal linkage, it is first necessary to remove cover as outlined in paragraph 175, or tilt cover back on draft control springs as outlined in paragraph 176. Both methods require the use of an overhead hoist. Refer to Figs. 207 and 208.

175. R&R LIFT COVER. To remove the hydraulic lift cover for service on any of the components or rear axle center housing, proceed as follows:

Disconnect lift cylinders from lift arms, and lift rods from lower links; then remove lift arms and rods from rockshaft as assemblies. Remove rear dress panel cover, draft control springs and draft control sensing linkage. Remove platform center panel and lower front panel from seat console. Disconnect pressure and vent hoses from seat cylinder and remove seat cylinder lower attaching pin; then unbolt and remove seat frame as a unit. Disconnect hydraulic lines from lift cover and remove cap screws securing cover to tractor center housing. Attach a suitable hoist to lift cover as shown in Fig. 207 and remove the cover.

When installing cover, make sure oscillator drive lever properly engages drive body on control valve. Aligning dowels help in positioning the cover. Complete the installation by reversing the removal procedure.

Fig. 206—Adjusting the response lever.

1. Dashpot needle valve
2. Setscrews
3. Lever

Fig. 207—View of lift cover removed as an assembly.

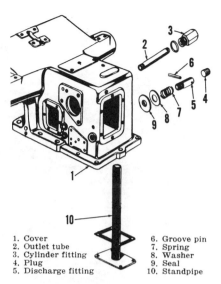

1. Cover	6. Groove pin
2. Outlet tube	7. Spring
3. Cylinder fitting	8. Washer
4. Plug	9. Seal
5. Discharge fitting	10. Standpipe

Fig. 209—Parts (2 through 10) must be removed from lift cover before control valve can be removed.

176. LIFT COVER PARTIAL REMOVAL. Completely lower the rockshaft. Move control lever to top of quadrant, turn selector knob to "PRESSURE" control and move control lever to bottom of quadrant. Remove platform center panel and lower front panel from seat console. Remove operator's seat and seat bracket. Remove seat cylinder and lower bracket. Disconnect pressure line from seat accumulator tee and lines leading from seat valve to lift cover.

Disconnect pressure line to transfer cap and remove lift cover front access plate. Disconnect hoses leading to rockshaft cylinders. Disconnect draft control sensing rod and unbolt and remove sensing linkage.

Fig. 208 — Hydraulic lift cover may be tilted back for service as shown.

1. Standpipe
2. Control valve

Attach a suitable hoist to top of lift cover and remove cap screws securing cover to tractor center housing, then carefully raise lift cover upward and back as shown in Fig. 208.

NOTE: Cover will try to jump rearward as front end is raised, and damage to oscillator mechanism can occur. Alignment dowels in front cap screw holes can help prevent damage as long as binding does not occur.

Install by reversing the removal procedure.

177. LINKAGE CONTROL VALVE. To remove the linkage control valve, first remove lift cover as outlined in paragraph 175 or tilt cover as in paragraph 176.

Drain cover reservoir and remove standpipe (10—Fig. 209). Remove plug (4). Working through plug hole and using special tool (MFN-767) or other suitable means, remove control valve discharge fitting (5) and associated parts.

Remove left hand cylinder hose fitting (3) and tube (2). Turn control quadrant selector knob to "PRESSURE" control and move lever to bottom of quadrant; then, working through front access opening, disconnect and remove pressure control piston rod.

Remove the two long cap screws securing transfer cap to front, right side of cover, leaving transfer cap in place. Support linkage control valve with hand and remove the other three cap screws securing valve to lift cover housing. Turn valve slightly and, working through front access plate opening, remove the slotted linkage guide plate from top of valve housing. Hold main control valve in discharge position and disconnect "Position" and "Draft" control links, then lift out the valve.

Fig. 211 shows an exploded view of main components of the valve. Shims (10—Fig. 212) control end clearance adjustment of intake flow control valve sleeve (7 or 8). Recommended adjustment is 0.001 clearance to 0.003 preload. Shims are available in thicknesses of 0.005, 0.008 and 0.010. To check required shim pack thickness, assemble the unit leaving "O"-ring off of plug (11) and using 0.026-0.030 thickness of shims (10). Tighten plug (11) finger tight, making sure parts are seated, then measure gap

Fig. 210—Removed view of linkage control valve showing proper assembly.

between flange of plug and surface of housing using a feeler gage. Remove the plug and deduct shims equal in thickness to the measured gap.

When valve is reassembled, check with Fig. 210 to be sure linkage is properly assembled, and adjust main valve as outlined in paragraph 171 before valve is reinstalled.

178. ROCKSHAFT. To remove the rockshaft, first remove hydraulic lift cover as outlined in paragraph 175 or tilt cover rearward as in paragraph 176. Disconnect lift cylinders from lift arms and remove lift arms from rockshaft if not previously removed.

Turn rockshaft until lock shaft (16—Fig. 213) engages notch in lock arm (7), then turn shaft (16) to "LOCK" position. Withdraw rockshaft (6) out right side of lift cover. Right hand bushing (9) will be removed with shaft, and left bushings (3) and spacer (4) can be removed at this time. Lock arm (7) can be removed after removing position control pickup rod. Rockshaft and arms have master splines for proper assembly. Assemble rockshaft by reversing the disassembly procedure.

179. OSCILLATOR DRIVE. Refer to Fig. 214 for and exploded view of main valve oscillator drive of the type used on late units. Early models are similar except shoe (1) was not used. The drive shoe (1) (or crank) rides on a cam machined on pto clutch drum.

To remove the unit, first remove hydraulic lift cover as outlined in paragraph 175 or tilt the cover as in paragraph 176. Remove front hydraulic cover as outlined in paragraph 147.

Remove hair pin (5) and drive crank (2). Remove pickup lever (7), spring (6) and key (4); then withdraw drive bushing (3) forward out of center housing bore. Only the late parts are available for service and parts (1, 2 and 3) must be individually ordered and all renewed.

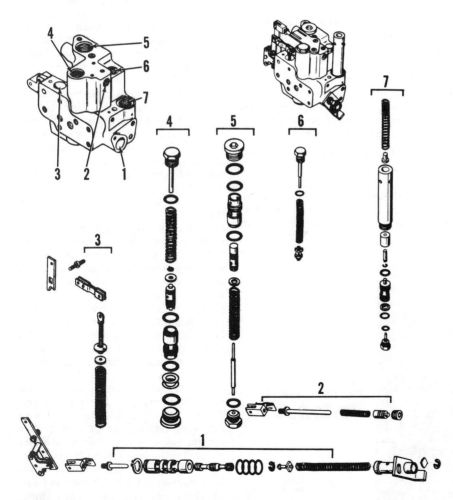

Fig. 211—Views of linkage control valve showing principal components. Location of valves in housing is indicated by corresponding numbers.

1. Main control valve
2. Pressure control pilot piston
3. Response dashpot
4. Inlet flow control valve
5. Discharge valve
7. Servo relief valve valve
6. Secondary relief

180. ROCKSHAFT CYLINDERS. Benton (Fig. 215) or Cessna (Fig. 216) rockshaft cylinders are used. The removed Benton cylinder can be disassembled by unscrewing and removing bearing nut (4—Fig. 215). All parts of cylinder are available individually.

The Cessna cylinder (Fig. 216) is of welded construction. To disassemble the cylinder, remove hose fitting and refer to Figs. 217 through 221.

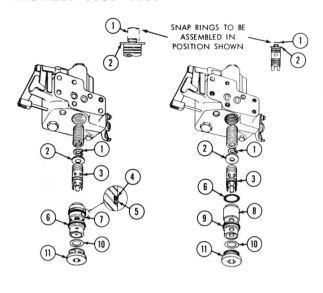

SNAP RINGS TO BE ASSEMBLED IN POSITION SHOWN

Fig. 212 — Partially exploded view of inlet flow control valves showing proper assembly. Early valve is shown on right, late valve on left.

1. Snap ring
2. Dampening washer
3. Flow control valve
4. Backup washer
5. O-ring
6. O-ring
7. Flow control sleeve
8. Flow control sleeve
9. O-ring
10. Shim pack
11. Plug

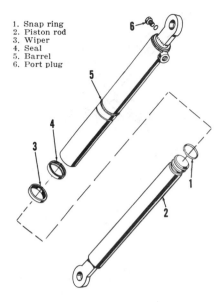

1. Snap ring
2. Piston rod
3. Wiper
4. Seal
5. Barrel
6. Port plug

Fig. 216—Exploded view of CESSNA rockshaft cylinder. Cylinder bore may be 2½, 2⅝ or 2⅞ inches.

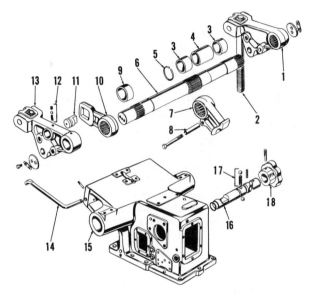

Fig. 213—Exploded view of lift cover cover, rockshaft and associated parts.

1. Lift arm, L.H.
2. Lift link
3. Bushing
4. Spacer
5. Snap ring
6. Rockshaft
7. Lock arm
8. Pickup tube
9. Bushing
10. Lock arm
11. Lock pin
12. Detent assembly
13. Lift arm, R.H.
14. Storage hook
15. Lift cover
16. Lock shaft
17. Detent
18. Lock knob

Fig. 217 — When disassembling CESSNA cylinder, align snap ring with hose port as shown.

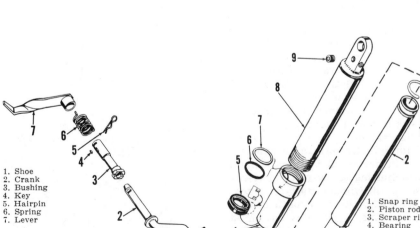

1. Shoe
2. Crank
3. Bushing
4. Key
5. Hairpin
6. Spring
7. Lever

Fig. 214—Exploded view of Main Valve oscillator drive and associated parts used on late models. Early tractors are similar except shoe (1) is not used.

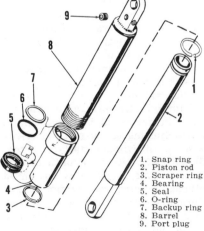

1. Snap ring
2. Piston rod
3. Scraper ring
4. Bearing
5. Seal
6. O-ring
7. Backup ring
8. Barrel
9. Port plug

Fig. 215—Exploded view of BENTON rockshaft cylinder used on some tractors.

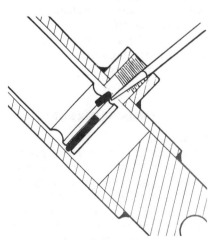

Fig. 218—Working through hose port, unseat snap ring into deep second groove.

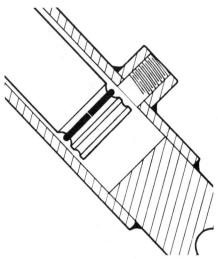

Fig. 219—Snap ring positioned for piston rod removal or installation.

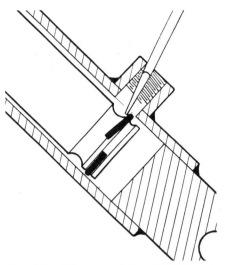

Fig. 220—When assembling CESSNA cylinder, work snap ring from deep groove back into end groove.

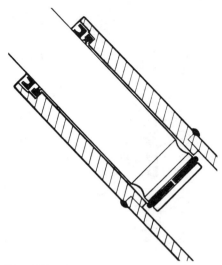

Fig. 221—Piston rod fully extended in CESSNA cylinder showing rod stop, bearing and seal configuration.

Align snap ring with cylinder port, with end of ring slightly to one side of port opening as shown. Insert a screwdriver or similar tool through port opening and work snap ring into deep, second groove in piston rod (Fig. 219). Piston can now be withdrawn. Assemble by reversing the disassembly procedure. Pull piston against stop (Fig. 221) after snap ring is repositioned, to be sure ring is fully seated.

TRANSFER CAP

All Models

181. The transfer cap (5—Fig. 224) is a one-way check valve used on all models with rockshaft and auxiliary valve. The unit mounts on right side of hydraulic lift cover. Check valve (4) prevents back flow from linkage valve to auxiliary valves when both units are actuated simultaneously or engine is shut off.

Check valve can be removed for service or cleaning after removing cap assembly and snap ring (1).

SENSING LINKAGE

All Models

182. Fig. 222 shows an exploded view of draft control springs and associated parts. Refer to paragraph 172 for adjustment if linkage is removed.

Fig. 223 shows an exploded view of pressure control coupler and drawbar frame. Use Fig. 223 as a guide when disassembling or assembling the unit.

AUXILIARY VALVE (POPPET TYPE)

Early Models

183. **ADJUSTMENT.** First center valve levers with neutral punchmarks on quadrant by loosening the clamp bolts (A—Fig. 224 or B—Fig. 225). Tighten clamp bolts to a torque of 12-15 ft.-lbs. when levers are correctly positioned.

The auxiliary valve body contains two high pressure (inlet) poppets (5) and two low pressure (outlet) poppets (9). The outer pair of poppets control fluid flow to and from the lower breakaway coupling, the inner pair control flow of upper coupling.

To adjust the valve, use a single-acting cylinder mounted on an implement or attached to a weight. Loosen the locknuts on outer pair of adjusting screws (6A & 8A) and connect the cylinder to lower breakaway coupling. With engine running at slow speed, move control lever to "RAISE" position and allow cylinder lever to extend about half way. Return control lever to neutral and observe the cylinder, which should neither raise nor lower. Any movement of cylinder indicates improper valve adjustment or leaking valve.

With cylinder partially extended, valve in neutral and engine running, turn adjusting screw (6A) in (clockwise) until cylinder starts to extend. Slowly back screw out until cylinder stops moving, back screw out an additional ¼ turn and tighten locknut. Turn lowering adjusting screw (8A) in until cylinder starts to lower, back screw out until cylinder stops, back screw out an additional ½ turn and tighten locknut. The slight overlap

with valve correctly adjusted allows inlet poppet to open first and pressurize the cylinder, preventing load from settling momentarily when lever is moved to raise position.

Adjust inner half of control valve in the same manner by connecting cylinder to upper breakaway coupling and moving control lever to "LOWER" to extend the cylinder.

Valve lever will stay in "RAISE" or "LOWER" position when lever is moved to end of travel and released. Lever must be manually returned to neutral when cylinder reaches the end of its stroke.

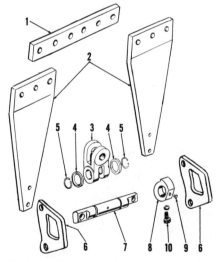

Fig. 222—Exploded view of draft control sensing springs and associated parts.

1. Clamp	6. Bracket
2. Sensing spring	7. Pin
3. Clevis	8. Pickup cam
4. Spacer	9. Drive plug
5. Snap ring	10. Lock screw

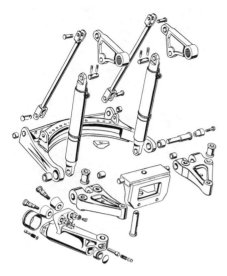

Fig. 223—Exploded view of pressure control linkage and drawbar support.

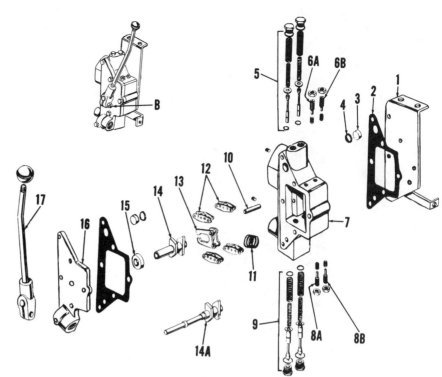

Fig. 225—Exploded view of single, poppet type auxiliary valve. Dual valve uses duplicate bodies, and camshaft (14A) in inner bank.

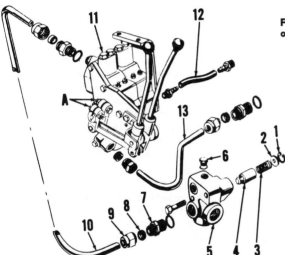

Fig. 224—Exploded view of transfer cap and auxiliary valve lines.

1. Snap ring
2. Washer
3. Spring
4. Valve plunger
5. Cap
6. Plug
7. Fitting
8. Sleeve
9. Nut
10. Pressure line
11. Auxiliary valve
12. Drain hose
13. Return hose
A. Clamp screws

B. Clamp screw
1. Side plate
2. Gasket
3. Retainer
4. Sealing ring
5. Upper (pressure) poppets
6A. Adj. screw (outer)
6B. Adj. screw (inner)
7. Body
8A. Adj. screw (outer)

8B. Adj. screw (inner)
9. Lower (return) poppets
10. Yoke guide
11. Centering spring
12. Actuating arm
13. Cam yoke
14. Cam (outer)
14A. Cam (inner)
15. Spacer
16. Valve cover
17. Lever

184. OVERHAUL. Tractors with rockshaft use remotely pivoted control levers and connecting links as shown in Fig. 224. On models without rockshaft the control lever attaches directly to valve camshaft as shown in Fig. 225.

On models with dual valve, the inner assembly uses camshaft (14A) which extends through the hollow outer cam (14). Mark the poppet assemblies (5 & 9) for reinstallation in same position from which removed. Adjust the valve after assembly as outlined in paragraph 183.

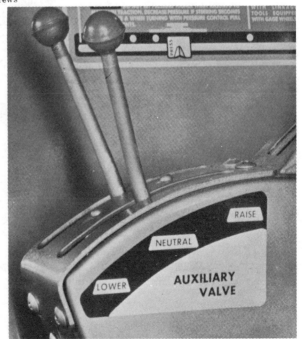

Fig. 226—Auxiliary valve levers used on models with 3-point lift.

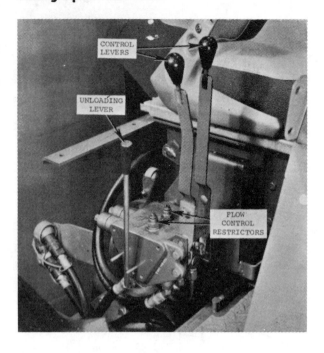

Fig. 227—Installed view of late type auxiliary valve with dress cover removed showing control levers, unloading lever and flow restrictors.

Fig. 229—Installed view of late type auxiliary valve on model without 3-point lift. Note the offset lowering stops in lever slots.

AUXILIARY VALVE (SPOOL TYPE)

Late Models

185. **OPERATION.** Fig. 227 shows an installed view of the Cessna Spool Type auxiliary valve and Fig. 228 a cross section of one bank.

Operating fluid is directed to and from cylinders by spool valve (12—Fig. 228). Poppet valve assemblies (11) seal the work ports (8 & 9) from transfer passages (2 & 6). When the spool valve is moved from neutral in either direction, the actuating passage (5) is connected to pressure passage (3) by internal drillings in valve spool. Fluid from actuating passage (5) enters the area between the two poppet valves, opening the valves and permitting fluid flow to work ports.

Feathering notches are located on return lands of spool and return is controlled to meter the flow. A mechanical detent position is provided at end of spool travel in each operating position. The spool will remain in detent until manually released.

A float detent position is provided by moving the lever beyond the lowering detent stop. In float position, both transfer passages are connected to return ports and actuating passage (5) is pressurized to open the poppet valves. Float position must be used to relieve pressure in cylinder and lines for connecting or disconnecting the cylinders, and if engine is not running, poppet valves must be manually unseated by moving the unloading lever to actuate plunger (10).

The spool valve is ported to permit fluid flow of up to 20 gpm, about double that of the earlier poppet type. The valve is equipped with a flow control restrictor (4) which may be adjusted to provide a regulated flow of 5-20 gpm. The factory adjusted setting is 10-12 gpm, which is obtained when distance (A) between end of adjusting screw and machined port boss on valve body is $1\frac{3}{32}$ inches.

186. **ADJUSTMENT.** To index the valve levers, move lever to "LOWER" detent position. Lever should just touch the offset lowering stop midway between neutral position and bottom of lever slot. If adjustment is incorrect, remove outer dress panel and loosen the clamp bolt securing lever extension to lever and reposition.

Connect a test gage to breakaway coupler and move control lever to pressurize the coupler. Gage should register system pressure. Pressure

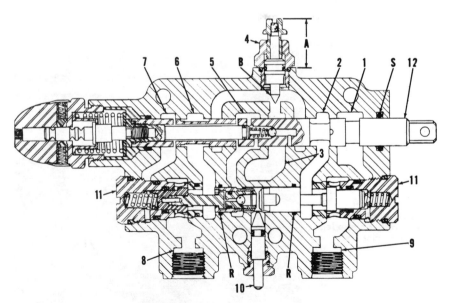

Fig. 228—Cross sectional view of CESSNA spool type auxiliary valve body. Valves may be stacked for multiple operation.

A. Adjustment	2. Transfer passage	7. Return port
B. Metering piston	3. Pressure port	9. Work port
R. O-rings	5. Actuating passage	10. Unloading plunger
S. O-ring	6. Transfer passage	11. Poppet assembly
1. Return port		

should not drop more than 150 psi when valve lever is returned to neutral, and pressure loss should not exceed 100 psi per minute with valve in neutral position. This leakage is approximately equivalent to 0.007 per minute movement of piston rod on a 3½ inch cylinder.

187. **OVERHAUL.** To disassemble the spool valve, first clean any paint, rust or burrs from eye end of valve spool, then unscrew the detent cap. Withdraw spool and cap as an assembly. Unscrew poppet plugs (11) and withdraw poppet plungers. Only the two $\frac{9}{16}$ inch O-rings (R) and spool seal ring (S) remain in stripped housing.

Examine centering spring and washers. If any parts are damaged, unscrew detent spool from valve spool to renew the parts. Install detent spool using Grade CV (blue) LOCTITE and tighten to 5-8 ft.-lbs. Tighten detent cap to a torque of 50-55 ft.-lbs. and valve throughbolts to 20-40 ft.-lbs.

NOTES

NOTES

MASSEY FERGUSON

Model ■ MF1150

Previously contained in I&T Shop Manual No. MF-30

SHOP MANUAL

MASSEY FERGUSON

MODEL
MF 1150

Tractor serial number stamped on instrument panel name plate.
Engine serial number stamped on side of engine.

INDEX (By Starting Paragraph)

CONDENSED SERVICE DATA

GENERAL

Torque Recommendations . . . . See End of Shop Manual
Engine Make . . . . Perkins
Engine Model AV8.510
Number of Cylinders 8
Bore, Inches 4 1/4
Stroke, Inches 4 1/2
Displacement, Cu. In. 510.7
Compression Ratio 17.5:1
Main Bearings, No. of 5
Cylinder Sleeves . . . . Dry
Forward Speeds 12
Reverse Speeds 4

TUNE UP

Firing Order 1-8-7-5-4-3-6-2
Valve Tappet Gap (Cold) Intake & Exhaust 0.012
Valve Face Angle . . . . 45°
Valve Seat Angle . . . . 45°
Injection Pump Spill Timing . . . . 30° BTDC
Timing Mark Location . . . . Flywheel
Battery
  Volts 12
  Capacity Amp/Hr 95
  Ground Polarity . . . . Negative
Injectors
  Opening Pressure . . . . 2575 psi
  Spray Hole Diameter 0.011
Governed Speed
  Engine
    Low Idle . . . . 800-850 rpm
    High Idle . . . . 2400 rpm
    Loaded . . . . 2200 rpm

Governed Speed Cont.
  1000 rpm PTO
    Low Idle . . . . 400-425 rpm
    High Idle . . . . 1200 rpm
    Loaded . . . . 1100 rpm
Horsepower @ PTO Shaft 135.6
Hydraulic System
  Maximum Pressure . . . . 2400 psi
  Rated Delivery . . . . 20 gpm

LIQUID CAPACITIES

Cooling System—U.S. Quarts 48
Crankcase—U.S. Quarts 21
Transmission, Differential and
  Hydraulic System—U.S. Gallons 22

SIZES—CLEARANCES

Crankshaft Journal Diameter 3.997
Crankpin Diameter 3.248
Camshaft Journal Diameter
  Front 2.372
  2nd 2.242
  3rd 2.232
  4th 2.222
  Rear 2.212
Crankshaft Bearing Clearance
  Main Bearings 0.004-0.0063
  Crankpin 0.0023-0.0043
Crankshaft End Play 0.002-0.017
Camshaft Bearing Clearance 0.002-0.006
Camshaft End Play 0.002-0.014
Piston to Cylinder Clearance 0.005-0.007

FRONT SYSTEM

Model 1150 tractors are available with Western Type (non-adjustable) front axle unit or Row Crop (wide adjustable) axle types. Refer to Figs. 1 and 2 for exploded views and to the following paragraphs for pertinent disassembly and overhaul data.

AXLE ASSEMBLY

1. Refer to Fig. 1 for an exploded view of Western Type axle assembly, front support, steering motor and associated parts. Fig. 2 shows Row Crop adjustable axle components.

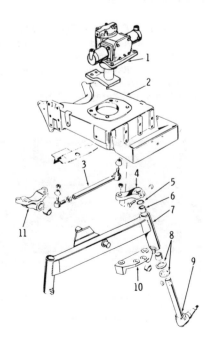

Fig. 1–Front support, steering motor and fixed-tread axle shown exploded. Refer to Fig. 2 for view of optional adjustable axle unit.

1. Steering motor
2. Front support
3. Drag link
4. Steering arm
5. Dust seal
6. Bushing
7. Axle
8. Thrust washers
9. Spindle
10. Pivot bracket
11. Pivot bracket

To remove either axle as an assembly, suitably support front of tractor. Disconnect drag links (3) from steering

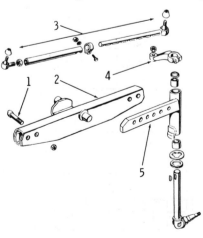

Fig. 2–Exploded view of adjustable axle.

1. Axle bolt
2. Center axle
3. Drag link
4. Steering arm
5. Axle half

arms (4). Unbolt and remove front pivot housing (10) from front support, then slide axle unit out of rear pivot support (11) and remove the unit from tractor. Install by reversing the removal procedure. Adjust toe-in if necessary as outlined in paragraph 2.

TIE RODS AND TOE-IN

2. Automotive type tie rod ends are used. Recommended toe-in is 0-¼ inch. Adjust both tie rods an equal amount to obtain the required setting.

POWER STEERING SYSTEM

All models are equipped with a closed center hydraulic system which provides standby power for operation of the hydrostatic steering system as well as the other hydraulic functions of operation and control of the tractor. This section covers only the Hydrostatic Power Steering Hand Pump (steering control valve), Steering Cylinder Assembly (steering motor), connecting lines and attaching parts. Refer also to HYDRAULIC SYSTEM Section for information on main hydraulic pump, priority valve, relief valve and operating fluid.

LUBRICATION AND BLEEDING

3. The hydrostatic steering hand pump and steering cylinder assembly (steering motor) are lubricated by the operating fluid. Refer to paragraph 100 for fluid type and checking procedure. The hand pump (P—Fig. 3) is lubricated by the high pressure oil as the system is operated. The shaft, bearings and gears of steering motor (C) are lubricated by return oil from the lubrication passages of the main hydraulic pump. The power steering system is self-bleeding.

ADJUSTMENT

4. No adjustments are required of steering hand pump or steering motor.

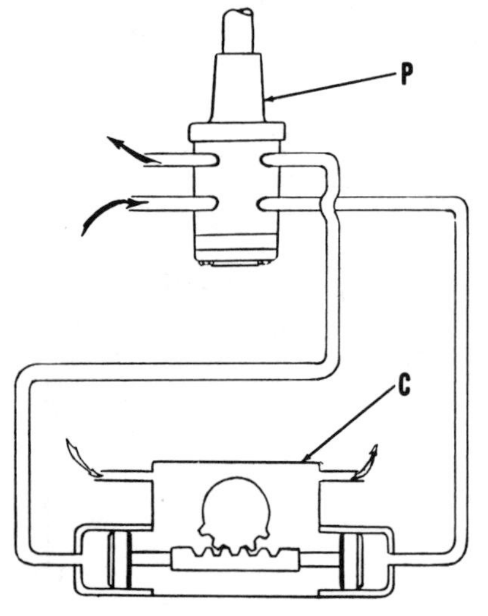

If clearance is excessive or steering action is erratic, overhaul is indicated. Refer to paragraph 5 for system diagnosis and the appropriate following paragraphs for removal and overhaul information.

TROUBLE SHOOTING

5. If malfunction of the steering system exists, first make an operational check of some other hydraulic unit to eliminate the hydraulic system as the cause of trouble. The procedure for completely checking the hydraulic system is given beginning with paragraph 101.

External leakage is not generally objectional from an operational standpoint and should be corrected only to prevent visible loss of fluid.

Internal leakage is objectional only to the extent performance is affected. Normal leakage of the steering system will permit approximately 2 revolutions per minute of the steering wheel with normal steering pressure applied and wheels fully turned to right or left steering stop. If considerable difference exists when turning in one direction only, the trouble is probably in steering cylinder. If slippage is approximately equal in either direction, the hand pump may be at fault.

STEERING CYLINDER (MOTOR)

Because of the welded center steering arm (T—Fig. 6), the steering motor cannot be removed without some disassembly, and most overhaul can be accomplished without removal. Paragraph 6 outlines the

procedure for removal and installation of the unit for other work or for renewal of steering arm bushings. If routine overhaul is planned, proceed as outlined in paragraph 7 and disassemble the unit only as far as required to accomplish the necessary repairs.

6. **REMOVE AND REINSTALL.** To remove the steering motor housing, first turn wheels to a straight ahead position and shut off engine. Remove front end weights if so equipped and front and right side grilles. Disconnect hydraulic lines and remove hydraulic system filter and oil cooler radiator. Disconnect both tie rods at inner ends.

Unbolt and remove right cylinder assembly. Reaching underneath the tractor, pull rear end of center steering arm to right as far as it will go. Remove the four retaining cap screws and twist top of steering motor to right as it is lifted out of front support.

Install by reversing the removal procedure. Tighten mounting cap screws to a torque of 275-285 ft.-lbs. and cylinder sleeve screws to 50-55 ft.-lbs. Cycle the system several times after completing assembly, to air-bleed the cylinder and lines, then refill hydraulic system reservoir if necessary, as outlined in paragraph 100.

7. **OVERHAUL.** Refer to Fig. 6 for an exploded view of steering motor unit. To overhaul the installed unit, first remove front end weights if so equipped, and front and right hand grilles. Disconnect hydraulic lines and remove hydraulic system filter and oil cooler radiator.

Remove right hand cylinder assembly and the four cap screws securing motor housing to front support. Unbolt and remove right cylinder, rotate motor housing until left cylinder can be withdrawn, then unbolt and remove left cylinder. Pry out expansion plug (A), remove snap ring (B) and

Fig. 3–Schematic view of hydrostatic power steering hand pump (P) and steering motor (C) showing direction of oil flow.

Fig. 4–Installed view of hydrostatic steering motor and hydraulic system oil filter with front grille screen removed.

Fig. 5–Reach through top opening to remove snap ring (Q–Fig. 6) then lift off housing unit.

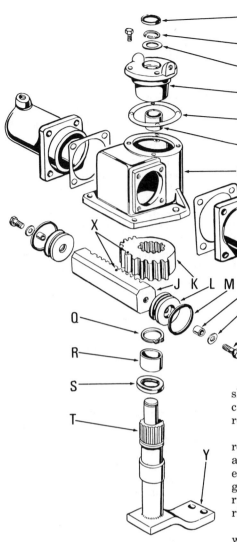

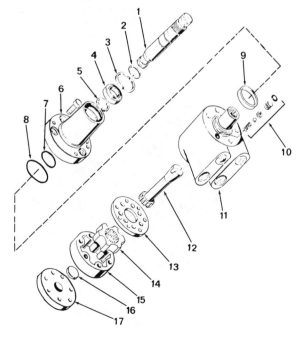

Fig. 6—Exploded view of hydrostatic power steering motor.

A. Expansion plug
B. Snap ring
C. Thrust washer
D. Top cover
E. Gasket
F. Bushing
G. Motor housing
H. Gasket
I. Cylinder
J. Rack
K. Pinion gear
L. Piston
M. Piston ring
N. Spacer
O. Washer
P. Cap screw
Q. Snap ring
R. Bushing
S. Seal
T. Steering shaft
X. Timing marks
Y. Steering arm

edge of chamfer in bore. Install seal (S) with lower edge flush to 1/64-inch below edge of housing boss. Install bushing (F) with upper edge flush to 1/64-inch below chamfer in cover (D).

Coat lip of seal (S) with lithium base bearing grease, then install housing over center steering arm shaft. Install snap ring (Q). Turn steering shaft until arm (Y) points straight to rear and install rack and pinion with timing punch marks (X) aligned as shown in Fig. 7.

Coat outer circumference of pistons (L—Fig. 6) with petroleum jelly and carefully install cylinders using new gaskets. Install top cover (D), thrust washer (C) and snap ring (B). Install a new expansion plug (A) using LOC-TITE to prevent oil leaks. Tighten piston cap screws (P) to a torque of 30-35 ft.-lbs., cylinder retaining screws to 50-55 ft.-lbs. and top cover cap screws to 33-38 ft.-lbs. Tighten motor housing mounting cap screws to 275-285 ft.-lbs.

Cycle the system several times after completing the assembly to air-bleed the cylinders and lines, then refill hydraulic reservoir if necessary as outlined in paragraph 100.

HYDROSTATIC HAND PUMP

The hydrostatic hand pump (Char-Lynn ORBITROL) is a remote hydraulic control and metering valve which directs pressurized fluid to the power steering motor. The unit also serves as a manual control valve and pump to provide steering effort when hydraulic pressure is not available. No mechanical linkage exists between the steering wheel and tractor front wheels.

8. **OPERATION.** The internal gear type hand pump (14 & 15—Fig. 8) con-

shaft unit (T) can be lifted out if indicated, after disconnecting steering tie rods.

If Teflon piston rings (M) are to be renewed, heat new rings in water to approximately 180° F. to soften and expand the rings. Install in piston grooves then clamp both pistons in a ring compresser while cooling, to allow rings to return to original size.

Bushings (F & R) are pre-sized and will not require reaming after installation. Install lower bushing (R) with open end of oil groove to inside, with top of bushing flush to 1/16-inch below

thrust washer (C); then unbolt and remove top cover (D).

Lift out pinion gear (K), then withdraw rack (J) with pistons attached, through either side opening. Reach through top opening and remove snap ring (Q), then unbolt and lift off motor housing (G). Center steering arm and

Fig. 7—Timing punch marks must be aligned and opposite center of steering arm when unit is reassembled.

Fig. 8—Exploded view of ORBITROL hydrostatic hand pump showing component parts. Body and spool assembly (11) is available only as a unit which contains check valve (10), but can be further disassembled for cleaning and inspection as outlined in text.

1. Input shaft
2. Snap ring
3. Snap ring
4. Bearing
5. Snap ring
6. Shaft housing
7. Quad ring
8. O-ring
9. Bushing
10. Check valve
11. Valve assembly
12. Drive link
13. Metering plate
14. Inner rotor
15. Outer stator
16. Disc
17. End plate

nects hydraulically to the opposing cylinders of the steering motor. The spring-loaded rotary valve (11) connects to hydraulic system standby pressure and system return lines.

Initial turning effort of the steering wheel meets resistance at the front wheels causing deflection of the rotary valve centering springs, metering a flow of pressurized hydraulic fluid through the rotary valve and hand pump to the steering motor. Return fluid from opposite cylinder of steering motor passes through return ports of rotary valve to the reservoir. Steering action and speed are thus in direct control of the operator at all times, steering EFFORT only, being supplied by hydraulic system standby pressure.

When hydraulic system pressure is not available, the hand pump operates manually to steer the tractor, with return fluid passing through check valve (10) to intake side of hand pump.

Fig. 9–Remove cap screws (1) and lift off metering pump components as shown in Fig. 10.

Fig. 10–Lift off metering pump.

9. **REMOVE AND REINSTALL.** Remove the hood, side panels and battery platform. Clean pump and surrounding area and disconnect the four

Fig. 11–Carefully invert the housing on a clean board to unbolt and remove input shaft and housing.

Fig. 12–Input shaft housing is positioned by a locator bushing (1).

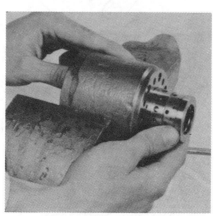

Fig. 13–Carefully push control valve spool and sleeve out bottom of housing as shown.

hydraulic lines, plugging valve ports as removed; then unbolt and remove hand pump unit.

Install by reversing the removal procedure. Cycle the system several times to air-bleed, then check and refill hydraulic system reservoir if necessary as outlined in paragraph 100.

10. **OVERHAUL.** Refer to Fig. 8 for a partially exploded view of the Char-Lynn ORBITROL hydrostatic hand pump. The rotary valve spool, sleeve and housing (11) is available only as an assembly which includes reaction springs, cross pin and check valve (10).

To disassemble the removed Orbitrol unit, first remove the cap screws (1—Fig. 9) retaining metering pump to bottom of housing and lift off pump components as shown in Fig. 10. Carefully invert the housing on a clean board as shown in Fig. 11 to prevent the rotary valve unit from sliding out, remove the four cap screws retaining top cover and lift off the cover as shown. Remove locator bushing (1—Fig. 12).

Carefully push control valve spool and sleeve assembly out bottom of housing as shown in Fig. 13.

NOTE: Be careful valve unit does not bind. Parts are fit to extremely close tolerance and a twisting motion may be required for withdrawal.

Insert a suitably bent wire through port nearest check valve plug and push out the plug as shown in Fig. 15. Re-

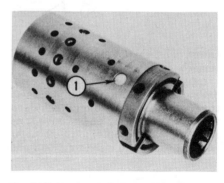

Fig. 14–Valve cross pin is positioned by two Nylon discs (1).

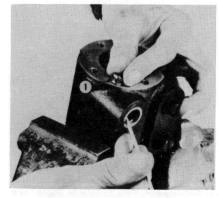

Fig. 15–Insert a suitably bent wire through valve port to push out check valve plug (1).

move the threaded check valve seat (1—Fig. 16) using an Allen wrench and remove check valve ball (2) and spring (3).

Remove nylon cross pin discs (1—Fig. 14) and cross pin, then separate spool and sleeve. Withdraw the six arched centering springs. Thoroughly clean all parts in a suitable mineral solvent and blow dry with air. Inspect lapped surfaces for scoring or other damage. Mating surfaces of cover (17—Fig. 8), rotor assembly (14 & 15) and metering plate (13) can be hand lapped to remove burrs, slight scratches or other imperfections. Use 600 grit abrasive paper or lapping compound. Rinse in clean solvent and blow dry after lap-

ping is complete. Rotor units are only available as a matched set.

Inspect centering springs (S—Fig. 17) for fractures or distortion. Springs should have a minimum arch of 7/32-inch when measured at center. Inspect spool, sleeve and body for nicks, scoring or wear. If any part is damaged, renew the housing assembly which includes spool, sleeve and check valve. Renew all O-rings when reassembling the unit.

Tighten the check valve seat (1—Fig. 16) to a torque of 150 inch-pounds. Install centering springs (S—Fig. 17) in sets of three with notched edge down. Insert the assembled spool and sleeve

assembly from bottom of valve body using a twisting motion. Do not allow sleeve to move up beyond flush with upper machined surface of body. Be sure pin slot in drive shaft aligns with valleys of inner rotor as shown in Fig. 18. Rotor has six gear teeth and 12 splines; if drive slot is improperly aligned, steering unit will operate in reverse of hand pressure. Tighten the seven lower cap screws evenly to a torque of 250 inch-pounds and the four cap screws in top cover to a torque of 220 inch-pounds.

Install the assembled unit as outlined in paragraph 9 and bleed by cycling the system until steering action is firm and response is immediate.

Fig. 16–Check valve seat (1) can be removed with an Allen wrench, then check ball (2) and spring (3) removed.

Fig. 17–Schematic end view of steering valve sleeves showing correct installation of drive pin (P) and centering springs (S).

Fig. 18–When properly assembled, slot in drive link will align with valley of inner gear as shown.

ENGINE AND COMPONENTS

All models are equipped with a Perkins Model AV8.510 Diesel Engine of V8 design, having a bore of 4¼ inches, a stroke of 4½ inches and a displacement of 510.7 cubic inches.

R&R ENGINE WITH CLUTCH

11. To remove the engine and clutch as a unit, first drain cooling system and if engine is to be disassembled, drain oil pan. Remove hood and nose piece as a unit. Remove batteries, battery carrier and steering support side cover plates.

Back out the four cap screws securing main hydraulic pump drive adapter to crankshaft pulley. Disconnect coolant hoses, muffler pipe, air cleaner hose and oil cooler hoses at front of cylinder block. Disconnect the six upper hydraulic system tubes at rear of engine.

Support tractor under transmission housing and front end assembly from a hoist, then unbolt and remove front axle, front support and radiator assembly as a unit as shown in Fig. 19.

Disconnect fuel supply and return lines, oil pressure gage line, throttle

control and fuel shut-off cable. Disconnect electrical wiring harness, temperature sending unit and tachometer drive cable. Remove the cap screws securing priority valve housing to battery support bracket. Unbolt and remove starter motor.

Support the engine with a hoist from all four lifting brackets as shown in Fig. 20; remove the retaining cap screws and swing the engine from its doweled position on clutch housing.

Install by reversing the removal procedure. Bleed fuel system as outlined in paragraph 46 and adjust throttle controls as in paragraph 56 after engine is installed.

CYLINDER HEAD

12. To remove either cylinder head, first drain cooling system and remove hood panel. Remove air induction manifold, coolant outlet and coolant bypass hose. Remove the interfering injector lines and injectors. Remove intake manifold then remove the banjo bolts connecting rocker arm shaft oil feed pipes to cylinder head.

Remove rocker arm covers, rocker arm assemblies and push rods. Discon-

Fig. 19–Removing front end as a unit for engine work.

Fig. 20–Attach hoist to all four lifting brackets when removing engine.

nect breather elbow at rear inside corner of head. Loosen then remove cylinder head stud nuts and cap screws, attach a hoist to the two lifting brackets and lift off cylinder head as shown in Fig. 21.

Cylinder heads are interchangeable from right to left providing necessary attachments are transferred. Cylinder head may be resurfaced providing no more than 0.015 is removed; nozzle protrusion does not exceed 0.148; or cylinder head thickness is not reduced to less than 3.970. If cylinder head is re-machined, test for leaks at 30 psi using warm water.

Cylinder head gasket is marked "TOP" for proper installation. Install gasket dry and tighten cylinder head cap screws and stud nuts progressively to a torque of 120-125 ft.-lbs. using the sequence shown in Fig. 22. Note that two different length cap screws are used to retain the head; the longer screws should be installed in center row, shorter screws at top of head. Install rocker arm assemblies as outlined in paragraph 18 and adjust tappet gap as in paragraph 19.

VALVES AND SEATS

13. Intake valves seat directly in the cylinder head while exhaust valves are equipped with hardened seat inserts. Original valves are numbered to correspond with a number stamped in combustion chamber area of head, each head being numbered individually. Intake valve must not protrude more than 0.054 nor exhaust valve more than 0.061. Intake valve (and/or seat) must be renewed if seated valve protrudes less than 0.001; exhaust valve (and/or seat) if protrusion is less than 0.008.

Valve and seat angle is 45° for all valves. The renewable exhaust valve seat inserts should have 0.0015-0.003 interference fit in cylinder head bore.

VALVES AND SEATS

14. Intake and exhaust valve guides are interchangeable. Refer to Fig. 23. Press all guides into head until distance (A) between top end of guide and spring seat of head measures ⅞ inch.

Inside diameter of a new guide is 0.375-0.376 and desired diametral clearance is 0.0015-0.0035 for intake valves or 0.002-0.004 for exhaust valves. Maximum permissible clearance for any valve is 0.0055, renew valves and/or guides if clearance is excessive.

VALVE SPRINGS

15. Springs, retainers, locks and oil deflectors are interchangeable for intake and exhaust valves. Inner and outer springs are used. Renew springs if they are distorted or heat discolored; or if they fail to meet the test specifications which follow. Renew other parts if they show signs of wear or damage.

INNER SPRING
Free length (approximate) . . 1 37/64 inches
Test (closed) length . . . . . 1 1/4 inches
Lbs. Test . . . . 30-34 lbs.
OUTER SPRING
Free length (approximate) . . 1 21/32 inches
Test (closed) length . . 1 21/64 inches
Lbs. Test . . . . 61-67 lbs.

VALVE ROTATORS

16. Intake and exhaust valves are equipped with positive type valve rotators which cause the valve to turn a slight amount each time the valve opens. Normal service consists of renewing the units, which are installed next to the cylinder head as shown in Fig. 24.

CAM FOLLOWERS

17. Mushroom type cam followers (Tappets) are contained in removable guide blocks as shown in Fig. 25. The guide block (containing the two cam followers for one cylinder) can be removed without major engine disassembly after removing injection pump and tappet cover. The 0.7475-0.7485 diameter tappets should have 0.006-0.0075 clearance in tappet bores. When installing guide blocks, tighten re-

Fig. 21–Attach hoist to two lifting brackets to remove cylinder head.

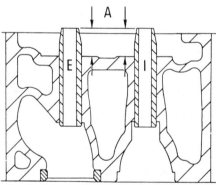

Fig. 23–Distance (A) from top of valve guides to spring seat should measure ⅞ inch.

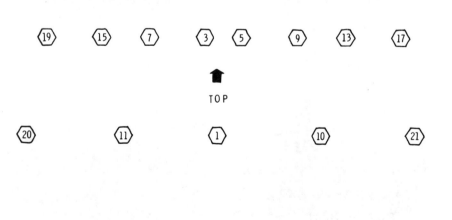

Fig. 22–Tighten cylinder head cap screws and stud nuts progressively to a torque of 120-125 ft.-lbs. using the sequence shown.

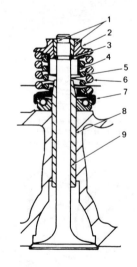

Fig. 24–Cross sectional view of installed valve showing component parts.

1. Valve locks
2. Retainer cap
3. Sealing O-ring
4. Oil deflector
5. Inner valve spring
6. Outer valve spring
7. Valve rotator
8. Valve guide
9. Valve

taining cap screws to a torque of 55 ft.-lbs.

ROCKER ARMS

18. Rocker arm covers can be removed without additional disassembly.

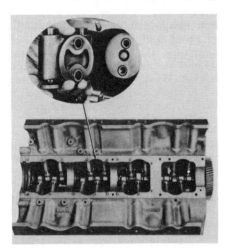

Fig. 25–Cam follower guide blocks are attached to cylinder block with cap screws and are removable in pairs as shown.

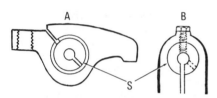

Fig. 26–Cross sectional views of rocker arm, shaft and bracket showing correct assembly. Refer to paragraph 18.

A. Rocker arm
B. Bracket

S. Shaft

The assembled rocker shafts will interchange between right-front and left-rear; or left-front and right-rear. Components of rocker shaft units are interchangeable in all positions. If rocker arms assemblies are removed for other work, make sure the bracket containing the shaft locating cap screw is positioned on attaching pads containing the lubricating pressure passages (P—Fig. 27 or 28). If shaft units are disassembled, the following conditions must be met when reassembling.

1. Bracket boss must point to right when facing head.

2. Bracket containing locating screw must align with mounting pad containing oil passage.

3. Locating screw must enter LARGE hole in rocker shaft (S—Fig. 26), thus aligning small hole with bracket oil passage.

4. Oil feed holes in rocker shaft must point toward valve stem end of rocker arms.

The cast iron rocker arms are interchangeable and are equipped with renewable bushings. Press bushings into rocker arm bores with oil holes aligned. Bushings are prefinished and should have 0.002-0.0045 clearance on the 0.7485-0.7495 diameter shaft when installed. Tighten rocker shaft bracket stud nuts to a torque of 20-25 ft.-lbs. and rocker cover screws to 45-50 inch pounds. Adjust tappet gap as outlined in paragraph 19.

VALVE TAPPET GAP

19. The recommended cold tappet gap setting is 0.012 for both the intake and exhaust valves. Cold (static) setting of all valves can be made from just two crankshaft positions, using the procedure shown in Figs. 27 and 28. Proceed as follows:

Remove rocker arm covers and turn crankshaft until TDC timing mark is aligned with notch in inspection hole as shown in Fig. 29.

Check the rocker arms for No. 1 and No. 4 cylinders. If both rocker arms on No. 1 cylinder are loose and both rocker arms on No. 4 cylinders tight, No. 1 cylinder is on compression stroke; adjust the eight tappets shown in Fig. 27. If

Fig. 29–View of flywheel housing showing TDC timing mark aligned.

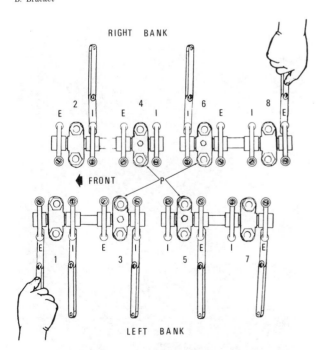

Fig. 27–With TDC timing mark aligned and No. 1 piston on compression stroke, adjust the indicated valves as outlined in paragraph 19. Numbers indicate cylinder numbering system and (P) denotes the rocker arm brackets containing oil pressure passages.

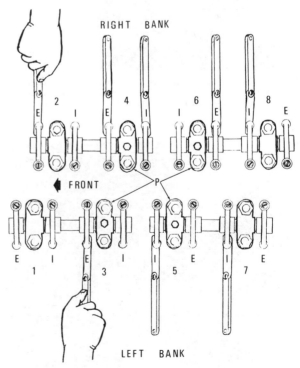

Fig. 28–With TDC timing mark aligned and No. 4 piston on compression stroke, adjust the indicated valves as outlined in paragraph 19.

both rocker arms on No. 4 cylinder are loose and rocker arms on No. 1 cylinder are tight, No. 1 cylinder is on the exhaust stroke; adjust the eight valves indicated in Fig. 28.

When the adjustment is completed, turn crankshaft one complete turn until TDC timing mark is again aligned, then adjust the remaining tappets.

Recheck the adjustment if desired, with engine running at slow idle speed. Recommended tappet clearance with engine at operating temperature is 0.010. Clearance may be adjusted with engine at slow idle speed or by following the procedure recommended for initial adjustment.

TIMING GEAR COVER

20. The timing gear cover consists of an upper section (Fig. 30) which contains mounting points for the water pump and alternator; and a lower section (Fig. 31) which houses the crankshaft oil seal.

To remove the timing gear cover, first remove hood and front end assembly as outlined in paragraph 11.

21. **UPPER COVER.** To remove the upper cover half after front end is re-

moved, first unbolt and remove fan blades. Loosen fan belt idler and remove fan belt. Remove remaining coolant hoses, remove retaining cap

Fig. 32–Crankshaft pulley (P) uses split, tapered retainer (R) to lock pulley on crankshaft. Refer to paragraph 22 for removal procedure.

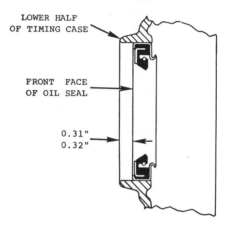

Fig. 33–Cross sectional view of crankshaft front oil seal showing installation dimensions.

screws and lift off upper timing gear cover section with alternator and water pump attached as shown in Fig. 30. Install by reversing the removal procedure.

22. **LOWER COVER.** To remove the lower half of timing gear cover it is first necessary to remove hood and front end assembly as outlined in paragraph 11 and upper cover half as in paragraph 21. Loosen oil pan screws. Remove the cap screw (S—Fig. 32) and washer (W) securing crankshaft pulley (P) to shaft. Using a center, attach a suitable puller to crankshaft pulley and apply slight pressure with pulley forcing screw; then strike forcing screw sharply with a hammer to bounce out the tapered locking ring (R). Pulley (P) can now be easily withdrawn.

The lip-type crankshaft oil seal should be installed flush with inside of cover bore (5/16-inch below front of bore) as shown in Fig. 33. Coat lip of seal with lithium type grease when assembling and tighten pulley retaining cap screw to a torque of 275-300 ft.-lbs.

TIMING GEARS

23. Fig. 34 shows a view of timing gear train with cover removed. Before attempting to remove any of the gears, first remove rocker arm covers and rocker arms to avoid the possibility of damage to pistons or valve train if camshaft or crankshaft should either one be turned independently of the other.

Fig. 30–Timing gear cover upper section can be lifted off as part of an assembly containing water pump, alternator and alternator support.

Fig. 31–Removing timing gear cover lower section.

Fig. 34–Timing gear train with timing marks aligned. Marks are aligned during assembly and will line up only occasionally during operation because of odd number of teeth in idler gears.

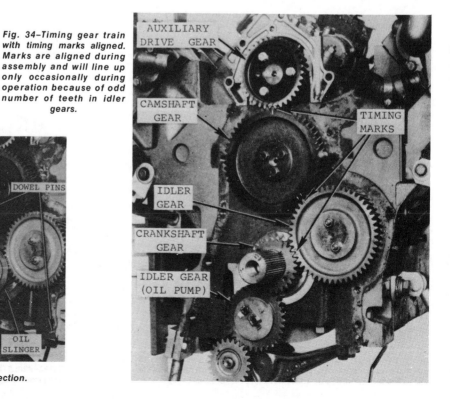

Timing gear backlash should be 0.006-0.009 between any of the gears in main timing gear train. Replacement gears are available in standard size only. If backlash is not within limits, renew gears, idler shafts, bushings or other items concerned.

NOTE: Because of the odd number of teeth in idler gear and auxiliary drive gears, all timing marks will not align with each revolution of the camshaft. Failure of marks to align does not necessarily indicate engine is mistimed.

To remove the timing gears or time the engine, refer to the appropriate following paragraphs.

24. **IDLER GEAR AND HUB.** The timing idler gear (Fig. 36) should have a diametral clearance of 0.001-0.004 on the 2.248-2.249 diameter hub. Sug-

gested end play is 0.012-0.023, with a wear limit of 0.030. Hub is a light interference fit in mounting bore of cylinder block and the two mounting studs are slightly offset, hub can only be installed in one position. The prefinished idler gear bushing is 0.002-0.005 press fit in gear bore and is renewable. Tighten retaining stud nuts to a torque of 27-30 ft.-lbs. when reinstalling.

25. **CAMSHAFT GEAR.** The camshaft is a transition fit in gear counterbore and is positively located by a dowel (1—Fig. 38) and locating slot (2). Gear is retained to camshaft by three cap screws. When installing the camshaft gear, use LOCTITE Stud N' Bearing Mount in capscrew threads and tighten screws to a torque of 27-30 ft.-lbs.

26. **CRANKSHAFT GEAR.** The crankshaft gear is a transition fit (0.0005 tight to 0.0015 loose) on shaft, and can usually be removed with a pry bar after timing gear cover is removed. When reinstalling crankshaft gear, remove upper idler gear and retime engine as outlined in paragraph 30.

27. **AUXILIARY HOUSING.** The auxiliary housing and drive mounts on

top of the engine block and drives the injection pump gear. The auxiliary housing can be removed independently of the timing gears by following the procedures outlined in paragraph 28; or during timing gear overhaul by following only those portions which apply.

28. **REMOVE AND REINSTALL.** To remove the auxiliary drive housing assembly, first drain cooling system and remove hood, air intake hoses and interfering air baffles. Remove coolant hoses and exhaust pipes. Disconnect alternator wiring, tachometer drive cable, fuel stop cable and throttle linkage.

Disconnect and remove fuel injection pump as outlined in paragraph 57. Remove alternator, water pump and timing gear cover upper half as a unit as shown in Fig. 30. Remove the retaining cap screws and lift off auxiliary drive housing as shown in Fig. 39.

When installing the auxiliary drive housing assembly, the injection pump must be correctly timed as follows: Make sure No. 1 cylinder is at TDC on compression stroke (NOTE: The single slash-marked tooth on timing gear should point up as shown in Fig. 40).

Fig. 35–Checking timing gear backlash using a feeler gage.

Fig. 37–Installing idler gear hub. Stud holes are slightly offset in hub center to assure proper alignment of idler gear oil hole.

Fig. 39–Removing auxiliary housing and shaft assembly.

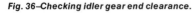

Fig. 36–Checking idler gear end clearance.

Fig. 38–Dowel (1) in camshaft gear must align with notch (2) in cam for proper timing.

Fig. 40–The single slash mark (circled) on timing gear is at the top as shown, when No. 1 piston is at TDC on compression stroke.

Install the auxiliary drive housing with the double punch-marked drive gear tooth space aligned with marked camshaft gear tooth as shown in Fig. 34. Tighten the retaining cap screws to a torque of 16 ft.-lbs. and install injection pump as outlined in paragraph 57.

29. OVERHAUL. The auxiliary drive shaft rotates in pressure lubricated, precision type bearings which attach to the housing with bearing caps. Diametral clearance of shaft in the bearings should be 0.0005-0.002, with a maximum permissible clearance of 0.003. Drive shaft end play of 0.001-0.011 is controlled by thrust washers which can be renewed after removing bearing caps. Drive shaft journal diameter is 1.374-1.3745.

Main bearing caps have stamped locating numbers as shown in Fig. 41 and are positioned by ring dowels which prevent incorrect installation. Gears are positioned to shaft by locating dowels and secured by cap screws. When reassembling the auxiliary drive shaft housing, tighten all cap screws to a torque of 16 ft.-lbs.

30. **TIMING THE GEARS.** Timing marks are affixed to crankshaft gear, idler gear, camshaft gear and auxiliary drive gear as shown in Fig. 34. Because of the odd number of teeth in idler gear and auxiliary drive gear, the marks on these gears will not align with every other crankshaft revolution and failure to align does not mean engine is improperly timed.

NOTE: The odd number of teeth prevents the development of wear patterns which coincide with stress loads of crankshaft and camshaft and therefore contribute to longer gear life with less gear noise.

When assembling the timing gears, align all timing marks as shown in Fig. 34. To check the timing when gears have not been removed, turn crankshaft until No. 1 piston is at TDC on compression stroke and the single punchmarked crankshaft gear tooth is meshed with idler gear; then count the

number of idler gear teeth between the marked crankshaft gear tooth and the first marked tooth on camshaft gear. The number should be nine. The auxiliary drive gear timing need not be checked if injection pump is correctly timed as outlined in paragraph 58.

CAMSHAFT

31. To remove the camshaft, first remove timing gear lower cover as outlined in paragraph 22, rocker arms as in paragraph 18 and camshaft gear as in paragraph 25. Remove tappet covers (paragraph 17) then remove or block up the cam followers (tappets). Remove the three cap screws retaining the camshaft thrust plate and withdraw the camshaft from block bore as shown in Fig. 42.

Normal camshaft bearing clearance is 0.002-0.006 for all journals, with a maximum permissible clearance of 0.0095. Normal camshaft end play is 0.002-0.014 with a maximum permissible end play of 0.020. End play is controlled by thickness of thrust washer and the annular thrust surface on rear face of camshaft gear. Pressure lubrication for rocker arms is metered by the third camshaft bearing journal.

When installing the camshaft, tighten the thrust plate retaining cap screws to a torque of 10-12 ft.-lbs. Install gear retaining cap screws using LOCTITE and tighten to a torque of 27-30 ft.-lbs. Camshaft journal diameters are as follows:

Front journal 2.371-2.373
No. 2 journal 2.241-2.243
No. 3 journal 2.231-2.233
No. 4 journal 2.221-2.223
No. 5 journal 2.211-2.213
Cam lift 0.330-0.333

ROD AND PISTON UNITS

32. Connecting rod and piston units are removed from above after removing cylinder head(s) and oil pan. Cylinder numbers are stamped on connecting rod and cap. On all cylinders except No's. 4 and 5, crankshaft must be turned after removing rod cap, to permit connecting rod to pass crankshaft counterweights. Bearings are available in undersizes of 0.010, 0.020

and 0.030 and crankshaft may be reground to these undersizes. When installing connecting rods, tighten retaining cap screws to a torque of 90-95 ft.-lbs.

PISTONS, SLEEVES AND RINGS

33. The aluminum alloy, cam ground pistons are supplied in standard size only. The toroidal combustion chamber is slightly offset in piston crown and top of piston contains relief pockets for valve heads. Because the same piston is used in both banks of the engine, there is no "Front" or "Rear" orientation. Rather, the piston is installed in cylinder with valve relief pockets (R—Fig. 43) up (toward center of vee).

Each piston is fitted with three chrome compression rings and one chrome oil ring, all located above the piston pin. The plain faced chrome top ring may be installed either side up in groove. The internally stepped chrome 2nd and 3rd compression rings are marked "TOP" for correct installation. The chrome faced oil control ring is equipped with an integral spring expander and may be installed either side up. Ring specifications are as follows:

End Gap
   All rings 0.017-0.024
Side Clearance
   Top ring 0.004-0.006
   All other rings 0.002-0.004
   Maximum wear limit, top
     ring 0.010

The production cylinder sleeves are 0.001-0.003 press fit in cylinder block. Service sleeves are a transition fit (0.001 tight to 0.001 clearance) in block

Fig. 43–Valve relief pockets (R) and long side of rod must be together and toward the top when rod and piston unit is assembled. Cylinder number (N) is also at top and right side up on factory assembled units.

Fig. 41–Assembled view of auxiliary housing and shaft unit.

Fig. 42–Installing camshaft in engine block.

bores and cylinder walls are finished. Use a heavy-duty sleeve puller capable of high pulling force to remove production sleeves. Thoroughly clean cylinder bore, giving particular attention to counterbore. Carefully examine sleeve contact area in cylinder block and remove any burrs, rust spots or other foreign material.

Thoroughly clean new service sleeves in a solvent and air day. Chill sleeves in dry ice or a deep-freeze and install dry. Push sleeve in completely by hand without stopping. When correctly installed, upper step of sleeve (A—Fig. 44) should extend 0.020-0.026 above gasket surface of block and flange area (B) should be flush to 0.004 below block surface. Maximum cylinder distortion after installation must not exceed 0.0025.

Cylinder sleeves should be renewed if scored, or if wear or taper exceeds 0.004. Crown of installed piston must be flush to 0.0085 BELOW gasket surface of block. Installed diameter of service sleeve is 4.251-4.252.

PISTON PINS

34. The 1.6248-1.6250 diameter, floating type piston pins are retained in piston bores by snap rings and are available in standard size only. The renewable connecting rod bushing must be final sized after installation to provide the recommended 0.0007-0.0018 diametral clearance for piston pin. Be sure the pre-drilled oil hole in bushing is properly aligned with hole in top of connecting rod and install bushing from chamfered side of bore. Piston pin is a transition fit (0.0000-0.0005 clearance) in piston bosses and piston should be heated to 120° F for pin removal or installation. Valve relief pockets (R—Fig. 43) and long leg (L) of rod must be on same side of assembly when pin is installed.

CONNECTING RODS AND BEARINGS

35. Connecting rod bearings are precision type, renewable from below after removing oil pan and rod bearing caps. When renewing bearing shells, make sure that the projection engages milled slots in rod and cap and that correlation marks on rod and cap are in register.

Connecting rod bearings should have a diametral clearance of 0.0023-0.0042 on the 3.248-3.2485 diameter crankpin. Recommended connecting rod side clearance is 0.015-0.022 with both rods fitted on crankpin journal. Bearings are available in undersizes of 0.010, 0.020 and 0.030 as well as standard. Tighten connecting rod cap screws to a torque of 90-95 ft.-lbs. when units are installed.

CRANKSHAFT AND BEARINGS

36. The crankshaft is supported in five precision type main bearings. To remove the rear main bearing cap, it is first necessary to remove the engine, clutch, flywheel and flywheel housing. The front main bearing cap carries the oil pump idler gear hub and removal of front bearing requires prior removal of oil pump and lines. All other main bearing caps can be removed from below after removing oil pan.

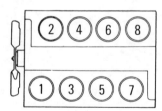

Fig. 45–Schematic view of block showing cylinder numbering sequence.

Main bearing cap cross bolts (Fig. 46) should be removed before loosening regular cap screws. Crankshaft end play is controlled by thrust washer halves located in block portion only, in front of and to the rear of center main bearing. A locating tab (Fig. 47) is machined on one end of thrust washer half and washer can be rolled out after removing main bearing cap.

Bearing inserts are available in undersizes of 0.010, 0.020 and 0.030 as well as standard; and thrust washer halves are available in standard thickness of 0.122-0.125, and oversize of 0.129-0.132. Recommended main bearing diametral clearance is 0.004-0.0063 and recommended crankshaft end play is 0.002-0.017. Tighten main bearing retaining cap screws to a torque of 190-200 ft.-lbs. and cross bolts to 45-50 ft.-lbs. Refer to paragraph 37 for installation of rear seal and oil pan bridge piece and to paragraph 41 for installation of oil pump and idler gear. Check the crankshaft against the values which follow:

Main journal diameter . 3.9967-3.9972
Crankpin diameter 3.248-3.2485

CRANKSHAFT REAR OIL SEAL

37. Most engines are equipped with an asbestos or Teflon rope type oil seal contained in a two-piece retainer. The seal can be removed after removing flywheel and flywheel housing.

Fig. 47–Identical thrust washer halves are located in block portion only of center main bearing.

Fig. 48–Crankshaft thrust washer halves can be pushed out as shown, without removing crankshaft.

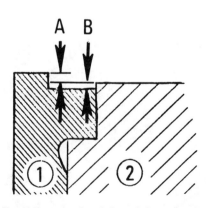

Fig. 44–Cross sectional view of sleeve counterbore showing installation dimensions. Upper step of sleeve (A) should extend 0.020-0.026 above gasket surface of block. Flange area (B) should be flush to 0.004 below block surface.

Fig. 46–Main bearing caps are provided with cross bolts for additional security.

Some late models have a one-piece lip type seal which seats on a wear ring pressed on crankshaft flywheel flange. The late type seal can be installed on earlier models with some modification.

The early, rope type seal is precision cut to length and must be installed in retainer halves with 0.025-0.035 of seal ends protruding from each end of retainer. Do not trim the seal. To install the rope half, clamp each half of retainer in a vise as shown in Fig. 49. Make sure seal groove is clean. Start each end of seal in groove with the specified amount of seal protruding. Allow seal rope to buckle in the center until about an inch of each end is bedded in groove, work center of seal into position then roll with a round bar as shown. Repeat the process with other half of seal.

On models with lip type seal, thread sealant must be used on those housing bolts which extend through block and on all flywheel cap screws, to eliminate the possibility of oil seepage.

When installing the cylinder block bridge piece, coat mating surfaces of block, bridge piece and seal retainer with a suitable non-hardening sealant and install using new end seals. Align gasket surfaces of bridge piece and cylinder block using a straightedge, and tighten retaining cap screws to a torque of 12 ft.-lbs.

FLYWHEEL

38. The engine flywheel can be removed after separating engine from clutch housing and removing clutch.

NOTE: Flywheel weighs about 150 pounds and is approximately 6 inches deep. To safely handle the flywheel during removal, a special handling tool is required. Tool can be shop constructed as shown in Fig. 51.

The flywheel is attached to crankshaft flange by twelve capscrews which are equally spaced. To assure correct alignment of flywheel timing marks, arrows (Fig. 52) on flywheel and crankshaft must align when flywheel is installed.

The starter ring gear can be renewed after flywheel is removed. Heat ring gear evenly to approximately 480° F. and install on flywheel with beveled end of teeth facing front of engine.

Install flywheel with alignment marks in register and tighten the twelve retaining cap screws to a torque of 75-80 ft.-lbs. as shown in Fig. 53. Maximum allowable face runout is 0.001 for each inch from flywheel centerline to point of measurement.

FLYWHEEL HOUSING

39. The cast flywheel housing is doweled to engine block and secured with cap screws. Housing can be removed after removing flywheel and must be removed for removal of crankshaft or crankshaft rear oil seal. When installing flywheel housing, tighten the retaining cap screws to a torque of 50 ft.-lbs.

Fig. 50—When installing cylinder block bridge piece, use a straight-edge to align gasket surfaces as shown.

Fig. 51—The heavy (150 lb.) flywheel is difficult to remove or install without the special handling tool which can be locally made as shown.

Fig. 49—The early model rope-type seal must be bedded in using a round bar as shown. Late units use a one-piece lip-type seal which seats on a wear ring on crankshaft flange.

OIL PAN

40. The heavy cast iron oil pan serves as part of the tractor frame and as an attaching point for tractor front support. If oil pan must be removed with engine installed in tractor, the following procedure is suggested, however, extreme care must be observed to prevent damage to tractor.

Drain the oil pan. Use a wood block on jack to distribute the weight, place jack underneath the oil pan and jack up the tractor until most of the weight is removed from front wheels. With tractor so supported, solidly block up beneath flywheel housing and front end of transmission, spreading the load over a wide housing area, then lower the jack underneath oil pan. Slightly loosen the six cap screws securing front support to oil pan and cylinder block as a check to make sure weight is properly supported.

With tractor secure, unbolt and remove oil pan, using a rolling floor jack or other means to lower and remove the pan unit. Install by reversing the removal procedure. Tighten oil pan cap screws to 15 ft.-lbs.

OIL PUMP

41. The gear type oil pump is mounted underneath the front main

Fig. 52—Crankshaft flange and flywheel are inscribed with arrows which must be aligned during installation to properly align flywheel timing marks.

Fig. 53—The flywheel adapter provides the drive for the IPTO shaft.

bearing cap and driven by the crankshaft gear through an idler. Oil pump can be removed as shown in Fig. 54 after removing oil pan.

The oil pump must maintain 30 psi of system pressure at normal operating speeds. Oil pump is serviced as an assembly only and individual parts are not available. Normal pump gear end play is 0.002-0.008 with cover gasket installed; pump must be renewed if end play exceeds 0.012. Radial gear clearance in housing should be 0.002-0.010 when measured as shown in Fig. 55. Backlash should be within the limits of 0.014-0.018 when measured as shown in Fig. 56. Drive gear backlash should not be less than 0.003 nor more than 0.012 when measured as shown in Fig. 57. Idler gear and idler gear hub are renewable individually. Diametral clearance of idler gear should be 0.001-0.004 and idler gear end play should be 0.015-0.022, with a maximum wear limit of 0.030.

RELIEF VALVE

42. The oil pressure relief valve (Fig.

Fig. 54–Removing engine oil pump and associated parts.

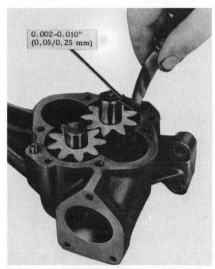

Fig. 55–Radial gear clearance should be 0.002-0.010 when measured with feeler gage as shown.

59) can be removed from engine block after removing oil pump as outlined in Fig. 41. The valve is pre-set to 60-65 psi; no attempt should be made to adjust the pressure except by renewal of worn or damaged parts.

OIL COOLER AND RELIEF VALVE

43. An engine oil cooler is located in bottom tank of coolant radiator. Full pressure engine oil is routed from oil pump and relief valve to an oil cooler

Fig. 56–Pump gear backlash can be measured with a feeler gage.

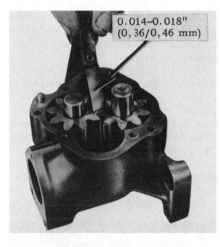

Fig. 57–Pump drive gear backlash should be 0.003-0.012 when measured with a feeler gage as shown.

Fig. 58–Idler gear hub has a drilled lubrication passage as shown.

relief valve housing assembly located on top of oil filter (Right side of Engine Block—Fig. 60), through the oil cooler; then to filter unit and engine oil gallery. The oil cooler relief valve housing contains a by-pass valve which operates on a 20 psi pressure differential to bypass the coolant radiator when oil is cold. The oil filter body contains a filter bypass valve which operates on a 15 psi pressure differential to provide engine lubrication in the event of cold oil or a plugged filter element.

AIR INTAKE SYSTEM

44. The engine air cleaner is of the dry element type protected by a mechanical type air restriction indicator.

IMPORTANT NOTE: Advise customers to observe indicator daily (or oftener in extreme dust conditions) and to service air cleaner ONLY WHEN "RED" WARNING SIGNAL IS LOCKED IN FULL VIEW WITH ENGINE STOPPED.

Restriction indicator should be tested periodically by installing a tee fitting in indicator line and connecting a water manometer to the fitting. With new elements installed and engine running at 2200 rpm, water manometer should read 10 inches. Remove air intake cap (pre-cleaner) and slowly slide a small steel plate over air intake

Fig. 59–Oil pressure relief valve housing attaches to engine block as shown.

Fig. 60–Right side of engine block showing dual engine oil filter and oil cooler by-pass valve.

15

opening while watching manometer. The air restriction indicator should "Lock" the red warning signal in place when manometer reads 25 inches of water.

If restriction indicator red warning fails to lock at 25 inches of water or locks at a substantially lower reading, indicator should be renewed. If manometer reading is materially lower than 10 inches of water at full load, inspect filter elements, canisters, hoses and manifold for air leaks which will admit unfiltered air.

Outer element can be cleaned with dry compressed air (not to exceed 100 psi) or by washing in clean lukewarm (100° F.) water and a non-foaming detergent. Outer element should be renewed after six washings or when paper element is ruptured. (Examine by placing a lighted bulb inside element in a darkened area). Inner element should be renewed at engine overhaul or whenever plugged, ruptured or otherwise damaged.

Fig. 61–View of air cleaner service door on left side of tractor hood, showing air restriction indicator and dry air filter element.

DIESEL FUEL SYSTEM

The diesel fuel system consists of three basic units; the fuel tank and filters, injection pump and injector nozzles. When servicing any unit associated with the diesel fuel system, the maintenance of absolute cleanliness is of utmost importance. Of equal importance is the avoidance of nicks, burrs or handling damage on any of the working parts.

Probably the most important precaution that service personnel can impart to owners of diesel powered tractors is to urge them to use an approved fuel that is absolutely clean and free from foreign materials. Extra precaution should be taken to make certain that no water enters the fuel storage tanks. Because of the high pressures and degree of control required of injection equipment, extremely

high precision standards are necessary in the manufacture and servicing of diesel components. Extra care in daily maintenance will pay big dividends in long service life and avoidance of costly repairs.

FUEL FILTERS AND LINES

45. **OPERATION AND MAINTENANCE.** All units are equipped with dual parallel fuel filters located just behind right rear corner of cylinder block as shown in Fig. 62.

Inspect glass bowl at bottom of both filters daily and drain off any water or dirt accumulation. Renew both elements at 500 hour intervals. Renew elements and clean the tank and lines if evidence of substantial water contamination exists.

46. **BLEEDING.** To bleed the system, open bleed plugs on top of filter units (1 & 3—Fig. 62) and actuate primer lever on fuel lift pump until air-free fuel flows from plug openings. Close bleed plugs while fuel is flowing. Open either bleed screw (3—Fig. 63) on injection pump and continue operating lift pump lever until air is eliminated from injection pump.

Fig. 62–Right rear corner of engine block showing dual parallel fuel filters. Items (1 & 3) are air bleed screws while items (2) are filter element cap screws.

Push in the fuel stop (shut-off) button, partially open the throttle and attempt to start the engine. If engine does not start, loosen compression nut at all injectors and turn engine over with starter until fuel escapes from all loosened connections. Tighten compression nuts and start the engine.

FUEL LIFT PUMP

47. The fuel lift pump is mounted on left side of fuel injection pump and driven by a lobe machined on injection pump camshaft. Pump parts are available separately or in a rebuilding kit. Output delivery pressure should be 5-8 psi.

INJECTOR NOZZLES

WARNING: Fuel leaves the injector nozzle with sufficient force to penetrate the skin. Keep exposed portions of your body clear of nozzle spray when testing.

48. **TESTING AND LOCATING A FAULTY NOZZLE.** If rough or uneven engine operation or misfiring indicates a faulty injector, the defective unit can usually be located as follows:

With engine running at the speed where malfunction is most noticeable (usually slow idle speed), loosen the compression nut on high pressure line for each injector in turn, and listen for a change in engine performance. As in checking spark plugs, the faulty unit is the one which, when its line is loosened, least affects the running of the engine.

If a faulty nozzle is found and considerable time has elapsed since the injectors have been serviced, it is recommended that all nozzles be removed and serviced or that new or reconditioned units be installed. Refer to the following paragraphs for removal and test procedure.

49. **REMOVE AND REINSTALL.** Before loosening any fuel lines, thoroughly clean the lines, connections,

Fig. 63–Installed view of injection pump showing points of adjustment and service.

1. Throttle rod
2. Baffle plate
3. Bleed screws
4. Timing plug
5. Shut-off cable
6. Low idle stop screw
7. High idle stop screw
8. Fuel line

injectors and surrounding area with air pressure and solvent spray. Disconnect and remove the leak-off line, disconnect pressure line and cap all connections as they are loosened, to prevent dirt entry into the system. Remove the stud nuts and withdraw injector unit from cylinder head.

Thoroughly clean the nozzle recess in cylinder head before reinstalling injector unit. It is important that seating surface be free from even the smallest particle of carbon or dirt which could cause the injector unit to fail to seat and result in blow-by. No hard or sharp tools should be used in cleaning. Do not re-use the copper sealing washer located between injector nozzle and cylinder head, always install a new washer. Each injector should slide freely into place in cylinder head bore without binding. Make sure that high pressure fuel line fits squarely at both ends and that leak-off line compression seals are in good condition. Tighten retaining stud nuts to a torque of 10-12 ft.-lbs. After engine is started, examine injectors for blow-by, making the necessary corrections before releasing the tractor for service.

50. **TESTING.** A complete job of testing and adjusting the injector requires the use of special test equipment. Only clean, approved testing oil should be used in tester tank. The nozzle should be tested for opening pressure, seat leakage, back leakage and spray pattern. When tested, the nozzle should open with a sharp popping or buzzing sound, and cut off quickly at end of injection with a minimum of seat leakage and a controlled amount of back leakage.

Before conducting the test, operate tester lever until fuel flows, then attach the injector. Close the valve to tester gage and pump tester lever a few quick strokes to be sure nozzle valve is not plugged, that four sprays emerge from nozzle tip, and that possibilities are good that injector can be returned to service without overhaul.

NOTE: Spray pattern is not symmetrical with centerline of nozzle tip, but forms a "tilted" pattern as shown in Fig. 65. Holes are positioned to provide the correct spray pattern in combustion chamber.

If adjustment is indicated in the preliminary tests, proceed as follows:

51. OPENING PRESSURE. Open the valve to tester gage and operate tester lever slowly while observing gage reading. Opening pressure should be 2575 psi; if not as specified, remove cap nut (1—Fig. 66) and turn adjusting sleeve (3) as required to obtain the recommended pressure.

NOTE: When adjusting a new injector or an overhauled unit with a new pressure spring (5), set the pressure at 2650 psi to allow for initial pressure loss as the spring takes a set.

52. SEAT LEAKAGE. The nozzle tip should not leak at a pressure less than 2400 psi. To check for leakage, actuate tester lever slowly and as the gage needle approaches the suggested test pressure, observe the nozzle tip. Hold the pressure for 10 seconds; if drops appear or nozzle tip is wet, the valve is not seating and injector must be disassembled and overhauled as outlined in paragraph 55.

53. BACK LEAKAGE. If nozzle seat as tested in paragraph 52 was satisfactory, check the injector and connections for wetness which would indicate external leakage. If no leaks are found, bring gage pressure to 2250 psi, release the lever and observe the time required for gage needle to drop from 2250 psi to 1500 psi. For a nozzle in good condition, this time should not be less than six seconds. A faster drop would indicate a worn or scored nozzle valve piston or body, and the nozzle assembly should be renewed.

NOTE: Leakage of the tester check valve or connections will cause a false reading, showing up in this test as excessively fast leakback. If all injectors tested fail to pass the test, the tester rather than the units should be suspected.

54. SPRAY PATTERN. If leakage and pressure are as specified when tested as outlined in paragraphs 51 through 53, operate the tester handle several times while observing the spray pattern. Four finely atomized, equally spaced, conical sprays should emerge from nozzle tip, with equal penetration into the surrounding atmosphere.

If pattern is uneven, ragged or not finely atomized, overhaul the nozzle as outlined in paragraph 55. NOTE: Spray pattern is not symmetrical with centerline of nozzle tip, but tilted as shown in Fig. 65.

55. **OVERHAUL.** Hard or sharp tools, emery cloth, grinding compound, or other than approved solvents or lapping compounds must never be used. An approved nozzle cleaning kit is available through a number of specialized sources.

Wipe all dirt and loose carbon from exterior of nozzle and holder assembly, refer to Fig. 66 and proceed as follows:

Secure the nozzle in a soft-jawed vise or holding fixture and remove cap nut

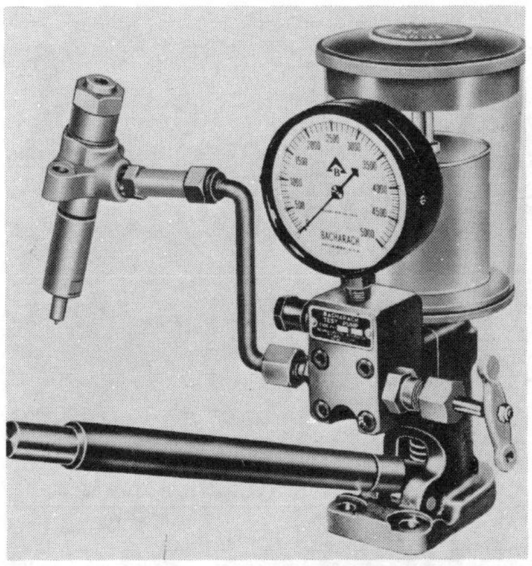

Fig. 64–A suitable injector tester is required to completely test and adjust the injector nozzles.

Fig. 65–Nozzle holes (arrows) are not located an equal distance from nozzle tip.

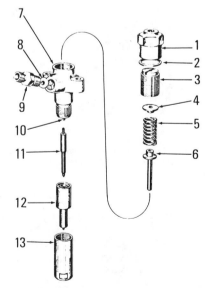

Fig. 66–Exploded view of the CAV / Simms nozzle of the type used.

1. Cap nut		
2. Gasket		
3. Adjusting sleeve	8.	Washer
4. Spring seat	9.	Inlet adapter
5. Spring	10.	Dowel
6. Valve spindle	11.	Nozzle valve
7. Nozzle holder	12.	Nozzle body
	13.	Nozzle nut

(1). Back off the adjusting sleeve (3) to completely unload pressure spring (5). Remove nozzle retaining cap nut (13) and nozzle body (12). Nozzle valve (11) and body (12) are matched and must never be intermixed. Place all parts in clean calibrating oil or diesel fuel as they are removed. Clean the exterior surfaces with a soft wire brush, soaking in an approved carbon solvent if necessary, to loosen hard carbon deposits. Rinse the parts in clean diesel fuel or calibrating oil immediately after cleaning, to neutralize the carbon solvent and prevent etching of polished surfaces. Clean the pressure chamber of nozzle tip using the special reamer as shown in Fig. 67. Clean the spray holes in nozzle with an 0.010 inch (0.24 mm) wire probe held in a pin vise as

shown in Fig. 68. Wire probe should protrude from pin vise only far enough to pass through spray holes (approximately 1/16-inch) to prevent bending and breakage. Rotate pin vise without applying undue pressure.

Clean valve seats by inserting small end of brass valve seat scraper into nozzle and rotating tool. Reverse the tool and clean upper chamfer using large end. Refer to Fig. 69. Use the hooked scraper to clean annular groove in top of nozzle body. Use the same hooked tool to clean internal fuel gallery as shown in Fig. 70.

With the above cleaning accomplished, back flush the nozzle by installing the reverse flusher adapter on injector tester and nozzle body in adapter, tip-end first. Secure with the

knurled adapter nut and insert and rotate the nozzle valve while flushing. After nozzle is back-flushed, seat can be polished using a small amount of tallow on end of polishing stick and rotating the stick while moderate pressure is applied.

Light scratches on valve piston and bore can be polished out by careful use of special injector lapping compound only, DO NOT use valve grinding compound or regular commercial polishing agents. DO NOT attempt to reseat a leaking valve using polishing compound. Clean thoroughly and back-flush if lapping compound is used.

Reclean all parts by rinsing thoroughly in clean diesel fuel or calibrating oil and assemble valve to body while immersed in the cleaning fluid. Reassemble the injector while still wet. With adjusting sleeve (3—Fig. 66) loose, reinstall nozzle body (12) to holder (7), making sure valve (11) is installed and locating dowel (10) properly aligned. Tighten nozzle retaining cap nut (13) to a torque of 50 ft.-lbs. Do not overtighten; distortion may cause valve to stick and no amount of overtightening can stop a leak caused by scratches or dirt.

Retest the assembled injector assembly as outlined in paragraphs 51 through 54.

INJECTION PUMP

The injection pump is a completely sealed unit. No service work of any kind should be attempted on the pump or governor unit without the use of special pump testing equipment and proper training. Inexperienced or unequipped service personnel should never attempt to overhaul an injection pump.

56. **ADJUSTMENT.** The slow idle stop screw (6—Fig. 72) should be adjusted with engine warm and running, to provide the recommended slow idle speed of 800-850 rpm. Hydraulic

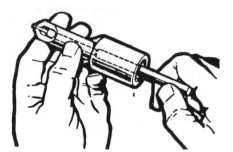

Fig. 67–Clean the pressure chamber in nozzle tip using the special reamer as shown.

Fig. 70–Using the hooked scraper to clean internal fuel gallery.

Fig. 68–Clean spray holes in nozzle tip using a pin vise and 0.010 inch wire probe.

Fig. 71–Clean the small fuel feed hole using a suitable tool.

Fig. 69–Clean the valve seat using brass scraper as shown.

Fig. 72–Installed view of injection pump showing points of adjustment and service.

1. Throttle rod
2. Baffle plate
3. Bleed screws
4. Timing plug
5. Shut-off cable
6. Low idle stop screw
7. High idle stop screw
8. Fuel line

system manual unloading valve must be in "down" (running) position and air cleaner must be clean. Some models may be equipped with an "Anti-Stall" device as shown in Fig. 73. If so equipped and the correct slow idle speed cannot be obtained, back out the "Anti-Stall" screw until it no longer affects idle speed. With idle speed correctly adjusted, turn in the anti-stall screw until idle speed begins to increase; then back out the screw ¼-turn and tighten the locknut.

The high speed stop screw (7—Fig. 72) is factory set and sealed. Governed speed under load should be 2200 rpm, with a high-idle (no-load) speed of 2375-2475 rpm. Refer to paragraph 58 for pump timing adjustment.

57. REMOVE AND REINSTALL. Before attempting to remove the injection pump, thoroughly wash pump and connections with clean diesel fuel or an approved solvent. Remove hood, air intake pipe and baffles (filler plates).

Disconnect throttle control rod and stop cable. Remove high pressure lines and cap all connections as lines are removed. Disconnect low pressure fuel lines at lift pump and injection pump main fuel gallery.

Turn crankshaft until No. 1 piston is at TDC on compression stroke. Remove injection pump rear support bracket, disconnect lubrication oil feed line,

then unbolt and lift off the pump as shown in Fig. 75.

Time the pump when installing as outlined in paragraph 58 and install by reversing the removal procedure. Injection pump plungers are numbered consecutively from front to rear and connect to cylinders as shown in Fig. 76. The injection pump lines are supported individually on vibration damper bosses cast into intake manifolds. Make sure lines rest squarely on vibration dampers after both ends are connected, then install and tighten line clamps (C—Fig. 77) and stud nuts (N). Remove oil filler plug (Fig. 74) from top of injection pump gear housing and pour in one pint of clean engine oil to provide initial lubrication. Bleed the pump and lines after installation as outlined in paragraph 46.

58. PUMP TIMING TO ENGINE. To time the injection pump during installation, first make sure that No. 1

piston is on compression stroke and that TDC timing mark on flywheel is aligned with notch in inspection hole as shown in Fig. 78. Remove the timing plug from right side of injection pump gear housing as shown in Fig. 79; if pump is being installed, turn pump gear until timing slot is centered in alignment hole then install the pump.

With pump installed and No. 1 piston at TDC on compression stroke timing slot should be centered as shown in inset. If timing slot does not perfectly center, remove tachometer drive housing from front of auxiliary housing as shown in Fig. 80. Loosen

Fig. 78–View of flywheel housing showing TDC timing mark aligned.

Fig. 75–Removing the injection pump.

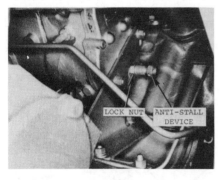

Fig. 73–Some models may be equipped with an "Anti-Stall" device. Refer to paragraph 56 for adjustment.

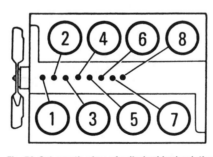

Fig. 76–Schematic view of cylinder block relating injection pump plungers to respective engine cylinders.

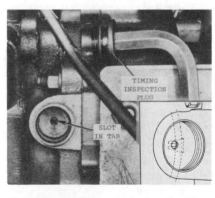

Fig. 79–With TDC timing mark aligned as shown in Fig. 78, slot in tab on injection pump camshaft should center in sight hole as shown.

Fig. 74–Installed view of injection pump.

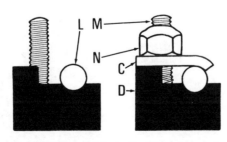

Fig. 77–Make sure injector lines (L) rest squarely on vibration dampers (D) before installing line clamps (C) and stud nuts (N).

M. Mounting stud

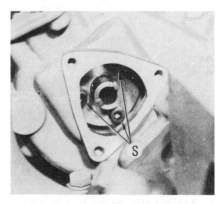

Fig. 80–Minor timing changes can be made by loosening the three cap screws (S) securing drive gear to injection pump camshaft.

the three socket head cap screws (S) and turn tachometer drive adapter until timing slot is correctly aligned. Retighten the screws (S) to a torque of 20 ft.-lbs.

IMPORTANT: The engine flywheel can be installed in any of 12 positions, only one of which will allow correct alignment of flywheel timing marks. In addition, the slotted timing tab (1—Fig. 81) may become misaligned due to accidental loosening of the retaining screw. It is therefore important that an alternate timing method be provided to serve as a double-check of the normal timing procedure. Refer to paragraph 59 for the suggested proceduure.

Fig. 81–Injection pump timing tab (1) is attached to pump camshaft with capscrew in slotted hole as shown.

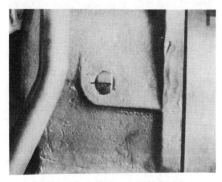

Fig. 82–Injection pump is spill timed with 30° (BTDC) flywheel timing mark aligned.

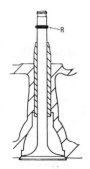

Fig. 83–When using engine valve to measure piston position, leave sealing O-ring (R) installed to prevent valve from accidentally dropping into cylinder.

59. SPILL TIMING THE INJECTION PUMP. To spill time the injection pump, first remove hood, air intake pipes and baffles. Thoroughly clean injection pump and surrounding area. Remove left, front rocker arm cover, No. 1 injector high pressure lines and No. 1 injector. Turn crankshaft until No. 1 piston is at TDC on compression stroke.

When it has been established that No. 1 piston is at TDC on compression stroke, remove timing plug on left hand side of flywheel housing and

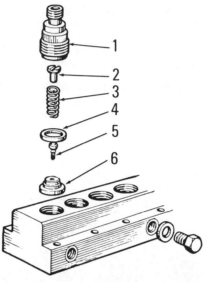

Fig. 84–Injection pump delivery valve components (2, 3 & 5) must be removed when spill timing the injection pump. A special wrench (Simms Part No. CT 2443) must be used to remove delivery valve adapter (1).

1. Adapter	4. Gasket
2. Volume reducer	5. Delivery valve
3. Spring	6. Valve seat

check the timing marks. If TDC timing mark is aligned as shown in Fig. 78, flywheel is correctly installed; turn crankshaft opposite normal direction of rotation until 30° BTDC timing mark aligns as shown in Fig. 82.

If "TDC" timing mark was not visible when checked, it may be assumed that flywheel is unmarked or incorrectly installed. It will be necessary to determine the 30° BTDC (Injection) position by measuring piston position as follows:

Remove left, front rocker arms assembly and push rods. Using a suitable fixture to depress the spring retainer cap, remove the valve locks; then lift off retainer cap, oil deflector, both valve springs and valve rotator from either of the valves in No. 1 cylinder. Leave the sealing O-ring (R—Fig. 83) in its groove in valve stem as a safety measure to keep the valve from dropping into cylinder if crankshaft should be turned. With valve head resting on piston crown, attach a dial indicator with pointer bearing on valve stem. Rock crankshaft back and forth through about ten degrees to find top of piston stroke and zero dial indicator at that point. Turn crankshaft opposite normal direction of rotation until a dial indicator reading of 0.373 is obtained, which is equivalent to 30° BTDC crankshaft position.

Remove the front (No. 1) delivery valve retainer (1—Fig. 84), extract volume reducer (2), spring (3) and delivery valve (5), then reinstall and tighten retainer (1). Install a suitable spill pipe to delivery valve retainer as shown at (B—Fig. 85). Disconnect fuel

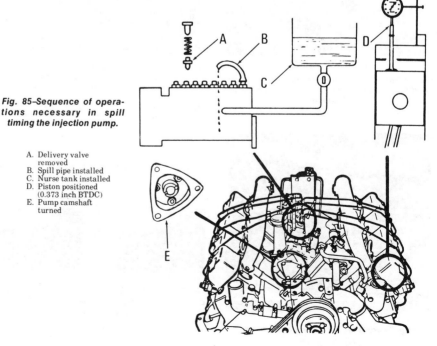

Fig. 85–Sequence of operations necessary in spill timing the injection pump.

A. Delivery valve removed
B. Spill pipe installed
C. Nurse tank installed
D. Piston positioned (0.373 inch BTDC)
E. Pump camshaft turned

delivery pipe from injection pump body and install a nurse tank with shut-off valve as shown at (C). Remove tachometer drive housing as shown in Fig. 80. Loosen the three socket head cap screws (S). Fill nurse tank reservoir with diesel fuel and turn shut-off valve to "ON" position. Turn tachometer drive adapter counter-clockwise to end of cap screw slots; a steady flow of fuel should now emerge from spill pipe (B—Fig. 85).

Now turn tachometer drive adapter

slowly clockwise until fuel flow just stops at spill pipe opening. Tighten the three cap screws (S—Fig. 80) at this point. Recheck injection timing by turning engine crankshaft opposite normal direction of rotation a few degrees until flow starts at spill pipe; then in normal direction of rotation until flow stops. Dial indicator (D—Fig. 85) should read 0.373 which corresponds to a beginning fuel delivery (injection) at 30° BTDC crankshaft position.

disconnect heat shield from upper, front of radiator. Unbolt radiator from right hand grille casting and slide radiator out left side of tractor as shown in Fig. 86. Install by reversing the removal procedure.

THERMOSTAT

61. A thermostat is located in the front end of each cylinder head and retained by outlet elbow as shown in Fig. 87. Jiggle pin (1) is installed in bypass passage to keep passage open. Make sure jiggle pin is free to move whenever elbow is removed for any reason. The thermostat should begin to open at 175°-182° F. and should be fully open at 208° F.

WATER PUMP

62. Fig. 88 shows an installed view of water pump and Fig. 89 shows an exploded view. To remove the water pump, drain cooling system and remove left-hand hood panel. Remove fan blades and sheet metal filler plates attached to rear of pump. Loosen both belts and remove alternator and support from top of water pump housing. Disconnect all hoses, then unbolt and remove water pump from timing gear upper cover.

When disassembling the pump it is first necessary to remove pulley (5) using a suitable puller. After removing snap ring (6), both shafts and the bearings can be pressed out from impeller end using a suitable mandrel.

When reassembling, pack area between the two bearings (7) half full of high melting-point grease. If bearings are removed, install with open sides together next to spacer (8).

COOLING SYSTEM

RADIATOR

60. A 7 psi pressurized cooling system is used. Capacity is approximately 40 U. S. quarts. An engine oil cooler is built into radiator core.

To remove the radiator, first drain cooling system and remove hood, grille screens and left side grille casting. Disconnect coolant hoses and engine oil cooler lines. Remove fan blades and

Fig. 86–Radiator can be removed from left side as outlined in paragraph 60.

Fig. 88–View of installed water pump with hoses connected.

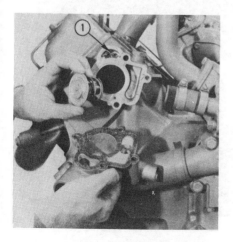

Fig. 87–Thermostat is located under outlet elbow in each cylinder head.

1. Wiggle pin

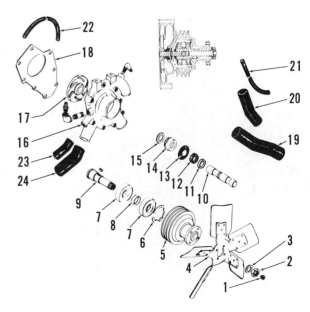

Fig. 89–Exploded view of water pump and associated parts.

1. Cap plug
2. Nut
3. Washer
4. Fan blades
5. Pulley
6. Snap ring
7. Bearing
8. Spacer
9. Front shaft
10. Rear shaft
11. Flange
12. Front seal
13. Slinger
14. Rear seal
15. Ceramic insert
16. Body
17. Impeller
18. Cover
19. Lower radiator hose
20. Left outlet hose
21. Left bleeder hose
22. Right bleeder hose
23. Bypass hose (2 used)
24. Right outlet hose

ELECTRICAL SYSTEM

ALTERNATOR & REGULATOR

63. ALTERNATOR. A "DELCO-TRON" generator (alternator) is used. All units are negative ground.

The only test which can be made without removal and disassembly of alternator is the output test. Maximum output of the standard duty alternator (DR-1100805) is 30 Amperes @ 5000 alternator rpm. Maximum output of the heavy duty alternator (DR-1100883) is 58 Amperes @ 5000 alternator rpm.

To disassemble the alternator, first put match marks (M—Fig. 90) on the two frame halves (5 and 16), then remove the four through-bolts. Pry frame apart with a screwdriver between stator frame (11) and drive end frame (5). Stator (11) must remain with slip ring end frame (16) when unit is disassembled.

NOTE: When frames are separated, brushes (B—Fig. 94) will jump out on rotor shaft at bearing area. Brushes MUST be cleaned of lubricant if they are to be reused.

Clamp the iron rotor lightly and carefully in a protected vise only tight enough to permit loosening of pulley nut (1—Fig. 90). Rotor and end frame can be separated after pulley is removed. Check the bearing surfaces of rotor shaft for visible wear or scoring. Examine slip ring for scoring or wear, and windings for overheating or other damage. Check rotor for shorted or open circuits using an ohmmeter as follows:

Refer to Fig. 91 and touch the ohmmeter probes to points (1-2) and (1-3); a reading near zero will indicate a ground. Touch ohmmeter probes to the two slip ring segments (2 & 3); reading should be 4.6-5.5 ohms, a higher reading will indicate an open circuit, a lower reading will indicate a short. If windings are satisfactory, mount the rotor between lathe centers and check runout at slip ring using a dial indicator. Runout should not exceed 0.002. Surface can be trued if runout is excessive or if surface is scored. Finish with 400 grit polishing cloth until scratches or machine marks are removed.

Disconnect the three stator leads and separate stator from slip ring end frame. The three leads (C—Fig. 92) have a common connection (Y) in the center of the windings. Check for continuity between the three leads (C). There should be no continuity between any of the leads and stator frame (F). Short circuits within the windings cannot be readily determined by test, because of the low resistance.

Three diodes (N—Fig. 93) are located in slip ring end frame (F) and three diodes (P) in heat sink (S). Diodes should test at or near infinity in one direction when tested with an ohmmeter, and at or near zero when meter leads are reversed. Renew any diode with approximately equal meter readings in both directions. Diodes must be removed and installed using an arbor press and suitable tool which contacts only the outer edge of the diode.

Brushes are available only in an assembly which includes the brush holder (H—Fig. 94). If brushes are re-

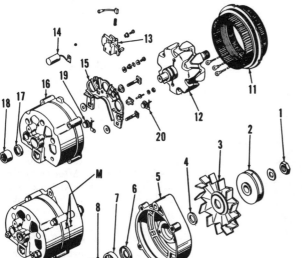

Fig. 90–Exploded view of DELCOTRON alternator of the type used.

1. Pulley nut
2. Drive pulley
3. Fan
4. Collar
5. Drive end frame
6. Slinger
7. Bearing
8. Gasket
9. Collar
10. Bearing retainer
11. Stator assembly
12. Rotor assembly
13. Brush holder
14. Capacitor
15. Heat sink
16. Slip ring end frame
17. Bearing retainer
18. Bearing
19. Negative diode
20. Positive diode
M. Match marks

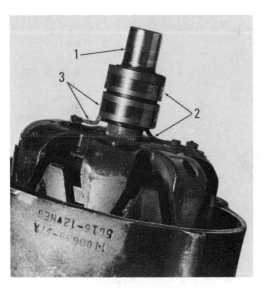

Fig. 91–View of rotor assembly showing probe points for testing.

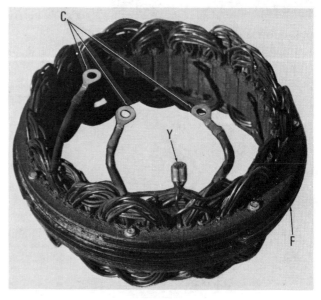

Fig. 92–Opposite ends of the three stator windings (C) are all connected at (Y). Windings must be insulated from frame (F).

used, make sure all grease is removed from surface of brushes before unit is reassembled. After degreasing, install both brushes and their springs in holder, push brushes up against spring pressure and insert a short piece of straight wire (W—Figs. 95 and 96) through brush holder and end frame to position brushes for assembly. Withdraw the wire from the outside ONLY after alternator is reassembled.

The capacitor (C—Fig. 93) connects to positive heat sink and is grounded to alternator frame. Capacitor protects the diodes from voltage surges. A shorted capacitor will cause a dead short in battery wiring and burn out the twin fuses located near the starter solenoid. Press the old capacitor out and new capacitor in, working from outside of slip ring end frame and

Fig. 95–To position brushes for assembly, insert a short, straight wire (W) through brush holder and end frame beneath the brushes. See Fig. 96 for other end of wire.

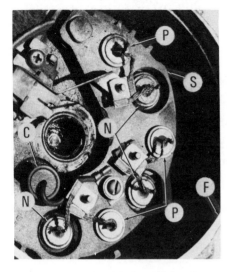

Fig. 93–Inside view of slip ring end frame showing component parts.

C. Capacitor
F. End frame
N. Negative diodes
P. Positive diodes
S. Heat sink

Fig. 94–Brushes (B) are only serviced in an assembly which includes brush holder (H). When alternator is separated, brushes will be contaminated with grease from end bearing and must be degreased before reassembly if brushes are being re-used.

Fig. 96–Wire (W) holds spring-loaded brushes away from contact with slip rings during assembly. Withdraw wire from outside with unit assembled. Refer also to Fig. 95.

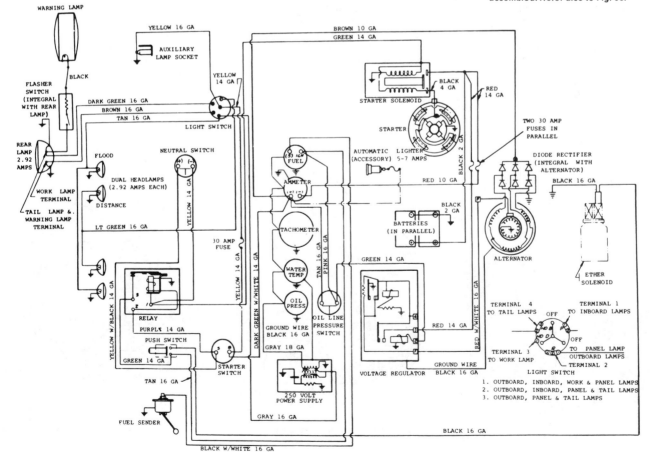

Fig. 97–Schematic wiring diagram typical of all models.

check unit for shorts after capacitor is connected.

Ball bearing (7—Fig. 90) and needle bearing (18) should be filled ½-full with Delco-Remy Bearing Lubricant when alternator is reassembled. Over-filling may cause lubricant to be thrown into alternator, resulting in malfunction. Assemble by reversing the disassembly procedure. Tighten pulley nut (1) to a torque of 50-60 ft.-lbs.

NOTE: A battery powered test light can be used instead of ohmmeter for all electrical checks except shorts in rotor winding.

64. **REGULATOR.** A DELCO-REMY standard two-unit regulator is used. Quick disconnect plugs are used at regulator and alternator. Production regulator is riveted to shock mount; service units are shipped less mount and are attached with screws. Test

specifications are as follows:
Regulator Model 1119513
Ground polarity .... Negative
Field Relay
  Air Gap 0.015
  Point Opening 0.030
  Closing Voltage Range 3.8-7.2
Voltage Regulator
  Air Gap 0.067*
  Point Opening 0.014
  Voltage Setting 13.9-14.8*
* The specified air gap setting is for bench repair only; make final adjustment to obtain specified voltage, with lower contacts opening not more than 0.4 volt less than upper contacts. The given voltage settings are for ambient temperature of 100° F. or less. Regulator is temperature compensated.

STARTING MOTOR

65. A DELCO-REMY 1114175 starting motor is used. Specifications are as follows:

Brush spring tension .... 35 oz.
No-Load Test
  Volts 9
  Amperes (Includes Solenoid) 140-190
  Minimum rpm 4000
Starter drive pinion clearance should measure 23/64-inch between edge of pinion and pilot housing wall with pinion in engaged position. Connect a 6-volt battery to solenoid terminals when checking pinion clearance, to keep armature from turning. Clearance may be adjusted by removing threaded plug in shift housing and turning the nut on outer end of solenoid plunger shaft.

CIRCUIT DESCRIPTION

66. Refer to Fig. 97 for a schematic wiring diagram. The two 95 Ampere Hour, 12 Volt batteries are connected in parallel and negative posts are grounded. Make sure both batteries are of equal capacity and condition.

ENGINE CLUTCH

All models are equipped with a dry type dual disc clutch mounted on the flywheel and released by a foot pedal. The clutch controls only tractor travel, the Independent Power Take-Off being controlled by a separate multiple disc wet clutch located in rear axle center housing.

ADJUSTMENT

67. Clutch pedal free play should be 2 inches when measured between pedal arm and platform as shown in Fig. 98. Adjust by shortening or lengthening linkage rod (R). Recheck the adjustment with engine running at high idle speed and readjust if required.

TRACTOR SPLIT

68. To detach engine from transmission assembly for clutch service, first remove hood, batteries, battery platform and side plates. Disconnect fuel

supply and return lines and the six hydraulic lines at the unions located underneath battery support.

Drain cooling system and disconnect temperature gage sending unit, oil gage line, tachometer cable, throttle and stop cable. Disconnect ether starting and electrical wiring, and remove starter motor. Unbolt priority valve from vertical battery support.

Wedge front axle to keep engine from tipping. Attach a hoist to all four engine lift brackets. If front end is not weighted, use spreader bars to distribute the load. If front end is heavily

weighted, it may be necessary to attach a fifth chain fall (or otherwise support) front of unit. Support transmission on a suitable stand or jack. Remove clutch housing cap screws and carefully separate the units.

Reconnect by reversing the disassembly procedure. Bleed fuel system as outlined in paragraph 46 and adjust throttle controls as in paragraph 56 after tractor is reconnected.

OVERHAUL

69. Refer to Fig. 99 for an exploded

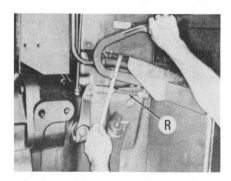

Fig. 98–Clutch pedal free play should be 2 inches, measured from platform stop as shown. Adjust by shortening or lengthening linkage rod (R).

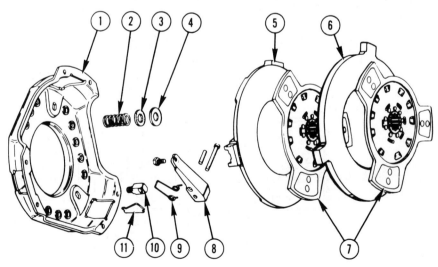

Fig. 99–Exploded view of clutch cover and associated parts used on all models. The two identical clutch discs (7) are separated by plate (6).

1. Cover	4. Retainer	9. Anti-rattle
2. Spring	5. Pressure plate	10. Pivot block
3. Insulator washer	6. Separator plate	11. Wear plate
	7. Disc	
	8. Release lever	

view of clutch cover, pressure plate and associated parts. Two identical clutch discs (7) are used, together with the floating separator plate (8) to increase friction surface.

To disassemble the cover, place the assembly in a suitable press as shown in Fig. 100. Remove the nuts from pivot blocks (10—Fig. 99) then release pressure slowly. The six wear plates (11) will be free to fall out as cover is removed, as will the fifteen springs (2), insulating washers (3) and retainers (4). Examine release levers (8) and pivot pins for wear. If release levers are removed from pressure plate, install retaining pins from right side as shown in Fig. 101 so that thrust will come against pin heads rather than cotter pins.

Springs (2—Fig. 99) are color coded with a light green stripe. A new spring has a free length of approximately 2 3/16 inches and should test 135-140 lbs. when compressed to a height of 1 9/32 inches. Insulating washers (3) and their retainers (4) must be placed on pressure plate end of springs as shown. Tighten pivot block nuts to a torque of 40 ft.-lbs. when assembling.

Flywheel, pressure plate or separator plate may be heat discolored but may not be warped more than 0.020 across clutch width. Flywheel may be resurfaced if required, provided clutch cover mounting pads are lowered an equal amount.

Install clutch discs (7—Fig. 99) with long hubs to rear. Clutch levers should be checked and adjusted if necessary using new clutch discs and Massey-Ferguson Special Tool MFN 766 as shown in Fig. 102. Usable worn clutch discs can be reinstalled after lever adjustment has been made.

RELEASE LINKAGE

70. Fig. 103 shows a view of clutch release linkage after clutch split. Release forks (1) clamp to release shaft (2), no keys, pins or splines being used. Release shaft can be withdrawn after loosening clamp bolts.

When assembling the linkage, make sure shaft (2) extends through outside of transmission housing 1⅜ inches. With pedal link disconnected, move release shaft arm forward until it contacts clutch housing casting and right clutch fork (1) rearward until it contacts front cover. Tighten clamp screw (65-70 ft.-lbs.) at this point. Adjust left fork (1) until 0.002-0.005 clearance exists between the fork and release bearing carrier (3). With clearance properly adjusted, tighten clamp bolt to 65-70 ft.-lbs. Recheck clearances and be sure linkage moves freely throughout range of travel without binding. Adjust pedal linkage after tractor is rejoined, as outlined in paragraph 67.

Fig. 102–Release lever height should be adjusted using new clutch discs and Special Tool MFN 766 as shown.

Fig. 103–View of release linkage after clutch split. Clutch fork arms (1) CLAMP to release shaft (2) and are not keyed. Make sure 0.002-0.005 clearance exists between forks and bearing carrier (3) when clamp screws are tightened.

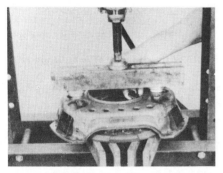

Fig. 100–Use a press to assist in clutch cover disassembly. Note method of support for pressure plate.

Fig. 101–Release lever retaining pins must be installed from right side as shown.

TRANSMISSION AND CONNECTIONS

All models are equipped with a basic three-speed, sliding gear transmission coupled to a manually selected, constant mesh dual range unit located in rear axle center housing at output end of main transmission unit.

Available as a factory installed option is a hydraulically actuated dual range unit located on the transmission input shaft which can be manually shifted while tractor is moving under load.

Transmissions with the hydraulically actuated dual range option are referred to as "Multi-Power" transmissions, and are capable of 12 forward and 4 reverse speeds. Transmissions without the option are referred to as "Dual-Range" transmissions and are capable of 6 forward and 2 reverse speeds.

Service procedures on the two transmissions are similar except where otherwise indicated.

REMOVE AND REINSTALL

71. To remove the transmission assembly from tractor, first drain transmission and rear axle center housing. Drain fuel tank. Split the tractor between engine and clutch housing as outlined in paragraph 68.

Remove operator's platform, brake pedals, brake control valve and lines, fuel supply line and the hydraulic line between priority valve and lift cover. Disconnect light wires, shift linkage (at lower end) and hydraulic lines at rear of transmission cover. Attach a hoist to steering console as shown in Fig. 104, remove the cap screws securing instrument panel and steering console to top of transmission and lift off the unit as shown. Remove transmission top cover as shown in Fig. 105.

Disconnect fuel tank crossover lines and remove the two tanks separately.

Suitably block the rear axle center housing to prevent it from tipping to front or rear. Swing transmission housing from the top and remove the attaching flange bolts, then lift off the transmission assembly.

The independent Power Take-Off spline coupling (Fig. 106) may remain with transmission or with rear axle

center housing when units are separated. When reinstalling the transmission unit, place coupling on rear unit with undercut end to rear. Complete the installation by reversing the removal procedure.

ADJUSTMENT

72. **SHIFT LINKAGE.** Except for the clutch release linkage, the only adjustment required is adjustment of link rods running from instrument panel mounted shift levers to the ac-

tuating levers on transmission housing. Refer to Figs. 107, 108, 109 and 110 for exploded views fo the complete shift linkage.

A suggested method of making the adjustment is to complete the assembly except for connecting the shift rod yokes to levers (7, 8, 9 & 11—Fig. 107). Shift the main transmission levers (19 & 21—Fig. 108) to neutral (center) detent, Hi-Lo shift lever (Fig. 109) to "LOW" position and park pawl (Fig. 110) to "DRIVE" position. Adjust the connecting rods (17 & 18—Fig. 108, 29 —Fig. 109 and 39—Fig. 110) until connecting pins can be easily inserted with main shift lever in neutral position and Hi-Lo shift lever in down position. With pins installed, check to be sure that shifting can be accomplished to all

Fig. 104–Lift off steering console as a unit using a hoist as shown.

Fig. 105–Removing transmission top cover.

Fig. 106–Place spline coupling on rear unit with undercut end to rear.

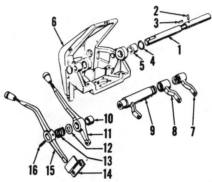

Fig. 107–Exploded view of instrument panel support showing shift levers and linkage.

1. Shaft	9. 2nd & 3rd hub
2. Pin	10. Bushing
3. Interlock key	11. Lever
4. Snap ring	12. Washer
5. Fiber bushing	13. Spring
6. Support	14. Bracket
7. Park hub	15. Lever
8. 1st & reverse hub	16. Pin

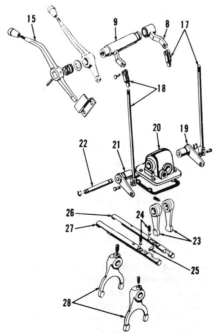

Fig. 108–Exploded view of shift tower, shift rails, forks and associated parts. Refer to Fig. 107 for parts identification not given below.

17. Shift rod	23. Shift finger
18. Shift rod	24. Detent assemblies
19. 1st & reverse lever	25. Interlock pin
20. Tower	26. Shift rail
21. 2nd & 3rd lever	27. Shift rail
22. Pivot shaft	28. Shift forks

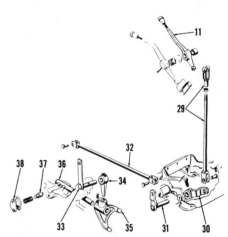

Fig. 109–Exploded view of Dual Range shift linkage.

11. Lever	34. Finger
29. Shift rod	35. Shift fork
30. Inner arm	36. Shift rail
31. Outer arm	37. Detent plunger
32. Link	38. Plate
33. Lever	

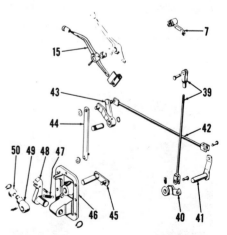

Fig. 110–Exploded view of park lock and linkage.

7. Park hub	44. Link
15. Shift lever	45. Lever
39. Shift rod	46. Park cover
40. Inner arm	47. Spring
41. Outer arm	48. Pawl
42. Link	49. Cam arm
43. Bellcrank	50. Roller

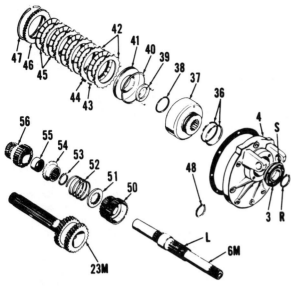

Fig. 111–Exploded view of transmission front cover, input gears, Multi-Power clutch and associated parts. Remainder of transmission front unit is similar to that shown in Fig. 119.

  L. Locating groove
  R. Snap ring
  S. Snap ring
  3. Bearing
  4. Front cover
  6M. Input shaft
23M. Countershaft
 36. Sealing rings
 37. Clutch hub
 38. Piston ring
 39. Thrust washer
 40. Piston ring
 41. Piston
 42. Separator plates
 43. Roller guides
 44. Separator springs
 45. Friction discs
 46. Pressure plate
 47. Snap ring
 48. Snap ring
 52. Spring
 53. Snap ring
 54. Jaw clutch
 55. Needle bearing
 56. Output gear

To disassemble the removed input shaft unit, first lift off the jaw clutch (54—Fig. 111) and spring (52). Unbolt and remove input bearing retainer (1—Fig. 119). Remove snap ring (R—Fig. 112) from front groove of input shaft then bump shaft and clutch assembly rearward out of front cover and bearing. Press the bearing forward out of front cover and remove snap ring (S) from retaining groove in bearing.

Remove snap ring (5—Fig. 114), clutch pressure plate (1), springs (S—Fig. 115) and their guide rollers. Remove the alternately installed five friction discs (45—Fig. 111) and separator plates (42). Remove snap ring (4—Fig. 114), thrust washer (3), clutch hub (2) and the doweled thrust washer from clutch drum hub. Invert the input shaft, remove snap ring securing clutch drum to input shaft splines, then remove clutch drum from input shaft. Place clutch drum open-end up on

detent positions without the levers bottoming or main shift lever binding through the neutral slot. If linkage cannot be properly adjusted, check for wear or damage and correct the trouble before proceeding further.

OVERHAUL
Multi-Power Models

73. **INPUT SHAFT AND FRONT COVER.**

Refer to Fig. 111 for an exploded view of input shaft, front cover and associated parts used on Multi-Power models. Fig. 119 shows main transmission shafts and gears used on Dual Range models. Except for input shaft, front cover and counter-shaft, parts are interchangeable between the two transmission types.

If only the input shaft, front cover, Multi-Power clutch and associated parts are to be removed, detach clutch housing from engine as outlined in paragraph 68. If transmission is to be disassembled, remove the unit as in paragraph 71.

Remove the clutch release shaft, forks and bearing. Remove Multi-Power inlet hydraulic tube and withdraw front IPTO shaft (11—Fig. 119). Remove the cap screws securing transmission front cover (4—Fig. 111) to housing, noting that top, center cap screw is machined to serve as a locating dowel. This special screw must be reinstalled in the same location during reassembly. Lift out the input shaft, front cover and Multi-Power clutch as an assembly as shown in Fig. 112.

The Multi-Power output gear (56—Fig. 111) and its bearing will remain in housing when input shaft is removed. If output gear must be removed, bump the gear and bearing rearward after removing mainshaft as outlined in paragraph 75 and rear snap ring (10—Fig. 119).

Fig. 114–Removed input shaft and Multi-Power clutch unit showing component parts.

1. Pressure plate
2. Clutch hub
3. Thrust washer
4. Snap ring
5. Snap ring

Fig. 112–Assembled view of transmission front cover and shaft unit. Refer to Fig. 111 for parts identification.

Fig. 113–Rear view of removed front cover showing jaw clutch and spring installed.

Fig. 115–Release springs (S) rest on splines of first separator plate.

bench and remove clutch piston using two pair of pliers as shown in Fig. 116.

Input shaft (6M—Fig. 111) has an inner tube at rear of shaft which is sealed at each end by an O-ring. The tube provides passage for lubricant to the Multi-Power jaw clutch. Removal is necessary only if leakage is suspected and a Massey Ferguson Special Tool, MFN 800HH is made available to facilitate removal.

Examine clutch plates (42) and discs (45) for heat discoloration, wear or other damage. Discs should be renewed if they measure less than 0.092 in thickness.

Lubricate all parts during assembly. Snap ring (R—Fig. 117) is available in four thicknesses. Select the proper thickness during assembly to limit clutch drum end play to 0.004. Available thicknesses are 0.109, 0.115, 0.120 and 0.125. Snap ring (R) is interchangeable with snap ring (48—Fig. 111) which fits in groove (L) in front of clutch drum.

When installing Multi-Power clutch friction discs and separator plates, proceed as follows: Note that separator plates have one external spline for each two internal splines in clutch drum. Drop one separator plate in clutch drum on top of piston; then alternate clutch discs and separator plates, posi-

tioning splines of remaining separator plates in the spline NOT occupied by original separator plate. Install springs (S—Fig. 115) and their guide rollers on top of splines of first separator plate as shown. Install pressure plate and large snap ring as shown in Fig. 118.

Install new bronze-teflon sealing rings (36—Fig. 111) in grooves on clutch hub (37), then carefully install the assembled shaft and clutch unit in front cover (4). Secure with snap ring (S) and install bearing retainer (1—Fig. 119).

Make sure clutch spring (52—Fig. 111) and jaw clutch (54) are properly positioned on input shaft before reinstalling front cover. Tighten cover re-

taining cap screws to a torque of 33-38 ft.-lbs. and complete the assembly by reversing the disassembly procedure. Service on the Multi-Power control valve is covered in paragraph 130.

Dual Range Models

74. **INPUT SHAFT AND FRONT COVER.** Refer to Fig. 119 for an exploded view of main transmission gears, shafts and associated parts.

Front cover (4), bearing (3) and associated parts can be removed after splitting engine from clutch housing as outlined in paragraph 68. To remove the input shaft (6), it is first necessary to remove the transmission as in paragraph 71 and transmission main shaft as in paragraph 75.

To remove the front cover after clutch split or transmission removal, first remove clutch release shaft, forks and bearings. Unbolt and remove bearing retainer (1) to protect the seal (2). Remove the cap screws retaining front cover (4) to transmission housing, noting that top, center cap screw is machined to serve as a locating dowel. Remove front cover and bearing (3) as an assembly, with input shaft (6) and rear bearing (9) remaining in transmission housing.

To remove input shaft (6) and bearing (9) with mainshaft and front cover removed, unseat rear snap ring (10) and bump input shaft and bearing rearward out of bearing bore. Install by reversing the removal procedure.

When reinstalling front cover (4), note that special (doweled) cap screw is

Fig. 118–Install pressure plate and retaining snap ring as shown.

Fig. 116–Work clutch piston out of drum using two pair of pliers as shown.

Fig. 117–Select thickest snap ring (R) which can be installed in groove during assembly. End play should be limited to 0.004.

1. Retainer
2. Seal
3. Bearing
4. Front cover
5. Snap ring
6. Input shaft
7. Seal
8. Snap ring
9. Bearing
10. Snap ring
11. IPTO input shaft
12. Snap ring
13. 2nd & 3rd gear
14. Needle bearing
15. Needle bearing
16. Main shaft
17. Snap rings
18. Needle bearings
19. Low/reverse gear
20. Bearing
21. Snap ring
22. Rear plate
23. Countershaft
24. Needle bearing
25. Spacer
26. Gear
27. Spacer
28. Gear
29. Spacer
30. Gear
31. Bearing
32. Snap ring
33. Reverse idler
34. Lock screw
35. Shaft

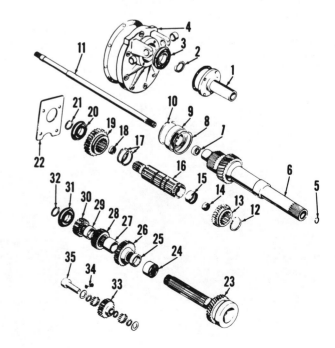

Fig. 119–Exploded view of main transmission shafts, gears and associated parts used on Dual Range Models. Most parts are similar or interchangeable on Multi-Power models.

installed in upper, center position and tighten all screws to a torque of 33-38 ft.-lbs.

All Models

75. **MAIN SHAFT AND SLIDING GEARS.** To remove the transmission main shaft and sliding gears, first remove transmission assembly as outlined in paragraph 71. Remove retainer plate (22—Fig. 119) from rear of transmission housing and slide IPTO front shaft (11) forward out of upper shafts. Remove the locking set screws in shift forks (28—Fig. 120) and with the other rail in neutral, slide either of the shift rails (26 to 27) rearward out of forks and housing. Retrieve detent balls and springs (24) and interlock pin (25) as rails are removed.

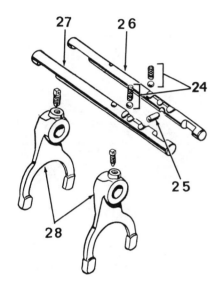

Fig. 120—Exploded view of main transmission shifter rails and forks. Refer also to Fig. 108.

24. Detent units
25. Interlock pin
26. Shift rail
27. Shift rail
28. Shift forks

Slide main shaft (16—Fig. 119) to the rear until 2nd-3rd sliding gear (13) can be lifted from shaft, unseat and remove snap rings (17), then slide shaft (16) rearward out of low-reverse sliding gear (19) and housing unit.

Rear bearing (20) can be pressed from shaft after removing snap ring (21). Inner bore contains three caged needle bearings (14, 15 and 18) which can be renewed using Massey-Ferguson Special Tools MFN 800P, MFN 800Q and MFN 800MM, or by carefully measuring the location of rear bearing before removal and installing new bearing in same position. The two front bearings should be installed just beyond flush with edge of bearing bores. The new shaft is shipped with bearings installed.

Reinstall the main shaft by reversing the removal procedure. Note that the two shift forks (28—Fig. 120) are interchangeable and that both are installed with long end of hub to rear.

76. **COUNTERSHAFT.** To remove the transmission countershaft (23M—Fig. 111 or 23—Fig. 119), first remove the mainshaft as outlined in paragraph 75 and transmission front cover as in paragraph 73 or 74.

Remove snap ring (32—Fig. 119) from rear of countershaft. Place suitable wood blocks (about 2 inches thick) between second gear (26) and housing wall and, using a suitable puller, remove countershaft forward out of transmission case, lifting off gears and spacers as shaft is removed.

When installing the countershaft, snap ring (32) is supplied in selective thicknesses of 0.093, 0.103, 0.113 and 0.125. Insert shaft from front, assembling spacers and gears on shaft as shown in Fig. 119 and 121. Block front of countershaft inside of front cover

flange and, using a short section of 2 inch ID pipe, tap bearing (31—Fig. 119) onto rear of shaft until it bottoms. Select the thickest snap ring (32) which can be installed. Countershaft end play should not exceed 0.010.

77. **REVERSE IDLER SHAFT.** To remove the reverse idler gear (33—Fig. 119) and shaft (35), first remove the main shaft as outlined in paragraph 75. Remove the locking bolt (34) from left side of transmission housing and tap the shaft (35) rearward, lifting gear and thrust washers out as shaft is withdrawn. Be careful not to tip the gear and spill the 56 loose needle rollers in transmission case as gear is withdrawn.

Assembly can be facilitated by using the Massey-Ferguson Special Tool MFN 800K or a piece of shaft about 0.010 shorter than reverse idler slot in housing. Use the shaft to retain the loose needle rollers and thrust washers, and install the gear with long hub forward. Shaft (35) will push the special tool out as it is inserted. Align the counterbore and install locking bolt (34), then complete the assembly by reversing the disassembly procedure.

Fig. 121—Installed view of countershaft showing correct positioning of gears.

REAR TRANSMISSION, CENTER HOUSING, MAIN DRIVE BEVEL GEARS AND DIFFERENTIAL

The rear axle center housing contains the Range Transmission, main drive bevel gears and differential, as well as power take-off gears, shafts and clutch. The hydraulic system charging pump is driven by gear teeth machined on outer edge of IPTO clutch drum.

TRACTOR SPLIT

78. To detach (split) the tractor between transmission and rear axle center housing, first drain transmission, center housing and fuel tanks.

Remove seat console lower front panel, operator's platform, lower step plates and tool box. Remove fuel tanks.

Disconnect park linkage rod and range shift link. Disconnect hydraulic lines and wiring running to front of tractor. Suitably support both sections and remove attaching flange bolts, then roll front of tractor forward away from rear axle center housing.

The IPTO spline coupling (Fig. 123) may remain with transmission or with rear axle center housing when units

are separated. When reconnecting the tractor, place coupling in rear unit as shown, with chamfered end of splines forward. Complete the assembly by reversing the disassembly procedure.

RANGE TRANSMISSION

79. **SHIFTER RAIL AND FORK.** Refer to Fig. 124 for an exploded view of range transmission shift rail and fork located in rear axle center housing. Actuating lever mounted on

instrument panel is shown in Fig. 109 and adjustment is covered in paragraph 72.

To remove the rails and forks located in center housing, first detach (split) transmission from center housing as outlined in paragraph 78. Disconnect remainder of hydraulic lines from center housing front, top cover; remove the cap screws and lift off the cover with hydraulic system charging pump attached.

Remove detent plate (38—Fig. 124), spring and plunger (37). Loosen the set screw in fork (35) and slide shift rail (36) forward out of housing bore. Install by reversing the removal procedure. Actuating lever (33) and arm (34) can be removed if desired, by loosening set screw in arm.

80. **INPUT CLUSTER GEAR.** To remove the input cluster gear shaft (6 —Fig. 125), first detach (split) transmission from rear axle center housing as outlined in paragraph 78. Disconnect remainder of hydraulic lines from center housing front, top cover and remove the cover and attached hydraulic system charging pump.

Remove IPTO output housing as outlined in paragraph 94. Withdraw rear IPTO shaft and sliding coupler (12—Fig. 126). Move IPTO drive shaft (10) rearward to expose snap ring (11), unseat and remove the snap ring; then slide IPTO drive shaft (10) rearward out of tractor.

Remove oil return line (R—Fig. 127) and oil suction tube (S—Fig. 125). Remove oscillator arm (A) by prying it forward. Remove snap ring (2—Fig. 126), insert a suitable puller through the hollow cluster gear (6) and pull the unit forward until large gear contacts

Fig. 122–Splitting tractor between transmission and rear axle center housing.

Fig. 123–Place spline coupling on rear unit with undercut end to rear.

Fig. 125–Top view of rear axle center housing with front cover removed, showing input cluster gear (6) and hydraulic pump intake tube (S).

Fig. 127–Rear axle center housing showing removal of charging pump return line (R). Oscillator arm (A) can be removed by prying it forward.

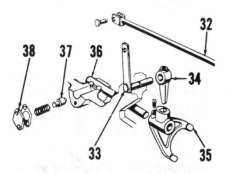

Fig. 124–Exploded view of range shift mechanism showing component parts.

32. Link
33. Lever
34. Finger
35. Shift fork
36. Shift rail
37. Detent plunger
38. Plate

1. Coupling
2. Snap ring
3. Bearing cup
4. Bearing cone
5. Needle bearing
6. Cluster gear
7. Bearing cone
8. Bearing cup
9. IPTO clutch
10. IPTO drive shaft
11. Snap ring
12. Coupler
13. Bearing cap
14. Lock pins
15. Shaft nut
16. Washer
17. Bearing cone
18. Bearing cup
19. Retaining ring
20. Thrust washer
21. Roller bearings
22. Spacer
23. Low gear
24. Shift collar
25. Hub
26. High gear
27. Thrust washer
28. Snap ring
29. Park gear
30. Snap ring
31. Bearing cup
32. Bearing cone
33. Bevel pinion

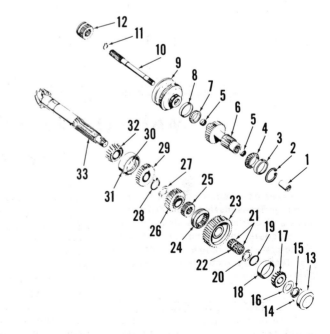

Fig. 126–Exploded view of bevel pinion shaft and range transmission gears located in rear axle center housing. IPTO clutch drum (9) drives hydraulic system charging pump and contains the actuating cam for hydraulic valve oscillating lever arm.

shift coupler, and bearing cup (3) is partially withdrawn from housing wall. Move shaft rearward until a slide puller can be installed as shown in Fig. 128, then remove the bearing cup. Pull front bearing cone as shown in Fig. 129, then lift input cluster gear from center housing.

Needle bearings (5—Fig. 126) can be installed in shaft (6) using Massey Ferguson Special Tools MFN 800P, MFN 800Q and MFN 800MM for correct positioning. If special tools are not available, note position of existing bearings and install new units in same location.

The rear bearing cup (8) is a slip fit in housing and can normally be removed by hand. If pulling is required, bearing cup and the press fit oil distributor ring (not shown) must be removed as a unit.

When installing the cluster gear, reverse removal procedure. Note that cluster gear bearing end play is controlled by a selective fit snap ring (2). Install bearing cup (3) and gently tap the cup rearward until shaft end play is eliminated, then install the thickest snap ring which can be installed. Pry the cup back against the installed snap ring, then measure end play using a dial indicator as shown in Fig. 131.

End play must be within the limits of 0.001-0.009. Snap rings (2—Fig. 126) are available in thicknesses of 0.077, 0.089, 0.099 and 0.109. IPTO interme-

Fig. 130–Front view of rear axle center housing showing IPTO shaft spline coupling and selective fit snap ring installed.

Fig. 131–Measuring cluster gear end play to determine correct thickness of selective fit snap ring shown in Fig. 130.

diate shaft end play must be checked as outlined in paragraph 94 after completion of assembly.

81. BEVEL PINION SHAFT. To remove the main drive bevel pinion shaft (33—Fig. 126), first detach (split) rear axle center housing from transmission as outlined in paragraph 78, remove input cluster gear as in paragraph 80 and differential assembly as in paragraph 83.

Remove pinion shaft bearing cap (13). Split the nut (15) using a sharp chisel, remove (and save) the two lock pins (14); then remove and discard the nut.

Remove range shift rail and fork

Fig. 133–View of bevel pinion shaft showing park gear correctly installed.

Fig. 134–Disassembling bevel pinion shaft.

Fig. 128–Using a slide hammer to remove range cluster gear front bearing cup.

Fig. 129–Removing front bearing cone from range cluster gear.

Fig. 132–IPTO clutch unit can be lifted out as shown after sliding intermediate shaft rearward and lifting out sliding coupler.

Fig. 135–Gears and associated parts correctly positioned for bevel pinion shaft installation.

(paragraph 79) and park lock cover. Unseat snap ring (28—Fig. 126) in front of park lock gear (29) and slide the gear forward on pinion shaft. Remove the tab washer (16) and bearing cone (17) then extract the spiral retaining ring (19). Move pinion shaft rearward in housing, lifting out the contained gears and associated parts as they are free.

Reassemble by reversing the disassembly procedure, making sure that

Fig. 136–Shift hub missing spline must align with oil hole in bevel pinion shaft during assembly.

Fig. 137–Reseat park gear retaining snap ring during reassembly.

Fig. 138–Installing low gear idler bearings during bevel pinion assembly.

missing tooth of shift hub spline aligns with oil hole in bevel pinion shaft as shown in Fig. 136.

Install a new nut (15—Fig. 126) and tighten the nut until a rolling torque of 10-15 inch pounds is provided for pinion shaft bearings. When the correct torque is obtained, drive the two lock pins between shaft and nut at counter-clockwise sides of key slots as shown in Fig. 139. Reinstall bearing cap (13—Fig. 126).

NOTE: Bearing cap (13) plays an important part in the centrifugal lubrication of gears (23 and 26). Make sure cap is installed.

PARK LOCK

82. The park lock cover mounts on left side of rear axle center housing. An exploded view is shown in Fig. 140 and an assembled view in Fig. 141. Park pawl (48) engages the gear (29—Fig. 126) which is keyed to bevel pinion shaft, thus locking pinion shaft in park position.

To remove the park lock cover, it is first necessary to drain rear axle center

Fig. 139–Pinion nut lock pins should be driven in at counter-clockwise end of key slots as shown.

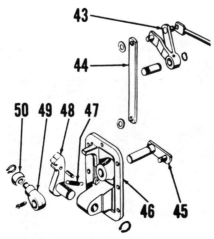

Fig. 140–Exploded view of park lock cover and associated parts. Park pawl (48) meshes with park gear (29–Fig. 126) on bevel pinion shaft. Refer also to Fig. 110.

housing and drain and remove interfering fuel tank.

DIFFERENTIAL AND BEVEL RING GEAR

83. **REMOVE AND REINSTALL.** To remove the differential assembly, first drain transmission and rear axle center housing. Remove both final drive units as outlined in paragraph 85 and IPTO rear housing as in paragraph 94.

Working through rear opening in center housing, block up between rim of main drive bevel gear and bottom of housing. Unbolt and remove both carrier bearing retainer plates and brake pressure plates as assemblies, allowing differential unit to rest on previously installed block. Roll differential and main drive bevel gear from housing as shown in Fig. 142. Overhaul the removed unit as outlined in paragraph 84.

Mesh position of main drive bevel

Fig. 141–Assembled view of park lock cover.

Fig. 142–Use a wood block as protection and roll differential assembly rearward out of center housing as shown.

gears is not adjustable. Backlash is not normally adjustable but should be checked if doubtful. If backlash exceeds 0.024, shims are available (in thicknesses of 0.003 and 0.007) which can be installed between ring gear and differential case half to reduce the backlash. Minimum recommended backlash is 0.006.

When installing the differential, use guide studs as shown in Fig. 143 to install retainer plates, and make sure the stamped word "TOP" is uppermost as shown in Fig. 144.

84. **OVERHAUL.** Refer to Fig. 145 for an exploded view of differential assembly. Differential case bolts are installed with LOCTITE and are not safety wired.

The main drive bevel ring gear and pinion (11) are available only as a matched assembly. Ring gear is riveted to differential case (12) at original assembly, but service parts are provided with bolts and nuts. Install using LOCTITE and tighten to a torque of 100-120 ft.-lbs. The ring gear may be pressed on case half, with up to 0.005 interference fit. A press may be required to remove the old ring gear or to install a new one;

The differential case is available only as a matched assembly which includes adapter (2), shim pack (3), case halves (4 & 12) and the two carrier bearing cups. The 0.004 thick shims (3) are used to obtain proper bearing adjustment at factory assembly and

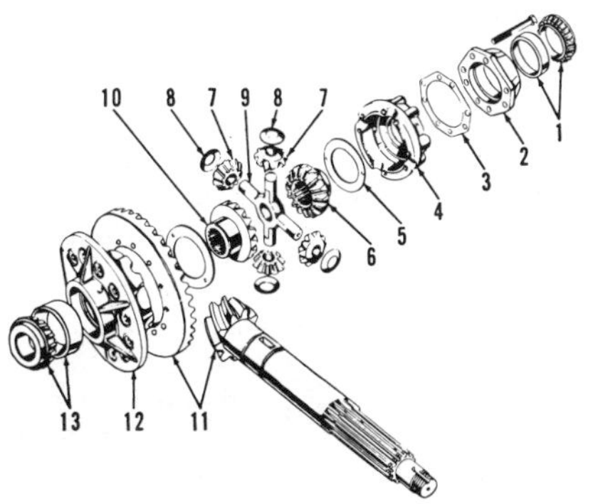

Fig. 145–Exploded view of main drive bevel gears, differential and associated parts of the type used.

1. Bearing
2. Adapter
3. Shim
4. Case half
5. Thrust washer
6. Side gear
7. Differential pinion
8. Thrust washer
9. Spider
10. Side gear
11. Ring gear & pinion
12. Case half
13. Bearing

thickness of shim pack should not be changed.

The two side gears (6 & 10) are not interchangeable, the one with the longer (2½ inch) hub being installed in gear half of case (12).

When assembling the differential unit, use LOCTITE on differential case bolts and tighten to a torque of 110-120 ft.-lbs.

Fig. 147–Make sure case halves are properly marked for reassembly before separating the units.

Fig. 143–Use three guide studs in alternate holes to align retainer plates when installing differential.

Fig. 146–Assembled view of differential unit.

Fig. 148–Spider and differential pinions installed for reassembly.

Fig. 144–Word "TOP" must be uppermost on retainer plate to properly position oil drain holes.

REAR AXLE AND
FINAL DRIVE

All models use a planetary final drive unit located in inner end of axle housing next to center housing. Final drive lubricant reservoir is common with transmission, differential and hydraulic reservoir. Axle stub shafts are free floating, with inner ends supported by differential side gears and outer ends serving as planetary sun gears. Service on any part of axle or final drive requires removal of unit as outlined in paragraph 85.

REMOVE AND REINSTALL

85. To remove either final drive unit as an assembly, first drain transmission and hydraulic system fluid, suitably support rear of tractor and remove rear wheel and fender. Remove drawbar and drawbar frame.

Attach a hoist to final drive housing as shown in Fig. 149. Remove the attaching stud nuts and cap screws and slide the complete final drive assembly away from rear axle center housing. Stub axle and brake disc may remain with center housing, or with final drive unit as shown in Fig. 150. When reinstalling final drive, place stub axle and brake in final drive unit as shown.

Install a new square-section seal (S—Fig. 151) on center housing and reassemble by reversing removal procedure. Tighten retaining stud nuts to a torque of 155 ft.-lbs. and flange cap screws to 170 ft.-lbs.

OVERHAUL

86. AXLE AND BEARINGS. With final drive unit removed as outlined in paragraph 85, withdraw stub axle, brake disc and cover plate as shown in Fig. 152. Remove capscrew retainer (Fig. 153) and capscrew, then withdraw planet carrier, retainer washer, shim pack, thrust washer and bearing cone.

Fig. 149–Removing right final drive unit.

Fig. 150–View of final drive housing with stub axle and brake disc installed.

Fig. 151–A square section seal (S) is used instead of a gasket to stop oil leak.

Use a suitable push-puller as shown in Fig. 154 and push axle shaft and outer seal from housing. (If push-puller is not available, remove outer seal, then withdraw axle assembly by hand.

Outer bearing cone and both bearing cups can be removed at this time. Outer bearing cone may require heavy pressure to fully seat on axle shaft. When assembling, pack outer bearing with lithium base multipurpose grease. Install inner seal as shown in Fig. 155 and carefully insert axle through seal. Install inner bearing cone and thrust washer as shown in Fig. 156. Install a

Fig. 152–Removing brake disc and stub axle from final drive unit.

Fig. 153–Planet carrier retaining cap screw is prevented from loosening by use of the retainer shown.

Fig. 154–Rear axle shaft and outer seal can be removed with a suitable push-puller after planet carrier is removed.

Fig. 155–Installing final drive housing inner seal.

Fig. 156–Install inner bearing cone and thrust washer after axle shaft is inserted.

Fig. 157–Using a guide stud to retain shim pack during planet carrier installation.

Fig. 158–Install planet carrier, temporarily install retaining cap screw, then check axle shaft end play using a dial indicator as shown in Fig. 159.

guide stud approximately 8-inches long in inner end of axle as shown in Fig. 157, then install the removed shim pack plus 0.010 in shims.

Install planet carrier, remove the guide stud and reinstall the retaining capscrew, tightening securely. Check axle end play as shown in Fig. 159. Remove cap screw and reinstall the guide stud. Lift off planet carrier then remove shims equal to the measured end play plus 0.001-0.005, to obtain the specified preload for axle bearings. Tighten the retaining cap screw to a

Fig. 159–Measuring axle shaft end play which must be finally adjusted to 0.001-0.005 preload during final assembly.

Fig. 160–Install outer seal after correct bearing preload is obtained.

Fig. 161–Planet pinion shafts are collectively retained by snap ring (R) which fits a groove in all three shafts and in carrier hub.

torque of 300-325 ft.-lbs. in final assembly and reinstall capscrew retainer. Install outer seal as shown in Fig. 160, then complete the assembly by reversing the removal procedure.

87. PLANET CARRIER. Planet pinion shafts are a slip fit in carrier and are retained by a snap ring (R—Fig. 161) which fits in a slot in all three shafts plus a retaining slot in carrier hub. To disassemble the carrier, place unit on a bench with inner (flat) side down and expand snap ring fully into shaft grooves as shown in Fig. 162. Work snap ring up out of carrier hub slot, withdrawing all three shafts until snap ring clears shaft grooves.

Slide the pinion, together with thrust washers and loose needle bear-

Fig. 162–To disassemble planet carrier, expand the snap ring into deeper pinion shaft grooves and lift all three shafts until clear of retaining groove in carrier hub.

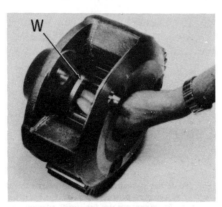

Fig. 163–Retainer washer (W) can be removed after one pinion is out as shown.

Fig. 164–All three planet pinion shafts and the retaining snap ring must be installed at the same time as shown.

ings, out of carrier as shown in Fig. 161. Retainer washer (W—Fig. 163) can be withdrawn after one planet pinion is out as shown.

When assembling, the snap ring and three planet pinion shafts must be installed together as shown in Fig. 164. Install planet carrier and adjust axle bearing preload as outlined in paragraph 86.

BRAKES

All models are equipped with individual hydraulic power brakes for each rear wheel, the power being supplied by the tractor central hydraulic system when engine is running and by a nitrogen filled accumulator when engine is not running. Brakes can be manually operated by depressing the brake pedals when main hydraulic system is inoperative, but manual power is not sufficient to stop a moving tractor.

The wet-type disc brakes are located on stub axle shafts between differential and final drive planetary gears. Brakes can be serviced after removing final drive housings as outlined in paragraph 85.

ADJUSTMENT AND BLEEDING

The only adjustment required of the brakes is adjustment and equalization of pedal height. Adjust and bleed the brakes as outlined in the appropriate following paragraphs. If brakes cannot be satisfactorily adjusted, remove and overhaul control valve as outlined in paragraph 128 or brake pistons as in paragraph 90.

88. BLEEDING. To bleed the hydraulic brakes, refer to Fig. 165 and proceed as follows:

To fill brake valve housing reservoir, start and run engine, initially depress both brake pedals; then release one pedal while holding the other pedal down for 10-15 seconds.

Loosen the locknuts and back out the two bleed screws (B—Fig. 165) two full

Fig. 165–Brake bleed screws (B) are located on lower side of rear axle center housing as shown. Bleed passage is internal (to main hydraulic reservoir) when bleed screw is opened.

turns. Depress each brake pedal in turn until resistance is felt and hold pedal down for 15 seconds. Release the pedals, tighten both bleed screws to a torque of 6-10 ft.-lbs. and secure with locknuts. Adjust pedal height as outlined in paragraph 89.

89. PEDAL ADJUSTMENT. Pedal height adjustment is correct when, with engine running, solid resistance is felt when pedals clear upper stop on platform by ⅝-inch with both pedals of equal height.

To make the adjustment, disconnect brake pedal interlock, refer to Fig. 166 and proceed as follows: Loosen locknuts (B) and turn brake rods (C) until adjustment (A) is correct.

Brake pedal interlock should engage with both pedals depressed when adjustment is correct.

Fig. 166–Brake pedal height adjustment is correct when distance (A) from upper stop to solid pedal feel (engine running) is ⅝ inch. Adjust by loosening locknuts (B–Inset) and turning brake rod (C). Pedal height should be equalized during adjustment.

Fig. 167–Brake actuating disc is located on stub axle shaft in final drive housing as shown.

OVERHAUL

90. BRAKE DISCS AND ASSOCIATED PARTS. Before any service can be performed on brake discs or actuating pistons, final drive unit must be removed as outlined in paragraph 85.

With final drive off, brake disc and cover plate can be removed as shown in Fig. 167. To remove the brake pressure plate, pistons and O-rings, first unbolt and remove differential cover plate. Remove the three cap screws (C—Fig. 168) and lift off the carrier plate containing pistons and return springs.

Brake pistons are marked "THIS SIDE OUT" for proper installation as shown at (M—Fig. 169). Back-up washer for piston O-ring goes to outside as shown in cross sectional view (Fig. 170).

91. ACCUMULATOR AND CHECK VALVE. A nitrogen filled accumulator and check valve are provided to store pressurized hydraulic fluid for the brake system with engine off or hydraulic system inoperative. Enough capacity is provided to apply the brakes 7 or 8 times.

Neither the accumulator unit nor the check valve can be overhauled, although the accumulator may be recharged. Service normally consists of checking to determine the faulty units, then renewing as required; proceed as follows:

Fig. 168–Brake pressure plate is attached to differential carrier plates by leaf type return springs. To disassemble the unit, remove the three cap screws (C).

Fig. 169–Brake pistons are marked "THIS SIDE OUT" for proper installation in differential carrier plate.

With system pressurized, shut off the engine, then count the number of times brakes apply before reserve power is exhausted. Seven or eight times is normal; a fewer number of times would indicate low hydraulic pressure, hydraulic leakage in brake cylinders or valve, defective accumulator or defective check valve.

If recharging equipment is available, accumulator nitrogen pressure can be checked by fully discharging the hydraulic fluid reserve (disconnect hydraulic line to brake valve to be sure) then installing a nitrogen test gage to charging port. Nitrogen pressure should be 1150-1250 psi at 70° F.

The brake system check valve is installed at either end of brake system supply line leading from main hydraulic system priority valve to brake accumulator. The in-line check valve cannot be disassembled and is serviced only as a complete unit. A rough test of

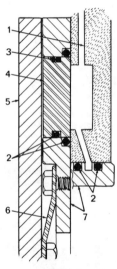

Fig. 170–Cross sectional view of rear axle center housing, differential carrier plate and associated parts showing brake actuating components.

1. Center housing
2. O-rings
3. Backup ring
4. Brake piston
5. Pressure plate
6. Return springs
7. Carrier plate

Fig. 171–Right hand view of tractor, showing brake accumulator (A).

the brake check valve can be made as follows:

Operate engine until brake hydraulic system is fully charged, shut off engine and wait ten seconds. Turn steering wheel; if residual hydraulic power provides steering assist, check valve is leaking and should be renewed. Install check valve with arrow

pointing TOWARD brake accumulator, in direction of fluid flow.

92. **BRAKE VALVE.** The brake valve is an integral part of the main hydraulic system and is covered in hydraulic system sections. Refer to paragraph 113 for brake valve tests and to paragraph 129 for overhaul procedure.

176) and detach top link clevis pin from draft control springs. Remove the pins securing ratio lever (R—Fig. 177) to top link clevis and draft control sensing rod and lift off the link. Disconnect draft control hand wheel from adjusting clevis and remove the clevis. Unbolt and remove drawbar rear support frame. Attach a hoist to draft control stop plate as shown in Fig. 177.

Remove one flange cap screw on each side of housing and install suitable

INDEPENDENT POWER TAKE-OFF

The IPTO input shaft is splined into an adapter attached to engine crankshaft along with flywheel, as shown in Fig. 172. The front shaft runs inside the hollow clutch shaft and transmission mainshaft, and is connected by a coupling to the IPTO drive shaft located in rear axle center housing.

The hydraulic pump drive gear is an integral part of the IPTO clutch drum. The multiple disc IPTO clutch is hydraulically actuated by the low volume, low pressure circuit of the main hydraulic system. Tests and service of pump and valve is covered in the HYDRAULIC SYSTEM Section of this manual.

OUTPUT SHAFT

93. The 540 and 1000 rpm IPTO output shaft drives consist of outer and inner flanges as shown in Fig. 173. Optional output shafts are shown in Fig. 174.

Any service on output shaft drive units requires removal of IPTO housing as outlined in paragraph 94 and overhaul of output shaft gear train as in paragraph 95.

IPTO OUTPUT HOUSING AND GEARS

94. **REMOVE AND REINSTALL.** To remove the IPTO output housing, first drain transmission and center housing and move IPTO coupling handle (H—Fig. 175) into engaged ("ON") position. Remove top link (Fig.

Fig. 175–View of control console showing IPTO coupling handle (H).

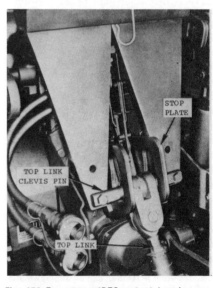
Fig. 176–To remove IPTO output housing, remove top link and disconnect top link clevis pin from sensing springs.

Fig. 172–View of flywheel with clutch removed, showing splined flywheel adapter which serves as the IPTO drive unit.

Fig. 173–View of IPTO output housing showing the dual speed drive flanges. The 1000 rpm output stub shaft is installed.

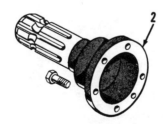

Fig. 174–View of IPTO output accessory shafts available for installation.

1. 1000 rpm ASAE shaft
2. 540 rpm ASAE shaft
3. 540 rpm 1¾-inch special shaft

Fig. 177–Remove drawbar rear support frame, ratio lever (R) and sensing linkage adjusting clevis.

guide studs; remove remainder of flange screws and lift off IPTO housing as shown.

IPTO rear drive shaft (Fig. 178) can be withdrawn after cover is removed, but be sure coupling is engaged to prevent coupler from dropping out of position. Also be sure shim pack is not lost or damaged when rear cover is removed. Shims control IPTO rear drive shaft end play which should be 0.010-0.026. Shims are available in thicknesses of 0.015, 0.030 and 0.062. Clearance can be checked with a feeler gage after removing center housing front, top cover or park lock cover.

When reinstalling IPTO output housing, tighten large flange cap screws to a torque of 50-55 ft.-lbs. and small cap screws to 15-18 ft.-lbs. Complete the assembly by reversing the removal procedures. Refer to paragraph 141 for draft control linkage adjustment.

95. **OVERHAUL.** To disassemble the removed output housing, remove the three cap screws (1—Fig. 180) and lift off gear cover (2). Refer to Fig. 181. Input gear (4) will be removed with cover while idler gear (3) may remain with cover or main housing. Note the

location and position of thrust washers as housing is disassembled.

When disassembling the 540 and 1000 rpm output drive flanges, unseat and remove the snap rings, gears and thrust bearings from inside of housing and withdraw both flange shafts as a

Fig. 180–To disassemble the removed output housing, remove the three cap screws (1) and lift off gear cover (2).

Fig. 181–Removed gear cover and associated parts. Refer also to Fig. 182.

2. Cover
3. Idler gear
4. Input gear
5. Shaft

unit as shown in Fig. 183. The shafts contain 66 loose bearing rollers assembled in two rows of 33 each, separated by a machined pipe spacer. Fig. 184 shows a cross sectional view of output flange shafts properly assembled. Install shafts carefully from rear of housing. The three thrust roller bearings have one thrust race each, the other race being machined on the 540 and 1000 rpm driven gears. The three thrust bearings are identical but thrust races are not. Place the thickest thrust race next to housing as shown in Fig. 185. Position the stepped thrust race/snap-ring retainer between the gears as shown in Fig. 186. Place the thin thrust race in cover recess. With cover installed, check to be sure that

Fig. 183–Assembled view of output drive flanges. Refer to Fig. 184 for cross section.

Fig. 178–Rear IPTO shaft can be withdrawn after IPTO output housing is removed. Be sure shift coupling is engaged and be careful not to lose shim pack.

Fig. 182–Output housing and gears with gear cover (2–Fig. 181) removed. Lift out cluster gear (6). Unseat snap rings and remove 1000 rpm output shaft gear (8) and 540 rpm output gear (7). Refer to Fig. 183.

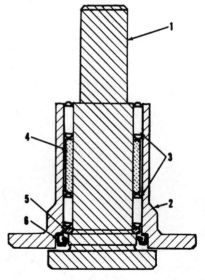

Fig. 184–Cross sectional view of output flange shafts showing bearing and seal installation.

1. 1000 rpm shaft
2. 540 rpm shaft
3. Needle rollers
4. Spacer
5. Thrust washer
6. Oil seal

Fig. 179–View of tractor showing IPTO output housing correctly installed.

gears and shafts turn freely with no interference. Reinstall the assembled output housing as outlined in paragraph 94.

IPTO CLUTCH

96. REMOVE AND REINSTALL. To remove the IPTO drive shaft, hydraulic clutch and pump drive gear, first remove IPTO output housing as outlined in paragraph 94 and center housing front, top cover as in paragraph 120.

Withdraw oscillator crank (O—Fig. 188) by prying forward. Unbolt and remove oil return tube (T). Withdraw IPTO coupler (Fig. 189). Slide intermediate IPTO shaft rearward a slight amount, extract the snap ring; then slide shaft rearward out of the way. With IPTO intermediate shaft re-

moved, lift out the IPTO clutch unit as shown.

Lubricate teflon bronze sealing rings with petroleum jelly and install clutch unit by reversing removal procedure.

Fig. 188—To remove the IPTO clutch, it is first necessary to remove return tube (T) and oscillator arm (O).

Fig. 185—When installing output flange shaft gears, place thickest thrust race next to housing and install one thrust bearing.

Fig. 186—Place the stepped thrust race over snap ring and install second thrust bearing.

Fig. 187—The third thrust bearing is installed over 1000 rpm gear, with thin thrust race in cover recess.

Fig. 189—Slide intermediate shaft rearward and lift out sliding coupler; then withdraw clutch assembly as shown.

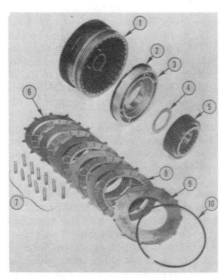

Fig. 190—Disassembled view of IPTO clutch showing component parts.

1. Clutch drum
2. Piston ring
3. Piston
4. Thrust washer
5. Clutch hub
6. Separator plates
7. Separator springs
8. Clutch discs
9. Pressure plate
10. Snap ring

97. OVERHAUL. With IPTO clutch removed as outlined in paragraph 96, disassemble as follows: Place clutch drum hub-end down on a bench and, applying slight pressure to plate (9—Fig. 190), unseat and remove snap ring (10). Remove items (4 through 9).

Place six dowel pins from Special Tool (MFN 768) in clutch housing splines (See Fig. 191) to protect the piston ring, and work piston (3—Fig. 190) from clutch drum by grasping stiffening ribs with pliers.

Examine the parts and renew any which are worn or damaged. Examine loose check ball in clutch housing to see that it is free in bore and seats without leakage against pressure from inside.

When assembling the clutch, work piston ring into drum using an ice pick or small screwdriver, install the eleven dowels as shown in Fig. 191; then push the piston fully into place by hand. Note that the externally splined separator plates (6—Fig. 190) are slightly dished. Sort the plates and stack so that all can be installed with concave in the same direction. Place one separator plate on top of piston, then alternate clutch discs and separator plates,

Fig. 191—Using the special dowels as ring compressors, push piston fully in place by hand as shown.

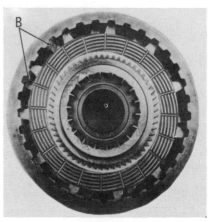

Fig. 192—External lugs of first separator plate should be one spline either way from remaining lugs as shown at (B).

HYDRAULIC SYSTEM

with lugs of separator plates one spline either way from lug of bottom plate as shown in Fig. 192. Lugs of bottom plate are to serve as pressure points for separator springs as shown in Fig. 193. Complete the assembly by reversing the disassembly procedure and install the assembled clutch as outlined in paragraph 96.

IPTO SHIFT LINKAGE

98. Fig. 194 shows and exploded view of IPTO shift linkage and associated parts. The sealing O-ring (7) fits in channeled groove of housing bore for shaft (6). No adjustment is required of linkage unit. In the event of improper shifting, check for bent or broken components.

Service on the IPTO control valve is contained in paragraph 131.

Fig. 193–Install separator springs on lugs of first separator plate as shown.

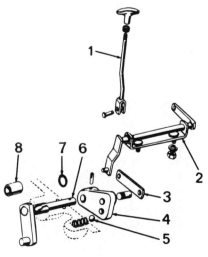

Fig. 194–Exploded view of IPTO shift linkage.

1. Link	5. Detent ball
2. Bellcrank	6. Shift crank
3. Link	7. O-ring
4. Lever	8. Bushing

All models have a closed center, constant pressure hydraulic system which can logically be divided into three circuits as follows:

LOW PRESSURE/LOW VOLUME Circuit. Operates at a volume of 6 gpm and a pressure of 225-275 psi to provide working fluid for Multi-Power, IPTO and Hydraulic Seat circuits. All valves are closed-center type and fluid discharges through combining valve into the Low Pressure/High Volume Circuit.

LOW PRESSURE/HIGH VOLUME Circuit, also referred to as "Charging Circuit". Supplies working fluid on demand to high pressure piston pump, and provides constant circulation to system filter and oil cooler. Circuit is supplied by a 17 gpm internal gear type pump and unused flow from low volume circuit. Charging circuit pressure is regulated at 90-100 psi.

HIGH PRESSURE (MAIN) Circuit. Supplies working fluid for power steering and power brakes on a priority basis, with sufficient reserve capacity for hydraulic working cylinders. Pressure is maintained at 2100-2400 psi, with available volume at 16-20 gpm.

The transmission lubricant is the operating fluid for the hydraulic system. Massey-Ferguson M-1127 Fluid is recommended. System capacity is 22 U.S. Gallons. Refill capacity is 17 U.S. Gallons, as 5 gallons will remain when reservoir is drained.

OPERATION

99. The hydraulic system operating fluid is drawn from reservoir in rear axle center housing through a screen-type filter. The internal gear type dual charging pump is engine driven by the IPTO input shaft. Fluid from the low volume side of dual pump flows under pressure to the hydraulic seat valve and to the Multi-Power and IPTO control valve. All units are equipped with closed-center valves; if no demand for fluid exists, the fluid passes through the combining valve where it is mixed with the flow from the high-volume side of charging pump.

The combined flow from the charging pump is open to two passages. One leads through the system filter and oil cooler to the inlet side of the main hydraulic pump, then to control port (C—Fig. 195) of regulating valve. Four to six gpm passes through the regulating valve spool orifice (O) to the sensing pressure regulating valve (P), where sensing pressure is maintained at 90-100 psi. If there are no other demands, the remainder of charging pump flow passes from bypass port (B) to dump

port (D) through the passage opened by movement of regulating valve spool (S).

The radial piston type main hydraulic pump is front mounted and driven by engine crankshaft. The main hydraulic pump attempts to maintain a pressure of 2100-2400 psi at all valves of the high pressure main hydraulic system. When this pressure is reached, flow from the main hydraulic pump ceases, the eight radial pistons are held away from the actuating cam and the pump drive shaft continues to rotate without performing work. The main pump fluid delivery is maintained at the flow necessary to satisfy system requirements up to a maximum delivery of 20 gpm.

LUBRICATION, TESTING AND ADJUSTMENT

100. **LUBRICATION.** The hydraulic system fluid serves as a lubricant for the power steering motor and hand pump, transmission, differential, final drive and IPTO gear train. The transmission and rear axle center housing serve as the system reservoir.

The only recommended fluid is MASSEY-FERGUSON M-1127 Fluid. System capacity is 22 U.S. Gallons; when system is not disassembled, however, only about 17 gallons will be drained from the system and only this amount is required for normal refill. To check the fluid level, be sure tractor is on level ground and remove dipstick (D—Fig. 196) from left side of transmission housing. Fill through dipstick opening if fluid is needed.

System filter (See Fig. 4) should be cleaned or renewed at 750 hour intervals or once a year, to coincide with lubricant change. At the same time, intake screen cover plate (bottom of center housing, Fig. 197), should be

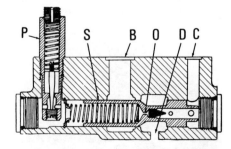

Fig. 195–Cross sectional view of hydraulic system regulating valve which controls charging pump pressure and flow.

B. Bypass port	P. Regulating pilot
C. Sensing control port	valve
D. Dump port	S. Valve spool
O. Control orifice	

removed and the screen cleaned. Do not attempt to flush the system as a considerable amount of the flushing solution cannot be drained. If fluid is badly contaminated, system must be disassembled for cleaning.

101. **QUICK TEST.** A quick test of the high pressure system can be performed using a pressure gage as follows:

Connect pressure gage to remote coupler and pressurize that coupler, with engine running first at 1000 rpm, then at 2000 rpm. Gage reading should be 2100-2400 psi on both tests.

Flow can be tested at breakaway coupling by using a flow meter and fully opening the flow control valve on top of auxiliary valve body. Connect flow meter inlet to breakaway coupling and direct flow meter outlet back to reservoir. A plug opening (return) is provided at rear, left side of rear axle center housing as shown in Fig. 198. Make sure all hydraulic functions are in neutral. Start and operate engine at 1000 rpm. Move auxiliary lever to pressurize the coupling containing flow meter and lock the lever in open position. Adjust load valve to 1000-1200 psi and allow fluid to reach an operating temperature of 90°-140° F., then check the measured flow which should be a MINIMUM of 9 gpm.

If flow and/or pressure are not as specified, refer to pump tests and adjustments as outlined in paragraph 103 to pinpoint the possible causes of trouble. For specific tests on system circuits and components, refer to the applicable paragraphs 110 through 113.

102. **CHARGING CIRCUIT PRESSURE AND FLOW.** To obtain access to front, top cover for service or testing, first remove the ring from manual unloading valve, remove operator's platform center panel as shown in Fig. 199 and lift off padding. All tests should be conducted with tractor hydraulic system fluid at operating temperature of 90°-140° F.

103. **PRESSURE TESTS.** Remove plug (S—Fig. 200) from sensing test port and install a suitable 0-400 psi pressure gage. With engine running at any speed above slow idle, sensing pressure should be 90-100 psi. Pressure should not drop below 80 psi at slow idle speed. If pressure is not correct, remove adjusting port plug (A) and, working through plug opening, turn the exposed slotted head plug clockwise to raise the pressure or counter-clockwise to lower pressure. Adjust the pressure to 95 psi @ 1000 engine rpm.

Remove plug (L) from low volume port and install 0-400 psi pressure gage. Low volume system pressure should not be less than 210 psi @ 1000 rpm nor more than 275 psi @ 2000 rpm.

Check the combined system pressure by installing pressure gage in high volume port plug (H). Pressures should be approximately the same as those given for low volume circuit.

Substantially higher pressures at low volume pressure port (L), may indi-

cate a sticking combining valve in pump body. If pressure is normal at 1000 rpm but higher than normal at 2000 rpm, checked for plugged main system filter or restricted lines. Low pressure at high rpm may indicate a sticking safety relief valve (behind port plug R).

104. **FLOW.** To check the combined flow from charging pump, disconnect charging supply tube from elbow (C— Fig. 200) and connect flow meter inlet line to elbow. Direct flow meter outlet to sump as shown in Fig. 198.

CAUTION: Operate tractor only long enough with charging supply tube disconnected to make the test.

Make sure IPTO valve lever is in "OFF" position, Multi-Power valve is in "LOW" position and seat valve lever is in neutral.

Start and run engine at the indicated speeds, maintaining a pressure of 150 psi by adjusting the flow meter load valve. Minimum flow should be as specified.

850 rpm . . . . 7 gpm
1000 rpm . . . . 11 gpm
2000 rpm . . . . 22 gpm

Low pump output can be caused by clogged intake screen, air leaks or restrictions in intake tube, faulty

Fig. 196–Hydraulic oil level dipstick (D) is on left side of transmission housing as shown. Dipstick opening is also filling port.

Fig. 199–For access to front top cover for service or testing, remove operator's platform center panel and padding as shown.

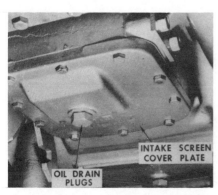

Fig. 197–Bottom view of rear axle center housing showing hydraulic system fluid plugs and intake screen cover plate.

Fig. 198–Left rear corner of rear axle center housing with fitting (1) and flow meter return line (2) attached.

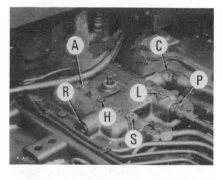

Fig. 200–View of front hydraulic cover showing test and adjusting ports.

A. Adjusting port
C. Charging port
H. High volume port
L. Low volume port

P. Low volume port
R. Relief valve plug
S. Sensing port

manual unloading valve, faulty safety relief valve or faulty pump.

To check the safety relief valve, leave flow meter connected and install a 0-400 psi pressure gage in plug port (H—Fig. 200). With engine running at 850 rpm and load valve on flow meter open, start closing load valve and note pressure at which a noticeable drop in flow occurs. Pressure should be 180 psi. Close flow meter load valve to block the flow. Pressure at plug port should be a minimum of 210 psi @ 1000 engine rpm and maximum of 350 psi @ 2000 rpm.

Low volume flow can be checked separately by connecting flow meter inlet to low volume port (P). Adjust flow meter load valve to maintain a pressure of 200 psi. Minimum flow should be 2.5 gpm @ 1000 rpm and 5 gpm @ 2000 rpm.

105. MAIN PUMP PRESSURE AND FLOW. To check the main hydraulic pump pressure and flow, disconnect outlet tube (O—Fig. 201) from right top of main pump housing. Cap the tube to prevent fluid loss. Attach flow meter inlet hose to pump body fitting and connect flow meter outlet to sump as shown in Fig. 198. Remove center plug (P—Fig. 202) from lower face of main pump front cover and attach a 0-2000 psi pressure gage to plug port.

Start and run engine at 1000 rpm and adjust flow meter load valve to obtain a pressure of 1900 psi. Allow fluid temperature to reach 140°-160° F.

106. RELIEF VALVE CHECK. With engine running and fluid at test temperature, adjust flow meter load valve until a flow of 5 gpm is obtained. Turn metering valve adjusting plug (M—Fig. 202) clockwise until pressure ceases to rise on FLOW METER pressure gage. Test gage reading should be 2650-2850 psi.

If maximum obtainable pressure is below 2650 psi, remove and renew relief valve cartridge (behind plug R), or

relief valve cartridge sealing O-ring. Cartridge may be accurately checked by hand after removal, using a hand test pump.

107. QUICK RECOVERY VALVE. After checking relief valve as outlined in paragraph 106, back out metering valve adjusting plug (M—Fig. 202) until flow meter pressure gage registers 1400 psi with engine running at 1000 rpm and flow adjusted to 5 gpm.

Flow meter pressure should be unstable. If it is not, turn quick recovery valve adjusting plug (Q) clockwise until pressure becomes unstable. Reset metering valve and check intermediate pressure as outlined in paragraph 108 after quick recovery valve is adjusted.

108. INTERMEDIATE PRESSURE. With quick recovery valve adjusted as outlined in paragraph 107 and engine running at 1000 rpm, adjust flow to 3 gpm and reset metering valve adjusting plug (M—Fig. 202), to obtain a flow meter pressure reading of 2050 psi. Check intermediate pressure gage installed in port (P), reading should be 375-425 psi. Increase engine speed to 2000 rpm; flow meter pressure should remain at 2050 psi and flow should remain at 3 gpm.

If any of the test results are not as stated, readjust settings as outlined in paragraphs 105 through 108 or overhaul pump as in paragraphs 117 through 119.

109. OUTPUT. With intermediate pressure checked as outlined in paragraph 108 and engine speed at 2000 rpm, adjust flow meter load valve to obtain an output pressure of 1900 psi on flow meter pressure gage. Main pump output flow should be a MINIMUM of 16 gpm.

Slowly close flow meter load valve and note gage pressure when flow stops. Shut-down pressure should be 2100-2400 psi at any speed. Minor adjustments of shutdown pressure can be made by turning metering valve adjusting plug (M—Fig. 202).

Slow engine speed to 1000 rpm and open flow meter load valve to obtain a flow of 5 gpm; then quickly close load valve while observing pressure. Gage should momentarily rise to 3200-3400 psi, then quickly drop to shutdown pressure and hold steady.

110. HIGH PRESSURE CIRCUIT TESTS. To check the high pressure system components, disconnect outlet tube (O—Fig. 201) from right top of main pump housing. Attach flow meter inlet hose to pump body port and flow meter outlet to tube (O).

Open load valve on flow meter and start and run engine with all hydraulic units inactive. Flow meter pressure gage registers shutdown pressure and flow gage registers high pressure circuit leakage, which should be 0-½ gpm.

111. PRIORITY VALVE. Connect both breakaway couplings of one valve bank with a diverter hose as shown in Fig. 203 and start engine with valve in neutral. With engine at slow idle speed, move valve lever to either operating position while watching flow meter pressure gage. Pressure should drop from standby pressure and hold steady at about 1300 psi.

A low or substantially higher reading indicates a malfunctioning priority valve, which may be disassembled and cleaned as outlined in paragraph 126, however parts are not available for overhaul.

112. POWER STEERING. Minor malfunction which does not materially affect steering operation will have no effect on hydraulic system.

As a quick hydraulic test, start and run engine at any speed. Turn steering wheel either way against steering stop. Continue to apply pressure at steering

Fig. 201–Front view of engine showing main hydraulic pump outlet tube (O).

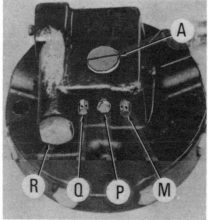

Fig. 202–Main hydraulic pump front cover showing points of adjustment and test.

A. Actuating piston plug
M. Metering valve adjustment
P. Pressure port plug
Q. Quick recovery adjustment
R. Relief valve plug

Fig. 203–Diverter hose installed for checking priority valve.

wheel. Leakage can be considered excessive if steering wheel can be turned one revolution with moderate pressure (36 inch pounds) applied; or if pump flow increases more than 0.25 gpm over reading with steering wheel released.

113. BRAKES. After bleeding brakes as outlined in paragraph 88 and adjusting linkage as outlined in paragraph 89, check for leakage and valve malfunction as follows:

Firmly depress brake pedals one at a time with engine running at 1000 rpm while watching flow meter gage. Flow should not increase more than 0.25 gpm when either pedal is depressed. If leakage is excessive, isolate the trouble by disconnecting the tube leading from valve to center housing and capping the connection. If leak continues, valve is at fault. If leak stops, brake unit is at fault. Refer to paragraph 90 for service on brake pistons, or paragraph 129 for service on brake valve.

Brake valve operation is based on a principle of proportional pressure, with 30 psi output pressure being transmitted for each 1 lb. pressure applied at brake pedal. Pressure should increase smoothly with increased pedal pressure until standby pressure of hydraulic system is reached.

The brake valve is equipped with equalizer piston assemblies which should open and equalize the output pressure when both brake pedals are applied at an unequal rate. Test equalizer action by installing a pressure gage in each brake line. With engine running, depress either pedal until about 1000 psi is obtained on that brake line. While holding that pedal steady, slowly depress the other pedal until about 400 psi is developed; then equalizer piston should open and both gages show equal pressure. Once equal pressure is attained, pressure should remain equal while either pedal is ap-

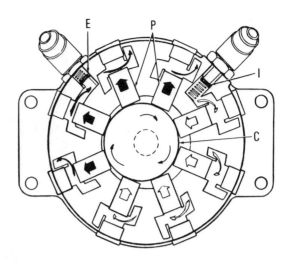

Fig. 205–Cross sectional schematic view of the radial piston type main hydraulic pump. Pistons are pushed outward by the rotating eccentric cam (C) and inward by hydraulic pressure from charging pump. Inlet valves (I) and exhaust valves (E) are interchangeable.

plied, even though the foot is completely removed from the other. Repeat all tests, depressing pedals in reverse order. Failure to perform as indicated could signify a malfunctioning equalizing valve.

With engine not running and accumulator reserve depleted, either or both pedals should develop and hold at least 500 psi pressure for a period of 30 seconds.

114. IPTO MODULATING VALVE. To check the IPTO modulating valve, remove left instrument panel cover and cap plug (T—Fig. 204) from IPTO outlet tee. Install a 0-500 psi pressure gage in plug opening.

With fluid at operating temperature and engine running at 1000 rpm, slowly move IPTO control lever toward "ON" position while watching gage pressure. The pressure should rise smoothly with movement of lever and hold at any selected pressure, until gage reading is 100 psi; then rise quickly to system pressure of 210-225 psi as lever is moved further toward "ON".

If IPTO pressure cannot be properly controlled, remove valve as outlined in paragraph 130 and overhaul as in paragraph 131.

MAIN PUMP

115. OPERATION. The radial piston type main hydraulic pump mounts on front support casting and is driven by engine crankshaft.

Fluid from the charging pump is provided at main pump inlet at moderate pressure. Unless flow is blocked by movement of the distributing valve, fluid passes through the inlet valves and pushes the radial pump pistons (P—Fig. 205) down into contact with pump camshaft (C). When demands for pressurized fluid are met, the distributing valve closes inlet valves off from charging pressure and pump pistons fail to follow cam. Pumping action thus stops until a drop in standby pressure signals a demand for additional fluid. The metering action of the distributing valve determines volume delivered, up to the maximum capacity of the system.

Pressurized fluid enters the pump cover at a charging pressure of up to 100 psi. The distributing valve piston is spring-loaded in the open position. Orifices lead from the charging pressure passages into bearing and camshaft area of pump to provide lubrication. Fluid from the main pump high pressure gallery passes through a metering valve to the area behind the distributing valve piston, then through fixed orifices to the charging passage, to balance the high pressure system requirements.

The control cover also contains a "Quick Recovery" valve which is spring loaded to the open position and closed by hydraulic pressure. The quick recovery valve connects the intermediate (control) pressure passage to charging passage. A sudden drop in standby pressure causes the quick recovery valve to open, dumping pressure from the control passage and allowing the distributing valve to completely open. A full supply of high pressure fluid is thus provided to meet system demands.

Fig. 204–To check IPTO modulating valve, remove cap plug (T) from IPTO outlet tee and install a suitable gage.

Fig. 206–To remove the main hydraulic pump, disconnect one tie rod and pivot axle support out of way as shown.

116. **REMOVE AND REINSTALL.** To remove the main hydraulic pump, first drain radiator and remove radiator hose. Suitably support front of tractor, remove axle pivot support bolts except right, rear bolt in front support, which must be loosened. Disconnect right tie rod and pivot axle and support out of the way as shown in Fig. 206. Remove the protector plate below the pump.

Disconnect inlet, outlet and lubrication lines. Unscrew the four cap screws (S—Fig. 207) retaining drive coupler (C) to crankshaft pulley, slide coupler against pump body and turn coupler until attaching bolts can be pushed into recesses provided in rear face of pump.

NOTE: The main hydraulic pump weighs more than 100 lbs. and means must be provided for lifting the pump down out of front support. The manufacturer provides a tool for pump removal as shown in Figs. 208 and 209.

Attach the tool or other suitable means to pump body and remove the mounting bolts, then swing pump away from tractor frame.

Install the pump by reversing removal procedure, making sure that drive coupler properly aligns with

Fig. 209–The special tool is designed to pivot pump from tractor as shown. Refer also to Fig. 208.

crankshaft pulley. Tighten all bolts securely.

117. **OVERHAUL.** With pump removed as outlined in paragraph 116, refer to Fig. 210 and proceed as follows:

Carefully withdraw pump coupling (1). Remove two of the 5/8-11 cap screws securing valve body (24) to piston housing (9) and install five-inch aligning studs. Loosen and remove the other two cap screws evenly, then carefully remove valve body and distributing valve springs (12 & 13).

118. **VALVE BODY.** Distributing valve piston (14—Fig. 210) must slide smoothly in the sleeve in valve body and actuating piston (25) must move smoothly in guide (15). Refer to Fig. 211 for cross sectional view.

Remove the snap ring from secondary relief valve (16—Fig. 210) and fish out the valve if necessary, using a

Fig. 207–Unscrew the four capscrews (S) securing drive coupler (C) to crankshaft pulley.

Fig. 208–Using the special tool (MFN-762) to remove the main hydraulic pump. Refer also to Fig. 209.

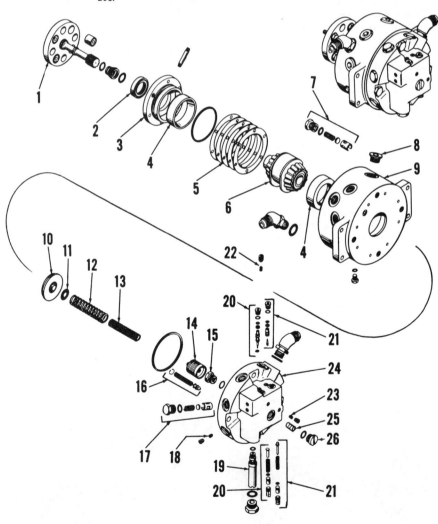

Fig. 210–Exploded view of crankshaft driven radial piston (main hydraulic) pump showing component parts.

1. Drive coupler	8. Piston plug	15. Guide		20. Quick recovery valve	
2. Oil seal	9. Piston housing	16. Secondary relief		21. Metering valve	
3. Carrier assembly	10. Orifice plate	valve		22. Orifice plug	
4. Bearing cup	11. Filter screen	17. Inlet valve		23. Orifice plug	
5. Shims	12. Outer spring	18. Orifice plug		24. Valve body	
6. Camshaft	13. Inner spring	19. Main relief valve		25. Actuating piston	
7. Discharge valve	14. Distributor valve			26. Plug	

small wire. Unscrew plug and remove main relief valve (19). Remove quick recovery valve (20) and metering valve (21), working from both ends of bore.

Remove the eight inlet valves (17), being careful the valve discs do not fall into inlet manifold and become lodged. Check the valve seats in body for damage.

The three orifice plugs (18, 22 and 23) will not need to be removed unless their condition is questionable. Actuating piston guide (15) is installed with LOCTITE and will not need to be removed unless O-ring or guide are to be renewed or trouble is suspected. Clean and inspect all parts and renew any which are damaged or questionable.

When assembling the metering valve, refer to Fig. 212 and proceed as follows: Working from bottom side of valve body, install valve guide (7), spring (8), spacer plug (10) and adjusting plug (11) in the order given. Make sure the O-ring is installed on spacer plug (10). Install O-ring (4) and backup ring (5) as shown, on valve (3), and install sleeve in top of bore; then drop valve needle (6) into sleeve (3), chamfered end up. Install and tighten plug (1).

When installing the quick recovery valve, refer to Fig. 213. Install valve guide (8), spring (9), spacer plug (11) and adjusting plug (12) in bottom of body bore. Install O-ring (4) and backup ring (5) as shown on valve sleeve (3), then install sleeve with smaller OD to top. Drop the piston (7) with small end down into sleeve bore. Install and tighten plug (1).

If piston guide (15—Fig. 210) has been removed, install with LOCTITE and tighten to a torque of 60-80 inch pounds. Tighten plug (26) to a torque of 20-25 ft.-lbs.

The eight intake valves, guides and springs (17) are interchangeable with each other and with outlet valve (7). The individual parts can be installed in valve body either side out. Tighten plugs to a torque of 45 ft.-lbs.

119. PISTON HOUSING. To disas-semble the piston housing (9—Fig. 210), first remove the eight chamber plugs (8) and pump pistons, keeping pistons in their proper order so they can be reinstalled in their mated bores. Remove the cap screws securing carrier (3) to pump body and remove the carrier by carefully prying it straight out. Remove and save shim pack (5). Shims control adjustment of the tapered camshaft bearings.

With pistons and carrier removed, shaft and bearing unit can be lifted out. Camshaft and bearings are available only as an assembly, renew the unit if any part is damaged.

Remove plugs and outlet valves (7), being careful that valve discs do not fall into discharge manifold of pump. Examine valve seats in piston housing for damage.

Pump pistons should have a clearance of 0.0003-0.0015 in their bores. Pistons and pump housing are factory matched and not available separately.

When reassembling the pump, first install discharge valves (7). Parts are interchangeable and can be turned either side out. Tighten plugs to a torque of 125 ft.-lbs.

Camshaft bearings should have an end play of 0.001-0.005. If in doubt, assemble camshaft and carrier (3), leaving out the seal (2). Install and tighten three alternate carrier cap screws and measure end play using a dial indicator. Shims (5) are available

in thicknesses of 0.003, 0.005, 0.006, 0.007 and 0.020. Tighten carrier cap screws to a torque of 30-35 ft.-lbs.

Install pump pistons in bores from which removed, with flat side next to camshaft. Install and tighten plugs to a torque of 225 ft.-lbs.

Install valve body on pump housing, making sure pump inlet and outlet ports are approximately 90 degrees apart and pointing up as shown in assembled view, Fig. 210. Tighten the four retaining cap screws evenly to a torque of 225 ft.-lbs.

Whenever the pump has been disassembled, renew the coupling shaft felt dust seal (D—Fig. 214) as follows:

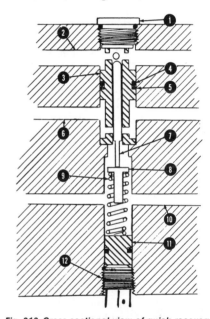

Fig. 213–Cross sectional view of quick recovery valve which is spring loaded in open position. When pressure in passage (2) drops below standby pressure, spring (9) moves valve (7) up, connecting the moderate pressure "Control" passage (6) and inlet passage (10). The rapid drop in control pressure allows the distributing valve to fully open. Radial pump will operate at full capacity until pressure balance is restored.

1. Plug	7. Valve plunger
2. High pressure passage	8. Valve guide
3. Valve sleeve	9. Spring
4. O-ring	10. Low pressure passage
5. Backup ring	11. Spacer
6. Control passage	12. Adjusting plug

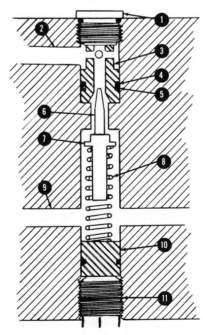

Fig. 212–Cross sectional view of metering valve and associated parts showing method of assembly.

1. Upper plug	6. Valve needle
2. High pressure passage	7. Valve guide
3. Valve sleeve	8. Valve spring
4. O-ring	9. Control passage
5. Backup ring	10. Spacer
	11. Adjusting plug.

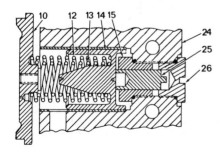

Fig. 211–Cross sectional view of distributing valve and associated parts. Refer to Fig. 210 for parts identification.

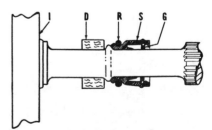

Fig. 214–Cross sectional view of pump coupler showing proper installation of shaft seal. The felt dust seal (D) is cemented in place as outlined in paragraph 119.

D. Dust seal	R. O-ring
G. Garter Spring	S. Rubber seal
I. Input flange	

Clean and degrease the area and cement the seal in the location shown and cement dust seal in position using Scotch Grip Industrial Adhesive No. 847 or equivalent. Position boot seal (S) and secure to shaft with O-ring (R), then install garter spring spreader (G). Lubricate outer portions of felt seal and boot seal, then reinstall shaft in pump. Reinstall pump as outlined in paragraph 116.

CHARGING PUMP.
120. REMOVE AND REINSTALL. The hydraulic system charging pump is mounted on front hydraulic cover as shown in Fig. 215. To remove the cover, first remove ring from manual unloading valve, then remove operator's platform center panel and padding.

Remove the pin securing seat cylinder to bracket mounted on cover and move the cylinder out of the way. Disconnect hydraulic tube assemblies and remove flange bolts, then lift off cover with charging pump and regulating valve attached.

Remove the two cap screws securing pump to lower side of top cover and lift off the pump. Renew O-rings and gaskets and install pump by reversing the removal procedure.

121. OVERHAUL. Refer to Fig. 216 for an exploded view of the rotor type dual charging pump. To disassemble the removed unit, clean outside of pump body and remove drive gear and key.

Firmly depress spring retainer (9) and remove snap ring (10); then remove retainer, spring (8) and combining valve spool (7). Remove cover (6) and small pumping rotors, then extract rotor key from exposed end of drive shaft.

NOTE: Inner rotor is a slip fit on shaft and key and both rotors are free to drop when cover is removed. DO NOT drop rotors when parts are disassembled.

Remove drive end cap (4) and withdraw shaft (2) and large internal rotor. Large external rotor may be with-

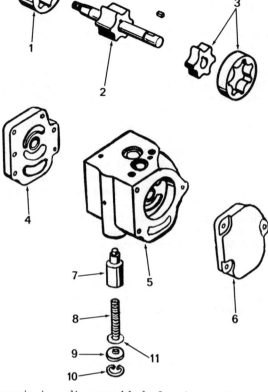

Fig. 216–Exploded view of rotor type dual charging pump.

1. High volume rotor
2. Shaft & gear
3. Low volume gears
4. End cover
5. Pump body
6. End cover
7. Combining valve
8. Valve spring
9. Spring retainer
10. Snap ring
11. Shim pack

drawn with shaft, or may remain in housing pocket.

Check the machined surfaces of end plates, body and rotors for scoring, wear or other damage. End clearance of rotors in housing bores should be 0.0008-0.0024. End plates should be absolutely flat. No gaskets are used. Assemble the pump by reversing the disassembly procedure. Tighten end plate retaining cap screws evenly to a torque of 10-15 ft.-lbs.

Assemble the combining valve using Fig. 216 as a guide. If low volume circuit pressure is to be corrected, add or remove shims (11) as required. Adding or removing one 0.020 shim will change combining valve pressure approximately 10 psi.

REGULATING VALVE
122. REMOVE AND REINSTALL. The regulating valve assembly is mounted on front hydraulic cover as shown in Fig. 215. Remove front cover as outlined in paragraph 120, remove attaching cap screws and lift off regulating valve.

Reinstall by reversing the removal procedure. After tractor is reassembled, adjust sensing pressure as outlined in paragraph 103.

123. OVERHAUL. The regulating valve is available only as an assembly and individual parts are not renewed. The regulating valve may, however, be

disassembled for inspection and cleaning. Refer to Fig. 217 for exploded view.

Spool (10) must slide freely in its bore. Center drilling contains a 0.014 diameter metering orifice and a conical screen.

Pilot valve (3) is prevented from seating by valve stop (5). If sensing pressure cannot be properly adjusted and no other trouble is encountered,

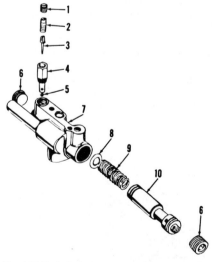

Fig. 217–Exploded view of regulating valve assembly. Component parts are not serviced.

1. Adjusting plug
2. Spring
3. Pilot valve
4. Valve body
5. Valve stop
6. End plug
7. Valve housing
8. Spring seat
9. Spring
10. Valve spool

Fig. 215–Removing front hydraulic cover with charging pump and regulating valve attached.

check the setting of pilot valve stop as follows: Clamp pilot valve body (4) lightly in a vise, right side up with both ends accessible. Note the setting of adjusting plug (1) and remove the plug and spring (2). Mount a dial indicator with button bearing on top end of needle valve (3). Back out the threaded stop (5) until dial indicator needle ceases to move; indicator reading should be 0.005-0.010. Adjust to those limits by turning stop (5) until needle is raised the required amount.

Assemble the regulating valve by reversing the disassembly procedure, using Fig. 217 as a guide. Install the valve as outlined in paragraph 122 and adjust as in paragraph 103.

PRIORITY VALVE

124. OPERATION. The priority valve is shown schematically in Fig. 218 and exploded in Fig. 219. Pressure supply ports (6 & 7—Fig. 218) leading to brake valve and power steering are located ahead of priority valve piston (4), while piston must unseat to supply fluid flow to hydraulic lift system port (5). When there is no flow, pressures are equal throughout the valve.

One end of the pilot valve piston (1) is open to high pressure while the other end is open to atmospheric pressure of bleed port (2). Before priority piston (4) can move to open lift system port (5), two things must happen in the priority valve. System pressure must overcome the pressure of spring (3), allowing pilot piston to move down and open the pilot passage to lift port (5), and lift port pressure must drop below system pressure. In normal operation, pilot flow (small arrows) through piston orifice and pilot piston, unbalances the system to allow pilot piston to open whenever movement of one of the control valves signals a fluid demand.

125. REMOVE AND REINSTALL. To remove the priority valve, first remove the hood, batteries, battery platform and voltage regulator. Remove right and left side covers from steering column support. Disconnect input line, brake line, steering line, bleed line and output line from valve housing. Remove the two mounting bolts and lift off the priority valve assembly. Install by reversing the removal procedure.

126. OVERHAUL. The priority valve is available as an assembly only. To disassemble the valve for cleaning or inspection, refer to Fig. 219 and proceed as follows:

Working through inlet port, unseat and remove snap ring (10), priority piston (9) and spring (8). Remove lower plug (1), spring discs (5), spring (6) and pilot pin (7). Jar valve body downward on a bench or wood block to dislodge the piston (3). If piston cannot be removed, valve must be renewed. Assemble by reversing the removal procedure.

BRAKE VALVE

127. OPERATION. A schematic view of the brake valve is shown in Fig. 220. Except for inlet pressure fitting (1) and inlet check valve (2), the parts are duplicated for right and left brakes. Initial movement of brake pedal depresses power piston (5) and closes off escape passage (4). Further movement of the pedal opens the modulating needle valve (3), allowing fluid pressure to enter through inlet check valve (2) and pass through connecting passage (8) to brake passage fitting (12). At the same time, fluid pressure reacts on the bottom side of power piston (5) and equalizing piston (9).

When pedal pressure is exceeded by pressure from the fluid, the pedal moves upward closing the modulating valve (3) and (if necessary) opening escape passage (4). The upward movement of equalizing valve (9) in response to actuating pressure, opens the connecting passage (10) which leads to the opposite equalizing valve. If both equalizing valves are opened simultaneously by depressing both pedals, actuating pressures are equalized through connecting passage (10) and both brakes will be applied with a pressure equal to that of the heaviest brake application.

One pound of pressure at brake pedal will generate 30 psi of actuating pressure in the power braking system.

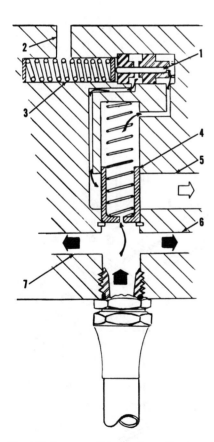

Fig. 218–Cross sectional schematic view of priority valve.

1. Pilot piston
2. Bleed port
3. Spring
4. Priority piston
5. Hydraulic lift port
6. Brake pressure port
7. Steering pressure port

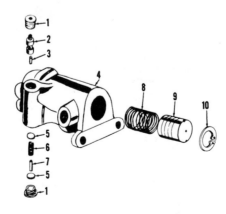

Fig. 219–Exploded view of priority valve showing component parts. Valve is serviced only as an assembly.

1. Plug
2. Valve sleeve
3. Pilot piston
4. Valve housing
5. Spring disc
6. Pilot spring
7. Pilot pin
8. Spring
9. Priority piston
10. Snap ring

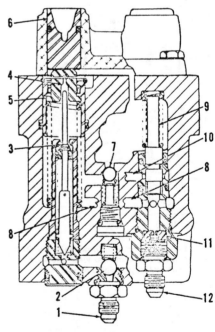

Fig. 220–Cross sectional schematic view of brake valve. Except for inlet pressure fitting (1) and check valve (2), parts are duplicated for right and left sides.

1. Inlet fitting
2. Check valve
3. Modulating valve
4. Escape passage
5. Power piston
6. Plunger
7. Manual check ball
8. Actuating passage
9. Equalizing piston
10. Equalizing passage
11. Orifice plate
12. Outlet fitting

In manual brake operation, a greater pedal travel depresses the power piston (5) and manual piston (33—Fig. 221), and trapped fluid is pumped through outlet port (12—Fig. 220) to actuating pistons. The reverse flow restrictors (11) slow the return flow of fluid, allowing piston area to refill through the reservoir check ball (7) when pedal is released. Three strokes of pedal should build up and hold pressure for manual operation.

128. REMOVE AND REINSTALL. To remove the brake valve, first remove cover panel and right foot rest. With engine not running, apply the brakes enough times to exhaust the pressure in standby system and accumulator.

NOTE: Pedal will move down into manual apply range when pressure is exhausted.

Thoroughly clean the valve and surrounding area. Disconnect lines from brake valve and remove snap ring from pedal pivot shaft. Remove the mounting cap screws securing valve to clutch housing. Rotate pedals rearward and slide off pivot shaft while pivoting the valve body around sump return line. Hold lines clear of valve and withdraw the unit.

Install by reversing the removal procedure. Bleed the brakes as outlined in paragraph 88 and adjust pedal linkage as in paragraph 89.

129. OVERHAUL. Refer to Fig. 221 for an exploded view of brake valve. To disassemble the removed valve, first unbolt and remove reservoir cover (5) and gasket (4). Remove valve cap (1) and gasket (2).

Using a C-clamp, press or other suitable tool, depress manual piston (33) and remove snap ring (36). Slowly re-

Fig. 222–Installed view of Multi-Power & IPTO control valve (V), with instrument panel side cover removed.

lease the pressure and remove manual piston (33), power piston (32), spring (31) and associated parts.

Remove snubber plug (24), snubber needle (26), cartridge (28) and associated parts by working from bottom of housing. Unscrew and remove equalizer valve spool (18), valve (16) and spring (14) as a unit.

Gaskets and O-rings are available in a seal kit which also includes snap rings (36). Renew any other parts which are questionable, and reassemble using Figs. 220 and 221 as a guide.

MULTI-POWER & IPTO CONTROL VALVE

130. REMOVE AND REINSTALL. To remove the Multi-Power & IPTO control valve (V—Fig. 222), first remove left instrument panel cover and disconnect all lines leading to valve. Disconnect lever yokes from valve spools. Remove the three bolts securing valve housing to instrument panel support and lift off the valve assembly.

When installing the valve, make sure lever links are adjusted to permit full movement of valve spools to both detent positions, and that lines are connected in proper order.

131. OVERHAUL. Refer to Fig. 223 for an exploded view. To disassemble the removed valve, first remove the three hex head cap screws, detent cover (10) and the two upper detent springs (1), balls (2) and shims (3) if used.

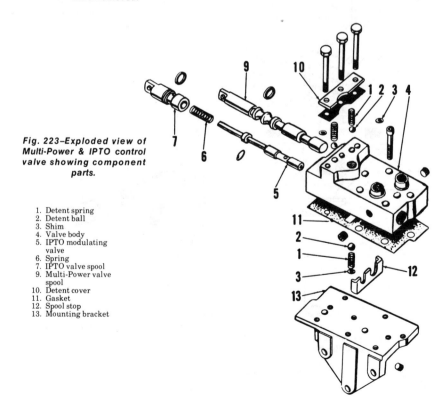

Fig. 223–Exploded view of Multi-Power & IPTO control valve showing component parts.

1. Detent spring
2. Detent ball
3. Shim
4. Valve body
5. IPTO modulating valve
6. Spring
7. IPTO valve spool
8. Multi-Power valve spool
9. Multi-Power valve spool
10. Detent cover
11. Gasket
12. Spool stop
13. Mounting bracket

TWO USED

Fig. 221–Exploded view of brake valve. Refer to Fig. 220 for cross sectional view and to paragraph 127 for principles of operation.

1. Cap
2. Gasket
3. Valve body
4. Gasket
5. Reservoir
6. Spring
7. Check ball
8. O-ring
9. Inlet pressure fitting
10. Check ball
11. Spring
12. O-ring
13. Plug
14. Spring
15. O-ring
16. Equalizing valve
17. O-rings
18. Valve spool
19. Orifice plate
20. O-ring
21. Outlet fitting
22. Backup washer
23. O-ring
24. Snubber plug
25. Spring
26. Snubber needle
27. O-ring
28. Cartridge
29. Backup washer
30. O-ring
31. Spring
32. Power piston
33. Manual piston
34. O-ring
35. Backup washer
36. Snap ring
37. Plunger
38. O-ring

Remove the six socket head cap screws and lift valve body (4) off of bracket (13). The IPTO valve lower detent ball, spring and shim will be free to fall out when parts are separated, as will spool stop (12). Do not lose any of the parts. The valve spools can be withdrawn from their bores after stop (12) is removed. Extract the snap ring and separate the modulating valve (5) and spring (6) from IPTO valve spool (7).

Examine all parts for binding, scoring or other wear. If IPTO valve cannot be properly modulated as outlined in paragraph 114, check for sticking of valve spool (5) in its bore or damage to valve spring (6). If IPTO valve will not stay in applied position, add additional detent shims (3) equally under upper and lower detent unit. Assemble the valve by reversing the disassembly procedure.

SEAT HYDRAULICS

The seat hydraulic system consists of a closed center, poppet type valve, a hydraulic positioning cylinder and an air (or nitrogen) filled accumulator teed to the hydraulic cylinder line. Refer to Fig. 224 for an exploded view of complete seat system and Fig. 225 for an exploded view of seat valve. The seat cylinder is shown exploded in Fig. 226. Refer to the appropriate following paragraphs for operation and service information.

132. **OPERATION.** The seat valve is connected to the hydraulic system Low Pressure/Low Volume circuit. Seat is raised or lowered by directing fluid to the single acting cylinder (11—Fig. 224), and cushioning obtained by transfer of fluid between the positioned cylinder and accumulator (10). Fluid leaking by the cylinder piston is returned to the cylinder through vent tube (12).

Settling of the seat where no fluid loss is visible would indicate a leaking cylinder piston or leaking return sealing washer in control valve. An increase in seat height would indicate a leaking inlet sealing washer in the valve. Loss of cushioning action would indicate failure of accumulator (10).

Remove and overhaul the system components as outlined in the appropriate following paragraphs.

133. **CONTROL VALVE.** Refer to Fig. 225 for and exploded view of seat control valve. To remove the valve, first remove knobs from seat valve and IPTO shift lever, then unbolt and remove dress cover panel. Disconnect hydraulic fitting connections and remove the valve through-bolts, then lift off valve assembly.

To disassemble the valve, remove wrap-around snap ring (7) and withdraw valve cam (11). Remove pressure fitting (1) and return fitting (8) and withdraw valve springs (3), poppet valves (4) and sealing washers (5). Renew sealing washers and O-rings, and any other parts which are damaged or questionable. Reassemble by reversing the disassembly procedure.

134. **ACCUMULATOR.** The seat accumulator (10—Fig. 224) is air or nitrogen filled to a pressure of 80-85 psi, and the compressibility of the air cushion serves as the seat spring. The line and fitting resistance to the transfer of trapped fluid between seat cylinder and accumulator serves as a damper and shock absorber for seat.

The nitrogen filled accumulator cannot be directly tested or filled without the aid of special adapter. Accumulator pressure of ALL accumulators can be indirectly checked, however, by teeing a low-pressure hydraulic gage in control line as follows: Fully lower the seat. Disconnect control line (6—Fig. 224) at either end, install a tee fitting and suitable gage, then reconnect the line. With engine running, and seat not occupied, move control lever handle to fill the cylinder. Release the lever and read the gage after cylinder has reached extended limit stop and initial pressure rise has ceased. The indicated pressure is approximately accumulator charge pressure.

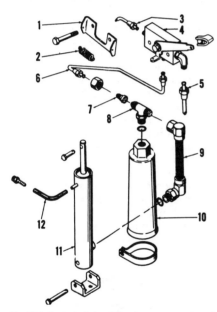

Fig. 224—Exploded view of main component parts of seat hydraulic system.

1. Bracket	7. Adapter
2. Centering spring	8. Tee
3. Inlet pressure fitting	9. Hose
4. Control valve	10. Accumulator
5. Return fitting	11. Cylinder
6. Control line	12. Vent tube

Fig. 225—Exploded view of seat control valve showing component parts.

1. Inlet pressure fitting
2. O-ring
3. Spring
4. Poppet
5. Sealing washer
6. Valve body
7. Wrap-around snap ring
8. Return fitting
9. Cylinder fitting
10. O-rings
11. Valve cam
12. Lever

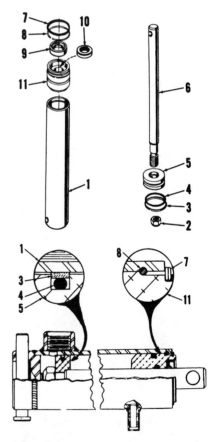

Fig. 226—Exploded view of seat cylinder showing component parts and details of construction.

1. Barrel	7. Snap ring
2. Rod nut	8. Snap ring
3. Teflon ring	9. Wiper
4. O-ring	10. Seal
5. Piston	11. Bearing
6. Piston rod	

NOTE: If pressure rises rapidly from minimal reading to low volume system pressure (210-275 psi) without hesitation, accumulator is probably discharged and should be renewed or recharged.

135. **SEAT CYLINDER.** The seat cylinder can be removed after removing the lower front panel from seat console. To disassemble the removed cylinder, refer to Fig. 226 and proceed as follows:

Remove outer snap ring (7) and push bearing (11) into barrel until inner snap ring (8) is exposed. Remove inner snap ring, then withdraw piston rod, piston and bearing as an assembly.

If piston is removed from piston rod (6), tighten nut (2) to a torque of 30-35 ft.-lbs. when reinstalling. A new locknut must be used whenever piston is disassembled. The piston contains a Teflon piston ring with an O-ring expander as shown in cross sectional view. The Teflon ring is not elastic but can be stretched when warmed in oil or water. A suggested method of installation is as follows:

Install O-ring in bottom of piston ring groove. Warm Teflon ring (90°-120° F.) in water or oil and stretch ONLY enough for installation. Compress with a suitable ring compressor for a few minutes until ring returns to original size. Remove the compressor, lubricate the ring and install piston and rod assembly carefully to prevent damage to piston ring. Complete the assembly by reversing the dissembly procedure.

HYDRAULIC LIFT SYSTEM (3-POINT LINKAGE)
All Models So Equipped

136. **OPERATION.** The three-point linkage hydraulic lift system offers three types of control for the lift arms, selected by turning the selector knob on top of control lever (C—Fig. 227).

DRAFT CONTROL. In "Draft Control", lift arm height is determined by the selected setting and controlled by top link pressure or tension on the leaf-type draft control springs. Draft control is the middle detent position on selector knob.

POSITION CONTROL. In "Position Control", lift arm height is maintained at a selected setting by an actuating lobe on rockshaft. Position control is obtained by turning selector knob counter-clockwise as far as possible after moving lever to top of quadrant.

PRESSURE CONTROL. In "Pressure Control", a selected pressure is maintained in rockshaft lift cylinders, the pressure being controlled by turning selector knob clockwise as far as possible. The master control valve is spring loaded in the "LIFT" position and is moved to neutral or lowering position by action of the control linkage.

137. **QUICK CHECKS.** For a quick check of hydraulic system, first be sure lift arm lock is "OPEN" as shown in Fig. 228 and flotation latch is locked as shown in Fig. 229.

Attach a heavy weight to lower links. Weight should allow rockshaft to travel through complete arc and should not be attached to upper link.

Fig. 228–Lift arm lock is located on left side of lift cover housing as shown.

Start and run engine at 1000-1600 rpm, then move control lever to top of quadrant. Weight should raise. With lever at top of quadrant, check to see that selector knob turns easily to all three detent positions.

Turn selector knob to "Position Control" and slowly move control lever downward on quadrant. When lever is ½-⅝ inch from top of slot, lift links should start to lower. Move lever down and up quadrant a little at a time. Lift links should respond to movement of control lever, completely lowering when lever is at bottom of quadrant and stopping at any selected intermediate point.

Lower the weight to ground and insert a ⅝ inch gage block between draft linkage ratio lever and lever guide at point shown at (A—Fig. 230), then turn draft control hand-wheel (B) until gage block is lightly held. Move control lever to top of quadrant and turn selector knob to "DRAFT" position; then move control lever about ⅔ the distance to bottom of quadrant, midway between the two sector marks (1—Fig. 231). Lower links should remain suspended in any selected position when control lever is within ¼-inch of midway between marks, raising when lever is moved upward slightly, or lowering when lever is moved downward.

138. **TROUBLE SHOOTING.** Relief pressure for the three-point hitch is controlled by maximum setting for pressure control. Instead of a relief valve opening when maximum setting is exceeded, control valve is returned to neutral by a pilot valve. If relief pressure malfunctions, check and/or adjust the setting of pressure control linkage.

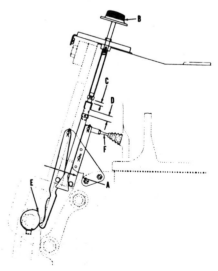

Fig. 230–Schematic view of draft control sensing linkage used on models so equipped.

A. Adjustment (5/8 in.)	D. Measurement (5/8 in.)
B. Handwheel	E. Sensing cam
C. Measurement (5/16 in.)	F. Pickup rod

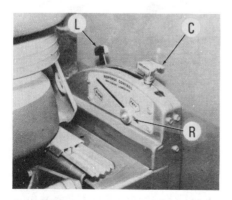

Fig. 227–Installed view of hydraulic control console showing three-point linkage controls.

C. Control lever
L. Locating stop   R. Response control

Fig. 229–Flotation latch (F) is in locked position when vertical as shown. When latch is turned 90°, some independent movement of each lift link is allowed.

Draft control linkage is returned to neutral in transport position by the "POSITION CONTROL" linkage. If "Position Control" linkage fails to return the main valve to neutral, valve will be neutralized by "Pressure Control" linkage when lift pistons reach the end of their stroke.

If lever must be moved ½-⅝ inch from top of quadrant before lift links start to lower, the position control linkage adjustment is probably satisfactory.

If lift hunts or searches in neutral position or is unduly noisy, leakage or improperly adjusted pressure control linkage is to be suspected.

1. Pickup rod
2. Clevis
3. Position control pickup
4. Position control link
5. Draft control link
6. Position control bellcrank
7. Adjusting screw
8. Locknut
9. Draft control bellcrank
10. Rubber plug
11. Stop screw
12. Pressure control bellcrank
13. Pressure control lever
14. Yoke
15. Adjusting rod
16. Valve lever
17. Clevis
18. Pressure control rod
19. Main valve rod
20. Drive link
21. Pivot stud
22. Dashpot beam
23. Dashpot piston rod
24. Nut

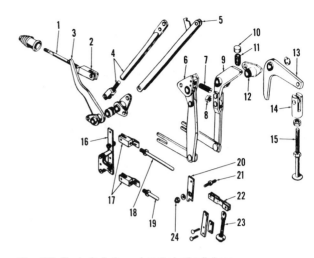

Fig. 233–Exploded view of control valve linkage.

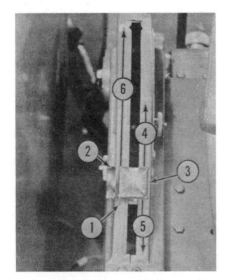

Fig. 231–View of control quadrant showing draft control positions.

1. Sector marks
2. Adjustable locator
3. Control lever
4. Decreasing depth
5. Increasing depth
6. Transport

Fig. 232–View of lift cover with front access plate removed.

1. Position control adj. screw
2. Draft control bellcrank
3. Pressure control adjusting rod

139. CHECK AND ADJUST. First be sure lift arm lock is "OPEN" as shown in Fig. 228 and flotation latch is locked as shown in Fig. 229.

Attach a heavy weight to lower links. Weight should allow links to travel through complete arc and should not be attached to upper link.

Remove upper port plug from either rockshaft cylinder and install a suitable pressure gage in port opening. Remove lower front dress panel from seat console and front access plate from hydraulic lift cover. Remove access plate from right side of cover housing.

Start and run engine at 1000-1600 rpm, move control lever to top of quadrant and note gage pressure. The gage will either show the maximum pressure setting (which should be 1800-2000 psi), or the pressure necessary to support the attached weight (which will be considerably less than maximum pressure setting).

Working through the opening in front of lift cover, back off the locknut on position control adjusting screw (1—Fig. 232). If gage reading is considerably less than maximum pressure setting, back out the adjusting screw (1) counter-clockwise until gage pressure rises sharply; then turn screw in until pressure drops. Tighten locknut while holding the adjusting screw. If gage reading was originally the maximum pressure setting, turn screw in (clockwise) until pressure drops then tighten locknut. After completing the initial adjustment, move control lever slowly forward on quadrant and measure lever position at which lift links first start to lower. Distance should be ½-⅝ inch. Readjust position control adjusting screw (1) if necessary. Turning screw clockwise will increase the distance, counter-clockwise will decrease distance.

With selector knob in "Position Control", move lever forward (downward) on quadrant a short distance at a time. Lift links should respond by moving a corresponding amount then stopping. Lift cylinders should be fully retracted when control lever reaches bottom of quadrant.

With lift arms lowered, turn lift arm lock knob to "LOCK" position, then move control lever back to top of quadrant. Gage pressure should now be 2000 psi. If it is not, loosen the locknut and adjust servo relief valve rod (3). Lengthen rod to increase pressure or shorten rod to decrease pressure.

Turn selector knob on control lever to "PRESSURE" position, then move lever slowly down the quadrant while watching gage reading. Pressure should drop steadily at a uniform rate as lever is moved; and should read 50-250 psi when lever reaches bottom of quadrant.

Move control lever to top of quadrant. Working through access plate opening on right side of lift cover, lengthen pressure control piston rod (18—Fig. 233) until pressure gage begins to fluctuate or cycle; then shorten rod until cycling stops. **NOTE: If piston rod (18) is too short, main control valve will not return to neutral and fluid entering through the main valve must escape through the servo relief valve, overworking and overheating the valve. Make sure adjustment of piston rod (18) is correct.**

Turn lift arm lock knob to "OPEN" position and check to see that lift arms raise and lower normally.

Move control lever to top of quadrant and turn selector knob to "DRAFT" position; then move control lever down the quadrant until aligned with "dimple" approximately ¼ the distance from top. Lower the lift links all the way by turning draft control hand wheel (Fig.

234), then check the clearance between draft link (9—Fig. 233) and stop screw (11). Clearance should be 0.002-0.004; if it is not, remove rubber access plug (10) from top of lift cover and readjust the screw (11).

140. MAIN VALVE ADJUSTMENT. If proper adjustment cannot be obtained as outlined in paragraph 139, check the main spool valve adjustment as follows:

Prepare the tractor as outlined in paragraph 139. Turn selector knob to "POSITION" control and move control lever to bottom of quadrant. With engine running and working through right access plate opening, press down on vertical dashpot link (20—Fig. 235) as shown by heavy arrow. Be certain lower notch engages spacer on actuating lever (16). Temporarily tighten locknut (24) to hold the link (20) in lowermost position.

Shorten control valve rod (19) until lower links begin to raise. Lengthen rod (19) when lower links are horizontal, until weight settles one inch in 10 seconds, measured at outer end of links.

Loosen locknut (24) until dashpot linkage is free to move; then readjust all the linkage as outlined in paragraph 139.

141. SENSING LINKAGE ADJUSTMENT. To adjust the draft control sensing linkage, first check to make sure than ⅜-7/16 inch clearance exists between top link clevis pin and top link bracket as shown in Fig. 236, with nothing attached to upper link and clevis pin in lower holes in springs.

If measurement is incorrect, remove the two bolts securing pin to sensing springs and turn the pin 180° until the other flats contact springs. (Attaching flats are offset on clevis pin as shown in inset). If measurement is still incorrect after rotating pin, check for bent or damaged springs or attaching parts.

Move control lever to top of quadrant and turn selector knob to "DRAFT"

position, then move the lever ⅔ the way down the quadrant until centered midway between the two sector marks. Upper link clevis must be hanging down and point of cam (E—Fig. 237) must be turned away from contact with pickup finger as shown. Turn draft control handwheel (B) counter-clockwise if necessary, until a ⅝-inch gage block can be inserted between ratio lever and bracket as indicated at (A); then turn handwheel (B) clockwise until gage block is just supported. With linkage thus positioned, distance (C) between clevis and upper stop nut should be

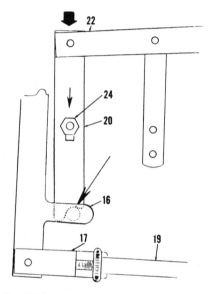

Fig. 235–Schematic view of main valve linkage showing adjustment procedure. Refer to Fig. 233 for parts identification and to paragraph 140 for details of adjustment.

approximately 5/16-inch and distance (D) between clevis and lower stop nut should be approximately ⅝-inch. Reposition the stop nuts if necessary, by loosening set screw and turning nut.

With engine running and linkage positioned as outlined, rockshaft should hold position without raising or lowering. Turning draft control handwheel (B) ¼-turn should cause rockshaft to raise or lower. If adjustment is incorrect, loosen the locknut and turn pickup rod (F) in or out as required.

142. RESPONSE ADJUSTMENT. With hydraulic fluid at operating temperature, quadrant lever in draft control, and approximately 1000 lbs. of weight attached to lower links, lowering time should not be more than 2 seconds with response control knob in "FAST" position, nor more than 5 seconds with control knob in "SLOW" position.

To adjust the response control, refer to Fig. 238. Loosen the clamp screw securing valve lever (2) to dashpot needle valve stem (1). Move response control lever (R—Fig. 239) fully to "SLOW" position and turn needle valve (1—Fig. 238) in until it bottoms; then retighten clamp screw. Readjust slightly if response is too slow; overhaul the valve if response is too fast.

143. **OVERHAUL.** The draft control sensing linkage, rockshaft arm and cylinders, or attaching links can be removed or installed without removing hydraulic lift cover. To obtain access to control valve, rockshaft or internal linkage, it is first necessary to remove cover as outlined in paragraph 144, or tilt cover back on draft control springs as outlined in paragraph 145. Both

Fig. 234–Draft control handwheel, located behind operator's seat. Turning handwheel clockwise decreases operating depth, counter-clockwise increases operating depth.

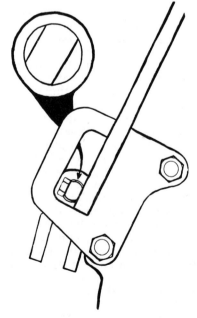

Fig. 236–Top link clevis pin attaching flats are slightly offset. Pin may be turned 180 degrees to establish initial setting as outlined in paragraph 141.

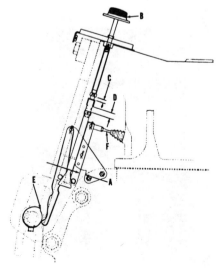

Fig. 237–Schematic view of draft control sensing linkage.

A. Adjustment (5/8 inch)
B. Handwheel
C. Measurement (5/16 inch)
D. Measurement (5/8 inch)
E. Sensing cam
F. Pickup rod

methods require the use of an overhead hoist.

144. R&R LIFT COVER. To remove the hydraulic lift cover for service on any of the components or rear axle center housing, proceed as follows:

Disconnect lift cylinders from lift arms and lift rods from lower links, then remove lift arms and rods from rockshaft as assemblies. Remove rear dress panel cover, draft control springs and draft control sensing linkage. Remove platform center panel and lower front panel from seat console. Disconnect pressure and vent hoses from seat cylinder and remove seat cylinder lower attaching pin; then unbolt and remove seat frame as a unit. Disconnect hydraulic lines from lift cover and remove cap screws securing cover to tractor center housing. Attach a suitable hoist to cover and lift off cover and control consoles as a unit.

When installing the cover, make sure oscillator drive lever properly engages drive body on control valve. Aligning dowels help in positioning the cover. Complete the installation by reversing the removal procedure.

145. LIFT COVER PARTIAL REMOVAL. Completely lower the rockshaft. Move control lever to top of quadrant, turn selector knob to "PRES-

Fig. 238–Dashpot needle valve (1) should be lightly seated when "Response" lever is in extreme slow position. Adjust by loosening clamp screw securing valve lever (2).

Fig. 239–View of control console showing "Response" control knob (R).

SURE" control and move control lever to bottom of quadrant. Remove platform center panel and the lower front panel from seat console. Remove operator's seat and seat bracket. Disconnect pressure line from seat accumulator tee and lines leading from seat valve to lift cover.

Disconnect pressure line to transfer cap and remove lift cover front access plate. Disconnect hoses leading to rockshaft cylinders. Disconnect draft control sensing rod and unbolt and remove sensing linkage.

Attach a suitable hoist to top of lift cover and remove cap screws securing cover to tractor center housing, then carefully raise lift cover upward and back as shown in Fig. 240.

NOTE: Cover will try to jump rearward as front end is raised, and damage to oscillator mechanism can occur. Alignment dowels in front cap screw holes can help prevent damage as long as binding does not occur.

The use of alignment dowels will also assist in installing the cover. Make sure oscillator drive lever properly engages drive body in control valve and reassemble by reversing the removal procedure.

146. LINKAGE CONTROL VALVE. To remove the linkage control valve, first remove lift cover as outlined in paragraph 144 or tilt cover as in paragraph 145.

Drain cover reservoir and remove standpipe (10—Fig. 241). Remove plug (4). Working through plug hole and using Special Tool MFN-767 or other suitable means, remove control valve discharge fitting (5) and associated parts.

Fig. 240–Hydraulic lift cover may be tilted back for service as shown.

S. Standpipe   V. Control valve

Remove left hand cylinder hose fitting (3) and tube (2). Turn control quadrant selector knob to "PRESSURE" position and move lever to bottom of quadrant; then working through front access opening, disconnect and remove pressure control piston rod.

Remove the two long cap screws securing transfer cap to front, right side of cover, leaving transfer cap in place. Support linkage control valve with hand and remove the other three cap screws securing valve to lift cover housing. Turn valve slightly and, working through front access plate opening, remove the slotted linkage guide plate from top of valve housing. Hold main control valve in discharge position and disconnect "Position" and "Draft" control links, then lift out the valve.

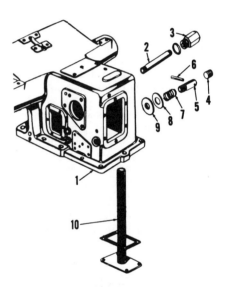

Fig. 241–Parts (2 through 10) must be removed from lift cover before control valve can be removed.

1. Cover	6. Groove pin
2. Outlet tube	7. Spring
3. Cylinder fitting	8. Washer
4. Plug	9. Seal
5. Discharge fitting	10. Standpipe

Fig. 242–Removed view of linkage control valve showing proper assembly.

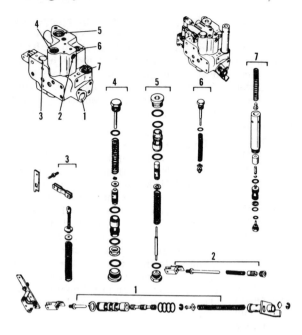

Fig. 243–Views of linkage control valve showing principal components. Location of valves in housing is indicated by corresponding numbers.

1. Main control valve
2. Pressure control pilot piston
3. Response dashpot
4. Inlet flow control valve
5. Discharge valve
6. Secondary relief
7. Servo relief valve

Fig. 243 shows an exploded view of main components of the valve. Shims (10—Fig. 244) control end clearance adjustment of intake control valve (7). Recommended adjustment is 0.001 clearance to 0.003 preload. Shims are available in thicknessess of 0.005, 0.008 and 0.010. To check required shim pack thickness, assemble the unit leaving O-ring off of plug (11) and using 0.026-0.030 thickness of shims (10). Tighten plug (11) finger tight, making sure parts are seated, then measure gap between flange of plug and surface of housing using a feeler gage. Remove the plug and deduct shims equal in thickness to the measured gap.

When valve is reassembled, check with Fig. 242 to be sure linkage is properly assembled, and adjust main valve as outlined in paragraph 140 before valve is installed.

147. ROCKSHAFT. To remove the rockshaft, first remove hydraulic lift cover as outlined in paragraph 144 or tilt cover rearward as in paragraph 145. Disconnect lift cylinders from lift arms and remove lift arms from rockshaft if not previously removed.

Turn rockshaft until lock shaft (16—Fig. 245) engages notch in lock arm (7), then turn shaft (16) to lock position. Withdraw rockshaft (6) out right side of lift cover. Right hand bushing (9)

will be removed with shaft, and left bushings (3) and spacer (4) can be removed at this time. Lock arm (7) can be removed after removing position control pickup rod. Rockshaft and arms have master splines for proper assembly. Assemble rockshaft by reversing the disassembly procedure.

148. OSCILLATOR DRIVE. Refer to Fig. 246 for an exploded view of main valve oscillator drive mechanism. The drive shoe (1) for crank rides on a cam machined on IPTO clutch drum.

To remove the unit, first remove hydraulic lift cover as outlined in paragraph 144 or tilt the cover as in paragraph 145. Remove front hydraulic cover as outlined in paragraph 120.

Remove hair pin (5) and drive crank (2). Remove pickup lever (7), spring (6) and key (4); then withdraw drive bushing (3) forward out of center housing bore. Assemble by reversing the disassembly procedure.

149. ROCKSHAFT CYLINDERS. The rockshaft cylinder unit is a welded assembly and is shown exploded in Fig. 247. To disassemble the cylinder, refer

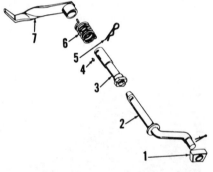

Fig. 246–Exploded view of main valve oscillator drive and associated parts.

1. Shoe
2. Crank
3. Bushing
4. Key
5. Hairpin
6. Spring
7. Lever

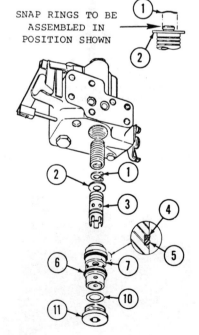

SNAP RINGS TO BE ASSEMBLED IN POSITION SHOWN

Fig. 244–Partially exploded view of inlet flow control valves showing proper assembly.

1. Snap ring
2. Dampening washer
3. Flow control valve
4. Backup washer
5. O-ring
6. O-ring
7. Flow control sleeve
10. Shim pack
11. Plug

Fig. 245–Exploded view of lift cover, rockshaft and associated parts.

1. Lift arm
2. Lift link
3. Bushing
4. Spacer
5. Snap ring
6. Rockshaft
7. Lock arm
8. Pickup tube
9. Bushing
10. Lock arm
11. Lock pin
12. Detent assembly
13. Lift arm
14. Storage hook
15. Lift cover
16. Lock shaft
17. Detent
18. Lock knob

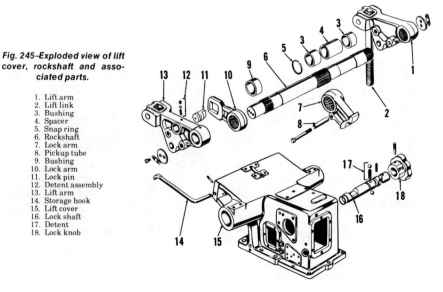

to Fig. 248. Remove hose fitting and align snap ring with cylinder port as shown in (A1). Insert a screwdriver or similar tool in port opening and work snap ring into deep, second groove in piston rod. Piston rod can now be withdrawn. Assemble by reversing the disassembly sequence as shown at (B). Pull piston against piston stop as shown in Fig. 249 after snap ring is repositioned, to be sure ring is fully seated.

TRANSFER CAP

150. The transfer cap (Fig. 250) is a one-way check valve used on all models with rockshaft and auxiliary valve. The unit mounts on right side of hydraulic lift cover. Check valve (4) prevents back flow from linkage valve to auxiliary valves when both units are actuated simultaneously or engine is shut off.

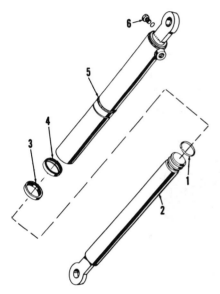

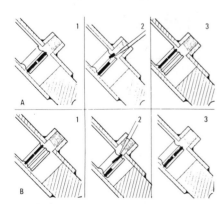

Fig. 247–Exploded view of rockshaft cylinder showing component parts. Refer to Fig. 248 for disassembly and assembly procedure.

Fig. 248–Cylinder is disassembled by displacing stop ring into deeper groove and reassembled by repositioning in stop groove. Work through hose port using a screwdriver or similar tool. Upper views (A1, 2 & 3) show disassembly procedure; lower views (B1, 2 & 3) show assembly procedure.

Check valve can be removed for service or cleaning after removing cap assembly and snap ring (1).

SENSING LINKAGE

151. Fig. 251 shows an exploded view of draft control springs and associated parts. Refer to paragraph 141 for adjustment if linkage is removed.

Fig. 252 shows an exploded view of pressure control coupler and drawbar frame. Use Fig. 252 as a guide when disassembling or assembling the unit.

AUXILIARY VALVE

152. **OPERATION.** Fig. 253 shows an installed view of the Cessna Spool Type auxiliary valve and Fig. 254 shows a cross section of one bank. Valves may be stacked for multiple operation.

Operating fluid is directed to and from cylinders by valve spool (12—Fig. 254). Poppet assemblies (11) seal the work ports (8 & 9) from transfer passages (2 & 6). When the spool valve is moved from neutral in either direction, the actuating passage (5) is connected to pressure passage (3) by internal drillings in valve spool. Fluid from ac-

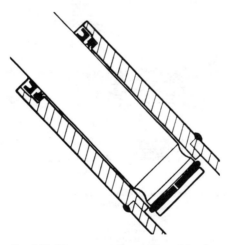

Fig. 249–When correctly positioned in stop groove, stop ring prevents withdrawal of cylinder rod as shown.

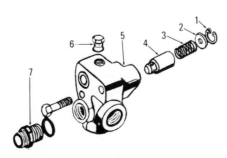

Fig. 250–Exploded view of transfer cap showing component parts.

1. Snap ring
2. Washer
3. Spring
4. Plunger
5. Body
6. Plug
7. Pressure fitting

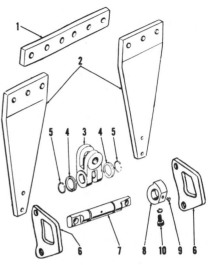

Fig. 251–Exploded view of draft control sensing springs and associated parts.

1. Clamp
2. Sensing spring
3. Clevis
4. Spacer
5. Snap ring
6. Bracket
7. Pin
8. Pickup cam
9. Drive plug
10. Lock screw

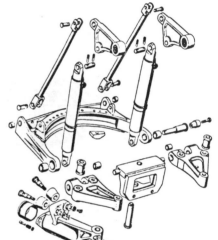

Fig. 252–Exploded view of pressure control linkage and drawbar support.

Fig. 253–Installed view of auxiliary valve with dress cover removed, showing control levers (L), unloading lever (U) and flow restrictors (R).

tuating passage (5) enters the area between the two poppet valves, opening the valves and permitting fluid flow to work ports.

Feathering notches are located on return lands of spool and return is controlled to meter the flow. A mechanical detent position is provided at end of spool travel in each operating position. The spool will remain in detent until manually released.

A float detent is provided by moving the lever beyond the lowering detent stop. In float position, both transfer passages are connected to return ports and actuating passage (5) is pressurized to open the poppet valves. Float position must be used to relieve the pressure in cylinder and lines for connecting or disconnecting the cylinders, and if engine is not running, poppet valves must be manually unseated by moving the unloading lever to actuate plunger (10).

The spool valve is ported to permit fluid flow of up to 20 gpm. The valve is equipped with a flow control restrictor (4) which may be adjusted to provide a regulated flow of 5-20 gpm. The factory adjusted setting is 10-12 gpm which is obtained when distance (A) between end of adjusting screw and machined port boss on valve body is 1 3/32 inches.

153. **ADJUSTMENT.** To index the valve levers, move lever to "LOWER" detent position. Lever should just touch the offset lowering stop, midway between neutral position and bottom of lever slot. If adjustment is incorrect, remove outer dress panel and loosen the clamp bolt securing lever extension to lever and reposition.

Connect a test gage to breakaway coupler and move control lever to pressurize the gage. Gage should register system pressure. Return valve lever to neutral while continuing to watch gage needle. Pressure should not drop more than 150 psi when valve lever is moved to neutral, and should not drop more than 100 psi per minute with valve in neutral position. This is the equivalent to a 3½-inch piston settling at the rate of 7/16-inch per hour.

154. **OVERHAUL.** To disassemble the spool valve, first clean any paint, rust or burrs from eye-end of valve spool, then unscrew the detent cap. Withdraw spool and cap as an assembly. Unscrew poppet plugs (11— Fig. 254) and withdraw poppet plungers. Only the two 9/16-inch O-rings (R) and spool seal ring (S) remain in stripped housing. Examine centering spring and washers. If any parts are damaged, unscrew detent spool from valve spool to renew the parts.

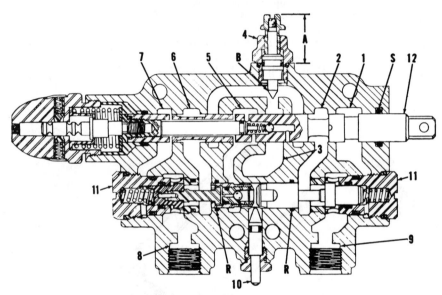

Fig. 254–Cross sectional view of spool type auxiliary valve body. Valves may be stacked for multiple operation.

A. Adjustment	1. Return port	7. Return port
B. Metering piston	2. Transfer passage	9. Work port
R. O-rings	3. Pressure port	10. Unloading plunger
S. O-ring	5. Actuating passage	11. Poppet assembly
	6. Transfer passage	

MASSEY-FERGUSON

Models ■ MF1105   ■ MF1135   ■ MF1155

Previously contained in I&T Shop Manual No. MF-31

SHOP MANUAL

MASSEY-FERGUSON

MODELS
MF1105   MF1135   MF1155

Tractor serial number is stamped on instrument panel name plate.
Engine serial number is stamped on side of engine.

INDEX (By Starting Paragraph)

CONDENSED SERVICE DATA

GENERAL

	MF1105	MF1135	MF1155
Torque Recommendations	See End of Shop Manual		
Engine Make	Perkins	Perkins	Perkins
Engine Model	ATC6.354	AT6.354	AV8.540
Number of Cylinders	6	6	8
Bore-Inches (mm)	3-7/8(98.5)	3-7/8(98.5)	4-1/4(108)
Stroke-Inches(mm)	5(127)	5(127)	4-3/4(120.7)
Displacement-Cu.-In. (cc)	354(5807)	354(5806)	539.1(8834)
Compression Ratio	16:1	16:1	17.5:1
Main Bearings, No. of	7	7	5
Cylinder Sleeves.	Dry	Dry	Dry
Forward Speeds	8 or 12	8 or 12	8 or 12
Reverse Speeds	2 or 4	2 or 4	2 or 4

TUNE-UP

	MF1105	MF1135	MF1155
Firing Order	1-5-3-6-2-4	1-5-3-6-2-4	1-8-7-5-4-3-6-2
Valve Tappet Gap—			
Cold—Inch (mm)	0.012(0.30)	0.012(0.30)	0.012(0.30)
Hot—Inch (mm)	0.010(0.25)	0.010(0.25)	. . . . . .
Valve Face Angle.	45°	45°	45°
Valve Seat Angle	45°	45°	45°
Injection Timing—			
Crankshaft Degrees.	26°	30°	28°
Piston Position—In. (mm)	0.325(8.26)	0.426(10.82)	0.352(8.94)
Injectors—			
Opening Pressure, psi	2645	2940	2720
Atmospheres	180	200	185
kg/cm²	186	206	191
Spray Hole Dia., Inch	0.011-0.0118	0.011-0.0118	0.0118-0.0122
mm	0.28-0.30	0.28-0.30	0.30-0.31
Governed Speeds—			
Engine rpm			
Low Idle.	800-850	800-850	800-850
High Idle	2325-2450	2325-2450	2325-2450
Loaded	2200	2200	2200
540 pto			
Low Idle.	216-230	216-230	216-230
High Idle	625-662	625-662	625-662
Loaded	594	594	594
1000 rpm pto			
Low Idle.	400-425	400-425	400-425
High Idle	1163-1225	1163-1225	1163-1225
Loaded	1100	1100	1100
Horsepower at pto Shaft*.	100.72	120.84	140.97
*According to Nebraska Test.			
Battery**			
Volts.	12	12	12
Ground Polarity.	Negative	Negative	Negative
Capacity Amp/hr.	95	95	95

**Each battery—two batteries are used, connected in parallel.

SIZES—CAPACITIES—CLEARANCES

	MF1105	MF1135	MF1155
Crankshaft Main Journal—			
Diameter, Inch.	2.9985	2.9985	3.9969
mm	76.17	76.17	101.52
Bearing Clearance,			
inch.	0.0025-0.0045	0.0025-0.0045	0.004-0.0063
mm	0.064-0.110	0.064-0.110	0.10-0.16
Crankshaft Crankpin—			
Diameter, inch	2.499	2.499	2.9983
mm	63.48	63.48	76.16
Bearing Clearance,			
inch	0.0015-0.0030	0.0015-0.0030	0.0018-0.0038
mm	0.038-0.076	0.038-0.076	0.05-0.10

CONDENSED SERVICE DATA CONT.

SIZES—CAPACITIES—CLEARANCES Cont.

Crankshaft End Play, Inch	0.002-0.0015	0.002-0.015	0.002-0.017
mm	0.05-0.38	0.05-0.38	0.05-0.43
Camshaft Journal Diameters—			
Front, Inch	1.997	1.997	2.372
mm	50.73	50.73	60.25
Second, inch	1.997	1.997	2.242
mm	50.73	50.73	56.95
Third, inch	1.997	1.997	2.232
mm	50.73	50.73	56.69
Fourth, inch	1.997	1.997	2.222
mm	50.73	50.73	56.44
Fifth, inch	.....	.....	2.212
mm	.....	.....	56.18
Camshaft Bearing Clearance—			
Inch	0.0025-0.055	0.0025-0.0055	0.002-0.006
mm	0.064-0.0140	0.064-0.0140	0.05-0.15
Camshaft End Play, inch	0.004-0.016	0.004-0.016	0.004-0.014
mm	0.1-0.41	0.1-0.41	0.1-0.36
Cooling System-Quarts	22	22	40
Liters	20.82	20.82	37.85
Crankcase Oil-Quarts	16 +	16 +	19 + +
Liters	15.14 +	15.14 +	17.98 + +
Transmission, Differential and Hydraulic Lift—			
Gallons	20 + + +	20 + + +	20 + + +
Liters	75.7 + + +	75.7 + + +	75.7 + + +

+ Add 1 quart (0.95 liter) if filter is changed.

+ + Add 2 quarts (1.9 liters) if filter is changed; 1½ quarts (1.4 liters) if the oil cooler is drained.

+ + + Capacity is given for initial fill. Drain and refill service will only remove approximately 17 gallons (64.4 liters).

FRONT SYSTEM

Refer to Figs. 1 through 4 for exploded views of the different front end types available, and to the following paragraphs for pertinent disassembly and overhaul data.

Model MF1155 tractors are available with Western Type (non-adjustable) front axle unit or Row Crop (wide adjustable) axle types only.

AXLE ASSEMBLY

1. Refer to Fig. 2 for an exploded view of Western Type axle assembly and to Fig. 3 for adjustable axle unit. Front support, steering motor and associated parts are shown in Fig. 1.

To remove either axle as an assembly, disconnect center steering arm (1—Fig. 2) from steering spindle, appropriately support the tractor and unbolt and remove axle pivot brackets (2 & 9 —Fig. 2 or 3). When reinstalling, align the center bolt hole of steering arm (1—Fig. 2) with "X" marked bolt hole on steering spindle flange. Use Grade "C" (Blue) LOCTITE on cap screws securing steering arm (1) and tighten

the screws to a torque of 175-200 ft.-lbs.

TIE RODS AND TOE-IN

2. Automotive type tie rod ends are used. Recommended toe-in is 0-3/8 inch.

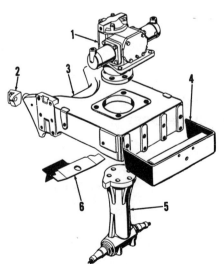

Fig. 1—Front support, steering motor and dual wheel tricycle pedestal shown exploded.

1. Steering motor
2. Engine support
3. Front support
4. Frame
5. Pedestal
6. Shield

Adjust both tie rods an equal amount to obtain the required setting.

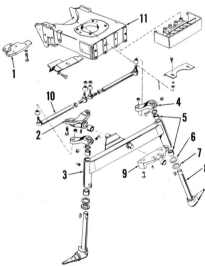

Fig. 2—Exploded view of Western Type front axle and associated parts. Bronze side of thrust washers (6) should be down against washer (7).

1. Center steering arm
2. Rear pivot bracket
3. Axle
4. Steering arm
5. Spindle bushing
6. Thrust washer
7. Thrust washer
8. Spindle
9. Front pivot bracket
10. Tie rod
11. Front support

POWER STEERING SYSTEM

All models are equipped with a closed center hydraulic system which provides standby power for operation of the hydrostatic steering system as well as the other hydraulic functions of operation and control of the tractor. This section covers only the Hydrostatic Power Steering Hand Pump (steering control valve), Steering Cylinder Assembly (steering motor), control connecting lines and attaching parts. Refer also to HYDRAULIC SYSTEM Section for information on main hydraulic pump, priority valve, relief valve and operating fluid.

LUBRICATION AND BLEEDING

3. The hydrostatic steering hand pump and steering cylinder assembly (steering motor) are lubricated by the operating fluid. Refer to paragraph 201 for fluid type and checking procedure. The hand pump (P—Fig. 5) is lubricated by the high pressure oil as the system is operated. The shaft, bearings and gears of steering motor (C) are lubricated by return oil from the lubrication passages of the main hydraulic pump. The power steering system is self-bleeding.

ADJUSTMENT

4. The only adjustment required is adjustment of cylinder rack clearance (backlash). If clearance is excessive, wear is usually indicated and cylinder unit should be removed as outlined in paragraph 6 and overhauled as in paragraph 7.

TROUBLE SHOOTING

5. If malfunction of the steering system exists, first make an operational check of some other hydraulic unit to eliminate the hydraulic system as the cause of trouble. The procedure for completely checking the hydraulic system is given beginning with paragraph 203.

External fluid leakage is not generally objectional from an operational standpoint and should be corrected only to prevent visible loss of fluid.

STEERING CYLINDER (MOTOR)

Two types of steering motor unit have been used on MF1105 and MF1135 Models. The type shown in Fig. 8 is equipped with a rack guide (R.G.) which is bolted to the front of housing as shown. Steering motor without the detachable rack guide (shown in Fig. 7) is used on MF1155 models and some MF1105 and MF1135 models. Some service procedures are different on the two units; therefore, refer to the appropriate following paragraphs.

Models Without Rack Guide

Many service procedures can be accomplished without removal; however, paragraph 6 outlines procedure for removal of the complete steering cylinder

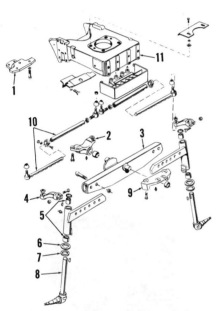

Fig. 3—Exploded view of adjustable axle. Bronze side of thrust washers (6) should be down against washer (7).

1. Center steering arm
2. Rear pivot bracket
3. Center axle
4. Steering arm
5. Spindle bushing
6. Thrust washer
7. Thrust washer
8. Spindle
9. Front pivot bracket
10. Tie rod
11. Front support

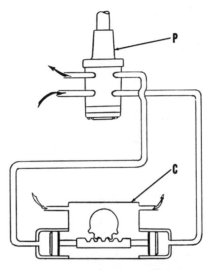

Fig. 5—Schematic view of hydrostatic power steering hand pump (P) and steering motor (C) showing direction of oil flow.

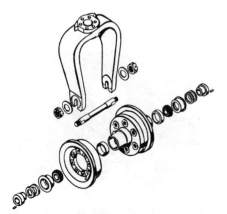

Fig. 4—Exploded view of single wheel tricycle pedestal and associated parts.

Fig. 6—Exploded view of hydraulic lines and associated parts running to front of tractor.

1. Filter to oil cooler tube
2. Oil cooler to regulating valve tube
3. Pump pressure tube
4. Steering motor bleed tube
5. Steering cylinder tube
6. Steering cylinder tube
7. Filter inlet tube
8. Pump inlet tube
9. Steering lubrication tube
10. Main hydraulic pump
11. Oil cooler
12. Bracket
13. Hydraulic filter
14. Steering motor

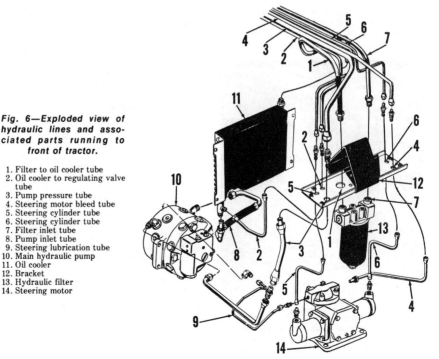

5

unit. If complete overhaul is indicated, refer to paragraph 6 for removal and paragraph 7 for overhaul procedures.

6. **REMOVE AND REINSTALL.** To remove the complete steering motor as a unit, first disconnect center steering arm from spindle. Remove front grille, disconnect steering cylinder lines, then unbolt and lift out steering motor assembly.

Install by reversing the removal procedure. Make sure that steering spindle (14—Fig. 9) and rack (18) are centered before installing center steering arm. Use Grade "CV" (Blue) LOCTITE on cap screws attaching center steering arm or pedestal to spindle (14) and tighten cap screws to a torque of 175-200 ft.-lbs. Steering motor is self-bleeding. Screws attaching steering motor to front housing should be tightened to 190-285 ft.-lbs. torque.

7. **OVERHAUL.** Refer to Fig. 9 for an exploded view of steering motor unit.

Disconnect cylinder lines, lube line and bleed line. Remove both cylinders

(1). Pry out welch plug (2), remove snap ring (3) and thrust washer (4); then unbolt and remove top cover (5). Lift out pinion gear (10), then withdraw rack (18) with pistons attached, through either side opening. Reach through top opening and remove snap ring (11) as shown in Fig. 10. Unbolt and lift housing (8—Fig. 9) from the front support. Center steering arm and shaft unit can be lifted out after disconnecting steering tie rods.

If Teflon piston rings (19) are to be renewed, heat rings in water to approximately 180 degrees F. to soften and expand the rings. Install in piston grooves then clamp both pistons in a ring compressor to allow seals to return to original size.

Bushings (6 & 12) are pre-sized and will not require reaming after installation. Install lower bushing (12) with open end of oil groove to inside, with outer edge 1/16 inch (1.6mm) below chamfer in bore. Install seal (13) with lower edge flush to 1/64 inch (0.40mm) below edge of housing bore.

Tighten cap screws retaining pistons (16) to rack (18) to a torque of 30-35 ft.-lbs. Tighten top cover retaining cap screws to 33-38 ft.-lbs. and cylinder retaining cap screws to 50-55 ft.-lbs.

After installing lower shaft snap ring (11) turn steering shaft until holes in steering shaft are aligned as shown in Fig. 11 and install rack and pinion with timing punch marks aligned as shown.

Install a new welch plug (2—Fig. 9) using Hydraulic Sealant (Brown) LOCTITE. Fill and bleed steering system as outlined in paragraph 3 after assembly is completed and lines connected.

Models With Rack Guide

Steering cylinders, pistons or piston rings can be renewed or backlash adjusted without removal of unit from the tractor. If complete overhaul is indicated, steering motor should be removed as outlined in paragraph 8.

8. **REMOVE AND REINSTALL.** To remove the complete steering motor as a unit, first disconnect center steering arm from spindle. Remove front grille, disconnect steering cylinder lines, then unbolt and lift out steering motor assembly.

Install by reversing the removal procedure. Make sure that steering spindle (9—Fig. 12) and rack (16) are centered

Fig. 7—Grille screens removed showing installed power steering motor. Unit shown (Arrow) is type without rack guide.

Fig. 8—Partially installed view of steering motor which has rack guide (RG). Refer to Fig. 13 for exploded view of this type.

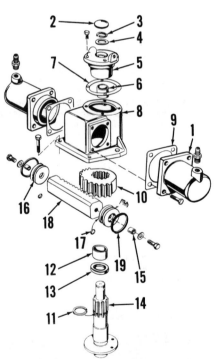

Fig. 9—Exploded view of hydrostatic steering motor showing component parts of type without rack guide.

1. Cylinder
2. Expansion plug
3. Snap ring
4. Thrust washer
5. Top cover
6. Upper bushing
7. Gasket
8. Housing
9. Gasket
10. Pinion gear
11. Snap ring
12. Lower bushing
13. Oil seal
14. Steering spindle
15. Spacer
16. Piston
17. "O" ring
18. Rack
19. Piston ring

Fig. 10—Reach through top opening to remove snap ring (11—Fig. 9) then remove steering spindle.

Fig. 11—Steering motor is correctly assembled when timing marks on rack and pinion are aligned as shown.

before installing center steering arm. Use "Grade CV (Blue) LOCTITE" on cap screws attaching center steering arm or pedestal to spindle (9) and tighten cap screws to a torque of 175-200 ft.-lbs. Steering motor is self-bleeding. Screws attaching steering motor to front housing should be tightened to 190-285 ft.-lbs. torque.

9. **OVERHAUL.** When disassembling the steering motor either on or off the tractor, unbolt and remove steering cylinders (23—Fig. 12) before removing the rack guide (21).

Piston can be removed by removing cap screw (10) after cylinder is off. Piston uses spacer (12) to free the installed piston or rack for alignment purposes. Piston should be free to turn with bolt (10) tightened. "O" ring (15) is fitted between piston (14) and rack (16) and seals inner hole in piston when hydraulic pressure is applied. Tighten bolt (10) to a torque of 30-35 ft.-lbs. when unit is reassembled.

Steering gear backlash should be 0.001-0.008 inch (0.03-0.20mm). Measure the backlash as follows: Assemble rack (16) and rack guide (21) without any shims (22). Tighten screws retaining rack guide to housing evenly to a torque of 30-35 ft.-lbs. Measure and record clearance between flange of rack guide and housing. Remove rack guide and install shims (22) equal in thickness to measured clearance plus 0.001-0.008 inch (0.03-0.20mm). Shims (22) are available in four thicknesses from 0.006 to 0.023 inch (0.15-0.58mm). Make sure rack is centered on gear (17) when unit is assembled.

To remove steering spindle (9), first remove cylinders, rack guide (21) and rack (16), then unbolt and remove housing cover. (1). Unseat and remove snap ring (3), then bump spindle (9) downward out of splines of gear (17).

Top edge of bearing (5) should be approximately 1/16 inch (1.6mm) below flush with top surface of housing. Lower bearing (7) should be approximately 1/4 inch (6.35mm) from seal counterbore. Seal (8) should be approximately 1/16 inch (1.6mm) above lower surface of housing. Large chamfer of washer (4) should be toward top.

An "X" mark is located on steering spindle (9) which is not visible when correctly assembled. Mark shaft across (180 degrees) from the stamped "X", then align this mark with timing mark on gear as shown in Fig. 13. Washers (18—Fig. 12) are not identical. The thin washer should be at top and thick washer below the thrust bearing (19) as shown in Fig. 13. Install rack with timing marks on gear and rack aligned as shown in Fig. 14. Check and adjust rack backlash as outlined in a previous paragraph.

HYDROSTATIC HAND PUMP

The hydrostatic hand pump (Char-Lynn ORBITROL) is a remote hydraulic control and metering valve which directs pressurized fluid to the power steering motor. The unit also serves as a manual control valve and pump to provide steering effort when hydraulic pressure is not available. No mechanical linkage exists between the steering wheel and tractor front wheels.

10. **OPERATION.** The internal gear type hand pump (14 & 15—Fig. 15) connects hydraulically to the opposing cylinders of the steering motor. The spring-loaded rotary valve (11) connects to hydraulic system stand-by pressure and system return lines.

Initial turning effort of the steering wheel meets resistance at the front wheels causing deflection of the rotary valve centering springs, metering a flow of pressurized hydraulic fluid through the rotary valve and hand pump to the steering motor. Return fluid from opposite cylinder of steering motor passes through return ports of rotary valve to the reservoir. Steering action and speed are thus in direct control of the operator at all times, steering EFFORT only, being supplied by hydraulic system standby pressure.

When hydraulic system pressure is not available, the hand pump operated manually to steer the tractor, with return fluid passing through check valve (10) to intake side of hand pump.

11. **REMOVE AND REINSTALL.** Remove the hood, side panels and battery platform. Clean pump and surrounding area and disconnect the four hydraulic lines, plugging valve ports as removed; then unbolt and remove hand pump unit.

Install by reversing the removal procedure. Cycle the system several times to air-bleed, then check and refill

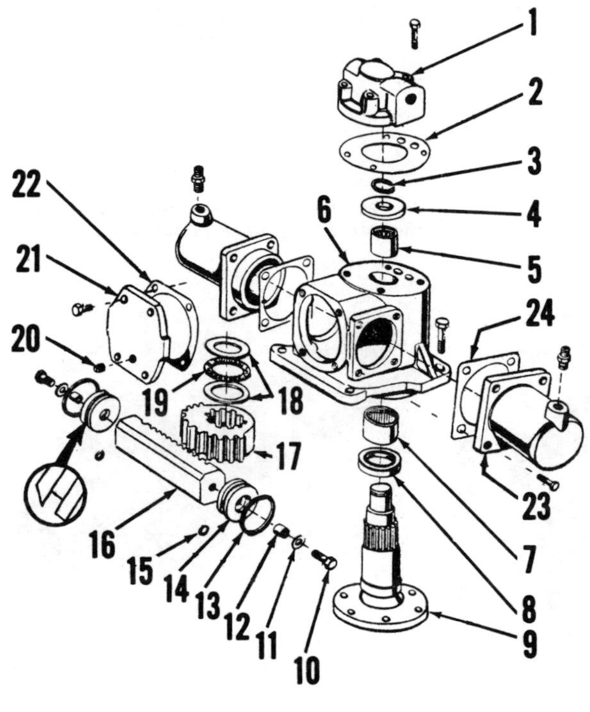

Fig. 13—The steering spindle shaft must be correctly timed to the gear, however "X" mark provided on shaft is not visible. Mark shaft opposite "X" mark and align this new mark with timing mark on gear.

Fig. 12—Exploded view of hydrostatic steering motor which has rack guide and needle bearings.

1. Cover
2. Gasket
3. Snap ring
4. Thrust washer
5. Needle bearing
6. Spindle housing
7. Needle bearing
8. Oil seal
9. Steering spindle
10. Cap screw
11. Washer
12. Spacer
13. Piston ring
14. Piston
15. "O" ring
16. Rack
17. Pinion
18. Bearing races
19. Thrust bearing
20. Drain plug
21. Rack guide
22. Shim
23. Cylinder
24. Gasket

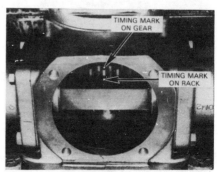

Fig. 14—Timing mark on gear and mark on rack must be aligned when assembling the steering rack.

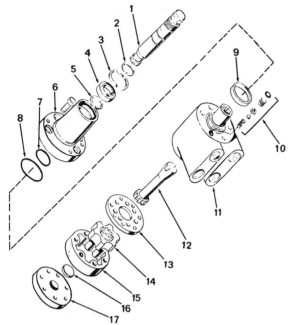

Fig. 15—Exploded view of hydrostatic hand pump showing component parts. Body and spool assembly (11) is available only as a unit which contains check valve (10), but can be further disassembled for cleaning and inspection as outlined in text.

1. Input shaft
2. Snap ring
3. Snap ring
4. Bearing
5. Snap ring
6. Shaft housing
7. Quad ring
8. "O" ring
9. Bushing
10. Check valve
11. Valve assembly
12. Drive link
13. Metering plate
14. Inner rotor
15. Outer stator
16. Disc
17. End plate

Carefully push control valve spool and sleeve assembly out bottom of housing as shown in Fig. 20.

NOTE: Be careful valve unit does not bind. Parts are fit to extremely close tolerance and a twisting motion may be required for withdrawal.

Insert a suitably bent wire through port nearest check valve plug and push out the plug as shown in Fig. 22. Remove the threaded check valve seat (1—Fig. 23) using an Allen wrench and

Fig. 19—Input shaft housing is positioned by a locator bushing (1).

hydraulic system reservoir if necessary as outlined in paragraph 201.

12. **OVERHAUL.** Refer to Fig. 15 for a partially exploded view of the Char-Lynn ORBITROL hydrostatic hand pump. The rotary valve spool, sleeve and housing (11) is available only as an assembly which includes reaction springs, cross pin and check valve (10).

To disassemble the removed Orbitrol unit, first remove the cap screws (1—Fig. 16) retaining metering pump to bottom of housing and lift off pump components as shown in Fig. 17. Carefully invert the housing on a clean board as shown in Fig. 18 to prevent the rotary valve unit from sliding out, remove the four cap screws retaining top cover and lift off the cover as shown. Remove locator bushing (1—Fig. 19).

Fig. 16—Remove cap screws (1) and lift off metering pump components as shown in Fig. 17.

Fig. 20—Carefully push control valve spool and sleeve out bottom of housing as shown.

Fig. 17—Lift off metering pump.

Fig. 18—Carefully invert the housing on a clean board to unbolt and remove input shaft and housing.

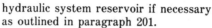

Fig. 21—Valve cross pin is positioned by two Nylon discs (1).

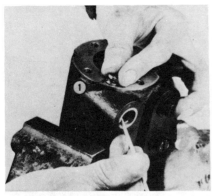

Fig. 22—Insert a suitably bent wire through valve port to push out check valve plug (1).

Fig. 23—Check valve seat (1) can be removed with an Allen wrench, then check ball (2) and spring (3) removed.

Fig. 24—Schematic end view of steering valve sleeves showing correct installation of drive pin (P) and centering springs(S).

Fig. 25—When properly assembled, slot in drive link will align with valley of inner gear as shown.

remove check valve ball (2) and spring (3).

Remove nylon cross pin discs (1—Fig. 21) and cross pin, then separate spool and sleeve. Withdraw the six arched centering springs. Thoroughly clean all parts in a suitable mineral solvent and blow dry with air. Inspect lapped surfaces for scoring or other damage. Mating surfaces of cover (17—Fig. 15), rotor assembly (14 & 15) and metering plate (13) can be hand lapped to remove burrs, slight scratches or other imperfections. Use 600 grit abrasive paper or lapping compound. Rinse in clean solvent and blow dry after lapping is complete. Rotor units are only available as a matched set.

Inspect centering springs (S—Fig. 24) for fractures or distortion. Springs should have a minimum arch of 7/32 inch (5.56mm) when measured at center. Inspect spool, sleeve and body for nicks, scoring or wear. If any part is damaged, renew the housing assembly which includes spool, sleeve and check valve. Renew all "O" rings when reassembling the unit.

Tighten the check valve seat (1—Fig. 23) to a torque of 150 inch-pounds. Install centering springs (S—Fig. 24) in sets of three with notched edge down. Insert the assembled spool and sleeve assembly from bottom of valve body using a twisting motion. Do not allow sleeve to move up beyond flush with upper machined surface of body. Be sure pin slot in drive shaft aligns with valleys of inner rotor as shown in Fig. 25. Rotor has six gear teeth and 12 splines; if drive slot is improperly aligned, steering unit will operate in reverse of hand pressure. Tighten the seven lower cap screws evenly to a torque of 250 in.-lbs. and the four cap screws in top cover to a torque of 220 in.-lbs.

Install the assembled unit as outlined in paragraph 11 and bleed by cycling the system until steering action is firm and response is immediate.

SIX CYLINDER ENGINE AND COMPONENTS

All MF1105 and MF1135 Models are equipped with a turbocharged, six cylinder Perkins Diesel engine having a bore of 3-7/8 inches (98.43mm) a stroke of 5 inches (127mm) and a displacement of 354 cubic inches (5801cc).

R&R ENGINE WITH CLUTCH

15. To remove the engine and clutch as a unit, first drain cooling system and if engine is to be disassembled, drain oil pan.

NOTE: Be careful to prevent entry of dirt or other foreign particles into any of the oil, hydraulic or fuel lines that must be disconnected for removal of the engine.

Disconnect batteries and remove hood and all hood side panels. Refer to Figs. 30 and 31. Remove the two oil

lines between engine oil cooler and oil filter adapter. Disconnect hydraulic oil lines and electrical connector at rear of engine. Disconnect or remove priority valve oil supply line. Disconnect air inlet and radiator hoses. Remove the

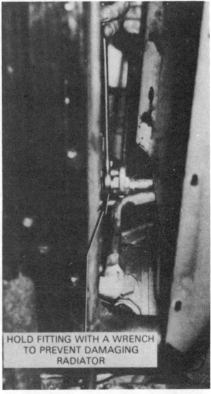

Fig. 31—View of oil cooler fittings. Use wrench on radiator hex to prevent damage to radiator.

Fig. 30—View of engine right side identifying parts which must be disconnected or removed in order to remove the six cylinder engine.

Fig. 32—View of left rear of engine showing parts which must be disconnected or removed before removing the six cylinder engine.

Fig. 35—View of six cylinder engine left side showing parts to be disconnected or removed before engine is removed.

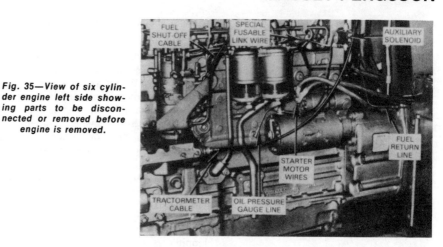

Fig. 33—The hydraulic pump coupler is attached to pulley with four screws as shown.

four screws (Fig. 33) which attach hydraulic pump drive coupling to crankshaft pulley. Block between front

Fig. 34 — View showing separation of front system from the engine.

axle support and axle to prevent tipping sideways and support front end and engine separately. Remove the 6 capscrews attaching front axle support to engine adapter plates and oil pan.

CAUTION: Carefully observe joint while loosening these six screws to be sure that front end and tractor are supported evenly. Change support as necessary if parting surfaces are not parallel.

Support tractor under forward end of transmission (NOT FLYWHEEL HOUSING). Disconnect electrical wires from temperature sending unit and alternator. Disconnect fuel supply line

from fuel lift pump, throttle control rod from bellcrank and priority valve from mounting plate. Refer to Fig. 35. Disconnect fuel return line near transmission, tractormeter (tachometer) and fuel shut-off cables at forward ends. Disconnect wires from starter motor and unbolt auxiliary solenoid from support plate. Unclamp oil lines from priority valve support plate and disconnect upper end of oil pressure gage line. Check all wires and tubes to be sure that they are disconnected and positioned so they will not be damaged. Support engine with an overhead hoist attached to the two lift brackets, then loosen capscrews and bolts attaching flywheel housing and oil pan to transmission.

CAUTION: Observe the joint while loosening bolts to be sure that tractor and engine are supported evenly. Adjust supports if mating surfaces are not parallel while separating.

When reinstalling engine observe the following: Install guide studs before sliding engine against transmission housing. Carefully move engine back against transmission housing while aligning splines of clutch with transmission input shaft and splines of flywheel adapter with pto shaft. If splines do not align, DO NOT force together, but turn flywheel slowly while sliding together. Tighten the large bolts attaching oil pan to transmission housing to 360-540 ft.-lbs. torque. Remainder of procedure for joining engine to transmission housing is reverse of disassembly.

The four lower screws attaching front support to the front of oil pan should be tightened first. Install adaptors onto studs in side of engine and start the upper front end retaining capscrews. Refer to Fig. 36. Tighten the two nuts on each adaptor enough to be sure that adaptor is tight against engine, then tighten the upper front end retaining capscrew. Install adaptor capscrew, then tighten both nuts and capscrews.

CYLINDER HEAD

16. To remove the cylinder head, first drain cooling system, disconnect batteries and remove hood and side panels. Disconnect turbocharger oil lines, air cleaner and inlet manifold hoses, unbolt and remove inlet manifold. Unbolt and remove exhaust manifold and turbocharger as a unit. Remove fuel injector pressure lines as an assembly. Remove fuel leak off pipe and all injector nozzles. Injector nozzles protrude from bottom of cylinder head and may be damaged if allowed to remain in head. Disconnect upper and lower coolant hoses and temperature indicator sending unit from water pump body. Unbolt and remove the fan blades. Disconnect belt tightener arm from water pump body and remove fan belt from water pump pulley. Remove breather pipe, rocker arm top cover, rocker arms and push rods; then remove rocker arm lower cover. Remove the cylinder head stud nuts, lift the head until clear of top of studs, then move head rearward until fan pulley is clear of radiator shroud.

The cylinder head can be resurfaced, providing the injector nozzle protrusion does not exceed 0.224 inch (5.69mm). Original thickness of cylinder head is 3.235-3.265 inches (82.17-82.93mm) and normally surfacing of less than 0.012 inch (0.3mm) will be possible. After resurfacing cylinder head, it will be necessary to reseat valves to correct depth as described in paragraph 17 and shown in Fig. 39.

Cylinder head gasket is marked "TOP FRONT" for proper installation. Head gasket should be installed dry. Tighten cylinder head stud nuts to a torque of 85 ft. lbs. using the sequence

shown in Fig. 37. Adjust valve tappet gap after head is installed as outlined in paragraph 22.

VALVES AND SEATS

17. On MF1105 models, both intake and exhaust valves seat directly in the cylinder head. Renewable inserts are used in all valve seats on MF1135 models. On all models valve heads should be recessed a specified amount into the cylinder head and inserts should be installed or renewed when specifications are exceeded. Clearance can be checked using a straight edge and feeler gage as shown in Fig. 38. Production clearances are held within the limits of 0.029-0.039 inch (0.74-0.99mm) for both inlet and exhaust valves. Preferred recess is close to minimum of 0.029-inch to allow for refacing valve and seat. Renew valve seat insert and/or valve if recess exceeds 0.060 inch (1.52mm). New valve seat inserts should be machined as shown in Fig. 39. Depth (A) of 0.094-0.099 inch (2.39-2.51mm) should provide correct amount of valve recess (Fig. 38) with new valve. All valve seats should be flared (30 degrees) and reseated (45 degrees) to correct depth and the nozzle protrusion should be checked after resurfacing cylinder head. Refer to paragraph 16. Valve face and seat angles should be 45 degrees for both inlet and exhaust valves.

VALVE GUIDES

18. Intake valve stem to guide wear limit is 0.005 inch (0.13mm) and exhaust valve stem to guide wear limit is 0.006 inch (0.15mm). The cast iron

valve guides are not interchangeable for intake and exhaust (Fig. 41). Exhaust guide is counterbored at valve head end as shown. Press new guide into head until distance (A) is 19/32 inch (15.08mm).

Inside diameter of new guides should be 0.375-0.376 inch (9.35-9.55mm) and provide desired diametral clearance of 0.0015-0.0035 inch (0.04-0.09mm) for intake valve stem, 0.002-0.004 inch (0.05-0.10mm) for exhaust valve stem.

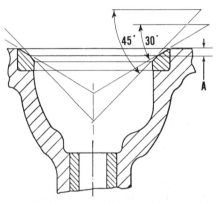

Fig. 39—Valve seat width and location is changed using 30 and 45 degree stones. Depth (A) should be 0.094-0.099 inch for new valves and seats.

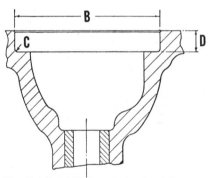

Fig. 40—Bore (B) in cylinder head for valve seat should be 2.0165-2.0175 inches for inlet valves; 1.678-1.679 inches for exhaust valves. Depth (D) should be 0.283-0.288 inch for inlet valves; 0.375-0.380 inch for exhaust valves. Radius (C) should be 0.015 inch for all valve seats.

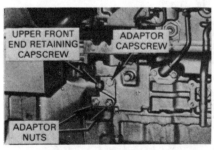

Fig. 36—The front adapters must be carefully installed. Refer to text.

Fig. 38—Using a straight edge and feeler gage to check valve head height. Refer to text.

Fig. 37—On six cylinder engines, tighten the cylinder head cap screws in sequence shown.

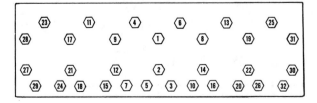

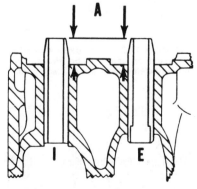

Fig. 41—Distance (A) should measure 19/32 inch for all guides. Difference in design of intake guide (I) and exhaust guide (E) is also shown.

Fig. 42—View of six cylinder rocker arms and shaft properly assembled.

Be sure to resurface valve seats after new guides are installed.

VALVE SPRINGS

19. Springs, retainers and locks are interchangeable for intake and exhaust valves; however, an umbrella type oil deflector is installed on intake valve stem. Springs should be installed with close (damper) coils toward cylinder head. Renew the springs if they are distorted, discolored, or fail to meet the test specifications which follow, or the other parts if they show signs of wear or damage.

INNER SPRING:
Lbs. Test @ 1.5625 inches . . . 14.63-16.17
Kg Test @ 39.69mm 6.64-7.33

OUTER SPRING:
Lbs. Test @ 1.78 inches 38-42
Kg Test @ 45.21mm 17.24-19.05

CAM FOLLOWERS

20. The mushroom type tappets (cam followers) operate in machined bores in the cylinder block and can be renewed after removing camshaft as outlined in paragraph 28. The 0.7475-0.7485 inch (18.99-19.01mm) diameter tappets should have 0.0015-0.0038 inch (0.04-0.10mm) diametral clearance in crankcase bores.

ROCKER ARMS

21. The rocker arm shaft assembly can be removed from left side after removing hood side panel and rocker arm cover.

NOTE: Push rods can drop into oil pan if dislocated at lower end. Be careful not to lose push rods into engine when removing or installing rocker arms.

Fig. 42 shows an assembled view of the rocker arm shaft. Right and left hand rocker arm units, mounting brackets and the oil feed tube must be positioned as shown. Oil feed holes in shaft must be installed toward valve stem side. Make sure the sealing "O" ring does not roll out of retaining grooves in feed tube as shaft unit is installed. Some mechanics prefer to position the "O" ring on tube below the positioning grooves and allow it to roll into proper position as shaft is installed.

Desired diametral clearance between rocker arms and shaft is 0.001-0.0035 inch (0.03-0.09mm). Renew shaft and/or rocker arms if clearance is excessive.

Tighten rocker arm support cap screws to a torque of 28-32 ft.-lbs.

VALVE CLEARANCE

22. The recommended cold valve clearance (tappet gap) is 0.012 inch (0.30mm) for both the intake and exhaust valves. Cold (static) setting of all valves can be made from just two crankshaft positions, using the procedure outlined in this paragraph and illustrated in Figs. 43 and 44.

Remove timing plug from left front side of flywheel adapter housing and turn crankshaft until "TDC" timing mark on flywheel is aligned with timing pointer. Check the rocker arms for front and rear cylinders. If rear rocker arms are tight and front rocker arms have clearance, No. 1 piston is on compression stroke; adjust the six tappets shown in Fig. 43. After adjusting the indicated tappets, turn crankshaft one complete turn until the "TDC" mark is again aligned and adjust the remaining tappets.

VALVE TIMING

23. Timing gears are keyed and valve timing will be correct if timing marks are properly aligned as outlined in paragraph 26.

TIMING GEAR COVER

24. The timing gear cover (housing) (12—Fig. 45) can only be removed after removing the camshaft gear (7) and

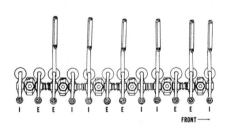

Fig. 43—With TDC timing marks aligned and No. 1 piston on compression stroke, adjust the indicated valves to 0.012 inch cold.

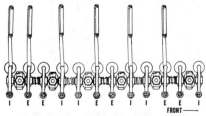

Fig. 44—With TDC timing marks aligned and No. 6 piston on compression stroke, adjust the indicated valves to 0.012 inch cold.

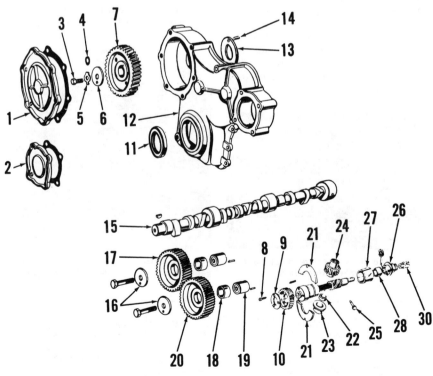

Fig. 45—Covers (1 & 2) and gears (7 & 10) must be removed before timing gear housing (12) can be removed for access to remainder of gears. Refer to text for details.

auxiliary drive shaft gear (10). The crankshaft front oil seal (11) seals against crankshaft pulley hub, and seal and seating surface can be inspected and renewed if necessary, after removing crankshaft pulley.

To remove the timing gear cover or renew the oil seal, first support tractor and remove front end as a unit as outlined in paragraph 15.

25. CRANKSHAFT PULLEY AND FRONT OIL SEAL. If front oil seal only is to be renewed, remove crankshaft pulley retaining cap screw (1—Fig. 46) and washer (2). Using a center, attach a suitable puller to crankshaft pulley (4) and apply slight pressure with puller forcing screw then strike forcing screw sharply with a hammer to bounce out the tapered locking ring (3). Pulley can now be easily withdrawn.

Coat outer edge of new seal with a hardening type gasket cement and install seal with lip toward crankshaft gear. Front face of seal must be 0.220-0.280 inch (5.6-7.1mm) below flush with front face of cover. Special tool (MFN-747B) can be used to press seal into correct depth without removing cover.

Thoroughly clean and degrease the splines on crankshaft and locking ring (3). Coat the splines with "Grade AV (Red) LOCTITE" and reinstall pulley and locking ring. Install washer (2) on capscrew (1) with inner chamfer next to capscrew head and install and tighten capscrew. If washer (2) is 3/8 inch thick, tighten capscrew to 300 ft.-lbs. torque; if washer (2) is ¼ inch thick, tighten capscrew to 250 ft.-lbs. torque.

26. COVER (HOUSING) AND GEARS. If the cover (12—Fig. 45) is to be removed, turn crankshaft until TDC

Fig. 47—Front view of engine block with timing marks properly aligned. Timing gear train will never appear as shown, because shaded gears (AS & CS) must be removed for access to timing gear housing.

AS. Accessory shaft gear
CS. Camshaft gear
CR. Crankshaft gear

timing mark on flywheel is aligned and No. 1 cylinder is on compression stroke. Remove crankshaft pulley as outlined in paragraph 25. Remove the two covers (1 and 2) and the gears (7 and 10), using a puller to remove camshaft gear (7). Cover can now be unbolted and removed.

The crankshaft gear and idler gears are now accessible for inspection or service. Allowable backlash between any two gears in timing gear train is 0.003-0.006 inch (0.08-0.15mm). Diametral clearance of idler gears to hubs is 0.001-0.003 inch (0.03-0.08mm) and installed end play of idler gears is 0.002-0.012 inch (0.05-0.30mm). Bushings in idler gears must be reamed after installation to an inside diameter of 1.3755-1.3771 inches (34.94-34.98mm). Oil holes are provided in gear hub and gear to supply lubrication to gear teeth.

Camshaft and auxiliary drive shaft can be withdrawn from cylinder block after timing gear housing is removed. Refer to paragraph 27 for details on auxiliary drive unit overhaul and to paragraph 28 for camshaft removal.

Before timing gear cover can be reinstalled, it is first necessary to remove the injection pump for proper retiming of auxiliary shaft. Remove right hand idler gear (17) and reinstall with timing punch marks on idler gears and crankshaft gear aligned as shown in Fig. 47.

NOTE: Left Hand Idler Gear may not have timing marks and marks may be ignored if present. Marks are not required for timing injection pump drive gear.

Because of the odd number of teeth in idler gears, complete timing marks will only occasionally align and cannot therefore be used as a timing check.

Make sure the camshaft, auxiliary drive shaft and their thrust washers are properly positioned, then reinstall the timing gear cover and loosely install the retaining cap screws. Center the crankshaft oil seal (or oil seal bore) in cover over the crankshaft, using the Special Tool MFN 747B or crankshaft

pulley and tapered locking ring; then tighten the cover retaining cap screws securely.

Install auxiliary drive shaft gear as outlined in paragraph 27 and camshaft gear as in paragraph 28.

27. AUXILIARY DRIVE SHAFTS AND GEARS. The auxiliary drive shaft is driven by the timing gear train and drives the engine oil pump and

Fig. 48—Timing idler gear withdrawn to show pressure oil passages (O) which provide lubrication to timing gear train.

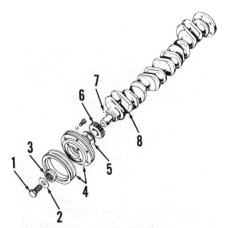

Fig. 46—Crankshaft front pulley (4) is retained by tapered lock ring (3). Refer to text for recommended removal procedure.

1. Cap screw
2. Washer
3. Locking ring
4. Pulley
5. Oil slinger
6. Crankshaft timing gear
7. Woodruff key
8. Crankshaft

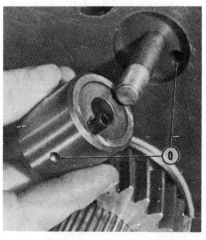

Fig. 49—Pressure oil passage (O) in idler gear hub and block. Refer also to Fig. 48.

injection pump. The auxiliary drive unit is shown exploded in Fig. 50 and in cross section in Fig. 51.

The auxiliary drive shafts and gears are lubricated by the engine oiling system. Renewable fitting (11) provides spray lubrication for drive worm.

To check the auxiliary shaft timing, remove injection pump as outlined in paragraph 57 and turn engine crankshaft until No. 1 piston is at TDC on compression stroke. Auxiliary shaft timing is correct if notches (N—Fig. 52) in pump drive shaft and adapter plate align as shown.

To remove the pump drive gear, unbolt and remove adapter (A). Remove thrust sleeve (16—Fig. 50) from top opening, then lift out gear (18).

If suitable equipment is available for finish honing concentric with housing, the bushing may be renewed in injection pump adapter housing. The lower thrust collar (20) may also be renewed. The adapter housing (upper) bushing should have 0.001-0.0042 inch (0.03-0.11mm) clearance and thrust collar (lower) bushing 0.0016-0.004 inch (0.04-0.10mm) clearance.

Auxiliary shaft thrust washers (5) fit in machined slots in shaft (7). Outer portion of washers fit a machined recess in front face of engine block and are retained by timing gear housing. Normal shaft end play of 0.0025-0.009 inch (0.06-0.23mm) is adjusted by renewing the thrust washers and/or shaft.

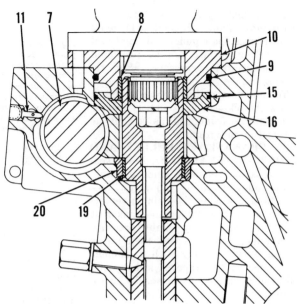

Fig. 51—Cross-sectional view of accessory drive housing and injection pump drive. Oil spray tube (11) directs lubricant to worm on accessory shaft (7). Refer to Fig. 50 for parts identification.

Auxiliary drive shaft should have a diametral clearance of 0.001-0.0042 inch (0.03-0.11mm) in front bushing (12) and 0.001-0.0036 inch (0.03-0.09mm) in rear bushing (13). Bushings are precision type and will not require reaming after installation. Make sure oil holes are aligned when bushings are installed.

Auxiliary shaft can be withdrawn from front after removing timing gear housing as outlined in paragraph 24. When reinstalling the auxiliary shaft, align engine timing marks as shown in Fig. 47 and pump drive shaft timing notch as shown in Fig. 52. The holes in auxiliary shaft drive gear (3—Fig. 50) are slightly elongated for proper alignment. Tighten cap screws (1) to a torque of 20-22 ft.-lbs. when alignment is correct.

CAMSHAFT

28. To remove the camshaft, first remove the rocker arms unit as outlined in paragraph 21 and timing gear housing as in paragraph 24. Unbolt and remove the fuel lift pump from right side of engine block. Remove the valve side covers and secure all cam followers

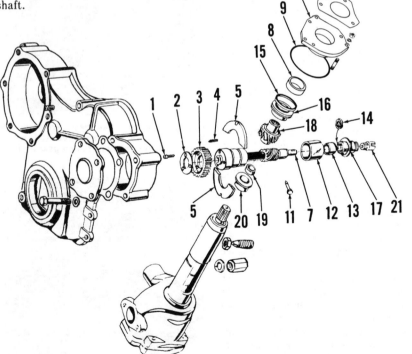

Fig. 50—View of accessory shaft, gears, injection pump drive and oil pump.

1. Screw	7. Accessory shaft	12. Bushing	17. Seal housing
2. Retainer plate	8. Bushing	13. Bushing	18. Worm gear
3. Drive gear	9. "O" ring	14. Oil seal	19. Bushing
4. Dowel pin	10. Pump adapter	15. "O" ring	20. Thrust collar
5. Thrust washer	11. Oil spray tube	16. Thrust sleeve	21. Connector

Fig. 52—Injection pump drive adapter with pump removed showing timing marks. Refer to text for shaft timing procedure.

A. Adapter
M. Pump timing mark
S. Master spline
N. Worm gear timing notches

in their uppermost position; withdraw camshaft forward out of engine block. Normal camshaft bearing clearance is 0.0025-0.0045 inch (0.06-0.11mm) for front bearing; 0.0025-0.0055 inch (0.06-0.14mm) for other journals. Cylinder block bores are unbushed. Camshaft end play of 0.004-0.016 inch (0.10-0.41mm) is controlled by a thrust washer located between front journal and camshaft gear. The thrust washer fits in a recess in front of engine block and is retained by timing gear housing and located by a dowel pin.

When reassembling the engine, reposition idler gear with timing marks aligned as outlined in paragraph 26 and shown in Fig. 47. With timing gear housing reinstalled, turn camshaft and gear until timing punch mark on idler gear is located between the two marks on camshaft gear and draw the gear into position using a suitable puller bolt. Tighten the camshaft gear retaining cap screw to a torque of 50 ft.-lbs. and lock in place by bending the tab washer.

ROD AND PISTON UNITS

29. Connecting rod and piston units are removed from above after removing the cylinder head and oil pan. Cylinder numbers are stamped on the connecting rod and cap. When reinstalling, make sure correlation numbers are in register and face away from camshaft side of engine. The connecting rod is drilled on turbocharged engines, and upper and lower connecting rod bearings are not interchangeable. Bearings are available in standard size and undersizes of 0.010, 0.020 and 0.030 inch. Some early crankshafts are "Nitrided" and can be identified by number 31322306, 31322307 or 31322317 stamped near the front end; usually on the number 1 web. Later crankshafts marked 31322309 and 31322311 are "Tufftrided". Regrinding any of these

crankshaft journals is recommended ONLY if the crankshaft can be retreated. DO NOT regrind shaft unless retreatment is possible. When installing connecting rod caps, use new self-locking nuts. Cadmium plated self-locking nuts should be torqued to 70 ft.-lbs.; plain self-locking nuts should be tightened to 90 ft.-lbs. torque.

PISTONS, SLEEVES AND RINGS

30. The aluminum alloy, cam ground pistons are supplied in a standard size only and are available in a kit consisting of piston, pin and rings for one cylinder. The toroidal combustion chamber is offset in piston crown as shown in Fig. 53 and piston is marked "FRONT" for proper assembly.

Each piston is fitted with a plain faced chrome top ring which may be installed either side up. The internally stepped 2nd & 3rd compression rings must be installed with groove toward top. Second and third compression rings are also chrome faced. Install the three piece oil control ring in the bottom (fourth) groove. Faces of rails are chrome plated. Specifications are as follows:

End Gap—
　Compression rings . . .0.016-0.021-inch
　　　　　　　　　　　　　0.41-0.53mm
Side Clearance in Groove—
　Top Ring 0.002-0.004-inch
　　　　　　　　　　　　　0.051-0.100mm
　Second and
　Third Rings 0.0019-0.0039-inch
　　　　　　　　　　　　　0.05-0.10mm
　Bottom Ring 0.0025-0.0045-inch
　　　　　　　　　　　　　0.06-0.11mm

The production cylinder sleeves are 0.001-0.003-inch press fit in cylinder block bores. Service sleeves are a transition fit (0.001 tight to 0.001 inch clearance). When installing new sleeves, make sure sleeves and bores are absolutely clean and dry, then chill the sleeves and press fully into place by hand. When properly installed, liner should extend 0.030-0.035 inch (0.76-0.89mm) above gasket surface of cylinder block as shown in Fig. 54.

PISTON PINS

31. The full floating piston pins are retained in piston bosses by snap rings. Piston pin diameter is 1.4998-1.50 inches (38.09-38.10mm) and are available in standard size only. On all models, the renewable connecting rod bushing must be final sized to provide a diametral clearance of 0.00075-0.0017 inch (0.0191-0.0432mm) for the pin. Be sure the predrilled oil hole (or holes) in bushings are aligned with holes in connecting rod and install bushing from chamfered side of rod bore. Piston pin should be a thumb press fit in piston after piston is heated to 160 degrees F.

CONNECTING RODS AND BEARINGS

32. Connecting rod bearings are precision type, renewable from below after removing oil pan and connecting rod bearing caps. When renewing bearing shells, be sure that the projection engages milled slot in rod and cap and that the correlation marks are in register and face away from camshaft side of engine. The upper half of bearing is equipped with an oil hole to provide pressure lubrication to the drilled rod, and upper and lower shells are not interchangeable.

Bearings are available in standard size and undersizes of 0.010, 0.020 and 0.030 inch. Resizing bearing journals for undersize bearings is recommended ONLY if specially treated ("Nitrided" or "Tufftrided") crankshaft can be retreated. Crankshafts with number 31322306, 31322307 or 31322317 stamped near the front end, usually on number 1 web, are "Nitrided". Crankshafts with number 31322309 or 31322311 are "Tufftrided". DO NOT regrind any crankshaft journal on these engines unless retreatment is possible.

Fig. 53—Top view of piston showing combustion chamber recess and piston markings.

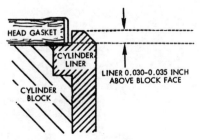

Fig. 54—Cross-sectional view of cylinder block showing cylinder liner properly installed.

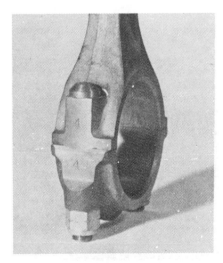

Fig. 55—Assembled view of serrated connecting rod and cap.

Connecting rod bearings should have a diametral clearance of 0.0015-0.003 inch (0.04-0.08mm) on the 2.499-2.4995 inch (63.47-63.49mm) diameter crankpin. Recommended connecting rod side clearance is 0.0095-0.0145 inch (0.24-0.37mm). Renew the self-locking connecting rod nuts. Cadmium plated nuts should be tightened to 70 ft.-lbs. torque; plain rod nuts should be tightened to 90 ft.-lbs. torque.

CRANKSHAFT AND BEARINGS

33. The crankshaft is supported in seven precision type main bearings. To remove the rear main bearing cap, it is first necessary to remove the engine, clutch, flywheel and rear oil seal. All other main bearing caps can be removed after removing the oil pan.

Upper and lower halves of bearing inserts are not interchangeable. The upper (block) half is slotted to provide pressure lubrication to crankshaft and connecting rods. Inserts are interchangeable in pairs for all journals except the center main bearing. When installing the thrust washers, make sure the steel back is positioned next to block and cap and the grooved, bronzed surface next to crankshaft thrust faces.

Bearing inserts are available in undersizes of 0.010, 0.020 and 0.030 inch as well as standard; and thrust washers are available in standard thickness of 0.089-0.091 inch (2.26-2.31mm). Regrinding is recommended ONLY if shaft journals can be retreated. Crankshafts with number 31322306, 31322307 or 31322317 stamped near front end, usually on number 1 web, are "Nitrided". Crankshafts with number 31322309 or 31322311 are "Tufftrided". DO NOT regrind any of the crankshaft journals unless retreatment is possible.

Recommended main bearing diametral clearance is 0.0025-0.0045 inch (0.06-0.11mm) and recommended crankshaft end play is 0.002-0.014 inch (0.05-0.36mm;. Tighten the main bearing retaining capscrews to a torque of 180 ft.-lbs. and secure by bending the tabs on locking washers. When renewing rear main bearing, refer to paragraph 34 for installation of rear seal and oil pan bridge pieces. Check the crankshaft against the values which follow:

Main Journal Standard Diameter—
   Inches 2.9985-2.999
   mm .76.16-76.17
Crankpin Journal Standard Diameter—
   Inches 2.499-2.4995
   mm .63.47-63.49

CRANKSHAFT REAR OIL SEAL

34. The lip type rear oil seal is contained in a one-piece housing attached to rear face of engine block as shown in Fig. 56. The seal retainer can be removed after removing flywheel. The lip seal is pressed into housing.

When installing the cylinder block bridge piece at front or rear, lightly coat metal-to-metal contact faces with a

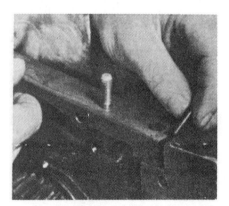

Fig. 57—Installing end seals in cylinder block bridge piece. Refer to text.

non-hardening gasket cement and insert the end seals as shown in Fig. 57; then use a straight edge as shown in Fig. 58 to make sure gasket faces are flush.

FLYWHEEL

35. To remove the flywheel, first separate engine from transmission housing and remove the clutch. Flywheel is secured to crankshaft mounting flange by six evenly spaced cap screws which also retain the pto shaft drive hub. The flywheel is not doweled. A seventh, unthreaded hole is drilled in crankshaft flange and a matching hole in flywheel rim; when these two holes are aligned, flywheel timing marks will be properly positioned.

The starter ring gear can be renewed after flywheel is removed. Heat ring gear evenly to approximately 475 degrees F. and install on flywheel with beveled end of teeth facing front of engine.

Check flywheel runout with a dial indicator after flywheel is installed. Maximum allowable flywheel runout is 0.001 inch for each inch (0.01mm for each cm) that measurement is taken from flywheel centerline. Tighten the flywheel retaining cap screws to a torque of 80 ft.-lbs. and secure with special tab washers.

ADAPTER PLATE

36. The engine adapter plate (Fig. 59) can be removed after removing flywheel as outlined in paragraph 35. The adapter plate is positively located on engine block by dowel pins and secured by six cap screws.

OIL PAN

37. The heavy cast-iron oil pan serves as part of the tractor frame and attaching point for the tractor front support. To remove the oil pan,

Fig. 56—Rear view of cylinder block and crankshaft showing rear oil seal housing.

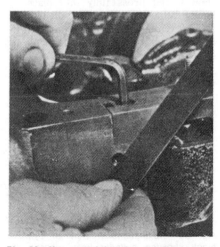

Fig. 58—Use a straight edge to align gasket faces of cylinder block and bridge piece.

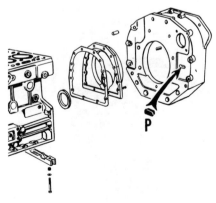

Fig. 59—View showing rear of block, oil seal housing and flywheel adapter. Flywheel timing marks are viewed through hole for plug (P).

Fig. 60—View showing blocking points when removing oil pan. Refer to text.

support tractor underneath the transmission housing and position a jack under front end to release pressure from ends of oil pan. Refer to Fig. 60. Drain oil from the oil pan, then remove pan retaining screws while using a rolling floor jack to lower the oil pan away from engine block. Install by reversing the removal procedure. Tighten the large screws at rear of pan to 360-540 ft.-lbs. torque.

OIL PUMP

38. The rotary type oil pump is mounted on lower side of engine block and driven by the injection pump drive shaft. Refer to Fig. 61 for an exploded view of oil pump and associated parts.

To remove the oil pump after oil pan is off, first unbolt and remove intake tube (9) and screen (11). Remove the cap screws retaining delivery tube (5) and housing (4) to cylinder block. Loosen or remove nut, then remove set screw (1) from outside of engine block. Withdraw oil pump (2), delivery housing (4) and associated parts as a unit.

Service parts for the oil pump are not available individually. Pump cover and rotors may be removed for inspection or cleaning. Inner rotor is retained

to shaft by pins and removal from shaft is not advised. Be sure to install outer rotor in pump body with chamfered outer edge to bottom of gear pocket, away from pump cover. Install oil pump by reversing the removal procedure. Make sure tapered end of set screw (1) fully locates the pump.

RELIEF VALVE

39. The piston type relief valve (6—Fig 61) is contained in oil delivery housing (4). The relief valve is set to open at 50-60 psi. Oil delivery housing and relief valve parts are available individually.

DIESEL FUEL SYSTEM (6 Cylinder Models)

The diesel fuel system consists of three basic units; the fuel tank and filters, injection pump and injector nozzles. When servicing any unit associated with the diesel fuel system, the maintenance of absolute cleanliness is of utmost importance. Of equal importance is the avoidance of nicks and burrs on any of the working parts.

Probably the most important precaution that service personnel can impart to owners of diesel powered tractors is

to urge them to use an approved fuel that is absolutely clean and free from foreign material. Extra precaution should be taken to be sure that no water enters the fuel storage tanks. Because of the high pressures and degree of control required of injection equipment, extremely high precision standards are necessary in the manufacture and servicing of diesel components. Extra care in daily maintenance will pay big dividends in extended service life and the avoidance of costly repairs.

FUEL FILTERS AND LINES

45. **OPERATION AND MAINTEN-ANCE.** A much greater volume of fuel is circulated within the system than is burned in the engine, the excess serving as a coolant and lubricant for the injection pump. Fuel enters the primary filter through inlet connection where it passes through the water trap and first stage filter element. Both lines leading to injection pump and the system return line enter a common passage in secondary filter and are separated from the filter inlet only by the filtering element. The greater volume of filtered fuel is thus recirculated between the secondary filter and injection pump, with the inlet line supplying little more fuel to the system than is actually burned in the engine.

Inspect the glass bowl at bottom of primary filter daily and drain off any water or dirt accumulation. Completely drain the primary filter at 100 hour intervals and renew the element each 500 hours. Renew element in secondary filter every 1000 hours. Renew both elements and clean the tank and lines if evidence of substantial water contamination exists.

46. **BLEEDING.** To bleed the system, make sure tank shut-off is open, have an assistant actuate the manual lever on fuel lift pump on right side of engine block, and proceed as follows:

Loosen the air vent (1—Fig. 65) on filter and continue to operate the lift

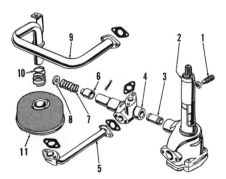

Fig. 61—Partially exploded view of oil pump and associated parts.

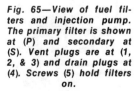

1. Set screw	
2. Oil pump	7. Valve spring
3. Delivery pipe	8. Valve cap
4. Delivery housing	9. Intake pipe
5. Pressure pipe	10. Spring
6. Relief valve piston	11. Intake screen

Fig. 65—View of fuel filters and injection pump. The primary filter is shown at (P) and secondary at (S). Vent plugs are at (1, 2, & 3) and drain plugs at (4). Screws (5) hold filters on.

pump until air-free fuel flows from vent plug hole. Tighten plug (1); then loosen vent plugs (2 and 3) in order, while continuing to operate the fuel lift pump. Tighten each plug as air is expelled, and proceed to the next.

NOTE: Air in governor housing relieved by bleed screw (3) will not prevent tractor from starting and running properly; however, condensation in the trapped air can cause rusting of governor components and eventual pump malfunction. Do not fail to bleed governor housing even though the tractor starts and runs properly.

Operate the manual lever on fuel lift pump for approximately ten extra strokes after tightening vent plug (3), to expel any air remaining in bleed-back lines to filter.

With the fuel supply system bled, push in the stop button, partially open throttle lever and attempt to start the tractor. If tractor fails to fire, loosen compression nut at all injector nozzles and turn engine over with starter until fuel escapes from all loosened connections. Tighten compression nuts and start engine.

FUEL LIFT PUMP

47. The fuel lift pump (Fig. 66) is mounted on right side of engine block and driven by the camshaft. Pump parts are available separately or in rebuilding kit. Output delivery pressure should be 5-8 psi.

INJECTOR NOZZLES

All models are equipped with C.A.V. multi-hole nozzles which extend through the cylinder head to inject the fuel charge into combustion chamber machined in crown of piston.

WARNING: Fuel leaves the injector nozzle with sufficient force to penetrate the skin. Keep exposed portions of your body clear of nozzle spray when testing.

48. **TESTING AND LOCATING A FAULTY NOZZLE.** If rough or uneven engine operation or misfiring indicates a faulty injector, the defective unit can usually be located as follows:

With engine running at the speed where malfunction is most noticeable (usually slow idle speed), loosen the compression nut on high pressure line for each injector nozzle in turn, and listen for a change in engine performance. As in checking spark plugs, the faulty unit is the one which, when its line is loosened, least affects the running of the engine.

If a faulty nozzle is found and considerable time has elapsed since the injectors have been serviced, it is recommended that all nozzles be removed and serviced or that new or reconditioned units be installed. Refer to the following paragraphs for removal and test procedure.

49. **REMOVE AND REINSTALL.** Before loosening any fuel lines, thoroughly clean the lines, connections, injectors and surrounding engine area with air pressure and solvent spray. Disconnect and remove the leak-off line, disconnect pressure line and cap all connections as they are loosened, to prevent dirt entry into the system. Remove the two stud nuts and withdraw injector unit from cylinder head.

Thoroughly clean the nozzle recess in cylinder head before reinstalling the injector unit. It is important that seating surface be free of even the smallest particle of carbon or dirt which could cause the injector unit to be cocked and result in blow-by. No hard or sharp tools should be used in cleaning. Do not re-use the copper sealing washer located between injector nozzle and cylinder head, always install a new washer. Each injector should slide freely into place in cylinder head without binding. Make sure that dust seal is reinstalled and tighten the retaining stud nuts to a torque of 10-12 ft.-lbs. After engine is started, examine injectors for blow-by, making the necessary corrections before releasing the tractor for service.

50. **TESTING.** A complete job of testing and adjusting the injector requires the use of special test equipment. Only clean, approved testing oil should be used in the tester. The nozzle should be tested for opening pressure, seat leakage, back leakage and spray pattern. When tested, the nozzle should open with a sharp popping or buzzing sound, and cut off quickly at end of injection with a minimum of seat leakage and controlled amount of back leakage.

Before conducting the test, operate tester lever until fuel flows, then attach the injector. Close the valve to tester gage and pump tester lever a few quick strokes to be sure nozzle valve is not plugged, that four sprays emerge from nozzle tip, and that possibilities are good that injector can be returned to service without overhaul.

NOTE: Spray pattern is not symmetrical with centerline of nozzle tip, but forms a "tilted" pattern as shown in Fig. 68. Hole location is engineered to

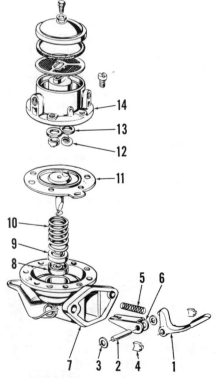

Fig. 66—Exploded view of fuel lift pump.

1. Actuating lever	8. Seal
2. Shaft	9. Retainer
3. Washer	10. Spring
4. Retainer	11. Diaphragm
5. Spring	12. Valve
6. Link	13. Gasket
7. Body	14. Cover

Fig. 67—A suitable injector tester is required to completely test and adjust the injector nozzles.

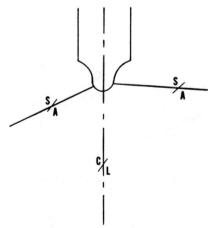

Fig. 68—Nozzle spray pattern is not symmetrical with centerline of nozzle tip.

provide the correct spray pattern in the combustion chamber.

If adjustment is indicated in the preliminary tests, proceed as follows:

51. OPENING PRESSURE. Open the valve to tester gage and operate tester lever slowly while observing gage reading. Opening pressure should be 2645 psi (180 atm) on MF1105 models; 2940 psi (200 atm) on MF1135 models. If opening pressure is not as specified, remove the injector cap nut (1—Fig. 69), loosen locknut (3) and turn adjusting sleeve (5) as required to obtain the recommended pressure.

NOTE: When adjusting a new injector or an overhauled injector with a new pressure spring (6), set the pressure at 2870 psi (195 Atm) for MF1105 Models; 3160 psi (215 Atm) for MF1135 Models. The additional pressure is to allow for initial pressure loss as the spring takes a set.

52. SEAT LEAKAGE. The nozzle tip should not leak at a pressure less than 2445 psi (165 atm) for MF1105 models; 2740 psi (185 atm) for MF1135 models. These pressures are 200 psi less than recommended opening pressure. To check for leakage, actuate tester lever slowly and as the gage needle approaches the suggested test pressure, observe the nozzle tip. Hold the pressure for 10 seconds; if drops appear or nozzle tip is wet, the valve is not seating and injector must be disassembled and overhauled as outlined in paragraph 55.

53. BACK LEAKAGE. If nozzle seat as tested in paragraph 52 was satisfactory, check the injector and connections for wetness which would indicate external leakage. If no leaks are noted, bring gage pressure to 2250 psi (150 atm), release the lever and observe the time for gage pressure to drop from 2250 psi (150 atm) to 1500 psi (100 atm). For a nozzle in good condition, this time should not be less than six seconds. A faster drop would indicate a worn or scored nozzle valve piston or body, and the nozzle assembly should be renewed.

NOTE: Leakage of the tester check valve or connections will cause a false reading, showing up in this test as excessively fast leakback. If all injectors tested fail to pass the test, the tester rather than the units should be suspected as faulty.

54. SPRAY PATTERN. If leakage and pressure are as specified when tested as outlined in paragraphs 51 through 53, operate the tester handle several times while observing the spray pattern. Four finely atomized, equally spaced, conical sprays should emerge from nozzle tip, with equal penetration into the surrounding atmosphere.

If pattern is uneven, ragged or not finely atomized, overhaul the nozzle as outlined in paragraph 55.

NOTE: Spray pattern is not symmetrical with centerline of nozzle tip, but tilted as shown in Fig. 68.

55. OVERHAUL. Hard or sharp tools, emery cloth, grinding compound, or other than approved solvents or lapping compounds must never be used. An approved nozzle cleaning kit is available through any C.A.V. Service Agency and other sources.

Wipe all dirt and loose carbon from exterior of nozzle and holder assembly. Refer to Fig. 69 and proceed as follows:

Secure the nozzle in a soft-jawed vise or holding fixture and remove the cap nut (1). Loosen jam nut (3) and back off the adjusting sleeve (5) to completely unload the pressure spring (6). Remove the nozzle cap nut (12) and nozzle body (11). Nozzle valve (10) and body (11) are matched assemblies and must never be intermixed. Place all parts in clean calibrating oil or diesel fuel as they are removed. Clean the exterior surfaces with a soft wire brush, soaking in an approved carbon solvent if necessary, to loosen hard carbon deposits. Rinse the parts in clean diesel fuel or calibrating oil immediately after cleaning, to neutralize the carbon solvent and prevent etching of polished surfaces. Clean the pressure chamber of nozzle tip using the special reamer as shown in Fig. 70. Clean the spray holes in nozzle with an 0.010 inch (0.24mm) wire probe held in a pin vise as shown in Fig. 71. Wire probe should protrude from pin vise only far enough to pass through spray holes (approximately 1/16 inch), to prevent bending and breakage. Rotate pin vise without applying undue pressure.

Clean valve seats by inserting small end of brass valve seat scraper into nozzle and rotating tool. Reverse the tool and clean upper chamfer using large end. Refer to Fig. 72. Use the hooked scraper to clean annular groove in top of nozzle body (Fig. 73.) Use the same hooked tool to clean the internal fuel gallery.

With the above cleaning accomplished, back flush the nozzle by installing the reverse flusher adapter on injector tester and nozzle body in adapter, tip end first. Secure with the knurled adapter nut and insert and rotate the nozzle valve while flushing. After nozzle is back flushed, seat can

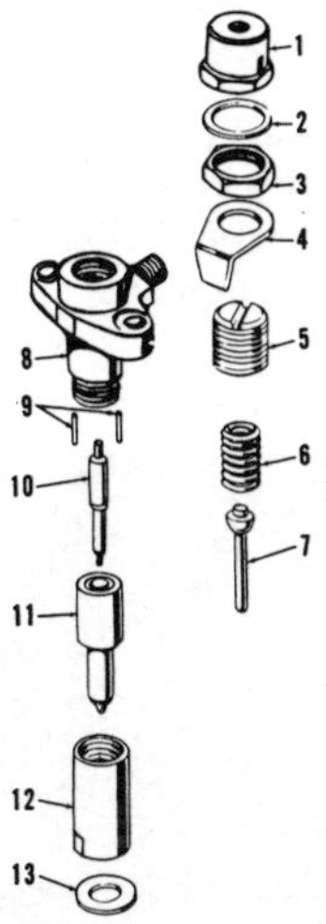

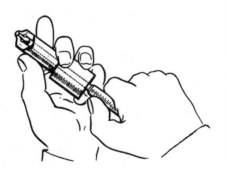

Fig. 70—Clean the pressure chamber in nozzle tip using the special reamer as shown.

Fig. 69—Exploded view of C.A.V. injector nozzle and holder assembly. Correct opening pressure is indicated on tab (4).

1. Cap nut	
2. Gasket	8. Nozzle holder
3. Jam nut	9. Dowels
4. Tab	10. Nozzle valve
5. Adjusting sleeve	11. Nozzle body
6. Spring	12. Nozzle nut
7. Valve spindle	13. Seat washer

Fig. 71—Clean spray holes in nozzle tip using a pin vise and 0.010 inch wire probe.

Fig. 72—Clean the valve seat using brass scraper as shown.

be polished using a small amount of tallow on end of polishing stick, rotating the stick as shown in Fig. 74.

Light scratches on valve piston and bore can be polished out by careful use of special injector lapping compound only, DO NOT use valve grinding compound or regular commercial polishing agents. DO NOT attempt to reseat a leaking valve using polishing compound. Clean thoroughly and back flush if lapping compound is used.

Reclean all parts by rinsing thoroughly in clean diesel fuel or calibrating oil and assemble valve to body while immersed in the cleaning fluid. Reassemble the injector while still wet. With adjusting sleeve (5—Fig. 69) loose, reinstall nozzle body (11) to holder (9), making sure valve (10) is installed and locating dowels aligned as shown in Fig. 75. Tighten nozzle cap nut (12—Fig. 69) to a torque of 50 ft.-

lbs. Do not overtighten; distortion may cause valve to stick and overtightening cannot stop a leak caused by scratches or dirt on the lapped metal surfaces of

Fig. 76—Installed view of injection pump showing slow idle stop screw (1). High speed stop screw is sealed when pump is adjusted on test stand.

Fig. 77—Pump should be properly timed when timing scribe lines (T) are aligned as shown. Timing window (C) can be removed and timing checked as outlined in text if conditions warrant.

valve body and nozzle holder.

Retest and adjust the assembled injector assembly as outlined in paragraphs 51 through 54.

NOTE: If overhauled injector units are to be stored, it is recommended that a calibrating or preservative oil rather than diesel fuel be used for the pre-storage testing. Storage of injectors containing diesel fuel for more than thirty days may result in the necessity of recleaning prior to use.

INJECTION PUMP

The injection pump is a completely sealed unit. No service work of any kind should be attempted on the pump or governor unit without the use of special pump testing equipment and training. Inexperienced or unequipped service personnel should never attempt to overhaul a diesel injection pump.

56. **ADJUSTMENT.** The slow idle stop screw (1—Fig. 76) should be adjusted with engine warm and running, to provide the recommended slow idle speed of 800-850 rpm. Also check to make sure that governor arm contacts both stop screws when hand lever is moved through full travel. Also check to make sure that stop lever arm moves fully to operating position when stop button is pushed in, and shuts off the fuel when stop button is pulled. The high speed stop screw is set at the factory and the adjustment is sealed. Governed speed under load should be 2200 rpm, with a high idle (no-load) speed of 2325-2450 rpm. Refer to paragraph 58 for pump timing adjustment.

57. **REMOVE AND REINSTALL.** Before attempting to remove the injection pump, thoroughly wash the pump and connections with clean diesel fuel or an approved solvent. Disconnect throttle control rod from governor arm and disconnect and remove stop cable and bracket. Disconnect high pressure

Fig. 73—Use the hooked scraper to clean annular groove in top of nozzle body.

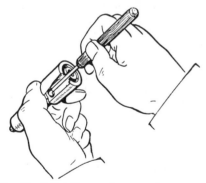
Fig. 74—Polish the nozzle seat using mutton tallow on a wood polishing stick.

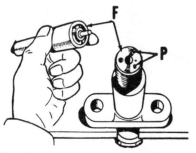
Fig. 75—Make sure locating dowels are carefully aligned when nozzle body is reinstalled.

Fig. 78—Timing marks are stamped on front face of engine flywheel as shown and are visible through hole in adapter housing in left side. The 26 degree mark is used for MF 1105 models; 30 degree mark is for MF 1135 tractors.

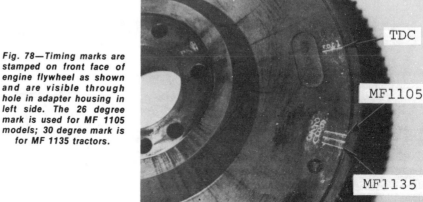
TDC

MF1105

MF1135

fuel lines at both ends, unbolt bracket from cylinder head, then lift off bracket and lines as a unit. Cap all connections as lines are removed. Disconnect and cap fuel inlet and return lines at pump, remove the remaining attaching stud nut and lift off the pump.

Normal installation of injection pump can be accomplished without reference to crankshaft timing marks or internal timing marks on injection pump. Be sure timing scribe lines (T—Fig. 77) are aligned when pump is installed. The master spline on pump drive shaft aligns with splines in pump to assure proper timing if auxiliary shafts are correctly installed as outlined in paragraph 27. Bleed the fuel system after installation as outlined in paragraph 46. If necessary to check injection pump timing, refer to paragraph 58.

58. PUMP TIMING TO ENGINE. Because of the master splines on pump drive shaft, pump timing will be correct if gears are properly installed and timing marks (T—Fig. 77) are aligned. To check gear installation, it is necessary to remove the injection pump as outlined in paragraph 57, then proceed as in paragraph 27.

Static timing (end of injection) should occur at 26 degrees BTDC for MF1105 models; 30 degrees BTDC for MF1135 models. The static timing mark for No. 1 cylinder is stamped on front face of flywheel as shown in Fig. 78. These marks are visible through the timing window in adapter housing on left side of engine block. Proceed as follows to check injection pump timing without removing the pump.

Turn engine over by hand until No. 1 piston is on compression stroke and the indicated "Static Timing" mark is aligned in timing window. Remove the timing cover (C—Fig. 77) from injection pump body; the "F" mark on pump rotor should be aligned with the square end of injection pump snap ring. Minor adjustment can be made by loosening the injection pump mounting stud nuts and shifting the pump mounting flange. To make a major timing adjustment, retime the auxiliary drive shaft as outlined in paragraph 27.

TURBOCHARGER
(6 Cylinder Models)

The six cylinder engines used in MF1105 and MF1135 models are equipped with exhaust driven turbochargers. The turbocharger used on MF1105 models is an Airesearch T-04 which provides 6-8 psi (12-16 inches of mercury) boost pressure at 2200 rpm with engine fully loaded. Model MF1135

tractors are equipped with a Holset 3LD41.6 turbocharger which provides 10-13 psi (20-26 inches of mercury) boost pressure at 2200 rpm with engine fully loaded. Refer to the appropriate following paragraphs for service.

TROUBLE SHOOTING

59. The following table can be used for locating difficulty with the turbocharger. Many of the probable causes are related to the engine air intake and exhaust systems which should be maintained in good condition at all times. Engine oil is used to cool and lubricate the turbocharger and damage can result from improper operation. The turbocharger should be allowed to cool by idling engine at approximately 1000 rpm for 2-5 minutes after engine has been operated under load. The engine should always be immediately restarted if it is killed while operating at full load. The oil in the turbocharger will be hardened by the intense heat if not allowed to cool and damage is sure to result. Use extreme care to avoid damaging any of the turbocharger moving parts when servicing.

SYMPTOMS	PROBABLE CAUSES
Engine Lacks Power	1,4,5,6,7,8,9,10 11,18,20,21,22,25,26,27,28,29,30
Black Exhaust Smoke	1,4,5,6,7,8,9,10 11,18,20,21,22,25,26,27,28,29,30
Blue Exhaust Smoke	1,4,8,9,19,21,22, 32,33,34,36
Excessive Oil Consumption	2,8,17,19, 20,33,34,36
Excessive Oil in the Turbine End	2,7,8, 16,17,19,20,22,32,33,34,36
Excessive Oil in the Compressor End	1,2,4,5,6,8,9,16,19, 20,21,33,36
Insufficient Lubrication	15,16,12,23, 24,31,36
Oil In Exhaust Manifold	2,7,19,20,22, 28,29,30,33,34

Damaged Compressor Wheel	3,6,8, 20,21,23,24,36
Damaged Turbine Wheel	7,8,18,20, 21,22,34,36
Drag or Bind in Rotating Assy	3,6,7,8, 13,14,15,16,20,21,22,31,34,36
Worn Bearings, Journals, Bearing Bores	6,7,8,12,13,14,15,16, 20,23,24,31,35,36
Noisy Operation	1,3,4,5,6,7,8,9,10, 11,18,20,21,22
Sludged or Coked Center Housing	2,15,17

KEY TO PROBABLE CAUSES

1. Dirty air cleaner element.
2. Plugged crankcase breathers.
3. Air cleaner element missing, leaking, not sealing correctly, loose connections to turbocharger.
4. Collapsed or restricted air tube before turbocharger.
5. Restricted (damaged) crossover pipe from turbocharger to inlet manifold.
6. Foreign object between air cleaner and turbocharger.
7. Foreign object in exhaust system from engine (check engine).
8. Turbocharger flanges, clamps or bolts loose.
9. Inlet manifold cracked, gaskets missing, connections loose.
10. Exhaust manifold cracked, burned, gaskets blown or missing.
11. Restricted exhaust system.
12. Oil lag (oil delay to turbocharger at start up).
13. Insufficient lubrication.
14. Lubricating oil contaminated.
15. Improper type lubricating oil used.
16. Restricted oil feed line.
17. Restricted oil drain line.
18. Turbine housing damaged or restricted.
19. Turbocharger seal leakage.
20. Worn journal bearings.
21. Excessive dirt buildup in compressor housing.

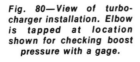

Fig. 80—View of turbocharger installation. Elbow is tapped at location shown for checking boost pressure with a gage.

22. Excessive carbon buildup behind turbine wheel.
23. Too fast acceleration at initial start (oil lag).
24. Too little warm-up time.
25. Fuel pump malfunction.
26. Worn or damaged injectors.
27. Valve timing.
28. Burned valves.
29. Worn piston rings.
30. Burned pistons.
31. Leaking oil feed line.
32. Excessive engine pre-oil.
33. Excessive engine idle.
34. Coked or sludged center housing.
35. Oil pump malfunction.
36. Oil filter plugged.

TURBOCHARGER UNIT

All Models

60. **REMOVE AND REINSTALL.** Remove hood and disconnect air inlet housing from turbocharger. Remove turbocharger air intake pipe, outlet tube and exhaust outlet elbow. Disconnect oil inlet and return lines from turbocharger, then unbolt and remove unit from exhaust manifold.

To reinstall, reverse removal procedure. Leave oil return line disconnected and crank engine with starter until oil begins to flow from return port, then tighten return line.

NOTE: Do not start engine until it is certain that turbocharger is receiving lubricating oil.

MF1105 Models (Airesearch)

61. **OVERHAUL.** Remove turbocharger unit as outlined in paragraph 60. Mark across compressor housing (Fig. 81), center housing and turbine housing to aid alignment when assembling.

CAUTION: Do not rest weight of any parts of impeller on turbine blades.

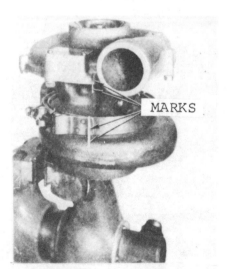

Fig. 81—Mark parts as shown to facilitate alignment when assembling.

Weight of only the turbocharger unit is enough to damage the blades.

Remove clamp (3—Fig. 82), compressor housing (1) and diffuser (2). Remove screws (T), lock plates (19L) and clamp plates (19C); then, remove turbine housing (21). Hold turbine shaft from turning using the appropriate type of wrench at center of turbine wheel (20) and remove locknut (4).

NOTE: Use a "T" handle to remove locknut in order to prevent bending turbine shaft.

On some turbine shafts, an allen wrench must be used at turbine end while others are equipped with a hex and can be held with a standard socket. Lift compressor impeller (5) off, then remove center housing from turbine shaft while holding shroud (18) onto center housing. Remove back plate retaining screw (C), then remove back plate (6), thrust bearing (11), thrust collar (8) and spring (9). Carefully remove bearing retainers (12) from ends and withdraw bearings (13 & 15).

CAUTION: Be careful not to damage bearings or surface of center housing when removing retainers. The center two retainers do not have to be removed unless damaged or unseated. Always renew bearing retainers if removed from grooves in housing.

Clean all parts in a cleaning solution which is not harmful to aluminum. A stiff brush and plastic or wood scraper should be used after deposits have softened. When cleaning, use extreme caution to prevent parts from being nicked, scratched or bent.

Inspect bearing bores in center housing (14—Fig. 82) for scored surface, out of round or excessive wear. Bearing bore diameter should be 0.6220-0.6223 inch (15.80-15.81mm) and must not exceed 0.6228 inch (15.82mm). Make certain bore in center housing is not grooved in area where seal (16) rides. Inside diameter of bearings (13 & 15) should be 0.4010-0.4014 inch (10.19-10.20mm) and outside diameter should be 0.6182-0.6187 inch (15.70-15.71mm). These bearings should be renewed each time unit is disassembled for service. Thrust bearing (11) should be measured at three locations around collar bore. Thickness should be 0.1716-0.1720 inch (4.36-4.37mm) and should be renewed each time unit is disassembled for service. Inside diameter of bore in backplate (6) should be 0.4995-0.5005 inch (12.69-12.71mm) and must not exceed 0.5010 inch (12.73mm). Thrust surface and seal contact area must be clean and smooth. Compressor impeller (5) must not show signs of rubbing with either the compressor housing (1) or the back plate (6). Make certain that impeller blades are not bent, chipped, cracked or eroded. Oil passage in

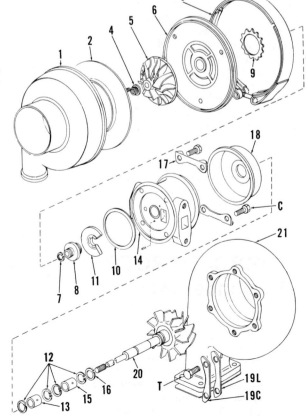

Fig. 82—Exploded view of Airesearch turbocharger used on MF 1105 models.

C. Backplate screws
T. Turbine housing screws
1. Compressor housing
2. Diffuser
3. Clamp
4. Locknut
5. Compressor impeller
6. Backplate
7. Seal ring
8. Thrust collar
9. Spring
10. Seal ring
11. Thrust bearing
12. Bearing retainers
13. Bearing
14. Center housing
15. Bearing
16. Seal ring
17. Lock plates
18. Turbine shroud
19C. Clamp plates
19L. Lock plates
20. Turbine wheel and shaft
21. Turbine housing

thrust collar (8) must be clean and thrust faces must not be warped or scored. Ring groove shoulders must not have step wear. Bearing area width should be 0.1740-0.1748 inch (4.42-4.44mm) and should not exceed 0.1752 inch (4.45mm). Width of groove for seal ring (7) should be 0.064-0.065 inch (1.63-1.65mm) and should not exceed 0.066 inch (1.68mm). Clearance between thrust bearing (11) and groove in collar (8) must be 0.001-0.004 inch (0.03-0.10mm), when checked at three locations. Inspect turbine shroud (18) for evidence of turbine wheel rubbing. Turbine wheel (20) should not show evidence of rubbing and vanes must not be bent, cracked, nicked, or eroded. Turbine wheel shaft must not show signs of scoring, scratching or overheating. Diameter of shaft journals should be 0.3997-0.4000 inch (10.15-10.16mm) and should not be less than 0.3994 inch (10.14mm). Groove in shaft for seal ring (16) must not be stepped and diameter of hub near seal ring should be 0.682-0.683 inch (17.34-17.35mm) and must not be less than

Fig. 83—View showing method of checking turbine shaft end play. Shaft end play should be checked after unit is cleaned to prevent false reading caused by carbon build-up.

Fig. 84—Turbine shaft radial play is checked with dial indicator through the oil outlet hole and touching shaft at (D—Fig. 85).

0.681 inch (17.30). Check shaft end play and radial clearance when assembling.

If bearing inner retainers (12) were removed, install new retainers using special Kent-Moore tools.

CAUTION: Bore in housing may be damaged if special retainer installing tool is not used.

Oil bearings (13 & 15) and install outer retainers using the special tool. Position the shroud (18) on turbine shaft (20) and install seal ring (16) in groove. Apply a light, even coat of engine oil to shaft journals, compress seal ring (16) and install center housing (14). Install new seal ring (7) in groove of thrust collar (8), then install thrust bearing so that smooth side of bearing (11) is toward seal ring (7) end of collar. Install thrust bearing and collar assembly over shaft, making certain that pins in center housing engage holes in thrust bearing. Install new rubber seal ring (10), make certain that spring (9) is positioned in back plate (6), then install back plate making certain that seal ring (7) is not damaged. Install lock plates (17) and screws (C). Tighten screws (C) to 75-90 in.-lbs. torque and bend lock plates up around heads of screws. Install compressor impeller (5) and make certain that impeller is completely seated against thrust collar (8). Install lock nut (4) to 18-20 in.-lbs. torque, then use a "T" handle to turn lock nut an additional 90 degrees.

CAUTION: If "T" handle is not used, shaft may be bent when tightening nut (4).

Install turbine housing (21) with clamp plates (19C) next to housing, tighten screws (T) to 100-130 in.-lbs. then bend lock plates (19L) up around screw heads.

NOTE: Coat threads of screws (T) with "FEL PRO" or similar high temperature compound before installing.

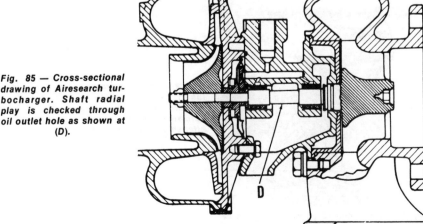

Fig. 85 — Cross-sectional drawing of Airesearch turbocharger. Shaft radial play is checked through oil outlet hole as shown at (D).

Check shaft end play and radial play at this point of assembly. If shaft end play (Fig. 83) exceeds 0.004 inch (0.10mm) thrust collar (8—Fig. 82) and/or thrust bearing (11) is worn excessively. End play of less than 0.001 inch (0.025mm) indicates incomplete cleaning (carbon not all removed) or dirty assembly and unit should be disassembled and cleaned. Refer to Figs. 84 and 85. If turbine shaft radial play exceeds 0.006 inch (0.15mm), unit should be disassembled and bearings, shaft and/or center housing should be renewed. Maximum permissible limits of all of these parts may result in radial play which is not acceptable.

Make certain that legs on diffuser (2—Fig. 82) are aligned with spot faces on back-plate (6) and install diffuser. Install compressor housing (1) and tighten nut of clamp (3) to 40-80 in.-lbs. torque. Fill reservoir with engine oil and protect all openings of turbocharger until unit is installed on tractor.

MF1135 Models (Holset)

62. **OVERHAUL.** Remove turbocharger unit as outlined in paragraph 60. Mark across compressor cover, center housing, turbine housing and clamp as shown in Fig. 81 to aid reassembly. Check axial play of turbine shaft. Normal end play is 0.003-0.005 inch (0.08-0.13mm). Remove clamp (6—Fig. 87) and lift core from the turbine housing (7). Remove the eight screws (3), then separate from compressor cover (20). Clamp a 5/8 inch 12 point wrench in vise, locate hub of turbine wheel (9) in this socket, then remove nut (16). Refer to Fig. 88. Carefully slide compressor wheel (17—Fig. 87) from shaft. The turbine wheel and shaft (9) and bearing (10) can be removed from housing. Use snap ring

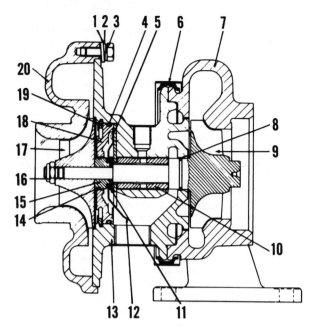

Fig. 87—Cross-section of Holset turbocharger showing location of parts.

1. Washer
2. Lock washer
3. Special screw
4. "O" ring
5. Thrust plate
6. "V" clamp
7. Turbine housing
8. Sealing ring
9. Turbine wheel and shaft
10. Bearing
11. Thrust ring
12. Oil deflector
13. Bearing (center) housing
14. Sealing ring
15. Spacer sleeve
16. Nut
17. Compressor wheel
18. Compressor insert
19. Snap ring
20. Compressor cover

meter, then slide insert (18) with spacer sleeve (15) and "O" ring installed into bore of housing.

NOTE: Use care to prevent spacer sleeve from falling out of insert when installing. Install snap ring (19) with BEVELED SIDE OUT. Slide compressor wheel (17) over shaft, coat threads with graphite grease and install nut (16). Use a wrench in vise as shown in Fig. 88 to hold turbine shaft. Tighten nut to 13 ft.-lbs. torque, then spin rotor wheels. The turbine and compressor wheels should spin freely with no rubbing or binding. Disassemble and recheck unit if any binding or rubbing is noted. Reassemble turbine housing (7), clamp (6) and compressor cover (20) making certain that the previously affixed alignment marks (Fig. 81) are in register. Coat threads and back face of screw and nut on clamp (6—Fig. 87) with graphite grease, then torque nut to 10 ft.-lbs. Coat machined flange of compressor cover (20) with graphite grease, and install using the eight special screws (3), lock washers (2) and flat washers (1). Tighten screws (3) evenly to 5 ft.-lbs. torque. Screws attaching oil

pliers to remove large retaining ring (19), then pry insert (18) from housing bore using two screwdrivers as shown in Fig. 89. Remainder of parts can be lifted out. Oil deflector (12), and thrust plate (5) are located by two groove pins which should not be removed.

Use only commercially approved solvents for cleaning and soak parts long enough to loosen deposits. Use plastic scraper or bristle type brush on aluminum parts. Glass bead dry blast may be used if pressure does not exceed 40 psi. Be sure to clean all parts with clean dry air blast before assembly.

Inspect all parts for scratches, wear,

cracks, bends or any other obvious damage. NEVER attempt to straighten any bent part. Very minor scratches can be tolerated, but install new parts if old part is in any way questionable. The small feed grooves in thrust plate (5—Fig. 87) must be clean and free from obstruction. Always renew "O" ring (4) when assembling.

Use care to not over-expand ring (8) when installing in groove. Lubricate turbine shaft (9) and ring (8), then assemble in bore of housing (13). Lubricate bearing (10) and install over turbine shaft and into housing bore. Lubricate thrust plate (5), then install in housing with bronze side up and holes engaging groove pins. Install thrust ring (11) over shaft and onto thrust plate, then position oil deflector over shaft and groove pins. Install ring (14) in groove of spacer sleeve (15), lubricate parts and slide spacer sleeve into bore of insert (18). Install "O" ring (4) in groove, lubricate outside dia-

Fig. 90—View showing left side of installed turbocharger.

Fig. 89—The compressor insert can be carefully withdrawn using two screwdrivers as shown after snap ring is removed.

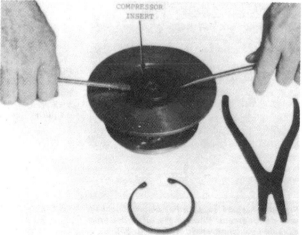

Fig. 88—A 5/8-inch 12 point wrench can be clamped in a vise to help remove and install nut as shown.

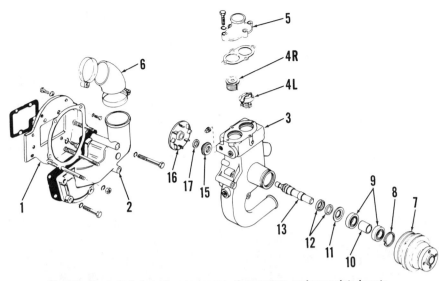

Fig. 91—Exploded view of water pump, thermostats and associated parts.

1. Mounting plate
2. Water connection
3. Water pump housing
4. Thermostats (left & right)
5. Water outlet
6. By-pass hose
7. Fan Pulley
8. Snap ring
9. Bearing
10. Spacer
11. Seal flange
12. Seal and retainer
13. Shaft
15. Water seal
16. Impeller
17. Collar

Fig. 92—The two thermostats are different and must be installed in correct locations as shown.

feed pipe adapter should be torqued to 25-30 ft.-lbs. and the oil drain pipe adapter should be torqued to 60-65 ft.-lbs. if they were removed.

COOLING SYSTEM (6 Cylinder Models)
RADIATOR

65. A 7 psi pressurized cooling system is used on all models. Cooling system capacity is approximately 22 U.S. quarts and an engine oil cooler is built into radiator core.

To remove the radiator, first drain cooling system and remove hood, grille and side panels from both sides. Disconnect both radiator hoses and detach cooling shroud from radiator. Move shroud back toward water pump as far as possible and disconnect the two engine oil cooler lines.

CAUTION: Use two wrenches to disconnect oil cooler lines to prevent damage to radiator.

Disconnect baffle from above radiator being sure to remember where foam pads belong. Detach radiator from brackets on both sides, remove the left side bracket, then withdraw radiator out left side.

When reinstalling be sure to hold engine oil cooler fittings on radiator with one wrench while tightening with another to prevent damage to radiator. Make sure that all of the foam pads are in place to keep out chaff and prevent recirculation of heated air.

THERMOSTAT

66. The two thermostats are contained in the water pump housing as

shown in Fig. 91 and retained by outlet connector (5). The two thermostats are different as shown in Fig. 92. Be sure that they are installed in correct positions as shown.

WATER PUMP

67. To remove the water pump, drain cooling system and remove hood side panels. Disconnect radiator hoses, bypass hose, throttle support (side cover) and temperature indicator sending unit

wire. Remove fan blades and belts. Remove the four nuts and two cap screws attaching pump to it's mounting plate, then slide pump forward off mounting plate and out left side of tractor.

NOTE: The screw on right side may not come completely out unless rocker cover is removed. In any case, be sure that this screw is in place before installing water pump and start this screw first.

Refer to Fig. 91 for exploded view of water pump. When installing, start the by-pass hose on water pump before the pump is completely located.

V-8 ENGINE AND COMPONENTS

All MF1155 models are equipped with a Perkins AV8.540 naturally aspirated, direct injection, V-8 diesel engine. Cylinder bore diameter of 4¼ inches (108mm) and stroke of 4¾ inches (120.7mm) provides 539.1 cubic inch (8834cc) displacement.

R&R ENGINE WITH CLUTCH

71. To remove the engine and clutch as a unit, first drain cooling system and if engine is to be disassembled, drain engine oil from pan. Remove hood, disconnect batteries, hydraulic oil lines (Fig. 95), electrical connector, air cleaner hose and radiator hoses. Remove priority valve oil supply line (Fig. 96) and disconnect the hydraulic pump drive coupling from the crankshaft pulley. Support tractor under engine oil pan until most of the front end weight is removed. Block between front support and axle to prevent tipping, then attach overhead hoist to front end so that front is safely supported (Fig. 97). Remove the six large cap screws

securing front axle support to the engine pan and adaptor plates.

CAUTION: Carefully observe the joint

Fig. 95—View showing oil lines and electrical connector which must be disconnected before removing V8 engine.

while loosening the six screws to be sure that front end and tractor are supported evenly. Change support as necessary if parting surfaces are not parallel.

Support tractor securely under forward end of transmission (NOT FLYWHEEL HOUSING). Disconnect tractormeter cable, fuel shut-off cable, throttle control rod, starter motor wires and oil pressure gage line. Detach rear end of fuel filter support plate from priority valve and disconnect fuel supply line at flexible hose connection. Refer to Figs. 98 and 99. Disconnect wire from the temperature indicator sending unit and electrical connections from alternator. Disconnect fuel return line from top of fuel filter and detach line from base of priority valve support plate. Remove hydraulic oil line retaining clamp (3—Fig. 100). Detach auxiliary solenoid from priority valve support plate, then relocate solenoid and starter wire out of the way. Support weight of engine at all four lift points (Fig. 101).

CAUTION: Balance point will be near rear lift straps because of flywheel, clutch and flywheel housing.

After engine is supported, loosen the bolts attaching engine flywheel housing to transmission. Check while loosening screws to be sure that weight is supported evenly. The flywheel housing is doweled on each side.

Observe the following when reinstalling engine. Install guide studs before sliding engine against transmission housing. Carefully move engine back against transmission housing while aligning splines of clutch with transmission input shaft and splines of flywheel adapter with pto shaft. If splines do not align, DO NOT force together, but turn flywheel slowly while sliding together. Locate the front end assembly, then install the radial piston pump protector plate and start all six screws. Tighten the four screws which attach the oil pan to the front support first. Loosen the screws attaching adaptors to engine slightly, then tighten the two screws attaching adaptors to front support. Tighten the six screws attaching adaptors to engine to 50-80 ft.-lbs. torque. Remainder of assembly is reverse of disassembly. Bleed fuel system as outlined in paragraph 106 and adjust throttle controls as in paragraph 116 after engine is installed.

CYLINDER HEAD

72. The cylinder heads can be removed with manifolds, fuel lines and injectors still attached to head. It is suggested that these items be removed

before removing cylinder head if any work is to be performed on head. Refer to Fig. 102.

To remove cylinder heads, drain cooling system, remove hood, mark the high pressure fuel lines for correct installation, then remove fuel lines and cover the openings to prevent entrance of dirt. Disconnect water outlet hose (1—Fig. 103), temperature gage wire (4) and by-pass hose from outlet elbow (3). Disconnect hose between air cleaner

and inlet manifold and remove compressor air inlet pipe. Remove breather assembly, inlet and exhaust manifolds and the fuel injectors. Remove rocker covers, rocker shaft oil feed pipes (Fig. 104), rocker arms and push rods. Remove cylinder head attaching cap screws and nuts in reverse of order shown in (Fig. 105). Lift head from the two studs.

Cylinder heads are interchangeable from right to left providing necessary

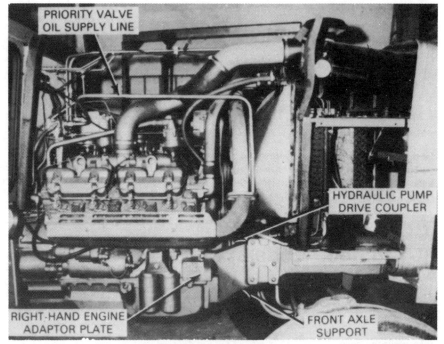

Fig. 96—View of right side of MF 1155 engine identifying some parts which must be disconnected for engine removal.

Fig. 97—View of left side of MF1155 showing the front system being separated from tractor.

attachments are transferred. Cylinder head may be resurfaced providing no more than 0.015 inch (0.38mm) is removed; nozzle protrusion (P—Fig. 107) does not exceed 0.143 inch (3.63mm); or cylinder head thickness (D) is not reduced to less than 3.970 inch (100.84 mm). If cylinder head is re-machined, test for leaks at 30 psi using warm water.

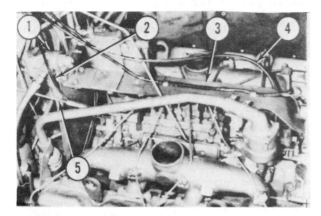

Fig. 98—Top right side view of injection pump controls and associated parts of V8 engine.

1. Priority valve
2. Fuel shut-off cable
3. Fuel filter support plate
4. Tachometer cable
5. Throttle control rod

Fig. 99—View showing components on right side of MF 1155 which must be disconnected before engine can be removed.

Fig. 102—The cylinder heads may be removed with manifolds, fuel lines and injectors in place as shown if cylinder head is not to be worked on. Use extreme care to prevent damaging nozzles and lines.

Fig. 100—View of left side of V8 engine showing wires and components which must be disconnected for removal of engine.

1. Alternator wires
2. Fuel return line
3. Hydraulic oil line clamp
4. Priority valve support plate
5. Auxiliary solenoid
6. Temperature indicator sending unit

Fig. 103—View of engine showing parts related to cylinder head removal.

1. Water outlet hose
2. Intake manifold
3. Water outlet elbow
4. Temperature sender
5. High pressure fuel line
6. Leak-off line
7. Exhaust manifold

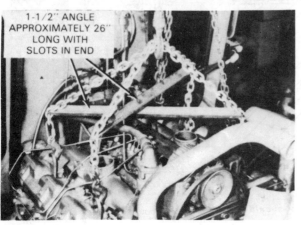

Fig. 101—Engine lifting chains should be supported as shown when removing engine.

Fig. 104—View showing rocker shaft oil feed pipe and injection pump oil feed pipe.

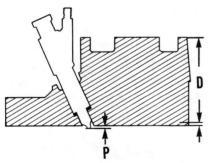

Cylinder head gasket is marked "TOP" for proper installation. Install gasket dry and tighten cylinder head cap screws and stud nuts progressively to a torque of 120-125 ft.-lbs. (16.6-17.3kgm) using the sequence shown in Fig. 105. Note that two different length cap screws are used to retain the head; the longest screws should be installed in center row, shorter screws at top of head and the shortest of the three lengths at bottom of head. Install rocker arm assemblies as outlined in paragraph 78 and adjust tappet gap as in paragraph 79.

VALVES AND SEATS

73. Intake valves seat directly in the cylinder head while exhaust valves are equipped with hardened seat inserts. Original valves are numbered to correspond with a number stamped in combustion chamber area of head, (Fig. 108). Intake valve must not protrude more than 0.054 inch (1.37mm) nor exhaust valve more than 0.0615 inch (1.56mm). Intake valve (and/or seat) must be renewed if seated valve protrudes less than 0.001 inch (0.03mm); exhaust valve (and/or seat) if protrusion is less than 0.0085 inch (0.22mm).

Valve and seat angle is 45 degrees for all valves. The renewable exhaust valve seat inserts should only be renewed if special "Churchill" tools or equivalent are available to roll edge over chamfer of seat insert.

VALVE GUIDES

74. Intake and exhaust valve guides are interchangeable. Refer to Fig. 111. Press all guides into head until distance (A) between top end of guide and spring seat of head measures 0.783-0.800 inch (19.89-20.32mm).

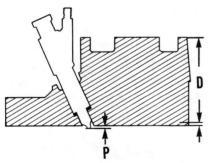

Fig. 107—Nozzle must not protrude (P) more than 0.143 inch and cylinder head thickness (D) must not be less than 3.970 inches.

Inside diameter of a new guide is 0.375-0.376 inch (9.53-9.55mm) and desired diametral clearance is 0.0015-0.0035 inch (0.04-0.09mm) for intake valves or 0.002-0.004 inch (0.05-0.10mm) for exhaust valves. Maximum permissible clearance for any valve is 0.0055 inch (0.14mm). Renew valves and/or guides if clearance is excessive.

VALVE SPRINGS

75. Springs, retainers, locks and oil deflectors are interchangeable for in-

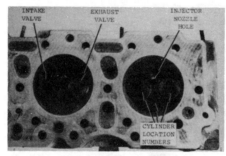

Fig. 108—Original valves are numbered to correspond with number stamped in combustion chamber area of head.

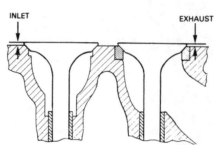

Fig. 109—The inlet valve must not protrude more than 0.054 inch and exhaust must not protrude more than 0.0615 inch. Refer to text for minimum protrusion limits.

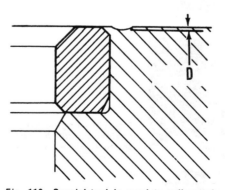

Fig. 110—Special tool is used to roll metal over edge of valve seat insert. Depth of roll (D) should be 0.004-0.005 inch, radius of depressed area should be 1/8 inch and diameter across rolled area should be 1.834-1.854 inches.

take and exhaust valves. Inner and outer springs are used. Renew springs if they are distorted or heat discolored; or if they fail to meet the test specifications which follow. Renew other parts if they show signs of wear or damage. Close coils of all springs should be toward cylinder head.

INNER SPRING
Free length (approximate) .1.983 inches
(50.37mm)
Test (closed) length . . . . . . .1.553 inches
(39.45mm)
Lbs. Test . . . . .30.3-33.5 lbs.
(13.73-15.2kg.)

OUTER SPRING
Free length (approximate) . .2.28 inches
(57.91mm)
Test (closed) length . . . . . . .1.833 inches
(46.56mm)
Lbs. Test . . . . .68.8-76 lbs.
(31.21-34.47kg.)

CAM FOLLOWERS

77. Mushroom type cam followers (Tappets) are contained in removable

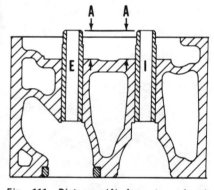

Fig. 111—Distance (A) from top of valve guides to spring seat should be 0.783-0.800 inch.

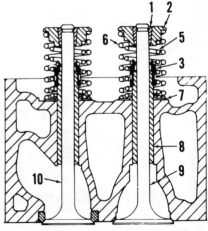

Fig. 112—Cross-sectional view of installed valve showing component parts.

1. Valve locks
2. Retainer cap
3. Stem seal
4. (not listed)
5. Inner valve spring
6. Outer valve spring
7. Seating washer
8. Valve guide
9. Intake valve
10. Exhaust valve

guide blocks as shown in Fig. 113. The guide block (containing the two cam followers for one cylinder) can be removed without major engine disassembly after removing injection pump and tappet cover. The 0.7475-0.7485 inch (18.99-19.01mm) diameter tappets should have 0.006-0.0085 inch (0.15-0.22mm) clearance in tappet bores. When installing guide blocks, tighten retaining cap screws to a torque of 52-58 ft.-lbs. (7.2-8.0kg.-m).

ROCKER ARMS

78. Rocker arm covers can be removed without additional disassembly. Rocker arms, brackets and shaft must be correctly assembled to provide proper lubrication for the rocker arms. Bushings (5—Fig. 114) in rocker arms are renewable and should be assembled with split (5S—Fig. 115) at 3 o'clock position toward valve side of arm (3) and oil hole (H) aligned with passage in rocker arm. Inside diameter of bushing should be 0.7515-0.7535 inch (19.09-19.14mm) and should provide 0.002-0.005 inch (0.05-0.13mm) diametral clearance between bushing and shaft.

Assemble each of the four sets of rocker arm assemblies as follows: Install both wire rings (10—Fig. 114) into grooves in shaft (12), then position one of the two retaining collars (9) over each wire ring. Examine the holes in shaft (12) and position the large locating hole (8H) toward left side as shown. Notice that a smaller oil delivery passage hole (P—Fig. 115) is below the locating hole and another oil delivery passage hole is at other end of shaft. The rocker arm lubrication holes (L) should be toward valve side of shaft and the large locating hole (8H) should be toward left as shown. Assemble one spring (6—Fig. 114) and one rocker arm (3) onto shaft. All of the valve contact surfaces should be toward you while

assembling. Slide one bracket (7) onto shaft (12). All brackets (7) must be assembled to shafts with boss on side toward right as shown. Position the bracket carefully, then install locating screw (8) through bracket (7) and into the correct hole (8H) of shaft. Position one washer (2), spring (6), rocker arm (3), another washer (2) and secure these parts to shaft with "E" clip (1). This completes assembly at left end of shaft. To assemble parts on right end of shaft, after clip (10) and retainer (9) are installed, proceed as follows: Slide one rocker arm (3C) onto shaft, followed by one spring (6), one washer (2) and bracket (7). Both brackets (7) are the same and the boss on each should be toward right, but the bracket used on right end of shaft is not fitted with a holding bolt. Install rocker arm (3D), spring (6), washer (2) and "E" ring (1).

When correctly assembled, the following conditions will be met:

1. Bosses of brackets (7) will be toward right.

2. Locating screw (8) will be in left

bracket and will correctly engage the large hole (8H) in rocker shaft (12).

3. Oil feed passages (L) in rocker shaft will be toward valve stem end of rocker arms.

When installing the rocker arm assemblies, install nut (24) on lower stud (23) with chamfered end of nut up. Studs (21) and nuts (22) are used on early models, later models are equipped with cap screw at these locations. Adjust valve clearance (tappet gap) as outlined in paragraph 79.

VALVE CLEARANCE

79. The recommended valve clearance (tappet gap) when engine is cold is 0.012 inch (0.30mm) for both intake and exhaust valves. Static setting of all valves can be accomplished with just two crankshaft positions using procedure shown in Figs. 116 and 117. Proceed as follows:

Remove rocker arm covers and turn crankshaft until TDC timing mark on pulley is aligned with indicator as shown in Fig. 118.

Fig. 114—Exploded view of valve operating parts. Refer to text for further explanation.

1. E ring
2. Washer
3. Rocker arms
4. Pin
5. Bushing
6. Spring
7. Bracket
8. Locating screw
8H. Hole for screw
9. Collar
10. Wire ring
11. Plug
12. Rocker shaft
13. Lock nut
14. Adjuster
15. Push rod
16. Cam follower
17. Guide block
18. Thimbles
19. Cap screws
20. Lock washer
21. Stud
22. Nut
23. Stud
24. Special Nut
25. Gasket
26. Cover
27. Retaining screw
28. Gasket
29. Oil feed pipe
30. Gasket
31. Banjo bolt
IP. Inj. Pump Connection

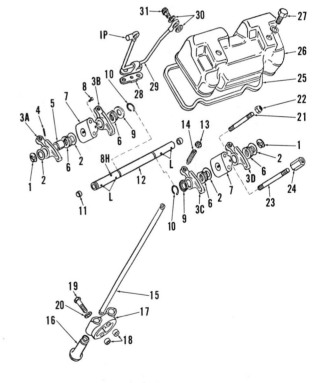

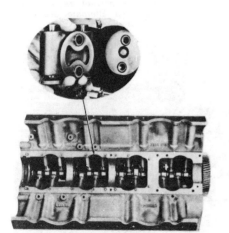

Fig. 113—Cam follower guide blocks are attached to cylinder block with cap screws and are removable in pairs as shown.

Fig. 115—Cross-section of rocker arm and bracket showing important features when assembling. Refer to text.

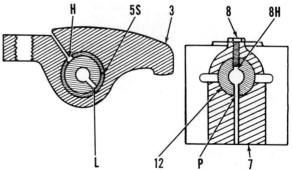

RIGHT BANK

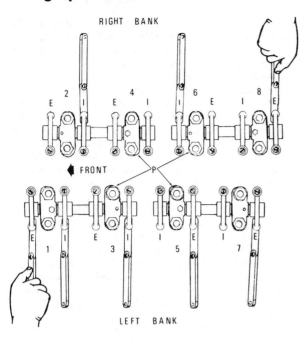

FRONT

P

LEFT BANK

Fig. 116—With TDC timing mark aligned and No. 1 piston on compression stroke, adjust the indicated valves as outlined in paragraph 79. Numbers indicate cylinder numbering system and (P) denotes the rocker arm brackets containing oil pressure passages.

RIGHT BANK

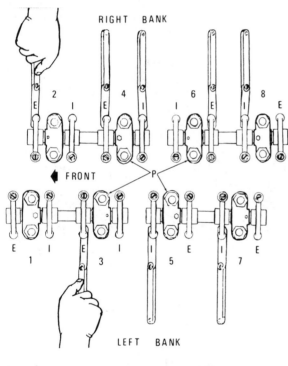

FRONT

P

LEFT BANK

Fig. 117—With TDC timing mark aligned and No. 4 piston on compression stroke, adjust the indicated valves as outlined in paragraph 79.

Check the rocker arms for No. 1 and No. 4 cylinders. If both rocker arms on No. 1 cylinder are loose and both rocker arms on No. 4 cylinders tight, No. 1 cylinder is on compression stroke; adjust the eight tappets shown in Fig. 116. If both rocker arms on No. 4 cylinder are loose and rocker arms on No. 1 cylinder are tight, No. 1 cylinder is on the exhaust stroke; adjust the eight valves indicated in Fig. 117.

When the adjustment is completed, turn crankshaft one complete turn until TDC timing mark is again aligned, then adjust the remaining tappets.

TIMING GEAR COVER

80. The timing gear cover consists of an upper section (Fig. 119) which contains mounting points for the water pump and alternator; and a lower section (Fig. 120) which houses the crankshaft oil seal.

To remove the timing gear cover, first remove hood and front end assembly as outlined in paragraph 71.

81. **UPPER COVER.** To remove the upper half after front end is removed, first unbolt and remove fan blades. Loosen fan belt idler and remove fan belt. Remove remaining coolant hoses, remove retaining cap screws and lift off upper timing gear cover section with alternator and water pump attached as

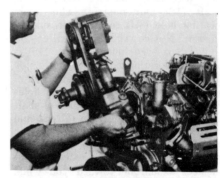

Fig. 119—Timing gear cover upper section can be lifted off as part of an assembly containing water pump, alternator and alternator support.

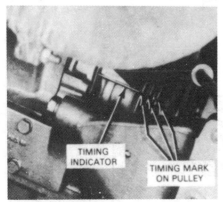

Fig. 118—View showing timing marks on crankshaft pulley and plate (indicator) aligned at number 1 cylinder TDC.

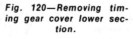

Fig. 120—Removing timing gear cover lower section.

shown in Fig. 119. Install by reversing the removal procedure.

82. **LOWER COVER.** To remove the lower half of timing gear cover it is first necessary to remove hood and front end assembly as outlined in paragraph 71 and upper cover half as in paragraph 81. Loosen oil pan screws. Remove the cap screw (S—Fig. 121) and washer (W) securing crankshaft pulley (P) to shaft. Attach a suitable puller to crankshaft pulley and apply slight pressure with pulley forcing screw; then strike forcing screw sharply with a hammer to bounce out the tapered locking ring (R). Pulley (P) can now be easily withdrawn.

The lip-type crankshaft oil seal should be installed 0.435-0.445 inch (11.05-11.30mm) below flush with front face of bore as shown in Fig. 122. Coat lip of seal with grease when assembling and be sure that slot in pulley aligns

Fig. 123—Timing gear train with timing marks aligned. Marks are aligned during assembly and will line up only occasionally during operation because of odd number of teeth in idler gears.

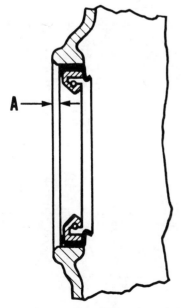

Fig. 121—Crankshaft pulley (P) uses split, tapered retainer (R) to lock pulley on crankshaft. Refer to paragrph 82 for removal procedure.

with roll pin in crankshaft. Install tapered locking ring (R—Fig. 121), washer (W) and retaining screw (S). Tighten the pulley retaining screw to a torque of 275-300 ft.-lbs.

TIMING GEARS

83. Fig. 123 shows a view of timing gear train with cover removed. Before attempting to remove any of the gears, first remove rocker arm covers and rocker arms to avoid the possibility of damage to pistons or valve train if camshaft or crankshaft should either

one be turned independently of the other.

Timing gear backlash should be 0.006-0.009 inch (0.15-0.023mm) between any of the gears in main timing gear train. Replacement gears are available in standard size only. If backlash is not within limits, renew gears, idler shafts, bushings or other items concerned.

NOTE: Because of the odd number of teeth in idler gear and auxiliary drive gears, all timing marks will not align with each revolution of the camshaft. Failure of marks to align does not necessarily indicate engine is mistimed.

To remove the timing gears or time the engine, refer to the appropriate following paragraphs.

84. **IDLER GEAR AND HUB.** The timing idler gear (Fig. 125) should have

Fig. 122—Cross-sectional view of crankshaft front oil seal which should be installed approximately 7/16 inch below flush with front of bore as shown at (A).

Fig. 124—Checking timing gear backlash using a feeler gage.

Fig. 125—Checking idler gear end clearance.

a diametral clearance of 0.001-0.004 inch (0.03-0.10mm) on the 2.248-2.249 inch (57.11-57.13mm) diameter hub. Suggested end play is 0.012-0.023 inch (0.31-0.58mm), with a wear limit of 0.030 inch (0.76mm). Hub is a light interference fit in mounting bore of cylinder block and the two mounting studs are slightly offset, hub can only be installed in one position. The prefinished idler gear bushing is 0.002-0.005 inch (0.05-0.13mm) press fit in gear bore and is renewable. Tighten retaining stud nuts to a torque of 27-30 ft.-lbs. when reinstalling.

85. **CAMSHAFT GEAR.** The camshaft is a transition fit in gear counterbore and is positively located by a dowel (1—Fig. 127) and locating slot (2). Gear is retained to camshaft by three cap screws. When installing the camshaft gear, use "Loctite Stud N'Bearing Mount" in cap screw threads and tighten screws to a torque of 27-30 ft.-lbs.

86. **CRANKSHAFT GEAR.** The crankshaft gear is a transition fit (0.0005 inch tight to 0.0015 inch loose) on shaft, and can usually be removed with a pry bar after timing gear cover is removed. When reinstalling crankshaft gear, remove upper idler gear and retime engine as outlined in paragraph 90.

87. **AUXILIARY HOUSING.** The auxiliary housing and drive mounts on top of the engine block and drives the injection pump gear. The auxiliary housing can be removed independently of the timing gears by following the procedures outlined in paragraph 88; or during timing gear overhaul by following only those portions which apply.

88. REMOVE AND REINSTALL. To remove the auxiliary drive housing assembly, first drain cooling system and remove hood, air intake hoses and interfering air baffles. Remove coolant hoses and exhaust pipes. Disconnect alternator wiring, tachometer drive cable, fuel stop cable and throttle linkage.

Disconnect and remove fuel injection pump as outlined in paragraph 117. Remove alternator, water pump and timing gear cover upper half as a unit as shown in Fig. 119. Remove the retaining cap screws and lift off auxiliary drive housing as shown in Fig. 128.

When installing the auxiliary drive housing assembly, the injection pump must be correctly timed as follows: Make sure No. 1 cylinder is at TDC on compression stroke.

NOTE: The single slash-marked tooth on timing gear should point up as shown in Fig. 129.

Install the auxiliary drive housing with the double punch-marked drive gear tooth space aligned with marked camshaft gear tooth as shown in Fig. 123. Tighten the retaining cap screws to a torque of 16 ft.-lbs. and install injection pump as outlined in paragraph 117.

89. OVERHAUL. The auxiliary drive shaft rotates in pressure lubricated, precision type bearings which attach to the housing with bearing caps. Diametral clearance of shaft in the bearings should be 0.0005-0.002 inch (0.013-0.051mm), with a maximum permissible clearance of 0.003 inch (0.08mm). Drive shaft end play of 0.001-0.011 inch (0.03-0.28mm) is controlled by thrust wash-

ers which can be renewed after removing bearing caps. Drive shaft journal diameter is 1.374-1.3745 inch (34.90-34.912mm).

Main bearing caps have stamped locating numbers as shown in Fig. 130 and are positioned by ring dowels which prevent incorrect installation. Gears are positioned to shaft by locating dowels and secured by cap screws. When reassembling the auxiliary drive shaft housing, tighten main bearing cap screws to 16-20 ft.-lbs. (2.2-2.8kg-m), gear retaining set screws to 16 ft.-lbs. (2.2kg-m) and fuel pump housing securing screws to 15-16 ft.-lbs. (2.1-2.3kg-m).

90. **TIMING THE GEARS.** Timing marks are affixed to crankshaft gear, idler gear, camshaft gear and auxiliary drive gear as shown in Fig. 123. Because of the odd number of teeth in idler gear and auxiliary drive gear, the marks on these gears will not align with every other crankshaft revolution and failure to align does not mean engine is improperly timed.

NOTE: The odd number of teeth prevents the development of wear patterns which coincide with stress loads of crankshaft and camshaft and therefore

Fig. 127—Dowel (1) in camshaft gear must align with notch (2) in cam for proper timing.

Fig. 129—The single slash mark (circled) on timing gear is at the top as shown, when No. 1 piston is at TDC on compression stroke.

Fig. 126—Installing idler gear hub. Stud holes are slightly offset in hub center to assure proper alignment of idler gear oil hole.

Fig. 128—Removing auxiliary housing and shaft assembly.

Fig. 130—Assembled view of auxiliary housing and shaft unit.

contribute to longer gear life with less gear noise.

When assembling the timing gears, align all timing marks as shown in Fig. 123. To check the timing when gears have not been removed, turn crankshaft until No. 1 piston is at TDC on compression stroke and the single punch marked crankshaft gear tooth is meshed with idler gear; then count the number of idler gear teeth between the marked crankshaft gear tooth and the first marked tooth on camshaft gear. The number should be nine. The auxiliary drive gear timing need not be checked if injection pump is correctly timed as outlined in paragraph 118.

CAMSHAFT

91. To remove the camshaft, first remove timing gear lower cover as outlined in paragraph 82, rocker arms as in paragraph 78 and camshaft gear as in paragraph 85. Remove tappet covers (paragraph 77) then remove or block up the cam followers (tappets). Remove the three cap screws retaining the camshaft thrust plate and withdraw the camshaft from block bore as shown in Fig. 131.

Normal camshaft bearing clearance is 0.002-0.006 inch (0.05-0.15mm) for all journals, with a maximum permissible clearance of 0.0095 inch (0.24mm). Normal camshaft end play is 0.004-0.014 inch (0.10-0.36mm) with a maximum permissible end play of 0.020 inch (0.51mm). End play is controlled by thickness of thrust washer and the annular thrust surface on rear face of camshaft gear. Pressure lubrication for rocker arms is metered by the third camshaft bearing journal.

When installing the camshaft, tighten the thrust plate retaining cap screws to a torque of 10-12 ft.-lbs. Install gear retaining cap screws using LOCTITE and tighten to a torque of 27-30 ft.-lbs. Camshaft journal diameters are as follows:

Front journal.......2.371-2.373 inches
60.23-60.27mm
No. 2 journal.......2.241-2.243 inches
56.92-56.97mm
No. 3 journal.......2.231-2.233 inches
56.67-56.71mm

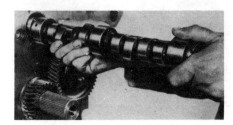

Fig. 131—Installing camshaft in engine block.

No. 4 journal........2.221-2.223 inches
56.42-56.46mm
No. 5 journal........2.211-2.213 inches
56.16-56.21mm
Cam lift............0.3325-0.3355 inch
8.45-8.52mm

ROD AND PISTON UNITS

92. Connecting rod and piston units can be removed from above after removing cylinder head(s) and oil pan. Cylinder numbers are stamped on connecting rod and cap. Connecting rod bearings are available in undersizes of 0.010, 0.020 and 0.030 inch as well as standard size. The crankshaft must be "RE-NITRIDED" if journal is reground to undersize.

When assembling, tighten the cap retaining screws to 100-105 ft.-lbs. (13.8-14.5kgm) torque.

PISTONS, SLEEVES AND RINGS

93. The aluminum alloy, cam ground pistons are supplied in standard size only. The toroidal combustion chamber is slightly offset in piston crown and top of piston contains relief pockets for valve heads. Because the same piston is used in both banks of the engine, there is no "FRONT" or "REAR" orientation. Rather, the piston is installed in cylinder with valve relief pockets (R—Fig. 132) up (toward center of vee).

Each piston is fitted with two compression rings and one oil control ring, all located above the piston pin. The barrel faced chrome top ring may be installed either side up in groove. The internally stepped chrome 2nd compression ring is marked "TOP" for correct installation with internal groove toward top of piston. Position oil ring spring in bottom groove and be sure that latch pin enters both ends of spring, then install the chrome plated oil control ring over the spring with gap of ring opposite (180 degrees from)

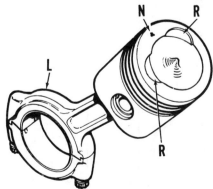

Fig. 132—Valve relief pockets (R) and long side of rod must be on same side and toward the top when rod and piston unit is assembled. Cylinder number (N) is also at top on factory assembled units.

the latch pin in the spring. Ring specifications are as follows:

End gap—
Top ring...........0.008-0.021-inch
0.20-0.53mm
Second ring &
oil ring..........0.017-0.024-inch
0.43-0.61mm
Side clearance in groove—
Top ring, desired....0.003-0.005-inch
0.08-0.13mm
Maximum.....0.008-inch (0.20mm)
Second ring &
oil ring..........0.002-0.004-inch
0.05-0.10mm

The production cylinder sleeves are 0.001-0.003 inch press fit in cylinder block bores. Service sleeves have 0.000-0.002 inch (0-0.05mm) clearance in block bores and cylinder walls are finished. Use a heavy-duty sleeve puller capable of high pulling force to remove production sleeves. Thoroughly clean cylinder bore, giving particular attention to counterbore. Carefully examine sleeve contact area in cylinder block and remove any burrs, rust spots or other foreign material.

Thoroughly clean new service sleeves in a solvent and air dry. Chill sleeves in dry ice or a deep-freeze and install dry. Push sleeve in completely by hand without stopping. When correctly installed, upper step of sleeve (A—Fig. 133) should extend 0.020-0.026 inch (0.51-0.66mm) above gasket surface of block and flange area (B) should be flush to 0.004 inch (0.10mm) below block surface. Make sure there is no distortion after installation.

Cylinder sleeves should be renewed if scored, or if wear or taper exceeds 0.002-inch (0.05mm). Crown of installed piston must be flush to 0.0085-inch (0.216mm) BELOW gasket surface of

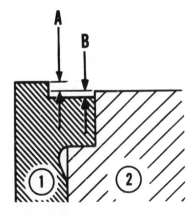

Fig. 133—Cross-sectional view of sleeve counterbore showing installation dimensions. Upper step of sleeve (A) should extend 0.020-0.026 inch above gasket surface of block. Flange area (B) should be flush to 0.004 inch below block surface.

block (D—Fig. 134). Installed diameter of service sleeve is 4.250-4.252 inches (107.95-108.00mm).

PISTON PINS

94. The 1.7498-1.7500 inches (44.445-44.450mm) diameter, floating type piston pins are retained in piston bores by snap rings and are available in standard size only. The renewable connecting rod bushing must be final sized after installation to provide the recommended 0.0007-0.0018 inch (0.018-0.046mm) diametral clearance of piston pin. Be sure the pre-drilled oil hole in bushing is properly aligned with hole in top of connecting rod and install bushing from chamfered side of bore. Piston pin is a transition fit of 0.0000-0.0005-inch clearance in piston bosses and piston should be heated to 100-120 degrees F for pin removal or installation. Valve relief pockets (R—Fig. 132) and long leg (L) of rod must be on the same side of assembly when pin is installed.

CONNECTING RODS AND BEARINGS

95. Connecting rod bearings are precision type, renewable from below after removing oil pan and rod bearing caps. When renewing bearing shells, make sure that the projection engages milled slots in rod and cap and that correlation marks on rod and cap are in register.

Connecting rod bearings should have a diametral clearance of 0.0018-0.0038 inch (0.05-0.10mm) on the 2.9980-2.9985 inches (76.149-76.162mm) diameter crankpin. Recommended connecting rod side clearance is 0.015-0.022 inch (0.38-0.56mm) with both rods fitted on crankpin journal. Bearings are available in undersizes of 0.010, 0.020 and 0.030 inch as well as standard. The crankshaft must be "RENITRIDED" if journal is reground undersize. Tighten connecting rod cap screws to a torque of 100-105 ft.-lbs. (13.8-14.5kg-m) when units are installed.

CRANKSHAFT AND BEARINGS

96. The crankshaft is supported in five precision type main bearings. To remove the rear main bearing cap, it is first necessary to remove the engine, clutch, flywheel and flywheel housing. The front main bearing cap carries the oil pump idler gear hub and removal of front bearing requires prior removal of oil pump and lines. All other main bearing caps can be removed from below after removing oil pan.

Main bearing cap cross bolts (Fig. 136) should be removed before loosening regular cap screws. Crankshaft end play is controlled by thrust washer halves located in block portion only, in front of and to the rear of center main bearing. A locating tab (Fig. 137) is machined on one end of thrust washer half and washer can be rolled out after

removing main bearing cap. Counterweights are stamped for proper location on crankshaft corresponding to web numbers. Oil weights and attaching screws lightly, install retaining screws. Torque weight retaining screws to 80-85 ft.-lbs. (11.1-11.8 kg-m), then loosen screws and retorque to 80-85 ft.-lbs. (11.1-11.8 kg-m). This procedure will properly seat weight and set the screws.

Bearing inserts are available in undersizes of 0.010, 0.020 and 0.030 inch as well as standard size. The crankshaft must be "RE-NITRIDED" if journals are reground to next undersize. The thrust washer halves are in standard thickness of 0.122-0.125 inch (3.10-3.18mm), and oversize of 0.1295-0.1325 inch (3.29-3.37mm). Recommended main bearing diametral clearance is 0.004-0.0063 inch (0.10-0.16mm) and recommended crankshaft end play is 0.002-0.017 inch (0.05-0.43mm). Tighten main bearing retaining cap screws to a torque of 190-200 ft.-lbs. (26.3-29.0 kg-m) and cross bolts to 70-75 ft.-lbs. (9.7-10.4 kg-m). Refer to paragraph 97 for installation of rear seal and oil pan bridge piece and to paragraph 101 for installation of oil pump and idler gear. Check the crankshaft against the values which follow:

Fig. 136—Main bearing caps are provided with cross bolts for additional security.

Fig. 137—Identical thrust washer halves are located in block portion only of center main bearing.

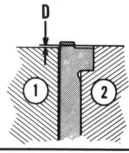

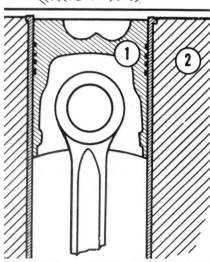

Fig. 134—Crown of piston should be 0.0085 inch below deck of block as shown at (D).

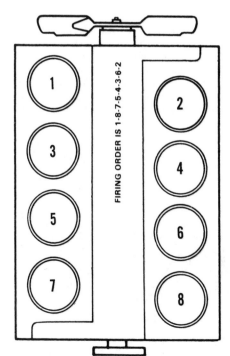

FIRING ORDER IS 1-8-7-5-4-3-6-2

Fig. 135—Schematic view of block showing cylinder numbering sequence.

Main journal diameter-
 Standard ....3.9967-3.9972 inches
       101.516-101.529mm
Crankpin journal diameter-
 Standard ....2.9980-2.9985 inches
       76.149-76.162mm

CRANKSHAFT REAR OIL SEAL

97. A spring loaded type lip seal is located in a retainer bolted to rear of engine (Fig. 139). The retainer and seal can be removed after splitting tractor, removing clutch and flywheel. During production, the oil seal is installed with its rear face flush with rear face of seal retainer. Check condition of seal contact surface of crankshaft before installing new seal in retainer. The seal can be pressed further into retainer to provide another contact surface if crankshaft is grooved. The recommended locations are: flush, 3/32 inch below flush and 3/16 inch below flush with rear face of retainer. Coat lip of seal and crankshaft surfaces with oil before assembling and use a seal protector when sliding seal over crankshaft. Use non-hardening sealer on gasket.

If cylinder block bridge piece if removed, fit the two small cylindrical rubber seals into recesses in ends of bridge piece as shown in Fig. 140. Use a straight edge to make sure that bottom (oil pan) surface of bridge piece

is aligned with gasket surface of block before tightening retaining screws.

FLYWHEEL

98. The engine flywheel can be removed after separating engine from clutch housing and removing clutch.

NOTE: The flywheel is heavy and approximately 6 inches deep. To safely handle the flywheel during removal, a special handling tool is required. Tool can be shop constructed as shown in Fig. 141. Also the use of guide studs is recommended.

The flywheel is attached to crankshaft flange by twelve capscrews which are equally spaced.

The starter ring gear can be renewed after flywheel is removed. Heat ring gear evenly to approximately 480 degrees F. (250 degrees C.) and install on flywheel with beveled end of teeth facing front of engine.

Install flywheel and tighten the twelve retaining cap screws to a torque of 74-80 ft.-lbs. (10.2-11.1 kg-m.). Maximum allowable face runout is 0.001 inch (0.025mm) for each inch from flywheel center-line to point of measurement.

FLYWHEEL HOUSING

99. The cast flywheel housing is

doweled to engine block and secured with cap screws. Housing can be removed after removing flywheel and must be removed for removal of crankshaft. When installing flywheel housing, tighten the retaining cap screws to a torque of 50 ft.-lbs. (6.9kg-m).

OIL PAN

100. The heavy cast iron oil pan serves as part of the tractor frame and as an attaching point for tractor front support. If oil pan must be removed with engine installed in tractor, the following procedure is suggested, however, extreme care must be observed to prevent personal injury or damage to tractor.

Drain oil pan and oil cooler (Fig. 143), then remove oil cooler. Raise tractor up slightly and block up solidly under flywheel housing.

NOTE: Use wood block to distribute weight and prevent slipping or breaking.

Position a jack under front end to release pressure from ends of pan. Position a rolling floor jack under oil pan and loosen all six screws attaching front support. Remove all of the oil pan retaining screws, then use jack to lower oil pan away from engine.

Install by reversing removal procedure. Tighten all screws securely.

Fig. 138—Crankshaft thrust washer halves can be pushed out as shown, without removing crankshaft.

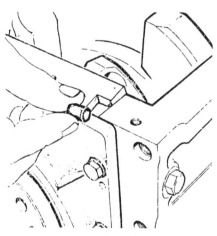

Fig. 140—Be sure to fit the two small cylindrical rubber seals into recesses at ends of bridge piece as shown.

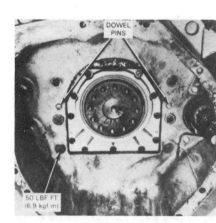

Fig. 142—View showing location of the two dowel pins which locate the flywheel housing (adapter) to engine block.

Fig. 139—View showing installation of rear oil seal housing. Location of the two dowel pins is indicated.

Fig. 141—The heavy flywheel is difficult to remove or install without the special handling tool which can be locally made as shown.

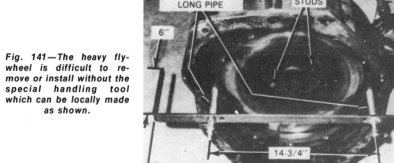

OIL PUMP

101. The gear type oil pump is mounted underneath the front main bearing cap and driven by the crankshaft gear through an idler. Oil is drawn up through a strainer and oil inlet tube mounted on the pump, and is delivered from the pump through a tube to the oil cooler mounted on side of oil pan. Oil enters cooler through the front tube and exits through rear tube to the oil pressure relief valve body. Oil pressure is regulated then the oil flows through the double filters before lubricating the engine bearings.

The oil pump can be removed after removing the oil cooler, withdrawing the two tubes used to direct oil to and from the oil cooler, then removing the oil pan. Refer to paragraph 100. Detach the oil inlet strainer brackets, then unbolt and remove the oil pump from bottom face of crankcase.

Refer to Fig. 146 for drawing of oil pan, pump, cooler, relief valve and associated parts. Refer to Fig. 147 for exploded view of the pump assembly. Normal pump gear end play is 0.002-0.012 inch (0.05-0.15mm) and maximum permissible wear limit 0.012 inch (0.30mm). Radial clearance between gear teeth and body should be 0.002-0.008 inch (0.05-0.20mm) and wear limit is 0.010 inch (0.25mm). Backlash be-

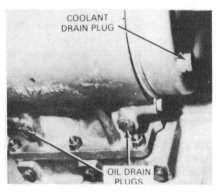

Fig. 143—View showing location of coolant and oil drain plugs on bottom and lower rear of oil cooler.

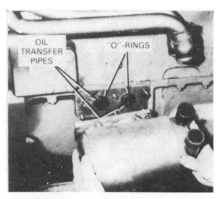

Fig. 144—The oil cooler must be removed and transfer pipes withdrawn before oil pan can be removed.

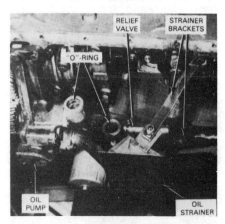

Fig. 145—View of installed oil pump and related parts with oil pan removed.

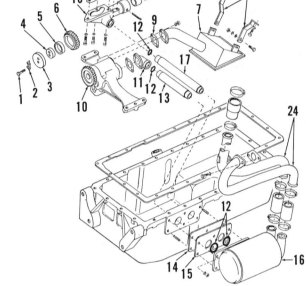

Fig. 146—Drawing of oil pan, pump, cooler, relief valve and related parts.

1. Cap screw
2. Lock plate
3. Plate
4. Hub
5. Bushing
6. Idler gear
7. Strainer
8. Brackets
9. Gasket
10. Oil pump
11. Elbow
12. "O" rings
13. Pipe from pump to cooler
14. Gasket
15. Plate
16. Oil cooler
17. Pipe from cooler to relief valve
18. Relief valve housing
19. Relief valve plunger
20. Spring
21. Cap
22. Oil deflector plate
23. Gasket
24. Water tubes to and from oil cooler

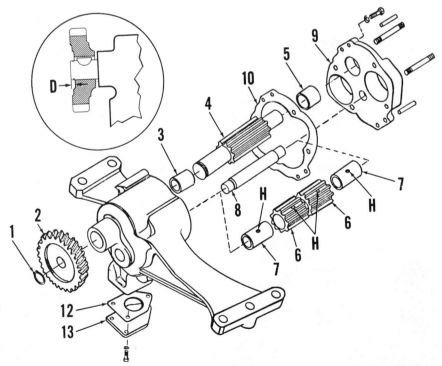

Fig. 147—Exploded view of oil pump. Inset shows distance (D) gear (2) should be pressed onto shaft (4).

1. Snap ring
2. Gear
3. Bushing
4. Drive shaft and gear
5. Bushing
6. Driven gears
7. Bushings
8. Idler shaft
9. Cover
10. Gasket
11. Body
12. Gasket
13. Plate

tween drive and driver gears should be 0.014-0.018 inch (0.36-0.46mm). Driver gear (2—Fig. 147) can be pressed from shaft after removing snap ring (1) if so equipped. Diametral clearance between bushings (3 & 5) and shaft (4) should be 0.001-0.003 inch (0.03-0.08mm). Diametral clearance between bushings (7) and shaft (8) should be 0.001-0.0025 inch (0.03-0.06mm).

When assembling, holes (H) in bushings (7) and gears (6) should be aligned. Press driver gear (2) onto shaft (over woodruff key if so equipped) until front face of gear is 0.16-0.18 inch (4.06-4.57mm) from end of shaft as shown at (D). Later models equipped with woodruff key should have snap ring (1) installed. Install driver gears (6) onto shaft (8) with shouldered ends out (away from each other) and holes (H) aligned.

The oil pump idler gear (6—Fig. 146) and associated parts can be removed after removing the timing gear cover as outlined in paragraph 82. Oil hole in hub (4) should be away from crankshaft and dowel located in oil passage of block. Bushing (5) should be 0.002-0.005 inch (0.05-0.13mm) interference fit in gear (6) and should have inside diameter of 2.000-2.002 inches (50.80-50.86mm) when installed. Diametral clearance between bushing (5) and hub (4) should be 0.001-0.004 inch (0.03-0.10mm). End play of installed gear should be 0.015-0.022 inch (0.38-0.56mm) with wear limit of 0.030 inch (0.76mm). Tighten screws (1) to 17-19 ft.-lbs. (2.4-2.6kg-m) torque and lock with plate (2).

RELIEF VALVE

102. The oil pressure relief valve Fig. 146) can be removed from engine block after removing oil pump as outlined in paragraph 101. The valve is pre-set to 60-65 psi (4.2-4.6kg-cm^2) and no attempt should be made to adjust the pressure except by renewal of worn or damaged parts. Hollow end of piston

(19) should be toward seat in body (18), away from spring (20).

OIL COOLER

103. An engine oil cooler (16—Fig. 146) is attached to the left side of the engine oil pan. Oil enters cooler through tube (13) from the oil pump (10) and leaves through longer tube (17). Coolant pipes (24) are used to circulate water from cylinder head through cooler and back to the water pump.

To remove oil cooler, first drain coolant and oil from the unit by removing plugs shown in Fig. 143. Disconnect hoses, then unbolt and remove cooler from side of oil pan. Use care to prevent damage to tubes (13 & 17—Fig. 146), plate (15) and "O" rings (12) at both ends of tubes. The tubes must be withdrawn before removing the oil pan.

No attempt should be made to disassemble oil cooler unless the unit can be properly pressure tested after reassembly. Mark body (26—Fig. 149) and one end of tube stack (25) before disassembly so that parts can be correctly aligned with assembling. Remove screws and washers locating tube stack in body, then slide the tube stack out toward the closed end until "O" ring (27) can be removed. Slide tube stack the other direction and out of body.

CAUTION: Do not pound on either end of the oil cooler. Slide the tube stack by supporting the stack and knocking the body mounting flange gently but firmly on a soft but firm surface such as a stout wooden bench.

Clean all parts thoroughly and use new "O" rings (27 & 27A). Oil or grease "O" rings, grooves in tube stack and bores in body before assembling. Install "O" ring (27A) at connection end of tube stack carefully enter closed end of stack through body being sure that previously affixed marks will align.

Slide tube stack through body and install "O" rings (27) in groove at closed end of tube stack. Slide the tube stack back into body aligning the marks. Install locating screws and washers and test for leakage. Fabricate a suitable adaptor to cover and seal the oil ports with provision to pressurize the oil passages. Fill coolant side of cooler with water, pressurize the oil side with air pressure of 90-150 psi (6.33-10.55kg-cm^2), then submerge the unit in water. The tube stack should be renewed if air escapes from coolant connections; new "O" rings should be installed if air escapes from ends.

AIR INTAKE SYSTEM

104. An air filter restriction gage located in the instrument panel is used to determine when air cleaner service is required. Advise customers to observe instrument and to service air cleaner only when the needle remains in the red zone with engine running. Proper maintenance is important and of particular importance is avoidance of leaks which would permit unfiltered air to enter the engine. Never operate the engine with any filters or cover missing. Never wash filter in gasoline, fuel oil or solvent. Use only solution of MF part number 1900 726 M1 mixed according to directions on the carton or lukewarm water and non-foaming detergent to clean the outer element. Install new outer element after six washings or once each year whichever occurs first. Be sure outer element is dry before installing. The inner filter element should not be cleaned, but new unit should be installed yearly or when restriction gage needle stays in red zone after outer element has been renewed.

DIESEL FUEL SYSTEM

(MF1155 Models)

The diesel fuel system consists of three basic units; the fuel tank and filters, injection pump and injector nozzles. When servicing any unit associated with the diesel fuel system, the maintenance of absolute cleanliness is of utmost importance. Of equal importance is the avoidance of nicks, burrs or handling damage on any of the working parts.

Probably the most important precaution that service personnel can impart to owners of diesel powered tractors is to urge them to use an approved fuel that is absolutely clean and free from foreign materials. Extra precaution

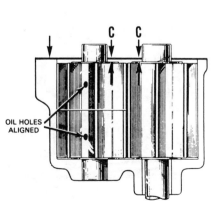

Fig. 148—Oil holes in idler gears and bushings must be aligned as shown. End clearance (C) of gears in body should be 0.015-0.022 inch.

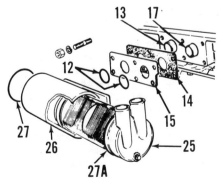

Fig. 149—Exploded view of engine oil cooler. Use extreme care when disassembling and assembling.

12. "O" rings	17. Exit tube
13. Delivery tube	25. Tube stack
14. Gasket	26. Cooler body
15. Plate	27 & 27A. "O" rings

should be taken to make certain that no water enters the fuel storage tanks. Because of the high pressures and degree of control required of injection equipment, extremely high precision standards are necessary in the manufacture and servicing of diesel components. Extra care in daily maintenance will pay big dividends in long service life and avoidance of costly repairs.

FUEL FILTERS AND LINES

105. **OPERATION AND MAINTENANCE.** All units are equipped with dual parallel fuel filters located just behind right rear corner of cylinder block as shown in Fig. 150.

Inspect glass bowl at bottom of both filters daily and drain off any water or dirt accumulation. Renew both elements at 500 hour intervals. Renew elements and clean the tank and lines if evidence of substantial water contamination exists.

106. **BLEEDING.** To bleed the system, open bleed plug on top of filter units (1—Fig. 150) and actuate primer lever on fuel lift pump until air-free fuel flows from plug openings. Close bleed plugs while fuel is flowing. Open either bleed screw (3—Fig. 151) on injection pump and continue operating lift pump lever until air is

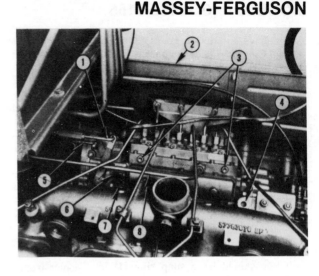

Fig. 151—Installed view of injection pump showing points of adjustment and service.

1. Throttle rod
2. Baffle plate
3. Bleed screws
4. Timing plug
5. Shut-off cable
6. Low idle stop screw
7. High idle stop screw
8. Fuel line

eliminated from injection pump.

Push in the fuel stop (shut-off) button, partially open the throttle and attempt to start the engine. If engine does not start, loosen compression nut at all injectors and turn engine over with starter until fuel escapes from all loosened connections. Tighten compression nuts and start the engine.

FUEL LIFT PUMP

107. The fuel lift pump is mounted on left side of fuel injection pump and driven by a lobe machined on injection pump camshaft. Pump parts are available separately or in a rebuilding kit. Output delivery pressure should be 5-8 psi.

INJECTOR NOZZLES

WARNING: Fuel leaves the injector nozzle with sufficient force to penetrate the skin. Keep exposed portions of your body clear of nozzle spray when testing.

108. **TESTING AND LOCATING A FAULTY NOZZLE.** If rough or uneven engine operation or misfiring indicates a faulty injector, the defective unit can usually be located as follows:

With engine running at the speed where malfunction is most noticeable (usually slow idle speed), loosen the compression nut on high pressure line for each injector in turn, and listen for a change in engine performance. As in checking spark plugs, the faulty unit is the one which, when its line is loosened, least affects the running of the engine.

If a faulty nozzle is found and considerable time has elapsed since the injectors have been serviced, it is recommended that all nozzles be removed and serviced or that new or reconditioned units be installed. Refer to the following paragraphs for removal and test procedure.

109. **REMOVE AND REINSTALL.** Before loosening any fuel lines, tho-

roughly clean the lines, connections, injectors and surrounding area with air pressure and solvent spray. Disconnect and remove the leak-off line, disconnect pressure line and cap all connections as they are loosened, to prevent dirt entry into the system. Remove the stud nuts and withdraw injector unit from cylinder head.

Thoroughly clean the nozzle recess in cylinder head before reinstalling injector unit. It is important that seating surface be free from even the smallest particle of carbon or dirt which could cause the injector unit to fail to seat and result in blow-by. No hard or sharp tools should be used in cleaning. Do not re-use the copper sealing washer located between injector nozzle and cylinder head, always install a new washer. Each injector should slide freely into place in cylinder head bore without binding. Make sure that high pressure fuel line fits squarely at both ends and that leak-off line compression seals are in good condition. Tighten retaining stud nuts to a torque of 10-12 ft.-lbs. After engine is started, examine injectors for blow-by, making the

Fig. 150—Dual parallel fuel filters are used on MF 1155 models. An air bleed screw (1) and filter element cap screw (2) is above each element. Drains are shown at (3).

Fig. 152—A suitable injector tester is required to completely test and adjust the injector nozzles.

necessary corrections before releasing the tractor for service.

110. **TESTING.** A complete job of testing and adjusting the injector requires the use of special test equipment. Only clean, approved testing oil should be used in tester tank. The nozzle should be tested for opening pressure, seal leakage, back leakage and spray pattern. When tested, the nozzle should open with a sharp popping or buzzing sound, and cut off quickly at end of injection with a minimum of seat leakage and a controlled amount of back leakage.

Before conducting the test, operate tester lever until fuel flows, then attach the injector. Close the valve to tester gage and pump tester lever a few quick strokes to be sure nozzle valve is not plugged, that four sprays emerge from nozzle tip, and that possibilities are good that injector can be returned to service without overhaul.

NOTE: Spray pattern is not symmetrical with centerline of nozzle tip, but forms a "tilted" pattern as shown in Fig. 153. Holes are positioned to provide the correct spray pattern in combustion chamber.

If adjustment is indicated in the preminary tests, proceed as follows:

111. OPENING PRESSURE. Open the valve to tester gage and operate tester lever slowly whole observing gage reading. Opening pressure should be 2720 psi (185 Atm); if not as specified, remove cap nut (1—Fig. 154) and turn adjusting sleeve (3) as required to obtain the recommended pressure.

NOTE: When adjusting a new injector or an overhauled unit with a new pressure spring (5), set the pressure at 3090 psi (210 Atm) to allow for initial pressure loss as the spring takes a set.

112. SEAT LEAKAGE. The nozzle tip should not leak at a pressure less than 2650 psi (180 Atm). To check for leakage, actuate tester lever slowly and as the gage needle approaches the suggested test pressure, observe the nozzle tip. Hold the pressure for 10 seconds; if drops appear or nozzle tip is wet, the valve is not seating and injector must be disassembled and overhauled as outlined in paragraph 115.

113. BACK LEAKAGE. If nozzle seat as tested in paragraph 112 was satisfactory, check the injector and connections for wetness which would indicate external leakage. If no leaks are found, bring gage pressure to 2400 psi (160 Atm) and observe the time required for gage pressure to drop to 1650 psi (110 Atm). For nozzle in good condition, this time should not be less than six seconds. A faster drop would indicate a worn or scored nozzle valve piston or body, and the nozzle assembly should be renewed.

NOTE: Leakage of the tester check valve or connections will cause a false reading, showing up in this test as excessively fast leakback. If all injectors tested fail to pass the test, the tester rather than the units should be suspected.

114. SPRAY PATTERN. If leakage and pressure are as specified when tested as outlined in paragraphs 111 through 113, operate the tester handle several times while observing the spray pattern. Four finely atomized, equally spaced, conical sprays should emerge from nozzle tip, with equal penetration into the surrounding atmosphere.

If pattern is uneven, ragged or not finely atomized, overhaul the nozzle as outlined in paragraph 115.

NOTE: Spray pattern is not symmetrical with centerline of nozzle tip, but tilted as shown in Fig. 153.

115. OVERHAUL. Hard or sharp tools, emery cloth, grinding compound, or other than approved solvents or lapping compounds must never be used. An approved nozzle cleaning kit is available through a number of specialized sources.

Wipe all dirt and loose carbon from exterior of nozzle and holder assembly, refer to Fig. 154 and proceed as follows:

Secure the nozzle in a soft-jawed vise or holding fixture and remove cap nut (1). Back off the adjusting sleeve (3) to completely unload pressure spring (5). Remove nozzle retaining cap nut (13) and nozzle body (12). Nozzle valve (11) and body (12) are matched and must never be intermixed. Place all parts in clean calibrating oil or diesel fuel as they are removed. Clean the exterior surfaces with a soft wire brush, soaking in an approved carbon solvent if necessary, to loosen hard carbon deposits. Rinse the parts in clean diesel fuel or calibrating oil immediately after cleaning, to neutralize the carbon solvent and prevent etching of polished surfaces. Clean the pressure chamber of nozzle tip using the special reamer as shown in Fig. 155. Clean the 0.0118-0.0122 inch (0.30-0.31mm) diameter

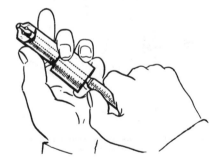

Fig. 155—Clean the pressure chamber in nozzle tip using the special reamer as shown.

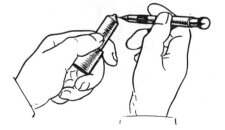

Fig. 156—Clean spray holes in nozzle tip using a pin vise and 0.010 inch wire probe.

Fig. 153—Nozzle holes (arrows) are not located an equal distance from nozzle tip.

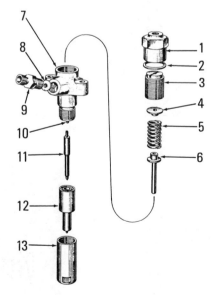

Fig. 154—Exploded view of the CAV/Simms nozzle of the type used.

1. Cap nut	
2. Gasket	
3. Adjusting sleeve	8. Washer
4. Spring seat	9. Inlet adapter
5. Spring	10. Dowel
6. Valve spindle	11. Nozzle valve
7. Nozzle holder	12. Nozzle body
	13. Nozzle nut

spray holes in nozzle with an 0.010 inch (0.24mm) wire probe held in a pin vise as shown in Fig. 156. Wire probe should protrude from pin vise only far enough to pass through spray holes (approximately 1/16 inch) to prevent bending and breakage. Rotate pin vise without applying undue pressure.

Clean valve seats by inserting small end of brass valve seat scraper into nozzle and rotating tool. Reverse the tool and clean upper chamfer using large end. Refer to Fig. 157. Use the hooked scraper to clean annular groove in top of nozzle body. Use the same hooked tool to clean internal fuel gallery as shown in Fig. 158.

With the above cleaning accomplished, back flush the nozzle by installing the reverse flusher adapter on injector tester and nozzle body in adapter, tip-end first. Secure with the knurled adapter nut and insert and rotate the nozzle valve while flushing. After nozzle is back-flushed, seat can be polished using a small amount of tallow on end of polishing stick and rotating the stick while moderate pressure is applied.

Light scratches on valve piston and bore can be polished out by careful use of special injector lapping compound only, DO NOT use valve grinding compound or regular commercial polishing agents. DO NOT attempt to reseat a leaking valve using polishing compound. Clean thoroughly and back-flush if lapping compound is used.

Reclean all parts by rinsing thoroughly in clean diesel fuel or calibrating oil and assemble valve to body while immersed in the cleaning fluid. Reassemble the injector while still wet. With adjusting sleeve (3—Fig. 154) loose, reinstall nozzle body (12) to holder (7), making sure valve (11) is installed and locating dowel (10) properly aligned. Tighten nozzle retaining cap nut (13) to a torque of 50 ft.-lbs. Do not overtighten; distortion may cause valve to stick and no amount of over-tightening can stop a leak caused by scratches or dirt.

Retest the assembled injector assembly as outlined in paragraphs 111 through 114.

INJECTION PUMP

The injection pump is a completely sealed unit. No service work of any kind should be attempted on the pump or governor unit without the use of special pump testing equipment and proper training. Inexperienced or unequipped service personnel should never attempt to overhaul an injection pump.

116. **ADJUSTMENT.** The slow idle stop screw (6—Fig. 160) should be adjusted with engine warm and running, to provide the recommended slow idle speed of 800-850 rpm. On some

models adjusting screw may be on opposite side of pump. Some models may be equipped with an "Anti-Stall" device as shown in Fig. 161. If so equipped and the correct slow idle speed cannot be obtained, back out the "Anti-Stall" screw until it no longer affects idle speed. With idle speed correctly adjusted, turn in the Anti-Stall screw until idle speed begins to increase; then back out the screw ¼ turn and tighten the locknut.

The high speed stop screw (7—Fig. 160) is factory set and sealed. Governed speed under load should be 2200 rpm, with a high-idle (no-load) speed of 2325-2450 rpm. Refer to paragraph 118 for pump timing adjustment.

117. **REMOVE AND REINSTALL.** Before attempting to remove the injection pump, thoroughly wash pump and connections with clean diesel fuel or approved solvent. Remove hood, air intake pipe and baffles (filler plates).

Disconnect throttle control rod and stop cable. Remove high pressure lines and cap all connections as lines are removed. Disconnect low pressure fuel lines at lift pump and injection pump main fuel gallery.

Turn crankshaft until No. 1 piston is at TDC on compression stroke. Remove injection pump rear support bracket, disconnect lubrication oil feed line, then

Fig. 159—Clean the small fuel feed hole using a suitable tool.

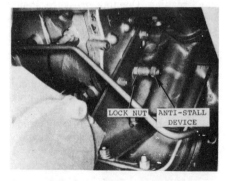

Fig. 161—Some models may be equipped with an "Anti-Stall" device. Refer to paragraph 116 for adjustment.

Fig. 157—Clean the valve seat using brass scraper as shown.

Fig. 158—Using the hooked scraper to clean internal fuel gallery.

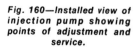

Fig. 160—Installed view of injection pump showing points of adjustment and service.

1. Throttle rod
2. Baffle plate
3. Bleed screws
4. Timing plug
5. Shut-off cable
6. Low idle stop screw
7. High idle stop screw
8. Fuel line

unbolt and lift off the pump as shown in Fig. 163.

Time the pump when installing as outlined in paragraph 118 and install by reversing the removal procedure. Injection pump plungers are numbered consecutively from front to rear and connect to cylinders as shown in Fig. 164. The injection pump lines are supported individually on vibration damper bosses cast into intake manifolds. Make sure lines rest squarely on vibration dampers after both ends are connected, then install and tighten line clamps (C—Fig. 165) and stud nuts (N). Remove oil filler plug (Fig. 162) from top of injection pump gear housing and pour in one pint of clean engine oil to provide initial lubrication. Bleed the pump and lines after installation as outlined in paragraph 106.

118. **PUMP TIMING TO ENGINE.** To time the injection pump during installation, first make sure that No. 1 piston is on compression stroke and that TDC timing mark on crankshaft pulley is aligned with pointer as shown in Fig. 166. Remove timing plug from right side of injection pump gear housing and align point of indicator screw with groove marked TDC as shown in Fig. 167, then install pump. With pump installed and No. 1 piston at TDC on compression stroke, timing marks should be perfectly aligned. If timing is not correct, remove the tachometer drive housing from front of auxiliary housing as shown in Fig. 168.

Loosen the three socket head screws (S) and turn pump shaft until timing groove is perfectly aligned with point of indicator screw, then tighten screws (S) to 24-28 ft.-lbs. (3.3-3.9kg-m). Check adjustment after changing timing by turning crankshaft two revolutions in normal direction (to remove backlash from gears) then recheck marks (Fig. 166).

119. **SPILL TIMING THE INJECTION PUMP.** To spill time the injection pump, first remove hood and thoroughly clean injection pump and surrounding area. Remove left, front rocker arm cover and No. 1 injection high pressure line. Turn crankshaft until No. 1 piston is at TDC on compression stroke.

Remove left front rocker arms assembly and push rods. Use a suitable fixture to depress the spring retainer cap, remove the valve locks; then lift

off retainer cap, and both valve springs from either of the valves in No. 1 cylinder. Leave the sealing "O" ring (R—Fig. 169) in its groove in valve stem as a safety measure to keep the valve from dropping into cylinder if crankshaft should be turned. With valve head resting on piston crown, attach a dial indicator with pointer bearing on valve stem. Rock crankshaft back and forth through about ten degrees to verify top of piston stroke and zero dial indicator at that point. Turn crankshaft opposite normal direction of rotation until a dial indicator reading of 0.352 inch (8.94mm) is obtained, which

Fig. 167—With TDC timing mark aligned as shown in Fig. 166, indicator screw pointer and TDC groove should be aligned in sight hole as shown.

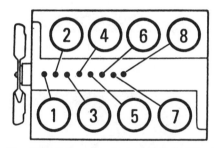

Fig. 164—Schematic view of cylinder block relating injection pump plungers to respective engine cylinders.

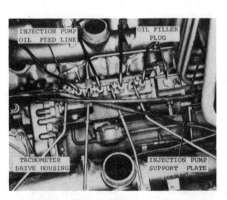

Fig. 162—Installed view of injection pump.

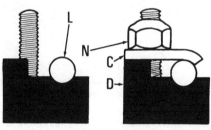

Fig. 165—Make sure injector lines (L) rest squarely on vibration dampers (D) before installing line clamps (C) and stud nuts (N).

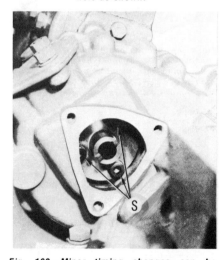

Fig. 168—Minor timing changes can be made by loosening the three cap screws (S) securing drive gear to injection pump camshaft.

Fig. 163—Removing the injection pump.

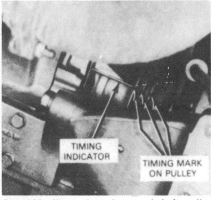

Fig. 166—View of engine crankshaft pulley (TDC) timing marks and indicator plate aligned.

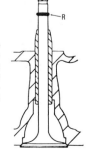

Fig. 169—When using engine valve to measure piston position, leave sealing "O" ring (R) installed to prevent valve from accidentally dropping into cylinder.

is equivalent to 28 degrees BTDC crankshaft position.

Remove the front (No. 1) delivery valve retainer (1—Fig. 170), extract volume reducer (2), spring (3) and delivery valve (5), then reinstall and tighten retainer (1). Install a suitable spill pipe to delivery valve retainer as shown at (B—Fig. 171). Disconnect fuel delivery pipe from injection pump body and install a auxiliary fuel tank with shut-off valve as shown at (C). Remove tachometer drive housing as shown in Fig. 168. Loosen the three socket head cap screws (S). Fill auxiliary fuel tank

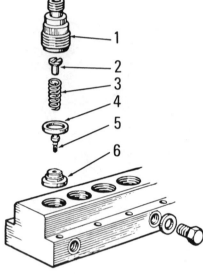

Fig. 170—Injection pump delivery valve components (2, 3 & 5) must be removed when spill timing the injection pump. A special wrench (MFN N2-5447) must be used to remove delivery valve adapter (1).

1. Adapter
2. Volume reducer
3. Spring
4. Gasket
5. Delivery valve
6. Valve seat

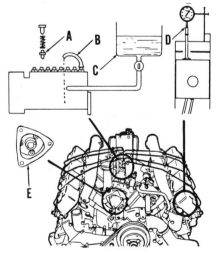

Fig. 171—Sequence of operations necessary in spill timing the injection pump.

A. Delivery valve removed
B. Spill pipe installed
C. Auxiliary fuel
D. Piston positioned (0.352 inch BTDC)
E. Pump camshaft turned

reservoir with diesel fuel and turn shut-off valve to "ON" position. Turn tachometer drive adapter counterclockwise to end of cap screw slots; a steady flow of fuel should now emerge from spill pipe (B—Fig. 171).

Fig. 172—View showing typical installation of auxiliary fuel tank and overflow pipe for checking spill timing.

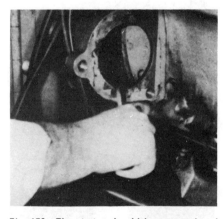

Fig. 173—The starter should be removed and crankshaft moved carefully by hand as shown when checking spill timing.

Now turn tachometer drive adapter slowly clockwise until fuel flow just stops at spill pipe opening. Tighten the three cap screws (S—Fig. 168) at this point. Recheck injection timing by turning engine crankshaft opposite normal direction of rotation a few degrees until flow starts at spill pipe; then in normal direction of rotation until flow stops. Dial indicator (D—Fig. 171) should read 0.352 inch (8.94mm) which corresponds to a beginning fuel delivery (injection) at 28 degrees BTDC crankshaft position.

COOLING SYSTEM (MF1155 Models)

RADIATOR

122. A 7 psi pressurized cooling system is used. Capacity is approximately 50 U.S. quarts. An engine oil cooler is attached to the side of the oil pan.

To remove the radiator, first drain cooling system and remove hood and side panels. Disconnect coolant hoses, remove fan blades and disconnect heat shield from upper, front of radiator. Unbolt radiator from right side mounting bracket, remove left side mounting bracket and slide radiator out left side. Install by reversing the removal procedure.

THERMOSTAT

123. A thermostat is located in the front end of each cylinder head and retained in outlet elbow by a snap ring. A jiggle pin is installed in bypass passage to keep passage open. Make sure jiggle pin is free to move whenever elbow is removed for any reason. The thermostat should begin to open at 157-168 degrees F. (69.4-75.6 degrees C.) and should be fully open at 188 degrees F. (86.7 degrees C.).

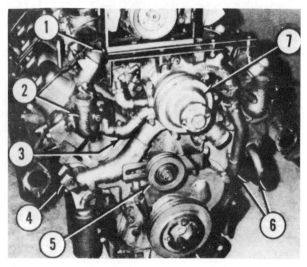

Fig. 175—View of installed water pump with hoses connected.

1. Alternator support channel
2. Right-hand outlet housing
3. Right-hand by-pass hose
4. Right-hand inlet connection
5. Idler pulley
6. Oil cooler pipes (coolant)
7. Water pump

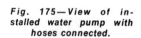

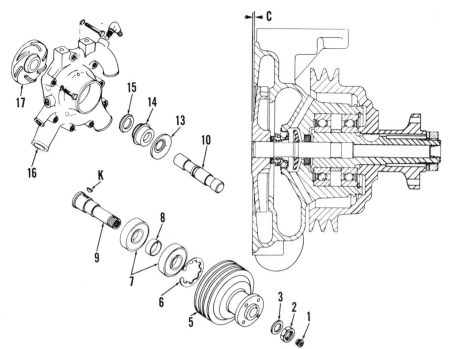

Fig. 176—Exploded and cross sectional views of water pump. Refer to Fig. 175 for installed view.

1. Cap plug	6. Snap ring	9. Front shaft	14. Rear seal
2. Nut	7. Bearing	10. Rear shaft	15. Ceramic insert
3. Washer	8. Spacer	13. Slinger	16. Body
5. Pulley			17. Impeller

WATER PUMP

124. Refer to Fig. 175 for installed view of water pump and hoses. Fig. 176 shows water pump cross section and exploded views. To remove the water pump, drain cooling system, remove fan belts and fan. Disconnect alternator bracket and hoses from pump, remove retaining nuts, then slide water pump forward off studs.

When disassembling the pump it is first necessary to remove pulley (5) using a suitable puller. After removing snap ring (6), both shafts and the bearings can be pressed out from impeller end using a suitable mandrel.

When reassembling, pack area between the two bearings (7) half full of high melting-point grease. If bearings are removed, install with open sides together next to spacer (8). Rear face of impeller center should be 0.015-0.020 inch (0.38-0.51mm) beyond flush with rear surface of housing as shown at (C).

ELECTRICAL SYSTEM (All Models)

ALTERNATOR

130. A Delco-Remy 10SI type 106 alternator with integral, solid state regulator is used on all MF1105, MF1135 and MF1155 models. Several different alternators have been used. Be sure of model when testing. Tractors without cabs use 37 or 42 amp units and models with cabs use alternators rated at 61 or 63 amps. Cold output rpm is alternator rpm (not engine speed).

1100578
Field Current (80°F.)
    Amperes 4.0-4.5
    Volts 12

Cold Output
    Amperes at 2000 rpm 22
    Amperes at 5000 rpm 33
Rated Hot Output (Amperes) at
    Maximum Operating Speed 37

1100579 & 1100583
Field Current (80°F.)
    Amperes 4.0-4.5
    Volts 12
Cold Output
    Amperes at 2000 rpm 25
    Amperes at 5000 rpm 37
Rated Hot Output (Amperes) at
    Maximum Operating Speed 42

1100581
Field Current (80°F.)
    Amperes 4.0-4.5
    Volts 12
Cold Output
    Amperes at 2000 rpm 30
    Amperes at 5000 rpm 55
Rated Hot Output (Amperes) at
    Maximum Operating Speed 61

1100585 & 1100589
Field Current (80°F.)
    Amperes 4.0-4.5
    Volts 12
Cold Output
    Amperes at 2000 rpm 30
    Amperes at 5000 rpm 57
Rated Hot Output (Amperes) at
    Maximum Operating Speed 63

To disassemble the alternator, first scribe matching marks (M—Fig. 180) on the two frame halves (4 and 16), then remove the four through bolts. Pry

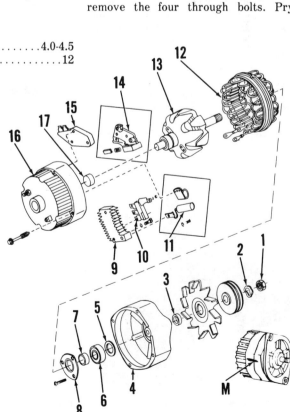

Fig. 180—Exploded view of alternator with internal mounted solid state regulator. Note match marks (M) on end frames.

1. Pulley nut
2. Washer
3. Spacer (outside drive end)
4. Drive end frame
5. Grease slinger
6. Ball bearing
7. Spacer (inside drive end)
8. Bearing retainer
9. Bridge rectifier
10. Diode trio
11. Capacitor
12. Stator
13. Rotor
14. Brush holder
15. Solid state regulator
16. Slip ring end frame
17. Bearing & seal assembly

frame apart with a screwdriver between stator frame (12) and drive end frame (4). Stator assembly (12) must remain with slip ring end frame (16) when unit is separated.

NOTE: When frames are separated brushes will contact rotor shaft at bearing area. Brushes MUST be cleaned of lubricant if they are to be reused.

Clamp the iron rotor (13) in a protected vise, only tight enough to permit loosening of pulley nut (1). Rotor end frame can be separated after pulley and fan are removed. Check bearing surface of rotor shaft for visible wear of scoring. Examine slip ring surface for scoring or wear, and rotor winding for overheating or other damage. Check rotor for grounded, shorted or open circuits using an ohmmeter as follows:

Refer to Fig. 181 and touch the ohmmeter probes to points (1-2 and 1-3); a reading near zero will indicate an open circuit and a lower reading will indicate an internal short. If windings are satisfactory, mount rotor in a lathe and check runout at slip rings using a dial indicator. Runout should not exceed 0.002 inch (0.05mm). Slip ring surfaces can be trued if runout is excessive or if surfaces are scored. Finish with 400 grit or finer polishing cloth until scratches or machine marks are removed.

Before removing stator, brushes or diode trio, refer to Fig. 182 and check for grounds between points A and C and B to C with an ohmmeter, using the lowest range scale. Then reverse the lead connections. If both A to C readings or both B to C readings are the same, the brushes may be grounded because of defective insulating washer and sleeve at the two screws. If the screw assembly is not damaged or grounded, the regulator is defective.

To test the diode trio, first remove the stator. Then remove the diode trio,

noting the insulator positions. With an ohmmeter, check between points A and D (Fig. 183) and then reverse the ohmmeter lead connections. If diode trio is good it will give one high and one low reading. If both readings are the same, the diode trio is defective. Repeat this test at points B and D and C and D.

The rectifier bridge (Fig. 184) has a grounded heat sink (A) and an insulated heat sink (E) that is connected to the output terminal. Connect ohmmeter to the grounded heat sink (A) and to the flat metal strip (B). Then reverse the ohmmeter lead connections. If both readings are the same, the rectifier bridge is defective. Repeat this test between points A and C, A and D, B and E, C and E, and D and E. Capacitor (11—Fig. 180) connects to the rectifier bridge and grounds to end frame, and protects the diodes from voltage surges.

Test the stator windings for grounded or open circuits as follows: Connect ohmmeter leads successively between each pair of leads. A high reading would indicate an open circuit.

NOTE: The three stator leads have a common connection in the center of the windings. Connect ohmmeter leads between each stator lead and stator frame. A very low reading would indicate a shorted circuit. A short circuit

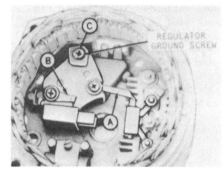

Fig. 182—Test points for brush holder. Refer to text.

within the stator windings cannot be readily determined by test because of the low resistance of the windings.

Brushes and springs are available only as an assembly which includes brush holder (14—Fig. 180). If brushes are reused, make sure all grease is removed from surface of brushes before unit is reassembled. When reassembling, first install regulator and then brush holder, springs and brushes. Push brushes up against spring pressure and insert a short piece of straight wire through hole and through end frame to outside. Be sure that the two screws at points A and B (Fig. 182) have insulating washers and sleeves.

NOTE: A ground at these points will cause no output, or controlled output. Withdraw the wire under brushes only after alternator is assembled.

Remove and inspect ball bearing (6—Fig. 180). If bearing is in satisfactory condition, fill bearing ¼ full with Delco-Remy lubricant No. 1948791 and reinstall. Inspect needle bearing (17) in slip ring end frame. This bearing should be renewed if its lubricant supply is exhausted; no attempt should be made to relubricate and reuse the bearing. Press old bearing out toward inside and press new bearing in from outside until flush with outside of end frame. Saturate felt seal with SAE 20 oil and install seal.

Reassemble alternator by reversing the disassembly procedure. Tighten pulley nut to a torque of 50 ft.-lbs.

STARTING MOTOR

131. A Delco-Remy starting motor is used. Model 1113139 is used on MF1105 tractors; model 1113683 is used on MF1135 tractors and model 1114190 is used on MF1155 tractors. Specifications are as follows:

1113139
Brush Spring Tension . . . . .80 oz. (min.)
No Load Test—
    Volts 11.5
    Amperes (w/solenoid) 57-70
    Rpm .5000-7400

Fig. 181—View of rotor showing test points for checking for grounds, shorts and opens.

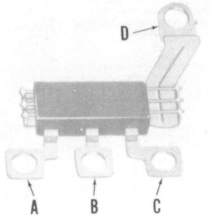

Fig. 183—Diode trio test points. Refer to text.

Fig. 184—Bridge rectifier test points. Refer to text.

Lock Test—
  Amperes 500
  Torque ft.-lbs. .... 22 (min.)
  Volts (approximate) 3.4

1113683
Brush Spring Tension ..... 82 oz. (min.)
No Load Test—
  Volts 9
  Amperes (w/solenoid) 75-105
  Rpm 5000-7000

1114190
No Load Test—
  Volts 9
  Amperes (w/solenoid) 140-190
  Rpm 4000-7000
Pinion Clearance-inch 21/64-25/64

CIRCUIT DESCRIPTION

132. Refer to Fig. 185 for a schematic wiring diagram. The two 95 Ampere Hour, 12 Volt batteries are connected in parallel and negative posts are grounded. Make sure both batteries are of equal capacity and condition. Quick disconnect couplings are used at all main connections.

ENGINE CLUTCH

All models are equipped with a dry type single disc clutch mounted on the flywheel and released by a foot pedal. The clutch controls tractor travel, the independent power take-off being controlled by a separate multiple disc wet clutch located in rear axle center housing.

ADJUSTMENT

140. Clutch pedal free play should be 11/16 inch (17.5mm) as measured between bottom of slot and clutch pedal arm as shown at (F—Fig. 190). Loosen lock nuts (1—Fig. 191) and turn the turnbuckle (2) as required to provide correct amount of pedal free play. Tighten lock nuts after adjustment is correct.

The safety start switch (S—Fig. 192) is located under the left hand foot rest and is actuated by the clutch pedal. This safety switch is designed to prevent starting unless clutch pedal is fully depressed. Adjust by turning the two lock nuts on switch so that switch is depressed 1/8 inch (3.2mm) when pedal is completely down.

TRACTOR SPLIT

141. Proceed as follows to detach (split) the engine from the transmission assembly for clutch service: Remove hood, disconnect all of the six hydraulic lines located above engine, detach priority valve from its mounting plate and disconnect batteries.

On MF1105 and MF1135 models,

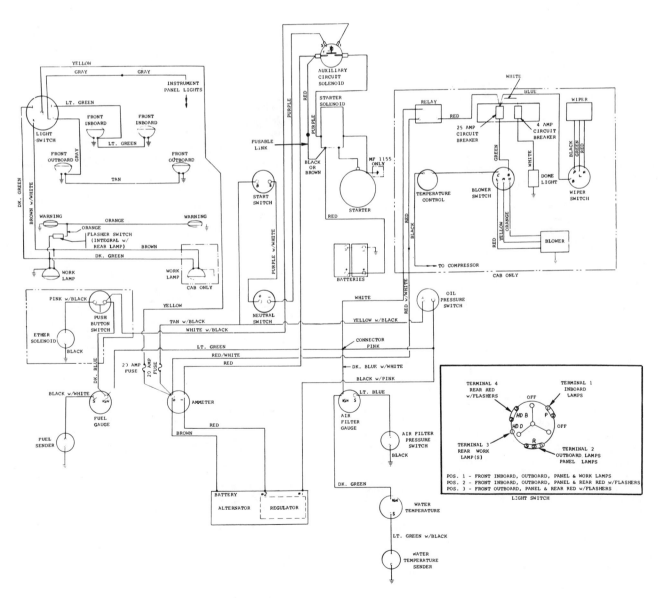

Fig. 185—Wiring diagram typical of all models.

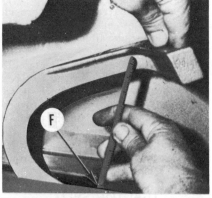

Fig. 190—Clutch pedal free play is measured at (F) between pedal and bottom of slot.

Fig. 191—Clutch pedal free play is adjusted at turnbuckle (2). Flat (F) is provided for removal of shaft.

Fig. 192—Clutch safety switch is located at (S).

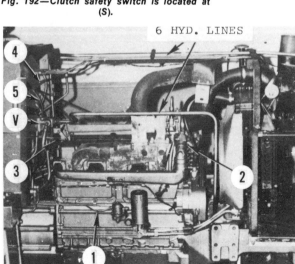

Fig. 193—View of right side showing items to be disconnected when splitting six cylinder engine and clutch. Refer to text for procedure.

refer to Figs. 193 and 194. Disconnect fuel line at lift pump (1—Fig. 193), electrical wires from alternator and temperature gage sending unit (2), throttle control rod from bellcrank (3), electrical connector (4) and oil pressure gage line (5). Disconnect tachometer cable (6—Fig. 194), fuel shut-off cable (7), fuel return line (8) and all electrical wires from starter. Detach auxiliary solenoid (9) from mount and move it out of the way without disconnecting wires.

On MF1155 models, refer to Figs. 195 and 196. Disconnect tachometer cable (1—Fig. 196), throttle control rod (2—Fig. 195), electrical connector (3), fuel shut-off cable (4), rear end of fuel filter support plate, fuel supply line (5) at flexible hose connection, electrical wires (6) from starter motor, oil pressure gage line (7), temperature indica-

Fig. 194—View of six cylinder engine left side. Refer to text for clutch split procedure.

Fig. 195—View of MF1155 tractor right side showing items to be disconnected for splitting tractor. Priority valve is shown at (V). Refer to text.

Fig. 196—View of MF1155 tractor left side. Connectors (C) are self sealing, couplers for air conditioning.

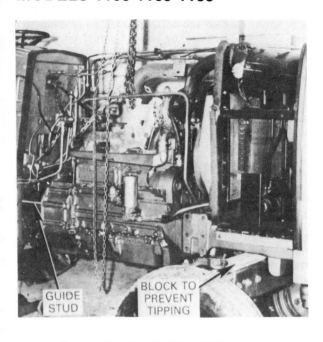

Fig. 197—View showing supports necessary when splitting. Six cylinder model is shown but blocking is similar for MF 1155.

Tighten the large bolts attaching oil pan to transmission housing to 360-540 ft.-lbs. torque. Remainder of procedure is reverse of disassembly.

OVERHAUL

142. Refer to Fig. 198 for exploded view of clutch cover, pressure plate, throwout bearing, linkage and associated parts. The clutch disc with four pads (6) is used in MF1105 models together with brown springs (2). Lavender springs (2) are used on other models with clutch disc (7) which has five pads.

To disassemble the cover, place the assembly in a suitable press as shown in Fig. 201 or use three threaded rods as shown in Fig. 199 to pull pressure plate (5—Fig. 198) into cover (1). Remove the three nuts from pivot blocks (10), then release the pressure plate

tor sending unit wire (8—Fig. 196), electrical wires from alternator and fuel return line (9) at top of fuel filter. Unhook fuel line from base of priority valve support plate and remove hydraulic oil line retaining clamp (10). Detach auxiliary solenoid (11) from mount and move it out of the way without disconnecting lead wires.

On all models, check to be sure that all wires, tubes and other equipment are positioned so that they will be free to separate without catching. Support tractor under transmission, block front axle to prevent tipping and support weight of engine so that front end and engine may be rolled forward. Refer to Fig. 197. Remove bolts and cap screws attaching engine to transmission and carefully separate by rolling front end and engine forward.

NOTE: It may be necessary to disconnect foot throttle plate and work through opening to remove the upper retaining bolts on models equipped with cabs.

CAUTION: Carefully observe the joint between engine and transmission while loosening bolts to be sure that front end and tractor are supported evenly. Change support as necessary if parting surfaces are not parallel. Dowel pins are installed between engine and transmission, but use of dowel guide studs is recommended when separating and rejoining.

When rejoining, install guide studs, then carefully move engine back against transmission housing while aligning splines of clutch with transmission input shaft and splines of flywheel adapter with pto shaft. If splines do not align, DO NOT force together, but turn flywheel slowly while sliding together.

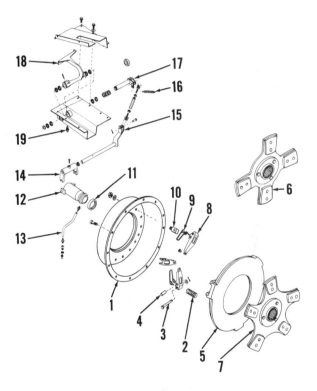

Fig. 198—Exploded view of clutch and linkage. Disc (6) with four pads is used with brown springs (2) on MF 1105 models only.

1. Clutch cover
2. Springs
3. Pin
4. Pin
5. Pressure plate
6. Disc (MF 1105)
7. Disc (MF 1135, MF 1155)
8. Release lever
9. Anti-rattle spring
10. Pivot block
11. Throw-out bearing
12. Collar
13. Lubrication hose
14. Fork
15. Release shaft
16. Spring
17. Shaft
18. Pedal
19. Safety switch

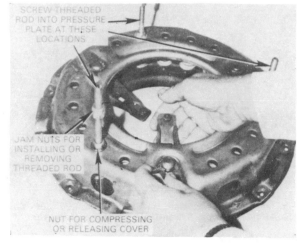

Fig. 199—Three threaded rods may be used as shown for relieving pressure while disassembling and reassembling clutch. Unit may be held by press as shown in Fig. 201.

SCREW THREADED ROD INTO PRESSURE PLATE AT THESE LOCATIONS

JAM NUTS FOR INSTALLING OR REMOVING THREADED ROD

NUT FOR COMPRESSING OR RELEASING COVER

slowly. Remainder of disassembly is self evident. Check springs (2) for over heating, breakage, wear or other damage and against the following pressure specifications:

Lavender (Models 1135 and 1155)

　Pressure at
　　1-13/16 inches . . . . 180-190 lbs.
　Pressure at 46.04mm . . 81.65-86.18kg.

Brown (Model 1105)

　Pressure at
　　1-13/16 inches . . . . 140-150 lbs.
　Pressure at 46.04mm . . 63.50-68.04kg.

When reassembling pressure plate and cover unit, observe the following: Coat the wear surfaces of release

Fig. 200—Heads of pins (3) must all be direction shown to prevent damage.

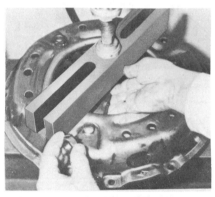

Fig. 201—A press may be used as shown to relieve spring pressure while disassembling and reassembling.

DAMPER SPRING
SIDE TOWARD REAR
OF TRACTOR

Fig. 202—Damper springs of clutch disc should be away from flywheel when assembling.

levers (8—Fig. 198), anti-rattle springs (9), pivot blocks (10) and pins (3 & 4) with extreme pressure and high temperature lubricant such as "Lubriplate 930-AA". Install pin (3) as shown in Fig. 200 with head toward side shown so that thrust will be against head of pin, not washer and cotter pin. Position the springs on pressure plate and place cover over the assembly with pivot block studs aligned with formed holes in cover. Compress springs with the three threaded rods (Fig. 199) or in press (Fig. 201) and install the three serrated washers and nuts. Tighten nuts to 40-45 ft.-lbs. torque.

The pads of clutch disc (6 or 7—Fig. 198) are 0.463 inch (11.76mm) thick when new. The friction surface of flywheel may be resurfaced; however, the mounting surface of clutch cover must also be machined the same amount. The disc contact surface must be 1.182-1.192 inches (30.02-30.28mm) below the cover mounting surface of flywheel.

When assembling clutch to flywheel use special tool (MFN531) to center the splines (Fig. 202) while cover is being installed. Damper spring side of clutch

Fig. 203—Release lever height (H) is measured from rear surface of adapter face to throw-out bearing contact surface of release levers.

INSTALL FORK WITH
CHAMFERED SIDE
TOWARD TRANSMISSION

Fig. 204—Chamfered side of clutch release fork should be toward transmission as shown.

disc should be to rear away from flywheel. The contact screw on each of the release levers should be exactly the same height so that all three will contact throwout bearing evenly. Height (H—Fig. 203) can be measured from a straight edge placed across machined surface of flywheel housing. Distance from surface of flywheel housing will vary from approximately 7/8 inch using a new disc to approximately ½-inch with an old, worn disc. The contact screws should be screwed into release levers as far as possible, then back out only the low screws only as far as necessary to make all three the same height.

RELEASE LINKAGE

143. The release linkage is shown in Fig. 198. The tractor must be split between the engine and transmission housing for removal of the release shaft (15) or fork (14). Disconnect linkage (Fig. 191), remove screw (S—Fig. 204), then withdraw release shaft. A flat (F—Fig. 191) is provided which permits withdrawing shaft.

Bores of bushings for release shaft should be coated with a dry lubricant before assembling. Chamfered side of fork should be toward transmission as shown in Fig. 204 and the hole in release shaft must be aligned with set screw in fork. Use safety wire to prevent screw(s) from loosening.

EIGHT SPEED TRANSMISSION

All models may be equipped with a main gear change transmission which has four forward and one reverse speed and is compounded by a range transmission with two speed which provides a total of eight forward speeds and two reverse speeds. Refer to the appropriate following paragraphs for service to this EIGHT speed transmission or to paragraphs 153 through 157 for service to the other available (Multi-Power) transmission.

OPERATOR'S PLATFORM

145. **REMOVE AND REINSTALL.** The operator's platform must be removed for several different operations, including removing the transmission. To remove, proceed as follows: Remove the seat assembly, disconnect the quadrant lever, control selector lever and response control lever. Disconnect hydraulic line from transfer cap and return line from lift cover. Detach auxiliary couplers from mounting plate and differential lock control from left of

seat. Disconnect left and right fender lights, then pull wire free from fenders. Remove the bolts attaching the cross brace to lift cover, bolts attaching platform to instrument console and bolts securing platform to braces at both sides of transmission. Attach hoist and lift platform as shown in Fig. 206. Be sure that all lines, wires and linkage are free.

NOTE: Platform is heavier on right rear because of auxiliary valve.

When reassembling, reverse removal procedure. Be careful to prevent damaging wires, linkage and hydraulic lines when lowering into position. Install all bolts securing platform to tractor.

INSTRUMENT CONSOLE

146. **REMOVE AND REINSTALL.** To remove the instrument console, first remove the platform as outlined in paragraph 145. Disconnect clutch linkage (1—Fig. 207) from cross shaft arm (2), lube line (3) from top of transmission housing and lines (4,5,6,7, & 8—Fig. 208). Disconnect high-low shift rod (9) and parking pawl rod (10) and locate out of the way. Disconnect the three shift levers (11, 12, & 13—Fig. 209) and master cylinder return line (14). Temporarily attach the priority valve to the tube assemblies at front of console and disconnect or remove tube (15—Fig. 210) between priority valve and transfer cap. Disconnect brake lines from master cylinder and wire

Fig. 206—View showing operator's platform being lifted from tractor.

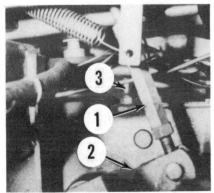

Fig. 207—Clutch release linkage is shown at (1), cross shaft release arm at (2) and lube line at (3).

(19—Fig. 208) from fuel tank sending unit. Tie the wires, tubes and rods together and secure to instrument console, then attach hoist to the console support bracket. Disconnect bracket

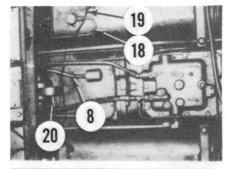

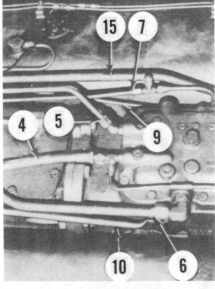

Fig. 208—Two views of transmission and center housings showing various items to be detached when removing instrument console. Refer to text.

Fig. 209—Views of shift levers (11, 12 & 13) and master cylinder return line (14).

Fig. 210—Two views of lines which must be disconnected when removing instrument console.

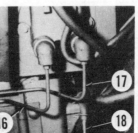

(20) from console, then carefully lift from top of transmission.

Reinstall by reversing removal procedure. Refer to Fig. 212 for routing battery cables.

MAIN TRANSMISSION REMOVAL

147. To remove the complete main transmission unit from the tractor, first detach (split) engine from transmission as outlined in paragraph 141, remove operator's platform as in paragraph 145 and instrument console as outlined in paragraph 146. Drain transmission housing, center housing and both fuel tanks, then remove the fuel tanks. Remove batteries, locate stand under the center housing and block up under drawbar at rear to prevent the center housing and axle from tipping backwards when transmission is being removed. Attach hoist to top of transmission then remove retaining bolts and separate transmission from center housing.

When reinstalling, reverse the removal procedure observing the following: Splined coupling between pto shafts should be installed with undercut end toward rear (center housing). Tighten side and bottom bolts securing transmission to center housing to 180-200 ft.-lbs. torque, then tighten the two top bolts to 130-140 ft.-lbs. torque.

MAIN TRANSMISSION OVERHAUL

148. **INPUT SHAFT AND FRONT COVER.** To remove only the input shaft and front cover, detach (split) engine from transmission as outlined in paragraph 141 and remove clutch link-

Fig. 211—Two views of instrument console being raised away from tractor.

Fig. 212—The left view shows correct routing of left battery cable and right view shows correct routing of right battery cables.

snap ring (9). Refer to Fig. 216. Leave front bearing support plate (16—Fig. 215) off until after housing (7) has been installed. Install retainer (4) and gasket (5) being careful not to damage seal (6). Position gasket (8) onto cover (7) with all holes carefully aligned, then install the cover with the special dowel screw in top center hole. Install remaining retaining screws and tighten to 35 ft.-lbs. torque.

Remove "O" ring (18) and shim gaskets (17) then temporarily locate the front bearing support (16) and bearing (19) in bore of cover (7). Install the three retaining screws by hand while tapping on the support (16) lightly. Measure clearance between flange of support and mating surface of housing as shown in Fig. 217. Select the necessary thickness of gaskets (17—Fig. 215) which will provide 0.004-0.008 inch preload on bearings. Remove the front bearing (19) and support plate (16), then reinstall using the correct gaskets (17) and "O" ring (18). Remainder of reassembly will be reverse of disassembly.

149. SHIFT FORKS AND SHAFTS. To remove the shift forks and shafts, it is necessary to remove the main transmission assembly as outlined in

Fig. 214—Views of hoist attached to main transmission for removal and installation.

Fig. 216—View of input shaft and front cover removed. Refer to Fig. 215 for legend.

age as outlined in paragraph 143. Withdraw front pto shaft (2—Fig. 215), remove retaining screws, then remove front cover assembly. Unbolt and remove input shaft retainer (4), remove snap ring (9), then pull shaft (13) forward out of gear (10) and bearing (12) out of bore of cover (7). Unbolt and remove countershaft front bearing support plate (16) and associated parts (17, 18 & 19).

Seal (14) is located in end of shaft (13) with lip toward rear and seal (6) is located in retainer (4) with lip toward rear. Press bearing and snap ring (12) onto shaft with snap ring and groove toward front, then install snap ring (11). Install new sealing rings (15) in grooves of shaft and coat sealing rings with petroleum jelly. Carefully slide shaft, sealing rings and bearing into bore of cover (7). Install gear (10) and

Fig. 215—Exploded view of input shaft and front cover for 8 speed transmission.

1. Snap ring
2. PTO input shaft
3. Tube
4. Retainer
5. Gasket
6. Seal
7. Cover
8. Gasket
9. Snap ring
10. Gear
11. Snap ring
12. Bearing and large snap ring
13. Input shaft
14. Seal
15. Rings
16. Support
17. Shims
18. "O" ring
19. Bearing
20. Bushings

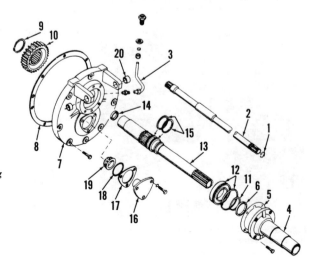

paragraph 147. Unbolt and remove the transmission top cover. Remove the detent springs (D—Fig. 220) and the three detent balls from under springs. Remove the retainer plate (P—Fig. 221) and loosen or remove set screws from shift forks. Pull the left (2 & R—Fig. 221) shift rail out of case toward rear and remove the shift fork. Pull the middle (4) shift rail out of case toward rear and remove the interlocking pin (Fig. 222). Pull the right (1 & 3—Fig. 221) shift rail out of case toward rear and remove shift fork. Use a wire as

Fig. 217—Bearing should be correctly set as described in text.

shown in Fig. 223 to move the two interlock balls out of case.

CAUTION: Be careful to remove these two balls. The two balls can be easily recovered if dropped into the front compartment, but not if allowed to fall into rear compartment.

When reassembling, start the three shift rails into rear of case as shown in Fig. 224. Lubricate and slide the two interlock balls (B—Fig. 219) into the case bores. Position the 1st and 3rd shift fork as shown in Fig. 225 and slide the rail into bore of shift fork. Move gear and shift rail to neutral then align set screw hole in rail and fork so that set screw can be installed. Coat threads of set screw with lock sealant and install set screw securely. Insert 2nd and reverse shift fork into gear and slide the rail into bore of shift fork. Move gear and shift rail to neutral then align set screw hole in rail and fork so that set screw can be installed. Coat threads with lock sealant and tighten securely. Install rear plate (P—Fig. 221) with flat smooth side away from main shaft rear bearing as shown. Remainder of assembly will be reverse of

disassembly procedure. Refer to paragraphs 147, 146, 145 and 141 for assembly notes.

150. MAIN SHAFT AND SLIDING GEARS. To remove the main shaft and sliding gears, first remove the main transmission as outlined in paragraph 147, the input shaft and front cover as in paragraph 148 and the shift rails and forks as in paragraph 149.

Bump the shaft to rear and remove the front (1st and 3rd) sliding gear. Relocate the shaft back into case and remove the rear snap ring (36—Fig. 226). Bump shaft forward out of rear bearing and withdraw the rear (2nd and

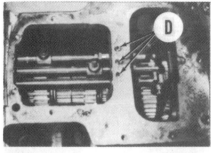

Fig. 220—View showing location of detent springs and balls (D).

Fig. 219—Exploded view of shift forks and related parts for 8 speed models.

1 & 3. Shift rail
2 & R. Shift rail
  4. Shift rail
10. Shaft and lever (4th)
11. "O" rings
12. Shaft and lever (2nd & Rev.)
13. "O" rings
14. Shaft and lever (1st & 3rd)
15. Snap ring
16. Lever (1st & 3rd)
17. Lever (4th)
18. Lever (2nd & Rev.)
19. Shift fork
20. Shift fork
21. Shift fork

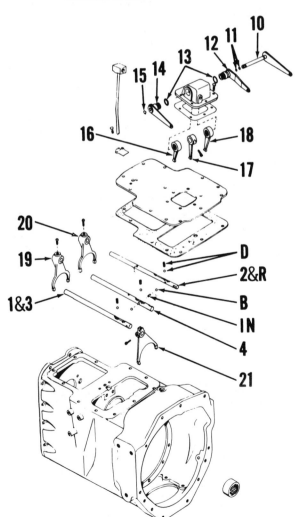

Fig. 221—View of rear of transmission showing plate (P) and shift rails.

Fig. 222—Be sure to catch the interlocking pin and don't let it fall into transmission. Pin is shown at (IN—Fig. 219).

reverse) sliding gear. The shaft, bearing and 4th gear can be withdrawn as an assembly through hole in front of case.

The reverse idler gear can be removed and at this time without removing the countershaft.

Inspect needle bearings (26 & 27) in main shaft and renew if necessary. Install both needle bearings with lettered end of case out toward ends of shaft.

Reassemble bearing (29), snap ring (30), gear (31) and snap ring (32) on shaft (25) and insert into position in transmission case.

NOTE: Shift grooves of all three gears (28, 31 & 33) should be toward front.

Install gear (33), then block front of shaft so that bearing (34) can be driven onto rear of shaft. Install snap ring (36) and snap ring (35). Install gear (28), then install shift forks as described in paragraph 149 and input shaft and front cover as outlined in paragraph 148. Refer to paragraphs

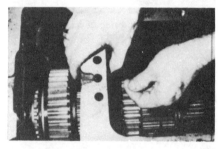

Fig. 223—A piece of wire can be used as shown to poke interlock balls out of housing. Be sure to catch balls as they fall out.

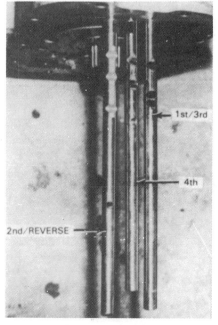

Fig. 224—View of the three shift rails inserted partially into the correct bores. Each rail is different.

147, 146, 145 and 141 for additional assembly notes.

151. COUNTERSHAFT. To remove the countershaft, first remove the mainshaft and sliding gears as outlined in paragraph 150. Remove snap ring (50—Fig. 227), then bump the shaft forward out of bearing (48). Slide countershaft (41) forward, while removing gear (47), wide spacer (46), gear (45), narrowest spacer (44), gear (43) and medium wide spacer (42).

The rear bearing (48) and the center needle bearing can be removed from case if renewal is required. Lettered

Fig. 225—View showing correct installation of shift fork (21—Fig. 219).

Fig. 226—Exploded view of main shaft and gears used in 8-speed transmission.

25. Main shaft
26. Bearing
27. Bearing
28. Gear (1st & 3rd)
29. Bearing
30. Snap ring
31. Gear (4th)
32. Snap ring
33. Gear (2nd & Rev.)
34. Bearing
35. Snap ring
36. Snap ring
P. Plate

side of needle bearing should be forward.

Insert the countershaft into housing while installing spacer (42—Fig. 228), gear (43) with hub to front, narrowest spacer (44), gear (45) with hub toward rear, widest spacer (46) and gear (47). The gear (47) may be installed either way. Insert shaft through case and block front of shaft so that bearing (48) can be bumped onto shaft far enough to remove all end play, then install the thickest snap ring (50) possible. End play must not be more than 0.010-inch and is adjusted by installing various thickness snap ring (50). Snap ring (50) is available in thicknesses of 0.093, 0.103, 0.113 and 0.123 inch.

Reverse removal procedure using assembly notes from paragraphs 150, 149, 148, 147, 146, 145 and 141. Be especially careful to adjust countershaft front support bearing (16, 19 & 40—Fig. 227) outlined in paragraph 148.

152. REVERSE IDLER GEAR. The reverse idler gear (54—Fig. 227) and shaft (55) can be removed after the main shaft and sliding gears are removed as outlined in paragraph 150. Remove the locking screw (51) and sealing washer, then withdraw shaft (55) out rear while lifting gear with bearing needles and spacer washers out the top.

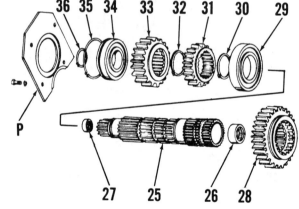

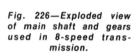

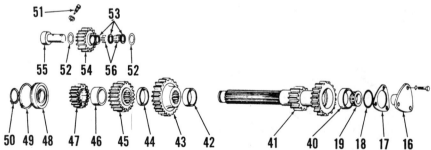

Fig. 227—Exploded view of countershaft and reverse idler assemblies.

16. Support
17. Shims
18. "O" ring
19. Bearing cone
40. Bearing cup
41. Countershaft

42. Spacer (7/8 in.)
43. Gear
44. Spacer (0.522 in.)
45. Gear
46. Spacer (1-1/8 in.)

47. Gear
48. Bearing
49. Snap ring
50. Snap ring
51. Locking screw

52. Thrust washers
53. Spacer washers
54. Reverse idler gear
55. Shaft
56. Loose rollers

NOTE: The 56 loose needle rollers (56), the three spacer washers (53) and two thrust washers (52) will fall out when shaft is removed. Be sure that all are accounted for before completing assembly.

When assembling, position one spacer washer (53) in bore of gear (54), then install one row of 28 loose rollers (56) followed by spacer washer (53), the remaining 28 rollers (56) and the last spacer washer (53). A special tool the correct diameter and length for holding the bearing rollers and spacer washers is available (MFN800K) from Massey-Ferguson. Insert the gear (54) with bearing needles (56), spacer washers (53) and thrust washers (52) into case with hub of gear toward front and slide shaft (55) into position. Be sure that shaft is correctly positioned and install screw (51) with sealing washer. Remainder of reassembly is reverse of disassembly.

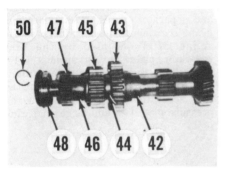

Fig. 228—View of countershaft showing location of parts.

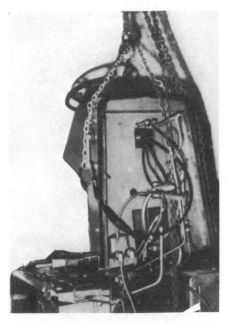

Fig. 230—After disconnecting all interfering parts, the instrument console can be lifted from tractor as shown for models with Multi-Power transmission.

MULTI-POWER (12 SPEED) TRANSMISSION

All models may be equipped with a gear change transmission having eight forward and two reverse speeds or a Multi-Power version providing twelve speeds forward and four reverse speeds. The Multi-Power transmission is a gear change transmission with six forward speeds and two reverse speeds which is additionally equipped with a hydraulically operated high-low range unit which may be shifted while tractor is moving under load. Many of the service procedures and even some parts may be similar. Refer to the appropriate following paragraphs for service of the Multi-Power (12 speed) TRANSMISSION or preceding paragraphs 145 through 152 for service to Eight Speed models.

MAIN TRANSMISSION REMOVAL

153. To remove the complete main transmission unit from tractor, first drain transmission and hydraulic system fluid. Remove the operator's platform, then detach (split) engine from transmission as outlined in paragraph 141. Disconnect shift linkage clutch linkage, brake lines, hydraulic line between priority valve and transfer cap, then remove fuel supply line and disconnect pto line at right side of case. Disconnect line from transmission top cover and lines from front cover assembly (on center housing). Disconnect light wires and remove instrument console as shown in Fig. 230. Drain and remove fuel tanks, place a stand under center housing and block under drawbar to prevent center housing and axle from tipping when transmission is removed. Attach hoist to transmission, remove retaining bolts and separate transmission from center housing.

Install by reversing removal procedure. Tighten side and bottom bolts securing transmission to center housing to 180-200 ft.-lbs. torque, then tighten top two bolts to 130-140 ft.-lbs.

MAIN TRANSMISSION OVERHAUL

154. **INPUT SHAFT AND FRONT COVER.** Refer to Fig. 231 for an ex-

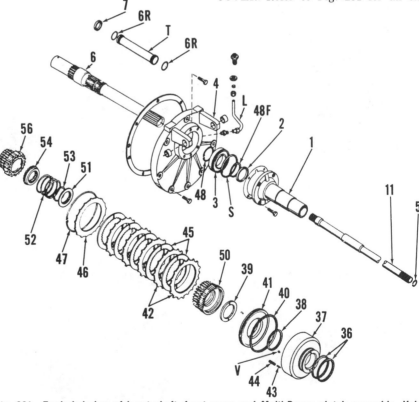

Fig. 231—Exploded view of input shaft, front cover and Multi-Power clutch assembly. Valve (V) is located in drum (37).

1. Retainer	11. PTO shaft	42. Separator plates	48F. Snap ring
2. Seal	36. Rings	43. Rollers (19 used)	50. Clutch hub
3. Bearing	37. Clutch drum	44. Springs (19 used)	51. Spacer
4. Cover	38. Piston seal	45. Clutch plates	52. Spring
5. Snap ring	39. Thrust washer	46. Pressure plate	53. Snap ring
6. Input shaft	40. Piston seal	47. Snap ring	54. Jaw clutch
6R. "O" rings	41. Piston	48. Snap ring	56. Constant speed gear
7. Seal			

ploded view of input shaft front cover and associated parts used on Multi-Power models. If only the input shaft, front cover, Multi-Power clutch and associated parts are to be removed, detach (split) engine from transmission housing as outlined in paragraph 141. Remove the main transmission as outlined in paragraph 153 if the unit is to be disassembled.

Remove the clutch release shaft, fork and bearing as outlined in paragraph 143 and withdraw the pto shaft (11). Remove hydraulic tube (L), then remove screws attaching transmission front cover (4) to housing.

NOTE: The top center retaining screw is machined to serve as a locating dowel for the front cover. Lift the input shaft, front cover and Multi-Power clutch out as an assembly.

To disassemble the removed unit, lift off jaw clutch (54) and spring (52). Unbolt and remove bearing retainer (1), remove snap ring (48F—Fig. 233) from front groove of input shaft, then bump shaft and clutch assembly rearward out of cover and bearing. Press bearing forward out of front cover and remove

snap ring (S) from groove in bearing.

Remove snap ring (47—Fig. 234), clutch pressure plate (46), spring (44—Fig. 235) and their guide rollers. Remove the five friction discs (45—Fig. 231) and separator plates (42). Remove snap ring (53—Fig. 234), thrust washer (51), clutch hub (50) and doweled thrust washer (39—Fig. 231) from clutch drum. Remove snap ring (48), then slide clutch drum (37) forward off of input shaft splines. Place clutch drum open-end up on bench and remove clutch piston using two pairs of pliers as shown in Fig. 236.

Input shaft (6—Fig. 231) has an inner tube (T) at rear of shaft which is sealed by an "O" ring (6R) at each end. The tube provides a passage for lubricant to the Multi-Power jaw clutch and removal is necessary only if leakage is suspected. Small end of tube (T) should be installed first and tube should be pushed into shaft (6) until it bottoms against flange in shaft. Lip of seal (7) should be outward (toward rear).

Examine all parts for wear, heat discoloration, warpage or other signs of damage. Renew all seals whenever unit

is disassembled for service.

Lubricate all parts during assembly. Coat piston rings (40 & 38) with petroleum jelly and rotate piston (41) while inserting into bore of drum (37). Position thrust washer (39) over dowel pins in drum (37) and slide drum assembly over splined end of input shaft (6). Install snap ring (48) and press bearing (3) into shaft with groove in outer diameter toward front end of shaft. Select correct thickness of snap ring (48F) which will limit end play of bearing to 0.004 inch (0.10mm) or less. Snap rings (48 & 48F) are alike; however, usually a 0.109 inch (thin) snap ring is installed at (48). Available thicknesses are 0.109, 0.115, 0.120 & 0.125 inch.

When installing the clutch friction discs (45) and separator plates (42), proceed as follows: Notice that the separator plates have one external lug for each two internal splines in clutch drum. Install one of the five separator plates (42), then alternate clutch discs and separator plates, positioning lugs of the remaining separator plates in the splines NOT occupied by the first separator plate. Install springs (44—Fig. 235) and their guide rollers (43—Fig. 231) on top of the lugs of the first separator plate, then install the pressure plate (46) and large snap ring (47).

Fig. 232—View of removed input shaft and cover assembly.

Fig. 234—View of Multi-Power clutch and input shaft. Refer to text for disassembly.

Fig. 236—Pliers can be used to withdraw piston.

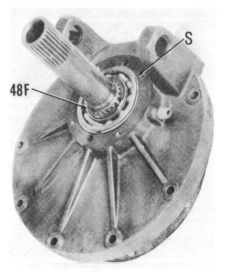

Fig. 233—View of input shaft and front cover with retainer removed.

Fig. 235—Nineteen springs (44) are installed as described in text.

Fig. 237—Refer to text for selection of snap ring (48F).

Install new sealing rings (36) in grooves of clutch drum (37), then carefully install the assembled shaft and clutch unit in front cover (4). Secure with snap ring (S) and install bearing retainer (1). Make sure clutch spring (52) and jaw clutch (54) are properly positioned on input shaft before reinstalling front cover. Be sure to install the special dowel screw at top center of cover and tighten all of the cover retaining screws to 33-38 ft.-lbs. torque. Complete reassembly by reversing removal procedure.

155. MAIN SHAFT AND SLIDING GEARS. To remove the transmission main shaft and sliding gears, first remove the main transmission assembly as outlined in paragraph 153. Remove the retainer plate (22—Fig. 239) from rear of transmission housing and slide pto shaft forward out of upper shafts. Remove the locking set screws in shift

Fig. 238—Install pressure plate and retaining snap ring as shown.

forks (28—Fig. 240) and with other shift rail in neutral, slide either rail (26 or 27) rearward out of forks and housing. Retrieve detent balls and springs (24) and interlock pin (25) as rails are removed.

Slide main shaft (16—Fig. 239) to rear until 2nd-3rd sliding gear (13) can be lifted from shaft, unseat and remove snap ring (17), then slide shaft (16) rearward out of low-reverse sliding gear (19) and housing. The constant speed gear (56) and bearing (9) can be removed from housing after removing front cover assembly as outlined in paragraph 154 and removing snap ring (10—Fig. 239A).

Rear bearing (20—Fig. 239) can be pressed from shaft after removing snap ring (21). Inner bore contains three caged needle bearings (14, 15 and 18) which can be renewed using Massey-Ferguson Special Tools MFN 800P, MFN 800Q and MFN 800MM, or by carefully measuring the location of rear bearing before removal and installing new bearing in same position. The two front bearings should be installed just beyond flush with edge of bearing bores. The new shaft is shipped with bearings installed.

Reinstall the main shaft by reversing the removal procedure. Note that the two shift forks (28—Fig. 240) are interchangeable and that both are installed with long end of hub to rear.

156. COUNTERSHAFT. To remove the transmission countershaft (23—Fig. 241), first remove the mainshaft as outlined in paragraph 155 and transmission front cover as in paragraph 154.

Remove snap ring (32) from rear of countershaft. Place suitable wood blocks (about 2 inches thick) between

second gear (26) and housing wall and, using a suitable puller, remove countershaft forward out of transmission case, lifting off gears and spacers as shaft is removed.

When installing the countershaft, snap ring (32) is supplied in selective thicknesses of 0.093, 0.103, 0.113 and 0.123 inch. Insert shaft from front, assembling spacers and gears on shaft

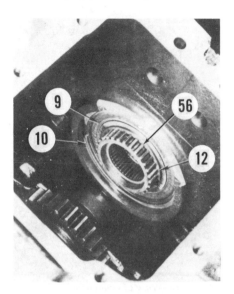

Fig. 239A—Constant speed gear (56) and bearing (9) can be removed after removing snap ring (10). Refer to text.

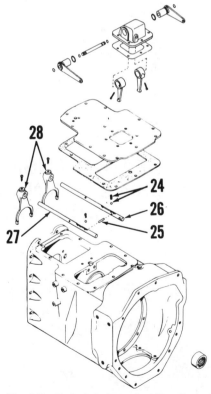

Fig. 239—Exploded view of countershaft and gears used in Multi-Power models.

R. Snap rings
S. Spacer
8. Snap ring
9. Bearing
10. Snap ring
12. Snap ring
13. Gear
14. Bearing
15. Bearing
16. Countershaft
17. Snap ring
18. Bearing
19. Gear
20. Bearing and snap ring
21. Snap ring
22. Plate

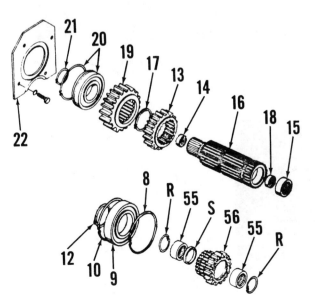

Fig. 240—Exploded view of shift rails and forks.

24. Detent spring and ball	26. Shift rail
25. Interlock pin	27. Shift rail
	28. Shift forks

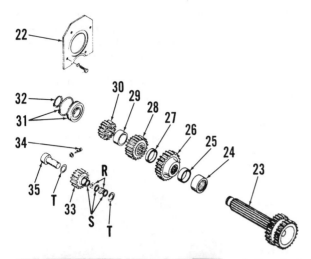

Fig. 241A—View of countershaft installed.

Fig. 241—Exploded view of countershaft and reverse idler assemblies.

R. Rollers
S. Spacer washers
T. Thrust washers
22. Plate
23. Countershaft
24. Bearing
25. Spacer (7/8 in.)
26. Gear
27. Spacer (0.522 in.)
28. Gear
29. Spacer (1-1/8 in.)
30. Gear
31. Bearing and snap ring
32. Snap ring
33. Reverse idler gear
34. Locking screw
35. Shaft

as follows: Medium spacer (25) which is 7/8 inch long, large gear (26) with hub forward, narrowest spacer (27) which is approximately ½-inch long, medium sized gear (28) with hub rearward, spacer (29) which is 1-1/8 inches long and small gear (30) can be installed either way. Block front of countershaft inside of front cover flange and, using a short section of 2 inch ID pipe, tap bearing (31) onto rear of shaft until it bottoms. Select the thickest snap ring (32) which can be installed. Counter-shaft end play should not exceed 0.010 inch (0.25mm).

157. **REVERSE IDLER SHAFT.** To remove the reverse idler gear (33—Fig. 241) and shaft (35), first remove the main shaft as outlined in paragraph 155. Remove the locking bolt (34) from left side of transmission housing and tap the shaft (35) rearward, lifting gear and thrust washers out as shaft is withdrawn. Be careful not to tip the gear and spill the 56 loose needle rollers in transmission case as gear is withdrawn.

Assembly can be facilitated by using the Massey-Ferguson Special Tool MFN800K or a piece of shaft about 0.010 inch (0.25mm) shorter than reverse idler slot in housing. Use the shaft to retain the loose needle rollers (R), spacers (S) and thrust washers (T). Install gear (33) with long hub forward. Shaft (35) will push the special tool out as it is inserted. Align counterbore and install locking bolt (34), then complete the assembly by reversing the disassembly procedure.

The hydraulic system charging pump is driven by gear teeth machined on outer edge of ipto clutch drum.

TRACTOR SPLIT

158. To detach (split) the tractor between transmission and rear axle center housing, first drain transmission, center housing and fuel tanks. Remove seat console lower front panel, operator's platform, lower step plates and tool box. Remove fuel tanks.

Disconnect park linkage rod and range shift link. Disconnect hydraulic lines and wiring running to front of tractor. Suitably support both sections and remove attaching flange bolts, then roll front of tractor forward away from rear axle center housing.

The ipto spline coupling (Fig. 243) may remain with transmission or with rear axle center housing when units are separated. When reconnecting the tractor, place coupling in rear unit as shown, with chamfered end of splines forward. Complete the assembly by reversing the disassembly procedure.

RANGE TRANSMISSION

159. **SHIFTER RAIL AND FORK.** Refer to Fig. 244 for an exploded view of range transmission shift rail and fork located in rear axle center housing. Actuating lever is mounted on instrument panel.

Fig. 243—Place spline coupling on rear unit with undercut end to rear.

REAR TRANSMISSION, CENTER HOUSING, MAIN DRIVE BEVEL GEARS AND DIFFERENTIAL

The rear axle center housing contains the Range Transmission, main drive bevel gears and differential, as well as power take-off gears, shafts and clutch.

Fig. 242—Views showing removal of final drive and axle. Be sure tractor is adequately supported and blocked.

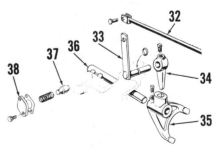

Fig. 244—Exploded view of range shift mechanism showing component parts.

32. Link
33. Lever
34. Finger
35. Shift fork
36. Shift rail
37. Detent plunger
38. Plate

To remove the rails and forks located in center housing, first detach (split) transmission from center housing as outlined in paragraph 158. Disconnect remainder of hydraulic lines from center housing front, top cover; remove the cap screws and lift off the cover with hydraulic system charging pump attached.

Remove detent plate (38—Fig. 244), spring and plunger (37). Loosen the set screw in fork (35) and slide shift rail (36) forward out of housing bore. Install by reversing the removal procedure. Actuating lever (33) and arm (34) can be removed if desired, by loosening set screw in arm.

160. **INPUT CLUSTER GEAR.** To remove the input cluster gear shaft (6—Fig. 245), first detach (split) transmission from rear axle center housing as outlined in paragraph 158 and remove shift rail and fork as outlined in paragraph 159. Disconnect remainder of hydraulic lines from center housing front, top cover and remove the cover and attached hydraulic system charging pump.

Remove ipto output housing as outlined in paragraph 186. Withdraw rear ipto shaft and sliding coupler (12—Fig. 246). Move ipto drive shaft (10) rearward to expose snap ring (11), unseat and remove the snap ring; then slide ipto drive shaft (10) rearward out of tractor.

Remove oil return line (R—Fig. 247) and oil suction tube (S—Fig. 245). Remove oscillator arm (A—Fig. 247) by prying it forward. Remove snap ring (2—Fig. 246), insert a suitable puller through the hollow cluster gear (6) and pull the unit forward until large gear contacts shift coupler, and bearing cup (3) is partially withdrawn from housing wall. Move shaft rearward until a slide

hammer puller can be installed as shown in Fig. 248, then remove the bearing cup. Pull front bearing cone as shown in Fig. 249, then lift input cluster gear from center housing.

Needle bearings (5—Fig. 246) can be installed in shaft (6) using Massey Ferguson Special Tools MFN 800P, MFN 800Q and MFN 800MM for correct positioning. If special tools are not available, note position of existing bearings and install new units in same location.

The rear bearing cup (8) is a slip fit in housing and can normally be removed by hand. If pulling is required, bearing cup and the press fit oil distributor ring (not shown) must be removed as a unit.

When installing the cluster gear, reverse removal procedure. Note that cluster gear bearing end play is controlled by a selective fit snap ring (2). Install bearing cup (3) and gently tap the cup rearward until shaft end play is eliminated, then install the thickest snap ring which can be installed. Pry the cup back against the installed snap ring, then measure end play using a dial indicator as shown in Fig. 251. End play must be within the limits of 0.001-0.009 inch (0.025-0.228mm). Snap rings (2—Fig. 246) are available in thicknesses of 0.077, 0.089, 0.099 and 0.109 inch. Ipto intermediate shaft end play must be checked as outlined in paragraph 186 after completion of assembly.

161. **BEVEL PINION SHAFT.** To remove the main drive bevel pinion shaft (33—Fig. 246), first detach (split) rear axle center housing from transmission as outlined in paragraph 158,

Fig. 247—Rear axle center housing showing removal of charging pump return line (R). Oscillator arm (A) can be removed by prying it forward.

Fig. 248—Using a slide hammer to remove range cluster gear front bearing cup.

Fig. 249—Removing front bearing cone from range cluster gear.

Fig. 245—Top view of rear axle center housing with front cover removed, showing input cluster gear (6) and hydraulic pump intake tube (S).

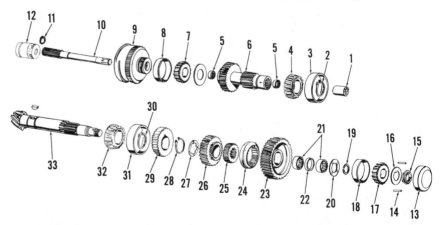

Fig. 246—Exploded view of bevel pinion shaft and range transmission gears located in rear axle center housing. IPTO clutch drum (9) drives hydraulic system charging pump and contains the actuating cam for hydraulic valve oscillating lever arm.

1. Coupling		
2. Snap ring	10. IPTO drive shaft	18. Bearing cup	26. High gear
3. Bearing cup	11. Snap ring	19. Retaining ring	27. Thrust washer
4. Bearing cone	12. Coupler	20. Thrust washer	28. Snap ring
5. Needle bearing	13. Bearing cap	21. Roller bearings	29. Park gear
6. Cluster gear	14. Lock pins	22. Spacer	30. Snap ring
7. Bearing cone	15. Shaft nut	23. Low gear	31. Bearing cup
8. Bearing cup	16. Washer	24. Shift collar	32. Bearing cone
9. IPTO clutch	17. Bearing cone	25. Hub	33. Bevel pinion

remove input cluster gear as in paragraph 160 and differential assembly as paragraph 163 or 166.

Remove pinion shaft bearing cap (13). Split the nut (15) using a sharp chisel, remove (and save) the two lock

pins (14); then remove and discard the nut.

Remove range shift rail and fork (paragraph 159) and park lock cover. Unseat snap ring (28—Fig. 246) in front of park lock gear (29) and slide the gear forward on pinion shaft. Remove the tab washer (16) and bearing cone (17) then extract the spiral retaining ring 19). Move pinion shaft rearward in housing, lifting out the contained gears and associated parts as they are free.

Reassemble by reversing the disassembly procedure, making sure that missing tooth of shift hub spline aligns with oil hole in bevel pinion shaft as shown in Fig. 256.

Install a new nut (15—Fig. 246) and tighten the nut until a rolling torque of 10-15 in.-lbs. is provided for pinion shaft bearings. When the correct torque is obtained, drive the two lock pins between shaft and nut at counterclockwise sides of key slots as shown in Fig. 259. Reinstall bearing cap (13—Fig. 246).

NOTE: Bearing cap (13) plays an important part in the centrifugal lubrication of gears (23 and 26). Make sure cap is installed.

Fig. 250—Front view of rear axle center housing showing IPTO shaft spline coupling and selective fit snap ring installed.

Fig. 251—Measuring cluster gear end play to determine correct thickness of selective fit snap ring shown in Fig. 250.

Fig. 252—IPTO clutch unit can be lifted out as shown after sliding intermediate shaft rearward and lifting out sliding coupler.

Fig. 253—View of bevel pinion shaft showing park gear correctly installed.

Fig. 254—Disassembling bevel pinion shaft.

Fig. 255—Gears and associated parts correctly positioned for bevel pinion shaft installation.

Fig. 256—Shift hub missing spline must align with oil hole in bevel pinion shaft during assembly.

Fig. 257—Reseat park gear retaining snap ring during reassembly.

Fig. 258—Installing low gear idler bearings during bevel pinion assembly.

PARK LOCK

162. The park lock cover mounts on left side of rear axle center housing. An assembled view is shown in Fig. 260A and an exploded view in Fig. 260. Park pawl (48) engages the gear (29—Fig. 246) which is keyed to bevel pinion shaft, thus locking pinion shaft in park position.

To remove the park lock cover, it is first necessary to drain rear axle center housing and drain and remove interfering fuel tank.

DIFFERENTIAL AND BEVEL RING GEAR

Two different ring and pinion sets have been used (7:36 and 8:41) on MF1105, MF1135 and MF1155 models. The reduction ratio is nearly the same (5.14:1 & 5.13:1) and either ratio may be installed, however, originally the 36 tooth ring gear is installed with 16 teeth side gears and differential pinions which are 2.95 inches (74.9mm) across teeth edges. The ring gear with 41 teeth was originally installed with 14 teeth side gears and differential pinions which are 3.33 inches (84.6mm) across teeth edges. On models with differential lock, both 3 and 5 plate type differential locks were used with 36 ring

Fig. 259—Pinion nut lock pins should be driven in at counter-clockwise end of key slots as shown.

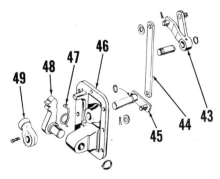

Fig. 260—Exploded view of park lock cover and associated parts. Park pawl (48) meshes with park gear (29—Fig. 246) on bevel pinion shaft.

43. Arm assy.	46. Cover
44. Link	47. Spring
45. Lever	48. Pawl

gear teeth; only 3 plate type was used with 41 teeth on ring gear. Refer to the appropriate following paragraphs for service.

Models Without Differential Lock

163. **REMOVE AND REINSTALL.** To remove the differential assembly, first drain transmission and center housing. Remove both final drive units as outlined in paragraph 170 and power take-off housing as in paragraph 186.

Working through rear opening in center housing, remove the oil return tube (Fig. 261) and block up between rim of main drive bevel gear and botton of housing. Unbolt and remove both carrier bearing retainer plates and brake pressure plates as assemblies, allowing differential unit to rest on previously installed block. Roll differential and ring gear rearward from center housing as shown in Fig. 262. Overhaul the removed unit as outlined in paragraph 164.

Fig. 260A—Assembled view of the park lock cover.

Fig. 261—The oil return tube shown above must be removed by removing nut and stud.

When installing differential, locate the assembly in center housing on block, positioning at approximate installed height. Install the right carrier bearing retainer plate first, then the left. Be sure stamped word "TOP" on carrier bearing retainer plates are up at that "O" rings (Fig. 263) remain in position while assembling. The 7/16 inch (11.1mm) hole and hex head screw is located at 10 o'clock position on right side as shown in Fig. 264 and at 2 o'clock position on left side. Install socket head screws in ½ inch (12.7mm)

Fig. 262—Use a wood block as protection and roll differential assembly rearward out of center housing.

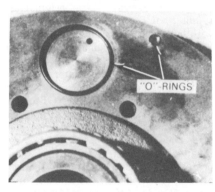

Fig. 263—Views showing the "O" rings located on inner surface of carrier bearing retainer plates. "O" rings seal brake fluid passages.

Fig. 264—View of right side showing correct position of the 7/16 inch hex head screw. Hex head screw should be at 2 o'clock position on left side.

holes and torque to 75-85 ft.-lbs. Install hex head screw in 7/16 inch (11.1mm) hole and tighten to 50-55 ft.-lbs. torque.

NOTE: If socket head screws are plated, discard and install new special unplated black screws.

Ring gear and pinion mesh position is not adjustable and normally it is not necessary to adjust backlash which should be 0.006-0.024 inch (0.15-0.61mm). Shims 0.003 inch thick (No. 190 027 M1) and 0.007 inch thick (No. 190 028M1) are available which may be installed between ring gear and differential housing to reduce backlash.

Install the square cut seal and the floating disc (Fig. 265) then reinstall axle assemblies. Tighten axle housing nuts and capscrews to 155-170 ft.-lbs. torque and drawbar bolts to 600-675 ft.-lbs. torque. Complete the assembly by reversing removal procedure and bleed brakes as outlined in paragraph 181.

164. OVERHAUL. Refer to Fig. 266 for an exploded view of differential assembly. Differential case bolts are installed with "LOCTITE" and are not safety wired.

Main drive bevel ring gear and pinion (11) are available only as a matched assembly. Ring gear is riveted to differential case (12) at original assembly, but service parts are provided with bolts and nuts. Center punch rivet heads and carefully remove heads using ½ inch drill bit. Use a 3/8 inch drill bit and carefully drill into center of rivet to depth of housing flange, then drive rivets out with punch. DO NOT chisel heads of rivets off. Ring gear may be 0.005 inch interference fit on case and may need to be pressed from case. Clean all parts carefully with a non-petroleum solvent such as "Loctite" primer, then install bolts and nuts using "Grade AV Loctite" (red) and tighten to 110-120 ft.-lbs. torque. The "Loctite" must cure for approximately

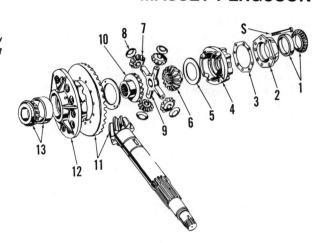

Fig. 266—Exploded view of standard non-locking differential.

1. Bearing
2. Adapter
3. Shim
4. Case half
5. Thrust washer
6. Side gear
7. Differential pinion
8. Thrust washer
9. Spider
10. Side gear
11. Ring gear & pinion
12. Case half
13. Bearing

2 hours at 72 degrees F. before load is applied.

Mark case halves before disassembling. The differential case is available only as a matched assembly which includes adapter (2), shim pack (3), case halves (4 & 12) and the two carrier bearing cups. If shims (3) are used to obtain proper bearing adjustment at factory assembly, thickness of shim pack should not be changed. Shims may not be used on some assemblies.

Be sure to carefully clean all parts thoroughly. Capscrews and threaded holes must be completely degreased, clean and dry before assembly. The center housing is hydraulic reservoir and must be cleaned and kept clean while assembling.

Assemble differential, coat threads of screws (S—Fig. 266) with "Loctite" grade AV (red) and tighten to 75-85 ft.-lbs. torque.

Models With Differential Lock

165. OPERATION. The differential lock assembly consists of a hydraulically actuated multiple disc clutch which selectively locks the left axle gear to differential cage, causing both rear wheels to rotate together as a unit. The differential lock clutch is automatically released if either or both brakes are applied, and can be released by pulling up on actuating valve knob or momentarily applying brake.

Power to operate the differential lock clutch is supplied by the low pressure, low volume circuit which also supplies working fluid to the Multi-Power and Power Take-Off hydraulic clutches.

Refer to Fig. 267 for an exploded view of differential assembly and to paragraph 237 for service procedures on hydraulic system components other than clutch.

166. REMOVE AND REINSTALL. To remove the differential assembly, first drain transmission and center housing. Remove both final drive units as outlined in paragraph 170 and power

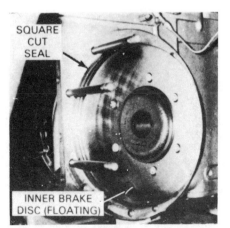

Fig. 265—View showing installation of the inner (floating) brake disc and the square cut seal ring.

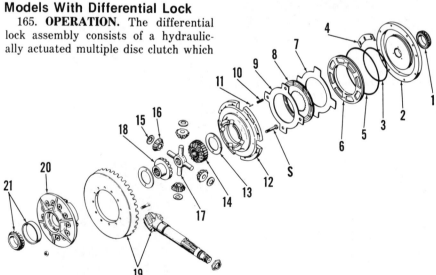

Fig. 267—Exploded view of locking type differential assembly. Refer to text for description of operation.

1. Bearing
2. Clutch cover
3. "O" ring
4. Shims
5. "O" ring
6. Piston
7. Separator plate (end)
8. Clutch disc
9. Separator plates
10. Separator springs
11. Roller guides
12. Case half
13. Thrust washer
14. Side gear
15. Thrust washers
16. Differential pinions
17. Spider
18. Side gear
19. Ring gear & pinion
20. Case half
21. Bearing

take-off housing as in paragraph 186.

Working through rear opening in center housing. remove the oil return tube (Fig. 261) and block up between rim of main drive bevel gear and bottom of housing. Unbolt and remove both carrier bearing retainer plates and brake pressure plates as assemblies, allowing differential unit to rest on previously installed block. Roll differential unit rearward from center housing as shown in Fig. 262. Overhaul the removed unit as outlined in paragraph 167.

When installing differential, locate the assembly in center housing on block, positioning at approximate installed height. Install the right carrier bearing retainer plate first. Be sure that stamped work "TOP" on bearing carrier plates are up and that "O" rings (Fig. 263) remain in position while assembling. The 7/16 inch (11.1mm) hole and hex head screw is located at 10 o'clock position on right side as shown in Fig. 264 and at 2 o'clock position on left side. Install socket head screws in ¹/₂ inch (12.7mm) holes and torque to 75-85 ft.-lbs. Install hex head screw in 7/16 inch (11.1mm) hole and tighten to 50-55 ft.-lbs. torque.

NOTE: If socket head screws are plated, discard and install new special unplated black screws. Be careful to prevent damage to differential lock seals (Fig. 268) when installing left carrier bearing retainer plate.

Ring gear and pinion mesh position is not adjustable and normally it is not necessary to adjust backlash which should be 0.006-0.024 inch (0.15-0.61mm). Shims 0.003 inch thick (No. 190 027 M1) and 0.007 inch thick (No. 190 028 M1) are available which may be installed between ring gear and differential housing to reduce backlash.

Install the square cut seal and the floating disc (Fig. 265) then reinstall axle assemblies. Tighten axle housing nuts and capscrews to 155-170 ft.-lbs. torque and drawbar bolts to 600-675 ft.-lbs. torque. Complete the assembly

Fig. 269—Views of three plate clutch and five plate clutch showing visual differences.

by reversing the removal procedure and bleed brakes as outlined in paragraph 181.

167. OVERHAUL. Refer to Fig. 267 for an exploded view of differential assembly. Both three and five plate differential lock clutches are used (Fig. 269).

Unbolt and remove cover (2—Fig. 267), separator plates (7 & 9), clutch discs (8) and springs (10). Differential lock piston (6) can be removed using puller screws as shown in Fig. 270.

NOTE: Be sure to save shims (4—Fig. 267) for reinstallation. Shims are used to obtain proper bearing adjustment at factory assembly and should not be changed. Each of the four stacks should be identical in thickness. Shims are included with new differential case if necessary. Some cases do not require shims.

Mark case halves before separating halves. Be sure to clean all parts thoroughly. Capscrews and threaded holes must be completely degreased, clean and dry before assembly. The

Fig. 270—Puller screws can be used as shown to remove the differential lock piston.

center housing is hydraulic reservoir and therefore must be cleaned and kept clean while assembling.

Main drive bevel ring gear and pinion (19) are available only as a matched assembly. Ring gear is riveted to differential case (20) at original assembly, but service parts are provided with bolts and nuts. Center punch rivet heads and carefully remove heads using ¹/₂ inch drill bit. Use a 3/8 inch drill bit and carefully drill into center of rivet to depth of housing flange, then drive rivets out with punch. DO NOT chisel heads of rivets off. Ring gear may be 0.005 inch interference fit on case and may need to be pressed from case. Clean all parts carefully with a non petroleum solvent such as "Loctite" primer, then install bolts and nuts using "Grade AV Loctite" (red) and tighten to 110-120 ft.-lbs. torque. The "Loctite" must cure for approximately 2 hours at 72 degrees F. before load is applied.

Assemble parts (13 thru 18) into case half (20), then install case half (12). Be sure that threaded holes are clean, degreased and dry then install screws (S) using "Grade AV Loctite". Tighten screws evenly to 110-120 ft.-lbs. Position thrust washer (13) on side gear (14) and the four springs (10) in seats. Install one clutch disc (8) followed by separator plate (9) which has holes for springs (10). The number of clutch discs (8) and separator plates (9) will be different for 3 and 5 plate assemblies, but install remaining parts (8 & 9) alternating discs and plates. Install the remaining plate (7) which has no holes for springs. Install new "O" ring (3) in groove of cover (2) and "O" ring (5) in groove of piston (6), then carefully assemble piston into bore of cover. Position the cover and piston

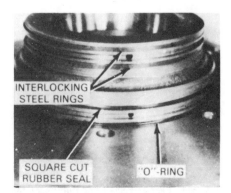

Fig. 268—View of seals on left carrier bearing retaining plate.

INTERLOCKING STEEL RINGS

SQUARE CUT RUBBER SEAL   "O"-RING

Fig. 271—Views showing location of springs (10), clutch disc (8), separator plates with holes (9) and plate without holes (7). Thrust washer is shown at (13).

assembly over the case half (12) and install retaining screws loosely. Insert the correct shims (4) into each of the four locations. Remove screws retaining cover to case one at a time, coat threads with "Grade AV Loctite" and reinstall. Tighten retaining screws evenly to 30-35 ft.-lbs. torque.

REAR AXLE AND FINAL DRIVE

All models use a planetary final drive unit located in inner end of axle housing next to center housing. Final drive lubricant reservoir is common with transmission, differential and hydraulic reservoir. Axle stub shafts are free floating, with inner ends supported by differential side gears and outer ends serving as planetary sun gear. Service on any part of axle or final drive requires removal of unit as outlined in paragraph 170.

REMOVE AND REINSTALL

170. To remove either final drive unit as an assembly, first drain transmission and hydraulic system fluid, suitably support rear of tractor and remove rear wheel and fender. Remove drawbar and drawbar frame.

Attach a hoist to final drive housing as shown in Fig. 279. Remove the attaching stud nuts and capscrews and slide the complete final drive assembly away from rear axle center housing. Stub axle and brake disc may remain with center housing, or with final drive unit as shown in Fig. 280. When reinstalling final drive, place stub axle and brake in final drive unit as shown.

Install a new square-section seal (Fig. 281) on center housing and reassemble by reversing removal procedure. Tighten retaining stud nuts and cap screws to 155-170 ft.-lbs. torque.

OVERHAUL

171. **AXLE AND BEARINGS.** With final drive unit removed as outlined in paragraph 170, withdraw stub axle, brake disc and cover plate as shown in Fig. 282. Remove capscrew retainer (Fig. 283) and capscrew, then withdraw planet carrier, retainer washer, shim pack, thrust washer and bearing cone.

Use a suitable push-puller as shown in Fig. 284 and push axle shaft and outer seal from housing. If push-puller is not available, remove outer seal, then withdraw axle assembly by hand.

Outer bearing cone and both bearing cups can be removed at this time. Outer bearing cone may require heavy pressure to fully seat on axle shaft. When assembling, pack outer bearing with lithium base multipurpose grease. Install inner seal as shown in Fig. 285 and carefully insert axle through seal.

Fig. 280—View of final drive housing with stub axle (10) and brake disc (9) installed.

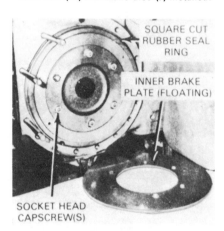

Fig. 281—View showing location of square cut seal ring.

Install inner bearing cone and thrust washer as shown in Fig. 286. Install a guide stud approximately 8 inches long in inner end of axle as shown in Fig. 287, then install the removed shim pack plus 0.010 inch (0.25mm) additional shims.

Install planet carrier, remove the guide stud and reinstall the retaining capscrew, tightening securely. Check axle end play as shown in Fig. 289. Remove capscrew and reinstall the guide stud. Lift off planet carrier then remove shims equal to the measured end play plus 0.001-0.005 inch (0.025-

Fig. 282—Removing brake disc and stub axle from final drive unit.

Fig. 283—Planet carrier retaining cap screw is prevented from loosening by use of the retainer shown.

Fig. 279—Views showing final drive units being removed.

Fig. 284—Rear axle shaft and outer seal can be removed with a suitable push-puller after planet carrier is removed.

0.127mm), to obtain the specified pre-load for axle bearings. Tighten the re-taining capscrew to a torque of 200-325 ft.-lbs. in final assembly and reinstall capscrew retainer. Install outer seal as shown in Fig. 290, then complete the assembly by reversing the removal procedure.

172. **PLANET CARRIER.** Planet pinion shafts are a slip fit in carrier and are retained by a snap ring (R—Fig. 291) which fits in a slot in all three shafts plus a retaining slot in carrier hub. To disassemble the carrier, place unit on a bench with inner (flat) side down and expand snap ring fully into shaft grooves as shown in Fig. 292. Work snap ring up out of carrier hub slot, withdrawing all three shafts until snap ring clears shaft grooves.

Slide the pinion, together with thrust washers and loose needle bearings, out of carrier as shown in Fig. 291. Retainer washer (W—Fig. 293) can be withdrawn after one planet pinion is out as shown.

When assembling, the snap ring the three planet pinion shafts must be installed together as shown in Fig. 294.

Install planet carrier and adjust axle bearing preload as outlined in paragraph 171.

Fig. 291—Planet pinion shafts are collectively retained by snap ring (R) which fits a groove in all three shafts and in carrier hub.

Fig. 288—Install planet carrier, temporarily install retaining cap screw, then check axle shaft end play using a dial indicator as shown in Fig. 289.

Fig. 285—Installing final drive housing inner seal.

Fig. 292—To disassemble planet carrier, expand the snap ring into deeper pinion shaft grooves and lift all three shafts until clear of retaining groove in carrier hub.

Fig. 289—Measuring axle shaft end play which must be finally adjusted to 0.001-0.005 inch preload during final assembly.

Fig. 286—Install inner bearing cone and thrust washer after axle shaft is inserted.

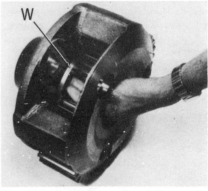

Fig. 293—Retainer washer (W) can be removed after one pinion is out as shown.

Fig. 287—Using a guide stud to retain shim pack during planet carrier installation.

Fig. 290—Installing outer seal after correct bearing preload is obtained.

Fig. 294—All three planet pinion shafts and the retaining snap ring must be installed at the same time as shown.

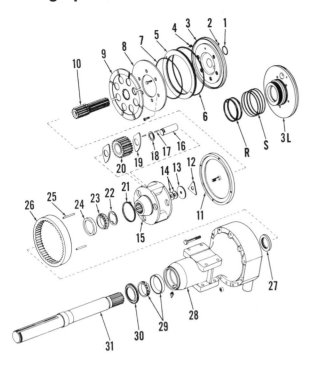

Fig. 295—Exploded view of right final drive assembly and brake disc. Differential carrier bearing retainer (3L), rings (R) and seals (S) are used on left side of models with locking differential.

1. "O" ring (pressure)
2. "O" ring (bleed passage)
3. Carrier retainer
4. Springs (4 used)
5. Brake piston
6. "O" ring
7. "O" ring
8. Inner brake plate
9. Brake disc
10. Stub shaft
11. Plate
12. Cap screw retainer
13. Retainer washer
14. Shims (0.002-0.005-0.010 & 0.018 in.)
15. Planet carrier
16. Shaft
17. Roller (25/row)
18. Spacer
19. Thrust washer
20. Planet pinion
21. Snap ring
22. Washer
23. Bearing cone
24. Bearing cup
25. Pins
26. Ring gear
27. Seal
28. Housing
29. Bearing cone and cup
30. Seal
31. Axle

The brake system should be bled as outlined in paragraph 181 after gage is removed.

181. FLUID AND BLEEDING. Lubricant from rear axle center housing is utilized as brake hydraulic fluid. A line connects brake reservoir to bleed-off from the main hydraulic system priority valve which supplies master cylinder with fluid. A second line returns excess fluid to the rear axle center housing. Whenever servicing brake master cylinder, or if cylinder has been drained, refill with proper lubricant, then bleed brakes as follows:

With engine running to keep master cylinder supplied with fluid, follow normal hydraulic brake bleeding procedure to remove all air from the system for both master cylinders and brake piston units. Bleeder fitting is located on rear axle center housing at top of rear axle housing on both sides of tractor as shown at (3—Fig. 301). On tractors equipped with differential lock, it will probably be necessary to bleed air from differential lock by loosening the fittings.

182. OVERHAUL MASTER CYLINDER. To remove the brake master cylinder, first disconnect the four lines (Fig. 303), remove cover plate from pedals, then unbolt and remove master cylinder.

BRAKES

180. ADJUSTMENT. The two rods (Fig. 300) between pedals and master cylinder pistons should be adjusted to provide 1/32-1/16 inch (0.8-1.6mm) free play. The brake interlock should be able to latch pedals together when pedals are depressed. The length of the rods will not usually need to be adjusted, but improper operation, damage to seals, possible difficulty in bleeding or leakage may be caused by rods incorrectly adjusted.

Fig. 300—View of brake pedals and operating rods with cover plate removed.

A pressure test is possible by connecting a gage into brake line, then bleeding air from lines as outlined in paragraph 181. On models without differential lock, a "T" fitting should be installed between housing and brake line. On Models with differential lock, disconnect line (2—Fig. 301) which runs to differential lock valve and connect pressure gage in its place. On all models, bleed system as outlined in paragraph 181.

Mark on line on each push rod (Fig. 300) and observe pressure when rod is moved ½ inch (12.7mm). If pressure is greater than 180 psi, the rod is too long and primary cup (10—Fig. 302) may be damaged as it passes across compensating port (C). Low pressure may be caused by wear, leakage or rod being too short.

Approximately 108-132 lbs. force on each pedal should move rod approximately 3 inches (76mm). Pedal should not creep within 30 seconds and pressure should be at least 500 psi.

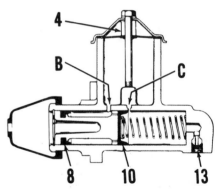

Fig. 302—Cross-section of brake master cylinder. Breather port is shown at (B); compensating port at (C).

Fig. 303—View of installed master cylinder. Line (1) is to right brake, line (2) to left brake, line (3) from priority valve and line (4) to transmission sump.

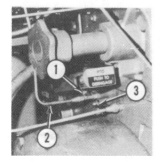

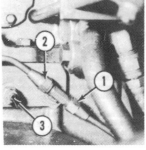

Fig. 301—Views of brake line (1), tube to differential lock valve (2) and bleeder valve (3). Left view is typical of left side, right typical of right side.

Refer to Fig. 304 for exploded view. Remove covers (6) and snap rings (7) then withdraw parts (8 thru 12). Seals (13) can be pulled out using tap or self tapping screw if renewal is necessary. Reassemble using Fig. 302 and Fig. 304 as guide.

Reinstall and connect lines (Fig. 303). Fill reservoir, start engine and bleed system as outlined in paragraph 181.

CAUTION: Use only Massey-Ferguson M1127 or equivalent oil as contained in center housing.

183. **BRAKE PISTONS AND DISCS.** Before any service can be performed on brake discs or actuating pistons, final drive unit must be removed as outlined in paragraph 170.

With final drive off, brake disc (Fig. 282) and cover plate can be removed. After the inner brake plate (8—Fig. 305) is removed, the piston (5) can be pulled from bore of carrier bearing retainer (3).

Install new "O" rings (6 & 7—Fig. 295). Sand any rough edges from bore

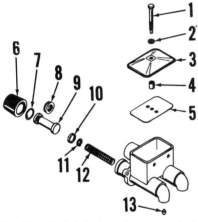

Fig. 304—Exploded view of master cylinder. Refer to Fig. 302 for cross-section.

1. Cap screw
2. Seal washer
3. Top cover
4. Spacer
5. Gasket
6. Boot
7. Snap ring
8. Secondary cup
9. Piston
10. Primary cup
11. Spring cup
12. Spring
13. Seat

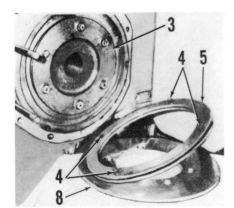

Fig. 305—View of brake operating piston (5) withdrawn from bore of carrier (3). Refer to Fig. 295 for exploded view.

of carrier (3) before inserting piston (5). Clean all parts thoroughly, lubricate and reinstall piston. Refer to Fig. 263 and paragraph 163 or 166 for renewal of "O" rings (1 & 2—Fig. 295).

Assembly is reverse of removal procedure. Be sure to bleed brakes as outlined in paragraph 181.

INDEPENDENT POWER TAKE-OFF

The ipto input shaft is splined into an adapter (A—Fig. 202) attached to engine crankshaft along with flywheel. The front shaft runs inside the hollow clutch shaft and transmission mainshaft, and is connected by a coupling (12—Fig. 246) to the ipto drive shaft located in rear axle center housing.

The hydraulic pump drive gear is an integral part of the ipto clutch drum (9). The multiple disc ipto clutch is hydraulically actuated by the low volume, low pressure circuit of the main hydraulic system. Tests and service of pump and valve is covered in

Fig. 308—View of IPTO output housing showing the dual speed drive flanges. The 1000 rpm output stub shaft is installed.

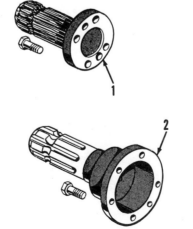

Fig. 309—View of IPTO output accessory shafts available for installation.

1. 1000 rpm ASAE shaft
2. 540 rpm ASAE shaft

the HYDRAULIC SYSTEM Section of this manual.

OUTPUT SHAFT

185. The 540 and 1000 rpm ipto output shaft drives consist of outer and inner flanges as shown in Fig. 308. Optional output shafts are shown in Fig. 309.

Any service on output shaft drive units requires removal of ipto housing as outlined in paragraph 186 and overhaul of output shaft gear train as in paragraph 187.

IPTO OUTPUT HOUSING AND GEARS

186. **REMOVE AND REINSTALL.** To remove the ipto output housing, first drain transmission and center housing and move ipto coupling handle (H—Fig. 310) into engaged ("ON") position. Remove top link (Fig. 311) and detach top link clevis pin from draft control springs. Remove the pins secur-

Fig. 310—View of control console showing IPTO coupling handle (H).

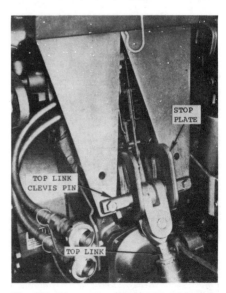

Fig. 311—To remove IPTO output housing, remove top link and disconnect top link clevis pin from sensing springs.

ing ratio lever (R—Fig. 312) to top link clevis and draft control sensing rod and lift off the link. Disconnect draft control hand wheel from adjusting clevis and remove the clevis. Unbolt and remove drawbar rear support frame. Attach a hoist to draft control stop plate as shown in Fig. 312.

Remove one flange capscrew on each side of housing and install suitable guide studs; remove remainder of flange screws and lift off ipto housing as shown.

Ipto rear drive shaft (Fig. 313) can be withdrawn after cover is removed, but be sure coupling is engaged to prevent coupler from dropping out of position. Also be sure shim pack is not lost or damaged when rear cover is removed. Shims control ipto rear drive shaft end play which should be 0.010-0.026 inch (0.25-0.66mm). Shims are available in thicknesses of 0.015, 0.030 and 0.062 inch. Clearance can be checked with a feeler gage after removing center housing front, top cover or park lock cover.

When reinstalling ipto output housing, tighten large flange capscrews to a torque of 50-55 ft.-lbs. and small capscrews to 15-18 ft.-lbs. Complete the

assembly be reversing the removal procedures. Refer to paragraph 251 for draft control linkage adjustment.

187. OVERHAUL. To disassemble the removed output housing, remove the three capscrews (1—Fig. 315) and lift off gear cover (2). Refer to Fig. 316. Input gear (4) will be removed with cover while idler gear (3) may remain with cover or main housing. Note the location and position of thrust washers as housing is disassembled.

When disassembling the 540 and 1000 rpm output drive flanges, unseat and remove the snap rings, gears and thrust bearings from inside of housing and withdraw both flange shafts as a

unit as shown in Fig. 318. The shafts contain 66 loose bearing rollers assembled in two rows of 33 each, separated by a machined pipe spacer. Fig. 319 shows a cross sectional view of output flange shafts properly assembled. Install shafts carefully from rear of housing. The three thrust roller bearings have one thrust race each, the other race being machined on the 540 and 1000 rpm driven gears. The three thrust bearings are identical but thrust races are not. Place the thickest thrust race next to housing as shown in Fig. 320. Position the stepped thrust race/snap ring retainer between the gears as shown in Fig. 321. Place the thin thrust race in cover recess. With cover installed, check to be sure that gears and shafts turn freely with no

Fig. 314—View of tractor showing IPTO output housing correctly installed.

Fig. 317—Output housing and gears with gear cover (2—Fig. 316) removed. Lift out cluster gear (6). Unseat snap rings and remove 1000 rpm output shaft gear (8) and 540 rpm output gear (7). Refer to Fig. 318.

Fig. 312—Remove drawbar rear support frame, ratio lever (R) and sensing linkage adjusting clevis.

Fig. 315—To disassemble the removed output housing, remove the three cap screws (1) and lift off gear cover (2).

Fig. 313—Rear IPTO shaft can be withdrawn after IPTO output housing is removed. Be sure shift coupling is engaged and be careful not to lose shim pack.

Fig. 316—Removed gear cover and associated parts. Refer also to Fig. 317.

2. Cover   4. Input gear
3. Idler gear   5. Shaft

Fig. 318—Assembled view of output drive flanges. Refer to Fig. 319 for cross-section.

interference. Reinstall the assembled output housing as outlined in paragraph 186.

IPTO CLUTCH

188. REMOVE AND REINSTALL. To remove the ipto drive shaft, hydraulic clutch and pump drive gear, first remove ipto output housing as outlined in paragraph 186 and center housing front, top cover as in paragraph 230.

Withdraw oscillator crank (O—Fig. 323) by prying forward. Unbolt and remove oil return tube (T). Withdraw ipto coupler Fig. 324. Slide intermediate ipto shaft rearward a slight amount, extract the snap ring; then slide shaft rearward out of the way.

Fig. 322—The third thrust bearing is installed over 1000 rpm gear, with thin thrust race in cover recess.

Fig. 323—To remove the IPTO clutch, it is first necessary to remove return tube (T) and oscillator arm (O).

With ipto intermediate shaft removed, lift out the ipto clutch unit as shown.

Lubricate teflon bronze sealing rings with petroleum jelly and install clutch unit by reversing removal procedure.

189. OVERHAUL. With ipto clutch removed as outlined in paragraph 188, disassemble as follows: Place clutch drum hub-end down on a bench and apply slight pressure to plate (9—Fig. 325), unseat and remove snap ring (10). Remove items (4 through 9).

Place six dowel pins from Special Tool (MFN768) in clutch housing splines (See Fig. 326) to protect the piston ring, and work piston (3—Fig. 325) from clutch drum by grasping stiffening ribs with pliers.

Examine the parts and renew any which are worn or damaged. Examine loose check ball in clutch housing to see that it is free in bore and seats without leakage against pressure from inside.

When assembling the clutch, work piston ring into drum using an ice pick or small screwdriver, install the eleven dowels as shown in Fig. 326; then push the piston fully into place by hand. Note that the externally splined separator plates (6—Fig. 325) are slightly dished. Sort the plates and stack so that all can be installed with concave in the same direction. Place one separator plate on top of piston, then alternate clutch discs and separator plates, with lugs of separator plates one spline either way from lug of botton plate as

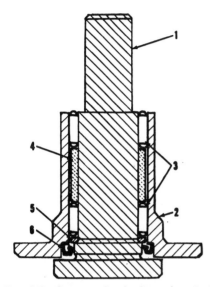

Fig. 319—Cross-sectional view of output flange shafts showing bearing and seal installation.

1. 1000 rpm shaft
2. 540 rpm shaft
3. Needle rollers
4. Spacer
5. Thrust washer
6. Oil seal

Fig. 320—When installing output flange shaft gears, place thickest thrust race next to housing and install one thrust bearing.

Fig. 321—Place the stepped thrust race over snap ring and install second thrust bearing.

Fig. 324—Slide intermediate shaft rearward and lift out sliding coupler; then withdraw clutch assembly as shown.

Fig. 325—Exploded view of IPTO clutch showing component parts.

1. Clutch drum
2. Piston ring
3. Piston
4. Thrust washer
5. Clutch hub
6. Separator plates
7. Separator springs
8. Clutch discs
9. Pressure plate
10. Snap ring
T. Teflon rings
V. Check valve

Fig. 326—Using the special dowels (D) as ring compressors, push piston (3) fully in place by hand as shown.

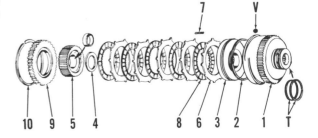

shown in Fig. 327. Lugs of bottom plate are to serve as pressure points for separator springs as shown in Fig. 328. Complete the assembly by reversing the disassembly procedure and install the assembled clutch as outlined in **paragraph 188.**

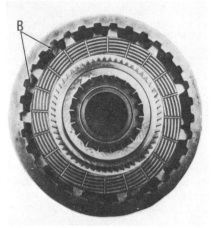

Fig. 327—External lugs of first separator plate should be one spline either way from remaining lugs as shown at (B).

Fig. 328—Install separator springs (7) on lugs of first separator plate as shown.

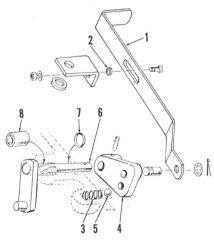

Fig. 329—Exploded view of IPTO shift linkage.

1. Handle
2. Bushing
3. Spring
4. Lever
5. Detent ball
6. Shift crank
7. "O" ring
8. Bushing

IPTO SHIFT LINKAGE

190. Fig. 329 shows an exploded view of ipto shift linkage and associated parts. The sealing "O" ring (7) fits in channeled groove of housing bore for shaft (6). No adjustment is required of linkage unit. In the event of improper shifting, check for bent or broken components.

Service on the ipto control valve is contained in paragraphs 240 and 241.

HYDRAULIC SYSTEM

All models have a closed center, constant pressure hydraulic system which can logically be divided into three circuits as follows:

LOW PRESSURE/LOW VOLUME Circuit. Operates at a volume of 6 gpm and a pressure of 225-275 psi to provide working fluid for Multi-Power, ipto and Differential Lock circuits. All valves are closed-center type and fluid discharges through combining valve into the Low Pressure/High Volume Circuit.

On models with 8-speed transmission instead of Multi-Power, the combination pto and Multi-Power control valve is still used. The Multi-Power valve is wired to "IN" (High) position directing oil to a metering hole between the transmission main input shaft and the front pto shaft. This oil is used for bearing lubrication and the size of the hole controls the amount of oil permitted to flow.

LOW PRESSURE/HIGH VOLUME Circuit, also referred to as "Charging Circuit". Supplies working fluid on demand to high pressure piston pump, and provides constant circulation to system filter and oil cooler. Circuit is supplied by a 17 gpm internal gear type pump and unused flow from low volume circuit. Charging circuit pressure is regulated at 90-100 psi.

HIGH PRESSURE [MAIN] Circuit. Supplies working fluid for power steering on a priority basis, with sufficient reserve capacity for hydraulic working cylinders. Pressure is maintained at 2100-2400 psi, with available volume at 16-20 gpm.

The transmission lubricant is operating fluid for the hydraulic system. Massey-Ferguson M-1127 Fluid is recommended. System capacity is 22 U.S. Gallons. Refill capacity is 17 U.S. Gallons, as 5 gallons will remain when reservoir is drained.

OPERATION

200. The hydraulic system operating fluid is drawn from reservoir in rear axle center housing through a screen-type filter. The dual, gerotor type charging pump is engine driven by the ipto input shaft. Fluid from the low volume side of dual pump flows under pressure to the hydraulic seat valve and to the Multi-Power and ipto control valve. All units are equipped with closed-center valves; if no demand for fluid exists, the fluid passes through the combining valve where it is mixed with the flow from the high-volume side of charging pump.

The combined flow from the charging pump is open to two passages. One leads through the system filter and oil cooler to the inlet side of the main hydraulic pump, then to control port (C—Fig. 335) of regulating valve. Four to six gpm passes through the regulating valve spool orifice (O) to the sensing pressure regulating valve (P), where sensing pressure is maintained at 90-100 psi. If there are no other demands, the remainder of charging pump flow passes from bypass port (B) to dump port (D) through the passage opened by movement of regulating valve spool (S).

The radial piston type main hydraulic pump is front mounted and driven by engine crankshaft. The main hydraulic pump attempts to maintain a pressure of 2100-2400 psi at all valves of the high pressure main hydraulic system. When this pressure is reached, flow from the main hydraulic pump ceases, the eight radial pistons are held away from the actuating cam and the pump drive shaft continues to rotate without performing work. The main pump fluid delivery is maintained at the flow necessary to satisfy system requirements up to the pumps maximum capacity.

LUBRICATION

201. **LUBRICATION.** The hydraulic system fluid serves as a lubricant for the power steering motor and hand pump, transmission, differential, final drive and ipto gear train. The trans-

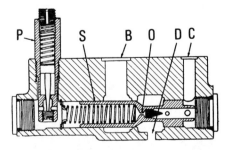

Fig. 335—Cross-sectional view of hydraulic system regulating valve which controls charging pump pressure and flow.

B. By-pass port
C. Sensing control port
D. Dump port
O. Control orifice

P. Regulating pilot valve
S. Valve spool

mission and rear axle center housing serve as the system reservoir.

The only recommended fluid is Massey-Ferguson M-1127 fluid. System capacity is 22 U.S. gallons; when system is not disassembled, however, only about 17 gallons will be drained from the system and only this amount is required for normal refill. To check the fluid level, be sure tractor is on level ground and remove dipstick (D—Fig. 336) from left side of transmission housing. Fill through dipstick opening if fluid is needed.

System filter (13—Fig. 6) should be cleaned or renewed at 750 hour intervals or once a year, to coincide with lubricant change. At the same time, intake screen cover plate (bottom of center housing (1—Fig. 337), should be removed and the screen cleaned. Do not attempt to flush the system as a considerable amount of the flushing solution cannot be drained. If fluid is badly contaminated, system must be disassembled for cleaning.

TESTS AND ADJUSTMENTS

202. **QUICK TEST.** A quick test of the high pressure system can be performed using a pressure gage as follows:

Connect pressure gage to remote

coupler and pressurize that coupler, with engine running first at 1000 rpm, then at 2000 rpm. Gage reading should be 2100-2400 psi on both tests.

Flow can be tested at breakaway coupling by using a flow meter and fully opening the flow meter and fully opening the flow control valve on top of auxiliary valve body. Connect flow meter inlet to breakaway coupling and direct flow meter outlet back to reservoir. A plug opening (return) is provided at rear, left side of rear axle center housing as shown in Fig. 338. Make sure all hydraulic functions are in neutral. Start and operate engine at 1000 rpm. Move auxiliary lever to pressurize the coupling containing flow meter and lock the lever in open position. Adjust load valve to 1000-1200 psi and allow fluid to reach an operating temperature of 140-160 degrees F., then check the measured flow which should be MINIMUM of 9 gpm.

If flow and/or pressure are not as specified, refer to pump tests and adjustments as outlined in paragraph 205 to pinpoint the possible causes of trouble. For specific tests on system circuits and components, refer to the applicable paragraphs 214 through 216.

203. **CIRCUIT TESTS.** To obtain access to front, top cover for testing,

remove center panel from operator's platform as shown in Fig. 339. All tests should be conducted with tractor hydraulic system fluid at operating temperature of 140-160 degrees F. A temperature gage can be installed in reservoir for checking fluid temperature. Be sure that reservoir is filled to proper level with Massey-Ferguson M-1127 fluid at all times during tests.

204. Move lift links to the lowered position before any plugs or lines are removed or disconnected. Always cover openings in lines and ports as soon as possible to prevent entrance of dirt and unnecessary loss of fluid. Attach approximately 1000 lbs. of weight to the lower link arms if 3-point linkage valve is to be checked. For most tests, the Multi-Power control valve must be in "LOW" position, pto valve in "OFF" position and differential lock in "OFF" position. On 8-speed models, remove the panel from left side of instrument console, remove wire from Multi-Power valve (1—Fig. 340) and move valve spool "OUT" to low position. On all models, refer to Fig. 341 and attach appropriate gages as required for specific tests.

205. CHARGING CIRCUIT PRESSURE. Remove plug (S—Fig. 341) from

Fig. 336—Hydraulic oil level dipstick (D) is on left side of transmission housing as shown. Dipstick opening is also filling port.

Fig. 337—Bottom view of rear axle center housing showing hydraulic system fluid drain plugs (2) and intake screen cover plate (1).

Fig. 338—Left rear corner of rear axle center housing with fitting (1) and flow meter return line (2) attached.

Fig. 339—For access to front top cover for service or testing, remove operator's platform center panel shown.

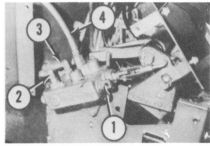

Fig. 340—View of IPTO and Multi-Power control valve. On models without Multi-Power transmission, spool (1) is wired in as shown.

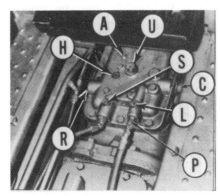

Fig. 341—View of front hydraulic cover showing test and adjusting ports.

A. Adjusting port
C. Charging port
H. High volume port
L. Low volume port
P. Low volume port

R. Relief valve plug
S. Sensing port
U. Automatic unloading valve

sensing test port and install a 0-400 psi pressure gage. Sensing pressure should be 90-100 psi with engine running at 1000 rpm. Pressure should not drop below 80 psi at slow idle speed. If pressure is not correct, remove adjusting port plug (A) and, working through plug opening, turn the exposed slotted head plug clockwise to raise the pressure or counter-clockwise to lower pressure. Adjust the pressure to 95 psi at 1000 engine rpm.

CAUTION: Do not completely remove plug (A) while engine is running. Stop engine, remove plug, adjust plug, reinstall plug temporarily, then retest pressure.

After pressure is adjusted to 95 psi at 1000 engine rpm, move control selector lever to "Position Control", run engine at 2000 rpm, raise links and observe pressure. Repeat test at 1000 rpm. If sensing line pressure drops and does not recover during tests, flow control and pressure regulator valve may be sticking. If pressure drops at idle, the unloading valve may be opening too soon.

Remove plug (L) from low volume port and install 0-400 psi pressure gage. Low volume system pressure should not be less than 225 psi at 1000 rpm nor more then 275 psi at 2000 rpm.

Check the combined system pressure by installing pressure gage in high volume port plug (H). Pressures should be approximately the same as those given for low volume circuit.

Substantially higher pressures at low volume pressure port (L), may indicate a sticking combining valve in pump body. Low pressure at 1000 engine rpm may indicate leakage in low pressure circuit. Be sure Multi-Power control valve is in "LOW", pto is "OFF" and differential lock is "OFF" when testing.

NOTE: On 8-speed models, valve spool (1—Fig. 340) must be unwired and pulled out.

On all models, if pressure is normal at 1000 rpm but higher than normal at 2000 rpm, check for plugged main system filter or restricted lines. Low pressure at high rpm may indicate a sticking safety relief valve (behind port plug (R—Fig. 341).

206. CHARGING CIRCUIT FLOW. To check the combined flow from charging pump, disconnect charging supply tube from elbow (C—Fig. 341) and connect flow meter inlet line to elbow. Direct flow meter outlet to sump as shown in Fig. 338. Connect a 0-400 psi gage to high volume port (H—Fig. 341).

CAUTION: Operate tractor only long enough with charging supply tube dis-

connected to make the test.

Make sure ipto valve lever is in "OFF" position, Multi-Power valve is in "LOW" position and differential lock is in "OFF" position.

NOTE: On 8-speed models, remove panel from left side of instrument console, remove wire from Multi-Power control valve and move valve spool out to "LOW" position.

Start and run engine at 850 rpm, maintaining 150 psi at port location (H—Fig. 341). Minimum flow is 7 gpm. Run engine at 1000 rpm maintaining 150 psi at port (H) by adjusting flow meter load valve. Minimum flow is 10 gpm at 1000 rpm. Low volume of flow may be caused by leakage in low pressure circuit. Also check safety relief valve and unloading valve.

To check for leakage in low pressure circuits, operate engine at 2000 rpm and adjust flow meter load valve to maintain 150 psi at port (H). Be sure that Multi-Power control valve is in "LOW", pto valve is "OFF" and differential lock is "OFF" and record the volume of flow.

Move the Multi-Power control to "HIGH", readjust flow load valve to maintain 150 psi at port (H) and observe volume of flow.

NOTE: Valve spool must be moved in to "HIGH" position on 8-speed models.

On all models, a maximum decrease of 1.6 gpm is permissible. Return Multi-Power control to low position after test is completed. If excessive loss of flow is indicated, the Multi-Power valve, Multi-Power clutch pack or sealing rings in 8-speed transmission may be at fault.

NOTE: On 8-speed models, be sure to wire valve spool in to "HIGH" position when tests are completed.

Move pto valve lever to "ON" position and adjust load valve on flow meter to maintain 150 psi at port (H—Fig. 341). The loss of volume from off to on position should not exceed 0.5 gpm. If loss is excessive, the pto clutch pack or pto control valve may be leaking. Return valve to off position after test is completed.

Fig. 342—Front view of engine showing main hydraulic pump outlet tube (O).

Move differential lock lever to "ON" position and adjust load valve on flow meter to maintain 150 psi at port (H). Loss of volume from off to on position should not exceed 0.5 gpm. If loss is excessive, piston seals, control valve, differential sealing rings or some other area of the differential lock circuit is leaking.

If volume of flow is low at each test, the low volume side of pump, safety relief valve or automatic unloading valve may be faulty.

207. SAFETY RELIEF VALVE. To check the safety relief valve, leave flow meter connected and install a 0-400 psi pressure gage in plug port (H—Fig. 341). With engine running at 850 rpm and load valve on flow meter open, start closing load valve and note pressure at which a noticeable drop in flow occurs. Pressure should be 180 psi. Close flow meter load valve to block the flow. Pressure at plug port should be a minimum of 210 psi at 1000 engine rpm and maximum of 350 psi at 2000 rpm. If pressure is incorrect, safety relief valve may be faulty. If relief valve cracking pressure is low, a loss of flow to the radial pump may incorrectly indicate a faulty main (radial) pump or faulty charging pump. Low cracking pressure can also be result of automatic unloading valve opening too soon.

208. AUTOMATIC UNLOADING VALVE. To check the automatic unloading valve, close the flow meter load valve and crank engine with starter (Fuel turned off to prevent starting) for approximately 5 seconds. Pressure indicated by gage at (H—Fig. 341) should be 20 psi maximum. Incorrect pressure indicates faulty unloading valve which is located under plug (U). Remove plug and withdraw spring and valve spool for inspection and cleaning.

209. MAIN PUMP PRESSURE AND FLOW. To check the main hydraulic pump pressure and flow, disconnect outlet tube (O—Fig. 342) from right top of main pump housing. Cap the tube to prevent fluid loss. It will be necessary to remove right hood panel and loosen supply line to priority valve inlet. Attach flow meter inlet hose to pump body fitting and connect flow meter outlet to sump as shown in Fig. 338. Tighten priority valve inlet supply line after flow meter is connected. Remove center plug (P—Fig. 343) from lower face of main pump front cover and attach a 0-2000 psi pressure gage to plug port.

Start and run engine at 1000 rpm and adjust flow meter load valve to obtain a pressure of 1900 psi. Allow fluid temperature to reach 140°-160°F.

210. To check the relief valve, proceed as follows: With engine running at 1000 rpm and fluid at test temperature, adjust flow meter load valve until a flow of 3-5 gpm is obtained. Turn metering valve adjusting plug (M—Fig. 343) clockwise until pressure ceases to rise on pressure gage. Test gage reading should be 2000-2100 psi.

If maximum obtainable pressure is below 2650 psi, remove and renew relief valve cartridge (behind plug R), or relief valve cartridge sealing "O" ring. Cartridge may be accurately checked by hand after removal, using a hand test pump.

211. To check the quick recovery valve, proceed as follows: First check the relief valve as outlined in paragraph 210, then back out metering valve adjusting plug (M—Fig. 343) until flow meter pressure gage registers 1400-1450 psi with engine running at 1000 rpm and flow adjusted to 3-5 gpm.

Flow meter pressure should be unstable. If it is not, turn quick recovery valve adjusting plug (Q) clockwise until pressure becomes unstable. Reset metering valve and check intermediate pressure as outlined in paragraph 212 after quick recovery valve is adjusted.

212. To check the intermediate pressure, proceed as follows: First adjust the quick recovery valve as outlined in paragraph 211 and with engine running at 1000 rpm, adjust flow to 3-5 gpm and reset metering valve adjusting plug (M—Fig. 343), to obtain a flow meter pressure reading of 2000-2100 psi. Check intermediate pressure gage installed in port (P), reading should be 375-425 psi. Increase engine speed to 2000 rpm; flow meter pressure should remain at 2050 psi and flow should remain at 3-5 gpm.

If any of the test results are not as stated, readjust settings as outlined in paragraphs 206 through 209 or overhaul pump as in paragraphs 227 through 229.

213. To check pump output, proceed as follows: First check intermediate pressure as outlined in paragraph 212, then with engine speed at 2000 rpm, adjust flow meter load valve to obtain an output pressure of 1900 psi on flow meter pressure gage. Main pump output flow should be a minimum of 16 gpm.

Slowly close flow meter load valve and note gage pressure when flow stops. Shut-down pressure should be 2100-2400 psi at any speed. Minor adjustments of shutdown pressure can be made by turning metering valve adjusting plug (M—Fig. 343).

Slow engine speed to 1000 rpm and open flow meter load valve to obtain a flow of 5 gpm; then quickly close load valve while observing pressure. Gage should momentarily rise to 3200-3400 psi, then quickly drop to shutdown pressure and hold steady.

214. HIGH PRESSURE CIRCUIT TESTS. To check the high pressure system components, disconnect outlet tube (O—Fig. 342) from right top of main pump housing. Attach flow meter inlet hose to pump body port and flow meter outlet to tube (O).

Open load valve on flow meter and start and run engine with all hydraulic

Fig. 344—Diverter hose installed for checking priority valve.

Fig. 350—Cross-sectional schematic view of the radial piston type main hydraulic pump. Pistons are pushed outward by the rotating eccentric cam (C) and inward by hydraulic pressure from charging pump. Inlet valves (I) and exhaust valves (E) are interchangeable.

units inactive, Flow meter pressure gage registers shutdown pressure and flow gage registers high pressure circuit leakage which should be 0-½ gpm.

215. To check the priority valve, proceed as follows: Connect both break-away couplings of one valve bank with a diverter hose as shown in Fig. 344 and start engine with valve in neutral. With engine at slow idle speed, move valve lever to either operating position while watching flow meter pressure gage. Pressure should drop from stand-by pressure and hold steady at about 1300 psi.

A low or substantially higher reading indicates a malfunctioning priority valve, which may be disassembled and cleaned as outlined in paragraph 236, however parts are not available for overhaul.

216. To check the power steering system, proceed as follows: Minor malfunction which does not materially affect steering operation will have no effect on hydraulic system.

As a quick hydraulic test, start and run engine at any speed. Turn steering wheel either way against steering stop. Continue to apply 36 in.-lbs. torque (approximately 5 lbs. at wheel rim) to steering wheel. Leakage can be considered excessive if steering wheel can be turned two revolutions per minute with moderate pressure (36 inch pounds) applied; or if pump flow increases more than 0.25 gpm over reading with steering wheel released.

MAIN PUMP

225. OPERATION. The radial piston type main hydraulic pump mounts on front support casting and is driven by engine crankshaft.

Fluid from the charging pump is provided at main pump inlet at moderate pressure. Unless flow is blocked by movement of the distributing valve, fluid passes through the inlet valves and pushes the radial pump pistons (P—Fig. 350) down into contact with

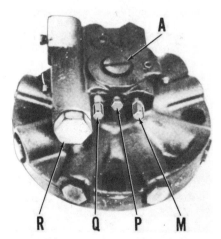

Fig. 343—Main hydraulic pump front cover showing points of adjustment and test.

A. Actuating piston   P. Pressure port plug
    plug   Q. Quick recovery
M. Metering valve   adjustment
    adjustment   R. Relief valve plug

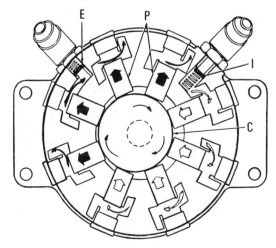

pump camshaft (C). When demands for pressurized fluid are met, the distributing valve closes inlet valves off from charging pressure and pump pistons fail to follow cam. Pumping action thus stops until a drop in standby pressure signals a demand for additional fluid. The metering action of the distributing valve determines volume delivered, up to the maximum capacity of the system.

Pressurized fluid enters the pump cover at a charging pressure of up to 100 psi. The distributing valve piston is spring-loaded in the open position. Orifices lead from the charging pressure passages into bearing and camshaft area of pump to provide lubrication. Fluid from the main pump high pressure gallery passes through a metering valve to the area behind the distributing valve piston, then through fixed orifices to the charging passage, to balance the high pressure system requirements.

The control cover also contains a "Quick Recovery" valve which is spring loaded to the open position and closed by hydraulic pressure. The quick recovery valve connects the intermediate (control) pressure passage to charging passage. A sudden drop in standby pressure causes the quick recovery valve to open, dumping pressure from the control passage and allowing the distributing valve to completely open. A full supply of high pressure fluid is thus provided to meet system demands.

226. **REMOVE AND REINSTALL.** To remove the main hydraulic pump, first drain radiator and remove lower radiator hose. Suitably support tractor under front of oil pan. Either remove axle assembly or move axle forward. To move axle forward, detach tie rod from right spindle, remove the rear support bracket (2—Fig. 2 or Fig. 3), loosen the right rear capscrew attaching front support bracket (9) and remove the other three attaching cap-

screws. The front axle can be rotated forward as shown in Fig. 351.

Remove protector plate from below pump. Loosen all four bolts securing pump drive shaft to crankshaft pulley. Move drive shaft into pump and align capscrews with holes in pump casting, then complete removal of the four capscrews. Disconnect all tubes from pump and cap all openings in tubes and in pump.

NOTE: The main hydraulic pump weighs more than 100 lbs. and means must be provided for lifting the pump down out of front support. The manufacturer provides a tool for pump removal as shown in Figs. 353 and 354.

Attach the tool or other suitable means to pump body and remove the mounting bolts, then swing pump away from tractor frame.

Install the pump by reversing removal procedure, making sure that drive coupler properly aligns with crankshaft pulley. Tighten all bolts securely.

227. **OVERHAUL.** With pump removed as outlined in paragraph 226, refer to Fig. 355 and proceed as follows:

Carefully withdraw pump coupling (1). Remove two of the 5/8-11 cap screws securing valve body (24) to piston housing (9) and install five-inch

aligning studs. Loosen and remove the other two cap screws evenly, then carefully remove valve body and distributing valve springs (12 & 13).

228. **VALVE BODY.** Distributing valve piston (14—Fig. 355) must slide smoothly in the sleeve in valve body and actuating piston (25) must move smoothly in guide (15). Refer to Fig. 356 for cross sectional view.

Remove the snap ring from secondary relief valve (16—Fig. 355) and fish out the valve if necessary, using a small wire. Unscrew plug and remove main relief valve (19). Remove quick recovery valve (20) and metering valve (21), working from both ends of bore.

Remove the eight inlet valves (17), being careful the valve discs do not fall into inlet manifold and become lodged. Check the valve seats in body for damage.

The two orifice plugs (18 and 23) will not need to be removed unless their condition is questionable. Actuating piston guide (15) is installed with "Loctite" and will not need to be removed unless "O" ring or guide are to be renewed or trouble is suspected. Clean and inspect all parts and renew any which are damaged or questionable.

When assembling the metering valve, refer to Fig. 357 and proceed as follows: Working from bottom side of

Fig. 352—View of pump and drive coupling (3). Pump inlet fitting is shown at (1) and pump drain line at (2).

Fig. 353—Using the special tool (MFN-762) to remove the main hydraulic pump. Refer also to Fig. 354.

1. Adjusting screw   3. Tension bar
2. Locking bar   4. Pump cradle

Fig. 351—To remove the main hydraulic pump, disconnect one tie rod and pivot axle support out of way as shown.

Fig. 354—The special tool is designed to lift pump from tractor and to lift it into position when installing. Refer also to Fig. 353.

valve body, install valve guide (7), spring (8), spacer plug (10) and adjusting plug (11) in the order given. Make sure the "O" ring is installed on spacer plug (10). Install "O" ring (4) and backup ring (5) as shown, on valve (3), and install sleeve in top of bore; then drop valve needle (6) into sleeve (3), chamfered end up. Install and tighten plug (1).

When installing the quick recovery valve, refer to Fig. 358. Install valve guide (8), spring (9), spacer plug (11) and adjusting plug (12) in bottom of body bore. Install "O" ring (4) and backup ring (5) as shown on valve sleeve (3), then install sleeve with smaller OD to top. Drop the piston (7) with small end down into sleeve bore. Install and tighten plug (1).

If piston guide (15—Fig. 355) has been removed, install with "Loctite" and tighten to a torque of 60-80 inch pounds. Tighten plug (26) to a torque of 20-25 ft.-lbs.

The eight intake valves, guides and springs (17) are interchangeable with each other and with outlet valve (7). The individual parts can be installed in valve body either side out. Tighten plugs to a torque of 45 ft.-lbs.

229. PISTON HOUSING. To disassemble the piston housing (9—Fig. 355), first remove the eight chamber plugs (8) and pump pistons, keeping pistons in their proper order so they can be reinstalled in their mated bores. Remove the cap screws securing carrier (3) to pump body and remove the carrier by carefully prying it straight out. Remove and save shim pack (5). Shims control adjustment of the tapered camshaft bearings.

With pistons and carrier removed, shaft and bearing unit can be lifted out. Camshaft and bearings are available only as an assembly, renew the unit if any part is damaged.

Remove plugs and outlet valves (7), being careful that valve discs do not fall into discharge manifold of pump. Examine valve seats in piston housing for damage.

Pump pistons should have a clearance of 0.0003-0.0015 inch (0.008-0.038mm) in their bores. Pistons and pump housing are factory matched and not available separately.

When reassembling the pump, first install discharge valves (7). Parts are interchangeable and can be turned either side out. Tighten plugs to a torque of 125 ft.-lbs.

Camshaft bearings should have an end play of 0.001-0.005 inch (0.025-

0.127mm). If in doubt, assemble camshaft and carrier (3), leaving out the seal (2). Install and tighten three alternate carrier cap screws and measure end play using a dial indicator. Shims (5) are available in thicknesses of 0.003, 0.005, 0.006, 0.007 and 0.020 inch. Tighten carrier cap screws to a torque of 30-35 ft.-lbs.

Install pump pistons in bores from which removed, with flat side next to camshaft. Install and tighten plugs to a torque of 225 ft.-lbs.

Install valve body on pump housing, making sure pump inlet and outlet ports are approximately 90 degrees apart and pointing up as shown in assembled view, Fig. 355. Tighten the four retaining cap screws evenly to a torque of 225 ft.-lbs.

Whenever the pump has been disassembled, renew the coupling shaft felt

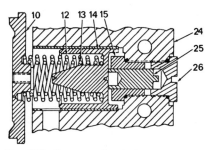

Fig. 356—Cross-sectional view of distributing valve and associated parts. Refer to Fig. 355 for parts identification.

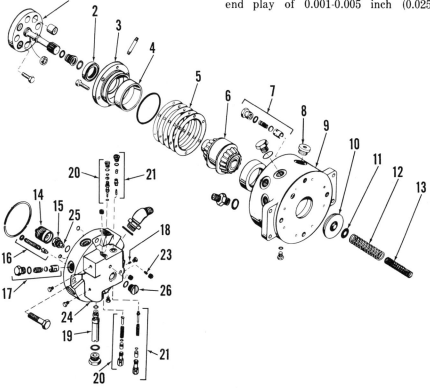

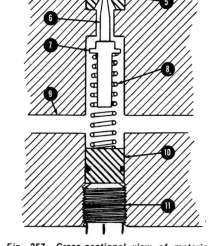

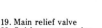

Fig. 355—Exploded view of crankshaft driven radial piston (main hydraulic) pump showing component parts.

Fig. 357—Cross-sectional view of metering valve and associated parts showing method of assembly.

1. Drive coupler	8. Piston plug	14. Distributor valve	19. Main relief valve
2. Oil seal	9. Piston housing	15. Guide	20. Quick recovery valve
3. Carrier assembly	10. Orifice plate	16. Secondary relief	21. Metering valve
4. Bearing cup	11. Filter screen	valve	23. Orifice plug
5. Shims	12. Outer spring	17. Inlet valve	24. Valve body
6. Camshaft	13. Inner spring	18. Orifice plug	25. Actuating piston
7. Discharge valve			26. Plug

1. Upper plug	6. Valve needle
2. High pressure	7. Valve guide
passage	8. Valve spring
3. Valve sleeve	9. Control passage
4. "O" ring	10. Spacer
5. Backup ring	11. Adjusting plug

dust seal (D—Fig. 359) as follows: Clean and degrease the area and cement the seal in the location shown and cement dust seal in position using Scotch Grip Industrial Adhesive No. 847 or equivalent. Position boot seal (S) and secure to shaft with "O" ring (R), then install garter spring spreader (G). Lubricate outer portions of felt seal and boot seal, then reinstall shaft in pump. Reinstall pump as outlined in paragraph 226.

CHARGING PUMP

230. REMOVE AND REINSTALL. The hydraulic system charging pump

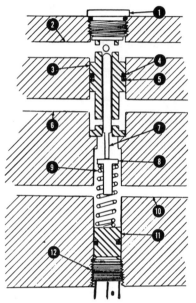

Fig. 358—Cross-sectional view of quick recovery valve which is spring loaded in open position. When pressure in passage (2) drops below standby pressure, spring (9) moves valve (7) up, connecting the moderate pressure "Control" passage (6) and inlet passage (10). The rapid drop in control pressure allows the distributing valve to fully open. Radial pump will operate at full capacity until pressure balance is restored.

1. Plug	7. Valve plunger
2. High pressure	8. Valve guide
passage	9. Spring
3. Valve sleeve	10. Low pressure
4. "O" ring	passage
5. Backup ring	11. Spacer
6. Control passage	12. Adjusting plug

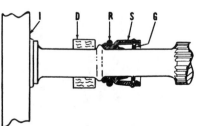

Fig. 359—Cross-sectional view of pump coupler showing proper installation of shaft seal. The felt dust seal (D) is cemented in place.

D. Dust seal	
G. Garter spring	R. "O" ring
I. Input flange	S. Rubber seal

(3–Fig. 360) is mounted on front hydraulic cover. To remove the cover (2–Fig. 360), first remove the center panel from operator's platform and remove the fuel tank cross brace. Disconnect the hydraulic tubes from the cover. Remove the pto cover to access the hydraulic tube (6). Remove the bolts retaining the cover to center housing, and carefully work the cover upward from the aligning dowels. Remove the cover with charging pump (3) and regulator valve (4) attached.

Remove the two cap screws securing pump to lower side of top cover and lift off the pump. Renew "O" rings and gaskets and install pump by reversing the removal procedure.

231. Overhaul. Refer to Fig. 361 for an exploded view of the rotor type dual charging pump. To disassemble the removed unit, clean outside of pump body and remove drive gear and key.

Drive out roll pin (10) then remove retainer (9), spring (8) and combining

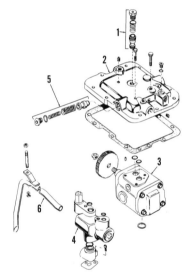

Fig. 360–The charging pump (3) and regulating valve are attached to the front hydraulic cover (2). The automatic unloader valve is shown at (1) and the safety relief valve at (5).

Fig. 361—Exploded view of the dual rotor type charging pump. Pump is shown at (3–Fig. 360).

1. High volume rotor assembly
2. Shaft
3. Low volume rotor assembly
4. End cover
5. Pump body
6. End cover
7. Combining valve
8. Valve spring
9. Retainer
10. Roll pin

valve spool (7). Remove cover (6) and small pumping rotors, then extract rotor key from exposed end of drive shaft.

NOTE: Inner rotor is a slip fit on shaft and key and both rotors are free to drop when cover is removed. DO NOT drop rotors when parts are disassembled.

Remove drive end cap (4) and withdraw shaft (2) and large internal rotor. Large external rotor may be withdrawn with shaft, or may remain in housing pocket.

Check the machined surfaces of end plates, body and rotors for scoring, wear or other damage. End clearance of rotors in housing bores should be 0.0008-0.0023 inch (0.020-0.058mm). End plates should be absolutely flat. No gaskets are used. Assemble the pump by reversing the disassembly procedure. Tighten end plate retaining capscrews evenly to a torque of 10-15 ft.-lbs.

Install the combining valve (7) and spring (8). Screw the retainer (9) in until it just touches spring (8). Turn retainer one additional turn to load spring and install roll pin.

REGULATING VALVE

232. REMOVE AND REINSTALL. The regulating valve assembly (4—Fig. 360) is mounted on front hydraulic cover. Remove front cover as outlined in paragraph 230, remove attaching capscrews and lift off regulating valve.

Reinstall by reversing the removal procedure. After tractor is reassembled, adjust sensing pressure as outlined in paragraph 205.

233. OVERHAUL. The regulating valve is available only as an assembly and individual parts are not renewed. The regulating valve may, however, be disassembled for inspection and cleaning. Refer to Fig. 362 for exploded view.

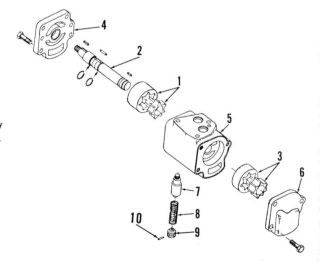

Spool (10) must slide freely in its bore with no perceptible clearance. Center drilling contains a small metering orifice and a conical screen as shown in Fig. 335.

Pilot valve (3) is prevented from seating by valve stop (5). If sensing pressure cannot be properly adjusted and no other trouble is encountered, check the setting of pilot valve stop as follows: Clamp pilot valve body (4) lightly in a vise, right side up with both ends accessible. Note the setting of adjusting plug (1) and remove the plug and spring (2). Mount a dial indicator with button bearing on top end of needle valve (3). Back out the threaded stop (5) until dial indicator needle ceases to move; indicator reading should be 0.005-0.010 inch (0.127-0.254mm). Adjust to those limits by turning stop (5) until needle is raised the required amount.

Assemble the regulating valve by reversing the disassembly procedure, using Fig. 362 as a guide. Install the valve as outlined in paragraph 232 and adjust as in paragraph 205.

PRIORITY VALVE

234. OPERATION. The priority valve is shown schematically in Fig. 363 and exploded in Fig. 365. Pressure supply ports (6 & 7—Fig. 363) leading to brake valve and power steering are located ahead of priority valve piston (4), while piston must unseat to supply fluid flow to hydraulic lift system port (5). When there is no flow, pressures are equal throughout the valve.

One end of the pilot valve piston (1) is open to high pressure while the other end is open to atmospheric pressure of bleed port (2). Before priority piston (4) can move to open lift system

port (5), two things must happen in the priority valve. System pressure must overcome the pressure of spring (3), allowing pilot piston to move down and open the pilot passage to lift port (5), and lift port pressure must drop below system pressure. In normal operation, pilot flow (small arrows) through piston orifice and pilot piston, unbalances the system to allow pilot piston to open whenever movement of one of the control valves signals a fluid demand.

235. REMOVE AND REINSTALL. To remove the priority valve, first remove the lower hood dress panels, disconnect lines (1, 4, 5 & 6—Fig. 364) from valve then remove mounting bolts (2) which also secure cross linkage for the engine throttle. Install by reversing the removal procedure.

236. OVERHAUL. The priority valve is available as an assembly only. To disassemble the valve for cleaning or inspection, refer to Fig. 365 and proceed as follows:

Working through inlet port, unseat and remove snap ring (10), priority piston (9) and spring (8). Remove lower plug (1), spring discs (5), spring (6) and pilot pin (7). Jar valve body downward on a bench or wood block to dislodge the piston (3). If piston cannot be re-

moved, valve must be renewed. Assemble by reversing the removal procedure.

DIFFERENTIAL LOCK VALVE

237. OPERATION. The differential lock control valve (Fig. 366) is located

Fig. 364—View of priority valve installed.

1. Line from radial pump	5. Line to steering control unit
2. Mounting bolts	6. Fitting at rear of priority valve
3. Throttle linkage	
4. Line to transfer cap	

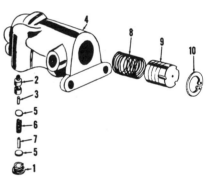

Fig. 365—Exploded view of priority valve showing component parts. Valve is serviced only as an assembly.

1. Plug	6. Pilot spring
2. Valve sleeve	7. Pilot pin
3. Pilot piston	8. Spring
4. Valve housing	9. Priority piston
5. Spring disc	10. Snap ring

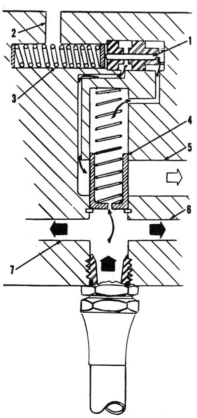

Fig. 363—Cross-sectional schematic view of priority valve.

1. Pilot piston	5. Hydraulic lift port
2. Bleed port	6. Brake pressure port
3. Spring	7. Steering pressure port
4. Priority piston	

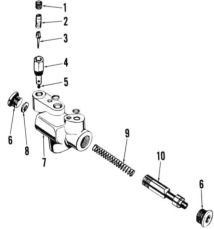

Fig. 362—Exploded view of regulating valve assembly. Component parts are not serviced.

1. Adjusting plug	6. End plug
2. Spring	7. Valve housing
3. Pilot valve	8. Spring seat
4. Valve body	9. Spring
5. Valve stop	10. Valve spool

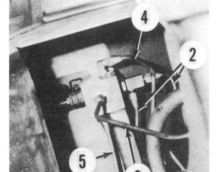

Fig. 366—View of differential lock valve installed. Lines from brakes are shown at (2). Line (4) is return to sump, line (5) is supply from pump and line (6) is to differential lock.

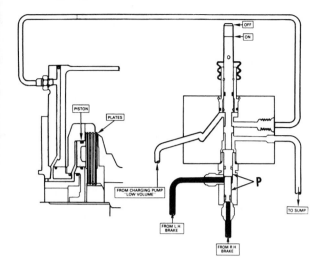

Fig. 367—Schematic drawing of differential lock hydraulic system. Pressure from either or both brakes pushes pistons (P) up which moves valve spool to OFF.

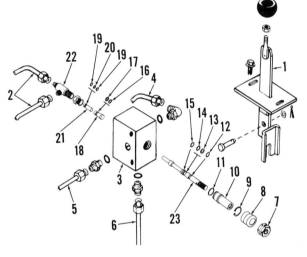

Fig. 368—Exploded view of differential lock valve.

1. Control lever
2. Brake lines
3. Body
4. Line to sump
5. Line from pump
6. Line to differential lock
7. Nut
8. Boot
9. Snap ring
10. Guide
11. "O" ring
12. Ring
13. Washer
14. "O" ring
15. Ring
16. Washer
17. "O" ring
18. Reset piston
19. Washers
20. "O" ring
21. Reset piston
22. Reset fitting
23. Spool

at left side of operator's platform. With spool pushed IN, oil from the charging pump low volume circuit is directed behind piston in differential assembly, which locks the differential locking clutch (Fig. 267). The valve may be disengaged by moving the control handle or automatically disengaged by engaging one or both of the tractor brakes. Oil from the brake circuit enters the reset fitting at lines (2—Fig. 366) and pushes either one or both of the pistons (P—Fig. 367) to move the

valve spool to OFF position.

238. **OVERHAUL.** To disassemble the removed valve, first remove reset fitting (22—Fig. 368). The inner piston (21) can be pushed out after pulling piston (18) out of bore. Remove rubber boot (8), snap ring (9), then pull spool (23) and guide (10) out of body. Refer to Fig. 368 for placement of "O" rings and back-up washers. Spool (23) should

move 5/16 inch (8mm) when correctly assembled.

MULTI-POWER AND IPTO CONTROL VALVE
240. **REMOVE AND REINSTALL.** To remove the Multi-Power & Ipto control valve (Fig. 369), first remove panel from left side of instrument console and disconnect all lines leading to valve. Disconnect lever yokes from valve spools. Remove the three bolts securing valve housing to support and lift off the valve assembly.

When installing the valve, make sure lever links are adjusted to permit full movement of valve spools to both detent positions, and that lines are connected in proper order.

241. **OVERHAUL.** Refer to Fig. 370 for an exploded view. To disassemble the removed valve, first remove the two detent plugs, "O" rings (11), shims (3) springs (1) and detent balls (2). Remove retaining rings (8) at both ends of Multi-Power spool, then carefully withdraw spool. Push spool (7) inward and remove retaining ring (12). Temporarily install plug in detent bore over ipto valve spool to keep ball from flying out. Turn spool (7) and withdraw assembly from bore. Remove plug, detent ball, spring and shims. Push valve (5) in and remove retaining ring (14).

Coat all parts with M-1127 oil or equivalent when assembling. If spools do not stay engaged in detent, it may be necessary to increase number of shims (3). Reverse disassembly procedure to reassemble.

HYDRAULIC LIFT SYSTEM (3-POINT LINKAGE)

All Models So Equipped
246. **OPERATION.** The three-point linkage hydraulic lift system offers three types of control for the lift arms,

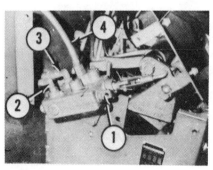

Fig. 369—View of IPTO and Multi-Power control valve installed. On models without Multi-Power transmission, valve spool (1) is wired in.

Fig. 370—Exploded view of IPTO and Multi-Power control valve showing component parts.

1. Detent spring
2. Detent ball
3. Shim
4. Valve body
5. IPTO modulating valve
6. Spring
7. IPTO valve spool
8. Snap ring
9. Multi-Power valve spool
10. "O" ring (3/4 in. ID)
11. "O" ring (11/16 in. ID)
12. Retaining ring
13. Nylon plug
14. Ring
15. "O" ring (7/16 in. ID)

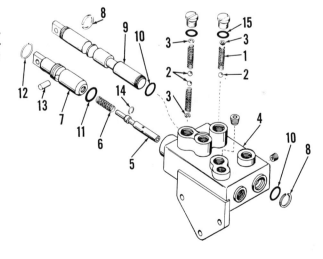

selected by turning the selector knob on top of control lever (C—Fig. 377).

DRAFT CONTROL. In "Draft Control", lift arm height is determined by the selected setting and controlled by top link pressure or tension on the leaf-type draft control springs. Draft control is the middle detent position on selector knob.

POSITION CONTROL. In "Position Control", lift arm height is maintained at a selected setting by an actuating lobe on rockshaft. Position control is obtained by turning selector knob counter-clockwise as far as possible after moving lever to top of quadrant.

PRESSURE CONTROL. In "Pressure Control", a selected pressure is maintained in rockshaft lift cylinders, the pressure being controlled by turning selector knob clockwise as far as possible. The master control valve is spring loaded in the "LIFT" position and is moved to neutral or lowering position by action of the control linkage.

247. **QUICK CHECKS.** For a quick check of hydraulic system, first be sure lift arm lock is "OPEN" as shown in

Fig. 378 and flotation latch is locked as shown in Fig. 379.

Attach a heavy weight to lower links. Weight should allow rockshaft to travel through complete arc and should not be attached to upper link.

Start and run engine at 1000-1600 rpm, then move control lever to top of quadrant. Weight should raise. With lever at top of quadrant, check to see that selector knob turns easily to all three detent positions.

Turn selector knob to "Position Control" and slowly move control lever downward on quadrant. When lever is 1/2-5/8 inch from top of slot, lift links should start to lower. Move lever down and up quadrant a little at a time. Lift links should respond to movement of control lever, completely lowering when lever is at bottom of quadrant and stopping at any selected intermediate point.

Lower the weight to ground and insert a 5/8 inch gage block between draft linkage ratio lever and lever

Fig. 379—Flotation latch (F) is in locked position when vertical as shown. When latch is turned 90 degrees, some independent movement of each lift link is allowed.

guide at point shown at (A—Fig. 380), then turn draft control hand-wheel (B) until gage block is lightly held. Move control lever to top of quadrant and turn selector knob to "DRAFT" position; then move control lever about 2/3 the distance to bottom of quadrant, midway between the two sector marks (1—Fig. 381). Lower links should remain suspended in any selected position when control lever is within 1/4-inch of midway between marks, raising when lever is moved upward slightly, or lowering when lever is moved downward.

248. **TROUBLE SHOOTING.** Relief pressure for the three-point hitch is controlled by maximum setting for pressure control. Instead of a relief valve opening when maximum setting is exceeded, control valve is returned to neutral by a pilot valve. If relief pressure malfunctions, check and/or adjust the setting of pressure control linkage.

Draft control linkage is returned to neutral in transport position by the "Position Control" linkage. If "Position Control" linkage fails to return the main valve to neutral, valve will be neutralized by "Pressure Control" linkage when lift pistons reach the end of their stroke.

If lever must be moved 1/2-5/8 inch from top of quadrant before lift links start to lower, the position control linkage adjustment is probably satisfactory.

If lift hunts or searches in neutral position or is unduly noisy, leakage or improperly adjusted pressure control linkage is to be suspected.

249. **CHECK AND ADJUST.** First be sure lift arm lock is "OPEN" as

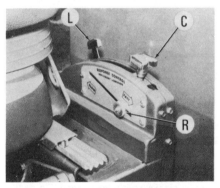

Fig. 377—Installed view of hydraulic control console showing three-point linkage controls.

C. Control lever
L. Locating stop
R. Response control

Fig. 378—Lift arm lock (L) is located on left side of lift cover housing as shown.

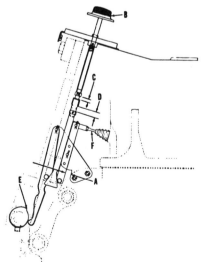

Fig. 380—Schematic view of draft control sensing linkage used on models so equipped.

A. Adjustment (0.625 inch)
B. Handwheel
C. Measurement (0.32 inch)
D. Measurement (0.62 inch)
E. Sensing cam
F. Pickup rod

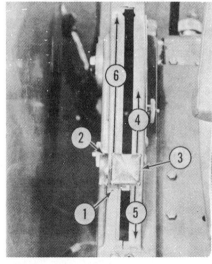

Fig. 381—View of control quadrant showing draft control positions.

1. Sector marks
2. Adjustable locator
3. Control lever
4. Decreasing depth
5. Increasing depth
6. Transport

shown in Fig. 378 and flotation latch is locked as shown in Fig. 379.

Attach a heavy weight to lower links. Weight should allow links to travel through complete arc and should not be attached to upper link.

Remove upper port plug from either rockshaft cylinder and install a suitable pressure gage in port opening. Remove lower front dress panel from seat console and front access plate from hydraulic lift cover. Remove access plate from right side of cover housing.

Start and run engine at 1000-1600 rpm, move control lever to top of quadrant and note gage pressure. The gage will either show the maximum pressure setting (which should be 1800-2000 psi), or the pressure necessary to support the attached weight (which will be considerably less than maximum pressure setting).

Working through the opening in front of lift cover, back off the locknut on position control adjusting screw

Fig. 382—View of lift cover with front access plate removed.

1. Position control adjust screw
2. Draft control bellcrank
3. Pressure control adjusting rod

(1—Fig. 382). If gage reading is considerably less than maximum pressure setting, back out the adjusting screw (1) counter-clockwise until gage pressure rises sharply; then turn screw in until pressure drops. Tighten locknut while holding the adjusting screw. If gage reading was originally the maximum pressure setting, turn screw in (clockwise) until pressure drops then tighten locknut. After completing the initial adjustment, move control lever slowly forward on quadrant and measure lever position at which lift links first start to lower. Distance should be $\frac{1}{2}$-5/8 inch (12.7-15.9mm). Readjust position control adjusting screw (1) if necessary. Turning screw clockwise will increase the distance, counterclockwise will decrease distance.

With selector knob in "Position Control", move lever forward (downward) on quadrant a short distance at a time. Lift links should respond by moving a corresponding amount then stopping. Lift cylinders should be fully retracted when control lever reaches bottom of quadrant.

With lift arms lowered, turn lift arm lock knob to "LOCK" position, then move control lever back to top of quadrant. Gage pressure should now be 2000 psi. If it is not, loosen the locknut and adjust servo relief valve rod (3). Lengthen rod to increase pressure or shorten rod to decrease pressure.

Turn selector knob on control lever to "PRESSURE" position, then move lever slowly down the quadrant while watching gage reading. Pressure should drop steadily at a uniform rate as lever is moved; and should read 50-250 psi when lever reaches bottom of quadrant.

Move control lever to top of quadrant. Working through access plate opening on right side of lift cover, lengthen pressure control piston rod (18—Fig. 383) until pressure gage begins to fluctuate or cycle; then shorten rod until cycling stops.

NOTE: If piston rod (18) is too short, main control valve will not return to neutral and fluid entering through the main valve must escape through the servo relief valve, overworking and overheating the valve. Make sure adjustment of piston rod (18) is correct.

Turn lift arm lock knob to "OPEN" position and check to see that lift arms raise and lower normally.

Move control lever to top of quadrant and turn selector knob to "DRAFT" position; then move control lever down the quadrant until aligned with "dimple" approximately ¼ the distance from top. Lower the lift links all the way by turning draft control hand wheel (Fig. 384), then check the clearance between draft link (9—Fig. 383) and stop screw (11). Clearance should be 0.002-0.004 inch (0.050-0.10mm); if it is not, remove rubber access plug (10) from top of lift cover and readjust the screw (11).

Fig. 384—Draft control handwheel, located behind operator's seat. Turning handwheel clockwise decreases operating depth, counter-clockwise increases operating depth.

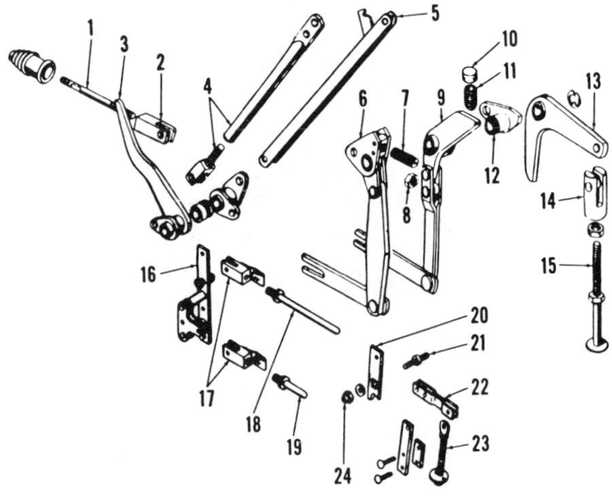

Fig. 383—Exploded view of control valve linkage.

1. Pickup rod
2. Clevis
3. Position control pickup
4. Position control link
5. Draft control link
6. Position control bellcrank
7. Adjusting screw
8. Locknut
9. Draft control bellcrank
10. Rubber plug
11. Stop screw
12. Pressure control bellcrank
13. Pressure control lever
14. Yoke
15. Adjusting rod
16. Valve lever
17. Clevis
18. Pressure control rod
19. Main valve rod
20. Drive link
21. Pivot stud
22. Dashpot beam
23. Dashpot piston rod
24. Nut

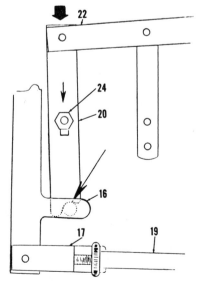

Fig. 385—Schematic view of main valve linkage showing adjustment procedure. Refer to Fig. 383 for parts identification and to paragraph 250 for details of adjustment.

250. MAIN VALVE ADJUSTMENT. If proper adjustment cannot be obtained as outlined in paragraph 249, check the main spool valve adjustment as follows:

Prepare the tractor as outlined in paragraph 249. Turn selector knob to "POSITION" control and move control lever to bottom of quadrant. With engine running and working through right access plate opening, press down on vertical dashpot link (20—Fig. 385) as shown by heavy arrow. Be certain lower notch engages spacer on actuating lever (16). Temporarily tighten locknut (24) to hold the link (20) in lowermost position.

Shorten control valve rod (19) until lower links begin to raise. Lengthen rod (19) when lower links are horizontal, until weight settles one inch in 10 seconds, measured at outer end of links.

Loosen locknut (24) until dashpot linkage if free to move; then readjust all the linkage as outlined in paragraph 249.

251. SENSING LINKAGE ADJUSTMENT. To adjust the draft control sensing linkage, first check to make sure that 3/8-7/16 inch (9.5-11.1mm) clearance exists between top link clevis pin and top link bracket as shown in Fig. 386, with nothing attached to upper link and clevis pin in lower holes in springs.

If measurement is incorrect, remove the two bolts securing pin to sensing springs and turn the pin 180 degrees until the other flats contact springs.

(Attaching flats are offset on clevis pin as shown in inset). If measurement is still incorrect after rotating pin, check for bent or damaged springs or attaching parts.

Move control lever to top of quadrant and turn selector knob to "DRAFT" position, then move the lever ⅔ the way down the quadrant until centered midway between the two sector marks. Upper link clevis must be hanging down and point of cam (E—Fig. 387) must be turned away from contact with pickup finger as shown. Turn draft control handwheel (B) counter-clockwise if necessary, until a 5/8 inch (15.9mm) gage block can be inserted between

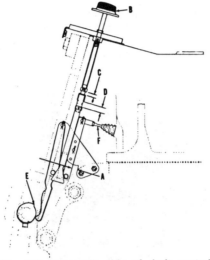

Fig. 387—Schematic view of draft control sensing linkage.

A. Adjustment (0.625 inch)
B. Handwheel
C. Measurement (0.32 inch)
D. Measurement (0.62 inch)
E. Sensing cam
F. Pickup rod

ratio lever and bracket as indicated at (A); then turn handwheel (B) clockwise until gage block is just supported. With linkage thus positioned, distance (C) between clevis and upper stop nut should be approximately 5/16 inch (8mm) and distance (D) between clevis and lower stop nut should be approximately 5/8 inch (15.9mm). Reposition the stop nuts if necessary, by loosening set screw and turning nut.

With engine running and linkage positioned as outlined, rockshaft should hold position without raising or lowering. Turning draft control handwheel (B) ¼-turn should cause rockshaft to raise of lower. If adjustment is incorrect, loosen the locknut and turn pickup rod (F) in or out as required.

252. RESPONSE ADJUSTMENT. With hydraulic fluid at operating temperature, quadrant lever in draft control, and approximately 1000 lbs. of weight attached to lower links, lowering time should not be more than 2 seconds with response control knob in "FAST" position, nor more than 5 seconds with control knob in "SLOW" position.

To adjust the response control, refer to Fig. 388. Loosen the clamp screw securing valve lever (2) to dashpot needle valve stem (1). Move response control lever (R—Fig. 389) fully to "SLOW" position and turn needle valve (1—Fig. 388) in until it bottoms; then retighten clamp screw. Readjust slightly if response is too slow; overhaul the valve if response is too fast.

253. OVERHAUL. The draft control sensing linkage, rockshaft arm and cylinders, or attaching links can be removed or installed without removing hydraulic lift cover. To obtain access to control valve, rockshaft or internal linkage, it is first necessary to remove cover as outlined in paragraph 254, or tilt cover back on draft control springs as outlined in paragraph 255. Both methods require the use of an overhead hoist.

254. R&R LIFT COVER. To remove the hydraulic lift cover for service on any of the components or rear axle center housing, proceed as follows:

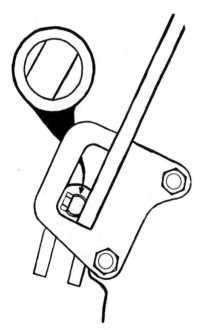

Fig. 386—Top link clevis pin attaching flats are slightly offset. Pin may be turned 180 degrees to establish initial setting as outlined in paragraph 251.

Fig. 388—Dashpot needle valve (1) should be lightly seated when "Response" lever is in extreme slow position. Adjust by loosening clamp screw securing valve lever (2).

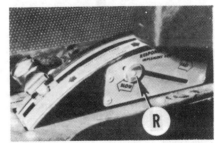

Fig. 389—View of control console showing "Response" control knob (R).

Disconnect lift cylinders from lift arms and lift rods from lower links, then remove lift arms and rods from rockshaft as assemblies. Remove rear dress panel cover, draft control springs and draft control sensing linkage. Remove platform center panel and seat frame. Disconnect hydraulic lines from lift cover to tractor center housing. Attach a suitable hoist to cover and lift off cover and control console as a unit.

When installing the cover, make sure oscillator drive lever properly engages drive body on control valve. Aligning

dowels help in positioning the cover. Complete the installation by reversing the removal procedure.

255. LIFT COVER PARTIAL REMOVAL. Completely lower the rockshaft. Move control lever to top of quadrant, turn selector knob to "PRESSURE" control and move control lever to bottom of quadrant. Remove platform center panel and the lower front panel from seat console. Remove operator's seat and seat bracket.

Disconnect pressure line to transfer cap and remove lift cover front access plate. Disconnect hoses leading to rockshaft cylinders. Disconnect draft control sensing rod and unbolt and remove sensing linkage.

Attach a suitable hoist to top of lift cover and remove cap screws securing cover to tractor center housing, then carefully raise lift cover upward and back as shown in Fig. 390.

NOTE: Cover will try to jump rearward as front end is raised, and damage to oscillator mechanism can occure. Alignment dowels in front cap screw holes can help prevent damage as long as binding does not occur.

The use of alignment dowels will also assist in installing the cover. Make sure oscillator drive lever properly engages drive body in control valve and reassemble by reversing the removal procedure.

256. LINKAGE CONTROL VALVE. To remove the linkage control valve, first remove lift cover as outlined in paragraph 254 or tilt cover as in paragraph 255.

Drain cover reservoir and remove standpipe (10—Fig. 251). Remove plug (4). Working through plug hole and using Special Tool MFN-767 or other suitable means, remove control valve discharge fitting (5) and associated parts.

Remove left hand cylinder hose fitting (3) and tube (2). Turn control quadrant selector knob to "PRESSURE" position and move lever to bottom of quadrant; then working through front access opening, disconnect and remove pressure control piston rod.

Remove the two long cap screws securing transfer cap to front, right side of cover, leaving transfer cap in place. Support linkage control valve with hand and remove the other three cap screws securing valve to lift cover

Fig. 390—Hydraulic lift cover may be tilted back for service as shown.

L. Shift linkage   T. Transfer cap
S. Standpipe   V. Control valve

Fig. 392—Removed view of linkage control valve showing proper assembly.

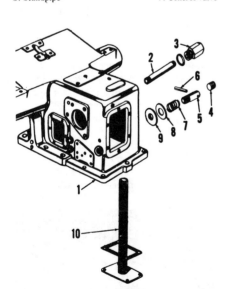

Fig. 391—Parts (2 through 10) must be removed from lift cover before control valve can be removed.

1. Cover   6. Groove pin
2. Outlet tube   7. Spring
3. Cylinder fitting   8. Washer
4. Plug   9. Seal
5. Discharge fitting 10. Standpipe

Fig. 393—Views of linkage control valve showing principal components. Location of valves in housing is indicated by corresponding numbers.

1. Main control valve
2. Pressure control pilot piston
3. Response dashpot
4. Inlet flow control valve
5. Discharge valve
6. Secondary relief
7. Servo relief valve
8. High pressure relief valve

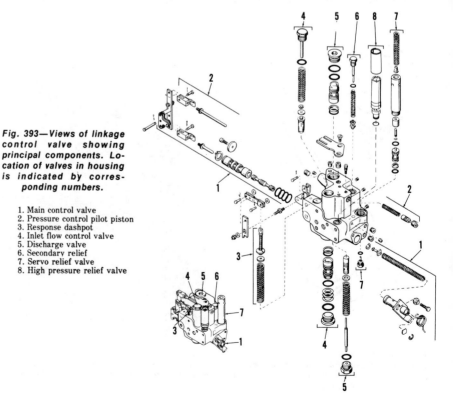

housing. Turn valve slightly and, working through front access plate opening, remove the slotted linkage guide plate from top of valve housing. Hold main control valve in discharge position and disconnect "Position" and "Draft" control links, then lift out the valve.

Fig. 393 shows an exploded view of main components of the valve. Shims (10—Fig. 394) control end clearance adjustment of intake control valve (7). Recommended adjustment is 0.001 inch (0.025mm) clearance to 0.003 inch (0.076mm) preload. Shims are available in thicknesses of 0.005, 0.008 and 0.010 inch. To check required shim pack thickness, assemble the unit leaving "O" ring off plug (11) and using 0.026-0.030 inch (0.66-0.76mm) thickness of shims (10). Tighten plug (11) finger tight, making sure parts are seated, then measure gap between flange of plug and surface of housing using a feeler gage. Remove the plug and deduct shims equal in thickness to the measured gap.

When valve is reassembled, check with Fig. 392 to be sure linkage is properly assembled, and adjust main valve as outlined in paragraph 250 before valve is installed.

257. ROCKSHAFT. To remove the rockshaft, first remove hydraulic lift cover as outlined in paragraph 254 or tilt cover rearward as in paragraph 255. Disconnect lift cylinders from lift arms and remove lift arms from rock-

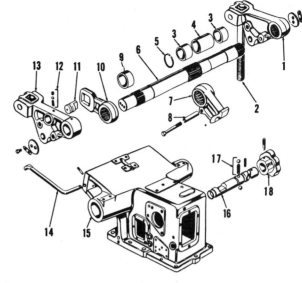

Fig. 395—Exploded view of lift cover, rockshaft and associated parts.

1. Lift arm
2. Lift link
3. Bushing
4. Spacer
5. Snap ring
6. Rockshaft
7. Lock arm
11. Lock pin
12. Detent assembly
13. Lift arm
14. Storage hook
15. Lift cover
16. Lock shaft
17. Detent
18. Lock knob

shaft if not previously removed.

Turn rockshaft until lock shaft (16—Fig. 395) engages notch in lock arm (7), then turn shaft (16) to lock posi-

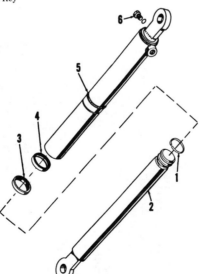

Fig. 396—Exploded view of main valve oscillator drive and associated parts.

1. Shoe
2. Crank
3. Bushing
4. Key

5. Hairpin
6. Spring
7. Lever

tion. Withdraw rockshaft (6) out right side of lift cover. Right hand bushing (9) will be removed with shaft, and left bushings (3) and spacer (4) can be removed at this time. Lock arm (7) can be removed after removing position control pickup rod. Rockshaft and arms have master splines for proper assembly. Assemble rockshaft by reversing the disassembly procedure.

258. OSCILLATOR DRIVE. Refer to Fig. 396 for an exploded view of main valve oscillator drive mechanism. The drive shoe (1) for crank rides on a cam machined on ipto clutch drum.

To remove the unit, first remove hydraulic lift cover as outlined in paragraph 254 or tilt the cover as in paragraph 255. Remove front hydraulic cover as outlined in paragraph 230.

Remove hair pin (5) and drive crank (2). Remove pickup lever (7), spring (6) and key (4); then withdraw drive bushing (3) forward out of center housing

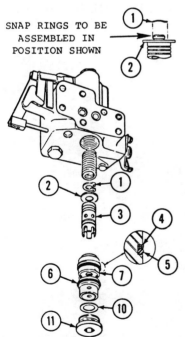

SNAP RINGS TO BE ASSEMBLED IN POSITION SHOWN

Fig. 394—Partially exploded view of inlet flow control valves showing proper assembly.

1. Snap ring
2. Dampening washer
3. Flow control valve
4. Backup washer
5. "O" ring

6. "O" ring
7. Flow control sleeve
10. Shim pack
11. Plug

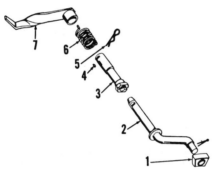

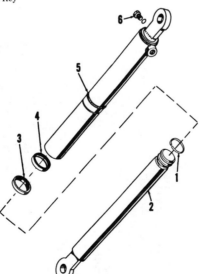

Fig. 397—Exploded view of rockshaft cylinder showing component parts. Refer to Fig. 398 for disassembly and assembly procedure.

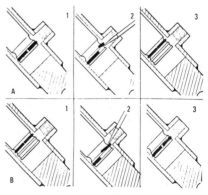

Fig. 398—Cylinder is disassembled by displacing stop ring into deeper groove and reassembled by repositioning in stop groove. Work through hose port using a screwdriver or similar tool. Upper views (A1, 2 & 3) show disassembly procedure; lower views (B1, 2 & 3) show assembly procedure.

bore. Assemble by reversing the disassembly procedure.

259. ROCKSHAFT CYLINDERS. The rockshaft cylinder unit is a welded assembly and is shown exploded in Fig. 397. To disassemble the cylinder, refer to Fig. 398. Remove hose fitting and

align snap ring with cylinder port as shown in (A1). Insert a screwdriver or similar tool in port opening and work snap ring into deep, second groove in piston rod. Piston rod can now be withdrawn. Assemble by reversing the dis-

assembly sequence as shown at (B). Pull piston against piston stop as shown in Fig. 399 after snap ring is repositioned, to be sure ring is fully seated.

TRANSFER CAP

260. The transfer cap (Fig. 400) is a one-way check valve used on all models with rockshaft and auxiliary valve. The unit mounts on right side of hydraulic lift cover. Check valve (4) prevents back flow from linkage valve to auxiliary valves when both units are actuated simultaneously or engine is shut off.

Check valve can be removed for service or cleaning after removing cap assembly and snap ring (1).

SENSING LINKAGE

261. Fig. 401 shows an exploded view of draft control springs and associated parts. Refer to paragraph 251 for adjustment if linkage is removed.

AUXILIARY VALVE

262. OPERATION. The auxiliary valve may be either single or dual spool type. The single spool valve (Fig. 405) is similar to dual spool valve (Fig. 406) except for plug (26—Fig. 410) in the unused spool bore and plugs (32) in the unused ports. Each spool has four positions (Raise, Neutral, Lower and Float), plus an adjustable flow restrictor.

263. ADJUSTMENT. The flow restrictors are normally adjusted to allow

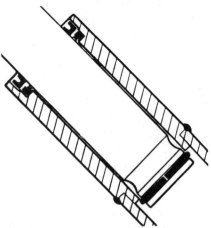

Fig. 399—When correctly positioned in stop groove, stop ring prevents withdrawal of cylinder rod as shown.

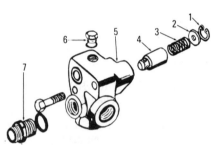

Fig. 400—Exploded view of transfer cap showing component parts.

1. Snap ring
2. Washer
3. Spring
4. Plunger
5. Body
6. Plug
7. Pressure fitting

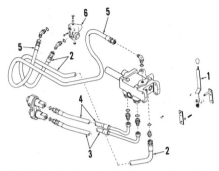

Fig 405—Drawing of single spool auxiliary control valve. Refer to Fig. 400 for exploded view of transport valve shown at (6).

1. Lever
2. Supply hose
3. Lower tube
4. Raise tube
5. Return line
6. Transport valve

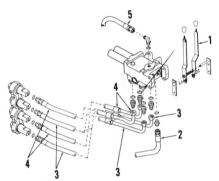

Fig. 406—Drawing of dual spool auxiliary control valve. Refer to Fig. 405 for legend.

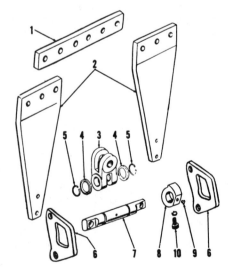

Fig. 401—Exploded view of draft control sensing springs and associated parts.

1. Clamp
2. Sensing spring
3. Clevis
4. Spacer
5. Snap ring
6. Bracket
7. Pin
8. Pickup cam
9. Drive plug
10. Lock screw

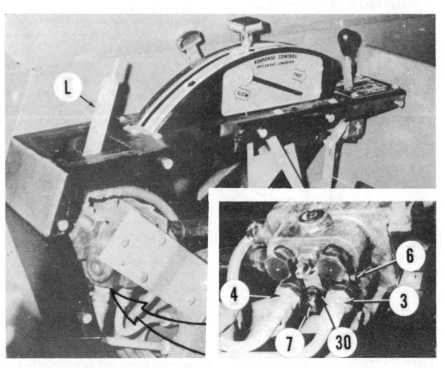

Fig. 407—View of auxiliary control valve with cover removed. Inset shows location of Lower tube (3), Raise tube (4), lock screw (6) and restrictor needle (7).

10-12 gpm of flow to regulate the speed of remote cylinder travel. Adjustment should be proportional to the size of the cylinder to prevent surge pressures and ease control of the remote cylinder. Turning restrictor needle (7—Fig. 407) in will decrease flow, out (counterclockwise) will increase flow. To check volume of flow, connect flow meter inlet hose to the raise coupler and connect outlet hose to sump as shown in Fig. 338. Start and run engine at 2000 rpm and move auxiliary control lever to Raise position. Adjust flow meter load valve to obtain 300 psi at meters pressure gage. If flow is not at desired rate of flow, turn restrictor needle (7—Fig. 407) as required, then

tighten lock nut. It will be necessary to readjust load valve of flow meter to provide 300 psi to check after changing restrictor needle setting. Check and adjust second spool using similar procedure.

An incorrectly adjusted lockout cartridge will cause high leakage of the cartridge and will result in remote cylinder leak-down. To check and adjust the lockout cartridge, proceed as follows: Connect a hand pump and pressure gage to a remote cylinder connector or to the cartridge as shown in Fig. 408 and move valve spool to Neutral position. Pressurize the output line to 200 psi with the hand pump and check for leak-down. Loosen lock screw (6) and if gage indicates that line was leaking down, turn cartridge (8) out until line will hold pressure without leaking. To set position, pressurize line to 200 psi and turn cartridge in slowly until pressure just begins to drop, then turn cartridge out until pressure stops falling, plus an additional 1/8-1/6 turn. Connect the hand pump and gage to each of the other output lines containing a cartridge (two lines for each valve spool) adjust each cartridge the same. Be sure to tighten lock screw (6) when position is correct.

Spool detent force can be adjusted by turning plug (9—Fig. 409) IN to increase or OUT to decrease as desired.

264. OVERHAUL. Refer to Fig. 410 or Fig. 411 for exploded view of auxiliary control valve. **When servicing,**

be sure to reinstall all parts in original locations. To disassemble loosen lock screws (6) and unscrew lockout cartridges (8). Disassembly of cartridges will not normally be necessary. Place spool in neutral, then carefully unscrew plug (9). Remove spring (12), cam (11), detent balls (10) and washer (13), then unscrew cap (14) from valve body. Remove spool by rotating slightly while pulling spring (16). "O" ring (21) and special washer (18) may not be withdrawn with spool. If necessary to separate parts (15 through 20), compress spring (16) and unscrew detent barrel (19) from end of valve spool (20).

NOTE: It will probably be necessary to clamp control lever end (flats) of spool (20) in a vise while turning detent barrel (19).

Blue Loctite is used on threads when assembling.

Clean all parts thoroughly and be especially careful not to install any parts in any location other than originally assembled. Use "Grade CV (Blue) Loctite" on threads of spool (20) and detent barrel (19). Compress retainer (17), spring (16) and retainer (15) and tighten detent barrel (19) to 9-10 ft.-lbs. torque. Complete setting for restrictor needles (7) is so that two threads are visible outside packing nut (30). Position of the lockout cartridges (8) can be set with 200 psi applied to port of cartridge with hand pump as described in paragraph 263. Reset restrictor needles after assembly to tractor as outlined in paragraph 263.

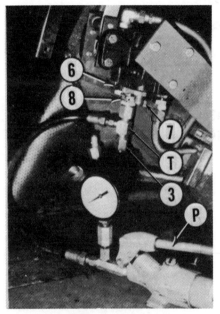

Fig. 408—View showing hand pump and gage (P) connected by "T" fitting (T) between line (3) and lockout cartridge (8).

Fig. 409—View showing location of the detent adjusting plug (9).

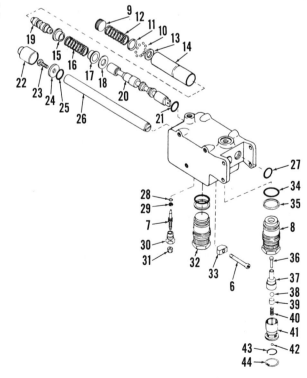

Fig. 410—Exploded view of single spool auxiliary valve.

6. Lock screw
7. Restrictor needle
8. Lockout cartridge
9. Detent adjust plug
10. Detent balls
11. Cam
12. Spring
13. Washer
14. Cap
15. Retainer (large opening)
16. Spring
17. Retainer (small opening)
18. Washer
19. Detent barrel
20. Spool
21. "O" ring
22. Cap
23. Screw
24. Washer
25. "O" ring
26. Plug
27. "O" ring
28. "O" ring
29. Packing
30. Packing nut
31. Nut
32. Plug
33. Clamp
34. "O" ring
35. Packing
36. Pin
37. Poppet
38. Steel ball
39. Spring guide
40. Spring
41. Guide
42. Steel ball
43. Clip
44. Snap ring

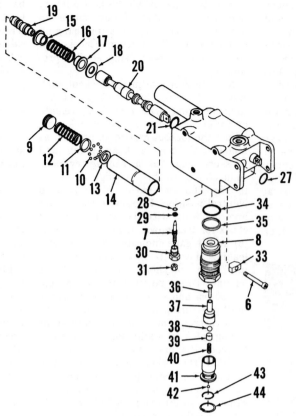

Fig. 411—Exploded view
of dual spool auxiliary
remote control valve. Parts
are duplicated for other
valve spool shown in-
stalled. Refer to Fig. 410
for legend.

General Torque Recommendations

Use the following torque *recommendations* as a guideline when a specification for a particular fastener is not available. In many cases manufacturers do not provide torque specifications, especially on older models.

Consider fastener condition carefully when referring to either a recommendation or a specification. If fastener reuse is appropriate, select the minimum value to account for fastener stretch. Softer fasteners or those securing softer materials, such as aluminum or cast iron, typically require less torque. In addition, lubricated or unusually long fasteners typically require less torque.

Determine fastener strength by referring to the grade mark on the bolt head. The higher the grade is, the stronger the fastener.

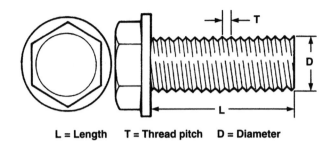

L = Length   T = Thread pitch   D = Diameter

Determine fastener size by measuring the thread diameter (D), fastener length (L) and thread pitch (T).

Size and Pitch	SAE grade 1 or 2 bolts	SAE grade 5 bolts	SAE grade 8 bolts
1/4—20	4-6 ft.-lbs.	6-10 ft.-lbs.	9-14 ft.-lbs.
1/4—28	5-7 ft.-lbs.	7-11 ft.-lbs.	10-16 ft.-lbs.
5/16—18	6-12 ft.-lbs.	8-19 ft.-lbs.	15-29 ft.-lbs.
5/16—24	9-13 ft.-lbs.	15-21 ft.-lbs.	19-33 ft.-lbs.
3/8—16	12-20 ft.-lbs.	19-33 ft.-lbs.	28-47 ft.-lbs.
3/8—24	17-25 ft.-lbs.	26-37 ft.-lbs.	36-53 ft.-lbs.
7/16—14	22-32 ft.-lbs.	31-54 ft.-lbs.	51-78 ft.-lbs.
7/16—20	27-36 ft.-lbs.	40-60 ft.-lbs.	58-84 ft.-lbs.
1/2—13	34-47 ft.-lbs.	56-78 ft.-lbs.	80-119 ft.-lbs.
1/2—20	41-59 ft.-lbs.	64-87 ft.-lbs.	89-129 ft.-lbs.
9/16—12	53-69 ft.-lbs.	69-114 ft.-lbs.	102-169 ft.-lbs.
9/16—18	60-79 ft.-lbs.	78-127 ft.-lbs.	115-185 ft.-lbs.
5/8—11	74-96 ft.-lbs.	112-154 ft.-lbs.	156-230 ft.-lbs.
5/8—18	82-110 ft.-lbs.	127-175 ft.-lbs.	178-287 ft.-lbs.
3/4—10	105-155 ft.-lbs.	165-257 ft.-lbs.	263-380 ft.-lbs.
3/4—16	130-180 ft.-lbs.	196-317 ft.-lbs.	309-448 ft.-lbs.
7/8—9	165-206 ft.-lbs.	290-382 ft.-lbs.	426-600 ft.-lbs.
7/8—14	185-230 ft.-lbs.	342-451 ft.-lbs.	492-665 ft.-lbs.
1—8	225-310 ft.-lbs.	441-587 ft.-lbs.	650-879 ft.-lbs.
1—14	252-345 ft.-lbs.	508-675 ft.-lbs.	742-1032 ft.-lbs.
1 1/8—7	330-480 ft.-lbs.	609-794 ft.-lbs.	860-1430 ft.-lbs.